—VNR ENVIRONMENTAL ENGINEERING SERIES—

Nelson L. Nemerow, Consulting Editor

Water Pollution Control

CONTROL AND TREATMENT OF COMBINED-SEWER OVERFLOWS, edited by Peter E. Moffa

TECHNOLOGIES FOR SMALL WATER AND WASTEWATER TREATMENT by E.J. Martin and E.T. Martin

INDUSTRIAL AND HAZARDOUS WASTE TREATMENT by Nelson Nemerow and Avijit Dasgupta

STREAM, LAKE, ESTUARY AND OCEAN POLLUTION, 2nd Ed., by Nelson Nemerow

Water Resources Development

CURRENT TRENDS IN WATER-SUPPLY PLANNING by David W. Prasifka

HANDBOOK OF PUBLIC WATER SYSTEMS by Culp, Wesner, and Culp

HANDBOOK OF CHLORINATION, 2nd Ed., by Geo. Clifford White

ANALYSIS OF WATER DISTRIBUTION SYSTEMS by Thomas M. Walski

WATER CLARIFICATION PROCESSES by Herbert E. Hudson, Jr.

DESIGN AND CONSTRUCTION OF WATER WELLS by the National Water Well Association

CORROSION MANAGEMENT IN WATER SUPPLY SYSTEMS by W. Harry Smith

DISINFECTION ALTERNATIVES FOR SAFE DRINKING WATER by Ted Bryant and George Fulton

Solid Waste Management

SMALL-SCALE MUNICIPAL SOLID WASTE ENERGY RECOVERY SYSTEMS by Gershman, Brickner and Bratton, Inc.

Hazardous Waste Treatment

GROUNDWATER TREATMENT TECHNOLOGY by Evan Nyer

HAZARDOUS WASTE SITE REMEDIATION by O'Brien and Gere, Inc.

SUBSURFACE MIGRATION OF HAZARDOUS WASTES by Joseph S. Divinny, Lorne G. Everett, James C.S. Lu and Robert L. Stollar

CONTINGENCY PLANNING FOR INDUSTRIAL EMERGENCIES by Piero M. Armenante

General Environmental

PROJECT PLANNING AND MANAGEMENT: AN INTEGRATED SYSTEM FOR IMPROVING PRODUCTIVITY by Louis J. Goodman

COMPUTER MODELS IN ENVIRONMENTAL PLANNING by Stephen I. Gordon

ENVIRONMENTAL PERMITS by Donna C. Rona

INDUSTRIAL
AND HAZARDOUS
WASTE TREATMENT

INDUSTRIAL AND HAZARDOUS WASTE TREATMENT

Nelson Leonard Nemerow
Avijit Dasgupta

ENVIRONMENTAL ENGINEERING SERIES

VNR VAN NOSTRAND REINHOLD
NEW YORK

Library of Congress Catalog Card Number 90-49704
ISBN 0-442-31934-7

Printed in the United States of America.

Van Nostrand Reinhold
115 Fifth Avenue
New York, New York 10003

Chapman and Hall
2-6 Boundary Row
London, SE1 8HN, england

Thomas Nelson Australia
102 Dodds Street
Ísouth Melbourne 3205
Victoria, Australia

Nelson Canada
1120 Birchmount Road
Scarborough, Ontario MIK 5G4, Canada

16 15 14 13 12 11 10 9 8 7 6 5 4 3 2 1

Library of Congress Cataloging-in-Publication Data

Nemerow, Nelson Leonard.
    Industrial and hazardous waste treatment / Nelson Leonard Nemerow.
  Avijit Dasgupta.
      p. cm.
  Includes index.
  ISBN 0-442-31934-7
    1. Factory and trade waste. 2. Hazardous wastes—Purification.
  I. Dasgupta, Avijit. II. Title.
  TD897.N377 1991
  628.4'2—dc20

                        90-49704
                        CIP

*To the grandchildren, Brian and Carolyn,
and the children, Debush, Brinda, and Binti—so that the environment
they inherit will be maintained in the highest quality.*

Senator John Glenn declared when describing hazardous wastes in December 1988:

We have a nuclear time bomb on our hands with the hazardous wastes in these plants and it's killing our people and must be taken care of now.

Richard Mahoney, Monsanto Chemical Company's CEO, said in 1988:

Society's telling us we have to change. We had better listen. Environmentalists today aren't fringe element tree-huggers. They're our neighbors, your students, our employees. You and me.

CONTENTS

PART 3 APPLICATIONS

This book is intended to meet the needs of many people: the college professor who teaches environmental engineering, the consulting engineer who seeks a solution to his client's problem, the municipal engineer who must understand the waste problem well enough to explain it to city officials and point out remedies, the industrial plant engineer who wants to prevent his company from polluting the water that receives his plant's wastes, the Environmental Protection Agency technical personnel charged with administering the Water Quality Act of 1970, and engineers at state and regional levels who are faced with the immediate and readily visible problem of pollution abatement.

A book to serve such diverse interests, in order to be useful, must naturally attack the subject from several viewpoints. The experience gained in using *Theories and Practices of Industrial Waste Treatment* (Addison-Wesley, 1963) over a period of five years and *Liquid Wastes of Industry* (Addison-Wesley, 1971) and *Industrial Water Pollution* (Krieger Publishing Co., 1978-1987) for another seven years proved invaluable to us when it came to writing this present text, which is divided into the following four parts.

Part 1 contains basic facts which the industrial waste engineer needs to know: effect of wastes on the surrounding environment, including the writing of environmental impact statements, ways to protect the stream from further pollution including the current E.P.A. industrial effluent guidelines, how to calculate the final treatment required before disposal of wastes into a receiving stream, how to sample the stream to ascertain the waste treatment required or the efficiency of existing treatment, and finally, how administrative decisions in pollution-abatement problems are influenced by the economics of waste treatment.

Part 2 delves into the theories and design of waste treatment, and talks about how wastes can be reduced by proper operation of manufacturing plants. Since no

waste problem is exactly similar to any other waste problem, students of this subject must have a coherent picture of the entire field of industrial waste treatment in order to decide which treatment process best suits the needs of a particular project. Section 2, therefore, differs from the conventional text on waste-water treatment in that it discusses not only removal of suspended and colloidal solids, but also the subjects of neutralization, equalization and proportioning, removal of inorganic dissolved salts, and private contract collection and treatment. The theories are similar to those expressed by other authors since theories have not changed very much. However, this section does present the latest new methods for the removal of dissolved organic solids. A novel method of the future for obtaining zero pollution by using environmentally balanced industrial complexes is also presented in Chapter 15 of this part.

Part 3 accentuates engineering practice and presents concrete examples of problems and their solutions. Theories are highly idealistic and seldom work in practice exactly as set forth on paper; quite often, in this field, they never become a reality. The reasons for this are numerous. Economics, public opinion, personality differences, local laws or customs, previous community experiences with certain industrial wastes, contradictory advice by consulting engineers, views of the local industrial development board, views of regulatory agencies—all these and many other factors help determine whether even well-conceived theories can be put into practice.

In each chapter in Part 3, we have reported on actual cases which we either know about or have executed personally. We have attempted to keep these case histories as contemporary as possible. For example, Chapters 19, 22 and 23 contain new cases and concepts in place of older ones used in the last text. The reader must realize, however, that it often takes as long as five years between the realization of a given waste problem and the initiation of work on it and the evaluation of the treatment plant used in solving it. Thus some cases described in the text originated several years ago, but we hope that they serve as typical examples and stand the test of time satisfactorily. The reader can follow step-by-step analyses and results, just as law students study legal cases. The basic question underlying this section is whether to treat industrial wastes separately or in conjunction with municipal sewage. In many instances theory calls for joint treatment, but practice demands separate treatment. One must know the reasons for recommending overall treatment systems, including production changes and closing of plants, as well as certain specific methods of waste treatment.

In these times of plant relocation due to business expansion and market changes, the separate chapter on site selection is especially useful. The more this country's land is developed with industries, cities, highways, parks, and reservoirs, the more important site selection becomes.

Part 4 gives separate treatises on all the major liquid industrial wastes—a subject which normally requires an entire book. We have classified all industries into six categories: apparel, food processing, materials, chemicals, energy, and non-point practices. We have found it desirable to divide the energy industry into two parts in order to specify the separate and different handling that must be accorded to radioactive wastes. Since hazardous wastes, in themselves, often differ from conventional industrial wastes and certainly represent such an immediate, dramatic, and dangerous impact on our environment, they are described in a separate and detailed Chapter 32. Treatments for these wastes, when different from conventional wastes, are given more extensive consideration. We do not attempt to present a comprehensive study of each waste; that in itself would require a separate text for each. Rather, we have given a condensed evaluation of the nature of each waste—its origin, characteristics, and more acceptable treatments. In addition, there is a recent bibliography, which presents the most readily available reference material for each type of waste. The bibliographies are largely those published since 1968. This should be invaluable to the person who needs to do efficient and rapid research on recent and older studies of a particular waste. Appendixes A and B contain provisions of the current Amended Federal Water Pollution Control Act (Sections 306 and 402). (1) National Standards of Performance and (2) National Pollutant Discharge Elimination System (NPDES). Appendix C contains a new useful reference list of EPA Effluent Guideline Documents.

Since no author could personally amass the extensive and varied data presented in this text, in writing this book, we have borrowed heavily—and gratefully—from source material on the subject by other writers. To these authors—as well as to our teachers, Dr. William Rudolfs, Dr. Hovhannes Heukelekian, and Dr. Harold Orford—we are sincerely indebted. We are also indebted to the hundreds of original researchers and to the journals which published their works. For their permission to quote excerpts in the book, we express our appreciation to the following publications, as well as to numerous others: *Journal of the Water Pollution Control Federation,* Washington, D.C.; *Wastes Engineering*, New York City; *Industrial Water and Wastes*, Chicago, Illinois; *Water and Sewage Works*, Chicago, Illinois; and *Proceedings of the Purdue University Industrial Waste Conference,* Lafayette, Indiana.

We are grateful to our many friends and colleagues who, after using both of the previous books on industrial wastes, offered their suggestions in a gracious manner. To our graduate students who were forced to study this book in great detail, let us say thank you for serving us and society so well. Without cooperative students on whom to try out one's ideas, it would be impossible to produce a meaningful textbook on this subject for graduate study.

N.L.N.
A.D.

INDUSTRIAL AND HAZARDOUS WASTE TREATMENT

Part 1 | BASIC KNOWLEDGE AND PRACTICES

CHAPTER 1

EFFECT OF WASTES ON STREAMS
AND WASTE-WATER TREATMENT PLANTS

1.1 Effects on Streams

All industrial wastes affect, in some way, the normal life of a stream [11].* When the effect is sufficient to render the stream unacceptable for its "best usage," it is said to be polluted. Best usage means just what the words imply: use of water for drinking, bathing, fishing, and so forth. A more detailed description of these uses is given in Chapter 2.

Streams can assimilate a certain quantity of waste before reaching a polluted state. Generally speaking, the larger, swifter, and more remote streams that are not much used are able to tolerate a considerable amount of waste, but too much of any type of polluting material causes a nuisance. To call a stream polluted, therefore, generally means that the stream contains an excessive amount of a specific pollutant or pollutants. Walter (17, 1975) describes pollution as originating basically from four sources in the economic process: 1—materials-source pollution involving environmental damage caused by extracting and transporting virgin renewable and non-renewable raw materials, and recycling used materials, required for production. 2—process-pollution originating in the production itself. 3—product pollution concerns damage done to the environment by products in their everyday use. 4—residual pollution involving the disposal of products once they have served their useful lives. The following materials can cause pollution:

Inorganic salts	Heated water
Acids and/or alkalis	Color
Organic matter	Toxic chemicals
Suspended solids	Microorganisms
Floating solids and	Radioactive materials
liquids	Foam-producing matter

*Numbers in brackets refer to the bibliographical references at the end of each chapter.

Inorganic salts, which are present in most industrial wastes as well as in nature itself, cause water to be "hard" and make a stream undesirable for industrial, municipal, and agricultural usage. We mention here just a few of the hundreds of difficulties arising from the use of hard water.

Salt-laden waters deposit scale on municipal water-distribution pipelines, increasing resistance to flow and lowering the overall capacity of the lines. Hard waters interfere with dyeing in the textile industry, brewing in the beer industry, and quality of the product in the canning industry. Magnesium sulfate, a particularly bothersome constituent in hard waters, has a cathartic effect on people. The chloride ion increases the conductance of electrical insulating paper; iron causes spots and stains on white goods manufactured by textile mills and on high-grade papers produced by paper mills; and carbonates produce a hard scale on peas processed in canneries. Most types of hard water encrust boiler tubes, so that transfer of heat to the water from the fire chamber is impaired. This condition, called "boiler scale," results in lowered boiler efficiency and increased cost of operation.

Another disadvantage is that, under proper environmental conditions, inorganic salts, especially nitrogen and phosphorus, induce the growth of microscopic plant life (algae) in surface waters. Although algae are really a secondary form of pollution, they can be of extreme importance. Their advantage is that of adding dissolved oxygen to the stream; their disadvantage is the organic loading they contribute after dying. Too little attention is given by industrial waste engineers to these inorganic products of waste liquors. The role of phosphorus is diverse and complicated, but it is known that in the absence of phosphorus there is practically total elimination of algae life.

There is another facet of the problem worth noting: a total absence of salts is apt to result in corrosive and/or tasteless water, whereas a certain degree of hardness enhances the development of a protective film on surfaces and renders water more palatable. Producers of baked goods, for instance, feel that some concentration of calcium sulfate helps to achieve a golden brown crust on bread. It is therefore desirable that *some* inorganic salts be present in the water supply. The amount, rather than the presence, is the important factor.

A rather different form of pollution may exist along the coasts of southern California and Florida, as well as in parts of Texas and Arizona, where excessive withdrawal of ground water has allowed subterranean intrusion of salt water into previously fresh-water aquifers.

Acids and/or alkalis discharged by chemical and other industrial plants make a stream unsuitable not only for recreational uses such as swimming and boating, but also for propagation of fish and other aquatic life. High concentrations of sulfuric acid, sufficient to lower the pH to below 7.0 when free chlorine is not present, have been reported to cause eye irritation to swimmers, rapid corrosion of ships' hulls, and accelerated deterioration of fishermen's nets. The toxicity of sulfuric acid for aquatic life is a function of the resulting pH; i.e., a dose that would be lethal in soft water may be quite harmless in hard or highly buffered water. It is generally agreed that the pH of a stream must be not less than 4.5 and not more than 9.5 if fish are to survive. Yet stream pH values from as low as 2 to as high as 11 may occur near industrial sources of pollution.

Sodium hydroxide—to cite an example of an alkali—is highly soluble in water and affects the alkalinity and pH. It appears in wastes from many industries, including soap manufacturing, textile dyeing, rubber reclaiming and leather tanning. Streams containing as little as 25 parts of sodium hydroxide per million have been reported deadly to fish. Alkali in boiler-feed water can, by its caustic action, cause caustic embrittlement of pipes. Water-treatment plants are also adversely affected by these pollutants; for example, treatment plants using alum as a coagulant often find shock loads of acid or alkali interfering with floc formation.

Some miscellaneous processes affected by using waters of certain pH values are the rate of industrial fermentation, quality of dough in baking, flavor in soft drinks, yeast activity in brewing of beer, taste of canned fruits, especially tomatoes, cleaning of industrial metals, and gelatin and glue manufacture. A low pH may cause corrosion in air-conditioning equipment, and a pH greater than 9.5 enhances laundering.

Organic matter exhausts the oxygen resources of rivers and creates unpleasant tastes, odors, and general septic conditions. Fish and most aquatic life are stifled by lack of oxygen, and the oxygen level, combined with other stream conditions, determines the life or death of fish. It is generally conceded that the critical range for fish survival is 3 to 4 parts per million (ppm) of dissolved oxygen. We know that some species of fish may not survive in water containing 3 ppm of dissolved oxygen, while other species may not be affected even slightly by the same low oxygen level. For example, trout are sensitive fish, requiring oxygen concentrations of at least 5 ppm, whereas carp are scavenger fish, capable of surviving in waters containing as little as 1 ppm of oxygen. This oxygen shortage, caused by organic matter, is often considered to be the most objectionable single factor in a stream's pollution.

Certain organic chemicals, such as phenols, affect the taste of domestic water supplies. If rivers containing phenols permeate nearby wells, they cause objectionable medicinal tastes, and in addition there is the less-obvious organic matter, which may cause discomfort or diseases.

Suspended solids settle to the bottom or wash up on the banks and decompose, causing odors and depleting oxygen in the river water. Fish often die because of a sudden lowering of the oxygen content of a stream, and solids that settle to the bottom will cover their spawning grounds and inhibit propagation. Visible sludge creates unsightly conditions and destroys the use of a river for recreational purposes. These solids also increase the turbidity of the water-course and enhance flooding by diminishing the stream-bed volume. Although each stream varies in the quantity of suspended solids it can safely carry away, most pollution-control authorities specify that suspended solids may be discharged into a stream only in amounts that will not impair the best usage of the stream.

Floating solids and liquids. These include oils, greases, and other materials which float on the surface; they not only make the river unsightly but also obstruct passage of light through the water, retarding the growth of vital plant food. Some specific objections to oil in streams are that it: (1) interferes with natural reaeration; (2) is toxic to certain species of fish and aquatic life; (3) creates a fire hazard when present on the water surface in sufficient amounts; (4) destroys vegetation along the shoreline, with consequent erosion; (5) renders boiler-feed and cooling water unusable; (6) causes trouble in conventional water-treatment processes by imparting tastes and odors to water and coating sand filters with a tenacious film; (7) creates an unsightly film on the surface of the water; and (8) lowers recreational, e.g. boating, potential.

Heated water. An increase in water temperature, brought about by discharging wastes such as condenser waters into streams, has various adverse effects. Stream waters which vary in temperature from one hour to the next are difficult to process effectively in municipal and industrial water-treatment plants, and heated stream waters are of decreased value for industrial cooling. Indeed, one industry may so increase the temperature of a stream that a neighboring industry downstream cannot use the water. Furthermore, warm water is lighter than cold, so that stratification develops, and this causes most fish life to retreat to stream bottoms. Since there may be less dissolved oxygen in warm water than in cold, aquatic life suffers, and less oxygen is available for natural biological degradation of any organic pollution discharged into these warm surface waters. Also, bacterial action increases in higher temperatures, resulting in accelerated depletion of the stream's oxygen resources.

Color, contributed by textile and paper mills, tanneries, slaughterhouses and other industries, is an indicator of pollution. Compounds present in waste waters absorb certain wavelengths of light and reflect the remainder, a fact generally conceded to account for color development of streams. Color interferes with the transmission of sunlight into the stream and therefore lessens photosynthetic action. It may also interfere with oxygen absorption from the atmosphere—although no positive proof of this exists.

Visible pollution often causes more trouble for industry than invisible pollution. Unseen pollution which does not create a nuisance will often be tolerated by state agencies, but the red and deep-brown colors of slaughterhouse wastes, the browns of paper-mill wastes, various intense colors of textile-mill wastes, and the yellows of plating-mill wastes will focus public indignation directly on those industries. It is only human to complain about visible pollution: property values decrease along a visibly polluted river, and fewer people will swim, boat, or fish in a stream highly colored by industrial wastes. Furthermore, municipal and industrial water plants have great difficulty, and scant success, in removing color from raw water.

Toxic chemicals. Both inorganic and organic chemicals, even in extremely low concentrations, may be poisonous to fresh-water fish and other, smaller, aquatic microorganisms. Many of these compounds are not removed by municipal treatment plants and have a cumulative effect on biological systems. Such insecticides as toxaphene, dieldrin, and dichlorobenzene have allegedly killed fish in farm ponds and streams. Insecticides used in cotton and tobacco dusting have their maximum effect following heavy rainfalls—i.e. they are more lethal in solution—but insecticides and rodenticides are hard to detect in a stream. However, newer techniques, e.g. electron-capture gas chromatography, can detect chlorinated hydrocarbon pesticides in concentrations of 0.001 micrograms per liter in one-liter samples of water.

New, highly complex, organic compounds produced by the chemical industry for textile and other companies have also proved extremely toxic to fish life. One example is acrylonitrile, a raw material used in the manufacture of certain new synthetic fibers.

Almost all salts, some even in low concentrations, are toxic to certain forms of aquatic life. Thus, chlorides are reportedly toxic to fresh-water fish in 4000 ppm concentration, as are hexavalent chromium compounds in concentrations of 5 ppm. Copper concentrations as low as 0.1 to 0.5 ppm are toxic to

Table 1.1   Limits set on contents of chemical elements or compounds in water supplies [1].

Characteristic	Natural mandatory limit, ppm	Recommended limit, ppm
Lead	0.1	0.05
Fluoride	1.5	0.7–1.2
Arsenic	0.05	0.01
Selenium	0.05	
Chromium (hexavalent)	0.05	
Copper		1.0
Iron		0.3
Magnesium		125
Zinc		5
Chloride		250
Sulfate		250
Phenolic compounds, in terms of phenol		0.001
Total solids		
Desirable		500
Permitted		1000
Normal carbonate ($CaCO_3$)		120
Excess alkalinity over hardness ($CaCO_3$)		35
pH (25°C)		10.6
Alkyl benzene sulfonate		0.5
Carbon chloroform extract (CCE)		0.2
Cyanide (Cn)		0.01
Manganese (Mn)		0.05
Nitrate (NO_3)		45
Strontium 90		10 $\mu\mu c$/liter
Radium 226		3 $\mu\mu c$/liter
Gross β radiation concentration		1000 $\mu\mu c$/liter

bacteria and other microorganisms. Although oyster larvae, for setting, *require* a copper concentration of about 0.05 to 0.06 ppm, concentrations above 0.1 to 0.5 ppm are toxic to some species. All three salts are often found in watercourses.

Accidental or intermittent discharge of certain toxic materials may go unnoticed and yet may completely disrupt stream life. Building-floor and storm-water drains that lead directly to the stream may convey contamination because of an upset in an industrial process or ignorance of the consequences. For example, the flushing of a chemical delivery tank at the unloading dock may carry dissolved toxic material into the stream through a storm drain.

Complex inorganic phosphates, such as P_2O_5, at levels as low as 0.5 ppm, perceptibly interfere with normal coagulation and sedimentation processes in water-purification plants. Increased coagulant dosages and/or increased settling times are required

[9] to solve the problem. Phenols in concentrations exceeding one part per billion have been found to be objectionable in a stream. Phenol reacts with chlorine and, even in extremely small quantities, gives the residual drinking water a noticeable medicinal taste. Table 1.1 lists water-quality limits recommended by the U.S. Public Health Service [1]. Toxic wastes are considered hazardous to humans, animals, and aquatic life in contact with such wastes. The reader is urged to read carefully the extensive coverage of these wastes in Chapter 32.

Microorganisms. A few industries, such as tanneries and slaughterhouses, sometimes discharge wastes containing bacteria. Vegetable and fruit canneries may also add bacterial contamination to streams. These bacteria are of two significant types: (a) Bacteria which assist in the degradation of the organic matter as the waste moves downstream. This process may aid in "seeding" a stream (deliberate inoculation with biological life for the purpose of degrading organic matter) and in accelerating the occurrence of oxygen sag in the water. (b) Bacteria which are pathogenic, not only to other bacteria, but also to humans. An example is the anthrax bacillus, originating in tanneries where hides from anthrax-infected animals have been processed.

Medical wastes discarded from hospitals, laboratories, and doctor's offices usually contain disease-causing bacteria and viruses. Indiscriminate and deliberate discharge by these users can put the entire society at risk of disease.

Radioactive materials. The manufacture of fissionable materials, the increasing peacetime use of atomic energy, and the projected development of atomic-power facilities have introduced new complications in the field of sanitary engineering. The problem of disposing of radioactive wastes is unique [2], since the effects of radiation can be immediate or delayed, and radiation is an insidious contaminant with cumulative damaging effects on living cells. Certain highly active radioisotopes such as Sr^{90} and Cs^{137} continue to release energy over long periods of time (several generations of the human race). This radiation is not readily detectable by the methods usually employed to determine the presence of contaminants in the environment. Furthermore, the biological and hydrological characteristics of a stream may have a profound influence on the uptake of radioactivity.

At present, the maximum safe concentration of mixed fission products for lifetime consumption, according to the Atomic Energy Commission, is 1×10^{-7} microcuries per milliliter. Therefore, regulatory agencies, as well as the public, are concerned about preventing contamination of surface streams by radioactive wastes [13].

The International Commission on Radiation Protection (The International Atomic Energy Agency, 1979) recommends that no member of the general society receive a dose of radiation to exceed a total of 500 millerems (radiation dose equivalents) per year over a lifetime. They also report that 67.6% comes from natural background, 30.7% from medical radiation, and only 0.6% from fallout, 0.45% from occupational exposure, 0.15% from nuclear industry releases, and 0.5% miscellaneous sources. This points out the major contribution of natural background, the next larger man-made

contribution from medical uses, and finally, the relatively minor contribution from the nuclear power industry (See Chapter 30).

Foam-producing matter, such as is discharged by textile mills, pulp and paper mills, and chemical plants, gives an undesirable appearance to the receiving stream. It is an indicator of contamination and is often more objectionable in a stream than lack of oxygen. More court cases have been fought and won on evidence about the appearance of a stream than about the unseen contents of the water. (This in itself should serve as a warning to industries discharging foam-producing wastes.)

1.2 Effects on Sewage Plants

It is only natural for industry to presume that its wastes can best be disposed of in the domestic sewer system, and municipal officials often feel that it is their responsibility to accept any wastes flowing into their city's disposal system. However, city authorities should not accept any waste discharges into the domestic sewer system without first learning the facts about the characteristics of the wastes, the sewage system's ability to handle them, and the effects of the wastes upon *all* components of the city disposal system. Institution of a sewer ordinance, restricting the types or concentrations of waste admitted in the sewer leading to a treatment plant, is one means of protecting the system.

To remove pollution from industrial wastes, a sewage-treatment plant must have sufficient capacity and of the proper type. Theoretically, a sewage-treatment plant could be designed to handle any type of industrial waste, but present plants fall short of this ideal. Joint treatment of municipal and industrial waste waters which are amenable to treatment may offer greater *removal* efficiencies, but economics will usually be the deciding factor.

The pollutional characteristics of wastes having readily definable effects on sewers and treatment plants can be roughly classed as follows: (1) biochemical oxygen demand (BOD); (2) suspended solids; (3) floating and colored materials; (4) volume; and (5) other harmful constituents. Table 1.2 presents a comparison of domestic-sewage pollutional characteristics and those of some industrial wastes.

Table 1.2 General comparison of pollutional loads in industrial wastes versus domestic sewage [14].

Origin of waste	Population equivalent*	
	Biochemical oxygen demand	Suspended solids
Domestic sewage	1	1
Paper-mill waste	16–1330	6100
Tannery waste	24–48	40–80
Textile-mill waste	0.4–360	130–580
Cannery waste	8–800	3–440

*Persons per unit of daily production.

Biochemical oxygen demand (BOD) is usually exerted by dissolved and colloidal organic matter and imposes a load on the biological units of the treatment plant. Oxygen must be provided so that bacteria can grow and oxidize the organic matter. An added BOD load, caused by an increase in organic waste, requires more bacterial activity, more oxygen, and greater biological-unit capacity for its treatment. This calls for an increase in both capital outlay and daily operating expense.

However, not all dissolved or colloidal organic matter oxidizes at the same rate, with the same ease, or to the same degree. Sugars, for example, are more readily oxidized than starches, proteins, or fats. The rate of decomposition for industrial organic matter may therefore be faster or slower than that for sewage organic matter, and this difference must be considered in the design and operation of biological units. Before private industry embarks on a joint disposal venture with the city, the oxidizability of industrial wastes should be determined by the use of Warburg or other similar respirometer tests, which instantaneously measure the oxygen utilized and the carbon dioxide evolved by various solutions.

Figure 1.1 illustrates one possible effect of a given industrial waste on a sewage plant. In this instance the industrial waste, with its constant rate of degradation, tends to smooth out the rate of decomposition of the sewage so that the result shows less upsurge due to nitrogenation. Also, the rate of decomposition of the industrial waste tends to slow down the initial rapid rate of domestic sewage.

After much experimentation, Ettinger [3] believes that there is still some doubt whether activated-sludge biological units are able to handle slug waste

discharges better than digesters, which have the advantages of inherent storage capacity and assumed complete mixing.

Suspended solids are found in considerable quantity in many industrial wastes, such as cannery and paper-mill effluents. They are screened and/or settled out of the sewage at the disposal plant. Solids removed by settling and separated from the flowing sewage are called *sludge*, which may then undergo an anaerobic decomposition known as digestion and be pumped to drying beds or vacuum filters for extraction of additional water. Certain settleable suspended solids from industrial wastes, e.g. fine grit and insoluble metal precipitates, may hinder sludge digestion.

Suspended solids in industrial waste may settle more rapidly or slowly than sewage suspended matter. If industrial solids settle faster than those of municipal sewage, sludge should be removed at shorter intervals to prevent excessive build-up. Quantities of stale sludge may be "scoured" (dislodged by physical means) off the bottom of the basin, with resultant increase of sludge in the effluent. A faster-settling industrial waste may accelerate the settling of sewage solids; a slower-settling one will require a longer detention period and larger basins and increase the likelihood of sludge decomposition, with accompanying nuisances, during slack sewage-flow periods. However, regardless of the settling rate, the quantity of sludge to be pumped to the sludge-disposal facilities at the treatment plant will be increased by the addition of such industrial waste. Since sludge digesters, drying beds, and filters are

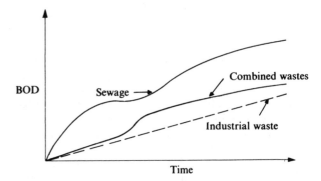

Fig. 1.1 Deoxygenation rates of sewage, a certain industrial waste, and a combination of the two.

designed to handle a certain number of pounds of solids per unit of capacity, any increased demands on the system usually require larger sludge-handling devices and may ultimately necessitate an increase in the plant's capacity, with resulting higher capital and operating expenses.

The settling characteristics of industrial wastes, alone and combined with municipal waste, should be determined before any disposal agreement between industry and city. Sludge consistency, percentage of total suspended solids removed, and weight of suspended solids removed are the criteria for evaluating settling characteristics.

Floating materials and colored matter, such as oil, grease, and dyes from textile-finishing mills, are disagreeable and visible nuisances. Visible pollution retards the development of a community or area, since it discourages camping, boating, swimming, and fishing—recreations indispensable to the vitality of a physically and mentally healthy community—and industry is reluctant to locate on a stream which is visibly polluted. Lack of industry further depresses the growth of city, county, and state, for less tax money means less progress. It is therefore imperative that nuisances such as color and floating matter be removed by the sewage-treatment plant.

A modern treatment plant will remove normal grease loads in primary settling tanks, but abnormally high loads of predominantly emulsified greases from laundries, slaughterhouses, rendering plants, and so forth, passing through the primary units (screens, grit chambers, and settling basins) into the biological units, will clog flow-distributing devices and air nozzles. A lengthy shutdown of these units may result in stream pollution and sudden loss of fish life.

Color removal by the treatment units of sewage plants is a knotty problem, and too little effort has been made so far to find an effective solution. The author found that trickling-filter plants in North Carolina were removing between 34 and 44 per cent of the dye color in the influent [12]. An overloaded primary plant, on the other hand, *added* 12 per cent to the color as waste passed through the plant. A knowledge of the character and measurement of color is essential. Since most colored matter is in a dissolved state, it is not altered by conventional primary devices, although secondary treatment units, such as activated sludge and trickling filters, remove a certain per-

centage of some types of colored matter. Sewage-treatment plants are generally not designed to remove color, so any reduction in this constituent is a fortunate coincidence, but, because of the previously mentioned detriment to streams, municipal disposal plants should in the future give increased consideration to removal of color. If an industry defines the type and quantity of colored matter in its waste, engineers can then make some prediction concerning the effectiveness of color removal by the treatment designed for domestic sewage.

Volume. A sewage plant can handle any volume of flow if its units are sufficiently large. Unfortunately, most sewage plants are already in operation when a request comes to accept the flow of waste from some new industrial concern. The hydraulic capacity of all units must then be analyzed; sewer lines must be examined for carrying capacity, bar screens for horizontal flow velocity, settling basins for detention periods and surface and weir overflow rates, trickling filters for excessive hydraulic loadings, and so forth.

An industry with a relatively clean waste such as condenser water can usually discharge it, after cooling, directly into the receiving stream and thus avoid overloading the sewage-treatment plant. This expedient saves capital and operating expenses at the disposal plant. However, before seemingly clean waters are accepted for direct disposal, they must be carefully examined for dissolved solids. Even a small concentration of solids in a large volume of waste water will sometimes result in a significant total-solids load.

Other harmful constituents. Industrial wastes may contain harmful ingredients in addition to the polluting load. These wastes can cause malfunctioning of the sewer system and/or the disposal plant. Some nuisances and their accompanying effects are:

(a) Toxic metal ions (Cu^{++}, Cr^{+6}, Zn^{++}, CN^-), which interfere with biological oxidation by tying up enzymes required to oxidize organic matter.

(b) Feathers, which clog nozzles, overload digesters, and impede proper pump operation.

(c) Rags, which clog pumps and valves and interfere with proper operation of bar screens or comminutors.

(d) Acids and alkalis, which may corrode pipes, pumps, and treatment units, interfere with settling, upset the biological purification of sewage, release odors, and intensify color.

(e) Flammables, which cause fires and may lead to explosions.

(f) Pieces of fat, which clog nozzles and pumps and overload digesters.

(g) Noxious gases, which present a direct danger to workers.

(h) Detergents, which cause foaming of aeration units.

(i) Phenols and other toxic organic material.

1.3 Differences Between Industrial and Municipal Wastes

Industry's primary objective is to produce the best possible product of its type at the lowest possible cost. Thus having to install waste treatment devices would contravene industry's primary objective. In order to compete in the world market, industry must hold production costs to a minimum. The most obvious area for cost-saving is in the area of waste treatment, since the product is least affected by reductions here. While a municipality may also be concerned with reducing its operating costs, it is more likely to be influenced by regulations requiring conformance to generally accepted practice. If this should result in higher costs, they will somehow be met by the larger and more distant "municipal fathers." A municipality is also concerned with providing a service in an orderly manner. It is significant to note that this service may provide additional employment for municipal constituents—a notion not inconsistent with the municipality's objective. The reader can comprehend rather readily that there are distinct and rather significant purposes for an industrial plant and its municipal counterpart.

Industry's operating schedule is one of extreme variability. One day an industrial plant might be closed while on another day it might be operating at two or three times its normal capacity in order to fill an urgent order. On one day a load of wastewater might be "dumped" at noon because of an error in operating schedules and on another day wastewater might not be discharged at that same time either because of a lack of orders or because of a malfunction in a production unit. There may be no wastewater from an industrial plant on weekends; how-

ever, the municipal wastewater system must and usually does operate 24 hours a day, 7 days a week. Further, wastewater flows from a municipality in a rather predictable pattern throughout each day. Therefore treatment plants must be designed and operated differently to handle the two systems.

Industry views wastewater treatment as an imposed necessity which it employs when it is compelled to, especially when wastewater's effect on the receiving watercourse is readily visible or when public approval and acclaim will be gained for the expenditure and effort. A municipality views wastewater treatment as a service to the community to be employed whenever the people are willing or can be convinced by higher authorities to pay the extra taxation required to implement it.

The number of contaminants in industrial waste can run the gamut of a scale from zero to about 100,000 parts per million. Industrial waste varies temporally and, in the normal course of operation, is unpredictable. The strengths and volumes of municipal sewage, on the other hand, are well established and occur within the rather narrow limits of 100 to 1000 parts per million for the contaminants that are generally measured in volumes of 50 to 150 gallons per person per day. Once again treatment systems for each type of contaminant must be designed and operated differently.

Industrial waste deoxygenates at rates which vary from negative values to about five times the rates at which normal domestic sewage deoxygenates. Some wastes have no organic matter and thus no deoxygenation rates or oxygen demands. Domestic sewage deoxygenates at a quite constant rate, seldom varies from a range of 0.07 to 0.20, and is usually discharged independently of the time of day or week.

Water-using industrial plants are generally located in specially zoned areas—that is, newer plants are located outside municipal limits and upstream from the sewage effluent discharges, while older plants are located nearer the city and many times within the city limits but still generally upstream from sewage outfalls. Often industrial plants are not conveniently located near municipal sewers or, if they are, these sewers may be of inadequate capacity to accept the industrial wastes. Municipal sewage-treatment plants are generally located in low-lying areas downstream from the municipality but near its boundaries.

Usually many miles may separate municipal plant and industrial operations.

Industrial plants are generally managed by personnel who have been centrally trained and brought in from some distant location. These managers frequently change location, thus they are seldom inbued with the area patriotism of a locally born and raised citizen. Industrial plant managers often attempt to overcome this barrier by purposely taking an active interest in community affairs. In some cases they are successful, but in many others their participation is primarily superficial and, in any event, is impeded by their temporary status at this plant location. Municipal officials are mostly "hometown" people who have gained the respect and confidence of their people over a great number of years. And further, they must act on the behalf of their people since they wish to remain respected citizens of the community after their official municipal duties are over.

These examples point out to the reader that one must not approach the solution to industrial-waste problems in a manner similar to those of municipal sewage problems. Now that you have seen the differences between industrial and municipal operations you are in a better position to absorb the remaining chapters of this text.

1.4 Environmental Impact Statements of Industrial Wastes

As you have read in the proceeding three sections, all wastes create a change and hence an impact on the specific environment into which it is discharged. Whereas domestic wastes such as sewage and municipal refuse exert a definite and predictable environmental impact, industrial wastes cause a less predictable and often less understood and extreme impact on all three of our environments (land, air, and water). One way to identify the potential impact of industrial wastes is to study the specific wastes and their environments into which they will be discharged in order to reply to many pertinent questions related to these impacts. The environment may even be considered by some, in addition to air, water, and land, as plants, animals, transportation systems, community structure and economic stability. In fact, Jain et al. (18, 1981) specify that "the environment is made up of a combination of our natural and physical surroundings and the relationship of people with the environment, which includes aesthetics, historic, cultural, economical and social aspects."

An environmental impact statement (EIS) is designed to disclose the environmental consequences of a proposed action—in our cases, usually the manufacture of a certain industrial product with its accompanying wastes—to assist the appropriate decision maker of the environmental risks involved.

An environmental impact statement generally is required and advisable whenever and wherever a new industrial product—with associated wastes—is manufactured. In the United States, however, an impact statement is only required legally if and when a preceding environmental impact assessment dictates its need.

The environmental impact statement should reveal any change which will occur either now or in the foreseeable future as a result of the industrial operation. The statement must point out precisely what is involved in the industrial manufacturing process, the nature of the entire environment into which the manufacturing is being imposed, the potential for future changes both in manufacturing and in the environment, and to present all this data in a manner which can easily be evaluated by any and all agencies reviewing the statement. The major United States agency which is required by law (National Environmental Policy Act—NEPA—of 1970) is the Council of Environmental Quality (CEQ). This council has offered a summarized outline for the contents of a typical impact statement which is shown in Figure 1–2. However, each government agency may use a somewhat different checklist for contents of an environmental impact statement. For example, the E.P.A., which reviews construction of sewage treatment plants and industrial waste treatment plants, has their own EIS content (slightly different than recommended by the CEQ shown in Fig. 1–2).

Goodman (19, 1988) presents a feasibility study checklist of six major areas of environmental concern. He points out that particular issues within these general guidelines are determined by the nature of individual projects. The six areas of study which Goodman describes are: (1) technical, (2) economic, (3) administrative/managerial, (4) environmental, (5) social/political, and (6) financial. Our major concern in this text is area number (4), in which Goodman includes 17 subjects, as contrasted to Jain's 49 given in the next pages. Obviously, the specific subjects to be considered as environmental concerns will vary with the reporter sup-

(1) COVER SHEET (1 page)
- Title of the action
- Its location
- EIS designation: Final _____ . Draft _____ . Supplement _____
- Lead agency and cooperating agencies
- Agency point of contact (name, address, and telephone number)
- Date by which comments must be received
- Abstract (one paragraph)

(2) SUMMARY (NTE 15 pages)
- Summarize EIS, use EIS format
- Conclusion
- Areas of controversy
- Issues to be resolved

(3) TABLE OF CONTENTS (suggest NTE 6 pages)
- Cover all headings and subheadings
- List of figures
- List of tables
- List of abbreviations
- List of scientific or Greek symbols

(4) PURPOSE OF AND NEED FOR THE ACTION (Sections 4–7; NTE 150 pages; 300 in special cases)
- Need or requirement
- Purpose or objective

(5) ALTERNATIVES INCLUDING THE PROPOSED ACTION
- Describe each alternative considered (including alternatives under lead agency jurisdiction and no action alternative)
- Alternatives not rigorously explored and reasons

- Environmental consequences of alternatives in a comparative form
- Preferred alternative
- Mitigation measures
- Sections 102 and 102(1); requirements of NEPA

(6) AFFECTED ENVIRONMENT
- Describe affected environment
- Necessary description relevant to impacts
- Summarize, consolidates, or reference information to minimize bulk

(7) ENVIRONMENTAL CONSEQUENCES
- Direct effects—significance
- Indirect effects—significance
- Conflicts with other federal, state, local, or native American tribe plans
- Energy requirements and conservation potential
- Natural or deplorable resource requirements and conservation potential
- Urban quality, historic and cultural resources, and the design of the built environment—reuse/conservation potential
- Mitigation measures not covered under "Alternatives" section

(8) LIST OF PREPARERS (NTE 2 pages)
- Names and qualifications of preparers
- Reference preparers of sections where possible

(9) APPENDIX
- Material prepared in support of EIS
- Analysis to support effects
- Analytic computations relevant to the decision

Fig. 1-2. Outline for CEQ-prescribed EIS content as adapted from Jain et al. (18, 1981).

plying the EIS. Goodman's environmental information needs are more general, while Jain's are rather specific.

The "heart" of an environmental impact statement resides in Section 6 and 7 of the contents of Figure 1–2. As mentioned previously, "the environment" has been and still is defined variously depending upon the subjective opinion of the particular sector of the public involved. Your authors believe that all aspects of our three natural environments—air, land, and water— should define amply "the environment." However,

Jain et al.(18, 1981) also include ecology, sound, human aspects, economics, and resources as categories to be added to the above. In Fig. 1–3 they list 49 "attributes" of the environment to be evaluated in Sections 6 and 7. In essence, the EIS contains subjective narrative discussion of these 49 characteristics of the environment as affected by the proposed industrial manufacturing operation. There appears to be a consensus among writers that the use of matrix analysis provides the most useful method of quantifying in an objective manner

Air
(1) Diffusion factor
(2) Particulates
(3) Sulfur oxides
(4) Hydrocarbons
(5) Nitrogen oxide
(6) Carbon monoxide
(7) Photochemical oxidants
(8) Hazardous toxicants
(9) Odors

Water
(10) Aquifer safe yield
(11) Flow variations
(12) Oil
(13) Radioactivity
(14) Suspended solids
(15) Thermal pollution
(16) Acid and alkali
(17) Biochemical oxygen demand (BOD)
(18) Dissolved oxygen (DO)
(19) Dissolved solids
(20) Nutrients
(21) Toxic compounds
(22) Aquatic life
(23) Fecal coliforms

Land
(24) Soil stability
(25) Natural hazard
(26) Land use patterns

Ecology
(27) Large animals (wild and domestic)
(28) Predatory birds
(29) Small game
(30) Fish, shell fish, and water fowl
(31) Field crops
(32) Threatened species
(33) Natural land vegetation
(34) Aquatic plants

Sound
(35) Physiological effects
(36) Psychological effects
(37) Communication effects
(38) Performance effects
(39) Social behavior effects

Human Aspects
(40) Life styles
(41) Psychological needs
(42) Physiological systems
(43) Community needs

Economics
(44) Regional economic stability
(45) Public sector review
(46) Per capita consumption

Resources
(47) Fuel resources
(48) Nonfuel resources
(49) Aesthetics

Figure 1-3. Environmental attributes (after Jain et al. 18, 1981).

the effects of the 49 characteristics. Luna Leopold (20, 1971) is generally conceded to be the originator of matrix analysis for EIS. He proposed 88 "existing characteristics (rather than 49) and conditions of the environment" and 100 "proposed actions which may cause environmental impact." He placed the former (88) on the vertical and the latter (100) on the horizontal of the matrix and gives each interaction a number from 1 to 10 depending upon the relative magnitude of the impact, 10 representing the greatest magnitude and 1 the least. He also uses a + sign to indicate a beneficial impact, and otherwise the box is left blank or given a − sign. The analyst must assume the responsibility of summing up the numbers and signs to quantify the results. Good-

man (21, 1984) presents a more detailed evaluation of the matrix system for the Salmon Falls Irrigation and Wildlife Enhancement Project in Upper Snake River Basin (Idaho-Wyoming, Jain et al. 18, 1981), and also gives an example matrix analysis for the construction of a 200 unit family housing project.

Your authors devised a EIS matrix analysis, Table 1–3, for the industrial problem project described in Chapter 19 of this text. We use the appropriate vertical components from Jain et al. (18, 1981), and alternatives open to the industrial plants as made available to the authors on the horizontal. Values of 1 to 10 are given for each with + for beneficial impact and − for detrimental impact. A zero (0) value indicates no effect

Table 1-3.   Alternative Actions to be Taken by Plants

		Moench					
					Wastewater Treatment		
	No Air Treatment	Partial Treatment		No Waste Treatment	DISCH to S.T.P.	Treat to 40% BOD Red	85% BOD Red.
		(Scrubbers)	(Backhouse)				
	1	2	3	4	5	6	7
*Air							
Sulfur Oxides	−2	+2	−2	−2	−1	0	+2
Odors	−4	+4	−4	−4	−3	−2	+4
Economics							
Moench Cost	0	−10	−8	0	−8	−5	−6
Glue plant Cost	—	—	—	—	—	—	—
Regional Econ. Stability	0	+2	0	−5	+5	+3	+5
Water							
Suspended Solids	—	—	—	−8	+4	+4	+8
Acid & Alkali	—	—	—	−8	+4	−4	+8
BOD	—	—	—	−10	+5	+2	+10
D.O.	—	—	—	−10	+10	+5	+10
Dissolved Solids	—	—	—	−8	+2	0	+6
Fecal Coliforms	—	—	—	−2	+2	0	+2
Toxic Compounds	—	—	—	−8	+2	0	+8
Land							
Land Use Patterns	−1	+1	0	−2	+2	+1	+2
Ecology							
Fish	—	0	0	−10	+5	+2	+10
Human Aspects							
Physiological Needs	−1	+1	0	−4	+4	+2	+4
Physchological Needs	−2	+2	0	−4	+4	+2	+4
Social Behavior Needs	−1	+1	0	−4	+4	+2	+4
					+53	21	+87
					−12	11	−6
Total	−11	+3	−14	−89	+41	+10	+81

*Selected from Jain's suggestions

or not pertinent to the problem. Addition of the numerical values in the vertical alternative columns gives the reviewer a ready quantitative comparison for decision making.

The reader will note that the authors have listed 28 potential alternative actions on the part of the industrial plants in the project, while limiting the environmental concerns (on the vertical of the matrix) to 17 characteristics. When referring to the totals in Table 1–3, it is apparent that for Moench Tannery the worst action that it could take would be no waste treatment (−89) and,

conversely, the best action would be treatment with 85% BOD reduction (+81). For the Peter Cooper glue plant, the same was true (−104 versus +84). For the tannery the most beneficial overall combined treatment alternative was computed to be a plant at the tannery (+100) using innovative combined beamhouse and tanhouse waste treatment. For the glue plant, the optimum alternative was to close the plant (+94) rather than completely treat its waste (−67). The reader should review Chapter 19 for the details leading to these impact evaluations.

PART 2

| | Moench | | | | | Peter Cooper Glue Plant | | | |
| Solid Waste Treatment | | | Wastewater Treatment | | | | | | |
None 8	Munic Collection 9	Private Collection 10	No Air Treatment 11	Partial Treatment (Scrubbers) 12	Partial Treatment (Backhouse) 13	No waste Treatment 14	DISCH to STP 15	Treat to 40% BOD red. 16	85% BOD red. 17
−2	+1	+1	−8	+8	0	−8	−5	−4	+7
−2	+1	+1	−10	+8	0	−10	−7	+4	+9
0	−2	−2	—	—	—	—	—	—	—
—	—	—	0	−8	−6	0	−10	−5	−10
+1	+1	+1	−5	+5	−3	−5	+5	+3	+8
—	—	—	—	—	—	−8	+5	+2	+7
—	—	—	—	—	—	−7	+5	+2	+6
—	—	—	—	—	—	−10	+5	+2	+8
—	—	—	—	—	—	−10	+6	+2	+8
—	—	—	—	—	—	−8	+2	+2	+6
—	—	—	—	—	—	−3	+2	+1	+2
—	—	—	—	—	—	−6	+1	+4	+4
−2	+2	+2	−6	+4	+1	−4	+2	+2	+4
0	0	0	−1	+1	0	−10	+8	+4	+10
−1	+1	+1	−5	+3	+1	−5	+3	+2	+5
−1	+1	+1	−5	+3	+1	−5	+3	+2	+5
−1	+1	+1	−2	+3	+1	−5	+3	+2	+5
+1	+8	+8		+35	+4		53	+38	+94
−9	−2	−2		−8	−9		−22	−5	−10
−8	+6	+6	−42	+27	−5	−104	+31	+33	+84

(continued)

PART 3

Peter Cooper Glue Plant

For Tannery

	Solid Waste Treatment		Overall Combined Treatment Alternatives			
None 18	Municipal Collection 19	Private Collection 20	A combined treatment plant and system servicing Moench Tannery and Peter Cooper Glue Company, and the Village of Gowanda at the site of the present sewage treatment plant. 21	The same system as No. 1 except that the plant at the site of the Indian Reservation downstream village plant and in a rather desolate, undeveloped creek segment. 22	Plant serving the tannery and glue plants only at the site of the tannery 23	Plant serving the tannery and village wastes only at the existing sewage treatment plant. 24
−10	+8	+8	+8	+10	+8	+2
−10	+9	+9	+8	+10	+8	+2
—	—	—	−8	−9	−7	−7
0	−10	−10	−10	−10	−8	0
−5	+5	+5	+10	+10	+10	+5
—	—	—	+7	+7	+7	+5
—	—	—	+6	+6	+6	+4
—	—	—	+6	+6	+6	+3
—	—	—	+8	+8	+8	+5
—	—	—	+4	+4	+4	+2
—	—	—	+4	+4	+4	+2
—	—	—	+5	+5	+5	+3
−8	+6	+6	+5	+7	+2	+6
0	0	0	+8	+8	+8	+6
−8	+6	+6	+5	+6	+4	+4
−10	+8	+8	+5	+6	+4	+4
−8	+6	+6	+5	+6	+4	+4
	+48		+94	+101	+86	+57
	−10		−18	−19	−15	−7
−58	+38	+38	+76	+82	+71	+50

PART 4

For Tannery		Glue Plant Production Alternatives	
Plant serving only the tannery at the tannery plant site using conventional treatment methods with preliminary separation of beamhouse and tanhouse wastes. 25	Plant serving only the tannery at the tannery plant site using combined treatment of equalized beamhouse and tanhouse wastes. 26	Build waste plant to achieve 85% reduction 27	Close plant rather than treat wastes 28
+6	+8	+4	+8
+6	+10	+5	+10
−10	−5	0	0
0	0	−10	0
+5	+5	+4	−5
+5	+8	+7	+8
+4	+8	+6	+7
+3	+10	+8	+10
+5	+10	+8	+10
+2	+6	+6	+8
+2	+2	+2	+3
+3	+8		
		+4	+6
+4	+4	+3	+4
+6	+10	+8	+10
+3	+5	+4	+5
+3	+5	+4	+5
+3	+5	+4	+5
+60	+105	+77	+99
−10	−5	−10	−5
+50	+100	−67	+94

(*continued*)

References

1. "Drinking water standards," U.S. Public Health Service, *Public Health Rept.* **61**, 371 (1946); (revised) U.S. Public Health Service Publication no. 956, Washington, D.C. (1962).
2. Dugan, P. R., R. M. Pfister, and M. L. Sprague, *Bibliography of Organic Pesticides—Publications Having Relevance to Public Health and Water Pollution Problems*, Prepared for the New York State Department of Health by Microbiology and Biochemical Center, Syracuse University Research Corporation, Syracuse, N.Y. (1963).
3. Ettinger, M. B., "Heavy metals in waste-recovery systems," Paper read at Interdepartmental Water Resources Seminar, March 1963, at Ohio State University, Columbus.
4. Gibbs, C. V., and R. H. Bothel, "Potential of large metropolitan sewers for disposal of industrial wastes," *J. Water Pollution Control Federation* **37**, 1417 (1965).
5. Gorman, A. E., "Waste disposal as related to site selected," Preprint 3, American Institute of Chemical Engineers Meeting, December 12–16, 1955.
6. Huet, M., "Water quality criteria for fish life," Paper read at Third Seminar on Biological Problems in Water Pollution, 13–17 August, 1962, U.S. Public Health Service Publication no. 999-WP-25, Washington, D.C. (1965), p. 160.
7. Jones, E., *Fish and River Pollution*, Butterworths, London (1964).
8. *Modern pH and Chlorine Control*, 19th ed., W. A. Taylor & Co., Baltimore (1966), pp. 57–103.
9. Moss, H. V., "Continuing research related to detergents in water and sewage treatment," *Sewage Ind. Wastes* **29**, 1107 (1967).
10. National Technical Advisory Committee, *Interior Report to the Federal Water Pollution Control Administration on Water Quality Criteria, June 30, 1967*, U.S. Department of the Interior, Washington, D.C. (1967).
11. Nemerow, N. L., *Water Wastes of Industry*, Bulletin no. 5, Facts for Industry Series, Industrial Experiment Program, North Carolina State College, Raleigh (1956).
12. Nemerow, N. L., and T. A. Doby, "Color removal in waste water treatment plants," *Sewage Ind. Wastes* **30**, 1160 (1958).
13. Palange, R. C., G. G. Robeck, and C. Henderson, "Radioactivity as a factor in stream pollution," Preprint 190, American Institute of Chemical Engineers Meeting, December 12–16, 1955.
14. "Survey of the Ohio river," in *Industrial Wastes Guides*, Supplement D, U.S. Public Health Service, Washington, D.C. (1943).
15. Tarzwell, C. M., "Water quality criteria for aquatic life," in *Biological Problems in Water Pollution*, U.S. Department of Health, Education and Welfare, Cincinnati (1957), pp. 246–272.
16. Tarzwell, C. M., "Dissolved oxygen requirement for fishes," in *Oxygen Relationships in Streams*, Report W58-2, U.S. Public Health Service, Cincinnati (1958), pp. 15–24.
17. Wurtz, C. B., "Misunderstandings about heated discharges," *Ind. Water Eng.* **4**, 28 (1967).
18. Walter, I., "International Economics of Pollution," Halstead Press Book, Chapter 1, John Wiley Co., New York City (1975).
19. "Radiation—A Fact of Life," International Atomic Energy, A-1400, Vienna, Austria (Sept. 1979).

Questions

1. When is a receiving stream considered to be "polluted" by industrial waste?
2. List and describe at least 11 contaminants originating from industrial waste.
3. What effects do each of the above contaminants have on a receiving stream; on a municipal sewage-treatment plant?
4. What are the major differences in characteristics between well-known municipal sewage effluents and industrial waste?
5. What are the essentials of the environmental impact statement? What purpose is it intended?

CHAPTER 2

STREAM PROTECTION MEASURES

Streams serve people in many ways, and the carrying away of pollution certainly is one of the chief services performed. However, there are other more important uses of stream waters: drinking, bathing, fishing, irrigation, navigation, recreation, and power. A stream must therefore be protected, so that it can serve the best interests of the people using it.

Since the competition for water uses has intensified dramatically during the '80s, in many instances we must satisfy several of these uses simultaneously. Proportioning the demand for varied uses of valuable water represents one of the greatest environmental challenges of the next decade.

2.1 Standards of Stream Quality

The methods of maintaining a stream in acceptable condition range from very flexible control, in which individuals make decisions about waste treatment, to rigid control by laws specifying stream or effluent standards. Because of the variations in procedure under the flexible control system, the method cannot be described in any detail. Normally, the state regulatory agency demands certain types and degrees of waste treatment based on the best interests of the persons living along the stream. An advantage of such a loose procedure is the freedom to make immediate decisions; a danger is the lack of representation of all the groups involved in the use of the stream. In addition, treatment may be different from one location to another, with resultant unfairness. However, nonuniformity can also be an advantage; for example, a plant located on a stream above the water-supply intake of a municipality would be expected to provide more complete treatment than a similar plant discharging waste into the same stream, but below the water-supply intake.

There are two schools of thought in the United States in regard to rigid protection: one group prefers "effluent standards" and the other "stream standards." The first system requires that, in all effluents from a certain type of industry, the waste discharged be kept below either a fixed percentage or a certain maximum concentration of polluting matter. A disadvantage to this approach is that there is normally no control over the total volume of polluting substance added to the stream each day. The large industry, although providing the same degree of waste treatment as the small one, may actually be responsible for a major portion of the pollution in the stream. It might be argued, however, that larger industries, by virtue of their value to the area, should be allocated a larger portion of the assimilating capacity of a stream.

The effluent-standard system is easier to control than the stream-standard system. No detailed stream analyses are needed to determine the exact amount of waste treatment required, and effluent standards can serve as a guide to a state in stream classification or during the organization of any pollution-abatement program. On the other hand, unless the effluent standards are upgraded, this system does not provide any effective protection for an overloaded stream. Standards for effluents are based more on economics and practicability of treatment than on absolute protection of the stream; the best usage of the stream is not the primary consideration. Rather the usage of the stream will depend on its condition after industrial-effluent standards have been satisfied. Upgrading and conservation of natural resources are somewhat neglected in favor of industrial economics.

The stream-standard system is based on establishing classifications or standards of quality for a stream and regulating any discharge into it to the extent necessary to maintain the established stream classification or quality. The primary motive of stream standards is to protect and preserve each stream for its best usage, on an equitable basis for both upstream and downstream users, although the upstream user often possesses a decided advantage because of location. A long and involved process of stream classification usually precedes any decision about waste treatment. Streams are classified in a manner set forth by state laws, sampled and analyzed for existing pollution, and surveyed for present and potential usages. The regulatory agency, after holding a public hearing to listen to comments from interested parties concerning the best usage, then decides on the highest usage of the stream, or section of stream. Naturally, the higher the classification, the cleaner the stream must be, with a resulting greater degree of waste treatment required. Formal notification is served on each polluter, giving a time limit for positive remedial action to maintain the stream in its classification. Implementation of the law becomes a matter of education, persuasion, public pressure, and sometimes action from the attorney general's office. It is left to each industry to decide the type and extent of treatment it must give to its waste to meet stream standards, although industries generally communicate directly with pollution-control authorities to determine what will be acceptable to them, since the pollution-control agency must review and approve the final construction plans for waste-treatment plants.

The main advantage of the stream-standard system is the prevention of excessive pollution, regardless of the type of industry or other factors such as the location of industries and municipalities. It also allows the public to establish goals for present and future water quality [11]. Loading is limited to what the stream can assimilate, and this may impose hardship on an industrial plant located at a critical spot along the stream. On the other hand, pollution abatement should be considered in decisions concerning the location of a plant, just as carefully as labor, transportation, market, and other conditions.

The difficulties in carrying to completion a classification system based on stream standards are:

1) Confusion when zones of different classifications are straddled by the waste: Does Waste 1 need to be treated to meet Class C or B stream standards?

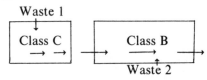

2) Controversy over the proportion of the stream to be reserved for future industrial, municipal, agricultural, and other uses.

3) Red tape and opposition from industry or the public to changes in the established classifications (upgrading or downgrading of a zone).

4) The need for a complex and thorough stream survey prior to classification. This can be costly, cumbersome, and result in delays.

On the brighter side is the security felt by industry, municipalities, and engineers, who know exactly the stream conditions for which they must design their plants. Industry will usually find a method of solving its problem *if* the extent of the situation and the precise degree of treatment required are known, although economics may still influence the action of the industry. Some states such as North Carolina have built in a degree of flexibility that would be unobtainable with effluent standards [5].

New York State has put a stream-standard method into practice [8] with the following classifications:

Fresh Water

Class AA—*Drinking water* after approved disinfection, with additional treatment, if necessary, to remove naturally present impurities.

Class A—*Drinking water* if subjected to approved treatment equal to a minimum of coagulation, sedimentation, filtration, and disinfection, plus additional treatment for natural impurities, if necessary.

Class B—*Bathing*, or any lesser use.

Class C—*Fishing*, or any lesser use.

Class D—*Agricultural, industrial cooling*, or *industrial process water*, or any lesser use.

Tidal Salt Waters

Class SA—*Shellfishing* for market purposes, and any other usages.

Class SB—*Bathing*, and any other usages except shellfishing for market purposes.

Class SC—*Fishing* and other usages, except bathing or shellfishing for market purposes.

Class SD—*Any usages* except fishing, bathing, and shellfishing.

Underground Waters

Class GA—*Drinking, culinary,* or *food processing*, and any other usages.

Class GB—*Industrial* or other water supply, and any other usages, except as used in GA.

Water-quality surveillance programs making use of monitoring networks have been used to measure such contaminants and ingredients of the air and water as dissolved oxygen, chlorides, pH, conductivity, oxidation-reduction potential, temperature, and solar radiation. In an increasing number of cases (such as ORSANCO) these programs have resulted in an effective improvement in the receiving water quality.

Once a stream has been classified, the regulatory agency attempts to maintain that classification, or better. Under this system, the burden of proof rests on the engineer designing the waste-treatment works. He must show that the plant's waste will not alter the classification of the stream into which the plant discharges. The regulatory agency is responsible to the public for reporting on the existing quality of streams and pointing out sources of significant pollution.

Cleary warns that "to strive for stream cleanliness is one thing but to insist on striving to make all waters as pure as holy water introduces further delays with what is now the obvious problem—the gross and visible pollution" [2].

2.2 Stream Quality Control

The Water Quality Act of 1965 has amended the 1948 Federal Water Pollution Control Act, to provide for establishment of water-quality standards for interstate waters. Most states have decided to establish their own standards which the federal government is expected to accept in turn. The act specifically states that standards shall be such as to protect the public health or welfare and enhance the quality of water. In establishing these standards consideration

will be given to the use and value of streams for public water supplies, propagation of fish and wildlife, recreational purposes, agricultural, industrial, and other legitimate uses. The act specifically states that "the discharge of matter into such interstate waters or portions thereof, which reduces the quality of such waters below the water quality standards established under this subsection (whether the matter causing or contributing to such reduction is discharged directly into such waters or reaches such waters after discharge into tributaries of such waters), is subject to abatement in accordance with provisions of . . ." the act. It also states that "Economic, health, esthetic, and conservation values which contribute to the social and economic welfare of an area must be taken into account in determining the most appropriate use or uses of a stream."

The federal government has established temporary quality guidelines for water used for (1) recreation, (2) public water supplies, (3) fish and wildlife, (4) agriculture, and (5) industry. A summary of the government's limits is presented in Tables 2.1 and 2.2.

Illustrations of two state quality standards are given by Grossman [4] for New York State in Table 2.3 and Rambow and Sylvester [11] for Washington State in Table 2.4. Stream water quality standards are also presented for the New England Interstate Water Pollution Control Commission in Table 2.5 and for the Ohio River Sanitation Commission in Table 2.6. (These four tables appear at the end of this section.) The reader can study these governmental agency standards and notice the tendency for more stringent controls from both federal and state governments.

The state of Pennsylvania, practicing "effluent standards," established (as an example) "raw waste" standards for pulp and paper mills (Table 2.7). These standards give the 5-day BOD and suspended solids per ton of product expected in wastes from well-run plants employing good housekeeping and recovery methods. The Sanitary Water Board has then applied the required effluent reduction for mills located on streams classified for primary or secondary treatment against the raw waste standard.

Four other procedures are prevalent for improving stream water quality: (1) stream specialization, (2) stream aeration, (3) low-flow augmentation, and (4) pumped storage. These are described briefly below.

Table 2.1 Water standards for various uses. National Technical Advisory Committee to F.W.P.C.A. on Water Quality Criteria, U.S. Dept. Interior, June 30, 1967, Wash. D.C.

Water quality	Recreation and aesthetic	Public water supply		Fish and aquatic wildlife			Agriculture		
		Permissible	Desirable	Fresh water organisms	Wild life	Marine and estuarine organisms	Farm water supplies	Livestock	Irrigation
Color, units		75	< 10	10% of light penetrating to bottom	10% of light penetrating 6 ft.				
Temperature, °F	< 85°	< 85°	< 85°	83–96° for 6 hr.					55–85°
Fecal coliform, no/100 ml	2000–200	2000	20						4000
Alkalinity (CACO₃), ppm		30–500	30–500	> 20	35–200	35–200			
Chloride, ppm		250	25						
Hexavalent chromium, ppm		0.05	Absent				0.05	0.05	5–20
Copper, ppm		1.0	Absent				1.0		0.2–5.0
Dissolved oxygen, ppm		> 3.0	Near to saturation	> 4.0	Bottom aerobic	> 4.0			
Hardness (CACO₃), ppm		300–500	60–120						
Iron, ppm		0.3	Virtually absent				0.3		
Manganese, ppm		0.05	Absent				0.05		2.0–20
Nitrates, ppm		10.0(N) Ind. NO₂	Virtually absent				45.0		
pH	5.0–9.0	6.0–8.5		6–9	7.0–9.2	6.5–8.5	6.0–8.5		4.5–9.0
Sulfate, ppm		250	50						
Total dissolved solids, ppm		500	200				500–5000	10,000	0–5000
Carbon chloroform extractable, ppm		0.15	0.04				0.0001–0.02		
Pesticide, ppm		0.001–0.1	Absent	Varies with organism	Varies with organism	Varies with organism			
Phenol, ppm		0.001	Absent						
Gross beta radioactivity, μμc/liter		1000	100	1000	1000	1000	1000	1000	1000
Cyanide, ppm		0.20	Absent				0.20		
Turbidity, ppm			Virtually absent	10–50					

Table 2.2 Water quality standards for industrial use.

Industry and process	Color, ppm	Alkalinity, ppm (CaCO₃)	Chloride, ppm	Hardness, ppm (CaCO₃)	Iron, ppm	Manganese, ppm
Textiles						
Size	5			25	0.3	0.05
Scouring	5			25	0.1	0.01
Bleaching	5			25	0.1	0.01
Dyeing	5			25	0.1	0.01
Paper						
Mechanical	30		1000		0.3	0.1
Chemical						
Unbleached	30		200	100	1.0	0.5
Bleached	10		200	100	0.1	0.05
Chemicals						
Alkali and chlorine	10	80		140	0.1	0.1
Coal tar	5	50	30	180	0.1	0.1
Organic	5	125	25	170	0.1	0.1
Inorganic	5	70	30	250	0.1	0.1
Plastic and resin	2	1.0	0	0	0.005	0.005
Synthetic rubber	2	2	0	0	0.005	0.005
Pharmaceutical	2	2	0	0	0.005	0.005
Soaps and detergents	5	50	40	130	0.1	0.1
Paints	5	100	30	150	0.1	0.1
Gum and wood	20	200	500	900	0.3	0.2
Fertilizer	10	175	50	250	0.2	0.2
Explosives	8	100	30	150	0.1	0.1
Petroleum			300	350	1.0	
Iron and steel						
Hot-rolled						
Cold-rolled						
Miscellaneous						
Fruit and vegetable canning	5.0	250	250	250	0.2	0.2
Soft drinks	10	85			0.3	0.05
Leather tanning	5		250	150	50	
Cement		400	250		25	0.5

(continued)

Stream specialization. This lets one stream become degraded so that others in the area are preserved in a relatively pristine state. The system is similar to that of stream classification except that it encourages the use of one stream as "an open sewer." Stream classification, on the other hand, is an attempt to label and use each stream for its best purpose and tends to upgrade all streams uniformly. Because social benefits are difficult to identify, all attempts to use streams in a discriminatory manner have been disappointing—most streams are found in a "great gray area."

Stream aeration. A novel approach to waste disposal for organic, decomposable-type wastes is to increase the amount of oxygen in the receiving streams by artificial means. Attempts are made to use vertical pumps (similar to those used in deep ponds) to re-

Table 2.2 (*continued*)

NO$_3$, ppm	pH	SO$_4$, ppm	Total dissolved solids, ppm	Suspended solids, ppm	SiO$_2$, ppm	Ca, ppm	Mg, ppm	HCO$_3$, ppm
	6.5–10		100	5.0				
	3.0–10.5		100	5.0				
	2.0–10.5		100	5.0				
	3.5–10		100	5.0				
	6–10							
	6–10			10	50	20	12	
	6–10			10	50	20	12	
	6–8.5			10		40	8	100
	6.5–8.3	200	400	5		50	14	60
	6.5–8.7	75	250	5		50	12	128
	6.5–7.5	90	425	5		60	25	210
0	7.5–8.5	0	1.0	2.0	0.02	0	0	0.1
0	7.5–8.5	0	2.0	2.0	0.05	0	0	0.5
0	7.5–8.5	0	2.0	2.0	0.02	0	0	0.5
		150	300	10.0		30	12	60
	6.5	125	270	10		37	15	125
5	6.5–8.0	100	1000	30	50	100	50	250
5	6.5–8.0	150	300	10	25	40	20	210
2	6.8	150	200	5	20	20	10	120
	6.0–9.0		1000	10		75	30	
	5–9							
	5–9			10				
10	6.5–8.5	250	500	10	50	100		
	6.0–8.0	250				60		
	6.5–8.5	250	600	500	35			

plenish the oxygen in the wastes. Some experiments have been carried out using pure oxygen rather than air, which appears to be rather costly at present. Although it apparently is less expensive to aerate the wastes than the streams which receive the wastes, a major exception occurs where a power dam is located across the stream. In this situation nearby industry plants can, and have, used the draft tubes of large turbines to draw oxygen into the water flowing through them by gravity.

Low-flow stream augmentation. Storage of water is generally considered beneficial for handling large amounts of pollution. Dams and reservoirs built on the upland portion of the main stream or on one of

Table 2.3 New York State classes and standards for fresh surface waters.

Class and best use*	Water standards†				
	Minimum dissolved oxygen, ml/liter	Coliform bacteria median, no/100 ml	pH	Toxic wastes, deleterious substances, colored wastes, heated liquids, and taste- and odor-producing substances‡	Floating solids, settleable solids, oil, and sludge deposits
AA—Source of un-filtered public water supply and any other usage	5.0 (trout) 4.0 (nontrout)	Not to exceed 50	6.5–8.5	None in sufficient amounts or at such temperatures as to be injurious to fish life or make the waters unsafe or unsuitable	None attributable to sewage, industrial wastes or other wastes
A—Source of filtered public water supply and any other usage	5.0 (trout) 4.0 (nontrout)	Not to exceed 5000	6.5–8.5		None which are readily visible and attributable to sewage, industrial wastes or other wastes
B—Bathing and any other usages except as a source of public water supply	5.0 (trout) 4.0 (nontrout)	Not to exceed 2400	6.5–8.5		
C—Fishing and any other usages except public water supply and bathing	5.0 (trout) 4.0 (nontrout)	Not applicable	6.5–8.5	None in sufficient amounts or at such temperatures as to be injurious to fish life or impair the waters for any other best usage	
D—Natural drainage, agriculture, and industrial water supply	3.0	Not applicable	6.0–9.5	None in sufficient amounts or at such temperatures as to prevent fish survival or impair the waters for agricultural purposes or any other best usage	

*Class B and C waters and marine waters shall be substantially free of pollutants that: unduly affect the composition of bottom fauna; unduly affect the physical or chemical nature of the bottom; interfere with the propagation of fish. Class D (marine) will be assigned only where a higher water use class cannot be attained after all appropriate waste-treatment methods are utilized. Any water falling below the standards of quality for a given class shall be considered unsatisfactory for the uses indicated for that class. Waters falling below the standards of quality for Class D, or SD (marine), shall be Class E, or SE (marine), respectively and considered to be in a nuisance condition.

†These Standards do not apply to conditions brought about by natural causes. Waste effluents discharging into public water supply and recreation waters must be effectively disinfected. All sewage-treatment plant effluents shall receive disinfection before discharge to a watercourse and/or coastal and marine waters. The degree of treatment and disinfection shall be as required by the state pollution control agency. The minimum average daily flow for seven consecutive days that can be expected to occur once in ten years shall be the minimum flow to which the standards apply.

‡Phenolic compounds cannot exceed 0.005 mg/liter; no odor-producing substances that cause the threshold-odor number to exceed 8 are permitted; radioactivity limits are to be approved by the appropriate state agency, with consideration of possible adverse effects in downstream waters from discharge of radioactive wastes, and limits in a particular watershed are to be resolved when necessary after consultation between states involved.

the tributaries of the main stream can store water during high-flow periods to be released on a programmed schedule when the stream flow diminishes below critical values. However, this can be somewhat dangerous to the water quality, since low-oxygen water may be released during low stream flow, scouring of attached growth may occur during any sudden increase in volume, tributary flow may be temporarily retarded (leading to quality deterioration) during the flow increase in the main stream, and the temperature of the stream may rise if the released water is also used as cooling water in power plants.

Pumped storage is a promising approach to drought control [14], which also has great potential for protecting streams from excessive contamination during critical low flows. The concept was originally developed as a means of storing electrical energy for subsequent re-use, thereby levelling demand. As Velz [14] pointed out, there are many river basins where conventional or even advanced treatment (tertiary) is not now, or will not be, adequate to meet desirable water-quality standards. Pollution control with the use of pumped storage for low-flow augmentation may provide an economical method of ensuring water quality, as well as having other benefits. Velz discussed four major advantages of pumped over conventional storage:

1) It minimizes the problem of locating reservoir sites on the main stream channel. Water may be pumped from a small on-channel pool to the pumped storage reservoir, located at a higher elevation and closer to the demand for augmentation. Simpler dams and elimination of costly flood spillways are obvious advantages.

2) The quality of stored water is improved, since it is pumped during maximum stream flow and stored away from the main stream.

3) There is a higher degree of flexibility and immediate response to deterioration of stream water.

4) Flow can be augmented incrementally along the course of a stream at or near points of greatest need, with high-quality water available at all times to augment the stream flow.

Pumped storage is practised by power plants to provide increased hydropower energy during peak periods. When these peak demand periods also correspond to increased contaminants downstream, the extra dilution provided by the powerplant discharge enchances water quality preservation. Pacific Gas and Electric's Helms pumped storage project (1984) represents a major example of this (Figure 2–1).

Proposed Maximum Contaminant Levels in Drinking Water

As directed by the "Safe Drinking Water Act" of December 1974, the EPA published its proposed interim standards for maximum levels of trace metals in public water systems.

The proposed standards are based on anticipated health effects of a lifetime exposure and an assumed consumption of two liters of water per day, and may well be revised to incorporate findings of new studies mandated under the Act.

Contaminant	Level mg/l
Arsenic	0.05
Barium	1.0
Cadmium	0.010
Chromium	0.05
Lead	0.05
Selenium	0.01
Silver	0.05
Mercury	0.002

2.3 Effluent Guidelines

On October 18, 1972 comprehensive federal water-pollution legislation was enacted under Public Law 92–500. Section 402 (the National Pollutant Discharge Elimination System) and Section 306 (the National Standards of Performance) are of major importance to the subject of industrial waste. Section 101(a) sets the following as its goals:

1) to eliminate the discharge of pollutants by 1985;
2) wherever possible, to have water quality suitable for sustaining fish, shellfish, and wildlife and for recreational purposes by July 1, 1983; and
3) to prohibit the discharge of toxic pollutants.

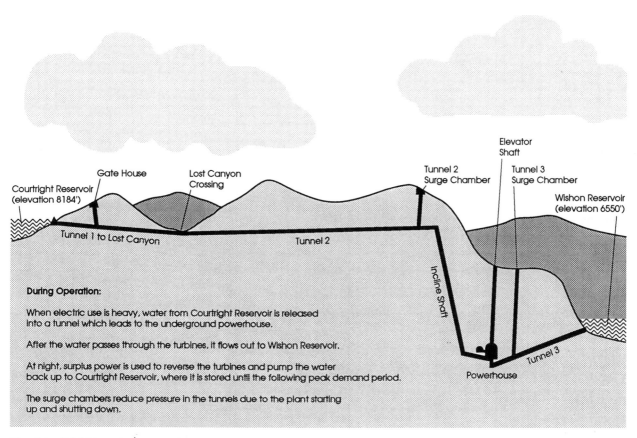

Fig. 2.1 PG&E's Helm Pumped Storage Project.

Other sections of Public Law 92-500 deal with permits. Here the law directs the EPA Administrator to establish guidelines within which individual states must operate their permit programs. Each state program must be approved by the EPA and is subject to takeover by the EPA if a state fails to fulfill its responsibility. The "best practicable control technology" was required for industrial use by July 1977. By July 1, 1983 the "best control technology economically available" was necessary. A no-discharge requirement can be imposed if it is both technologically and economically achievable.

To enforce the July 1977 regulation, the EPA would take into account the total impact of the action taken by industry within a given category (e.g., steel, chemical, paper, etc.) the overall financial ability of industry to comply, and the national impact of such compliance on communities and workers. To enforce the 1983 regulations, the EPA considered whether such application would be economically achievable by the category or class of industries affected and would result in reasonable progress toward the national goal of eliminating all water pollution. The law also requires that a study be

Table 2.4 Summary of surface water quality limit proposals for the state of Washington.* (After Rambow and Sylvester [11].)

Characteristic	Fresh water		Salt water	
	Goal	Standard	Goal	Standard
Alkalinity (phenolphthalein and total)[1]				
Ammonia nitrogen	0.3	0.5	0.0025	0.003
Arsenic	0.003	0.005	0.003	0.004
Bacteria[2]				
Barium	0.01	0.05	0.05	0.06
Bicarbonate[3]				
BOD	1.0	2.0	1.0	2.0
Boron	0.1	0.3	4.7	5.5
Bottom deposits from waste-water discharge	None	None	None	None
Cadmium	0.0005	0.001	0.00011	0.00013
Calcium[4]				
Carbonate[5]				
CCE (carbon chloroform extract)	0.00	0.10	0.05	0.10
Chloride	10	20	Natural	120% of natural
Chromium	Trace	0.01	0.00005	0.00006
COD[6]				
Coliforms (domestic sewage origin)	50/100 ml	240/100 ml	50/100 ml	240/100 ml
Color	5 units	5 units over natural	None	5 units
Conductivity	110% of natural	125% of natural	Natural	120% of natural
Copper	0.05	0.02 above background	0.05	0.06
Cyanide	0.005	0.01	None	0.01
Dissolved oxygen	95% saturation	85% saturation	95% saturation	85% saturation
Fecal streptococci[7]				
Floating solids	None	None	None	None
Fluoride	0.5	1.0	1.3	1.5
Hardness (as CaCO₃)	20 to 75	20 to 125	—	—
Hydroxide	None	None	None	None
Iron	0.0 above natural	0.1 above natural	0.01 above natural	0.2
Lead	Limit of detection	0.02	Limit of detection	0.004
Magnesium[8]				
Manganese	Trace	0.01	0.002	0.04
Nitrate	0.1 above natural	1.0 above natural	0.5	0.6
Nitrogen (total)	0.4 above natural	1.0 above natural	0.5	0.6
Threshold-odor number	1.0	3	1.0	3
Oil and tars	None	None	None	None
Pesticides[9]				
pH	7.0–8.0	6.5–8.5	7.5–8.4	7.5–8.4

Table 2.4 (*continued*)

Characteristic	Fresh water		Salt water	
	Goal	Standard	Goal	Standard
Phenol	Limit of detectability	0.0005	0.04	0.05
Phosphate (total)	0.03	0.15	0.3	0.4
Potassium	2.5	5.0	380	450
Radioactivity	None	USPHS DWS	None	USPHS DWS
Selenium	Limit of detectability	0.002	0.004	0.005
Silica[10]				
Silver	Limit of detectability	0.003	0.0003	0.0004
Sodium	10 over natural	35 over natural	10,500	12,500
Spent sulfite liquor[11]				
Sulfate	15	30	2700	3200
Surfactants	Trace (LAS)	0.10 (LAS)	Trace (LAS)	0.10 (LAS)
Temperature	Natural + 1°C	Natural + 2°C	Natural + 1°C	Natural + 2°C
Total dissolved solids[12]				
Toxicants, miscellaneous	None detectable	None detectable	None detectable	None detectable
Turbidity	5 units	Natural	3 units	5 units
Viruses[13]				
Zinc	Limit of detectability	Limit of detectability	0.01	0.012

*All values in mg/liter unless otherwise specified.

[1] No specific limits. A waste discharge is not to increase the natural total alkalinity by more than 10 per cent or to impart phenolphthalein (CO_3 and OH^-) alkalinity to a receiving water.

[2] No limit specified: refer to coliform organisms.

[3] No limit specified: relates to conductivity and pH.

[4] No limit specified: see hardness.

[5] Although carbonate itself in moderate concentrations is not particularly detrimental, it is associated with high pH values (greater than 8.3). Any carbonate discharge is not to be detectable below the point of discharge.

[6] Since the chemical oxygen demand is related to the BOD, DO, and CCE, it has no limit specified herein.

[7] Fecal bacteria are represented by the coliform group standard.

[8] Controlled by hardness content: no limit specified.

[9] Insufficient data.

[10] No standard proposed. Turbidity will include colloidal silica.

[11] Effect covered by other parameters.

[12] No standard proposed. Conductivity standards are related.

[13] None proposed at present.

Table 2.5 New England Interstate Water Pollution Control Commission: classification and standards of quality for interstate waters (as revised and adopted April 18, 1967).

Water-use class and description*	Standards of water quality		
	Dissolved oxygen	Sludge deposits, solid refuse, floating solids, oils, grease, and scum	Color and turbidity
A—Suitable for water supply and all other water uses; character uniformly excellent†	75% saturation, 16 hr/day; 5 mg/liter at any time	None allowable	None other than of natural origin
B—Suitable for bathing, other recreational purposes, agricultural uses; industrial processes and cooling; excellent fish and wildlife habitat; good aesthetic value; acceptable for public water supply with appropriate treatment	75% saturation, 16 hr/day; 5 mg/liter at any time	None allowable	None in such concentrations that would impair any usages specifically assigned to each class
C—Suitable for fish and wildlife habitat; recreational boating; industrial processes and cooling; under some conditions acceptable for public water supply with appropriate treatment; good aesthetic value	5 mg/liter, 16 hr/day; not less than 3 mg/liter at any time. (For cold-water fishery not less than 5 mg/liter at any time)	None§	
D—Suitable for navigation, power, certain industrial processes and cooling, and migration of fish; good aesthetic value	Minimum of 2 mg/liter at any time.	None§	

*Waters shall be free from chemical constituents in concentrations or combinations which would be harmful to human, animal, or aquatic life for the appropriate, most sensitive, and governing water-class use. In areas where fisheries are the governing considerations and approved limits have not been established, bioassays shall be performed as required by the appropriate agencies. For public drinking-water supplies the limits prescribed by the U.S. Public Health Service may be used where not superseded by more stringent signatory state requirements.
†Class A waters reserved for water supply may be subject to restricted use by state and local regulation.

Table 2.5 (*continued*)

	Standards of water quality		
Coliform bacteria, no./100 ml	Taste and odor	pH	Allowable temperature increase
Not to exceed a median of 100/100 ml nor more than 500 in more than 10% of samples collected	None other than of natural origin	As naturally occurs	None other than of natural origin
Not to exceed a median of 1000/ml nor more than 2400 in more than 20% of samples collected	None in such concentrations that would impair any usages specifically assigned to each class or cause taste and odor in edible fish	6.5–8.0	Only such increases that will not impair any usages specifically assigned to each class‡
None in such concentrations that would impair any usages specifically assigned to each class		6.0–8.5	
	None in such concentrations that would impair any usages specifically assigned to this class	6.0–9.0	None except where the increase will not exceed the recommended limits on the most sensitive water use and will in no case exceed 90° F

‡The temperature increase shall not raise the temperature of the receiving waters above 68°F for waters supporting cold-water fisheries and 83°F for waters supporting a warm-water fishery. In no case shall the temperature of the receiving water be raised more than 4°F.
§Sludge deposits, floating solids, oils, grease, and scum shall not be allowed except for such small amounts that may result from the discharge of appropriately treated sewage or industrial waste effluents.

Table 2.6   ORSANCO quality criteria.

ORSANCO resolution no. 16–66 (adopted May 12, 1966; amended September 8, 1966, and May 11, 1967)

Whereas: The assessment of scientific knowledge and judgments on water-quality criteria has been a continuing effort over the years by the Commission in consultation with its advisory committees; and

Whereas: The Commission now finds it appropriate to consolidate viewpoints and recommendations relating to such criteria;

Now, therefore, be it resolved: That the Ohio River Valley Water Sanitation Commission hereby adopts the following statement and specifications:

Criteria of quality are intended as guides for appraising the suitability of interstate surface waters in the Ohio Valley for various uses, and to aid decision-making in the establishment of waste-control measures for specific streams or portions thereof. Therefore, the criteria are not to be regarded as standards that are universally applicable to all streams. What is applicable to all streams at all places and at all times are certain minimum conditions, which will form part of every ORSANCO standard.

Standards for waters in the Ohio River Valley Water Sanitation District will be promulgated following investigation, due notice and hearing. Such standards will reflect an assessment of the public interest and equities in the use of the waters, as well as consideration of the practicability and physical and economic feasibility of their attainment.

The ORSANCO criteria embrace water-quality characteristics of fundamental significance, and which are routinely monitored and can be referenced to data that are generally available. The characteristics thus chosen may be regarded as primary indicators of water-quality, with the understanding that additional criteria may be added as circumstances dictate. Unless otherwise specified, the term average as used herein means an arithmetical average.

Minimum Conditions Applicable to All Waters at All Places and at All Times

1. Free from substances attributable to municipal, industrial or other discharges or agricultural practices that will settle to form putrescent or otherwise objectionable sludge deposits
2. Free from floating debris, oil, scum and other floating materials attributable to municipal, industrial or other discharges or agricultural practices in amounts sufficient to be unsightly or deleterious
3. Free from materials attributable to municipal, industrial or other discharges or agricultural practices producing color, odor or other conditions in such degree as to create a nuisance
4. Free from substances attributable to municipal, industrial or other discharges or agricultural practices in concentrations or combinations which are toxic or harmful to human, animal, plant or aquatic life

Stream-Quality Criteria

For public water supply and food-processing industry

The following criteria are for evaluation of stream quality at the point at which water is withdrawn for treatment and distribution as a potable supply:

1. *Bacteria:* Coliform group not to exceed 5000/100 ml as a monthly average value (either MPN or MF count); nor exceed this number in more than 20 per cent of the samples examined during any month; nor exceed 20,000 per 100 ml in more than five per cent of such samples
2. *Threshold-odor number:* Not to exceed 24 (at 60°C) as a daily average
3. *Dissolved solids:* Not to exceed 500 mg/l as a monthly average value, nor exceed 750 mg/l at any time. (For Ohio River water, values of specific conductance of 800 and 1200 micromhos/cm (at 25°C) may be considered equivalent to dissolved-solids concentrations of 500 and 750 mg/l)
4. *Radioactive substances:* Gross beta activity not to exceed 1000 picocuries per liter (pCi/l), nor shall activity from dissolved strontium 90 exceed 10 pCi/l, nor shall activity from dissolved alpha emitters exceed 3 pCi/l
5. *Chemical constituents:* Not to exceed the following specified concentrations at any time:

Table 2.6 (*continued*)

Constituent	Concentration (mg/l)
Arsenic	•0.05
Barium	1.0
Cadmium	0.01
Chromium (hexavalent)	0.05
Cyanide	0.025
Fluoride	1.0
Lead	0.05
Selenium	0.01
Silver	0.05

For industrial water supply

The following criteria are applicable to stream water at the point at which the water is withdrawn for use (either with or without treatment) for industrial cooling and processing:

1. *Dissolved oxygen:* Not less than 2.0 mg/l as a daily-average value, nor less than 1.0 mg/l at any time
2. *pH:* Not less than 5.0 or greater than 9.0 at any time
3. *Temperature:* Not to exceed 95°F at any time
4. *Dissolved solids:* Not to exceed 750 mg/l as a monthly average value, nor exceed 1000 mg/l at any time. (For Ohio River water, values of specific conductance of 1200 and 1600 micromhos/cm (at 25°C) may be considered equivalent to dissolved-solids concentrations of 750 and 1000 mg/l)

For aquatic life

The following criteria are for evaluation of conditions for the maintenance of a well-balanced, warm-water fish population. They are applicable at any point in the stream except for areas immediately adjacent to outfalls. In such areas cognizance will be given to opportunities for the admixture of waste effluents with river water.

1. *Dissolved oxygen:* Not less than 5.0 mg/l during at least 16 hours of any 24-hour period, or less than 3.0 mg/l at any time
2. *pH:* No values below 5.0 or above 9.0; daily average (or median) values preferably between 6.5 and 8.5
3. *Temperature:* Not to exceed 93°F at any time during the months of May through November and not to exceed 73°F at any time during the months of December through April
4. *Toxic substances:* Not to exceed one-tenth of the 48-hour median tolerance limit, except that other limiting concentrations may be used in specific cases when justified on the basis of available evidence and approved by the appropriate regulatory agency

For recreation

The following criterion is for evaluation of conditions at any point in waters designated to be used for recreational purposes, including such water-contact activities as swimming and water skiing:

Bacteria: Coliform group not to exceed 1000/100 ml as a monthly average value (either MPN or MF count); nor to exceed this number in more than 20 per cent of the samples examined during any month or exceed 2400/100 ml (MPN or MF count) on any day

For agricultural use and stock watering

Criteria are the same as those shown for minimum conditions applicable to all waters at all places and at all times

Table 2.7 Pennsylvania raw-waste standards for pulp and paper mills.

Type of product or process	Population equivalent, per ton of product, based on 5-day BOD		Pounds of suspended solids per ton of product	
	3-day average*	8-hr average	3-day average	8-hr average
Group A				
Tissue paper	75	80	40	50
Glassine paper	25	30	15	20
Parchment paper	40	45	20	30
Miscellaneous papers	25	30	5	10
Flax papers—condenser	375	415	300	350
Group B (speciality group)				
Fiber paper	800	850	200	235
Asbestos paper	125	185	290	350
Felt paper	210	230	60	65
Insulating paper	2250	2500	325	350
Speciality papers	1000	1200	135	160
Group C (coarse paper)	90	120	35	50
Group D (integrated mills)				
Wood preparation	80	100	40	50
Pulp (sulfite)	3000	3500	35	40
Pulp (alkaline)	300	350	20	35
Pulp (groundwood)	115	130	80	85
Pulp (deinked unfilled stock)	500	650	375	500
Pulp (deinked filled stock)	400	500	600	800
Pulp (rag cooking)	1400	1550	475	500
Bleaching, (long-fiber stock, multi- or single-stage bleaching and short-fiber stock, single-stage bleaching)	60	70	3	6
Bleaching (short-fiber stock, multi-stage bleaching)	165	185	30	35
Paper-making	100	125	75	85

*All averages are for periods of consecutive operation.

undertaken into the environmental, technological, economic, and social effects that would result from enforcing the 1983 regulation on industry and attaining the 1985 goal of pollution-free water.

Section 306 (National Standards of Performance) is presented in its entirety in Appendix A. Section 306(a) is a definitive section. The phrase "Standards of performance" is defined to include process change as well as control through the use of technology. The "best available demonstrated control technology" must be used. Section 306(b) provides a list of industries to be regulated. It also requires that within 1 year and 90 days proposed standards for new sources be promulgated. Within 120 days final regulations are required. The Administrator shall consider costs and other nonwater environmental impacts. This section also applies to new sources operated by the United States government. Section 306(c) allows the Administrator to delegate this authority to the states except for sources operated by the United States government.

Section 306(d) states that any point source whose construction begins after October 18, 1972, if it meets all applicable standards of performance, shall not be subject to any more stringent standards of performance for 10 years or its depreciation period, which ever occurs first.

Section 402 (National Pollutant Discharge Elimination System) is presented in its entirety in Appendix B. This section provides for a permit system. All permits that were in force under the 1899 Rivers and Harbors Act would continue until modified. No new permits would be issued under that Act. Pending permits were to be deemed applications under the new Act. This section also requires that the Administrator shall turn the program over to the states subject to the guidelines of previous sections, unless he or she determines that the state permit program does not meet Federal requirements. Section 402(d) gives the Administrator limited veto over the granting of a permit if it affects another state or is outside the guidelines of this Act.

The discharge limits for contaminant levels in the treated effluent were specified by EPA for the different industry categories. The regulated contaminants generally can be divided into two groups: namely, (1) conventional constituents such as BOD, COD, T.S.S., pH, oil and grease, etc., and (2) non-conventional and toxic constituents such as heavy metals, toxic organics, etc. applicable to the particular industry category. It would be too voluminous to include all the discharge limits for the different industrial categories. The following tables (Table 2.8) present some of the constituents for each of the industries for comparison purposes. These discharge limits are revised periodically by EPA. A reference list of the Effluent Guidelines document for the different industries published by EPA is included in Appendix C.

Table 2-8.   1977 Effluent Guidelines (USEPA)

Chapter code and industrial category	Contaminant levels							
	BOD		TSS				Oil and grease	
	maximum daily	average 30 days	maximum daily	average 30 days	pH		maximum daily	average 30 days
26.5 Dairy (#/1000s BOD inputs)								
Large plants								
Receiving stations	0.048	0.0190	0.0713	0.029	6–9			
Fluid producing plants	0.338	0.135	0.551	0.203	6–9			
Cultured products	0.338	0.135	0.506	0.203	6–9			
Butter	0.138	0.0550	0.206	0.083	6–9			
Cottage cheese	0.670	0.268	1.005	0.402	6–9			
Natural and processed cheese	0.073	0.029	0.109	0.044	6–9			
Fluid ice cream mix	0.220	0.088	0.330	0.132	6–9			
Ice cream frozen desserts	0.460	0.184	0.69	0.276	6–9			
Condensed milk	0.345	0.138	0.518	0.207	6–9			
Dry milk	0.163	0.065	0.244	0.098	6–9			
Condensed whey	0.100	0.040	0.150	0.060	6–9			
Dry whey	0.100	0.040	0.150	0.060	6–9			
Grain mills (#/100 standard bushels)								
Corn wet milling	150.0	50.5	150.0	50.0	6–9			
Corn dry milling	12.0	4.0	10.5	3.5	6–9			
Bulgur wheat flour mills	1.5	0.5	1.5	0.5	6–9			
Parboiled rice (#/100# rice)	0.042	0.014	0.024	0.008	6–9			
Ready-to-eat cereal (#/1000# cereal)	1.2	0.40	1.2	0.40	6–9			
Wheat starch gluten (#/1000# wheat flour)	6.0	2.0	6.0	2.0	6–9			
26.4 Canned and preserved fruits and vegetables (#/1000# raw materials)								
Apple juice	0.60	0.30	0.80	0.40	6–9			
Apple products	1.10	0.55	1.40	0.70	6–9			
Citrus products	0.80	0.40	1.70	0.85	6–9			
Frozen potatoes	2.80	1.40	2.80	1.40	6–9			
Dehydrated potatoes	2.40	1.20	2.80	1.40	6–9			
26.18 Canned and preserved seafood processing (#/1000# seafood)								
Farm-raised catfish			28.0	9.2	6–9		10.0	3.4
Blue crab (conventional)			2.2	0.74	6–9		0.6	0.2
Blue crab (mechanized)			36.0	12.0	6–9		13.0	4.2
Alaskan crabmeat (remote) 1983			16.0	5.3	6–9		1.6	0.52
Alaskan crabmeat (nonremote)			19.0	6.2	6–9		1.8	0.61
Alaskan whole crab (nonremote)			12.0	3.9	6–9		1.3	0.42
Alaskan whole crab (remote) 1983			9.9	3.3	6–9		1.1	0.36
Dungeness and Tanner crab			8.1	2.7	6–9		1.8	0.61
Alaskan shrimp process (nonremote)			320.0	210.0	6–9		51.0	17.0
Alaskan shrimp process (remote)			270.0	180.0	6–9		45.0	15.0
Northern shrimp processing			160.0	54.0	6–9		126.0	42.0

1977 Effluent Guidelines (USEPA)

Chapter code and industrial category	Contaminant levels							
	BOD		TSS				Oil and grease	
	maximum daily	average 30 days	maximum daily	average 30 days	pH		maximum daily	average 30 days
26.18 (Continued)								
Southern (nonbreaded) shrimp			110.0	38.0	6–9		36.0	12.0
Breaded shrimp processing			280.0	93.0	6–9		36.0	12.0
Tuna processing	230.0	9.0	8.3	3.3	6–9		2.1	0.84
Fish meal	4.7	3.5	2.3	1.3	6–9		0.8	0.63
Alaskan hand-butchered salmon			1.7	1.4	6–9		0.2	0.17
Alaskan mech. processed salmon			27.0	22.0	6–9		27.0	10.0
West Coast hand-butchered salmon			1.7	1.4	6–9		0.2	0.17
West Coast mech. processed salmon	41.0	34.0	8.2	6.7	6–9		4.0	1.6
Alaskan bottom fish processing			1.9	1.7	6–9		0.11	0.09
Non-Alaskan conventional bottom-fish processing			14.0	10.0	6–9		5.7	3.3
Hand-shucked clam processing			30.0	19.0	6–9		0.29	0.19
Mechanized clam processing			7.7	6.1	6–9		0.55	0.48
Pacific Coast hand-shucked oyster processing			37.0	35.0	6–9		1.7	1.6
Atlantic and Gulf Coast hand-shucked oyster processing			19.0	15.0	6–9		0.77	0.70
Steamed/canned oyster processing			54.0	36.0	6–9		1.6	1.3
Sardine processing			4.2	3.3	6–9		2.9	1.6
Alaskan scallop processing			0.82	0.62	6–9		0.63	0.32
Non-Alaskan scallop processing			0.82	0.62	6–9		0.63	0.32
Alaskan herring fillet processing			25.0	24.0	6–9		8.4	6.9
Non-Alaskan herring fillet processing			25.0	24.0	6–9		8.4	6.9
26.14 Sugar processing (#/1000# product)								
Beet sugar processing	3.3	2.2	—	—	6–9	Temperature	90°F	
26.15 Crystalline cane	2.38	0.86	0.54	0.18	6–9			
Liquid cane sugar	1.56	0.63	0.99	0.33	6–9			
Louisiana cane sugar	1.14	0.63	1.41	0.47	6–9			
Hawaii raw cane sugar (Hilo-Hamakua coast)	—	—	4.2	2.1	6–9			
Hawaii raw cane sugar	1.11	0.63	1.11	0.47	6-9			
Other inorganic chemicals (parts per million)								
Aluminum sulfate			50.0	25.0	6–9			
Calcium chloride (#/1000 product)			0.016	0.0082	6–9			
Calcium oxide and lime (ppm)			50.0	25.0	6–9			
Chlorine and sodium or potassium hydroxide (#/1000# product)								
Mercury cell process			0.64	0.32	6–9	Hg	0.00028	0.00014

(*continued*)

1977 Effluent Guidelines (USEPA)

Chapter code and industrial category	BOD		TSS				Oil and grease	
	maximum daily	average 30 days	maximum daily	average 30 days	pH		maximum daily	average 30 days
(Continued)								
Diaphragm cell process			0.64	0.32	6–9	Pb	0.005	0.0025
Hydrogen fluoride (ppm)			50.0	25.0	6–9	Fluoride	30.0	15.0
Hydrogen peroxide (#/1000# product)			0.80	0.40	6–9	TOC	0.44	0.22
Potassium sulfate (ppm)			50.0	25.0	6–9			
Sodium carbonate (#/1000# product)			0.20	0.10	6–9			
Sodium chloride			0.34	0.17	6–9			
Sodium chromate and sodium sulfate (#/1000# product)			0.44	0.22	6–9	Hex. chrome	0.009	0.0005
						Total chrome	0.0088	0.0044
Sodium metal (#/1000# product)			0.46	0.23	6–9			
Sodium silicate			0.01	0.005	6–9			
Sodium sulfite			0.032	0.016	6–9	TOC	3.4	1.7
Titanium oxide								
Chloride process			4.6	2.3	6–9	Iron	1.7	0.36
Sulfate process			21.0	10.5	6–9	Iron	1.7	0.84
Aluminum fluoride			0.86	0.43	6–9	Fluoride	0.68	0.34
						Aluminum	0.34	0.17
Ammonium chloride					6–9	Ammonia-N	8.8	4.4
Boric acid			0.14	0.07	6–9	Arsenic	0.0028	0.0014
Calcium carbonate (#/1000# product)								
Milk of line process			0.56	0.28	6–9			
Solvay process			1.16	0.58	6–9			
Carbon monoxide and hydrogen			0.12	0.06	6–9	COD	0.50	0.25
Chrome pigments			5.1	1.7	6–9	Total chrome	0.1	0.034
						Hex chrome	0.01	0.0034
						Lead	0.42	0.14
						Zinc A	0.72	0.27
						Cyan. A	0.01	0.0034
						Cyan.	0.1	0.034
						Iron	0.72	0.27
Copper sulfate								
Pure raw material					6–9	Copper	0.0006	0.0002
Recovery process			0.069	0.023	6–9	Copper	0.003	0.001
						Nickel	0.006	0.002
						Selenium	0.0015	0.0005
Hydrogen cyanide (Andrussow process)	3.6	1.8	2.4	1.2	6–9	Cyanide	0.05	0.025
						Cyanide A	0.005	0.0025
						Ammonia-N	0.36	0.18
Lithium carbonate			2.7	0.9	6–9			
Manganese sulfate			0.0062	0.032	6–9	Nickel	0.006	0.002
Oxygen and nitrogen					6–9	Oil & Grease	0.002	0.001
Potassium iodide			0.09	0.03	6–9	Iron	0.015	0.005
						Sulfide	0.015	0.005
						Barium	0.009	0.003

1977 Effluent Guidelines (USEPA)

Chapter code and industrial category	Contaminant levels							
	BOD		TSS					
	maximum daily	average 30 days	maximum daily	average 30 days	pH		maximum daily	average 30 days
(Continued)								
Silver nitrate			0.069	0.023	6–9	Silver	0.009	0.003
Sodium silicofluoride			0.6	0.3	6–9	Fluoride	0.5	0.25
Zinc sulfate, sodium fluoride, potassium chloride, lead oxide, iodine, hydrogen, fluorine, ferric chloride, chromic acid, lime, bromine, borax, sulfuric acid, sodium bicarbonate, potassium dichromate, potassium, nitric acid, and hydrochoric acid	There shall be no discharge from these industries to navigable waters...							
28.8 Plastic and synthetics (#/1000# product)								
Polyvinyl chloride								
Suspension polymerization	0.70	0.36	1.8	0.99	6–9			
Emulsion polymerization	0.26	0.13	0.65	0.36	6–9			
Bulk polymerization	0.12	0.06	0.29	0.16	6–9			
Polyvinyl acetate	0.39	0.20	1.0	0.55	6–9			
Polystyrene								
Suspension polymerization	0.43	0.22	1.1	0.61	6–9	Total chrome	0.0046	0.0023
Bulk polymerization	0.08	0.14	0.2	0.11	6–9			
Polypropylene	0.81	0.42	2.1	1.16	6–9			
Polyethylene								
Low density	0.39	0.20	1.0	0.55	6–9			
High density	0.58	0.30	1.5	0.83	6–9	Total chrome	0.0062	0.0031
Cellophane	17.8	8.7	29.1	16.0	6–9			
Rayon	10.0	4.8	16.0	8.8	6–9	Zinc	0.91	0.534
Acrylonitrile-butadiene-stryene (ABS) and stryene-acrylonitrile (SAN)	1.3	0.63	2.1	1.16	6–9	Chromium	0.0088	0.0044
Polyester of fiber (batch process) by resin and continuous processing	1.4	0.78	0.95	0.52	6–9			
Nylon 66								
Resin	1.2	0.66	0.80	0.44	6–9			
Fiber	1.1	0.58	0.70	0.39	6–9			
Resin and fiber	2.3	1.24	1.5	0.83	6–9			
Nylon 6								
Resin	6.8	3.71	4.5	2.48	6–9			
Fiber	3.5	1.90	2.3	1.27	6–9			
Resin and fiber	10.3	5.61	6.8	3.75	6–9			
Cellulose acetate								
Resin	7.5	4.13	5.0	2.75	6–9			
Fiber	7.5	4.13	5.0	2.75	6–9			
Resin and fiber	15.0	8.26	10.0	5.5	6–9			
Acrylics	5.0	2.75	2.0	1.1	6–9	Phenol	0.017	0.0083
						COD	25.0	13.8
Ethylene-vinyl acetate	3.9	0.20	1.0	0.55	6–9			

(continued)

1977 Effluent Guidelines (USEPA)

Chapter code and industrial category	Contaminant levels							
	BOD		TSS					
	maximum daily	average 30 days	maximum daily	average 30 days	pH		maximum daily	average 30 days
28.8 (Continued)								
Polytetrafluorethylene	7.0	3.6	1.8	9.9	6–9	Fluoride	1.2	0.60
Polypropylene fiber	0.78	0.40	2.0	1.1	6–9	Oil & Grease	1.0	0.50
Alkyds and unsaturated polyester resins	0.60	0.33	0.40	0.22	6–9			
Cellulose nitrate	26.0	14.0	17.0	9.4	6–9			
Polyamide (Nylon 6/12)	1.2	0.66	0.80	0.44	6–9			
Polyester (thermo-plastic resins)	1.4	0.78	0.95	0.52	6–9			
Silicones	1.9	1.0	1.25	0.69	6–9	Copper	0.010	0.005
28.4 Soap and detergent								
Soap manufacturing by kettle (#/1000# product)	1.80	0.60	1.20	0.40	6–9	Oil & Grease	0.30	0.10
						COD	4.5	1.5
Fatty acids								
Fat splitting	3.6	1.20	6.6	2.2	6–9	Oil & Grease	0.90	0.30
						COD	9.9	3.0
Hydrogenation	0.45	0.15	0.3	0.1	6–9	Oil & Grease	0.30	0.10
						COD	0.75	0.25
Soap manufacturing by fatty acid neutralization	0.03	0.01	0.10	0.2	6–9	Oil & Grease	0.13	0.01
						COD	0.15	0.05
Glycerine conc.	4.5	1.5	0.60	0.20	6–9	Oil & Grease	0.30	0.10
						COD	13.5	4.5
Glycerine distillation	1.5	0.50	0.60	0.20	6–9	Oil & Grease	0.30	0.10
						COD	4.5	1.5
Manufacturing of soap flakes and powders (#/1000# anhydrous products)						Oil & Grease	0.03	0.01
	0.03	0.01	0.10	0.01	6–9	Oil & Grease	0.03	0.01
						COD	0.15	0.05
Manufacturing of bar soaps	1.02	0.34	1.74	0.58	6–9	Oil & Grease	0.12	0.04
						COD	2.55	0.85
Manufacturing of liquid soaps	0.10	0.01	0.03	0.01	6–9	Oil & Grease	0.03	0.01
						COD	0.15	0.05
Oleum sulfonation and sulfation	0.09	0.02	0.15	0.03	6–9	Oil & Grease	0.25	0.07
						COD	0.40	0.09
						Surfactants	0.15	0.03
Air SO₃ sulfation and sulfonation	0.09	0.03	0.09	0.03	6–9	Oil & Grease	0.15	0.05
						COD	4.05	1.35
						Surfactants	0.90	0.3
SO₃ solvent and vacuum sulfonation	0.09	0.03	0.09	0.03	6–9	Oil & Grease	0.10	0.05
						COD	3.05	1.35
						Surfactants	0.90	0.03
Sulfamic acid sulfation	0.90	0.30	0.09	0.03	6–9	Oil & Grease	0.15	0.05
						COD	4.05	1.35
						Surfactants	0.90	0.30
Chlorosulfonic acid sulfation	0.90	0.30	0.09	0.03	6–9	Oil & Grease	0.15	0.05
						COD	4.05	1.35
						Surfactants	0.90	0.30

1977 Effluent Guidelines (USEPA)

Chapter code and industrial category	Contaminant levels							
	BOD		TSS					
	maximum daily	average 30 days	maximum daily	average 30 days	pH		maximum daily	average 30 days
28.4 (Continued)								
Neutralization of sulfuric acid esters and sulfonic acids	0.03	0.01	0.09	0.03	6–9	Oil & Grease	0.03	0.01
						COD	0.15	0.05
						Surfactants	0.06	0.02
Spray dried detergents	0.03	0.01	0.03	0.01	6–9	Oil & Grease	0.015	0.005
						COD	0.15	0.05
						Surfactants	0.06	0.02
Liquid detergents (normal)	0.60	0.20	0.015	0.005	6–9	Oil & Grease	0.015	0.005
						COD	0.8	0.60
						Surfactants	0.30	0.13
Detergents by dry blending	0.03	0.01	0.03	0.01	6–9	Oil & Grease	0.015	0.005
						COD	0.21	0.07
						Surfactants	0.03	0.01
Drum dried detergents	0.03	0.01	0.03	0.01	6–9	Oil & Grease	0.03	0.01
						COD	0.15	0.05
						Surfactants	0.03	0.01
Detergent bars and cakes	2.10	0.70	0.60	0.20	6–9	Oil & Grease	0.06	0.02
						COD	9.9	3.3
						Surfactants	1.5	0.50
28.9 Fertilizer manufacturing								
Phosphate (ppm)	50.0	25.0			8–9.5	Phosphate	70.0	35.0
						Fluoride	30.0	15.0
Ammonia (#/1000# product)					6–9	Ammonia-N	0.1875	0.0625
Urea (#/1000# product)					6–9	Ammonia-N	0.075	0.0375
						Organic-N	0.125	0.0625
Ammonium nitrate								
Aqueous solution					6–9	Ammonia-N	0.075	0.375
						Organic-N	0.10	0.03
Prilled or granulated					6–9	Ammonia-N	0.20	0.10
						Organic-N	0.22	0.11
Nitric acid	No discharge permitted							
Ammonium sulfate	No discharge permitted							
Mixed and blended fertilizer	No discharge permitted							
27.9 Petroleum refining								
Topping* (#/1000# bbl of feedstock)								
Point source	8.0	4.25	5.6	3.6	6–9	Ammonia-N	0.99	0.45
						Oil & Grease	2.5	1.3
						COD	41.2	21.3
						Phenolic	0.060	0.027
						Sulfide	0.053	0.024
						Total chrome	0.122	0.071
						Hex.chrome	0.10	0.0044
Runoff	0.4	0.21	0.26	0.17	6–9	Oil & Grease	0.126	0.067
						COD	3.1	1.6

*Process and size factors are also used and given in the published guidelines—greater size and greater process configurations permit increased contaminants in effluent.

(continued)

1977 Effluent Guidelines (USEPA)

Chapter code and industrial category	BOD maximum daily	BOD average 30 days	TSS maximum daily	TSS average 30 days	pH		Oil and grease maximum daily	Oil and grease average 30 days
27.9 (Continued)								
Ballast	0.4	0.21	0.26	0.17	6–9	Oil & Grease	0.126	0.067
						COD	3.9	2.0
Cooling water—once through	TOC less than 5 mg/l							
Cracking* (same as for topping, runoff, ballast, and cooling water)	9.9	5.5	6.9	4.4	6–9	Ammonia-N	6.6	3.0
						Oil & Grease	3.0	1.6
						COD	74.0	38.4
						Phenolic	0.074	0.036
						Sulfide	0.065	0.029
						Total chrome	0.15	0.088
						Hex.chrome	0.012	0.0056
27.10 Petrochemicals (same as for topping, runoff, ballast, and cooling water) /1000 bbl feedstock	12.1	6.5	8.3	5.25	6–9	Ammonia-N	8.25	3.8
						Oil & grease	3.9	2.1
						COD	74.0	38.4
						Phenolic	0.088	0.0425
						Sulfide	0.078	0.035
						Total chrome	0.183	0.107
						Hex.chrome	0.016	0.0072
27.9 Lube (same as for topping, runoff, ballast, and cooling water)	17.9	9.11	12.5	8.0	6–9	Ammonia-N	8.3	3.8
						Oil & grease	5.7	2.1
						COD	127.0	66.0
						Phenolic	0.133	0.065
						Sulfide	0.118	0.053
						Total chrome	0.273	0.160
						Hex.chrome	0.024	0.011
Integrated	19.2	10.2	13.2	8.4	6–9	Ammonia-N	8.3	3.8
						Oil & grease	6.0	3.2
						COD	136.0	70.0
						Phenolic	0.14	0.068
						Sulfide	0.124	0.056
						Total chrome	0.29	0.17
						Hex. chrome	0.025	0.01
27.4 Iron and steel manufacturing (#/1000# product)								
By product code*			0.1095	0.0365		Ammonia-N	0.2736	0.0912
						Cyanide	0.0657	0.0219
						Oil & grease	0.0327	0.0109
						Phenol	0.0045	0.0015

*Process and size factors are also used and given in the published guidelines—greater size and greater process configurations permit increase contaminants in effluent.

1977 Effluent Guidelines (USEPA)

Chapter code and industrial category	Contaminant levels							
	BOD		TSS				Oil and grease	
	maximum daily	average 30 days	maximum daily	average 30 days	pH		maximum daily	average 30 days
27.4 (Continued)								
Beehive	No discharge allowed							
Sintering			0.0312	0.0104		Oil & grease	0.0063	0.0021
Blast furnace								
Iron						Ammonia-N	0.1953	0.0651
						Cyanide	0.0234	0.0078
						Phenol	0.0063	0.0021
Ferro manganese			0.3129	0.104		Ammonia-N	1.56	0.5212
						Cyanide	0.4689	0.1563
						Phenol	0.0624	0.0208
Basic oxygen furnace								
Semiwet air-pollution control	No discharge allowed							
Wet air-pollution control			0.0312	0.0104				
Open hearth furnace			0.0312	0.0104				
Electric arc furnace	No discharge allowed							
Electric arc furnace with wet air-pollution control			0.0312	0.0104				
Vacuum degassing			0.0156	0.0052				
Continuous casting			0.0780	0.0260		Oil & grease	0.0234	0.0078
27.5 Non-ferrous metals								
Bauxite refining by Bayer process	No discharge to navigable waters							
Primary Al smelting by Hall-Heroult process (#/1000# product)			3.0	1.5	6–9	Fluoride	2.0	1.0
Aluminium smelting	No discharge allowed							
AlF_3 for Mg removal	No discharge allowed							
Cl_2 in Mg removal (#/1000 # Mg removed)			175.0		7.5–9	COD	6.5	
Wet methods for processing residuals (#/1000#product)			1.5		7.5–9	COD	1.0	
						NH_3-N	0.01	
						Al	1.0	
						Cu	0.003	
Primary copper smelting on site†	No discharge allowed							
Process water impoundment designed to contain 10-year, 24-hour rainfall at smelter (ppm)			50.0	25.0	6–9	Arsenic	20.0	10.0
						Lead	1.0	0.5
						Copper	0.05	0.25
						Cadmium	1.0	0.5
Primary copper refining not on site† Process wastewater impoundment designed to contain 10-year, 24-hour	No discharge allowed							

*May be exceeded by 15 percent for plants using gas desulfurization units and variations for plants using indirect NH_3 recovery systems.
†Selenium and zinc limits are 10 and 5 ppm, respectively, for maximum and average values.

(continued)

1977 Effluent Guidelines (USEPA)

Chapter code and industrial category	Contaminant levels							
	BOD		TSS					
	maximum daily	average 30 days	maximum daily	average 30 days	pH		maximum daily	average 30 days
27.5 (Continued)								
rainfall at smelter			50.0	25.0	6–9	Arsenic	20.0	10.0
						Lead	20.0	10.0
						Copper	0.5	0.25
Secondary copper processing of new and used copper*	No discharge of process waters allowed							
Process wastewater impoundment designed to contain 10-year, 24-hour rainfall (ppm)			50.0	25.0	6–9	Lead	20.0	10.0
						Copper	0.5	0.25
Primary lead	No discharge of process waters allowed							
Process wastewaters from impoundments designed to contain 10-year, 24-hour rainfall (ppm)			50.0	25.0	6–9	Lead	1.0	0.5
						Zinc	10.0	5.0
						Cadmium	1.0	0.5
Primary zinc by electrolytic or pyrolytic methods (#/1000# product)			0.42	0.21	6–9	Arsenic	0.0016	0.0008
						Cadmium	0.008	0.004
						Zinc	0.08	0.04
						Selenium	0.08	0.04
27.5 Ferro alloy manufacturing								
Open electric furnace with wet air-pollution control (#/megawatt hour)			0.703	0.352	6–9	Total chrome†	0.014	0.007
						Manganese	0.11	0.070
Covered electric furnaces and smelters with wet air-pollution control ‡ (#/megawatt hour)			0.922	0.461	6–9	Total chrome†	0.018	0.009
						Manganese	0.184	0.092
						Total cyanide	0.009	0.005
						Phenol	0.013	0.009
Slag processing (#/z ton processed)			5.319	2.659	6–9	Total chrome†	0.016	0.053
						Manganese	1.064	0.532
Covered calcium carbide furnace with wet air-pollution control devices (#/1000# product)			0.380	0.190	6–9	Total cyanide	0.0056	0.0028
Other calcium carbide furnaces	No discharge of process waste waters allowed							
Electrolytic manganese products								
Electro. Mn #/1000# product			6.778	3.389	6–9	Manganese	2.771	1.356
						NH_3-N	40.667	20.334
Electro. MnO_2			1.762	0.881	6–9	Manganese	0.705	0.352
						NH_3-N	10.574	5.277

* Selenium and zinc limits are 10 and 5 ppm, respectively, for maximum and average values.
† 10 percent of these limits for Chrome VI.
‡ For nonelectric furnace smelting multiply all values by 3.3 (for # per tone of product guidelines limit.

1977 Effluent Guidelines (USEPA)

Chapter code and industrial category	Contaminant levels							
	BOD		TSS					
	maximum daily	average 30 days	maximum daily	average 30 days	pH		maximum daily	average 30 days
27.4 (Continued)								
Electrolytic chromium metal (#/1000# product)			5.276	2.638	6–9	Total chrome* Manganese NH₃-N	0.106 2.111 10.553	0.053 1.055 5.276
25.4 Leather tanning and finishing								
Hair pulp with chrome tanning† (#/1000# product)	8.0	4.0	10.0	5.0	6–9	Total chrome* Oil & grease	0.20 1.5	0.10 0.75
Hair save with chrome tanning	9.2	4.6	11.6	5.8	6–9	Total chrome* Oil & grease	0.24 1.8	0.12 0.90
Unhairing with vegetable or alum tanning	7.6	3.8	9.6	4.8	6–9	Total chrome* Oil & grease	0.10 1.5	0.05 0.75
Finishing of tanned hides (#/1000# raw material)	3.2	1.6	4.0	2.0	6–9	Total chrome Oil & grease	0.20 0.5	0.10 0.25
Vegetable or chrome tanning of unhaired hides (#/1000# raw material)	9.6	4.8	12.0	6.0	6–9	Total chrome Oil & grease	0.12 1.8	0.06 0.90
Unhairing with chrome tanning, no finishing	5.6	2.8	6.8	3.4	6–9	Total chrome Oil & grease	0.2 0.7	0.10 0.35
27.13 Glass manufacturing								
Insulation fiberglass Wastewater from advanced air-emission controls (#/1000# product)	No discharge of wastewaters to navigable waters							
	0.024	0.012	0.03	0.015	6–9	COD	0.33	0.165
Sheet glass manufacturing	No discharge of process wastewater							
Rolled glass	No discharge of process wastewater							
Plate glass (#/ton product)			0.09	0.09	6–9			
Float glass (#/ton product)			0.004	0.004	6–9	Phosphate Oil & grease	0.0001 0.0028	0.0001 0.0028
Automotive glass tempering (#/1000/Ft² product)			0.40	0.25	6–9	Oil & grease	0.13	0.13
Automotive glass laminating (#/1000/Ft² product)			0.90	0.90	6–9	Total chrome Oil & grease	0.22 0.36	0.22 0.36
Glass container manufacturing (#/1000# furnace pull)			0.14	0.07	6–9	Oil & grease	0.06	0.03
Glass tubing (Danner) (#/1000# furnace pull)			0.46	0.23	6–9			
Television tube (#/1000# furnace pull)			0.30	0.15	6–9	Fluoride	0.14	0.07

*10 percent of these limits for Chrome VI.

†Additional allocations equal to 1/2 these limitations for BOD and TSS and for plants which produce less than 17,000 kg hides per day.

(*continued*)

1977 Effluent Guidelines (USEPA)

Chapter code and industrial category	BOD maximum daily	BOD average 30 days	TSS maximum daily	TSS average 30 days	pH		maximum daily	average 30 days
27.13 (Continued)								
Television tube						Oil & grease	0.26	0.13
						Lead*	0.009	0.0045
Incandescent lamp envelope								
(#/1000# furnace pull)								
Clear			0.23	0.115	6–9	Oil & grease	0.23	0.115
Frosted			0.46	0.23	6–9	Fluoride	0.23	0.115
Hand-pressed and blown	No limitations for 1977							
27.23 Asbestos manufacturing								
Asbestos cement pipe (#/ton product)			1.14	0.38	6–9			
Asbestos cement sheet			1.35	0.45	6–9			
Asbestos paper (starch binder)			1.10	0.70	6–9			
Asbestos paper (electromeric binder)			1.10	0.70	6–9			
Asbestos millboard	No discharge to waters							
Asbestos roofing			0.02	0.02	6–9	COD	0.029	0.016
Asbestos floor tile (#/1000/pieces tile)			0.13	0.08	6–9	COD	0.30	0.18
Coating or finishing of asbestos textiles	No discharge							
Solvent recovery (#/1000# finished asbestos product)			0.18	0.09	6–9	COD	0.30	0.15
Vapor adsorption	No discharge to waters							
Wet dust collection (#/MM std.cu.ft. of air-scrubbed)			5.0	2.5	6–9			
27.12 Rubber manufacturing								
Tire and inner tubes (#/1000# raw material)			0.096	0.064	6–9	Oil & grease	0.024	0.016
Emulsion crumb rubber (#1000# product)	0.60	0.40	0.98	0.65	6–9	COD	12.0	8.0
Solution crumb rubber	0.60	0.40	0.98	0.65	6–9	Oil & grease	0.24	0.16
						COD	5.91	3.64
Latex rubber	0.51	0.34	0.82	0.55	6–9	COD	10.27	6.85
						Oil & grease	0.21	0.14
Small-sized, general molded, extruded and fabricated rubber (# /1000# raw material)								
Lead-sheathed hose			1.28	0.64	6–9	Lead	0.0017	0.0007
						Oil & grease	0.70	0.25
Wet scrubbers			0.58	2.9				
Medium-sized, general molded, extruded, and fabricated rubber plants (#/1000# raw material)			0.8	0.40	6–9	Oil & grease	0.42	0.15

*Applicable only to abrasive and acid polishing wastewater.

1977 Effluent Guidelines (USEPA)

Chapter code and industrial category	Contaminant levels							
	BOD		TSS					
	maximum daily	average 30 days	maximum daily	average 30 days	pH		maximum daily	average 30 days
27.12 (Continued)								
Wet scrubbers			5.8	2.90				
Lead-sheathed hose						Lead	0.0017	0.0007
Large-sized, general molded, extruded, and fabricated rubber plants			0.5	0.25	6–9	Lead	0.093	
Lead-sheathed hose						Lead	0.0017	0.0007
Wet digestion, reclaimed rubber	14.7	6.11	1.04	0.52	6–9	Lead	0.40	0.144
Pan, dry digestion, and medicenial reclaimed rubber (#/1000# product)	6.7	2.8	0.384	0.192	6–9	Lead	0.40	0.144
Latex dipeed, extruded, and molded rubber (#/1000# raw materials)	3.72	2.20	6.96	2.90	6–9	Lead	2.0	0.73
Chromic acid foam-cleaning operation						Chromium	0.0086	0.0036
Latex foam—for integrated wet and dry digestion processing	2.4	1.40	2.26	1.4	6–9	Zinc	0.058	0.024
28.3 Phosphate manufacturing								
Phosphorous production (#/1000# product)			1.0	0.5	6–9	Fluoride	0.10	0.05
						Total P	0.30	0.15
						Elem. P	Not detectable	
Phosphate consuming	No discharge from manufacture of H_3PO_4, P_2O_5, P_2S_5							
From POl_3 manufacture			1.4	0.7	6–9	Total P	1.6	0.08
						Elm. P	Not detectable	
						Arsenic	0.0001	0.00005
From $POCl_2$			0.3	0.15	6–9	Total P	0.34	0.17
Phosphate	No discharge from manufacture of Na Tripolyphosphate or animal grade feed $Ca_3(PO_4)_2$							
From human food grade $Ca_3(PO_4)_2$			0.12	0.06	6–9	Total P	0.06	0.03
Defluorinated phosphate rock*	No discharge to navigable waters							
Wastewater from impoundment designed to retain 10-year, 24-hour rainfall in (mg/l)			50.0	25.0	6–9	Fluoride	30.0	15.0
						Total P	70.0	35.0
Same for manufacture of defluorinated phosphoric acid*								
Na_3PO_4 manufacturing (#/1000# product)			0.50	0.25	6–9	Fluoride	0.30	0.15
						Total P	0.80	0.40
29.1 Steam electric power†								
Generating unit* (ppm)			100.0	30.0	6–9	PCB	none	
						Oil & grease	20.0	15.0
Ash transport* (ppm)			100.0	30.0	6–9	Oil & grease	20.0	15.0
Metal cleaning* (ppm)			100.0	30.0	—	Oil & grease	20.0	15.0

*Neither free available chlorine nor total residual chlorine can be discharged for more than two hours in any day and not more than one unit at a time.
†Includes both small units and old units.

(*continued*)

1977 Effluent Guidelines (USEPA)

Chapter code and industrial category	Contaminant levels							
	BOD		TSS				Oil and grease	
	maximum daily	average 30 days	maximum daily	average 30 days	pH		maximum daily	average 30 days
29.1 (Continued)								
Metal cleaning (ppm)						Copper	1.0	1.0
						Iron (total)	1.0	1.0
Boiler blowdown* (ppm)			100.0	30.0	—	Oil & grease	20.0	15.0
						Copper	1.0	1.0
						Iron (total)	1.0	1.0
Once through cooling water* (ppm)						Free available		
Cooling tower blowdown* (ppm)						chlorine	0.5	0.2
Area runoff from 10-year, 24-hour rainfall (ppm)—less than 50 TSS					6–9			
Timber products								
Barking (#/Ft³ products)	No discharge into navigable waters except from hydraulic operations							
Barking hydraulic	0.09	0.03	0.431	0.144	6–9			
Veneer	No discharge							
Softwood veneer (steaming) (#/Ft³ products)	0.045	0.015			6–9			
Hardwood veneer (steaming)	0.10	0.034			6–9			
Plywood	No discharge							
Hardboard dry processing	No discharge							
Hardboard wet process (#/2000 # product)	1.56	5.2	33.0	11.0	6–9			
Wood preserving	No discharge							
Wood preserving—steam (#/1000 ft³ product)					6–9	COD	68.5	34.5
						Oil & grease	1.5	7.5
						Phenols	0.14	0.04
Wood preserving-boultonizing	No discharge							
Wet storage	No debris discharge							
Log washing (ppm)			equal to or less than 50 TTS					
Saw mills and planing	No discharge							
Finishing	No discharge							
Particleboard	No discharge							
27.22 Wood furniture and fixture production								
without water wash, spray booths, or laundry facilities	No discharge							
with water wash, spray booths, or with laundry facilities—0.2 ml/l TSS					6–9			

*Neither free available chlorine nor total residual chlorine can be discharged for more than two hours in any day and not more than one ur at a time.

1977 Effluent Guidelines (USEPA)

Chapter code and industrial category	Contaminant levels							
	BOD		TSS					
	maximum daily	average 30 days	maximum daily	average 30 days	pH		maximum daily	average 30 days
25.1 Textiles								
Wool scouring (#/1000# wool)	10.6	5.3	32.2	16.1		COD	110.0	50.0
						Total chrome	0.05	0.05
						Phenols	0.05	0.05
Wool finishing (#/1000# fiber)	22.4	11.2	35.2	17.6		COD	163.0	81.5
						Total chrome	0.14	0.07
						Phenols	0.14	0.07
Dry process (#/1000# product)	1.4	0.7	1.4	0.7		COD	2.8	1.4
Woven fabric finishing (natural or synthetic fiber) (#/1000# products)	6.6	3.3	17.8	8.9		COD*	60.0	30.0
						Total chrome	0.10	0.05
						Phenols	0.10	0.05
Knit fabrics finishing	5.0	2.5	21.8	10.9		COD*	60.0	30.0
						Total chrome	0.10	0.05
						Phenols	0.10	0.05
Carpet mills	7.8	3.9	11.0	5.5		COD	70.2	35.1
						Total chrome	0.04	0.02
						Phenols	0.04	0.02
Complex manufacturing						COD	20.0	10.0
Stock and yarn dyeing (#/1000# product)	6.8	3.4	17.4	8.7		COD	84.6	42.3
						Total chrome	0.12	0.06
						Phenols	0.12	0.06
Wool scouring (#/1000# wool)					6–9	Sulfide	0.20	0.10
Wool finishing					6–9	Sulfide	0.28	0.14
Dry process					6–9	Total coliform bacteria	400/100 ml.	
Woven fabric finishing					6–9	Sulfide	0.20	0.10
Knit fabric finishing					6–9	Sulfide	0.20	0.10
Carpet mills					6–9	Sulfide	0.08	0.04
Stock and yarn dyeing					6–9	Sulfide	0.24	0.12
27.21 Cement manufacturing								
Nonleaching (#/1000# product)	0.005				6–9	Temperature	3°	
Leaching	0.4				6–9	Temperature	3°	
Materials storage pile runoff	≤ 50 mg/l				6–9			
26.12 Feedlots								
All categories except ducks	No discharge to navigable waters except the wastewater whenever rainfall occurs in excess of the 10-year, 24-hour rainfall.							
Ducks (#/1000# ducks)	3.66	2.00				Coliforms	≤ 400/100 ml.	

*Varies according to fiber and blend.

(*continued*)

1977 Effluent Guidelines (USEPA)

Chapter code and industrial category	Contaminant levels							
	BOD		TSS					
	maximum daily	average 30 days	maximum daily	average 30 days	pH		maximum daily	average 30 days
28.15 Organic chemicals manufacturing								
Nonaqueous process	0.045	0.02	0.067	0.03	6–9			
Processes with process-water contact as steam dilutent or absorbent (#/1000# product), acetone, butadiene, ethylbenzene, ethylene, propylene, ethylene dichloride, ethylene oxide, formaldehyde, methanol, methylamines, vinyl acetate, or vinyl chloride	0.013	0.058	0.20	0.088	6–9			
Acetaldehyde, acetylene, butadiene by three other processes	0.95	0.42	1.42	0.64	6–9			
Aqueous liquid phase reaction systems (acetic acid, acrylic acid, coal tar, ethylene glycol, terephthalate acid by two processes)	0.28	0.12	0.42	0.19				
Acetaldehyde, caprolactum, coal tar, oxo chemicals, phenol, acetone by one process	0.55	0.25	0.56	0.25				
Acetaldehyde by one process, analine, bisphenol A, dimethyl terephthalate)	1.15	0.51	0.15	0.068				
Acrylates, p cresol, methyl methacrylate, terephthalic acid by one process, or tetraethyl lead	3.08	1.37	2.80	1.25				
Phenol and acetone, bisphenol A or p cresol by a new process						Phenols	0.045	0.020
27.6 Electroplating								
Common metals (#/million Ft² operation)						Cu	32.7	16.4
						Ni	32.7	16.4
						Cr total	32.7	16.4
						Cr_{VI}	3.32	1.6
						Zn	32.7	16.4
						Cn total	32.7	16.4
						Cn_A	3.3	1.6
						Fl	1308.0	654.0
						Cadium	19.2	9.6
						Pb	32.7	16.4
						Fe	65.4	32.7
						Sn	65.4	32.7
	1308.0	654.0			6–9.5	P total	65.4	32.7
Precious metals (#/million Ft² operation)						Ag	3.3	1.6
						Au	3.3	1.6
						Cr total	32.7	16.4
						Cr_{VI}	3.3	1.6

1977 Effluent Guidelines (USEPA)

Chapter code and industrial category	BOD maximum daily	BOD average 30 days	TSS maximum daily	TSS average 30 days	pH		maximum daily	average 30 days
27.6 (Continued)						Iridium	3.3	1.6
						Cn total	32.7	16.4
						Cn_A	3.3	1.6
						Osmium	3.3	1.6
						Palladium	3.3	1.6
						Platinum	3.3	1.6
						Rhodium	3.3	1.6
						Ruthenium	3.3	1.6
Anodizing (#/million Ft² operation)			1308.0	654.0	6–9.5	P total	65.4	32.7
						Cu	18.4	9.2
						Ni	18.4	9.2
						Cr total	18.4	9.2
						Cr_{VI}	1.8	0.92
						Zn	18.4	9.2
						Cn	18.4	9.2
						Cn_A	1.8	0.92
						Fl	738.0	369.0
						Cadium	8.8	4.4
						Fe	36.8	18.4
						Cn total	36.8	18.4
Coatings (#/million Ft² operation)			738.0	369.0	6–9.5	P total	36.8	18.4
						Cu	16.4	8.2
						Ni	16.4	8.2
						Cr total	16.4	8.2
						Cr_{VI}	1.6	0.82
						Zn	16.4	8.2
						Cn	16.4	8.2
						Cn_a	1.6	0.82
						Fl	646.0	323.0
						Cadium	9.8	4.9
						F	32.8	16.4
						Sn	32.8	16.4
Chemical etching and milling (#/million Ft² operation)			644.0	323.0	6–9.5	P total	32.8	16.4
						Cu	24.6	12.3
						Ni	24.6	12.3
						Cr total	24.6	12.3
						Cr_{VI}	2.4	1.2
						Zn	24.6	12.3
						Cn total	24.6	12.3
						Cn_A	3.8	1.9
						Fl	981.0	492.0
						Cadium	14.8	7.4
						Pb	49.2	24.6
						Fe	49.2	24.6
						Sn	49.2	24.6
			981.0	492.0	6–9.5	P total	49.2	24.6

(*continued*)

1977 Effluent Guidelines (USEPA)

Chapter code and industrial category	Contaminant levels							
	BOD		TSS				Oil and grease	
	maximum daily	average 30 days	maximum daily	average 30 days	pH		maximum daily	average 30 days
27.1 Pulp, paper and paperboard								
Unbleached kraft (#/ton product)	11.2	5.6	24.0	12.0	6–9			
Na based neutral sulfite semi-chemical	17.4	8.7	22.0	11.0	6–9			
NH₄ based neutral sulfite semi-chemical	16.0	8.0	20.0	10.0	6–9			
Unbleached kraft neutral sulfite semichemical	16.0	8.0	25.0	12.5	6–9			
Paperboard from wastepaper	6.0	3.0	10.0	5.0	6–9			
27.2 Builders paper and roofing felt								
Builders paper and roofing (#/ton product)	10.0	6.0	10.0	6.0	6–9	Settleable solids	≤ 0.2 ml/l	
26.10 Meat products								
Simple slaughterhouse								
(a) On-site slaughtering (#/1000# LWK)	0.24	0.12	0.40	0.20	6–9	Oil & grease	0.12	0.06
						Fecal coliform	≤ 400/100 ml	
(b) Off-site slaughtering	0.04	0.02	0.08	0.04				
(c) Blood processing off site	0.04	0.02	0.08	0.04				
(d) Wet or low temperature rendering of material derived from off-site slaughtering (#/1000# ELWK)	0.06	0.03	0.12	0.06				
(e) Dry rendering of material derived from off-site slaughtering	0.02	0.01	0.04	0.02	6–9			
Complex slaughterhouse								
(a)	0.42	0.21	0.50	0.25	6–9	Oil & grease	0.16	0.08
						Fecal coliform	≤ 400/100 ml	
(b)	0.42	0.02	0.08	0.04				
(c)	0.04	0.02	0.08	0.04				
(d)	0.06	0.03	0.12	0.06				
(e)	0.02	0.01	0.04	0.02				
Low processing packinghouse						Fecal coliform	≤ 400/100 ml	
(a)	0.34	0.17	0.48	0.24	6–9	Oil & grease	0.16	0.08
(b)	0.04	0.02	0.08	0.04				
(c)	0.04	0.02	0.08	0.04				
(d)	0.06	0.03	0.12	0.06				
(e)	0.02	0.01	0.04	0.02				
High processing packinghouse					6–9	Oil & grease	0.26	0.13
(a)	0.48	0.24	0.62	0.31		Fecal coliform	≤ 400/100 ml	
(b)	0.04	0.02	0.08	0.04				
(c)	0.04	0.02	0.08	0.04				
(d)	0.06	0.03	0.12	0.06				
(e)	0.02	0.01	0.04	0.02				

1977 Effluent Guidelines (USEPA)

Chapter code and industrial category	Contaminant levels							
	BOD		TSS					
	maximum daily	average 30 days	maximum daily	average 30 days	pH		maximum daily	average 30 days
26.10 (Continued)								
Small processor (up to 3 tons product/day) (#/1000# product)	2.0	1.0	2.4	1.2	6–9	Oil & grease Fecal coliform	1.0 no limit	0.5
Meat cutter (#/1000# finished product)	0.036	0.018	0.044	0.022	6–9	Oil & grease Fecal coliform	0.012 ≤ 400/100 ml	0.006
Sausage and luncheon meats	0.56	0.28	0.68	0.34	6–9	Oil & grease Fecal coliform	0.20 ≤400//100 ml	0.10
Ham processor	0.62	0.31	0.74	0.37	6–9	Oil & grease Fecal coliform	0.22 ≤ 400/100 ml	0.11
Canned meat processor	0.74	0.37	0.90	0.45	6–9	Oil & grease Fecal coliform	0.26 ≤ 400/100 ml	0.13
Rendering	0.14	0.07	0.20	0.10	6–9	Oil & grease Fecal coliform NH_3	0.10 ≤ 400/100 ml 0.04	0.05 0.02
29.3 Coal mining								
Coal preparation —any untreated overflow from facilities designed to contain process-generated wastewater and surface runoff from 10-year, 24-hour expected rainfall					6–9			
Coal storage, refuse storage, coal preparation plant ancillary area; Fe total, Fe dis., Al total, Man.$_t$, Ni.$_t$, and TSS to be determined					6–9			
Acid or ferruginous mine drainage	To be determined				6–9			
Alkaline mine drainage	To be determined							
27.9 Oil and gas extraction								
Near offshore (mg/l) (produced water)						Oil & grease	72.0	48.0
Near offshore (mg/l) (deck drainage)						Oil & grease	72.0	48.0
Far offshore (produced water)						Oil & grease	72.0	48.0
Far offshore (deck drainage)						Oil & grease	72.0	48.0
27.23 Mineral mining and processing								
Gypsum (impoundment)	No discharge of process waters for operations not employing wet scrubbers; only that volume of water resulting from rain that exceeds maximum safe surge capacity							
Asphaltic mineral	Same as for gypsum							
Asbestos and wollastonite	Same as for gypsum							
Barite	No discharge of process wastewater not employing wet operations or flotation processes							
Fluorspar	No discharge of process wastewater not employing heavy media separation or flotation							

(continued)

1977 Effluent Guidelines (USEPA)

Chapter code and industrial category	Contaminant levels							
	BOD		TSS				Oil and grease	
	maximum daily	average 30 days	maximum daily	average 30 days	pH		maximum daily	average 30 days
27.23 (Continued)								
Salines from brine lakes	No discharge of process wastewater							
Borax	Same as for gypsum							
Potash	Same as for gypsum							
Na$_2$SO$_4$	Same as for gypsum							
Frasch sulfur	Same as for gypsum							
Bentonite	No discharge of process generated wastewater							
Magnesite	Same as for gypsum							
Diatomite	Same as for gypsum							
Jade	Same as for gypsum							
Novaculite	Same as for gypsum							
Tripoli	No discharge of process wastewater							
Graphite (mg/l)			20.0	10.0	6–9	Total Fe	2.0	1.0

References

1. Bloodgood, D. E., "1953 Industrial wastes forum," *Sewage Ind. Wastes* **26**, 640 (1954).
2. Cleary, E. J., "Water pollution control—Gearing performance to promise," *Civil Eng.* **38**, 63 (1968).
3. Dappert, A. P., "Pollution control through the mechanism of classes and standards," *Sewage Ind. Wastes* **24**, 313 (1952).
4. Grossman, I., "Experiences with surface water quality standards," *J. Sanit. Eng. Div. Am. Soc. Civil Engrs.* **94**, (SAI) 13 (1968).
5. Hubbard, E. C., "Stream standards," *J. Water Pollution Control Federation* **37**, 308 (1965).
6. Ingols, R. S., "Surface water pollution and natural purification," *Municipal South*, p. 31 (January 1956).
7. Kittrell, F. W., "Effects of impoundments on dissolved oxygen resources," *Sewage Ind. Wastes* **31**, 1965 (1959).
8. New York State Water Pollution Control Board, "Classification and standards of water quality and purity," *J. Am. Water Works Assoc.* **42**, 1137 (1950).
9. *Public Health Service Drinking Water Standards*, No. 956, Wash. D.C. (1962).
10. "ORSANCO's success story," *Public Works*, **19**, 66 (1965).
11. Rambow, C. A., and R. O. Sylvester, "Methodology in establishing water quality standards," *J. Water Pollution Control Federation* **39**, 1155 (1967).
12. Sheets, J. L., *Evaluation of Pollution Abatement Benefits from Low-Flow Augmentation*, Department of Civil Engineering, University of Illinois, Urbana (July 1964) p. 91.
13. Streeter, H. W., "Standards of stream sanitation," *Sewage Ind. Wastes* **21**, 115 (1949).
14. Velz, C. J., J. D. Calvert, Jr., R. A. Deininger, W. L. Heilman, and J. Z. Reynolds, *J. Sanit. Eng. Div. Am. Soc. Civil Engrs.* **SAI**, 159 (1968).
15. Wendell, M., "Intergovernmental relations in water quality control," *J. Water Pollution Control Federation* **39**, 278 (1967).
16. Worley, J. L., F. J. Burgess, and W. W. Towne, "Identification of low flow augmentation requirements for water quality control by computer techniques," *J. Water Pollution Control Federation* **37**, 659, (1965).
17. "Helms Now Operating," Pacific Gas and Elec. Co., 2nd Qtr. Report, San Francisco, CA, 4 (June 30, 1984).
18. Jain, R.K., L.V.Urban and G.S. Stacey, "Environmental Impact Analysis," Van Nostrand Reinhold Publishing Company, New York City (1981).
19. Goodman, L.J., "Project Planning and Management," Van Nostrand Reinhold Publishing Company, New York City (1988).
20. Leopold, L.B. et al., "A Procedure for Evaluating Environmental Impact." U.S. Geological Survey Circulat, 655 (1971).

21. Goodman, A. "Principles of Water Resources Planning," Prentice Hall Publishing Company, Englewood Cliffs, New Jersey (1984).

Questions

1. What are the major uses of rivers, streams, and lakes?

2. Describe the two predominant systems of receiving-water protection.

3. Which system is in current use in your region of the country?

4. List and describe four procedures for improving receiving-water quality exclusive of waste treatment.

5. What is the "Effluent Guideline System?"

CHAPTER 3

COMPUTATION OF ORGANIC WASTE LOADS ON STREAMS

Despite an industrial plant's efforts to reduce the volume and strength of its wastes to a minimum (to be described in Chapters 6 and 7), some waste will still remain to be disposed of. A detailed analysis is needed, to show the volume and character of the remaining waste and to determine how much treatment the waste requires, before an industry can satisfy either laws or public opinion.

The degree of treatment necessary depends primarily on the condition and best usage of the receiving stream. Overtreatment of wastes results in unnecessary, burdensome expense; undertreatment is only a waste of effort and money, since it does not abate the pollution problem. Thus one can readily comprehend the value of calculating as closely as possible the amount of pollution that can safely be discharged into a stream.

Although this text does not propose to cover the entire field of stream sanitation, the major methods of determining the required degree of waste treatment will be described here in some detail. The Streeter-Phelps formulation [6] has been used with various degrees of success over the past thirty years. Thomas simplified the Streeter-Phelps formulations in actual practice. Another method (Churchill) has come into common usage only within the last thirteen years. Other methods are available for determining stream reaeration and stream purification coefficients (such as Fair's factor), but these will not be discussed in this chapter because of space limitations.

3.1 Streeter-Phelps Formulations

We first present a list of the symbols used to denote the various parameters to be calculated:

K_1 = deoxygenation rate/day
Δ_t = time of travel (days)
L_A = ultimate upstream BOD (ppm or lb)
L_B = ultimate downstream BOD (ppm or lb)
K_2 = reaeration rate/day
$\overline{L}$ = average ultimate oxygen demand in stream section (ppm or lb)
$\overline{D}$ = average oxygen deficit in stream section (ppm or lb)
ΔD = change in oxygen deficit from upstream to downstream sampling points (ppm or lb)
t = time of stream flow from upstream to downstream points of sampling (days)
D_t = dissolved-oxygen deficit downstream (ppm or lb), at time t
D_A = dissolved-oxygen deficit upstream (ppm or lb).

The following formulas are used to compute the deoxygenation rate (K_1),* reaeration rate (K_2), and the dissolved-oxygen deficit (D_t) at a downstream location:

$$K_1 = \frac{1}{\Delta t} \log \frac{L_A}{L_B} = \text{deoxygenation rate}, \qquad (1)$$

$$K_2 = K_1 \frac{\overline{L}}{\overline{D}} - \frac{\Delta D}{2.3 \Delta t \overline{D}} = \text{reaeration rate}, \qquad (2)$$

$$D_t = \frac{K_1 L_A}{K_2 - K_1}[10^{-K_1 t} - 10^{-K_2 t}] + D_A \cdot 10^{-K_2 t}$$
$$= \text{dissolved oxygen deficit downstream.} \quad (3)$$

We now apply the Streeter-Phelps formulations to calculate the allowable organic loading for the stream situation sketched in Fig. 3.1. Using the stream data

---

*K_1 includes K_3 (deoxygenation due to bottom deposits), unless computed separately.

K_3 is the constant of proportionality which reflects the composition of the waste and the characteristics of the receiving water, as well as the quiescence of the stream at the point under consideration. It represents the amount of BOD which is removed by sedimentation. Therefore, in regions of considerable turbulence, K_3 is usually zero and under scouring conditions may even be negative. Since it is so difficult to compute, it is generally considered as an integral part of the deoxygenation rate, K_1.

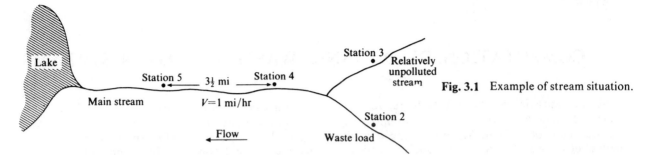

Fig. 3.1 Example of stream situation.

Table 3.1 Data for the stream illustrated in Fig. 3.1. Two samples, collected on different days, were taken at each station.

| | Temperature, °C | Flow | | 5-day BOD | | | Dissolved oxygen | | |
		cfs*	mgd	ppm	lb/day	lb/day	Saturated, ppm	ppm	Deficit, ppm
Station 4	20	60.1	38.82	36	11,650	1325	9.2	4.1	5.1
	17	54	34.88	10.35	3,015	1512	9.8	5.2	4.6
				Total	14,665				
				Average	7,332				
Station 5	20	51.7	33.40	21.2	5,905	725	9.2	2.6	6.6
	16.5	66.6	43.02	5.83	2,095	1050	9.9	2.8	7.1
				Total	8,000				
				Average	4,000				

*Cubic feet per second.

given in Table 3.1, as representing the stream prior to any waste treatment, we obtain the following quantitative results:

$$K_1 = \frac{1}{3.5/24} \log \frac{7332 \times 1.46\dagger}{4000 \times 1.46\dagger},$$
$$= 1.8; \tag{4}$$

$$L = \frac{\text{BOD (Station 4)} + \text{BOD (Station 5)}}{2}$$
$$= \frac{7332 \times 1.46 + 4000 \times 1.46}{2}$$
$$= 8273 \text{ lb/day};$$

$$\overline{D} = \frac{\left\{ \begin{array}{c} 6.6 \times 8.34 \times 33.4 + 5.1 \times 8.34 \times 38.8 \\ + 7.1 \times 8.34 \times 43 + 4.6 \times 8.34 \times 34.9 \end{array} \right\}}{4}$$
$$= 1843 \text{ lb/day};$$

$$\Delta D = 1/2 \, (6.6 \times 8.34 \times 33.4 + 7.1 \times 8.34 \times 43)$$
$$-1/2 \, (5.1 \times 8.34 \times 38.8 + 4.6 \times 8.34 \times 34.9)$$
$$= 698 \text{ lb/day};$$

$$\Delta t = \frac{3.5}{24} = 0.146 \text{ day};$$

$$K_2 = 1.8 \cdot \frac{8273}{1843} - \frac{698}{2.3(0.146)(1843)}$$
$$= 8.08 - 1.12 = 6.96. \tag{5}$$

---

†Multiplier used to convert 5-day 20°C BOD to ultimate first-stage BOD, assuming normal domestic sewage deoxygenation rates.

Using Fair's f formula, we have

$$f = \frac{K_2}{K_1} = \frac{6.96}{1.80} = 3.8.$$

According to Fair's classification, the result is characteristic of streams falling in his group D, i.e., "streams with normal velocity that can almost be considered as swift streams."

With these stream reaction rates (K_1 and K_2) and the initial condition at Station 4 (D_A and L_A), it is possible to plot the dissolved-oxygen deficit (sag curve) at any point in the stretch between Station 4 and Station 5, assuming that the reaction rates remain constant in the stretch. Figure 3.2 is a graphic plot of the actual sag curve.

If we assume that the waste load entering at Station 2 has undergone primary treatment resulting in a 30 per cent reduction of BOD and that the distance between Stations 2 and 4 does not change the effect of the pollution load or dissolved-oxygen deficit, we obtain the following BOD values:

$$90 \text{ (ppm BOD at Station 2)}$$
$$\times \ 8.34 \times 8.1 \text{ cfs}$$
$$\times \ 0.65 \text{ (mgd/cfs)} = 3951 \text{ BOD lb/day},$$

$$92 \text{ (ppm BOD at Station 2)}$$
$$\times \ 8.34 \times 8.22 \text{ cfs}$$
$$\times \ 0.65 \text{ (mgd/cfs)} = 4099 \text{ BOD lb/day},$$

$$\text{Total} = 8050,$$
$$\text{Average} = 4025 \text{ BOD lb/day},$$
$$L = 4025 \times 1.46$$
$$= 5876 \text{ BOD lb/day}.$$

Hence, at Station 4, we have

$$L_4 = 7332 \times 1.46 - 0.30 \ (5876)$$
$$= 8942.3 \text{ lb/day}.$$

The dissolved-oxygen deficit at Station 4 is obtained by the following calculation:

$$D_4 = \frac{4.6 \times 8.34 \times 34.9 + 5.1 \times 8.34 \times \overline{38.8}}{2}$$

$$1494 \text{ lb/day},$$

and the dissolved-oxygen deficit downstream at Station 5 is obtained from Eq. (3):

$$D_5 = 1.8 \times \frac{8942}{6.96 - 1.80} [10^{-1.8(0.146)} - 10^{-6.96(0.146)}]$$
$$+ \ 1494 \times 10^{-6.96(0.146)}$$
$$= 3118 \ [10^{-0.263} - 10^{-1.016}] + 1494 \times 10^{-1.016}$$
$$= 3118 \ [0.5458 - 0.0964] + 1494 \ (0.0964)$$
$$= 3118 \ [0.4494] + 144$$
$$= 1400 + 144$$
$$= 1544 \text{ lb/day}.$$

At the lowest flow (33.4 mgd) and the highest temperature (20°C) observed at Station 5, we have 9.2 (ppm dissolved oxygen (DO) at saturation at

$$\text{Station 5)} - \frac{1544}{33.4 \times 8.34},$$

or

$$9.20 - 5.55 = 3.65 \text{ ppm DO};$$

and at 25°C we have

$$8.40 - 5.55 = 2.85 \text{ ppm DO}.$$

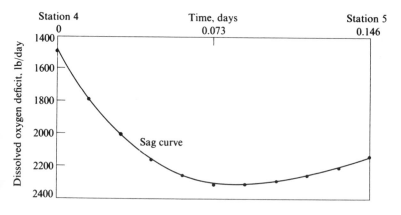

Fig. 3.2 Plot of sag curve.

Therefore, primary treatment would be insufficient to keep the stream at a minimum of 4 ppm DO during summer low flows and high temperatures.

The actual dissolved-oxygen deficit at the bottom of the sag at $t = 0.073$ days (see Fig. 3.2) can be computed as follows:

$$D_{0.075 \text{ days}} = 1.8 \times \frac{8942}{6.96-1.80}(10^{-1.86 \times 0.146}$$
$$-10^{-6.96 \times 0.146}) + 1494 \times 10^{-6.96 \times 0.146}$$

Cohen and O'Connell [2] have found the analog computer valuable in providing rapid information on the effect of variations in initial stream conditions. Because it is easy to modify the data fed in, many more conditions of potential interest can be simulated and thus investigated with little additional effort. It will be apparent to the reader that more rapid, automatic methods are essential whenever the Streeter-Phelps equations are used and especially when the engineer is attempting to ascertain the effects of industrial wastes under a variety of possible stream conditions.

3.2 Thomas Method for Determining Pollution-Load Capacity of Streams

Thomas [7] developed a useful simplification of the Streeter-Phelps equations for computing stream capacity. In his method the stream-reaction constants K_1 and K_2 are computed as in Section 3.1. However, he proposes using a nomograph to compute the dissolved-oxygen deficit at any time t downstream from a source of pollution load. Conversely, a pollution load producing a critical dissolved-oxygen deficit downstream can be calculated from the same nomograph. The nomograph which plots D/L_A versus $K_2 t$ for various ratios of K_2/K_1 is shown in Fig. 3.3. Thomas recognizes that Eq. (3) is unwieldy and in most practical applications can be solved only by tedious trial-and-error procedures. He believes that this disadvantage can be overcome by the use of his nomogram. Before the nomogram is used, K_1, K_2, D_A, and L_A must be computed. By means of a straight-edge, a straight line (isopleth) is drawn, connecting the value of D_A/L_A at the left with the point presenting the appropriate day multiplied by the reaeration constant ($K_2 t$) on the appropriate (K_2/K_1)-curve. The value of D/L_A is then read at the intersection with the isopleth. Finally, the value of the deficit at the end of the appropriate day is obtained by multiplying L_A and the intersection value.

The author has used the Thomas nomogram on many occasions and found it very convenient, accurate, and time-saving. The following problem illustrates the use of the method.

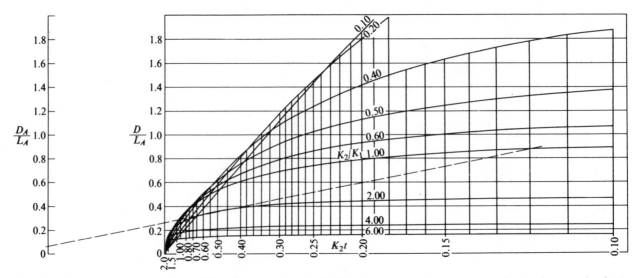

Fig. 3.3 Nomogram for the dissolved-oxygen sag. Oxygen deficits downstream from a point of pollution may be determined from initial BOD and DO and stream self-purification constants. (After Thomas [7].)

N ←——————

Station 6

Cape Fear River

Carvers

33

Lake Singletary

Station 5

87

White Lake

Station 4

U.S. lock no. 2

Elizabethtown

Station 3

Harrison Creek

33

87

Station 2

Dublin

Station 1

Tarheel

Fig. 3.4 Section of Cape Fear River from ferry at Tarheel, N.C., to ferry near Carvers, N.C. (N.C. State Board of Health, Sanitary Eng. Div., 8-2-51.)

An industrial plant planning to discharge a 5-day 20°C BOD load of 5000 pounds was proposed at Station 1 in the Cape Fear River in North Carolina (Fig. 3.4). The North Carolina Stream Sanitation Committee was concerned about both the present condition of the stream stretch below Station 1 and the future condition, if and when the site was approved.

The river was sampled at six locations from Station 1 to Station 6, a distance of 38.1 miles. Ultimate BOD, deoxygenation, and reaeration rates were computed in each stretch of the river and used to obtain the oxygen-sag curve. A hypothetical load of 5000 pounds of 5-day BOD was imposed on the river at Station 1, and the Thomas method was used to draw the new oxygen-sag curve based on similar reaction rates.

The calculations are shown in Tables 3.2 through 3.9. The subscript B refers to laboratory bottle values and Str is the value at existing stream temperature. The sag curves (existing and projected), with present flow and 5-year minimum flow, are shown in Fig. 3.5.

Calculations of oxygen deficits in stream. We first present the calculations of the oxygen deficit existing in the stream at the time of the investigation. To obtain the K_2-value for the stream stretch between stations, we proceeded as follows, using the K_1-value (0.331) as the most representative, realistic, and appropriate value. (Negative K_1 values must be erroneous.)

Table 3.2 Cape Fear velocity measurements.

Floats: 4 miles/280 min = 0.86 miles/hr × 0.85* = 0.69 miles/hr

Station	Mile	Distance between stations, miles	Flow time		Total flow time, days	L at 20°C, ppm	K_B	DO, ppm	T,°C	L_R, ppm	$K_{B_{Str}}$	K_1	K_2
			hr	days									
1	0	0	0	0	0	1.71	0.101	5.8	26.0	1.92	0.133	−0.95	−0.139
2	5.33	5.33	7.73	0.322	0.322	3.40	0.054	5.4	26.5	3.84	0.073	0.585	0.159
3	13.33	8.00	11.6	0.483	0.805	1.75	0.098	5.0	27.2	2.00	0.135	−2.21	−1.73
4	16.27	2.94	4.25	0.177	0.982	4.18	0.035	5.9	29	4.93	0.052	0.331	0.463
5	31.17	14.90	21.65	0.902	1.884	2.09	0.100	5.4	28.9	2.47	0.149	0.332	0.105
6	38.11	6.94	10.06	0.419	2.303	1.54	0.151	5.1	28.2	1.79	0.216		

*To correct for wind and other surface effects.

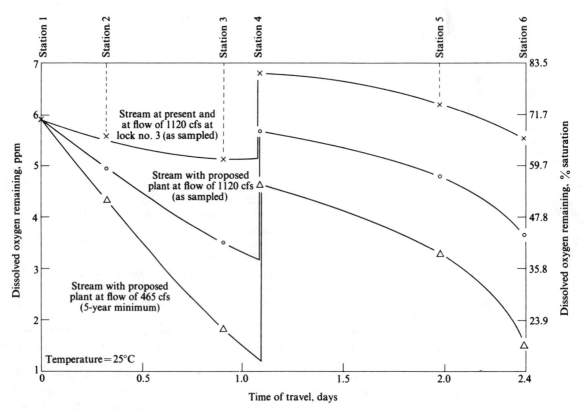

Fig. 3.5 Oxygen-sag curves of Cape Fear River below Tarheel, N.C. (N.C. State Board of Health, Sanitary Eng. Div., 8-2-51, N.L.N.)

Table 3.3 Calculations of deoxygenation rate (K_1) of stream.

$$K_1 = \frac{1}{t} \log \frac{L_A}{L_B}$$

From Station 1 to Station 2:

$$K_1 = \frac{1}{0.322} \log \frac{1.92}{3.84}$$
$$= \frac{1}{0.322} (0.282 - 0.584)$$
$$= \frac{1}{0.322} (-0.302)$$
$$= -0.95$$

From Station 2 to Station 3:

$$K_1 = \frac{1}{0.483} \log \frac{3.84}{2.00}$$
$$= \frac{0.282}{0.483} = 0.585$$

From Station 3 to Station 4:

$$K_1 = \frac{1}{0.177} \log \frac{2.00}{4.93}$$
$$= \frac{1}{0.177} (0.301 - 0.692)$$
$$= -\frac{0.391}{0.177}$$
$$= -2.21$$

From Station 4 to Station 5:

$$K_1 = \frac{1}{0.902} \log \frac{4.93}{2.47}$$
$$= \frac{0.299}{0.902} = 0.331$$

From Station 5 to Station 6:

$$K_1 = \frac{1}{0.419} \log \frac{2.47}{1.79}$$
$$= \frac{0.139}{0.419} = 0.332$$

Between Stations 1 and 2:

$$K_2 = \frac{L}{D} K_1 - \frac{\Delta D}{2.3 \, \Delta t \, \overline{D}},$$

$$\overline{L} = \frac{11,600 + 25,800}{2} = 18,700,$$

$$\overline{D} = \frac{18,500 + 14,600}{2} = 16,550,$$

$$D = 18,550 - 14,600 = 3900,$$

$$t = 0.322;$$

hence

$$K_2 = 0.331 \frac{18,700}{16,550} - \frac{3900}{2.3 \times 0.322 \times 16,550}$$
$$= 0.374 - \frac{3900}{12,257} = 0.056.$$

Again, using 0.331 as K_1 of the stream section, one obtains

K_2	K_1	K_2/K_1	D_A	L_A	D_A/L_A	$K_2 t$
0.056	0.331	0.202	2.42	1.92	1.26	0.067 × 0.322 = 0.022

From extension of the nomograph, we obtain

$$\frac{D}{L_A} = 1.46,$$

and therefore

$$D = 1.46 \times 1.92 = 2.80 \text{ (calculated deficit at Station 2)}$$
$$(2.75 = \text{observed deficit}).$$

Between Stations 2 and 3:

K_2	K_1	K_2/K_1	D_A	L_A	D_A/L_A	$K_2 t$
0.159	0.331	0.48	2.75	3.84	0.716	0.159 × 0.483 = 0.077

From extension of the nomograph, we obtain

$$\frac{D}{L_A} = 0.9,$$

and therefore

$$D = 0.9 \times 3.84 = 3.25 \text{ (calculated deficit at Station 3)}$$
$$(3.04 = \text{observed deficit}).$$

Between Stations 3 and 4:

$$K_2 = 0.331 \frac{31,150}{22,550} - \frac{(-12,100)}{(2.3 \times 0.177 \times 22,550)}$$
$$= 0.457 + \frac{12,100}{9,180} = 0.457 + 1.320 = 1.777.$$

K_2	K_1	D_A	L_A	K_2/K_1	D_A/L_A	$K_2 t$
1.777	0.331	3.04	2.00	5.36	1.53	1.777 (0.177) = 0.314

From nomograph, we obtain

$$\frac{D}{L_A} = 0.85,$$

and therefore

$D = 0.85 \times 2.00 = 1.70$ (calculated deficit at Station 4)
(1.87 = observed deficit).

Between Stations 4 and 5:

K_2	K_1	K_2/K_1	D_A	L_A	D_A/L_A	K_2t
0.463	0.331	1.4	1.87	4.93	0.38	0.463 (0.902) = 0.418.

From nomograph, we obtain

$$\frac{D}{L_A} = 0.45,$$

and therefore

$D = 0.45 (4.93) = 2.22$ (calculated deficit at Station 5)
(2.38 = observed deficit).

Between Stations 5 and 6:

K_2	K_1	K_2/K_1	D_A	L_A	D_A/L_A	K_2t
0.105	0.322	0.316	2.38	2.47	0.96	0.105 (0.42) = 0.044

From extension of the nomograph, we obtain

$$\frac{D}{L_A} = 1.18,$$

and therefore

$D = 1.18 \times 2.47 = 2.91$ (calculated deficit at Station 6)
(2.79 observed deficit).

Imposition of 5000 pounds of 5-day BOD at Station 1. We next present calculations which establish what the new oxygen-sag curve would be after imposition of 5000 pounds of 5-day BOD on the river at Station 1. The Thomas nomograph was used, and it was assumed that reaction rates would be similar to those already observed in the river:

724 mgd $\times$ 5.8 + 1.5 $\times$ 0 = 725.5X (initial river DO with industrial waste)

$$\frac{4200}{725.5} = X = 5.79$$

$8.22 - 5.79 = 2.43 =$ oxygen deficit (D)
5000 lb/day = ppm $\times$ 8.34 $\times$ 1.5 mgd
ppm = 400 of 5-day BOD.

(a)   (b)

(724 mgd $\times$ 1.23 + 1.5 $\times$ 400) $\dfrac{1.92}{1.23}$ = 725.5X

(890 + 600) $\dfrac{1.92}{1.23}$ = 725X

ppm L $\dfrac{2325}{725.5} = X = 3.21$

(a) = 5-day BOD of river at Station 1 at 20°C;
(b) = increase due to conversion to L-value and stream temperature.

Table 3.4   Calculations of stream oxygen deficits.

Station	Temperature, °C	DO, ppm			Assumed flow, mgd	Downstream minus upstream oxygen deficit, lb/day
		Saturated	Observed	Deficit		
1	26	8.22	5.8	2.42	724	(807 $\times$ 8.34 $\times$ 2.75) − (724 $\times$ 8.34 $\times$ 2.42)
						18,500 − 14,600 = +3,900
2	26.5	8.15	5.4	2.75	807	(1130 $\times$ 8.34 $\times$ 3.04) − (807 $\times$ 8.34 $\times$ 2.75)
						28,600 − 18,500 = +10,100
3	27.2	8.04	5.0	3.04	1130	(1060 $\times$ 8.34 $\times$ 1.87) − (1130 $\times$ 8.34 $\times$ 3.04)
						16,500 − 28,600 = −12,100
4	29	7.77	5.9	1.87	1060	(1060 $\times$ 8.34 $\times$ 2.38) − (1060 $\times$ 8.34 $\times$ 1.87)
						21,000 − 16,500 = +4,500
5	28.9	7.78	5.4	2.38	1060	(1060 $\times$ 8.34 $\times$ 2.79) − (1060 $\times$ 8.34 $\times$ 2.38)
						24,700 − 21,000 = +3,700
6	28.2	7.79	5.1	2.69	1060	

Table 3.5 Calculations of stream ultimate biochemical oxygen demand (BOD).

Station	$L_{20°C}$, ppm	L_R, ppm	L, lb/day
1	1.71	1.92	$\left\{\begin{array}{c}1.92 \times 724 \times 8.34 \\ + 3.84 \times 807 \times 8.34\end{array}\right\}$ / 2 $=\dfrac{11,600 + 25,800}{2}$ $= 18,700$
2	3.40	3.84	$\left\{\begin{array}{c}3.84 \times 807 \times 8.34 \\ + 2.00 \times 1130 \times 8.34\end{array}\right\}$ / 2 $=\dfrac{25,800 + 18,800}{2}$ $= 21,800$
3	1.75	2.00	$\left\{\begin{array}{c}2.00 \times 1130 \times 8.34 \\ + 4.93 \times 1060 \times 8.34\end{array}\right\}$ / 2 $=\dfrac{18,800 + 43,500}{2}$ $= 31,150$
4	4.18	4.93	$\left\{\begin{array}{c}4.93 \times 1060 \times 8.34 \\ + 2.47 \times 1060 \times 8.34\end{array}\right\}$ / 2 $=\dfrac{43,500 + 21,800}{2}$ $= 32,650$
5	2.09	2.47	$\left\{\begin{array}{c}2.47 \times 1060 \times 8.34 \\ + 1.79 \times 1060 \times 8.34\end{array}\right\}$ / 2 $=\dfrac{21,800 + 15,800}{2}$ $= 18,800$
6	1.54	1.79	

Between Stations 1 and 2:

K_2	K_1	K_2/K_1	D_A	L_A	D_A/L_A	K_2t
0.067	0.331	0.202	2.43	3.21	0.76	0.067×0.322 $= 0.022$

From nomograph, we obtain

$$\frac{D}{L_A} = 1.06,$$

and therefore

$$D = 1.06(3.21) = 3.40 \text{ (calculated deficit at Station 2).}$$

Between Stations 2 and 3:

K_2	K_1	K_2/K_1	D_A
0.159	0.331	0.48	3.40

L_A	D_A/L_A	K_2t
$\left(3.21 \times \dfrac{3.84}{1.92}\right)$ 6.42^*	0.53	0.159×0.483 $= 0.077$

From nomograph, we obtain

$$\frac{D}{L_A} = 0.76,$$

and therefore

$$D = 0.76(6.42) = 4.88 \text{ (calculated deficit at Station 3).}$$

Between Stations 3 and 4:

K_2	K_1	K_2/K_1	D_A
1.777	0.331	5.36	4.88

L_A	D_A/L_A	K_2t
$\left(6.42 \times \dfrac{2.00}{3.84}\right)$ 3.34	1.46	1.777×0.177 $= 0.314$

From nomograph, we obtain

$$\frac{D}{L_A} = 0.81,$$

and therefore

$$D = 0.81(3.34) = 2.71 \text{ (calculated deficit at Station 4).}$$

$^*\dfrac{\text{New } L_{R^2}}{\text{New } L_{R^1}} = \dfrac{\text{Old } L_{R^2}}{\text{Old } L_{R^1}}$

and therefore

New $L_{R^2} = $ New $L_{R^1} \times \dfrac{\text{Old } L_{R^2}}{\text{Old } L_{R^1}}$.

(R^1 and R^2 refer to station locations on the river, R.)

Between Stations 4 and 5:

K_2	K_1	K_2/K_1	D_A
0.463	0.331	1.4	2.71

L_A	D_A/L_A	$K_2 t$
$\left(3.34 \times \dfrac{4.93}{2.00}\right)$	0.33	0.463×0.902
8.23		$= 0.418$

From nomograph, we obtain

$$\frac{D}{L_A} = 0.44,$$

and therefore

$$D = 0.44(8.23) = 3.62 \text{ (calculated deficit at Station 5).}$$

Between Stations 5 and 6:

K_2	K_1	K_2/K_1	D_A
0.105	0.332	0.316	3.62

L_A	D_A/L_A	$K_2 t$
$\left(8.23 \times \dfrac{2.47}{4.93}\right)$	0.88	$0.105(0.42)$
4.13		$= 0.044$

From nomograph, we obtain

$$\frac{D}{L_A} = 1.16,$$

and therefore

$$D = 1.16 \times 4.13 = 4.80 \text{ (calculated deficit at Station 6).}$$

Table 3.6   Calculation of stream average oxygen deficits.

Station	$\bar{D}$, lb/day
1	$\dfrac{18,500 + 14,600}{2} = \dfrac{33,100}{2} = 16,555$
2	$\dfrac{28,600 + 18,500}{2} = \dfrac{47,100}{2} = 23,550$
3	$\dfrac{16,500 + 28,600}{2} = \dfrac{45,100}{2} = 22,550$
4	$\dfrac{21,000 + 16,500}{2} = \dfrac{37,500}{2} = 18,750$
5	$\dfrac{23,800 + 21,000}{2} = \dfrac{44,800}{2} = 22,400$

Table 3.7   Calculations of stream reaeration rate (K_2).

Between Stations	
1 and 2	$K_2 = K_1 \cdot \dfrac{L}{D} - \dfrac{\Delta D}{2.3\, \Delta t\, \bar{D}}$
	$= -0.95 \cdot \dfrac{18,700}{16,555} - \dfrac{(3900)}{2.3(0.322)16,555}$
	$= -1.07 - \dfrac{3900}{12,265} = -1.07 - 0.318$
	$= -1.390$
2 and 3	$K_2 = 0.585 \cdot \dfrac{21,800}{23,550} - \dfrac{10,100}{2.3 \times 0.483 \times 23,550}$
	$= 0.541 - \dfrac{10,100}{26,200} = 0.541 - 0.382$
	$= 0.159$
3 and 4	$K_2 = -2.21 \cdot \dfrac{31,150}{22,550} - \dfrac{-12,100}{2.3 \times 0.177 \times 22,550}$
	$= -3.05 + \dfrac{12,100}{9160} = -3.05 + 1.32$
	$= -1.73$
4 and 5	$K_2 = 0.331 \cdot \dfrac{32,650}{18,750} - \dfrac{4500}{2.3 \times 0.902 \times 18,750}$
	$= +0.579 - \dfrac{4500}{38,850} = +0.579 - 0.116$
	$= +0.463$
5 and 6	$K_2 = 0.332 \cdot \dfrac{18,800}{22,850} - \dfrac{3700}{2.3 \times 0.419 \times 22,850}$
	$= +0.273 - 0.168 = 0.105$

Table 3.8   Summary of deficits and reaction rates used.

Between Stations	K_1	K_2	Downstream oxygen deficit	
			Calculated, ppm	Observed, ppm
1 and 2	0.331	0.067	2.80	2.75
2 and 3	0.331	0.159	3.25	3.04
3 and 4	0.331	1.777*	1.70	1.87
4 and 5	0.331	0.463	2.22	2.38
5 and 6	0.332	0.105	2.91	2.69

*City of Elizabethtown plus Lock No. 2 caused this high value.

Table 3.9 Oxygen left in water at Stations 1 to 6 at 25°C.

Station	DO satu-rated, ppm	Deficits (calculated), ppm			O_2 remaining, ppm			O_2 remaining, % saturation		
		No plant	With plant	Plant at low flow	No plant	With plant	Plant at low flow	No plant	With plant	Plant at low flow
1	8.38	2.42	2.43	2.45	5.96	5.95	5.93	71	71	70.5
2	8.38	2.80	3.40	4.07	5.58	4.98	4.31	66.7	59.5	51.4
3	8.38	3.25	4.88	6.54	5.13	3.50	1.84	61.3	41.8	22.0
4	8.38	1.70	2.71	3.71	6.68	5.67	4.67	79.8	67.7	55.8
5	8.38	2.22	3.62	5.16	6.16	4.76	3.22	73.6	57.0	38.4
6	8.38	2.91	4.80	6.97	5.47	3.58	1.41	65.3	42.8	16.8

Imposition of 5000 pounds 5-day BOD at 5-year minimum flow. We shall now examine the situation if the plant's waste load were imposed on the river at its 5-year minimum flow. Again we use the Thomas nomograph to calculate the oxygen deficit:

$$300 \times 5.8 + 1.5 + 0 = 301.5X$$

$$\frac{1740}{301.5} = X = 5.77 \text{ (initial DO at low flow)}$$

$$8.22 - 5.77 = 2.45 = \text{oxygen deficit } (D) \text{ at low flow.}$$

When $y = $ ppm L at low flow, then

$$(300 \times 1.23 + 1.5 \times 400) \frac{1.92}{1.23} = 301.5y$$

$$(369 + 600) \frac{1.92}{1.23} = 301.5y$$

$$\frac{1510}{301.5} = y = 5.02 \text{ ppm } L \text{ at low flow.}$$

Between Stations 1 and 2:

K_2	K_1	K_2/K_1	D_A	L_A	D_A/L_A	K_2t
0.067	0.331	0.202	2.45	5.02	0.488	0.067×0.322 = 0.022

From nomograph, we obtain

$$\frac{D}{L_A} = 0.81,$$

and therefore

$$D = 0.81(5.02) = 4.07 \text{ (calculated deficit at Station 2).}$$

Between Stations 2 and 3:

K_2	K_1	K_2/K_1	D_A
1.59	0.331	0.48	4.07

L_A	D_A/L_A	K_2t
$\left(5.02 \times \dfrac{3.84}{1.92}\right)$ 10.05	0.405	0.159×0.483 = 0.077

From nomograph, we obtain

$$\frac{D}{L_A} = 0.65,$$

and therefore

$$D = 0.65(10.05) = 6.54 \text{ (calculated deficit at Station 3).}$$

Between Stations 3 and 4:

K_2	K_1	K_2/K_1	D_A
1.777	0.331	5.36	6.54

L_A	D_A/L_A	K_2t
$\left(10.05 \times \dfrac{2.00}{3.84}\right)$ 5.23	1.25	1.777×0.177 = 0.314

From nomograph, we obtain

$$\frac{D}{L_A} = 0.71,$$

and therefore

$$D = 0.71(5.23) = 3.71 \text{ (calculated deficit at Station 4).}$$

Between Stations 4 and 5:

K_2	K_1	K_2/K_1	D_A
0.463	0.331	1.4	3.71

L_A	D_A/L_A	$K_2 t$
$\left(5.23 \times \dfrac{4.93}{2.00}\right)$	0.288	0.463×0.902
12.9		$= 0.418$

From nomograph, we obtain

$$\frac{D}{L_A} = 0.4,$$

and therefore

$D = 0.4(12.9) = 5.16$ (calculated deficit at Station 5).

Between Stations 5 and 6:

K_2	K_1	K_2/K_1	D_A
0.105	0.332	0.316	5.16

L_A	D_A/L_A	$K_2 t$
$\left(12.9 \times \dfrac{2.47}{4.93}\right)$	0.8	0.105×0.42
6.45		$= 0.044$

From nomograph, we obtain

$$\frac{D}{L_A} = 1.08,$$

and therefore

$D = 1.08(6.45) = 6.97$ (calculated deficit at Station 6).

Drainage from Harrison Creek was evidently the cause of the existing oxygen sag in the river stretch between Station 1 and the dam at Station 4.

The critical oxygen concentration at 25°C in the river stretch sampled was 5.1 ppm just above the dam at U.S. Lock No. 2 (Station 4), as illustrated in Fig. 3.5. However, aeration of the water as it passed over this dam brought the oxygen level up to 6.8 ppm (the highest level obtained in the entire 38-mile reach). The pollution from Elizabethtown caused the oxygen to sag once again to 6.1 ppm at Station 5, and the relatively low reaeration rate in the stretch from Station 5 to Station 6 caused a continued sag to 5.75 ppm at the end of the reach at Station 6 (Table 3.9).

Rather drastic results occurred when the plant load was superimposed on the river at Station 1. The oxygen sagged from 5.9 ppm at Tarheel to about 3.2 ppm just above the dam at Station 4. Aeration at this point brought the level back to 5.7 ppm, from which point it sagged once again to 3.6 ppm in the stretch between Stations 4 and 6.

Thomas [7] also made available a formulation which allows the industrial-waste analyst to approximate L_A, the maximum BOD load that may be introduced into the stream without causing the oxygen concentration downstream to fall below a specified value:

$$\log L_A = \log Dc + \left[1 + \frac{K_1}{K_2 - K_1}\left(1 - \frac{D_A}{D_C}\right)^{0.418}\right]\log\frac{K_2}{K_1}.$$

By substituting in the above equation the values of the oxygen deficit to be maintained downstream, the stream-reaction constants (K_1 and K_2), and the initial oxygen deficit at the point of pollution, it is possible to calculate the maximum ultimate first-stage BOD which can be added to the stream.

When the plant waste load was imposed on the river at its 5-year minimum low flow, the oxygen sagged to about 1.4 ppm just above the dam. The greater the oxygen deficit, the greater the reaeration rate; thus, aeration at the dam brought the oxygen level to 4.7 ppm. From this point (Station 4), the oxygen level began to sag once again to a value of only 1.4 ppm at Station 6.

It was therefore recommended that the proposed plant be located just below U.S. Lock No. 2 (Station 4) where stream assets are at a maximum. With this location, some damage would be done to the river, but the oxygen level would not drop as low as if the plant were located at Station 1. Because the river already had an oxygen sag from Station 4 to Station 6, it was also recommended that a primary-treatment installation be a minimum condition of building the plant at this acceptable site. Furthermore, it was the author's opinion that this plant should not be allowed to locate at Station 1 unless it installed efficient secondary treatment.

3.3 Churchill Method of Multiple Linear Correlation

From 24 stream samples taken at appropriate points Churchill and Buckingham [1] found that a good correlation generally exists between BOD, DO, temperature, and stream flow. In other words, they found that the dissolved-oxygen sag in a stream depends upon only three variables: BOD, temperature, and flow. By means of the least-squares method [2], the line of regression can be computed, so as to predict the dissolved-oxygen sag for any desired BOD loading. This method eliminates the often questionable and always cumbersome procedure for determining times of flow between stations and the resulting stream-reaction rates (K_1, K_2, and K_3).

The author [4] found that the Churchill and Buckingham method provides a good correlation, if each stream sample is collected and observed under maximum or minimum conditions of one of the three stream variables. Only six samples were required in one study [4] to produce practical and dependable results. Additional samples may add some small degree of refinement to the results, but the refinement probably would not offset the effort of planning, collecting, and analyzing the samples and calculating the results.

The following data, which refer to the stream situation sketched in Fig. 3.6, illustrate the use of the Churchill method to obtain the line of best fit and the resulting oxygen sag. Data collected on four different days during various low- to medium-flow periods gave the sag values between Stations 3 and 5 shown in Table 3.10.

For our calculations, we use the following three normal equations, based on the principle of least squares [3], and the Doolittle method [5] of solving three simultaneous equations:

$$b_1\Sigma X_1^2 + b_2\Sigma X_1X_2 + b_3\Sigma X_1X_3 = \Sigma X_1Y, \qquad (1)$$
$$b_1\Sigma X_1X_2 + b_2\Sigma X_2^2 + b_3\Sigma X_2X_3 = \Sigma X_2Y, \qquad (2)$$
$$b_1\Sigma X_1X_3 + b_2\Sigma X_2X_3 + b_3\Sigma X_3^2 = \Sigma X_3Y. \qquad (3)$$

The form of the equation for the dissolved oxygen drop is

$$Y = a + b_1X_1 + b_2X_2 + b_3X_3,$$

where a, b_1, b_2, and b_3 are constants and

X_1 = BOD at sag (ppm)
X_2 = temperature at sag (°C)
X_3 = flow at sag (1000/cfs)
Y = DO drop (ppm).

Table 3.10 Multiple linear correlation of DO drop, BOD, temperature and stream discharge (four samples).

Date	Dissolved oxygen At Sta. 3, ppm	At Sta. 5, ppm	Drop in O_2, ppm (Y)	BOD at sag, ppm (X_1)	Temp., °C (X_2)	Flow, 1000/cfs (X_3)	Y^2	YX_1	YX_2	YX_3	X_1^2	X_1X_2	X_1X_3	X_2^2	X_2X_3	X_3^2
6/18/58	10.0	7.9	2.1	8.4	11.5	7.14	4.41	17.64	24.15	15.00	70.50	96.60	59.98	132.25	82.11	50.98
7/1/58	8.2	7.2	1.0	2.6	20.5	11.71	1.00	2.60	20.50	11.71	6.76	53.30	30.45	420.25	240.05	137.12
7/22/58	7.2	2.6	4.6	21.2	20.0	19.35	21.15	97.52	92.00	89.0	449.44	424.00	410.22	400.00	387.00	374.42
8/4/58	8.3	2.8	5.5	5.83	16.5	15.02	30.25	32.06	90.75	82.61	33.99	96.19	87.57	272.25	247.83	225.60
Total			13.2	38.03	68.5	53.22	56.81	149.82	227.40	198.33	560.70	670.10	588.20	1224.75	956.95	788.12
			$\bar Y$	$\bar X_1$	$\bar X_2$	$\bar X_3$										
Means			3.30	9.51	17.12	13.31										
							*$n\bar Y^2$	$n\bar Y\bar X_1$	$n\bar Y\bar X_2$	$n\bar Y\bar X_3$	$n\bar X_1^2$	$n\bar X_1\bar X_2$	$n\bar X_1\bar X_3$	$n\bar X_2^2$	$n\bar X_2\bar X_3$	$n\bar X_3^2$
Corrected items							43.50	125.54	225.98	175.69	361.77	651.25	506.30	1172.31	911.47	708.62
Corrected sums							13.25	24.29	1.42	22.64	199.10	18.86	81.58	52.48	44.48	79.50

*n = number of terms.

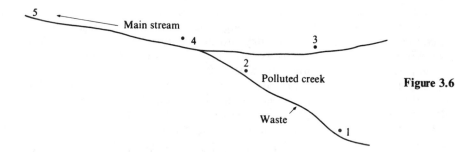

Figure 3.6

We now substitute the numerical data given in Table 3.10 and proceed step by step as follows. From Eq. (1) we obtain

$$198.93b_1 + 18.85b_2 + 81.90b_3 = 24.28. \quad (4)$$

Dividing Eq. (4) by 198.93 yields

$$b_1 + 0.0947b_2 + 0.4117b_3 = 0.1220. \quad (5)$$

Multiplying by 0.0947 gives

$$18.85b_1 + 1.785b_2 + 7.65b_3 = 2.27. \quad (6)$$

We next apply Eq. (2) and obtain

$$18.85b_1 + 52.44b_2 + 44.48b_3 = 1.42. \quad (7)$$

Subtracting Eq. (7) from (6) yields

$$-50.66b_2 - 36.72b_3 = 0.82. \quad (8)$$

Dividing Eq. (8) by -50.66, we obtain

$$b_2 + 0.727b_3 = -0.01760. \quad (9)$$

We now multiply Eq. (4) by -0.412 and have

$$-81.96b_1 - 7.76b_2 - 33.74b_3 = -10.00. \quad (10)$$

Next we multiply Eq. (8) by 0.725:

$$-36.72b_2 - 26.62b_3 = +0.6179. \quad (11)$$

Finally, we apply Eq. (3) and obtain

$$81.90b_1 + 44.48b_2 + 79.50b_3 = 22.64. \quad (12)$$

Adding Eqs. (10), (11), and (12), we have

$$19.14b_3 = 13.28, \quad (13)$$

$$b_3 = 0.6938. \quad (14)$$

From Eq. (9),

$$b_2 = -0.5198. \quad (15)$$

From Eq. (5),

$$b_1 = -0.1160. \quad (16)$$

As a check, we substitute all values in Eq. (12):

$$22.64 = 22.64,$$
$$a = \overline{Y} - (b_1\overline{X}_1 + b_2\overline{X}_2 + b_3\overline{X}_3),$$
$$a = 4.058.$$

The preceding computations yield the following equation for DO drop:

$$Y = a + b_1X_1 + b_2X_2 + b_3X_3,$$
$$Y = 4.058 - 0.1160X_1 - 0.5198X_2 + 0.6938X_3,$$

where
$$Y = \text{DO drop (ppm)}$$
$$X_1 = \text{BOD at sag (ppm)}$$
$$X_2 = \text{temperature at sag (°C)}$$
$$X_3 = \text{flow at sag (1000/cfs)}.$$

The results obtained from the dissolved-oxygen equation are summarized in Table 3.11 to show that this equation can predict with an acceptable degree of accuracy the sag occurring below a source of pollution. Another correlation procedure must be used to compute the allowable BOD loading at the source of

Table 3.11 Comparison of calculated oxygen drops with observed values of oxygen drop at the sag.

| | Oxygen drop | |
Date	Calculated from formula, ppm	Observed, ppm
6/18/58	2.07	2.10
7/1/58	1.22	1.00
7/22/58	4.66	4.60
8/4/58	5.23	5.50

pollution. The BOD equation can be derived from the same least-squares method by correlating the upstream BOD load with the temperature, discharge, and resulting BOD at the sag point in the stream. The necessary data for the development of the BOD equation are given in Table 3.12.

Using the three normal equations, (1), (2), and (3), we obtain from Eq. (1)

$$4.65b_1 - 15.06b_2 + 13.47b_3 = 4.99. \quad (17)$$

Dividing Eq. (17) by 4.65 yields

$$b_1 - 3.24b_2 + 2.90b_3 = 1.07. \quad (18)$$

We now use Eq. (2) and get

$$-15.06b_1 + 51.00b_2 - 39.23b_3 = -15.92. \quad (19)$$

If we multiply Eq. (17) by 3.24, we get

$$+15.06b_1 - 48.79b_2 + 43.64b_3 = 16.17. \quad (20)$$

Adding Eqs. (19) and (20) yields

$$2.21b_2 + 4.41b_3 = 0.25; \quad (21)$$

dividing Eq. (21) by 2.21 leads to

$$b_2 + 2.00b_3 = 0.113. \quad (22)$$

Multiplying Eq. (17) by -2.897, we have

$$-13.47b_1 + 43.63b_2 - 39.02b_3 = -14.46. \quad (23)$$

Multiplying (22) by -4.40 yields

$$-4.40b_2 - 8.80b_3 = -0.497. \quad (24)$$

Using Eq. (3), we obtain

$$13.47b_1 - 39.23b_2 + 44.87b_3 = 10.13. \quad (25)$$

We then add Eqs. (23), (24), and (25):

$$-2.95b_3 = -4.827,$$
$$b_3 = +1.64.$$

Substitution of this result in Eq. (22) yields

$$b_2 = -3.167,$$

and substitution in Eq. (25) gives

$$b_1 = -13.93.$$

We use Eq. (20) to check these results:

$$-209.79 = -209.92,$$
$$a = Y - b_1X_1 - b_2X_2 - b_3X_3,$$
$$a = 114.96.$$

Table 3.12   Multiple linear correlation of BOD loads, temperature, and discharge from the source of pollution to the sag (Station 5).

Date	BOD at Sta. 5, ppm (Y)	Applied BOD load (1000 lb/day) at Sta. 2 + 3 (X_1)	At Station 5 Temp., °C (X_2)	At Station 5 Discharge, 10 cfs (X_3)	Y^2	YX_1	YX_2	YX_3	X_1^2	X_1X_2	X_1X_3	X_2^2	X_2X_3	X_3^2
6/18/58	6.37	6.91	11.5	14.0	40.58	44.02	73.26	89.18	47.75	79.47	96.74	132.25	161.00	196.00
7/1/58	1.20	4.41	20.5	8.54	1.44	5.29	24.60	10.25	19.45	90.41	37.66	420.25	175.07	72.93
7/22/58	5.94	4.19	20.0	5.17	35.23	24.89	118.80	30.71	17.56	83.80	21.66	400.00	103.40	26.73
8/4/58	2.10	4.85	16.5	6.66	4.41	10.19	34.65	13.99	23.52	80.03	32.30	272.25	109.89	44.36
Sums	15.61	20.36	68.5	34.37	81.66	84.39	251.31	144.13	108.28	333.71	188.36	1224.75	549.36	340.02
	$\bar{Y}$	$\bar{X}_1$	$\bar{X}_2$	$\bar{X}_3$										
Means	3.90	5.09	17.13	8.59										
					$n\bar{Y}^2$	$n\bar{Y}X_1$	$n\bar{Y}X_2$	$n\bar{Y}X_3$	$n\bar{X}_1^2$	$n\bar{X}_1\bar{X}_2$	$n\bar{X}_1\bar{X}_3$	$n\bar{X}_2^2$	$n\bar{X}_2X_3$	$n\bar{X}_3^2$
Corrected items					60.79	79.40	267.23	134.0	103.63	348.77	174.89	1173.75	588.59	295.15
Corrected sums					20.87	4.99	−15.92	10.13	4.65	−15.06	13.47	51.00	−39.23	44.87

Therefore, the equation yielding the allowable BOD load becomes

$$Y = 114.96 - 13.93X_1 = 3.167X_2 + 1.64X_3,$$

where Y = BOD load at sag (Station 5) (1000 lb/day)
  X_1 = combined BOD loads of upstream Stations 2 and 3 (1000 lb/day)
  X_2 = temperature at sag (Station 5) (°C)
  X_3 = flow at sag (Station 5) (10 cfs).

Applying the 30 per cent BOD reduction (primary treatment) to the pollution load added at Station 2 only, we have

$$0.70 \times 3692^* = 2584 \text{ lb/day at Station 2}$$
$$159 \text{ lb/day at Station 1} \left. \right\} \text{Natural}$$
$$1159 \text{ lb/day at Station 3} \left. \right\} \text{pollution}$$
$$\text{Total} = 3902 \text{ lb/day.}$$

If the above result (a somewhat reduced load) is inserted in the BOD equation at 20.5°C and 51.7 cfs at Station 5 (most critical temperature and flow), we should arrive at the BOD load remaining at the sag:

$$\begin{aligned} \text{BOD} &= 114.96 - 13.93(3.902) - 3.167(20.5) \\ &+ 1.64(5.17) \\ &= 4.17 \text{ (1000 lb/day)} \\ &= 4170 \text{ lb/day, or } 14.94 \text{ ppm at } 51.7 \text{ cfs.} \end{aligned}$$

Using this BOD in the dissolved-oxygen-sag formula, we obtain the DO drop that would occur if the reduced effects from bottom deposits were ignored. (Less sewage pollution results in less bottom deposits, and this factor may change the deoxygenation rate of a flowing river.) We thus have

$$\begin{aligned} \text{DO drop} &= 4.555 - 0.1371(14.94) - 0.5931(20.5) \\ &+ 0.7639\left(\frac{1000}{51.7}\right) \\ &= 5.095 \text{ ppm.} \end{aligned}$$

At saturation (20°C),

$$\text{DO} = 9.2 \text{ ppm}$$
$$\text{Dissolved oxygen} = 9.2 - 5.095 = 4.105 \text{ ppm.}$$

Thus, Churchill's method (with only four samples at critical conditions) shows that primary treatment would be sufficient to sustain 4 ppm DO in the stream during critical conditions, providing the upstream station is fully saturated with dissolved oxygen.[†] The somewhat higher level of DO obtained by the Churchill method (3.65 according to the Streeter-Phelps formula) might be attributed to the omission of the slime effect when the BOD load is reduced. Slime growths tend to accelerate the removal of dissolved organic matter from the flowing water by increasing biological action. On the other hand, when sludge deposits exist or the slime growths become dense and voluminous, it is safe to conclude that the anaerobic decomposition products formed will require oxygen from the overlying water layers. The increased removal of organic matter in the slime counteracts to some degree the demand for oxygen from the water. The net result from slime growths and bottom deposits is accelerated local oxygen demand with subsequent (downstream) reduced oxygen demand.

References

1. Churchill, M. A., and R. A. Buckingham, "Statistical method for analysis of stream purification capacity," *Sewage Ind. Wastes* **28**, 517 (1956).
2. Cohen, J. B., and R. L. O'Connell, "The analog computer as an aid to stream self-purification computations," *J. Water Pollution Control Federation* **35**, 951 (1963).
3. Croxton, F. E., and D. J. Cowden, *Applied General Statistics*, Prentice-Hall, Englewood Cliffs, N. J. (1955), pp. 261–280.
3a. Fair, G. M., "The dissolved oxygen sag—An analysis," *Sewage Works J.* **11**, 451 (1939).
4. Simmons, J. D., N. L. Nemerow, and T. F. Armstrong, "Modified river sampling for computing dissolved oxygen sag," *Sewage Ind. Wastes* **29**, 936 (1957).
5. Steel, R. G. D., and J. H. Torrie, *Principles and Procedures of Statistics*, McGraw-Hill Book Co., New York (1960), p. 290.
6. Streeter, H. W., and E. B. Phelps, *A Study of the Pollution and Natural Purification of the Ohio River*, Bulletin no. 146, U.S. Public Health Service, Washington, D.C. (1925).
7. Thomas, H. A., "Pollution load capacity of streams," *Water Sewage Works*, **95**, 409 (1948).

---

*Average BOD added at Station 2.
†See result obtained by Streeter-Phelps formulations on p. 39.

Suggested Additional Reading

Camp, T. R., "Field estimates of oxygen balance parameters," *J. Sanit. Eng. Div. Am. Soc. Civil Engrs.* **92**, 115 (1966).

Fair, G. M., "The dissolved oxygen sag—An analysis," *Sewage Works J.* **11**, 445 (1939).

Kittrell, F. W., and O. W. Kochtitsky, "Shallow turbulent stream, self purification characteristics," *Sewage Works J.* **19**, 1032 (1947).

LeBosquet, M., Jr., and E. C. Tsivoglou, "Simplified dissolved oxygen computations," *Sewage Ind. Wastes* **22**, 1054 (1950).

Liebman, J. C., and Loucks, D. P., "A note on oxygen sag equations," *J. Water Pollution Control Federation* **38**, 1963 (1966).

Streeter, H. W., "A nomograph solution of the oxygen sag equation," *Sewage Ind. Wastes* **21**, 884 (1949).

Questions

1. What stream factors are required in order to compute the treatment required for industrial wastewater when using the Streeter Phelps formulation?
2. How is each of these stream factors obtained and what are the problems associated with each of them?
3. What stream factors are required for computing the same treatment when using the Churchill method?
4. How is each stream factor obtained and what are the problems associated with each of them?
5. When would you use the Streeter Phelps formulation and when would you select the Churchill method?
6. How would you determine from the ultimate first stage BOD's derived by either method the degree of treatment required?

CHAPTER 4

STREAM SAMPLING

Any decision on industrial waste treatment is only as dependable as the stream-sampling program on which it is based. Any time spent in planning a comprehensive stream-sampling program will be well rewarded when engineers design treatment plants which operate efficiently. Among the many factors to be considered when one designs such a sampling program [8] are:

Overall objectives of the program	Frequency of sample collection
Total number of samples	Time of year for sampling
Points of collection	Statistical handling of data
Method of collection	
Data to be obtained	Care of samples prior to analysis

Overall objectives of the program. Programs may vary considerably in their objectives. In one instance the engineer may be concerned with the effect of an upstream industry on the water quality downstream; of special interest in this case might be the color of the receiving stream. In another instance he may be attempting to ascertain the dissolved-oxygen-sag characteristics of the stream during the summer season only. In still another case, he could be concerned that the stream characteristics comply with the classification standards established by the state pollution-control authorities. These are but a few examples. The importance of the other eight factors will depend to some degree on the overall objectives of the survey.

Total number of samples. The number of samples required depends on the objectives of the program and the amount of time and effort being devoted to the survey. The use of a few locations and enough samples to define the results in terms of statistical significance is usually much more reliable than using many stations with only a few samples from each. Also, samples are frequently taken over a long time interval, during which the condition of the watercourse is subject to variation. In many instances, an attempt to test all conditions by infrequent, random sampling produces no definite pattern and, in fact, may be misleading. It may be better to concentrate on well-defined, frequent, and intensive sampling.

A well-planned survey with a specific objective will require a minimum of samples. If, for example, someone wishes to determine the river characteristics—especially the dissolved-oxygen profile—during low-flow periods, two or three samples, collected at the proper time, will suffice. These samples must be collected, however, during extreme drought periods, if factors influencing the character of the stream are to be clearly established. The number of samples required often depends on the ability of the engineer to collect samples which include all significant factors. To illustrate, if a stream is known to contain phenols in addition to other organic matter but only in the fall season, the engineer must obviously collect one or more samples during the fall, despite the fact that the lowest flows and highest temperatures occur in July. Four to six river samples are the generally accepted minimum for reliable analysis and predictions. Factors such as sporadic influx of a toxic metal, flooding of the banks during certain seasons, or a peculiar flow pattern during droughts may warrant the collection of more than the minimum number of samples. It should be quite clear, however, that industrial processing is so varied, and usually unpredictable, that many samples of the receiving water under all conditions must be collected in order to evaluate truly the effect of a waste upon the stream.

Points of collection. Sampling points should be selected with great care and special consideration given to sources of pollution, dilution by branch streams, changes in surrounding topography, and slope of the river. Significant riverside features should also influence the choice of sampling points: a municipal water intake, a state park, an industrial area, a good fishing spot, a hotel, or a camping site would each have a definite bearing on the usage of the stream. Since the acceptable pollution limits for waters vary according to usage, samples should be collected and a record made of the condition of the stream just above and just below all such points of stream use or change.

As in the case history presented in Chapter 3, a minimum of four stream stations (sampling points) is recommended: (a) an upstream site, where the water is uncontaminated; (b) just below the source of pollution or dilution; (c) where the stream is in the worst condition due to a specific source of pollution (bottom of oxygen sag); (d) a point midway between bottom of oxygen sag and recovery of oxygen level. A rapid method of locating these points is to run a small boat up and down the stretch in question, using a dissolved-oxygen probe to locate sag and recovery points. This can be done from bridges but only with approximate location accuracy.

Whichever of the methods described in Chapter 3 is utilized to ascertain acceptable stream loadings, these four stations will be adequate for subsequent analysis. The necessity of additional sampling depends on various local conditions mentioned above or on unforeseen abnormalities in the characteristics of a stream, such as areas of immediate oxygen demand, sludge deposits, biological absorption, and algae growth and decay. Sampling stations should be located, as nearly as possible, at points of uniform cross-section, nonshifting bottom, minimum stream width (to facilitate sampling and increase accuracy of flow measurements), and average velocity. In addition, one should consider ease of approach to the station and ease of obtaining a representative sample. Bridges over the stream are of considerable assistance in collecting uniform samples and measuring cross-sections and stream velocity.

Sampling in tidal estuaries is a difficult and controversial problem, and engineers' answers to it have ranged from complete mathematical formulations to simple grab samples. The problem arises from the fact that pollution in tidal streams ebbs and floods with the tide, so that a portion often remains in the reach below the source of pollution for many days, rather than hours. One method of sampling which the author has found effective is to determine the tide cycle of the particular stream, then to collect samples on the high and low, as well as the mean, tide. This method provides the analyst with a consistent and overall picture of the tidal pollution situation.

Method of collection. Samples should be taken from a 0.6 depth in streams less than 2 feet deep (i.e., 1.2 feet in a 2-foot-deep stream). In deeper streams, it is necessary for the sampler to composit portions taken from depth levels of 0.2 and 0.8. When the stream flow remains quite steady, equal portions of each sample may be composited for analysis. The volume of samples depends on the number and type of analyses to be carried out on the individual and/or the composite samples. A standard-type dissolved-oxygen sampler is recommended for collecting most samples. Glass bottles with glass caps or polyethylene containers are most widely used. Any doubt about the cleanliness of the sample bottle can usually be dispelled by rinsing it first with some of the actual stream water, but special sterile bottles are required for bacteriological samples.

Data to be obtained. The scientific collection of data for stream analysis may be divided into three major categories: hydrologic factors, sources of pollution, and watercourse sampling. The type of data to be obtained depends on the objectives of the survey and the amount of time and money available for the investigation. For example, if the oxygen resources of a section of a stream are the main concern of the regulatory agency, dissolved oxygen, water temperature, and stream flow should be measured over as long and as critical a period as possible. If the survey is of a general nature, the stream analyst should undertake as many chemical, physical, and biological tests as possible, to assist him in later interpretation and evaluation of the data. The writer has found that many surveys containing only four measurements—rate of flow, temperature, BOD, and dissolved oxygen—can supply information sufficient for design of waste-treatment units. In addition, data on pH, color, and turbidity may indicate the general physical condition of the stream. Biological analyses are required when the stream water is used for drinking, bathing, or fishing. In this case, the coliform count is usually determined.

Frequency. Samples should be collected as frequently as is necessary to provide a representative total sample. The master sample should contain individual constituents of every variation expected. For example, if the pH is known to vary from 4 to 10, individual samples with pH values of 4, 5, 6, 7, 8, 9, and 10 should appear in the composite at least once during each sampling period. If the situation requires instantaneous analysis, more individual samples, with little or no compositing, are collected. The former method is practiced when one seeks to determine average existing stream conditions.

Time of year. The time of year is of utmost importance when there is a deadline for producing results. In stream studies dealing with industrial-waste treatment one is concerned primarily with critical conditions of pollution, which generally exist when the environment is at its warmest, the stream flow slowest, and the man-made pollution greatest. In most parts of the United States, under normal conditions, these critical situations occur in the summer months, so that the ideal time for stream survey is during the summer. But many studies must be undertaken in the spring and fall seasons, owing to stream conditions or the manpower situation. Indeed, because of the urgency of the problem or unusual conditions of runoff or pollution, investigations may be carried out in any season. However, the objective of every stream analyst and industrial waste engineer should be to collect his data during critical stream conditions of temperature, flow, and pollution load, when the probability of error is less.

It may be necessary to project stream analyses to conditions which might obtain during future critical periods. This is sometimes required even when a survey is made during the summer, because the intensity of the problem varies to some degree from year to year and even from one period of years to another.

Statistical handling of data. It is a well-known fact that data can be manipulated to emphasize that aspect of the survey which the analyst deems most important. This can be an ethical practice, and it does not preclude other conditions or phenomena which may exist. However, the engineer must have a working knowledge of statistics and mathematics in order to convey this information in the best form to the layman. For example, when studying the "most probable number" (MPN) of coliform bacteria present in a stream, the arithmetic mean would not clearly describe and emphasize this number, whereas the geometric mean, or mode, may well illustrate it accurately. In addition, considerable variation can, and often does, exist between the arithmetic mean and the mode in biological systems. The following example illustrates the difference between the arithmetic mean and mode for a series of coliform bacteria counts:

Sample number	Coliform count, MPN/100ml
1	0
2	0
3	0
4	9.0
5	3.6

$$\text{Arithmetic mean} = \frac{12.6}{5} = 2.52$$

$$\text{Mode} = 0$$

The rate of flow during critical conditions is of importance to the industrial waste engineer. If he incorporates in his figures the minimum flow ever recorded in a stream, he may end up with an unrealistic evaluation. On the other hand, the use of the mean summer or low-flow value can be dangerous, since lower flows than this occur quite often. Some state regulatory agencies use the minimum seven-day flow likely to recur once in ten years as the criterion for designing waste-treatment facilities.

Statistical handling of industrial-waste data is just as important as statistical handling of river data. The waste engineer should realize, for example, that figures on peak waste flows are significant under certain conditions, but the arithmetic mean BOD values are the figures required when one is designing facilities for treating these wastes. In computing treatment-plant efficiencies, the engineer can obtain a more complete picture of the plant's operation by using the standard deviation from the arithmetic mean than by using the mean value alone. Also, when comparing the efficiency of one treatment plant with another, he may find it desirable to use a coefficient of variation.

Care of samples prior to analysis. All samples should be analyzed as soon as possible after collection. The writer would recommend on-the-bank analysis whenever this is possible. With modern portable testing equipment available, there should be little reason (except convenience) for bringing samples back to the laboratory for every analysis. However, it is impractical to carry out detailed tests such as coliform counts and determinations of phenol concentration and suspended-solids quantity on the stream site. All samples subject to even the slightest chemical, physical, or biological change should be chilled immediately and kept at a temperature from 0°C to 10°C (4°C is optimum) until analyses are carried out. Dissolved-oxygen samples should be carried through the acidification stage on the stream site. Phenol samples should be preserved with copper sulfate. (For preserving samples for other analyses, see reference 7.) Plastic sampling bottles should be avoided when a reaction is possible between constituents of the waste, such as organic solvents, and plastic. Likewise, metal containers and caps should not be used to hold wastes on which metals are to be determined.

Porges [6] concerned himself with measuring the true concentration of dissolved oxygen in a stream water. He suggests: (1) whenever possible, dissolved-oxygen determinations should be made immediately after a sample is collected, with minimum exposure to light; (2) fixing a sample with acid-azide or processing it to the iodine stage and then leaving it exposed to sunlight may result in unreliable values; (3) for the particular waters sampled, untreated samples, samples processed to the iodine stage, and samples fixed by acid-azide, when stored not more than six hours on ice in the dark, all gave reasonably accurate dissolved-oxygen results; (4) icing of untreated samples under dark conditions merits consideration for stream surveys since the collection time is minimal and the samples are brought to the laboratory sooner, where analyses may be performed under optimum conditions; (5) samples should be exposed to a minimum of light during the time necessary for collection, transportation, and analysis. Porges' recommendations, though not always practicable, are significant, since the dissolved-oxygen measurement is one of the most vital in any stream survey.

Kittrell and West [5] suggest twelve "commandments" on stream survey procedures:

1. Develop a specific objective and define it clearly before undertaking a stream-pollution study

2. Review all available reports and records (essential to sound planning)

3. Do not assume that a previous occurrence will necessarily be duplicated but be alert to possible variations in the patterns of events; because of the complexities of a river system almost anything can happen, and frequently does

4. Make a thorough reconnaissance of the stream, sources of wastes, and water uses and prepare a detailed plan of operation before sending a crew into the field

5. Observe and take into account all the characteristics of the stream

6. Select stream sampling stations that will give accurate measurements of waste loads and of the orderly course of natural purification, not those that primarily save time or trouble in sampling

7. When waste discharges, stream flow, or other stream conditions vary diurnally, around-the-clock sampling is highly desirable, if not imperative; in any event, sampling times should be varied throughout the daylight hours as much as is feasible

8. Carefully evaluate wastes that may cause water quality degradation and make all analyses necessary to show damage, but avoid wasting energy and money on analyses that do not contribute to proof of pollution. One or two sets of preliminary samples will indicate whether certain analyses will be productive in marginal cases

9. Always obtain agreement with the laboratory chief on the number of samples, for the specified determinations, that can be processed daily by the personnel and facilities available

10. Maintain a continuing review of the analytical data produced by the laboratory to detect any need for revision of the original study plan

11. Always be alert for any significant clue to some important facet of the survey that could not be anticipated in the original planning of the study

12. Consider all possible interpretations of data and beware of pat answers that on first consideration appear to fit the findings

References

1. Baily, T. E., "Fluorescent-tracer studies of an estuary," *J. Water Pollution Control Federation* **38**, 1986 (1966).
2. Clark, R. N., "Discussion on sampling for effective evaluation of stream pollution," *Sewage Ind. Wastes* **22**, 683 (1950).
3. Gunnerson, C. G., "Hydrologic data collection in tidal estuaries," *Water Resources Res.* **2**, 491 (1967).
4. Haney, P. D., and J. Schmidt, *Representative Sampling and Analytical Methods in Stream Studies*, Technical Report W58-2, U.S. Public Health Service, Washington, D.C. (1958), pp. 133–142.
5. Kittrell, F. W., and A. W. West, "Stream survey procedures," *J. Water Pollution Control Federation* **39**, 627 (1967).
6. Porges, R., "Dissolved oxygen determination for field surveys," *J. Water Pollution Control Federation* **36**, 1247 (1964).
7. *Standard Methods for the Examination of Water, Sewage, and Wastes,* 10th ed., American Public Health Association, New York (1955).
8. Velz, C. J., "Sampling for effective evaluation of stream pollution," *Sewage Ind. Wastes* **22**, 666 (1950).
9. Weaver, L., "Stream surveillance programs," *J. Water Pollution Control Federation* **38**, 1334 (1966).

Questions

1. What are the nine factors to be considered when conducting a stream-sampling program?
2. Discuss the reasons why each of these factors play an important role in the design of industrial waste-treatment facilities.

CHAPTER 5

ECONOMICS OF WASTE TREATMENT

At this point the author feels it appropriate to express his views on the choices open to an industry faced with the cost of pollution abatement. One often hears industry threaten to move to another part of the country or to close its doors completely, and, unfortunately, some of these threats have been carried out. The author believes rather that industry has a moral, legal, and economic responsibility to consider waste treatment as one of the variable costs of doing business, akin to labor, marketing, and raw materials. A company should seek to minimize its production costs by selecting a plant site where the total costs (labor, taxes, raw materials, water and other utilities, marketing, *and* waste treatment) are lowest, regardless of the existence or absence of a municipal plant, tax aid, or subsidy.

Industry should attempt to treat its waste at the lowest cost that will yield a satisfactory effluent for the particular receiving stream, which may necessitate considerable study, research, and pilot investigations. Planning ahead will provide time to make appropriate decisions. Conversely, lack of planning on minimizing waste-treatment costs may mean that a sudden demand for an immediate solution will cause industry to decide to cease production. We can expect industrial production costs to rise all over the U.S. as pollution-abatement costs are accepted as part of production costs. However, the advantages of particular locations and ingenuity in waste-treatment methods can have a considerable effect and industry is rapidly becoming aware of the importance of these factors.

Nevertheless, among the remarks most often made by industry when faced with abatement demands are: "We can't afford it," or "It'll put us out of business." The viewpoint of the federal government is typified by a statement by former Secretary of the Interior Stewart Udall, "It's not pollution abatement that we cannot afford—it's pollution." Thus, we have apparently reached an impasse in communication between these groups. The purpose of this chapter is to elucidate the actual benefits to industry of pollution abatement, to illustrate means of putting a dollar value on these benefits, and finally to relate the dollar value of benefits to the quality of stream water through the use of a new pollution index being developed.

Industrial waste treatment has been encouraged, promoted, and even dictated by state and federal governments on the premise that industrial wastes are similar to municipal wastes and constitute a public health menace. It has long been recognized, but never openly, that the public health hazard of industrial waste is unlike that of municipal wastes and is often nonexistent. However, there are benefits from waste treatment which exist but are difficult to describe in quantifiable terms. Since both the polluter and the public are aware that industrial waste treatment to prevent stream pollution is "a good thing," few have questioned the true reasons why their wastes should be treated.

Pollution of our watercourses has increased during the last ten years despite an increase in treatment plants. The U.S. Public Health Service reported that industrial waste-water effluents carry more than twice as much degradable organic matter into U.S. watercourses as the sewage effluents of all our municipalities combined. Population growth and increased industrial production have surpassed the elimination of wastes by proper treatment. We are faced with the dilemma of demanding expanded industrial waste treatment at more cost to industry without being able to present the benefits to be derived from such expenditure.

Most of the real benefits which result from industrial waste treatment are considered "irreducibles." This is primarily because no one has been willing or able to put a dollar value on them. If we can identify these benefits and indicate means to quantify them (whenever possible), both industry and the public will be in a position to examine the economics of a given waste-treatment situation. It then becomes important to devise a workable system for determining how pollution capacity resources in our streams should be equitably proportioned and distributed among the various competing consumers.

Streams are no longer just a means of waste conveyance to the oceans but valuable resources which can be used and reused for many purposes during their passage. Their pollution-carrying capacity must, however, be protected and preserved by all consumers and utilized in ways which are most beneficial to all society. Determination of the optimum beneficial uses of our rivers and streams and their true total costs must be a major concern of scientists and administrators.

Some of these pollution-carrying stream resources include tolerance levels for accepting quantities of:

Organic matter (decomposable)	Grease and oils (floating matter)
Salt (chloride, sulfate. etc.)	Surface-active materials
Toxic materials	Bacteria, protozoa, and viruses
Color and turbidity	Odors or odoriferous matter
Suspended solids	
Heat (temperature)	Other contaminants such as pesticides and nondegradable organics (CCE)
Algae nutrients	
Radioactive matter	

J. G. Moore, Jr., the former Commissioner of the Federal Water Pollution Control Administration, reveals that the water pollution control policy is moving toward determining what degree of waste treatment is feasible and accepting that limit [5]. He gives the following principal reasons for this new policy:

Continued growth of population and industry, with the dual effect of generating both more wastes and greater demand on limited water resources

Growing awareness that the assimilative capacity of many lakes and streams would still be severely taxed by natural and other diffuse sources of pollution even if all point-sources of pollution were brought under complete control

The knowledge that feasible means are available, from both a technical and economic standpoint,

to provide a high degree of treatment for industrial as well as municipal wastes

A growing conviction that effective water pollution control should be looked upon as one of the normal costs of running a government or a business

Increasing interest in the aesthetic value of clean water

Rising concern over the accelerated eutrophication* of lakes

Mounting public disgust over the wholesale killing of fish, birds, and other wildlife, and the growing belief that whatever effective water pollution control ultimately costs, it will be worth it

The increasing sophistication and effectiveness of conservation and other special interest groups in assessing both the strengths and weaknesses of specific water pollution control programs and policies.

The list could be continued. But these few points are enough to suggest that the current trend toward maximum feasible treatment of both municipal and industrial wastes is not likely to be reversed, now or in the future.

As inevitable as day follows night, someone or some group of persons will be forced to make decisions concerning the specific quality of receiving water which must be maintained in each separate instance. In some cases it may be the lowest quality level which can be tolerated by society: in other cases it may be a higher quality mandated by society. The decisions can be made by the Federal government, the state government, a local drainage basin authority or agency, or by a coalition of one or more of the three potential administrative units. Although who makes these decisions is of considerable interest and quite important to all of us, how the decisions are made is more germane to this discussion.

Essentially there are two major methods of arriving at these decisions: (1) by the arbitrary conclusion that no degradation of our streams is allowable and strict regulation will be used to enforce this conclusion or (2) by determining the economic benefits and resulting

---

*This term refers to an accelerated dying of bodies of water caused by the presence of excessive biological nutrient material.

costs to society of alternate water-quality levels and choosing the level yielding maximum benefit at minimum cost. We have been trying the first system for many years and our pollution growth has exceeded our pollution abatement with an accompanying net loss in water pollution carrying-capacity resources. Further, a regulation system requires predetermined knowledge of how much money a polluter can actually spend on wastewater treatment and still remain solvent. The obvious solution is that of cost-benefit analysis.

5.1 Benefits of Pollution Abatement

Most economists agree that there are three categories of benefit in any project: primary, secondary, and intangible. A notable exception is the system of categorizing benefits into "technological" and "pecuniary" types. Difficulties arise in this method when there is overlap from one type of benefit to another. The writer is inclined to prefer a system which separates benefits on the basis of the recipients and measurability.

Primary benefits may be defined as the accumulated worth of products and services originating directly from the project. Although many industrial firms claim that there are no primary benefits associated with pollution abatement, their proclamations may be intentionally misleading. What industries really wish to say is that the direct costs of waste treatment *to them* exceed any measurable good to them which results from the treatment. If, on the other hand, more of these benefits were measurable, industry might be inclined to think and act differently.

Secondary benefits are often called "indirect benefits," since they tend to occur to those who do not use the output of the product and services directly. Many readily understood benefits of waste treatment fit into this category, such as a community's recreational use of clean water downstream after waste treatment by an industrial firm. The people of the community benefit *indirectly* from the industrial firm treating its waste. This situation can be referred to as a technical external economy for the community.

Intangible benefits are irreducible since no dollar value can be easily assigned to them, although it is readily apparent that they exist. For example, waste treatment might improve the morale of the community by virtue of its possession of a clean river: a sort of mental well-being exists among the inhabitants which, although real, defies quantification.

From a practical engineering standpoint the benefits of waste treatment are directly related to the value of the water and associated land downstream and should include (1) the lowered true cost of the water downstream, (2) the lessened damages for consumers utilizing contaminated water downstream, and (3) increased opportunities of associated land and water use downstream.

Commissioner Moore supports the contention that a minimum of secondary treatment for all domestic, commercial, and industrial wastes discharged to fresh water and for most wastes discharged to salt water should be required, regardless of the benefits:

1. In the vast majority of instances, secondary treatment is the minimum needed either to enhance or maintain the quality of the receiving waters.

2. Secondary treatment is economically and technologically feasible for municipal wastes; its equivalent is also feasible, through treatment or process changes or both, for industrial wastes. This fact has been abundantly demonstrated by both cities and industries.

What you have, in other words, is a virtually universal need and practicable means for meeting it. There is a lot of thrust in that combination.

Mr. Moore believes that using a true benefit-cost ratio varies from "the easy to the impossible." Where aesthetic values are concerned, he says:

One of the major long-term benefits of water pollution prevention and control—to use the word "benefit" in its broad sense rather than in the technical sense of the word as used in the term benefit-cost analysis—is the aesthetic value of clean lakes and streams. It is not the function of benefit-cost analysis to set water quality goals or to provide economic justification for one level of water quality against another. The function of benefit-cost analysis or cost-effectiveness analysis is to determine the most practicable means of *achieving agreed-upon water quality goals*. Determining water quality *goals* is a matter of public policy. And in water quality management, as in a growing number of other areas, public policy is taking into account the indirect as well as the direct benefits of water pollution control.

5.2 Measurement of Benefits

Benefits even more than costs must indicate the recipient of the services. Whom are we benefiting: the local persons, the regional inhabitants, the entire country, or civilization as a whole? Our answer to this question will affect the decisions of administrations.

Any pollution abatement action undertaken on a river basin will have some reverberating effects on all the peoples of the earth. In most cases the effects will be greatly reduced as the distance increases from the point of abatement—like the ripples produced by a pebble thrown into a large, still lake. In some instances the ripples will be helpful or beneficial to some persons and harmful or costly to others. The total picture is complex indeed when we consider that each ripple represents many benefits and costs, some of which are intangible and most are of the secondary type. We must presume that the sum of benefits exceeds the total costs to all persons.

The question of whether the benefits exceed the costs of waste treatment on anything less than an entire civilization basis is one worth pondering. One would imagine that the smaller the area selected surrounding the abatement action the greater the excess of benefits over costs. This premise is based on the theory that people are basically selfish and will tend to do more good for themselves and immediate neighbors than harm; whereas they might do less good and more harm to persons situated far from the pollution abatement. Some would argue, on the other hand, that local people must share the major portion of the costs of pollution abatement (and thus deserve the major benefits thereof) while much of the benefit accrues to the larger segments of society. In other words, Industry A treats its wastes and incurs most of the total costs while Industry B 20 miles downstream receives most of the benefits from this treatment and pays little of the costs.

If all the true benefits (Table 5.1) could be quantified, the author believes that they would not only exceed the total costs, but also would fall largely upon persons living and working in the local river basin and using the facilities provided by the local governments. These premises cannot be verified, however, until one can truly measure all the benefits. It is suggested that engineers proceed on faith that this premise is valid until and unless disproven by factual evidence. In view of these statements the author has decided to limit the sphere of benefits of waste treatment to the area of local river basin influence.

W. David Slawson in a "letter to the editor" in the *Wall Street Journal* (July 30, 1971) made the realistic statement that "Since no sensible person has suggested controlling pollution in any instance in which the costs of control would exceed the damage the pollution inflicts, controlling pollution would cost less than our allowing it to continue uncontrolled. As for our competitive position in the world markets, that is of significance only as it affects our standard of living, and our standard of living will be relatively improved to the extent we control pollution at less cost than the damage the pollution would inflict if not controlled."

5.3 A Proposed Method for Resource Allocation

Pollution cannot be continued without seriously affecting our country's progress as measured by the gross national product and without placing rather severe mental and physical limitations on the existing population using stream resources. Two solutions of the pollution dilemma are open to us: the first involves regulation—policing and the courts—while the second allows the free market to solve the problem of supply and demand which is now becoming unbalanced.

Regulation with the aid of education and gentle persuasion has been practiced in the United States since the federal and state governments recognized the problem. Progress has been too slow; roadblocks have been placed in many critical areas; and politics have often replaced pollution abatement. Society seems reluctant to enforce the regulation of a resource which it does not really believe is exhaustible or limited. People are still slow to realize that pollution will finally overtake all our streams unless measures are taken to protect them now. Too many responsible

Table 5.1 Specific benefits of industrial waste treatment.

Primary benefits

a) Savings in dollars to the industrial firm by reuse of treated effluent instead of fresh water
b) Savings in dollars resulting from compliance with regulatory agencies, i.e. avoidance of legal and expert fees and time of management involved in court cases
c) Savings in dollars from increased production efficiency, made possible by improved knowledge of the waste-producing processes and practices

Secondary benefits

a) Saving in dollars to downstream consumers from improved water quality and hence lowered operating and damage costs
b) Increase in employment, higher local payroll, and greater economic purchasing power of labor force used in construction and operation of waste-treatment facilities
c) Increased economic growth of the area due to the commitment of industry to waste treatment and potential for expansion at the existing plant
d) Increased economic growth of area with more clean water available for additional industrial operations, which in turn yield more employment and money for the area.
e) Increased value of adjacent properties as a result of a cleaner, more desirable, receiving stream
f) Increased population potential for the area since cleaner water will be available at a lower cost; the limiting factors of water cost and quantity have been pushed back further into the future

g) Increased recreational uses, such as fishing, boating, swimming, as a result of increased purity of water; recreational opportunities previously eliminated are available again

Intangible benefits

a) Good public relations and an improved industrial image after installation of pollution abatement devices
b) Improved mental health of citizens in the area confident of having adequate waste treatment and clean waters
c) Improved conservation practices, which will eventually yield payoffs in the form of more clean water for more people for more years
d) Renewal and preservation of scenic beauty and historical sites
e) Residential development potential for land areas nearby because of the presence of clean recreational waters
f) Elimination of relocation costs (of persons, groups, and establishments) because of impure waters
g) Removal of potential physical health hazards of using polluted water for recreation
h) Industrial capital investment assures permanence of the plant in the area thus lending confidence to other firms and citizens depending on the output produced by the industry
i) Technological progress, resulting from the conception, design, construction, and operation of industrial waste treatment facilities

---

citizens really believe that it will be time enough to correct the situation if and when our streams become unusable.

The economists advocate the use of the free market as a means of limiting the overdraft of water resources. In essence, they say, "Let the user pay the price of polluting a stream." If and when the price becomes too high he will be forced by economics to reduce his pollution and thus make more resources available for other consumers. There is little experience with this theory in the United States, although Kneese [4] and Fair [2] have reported and discussed the Ruhr River Valley Genossenschaften system in Germany.

However, there would be certain indisputable dangers involved in depending solely on the free

market system in this country. The major one, ironically, is that aspect of the theory which makes it so desirable—making pollution-carrying resources available to those who can contribute most to the growth of the gross national product. In other words, those consumers most able to pay for the resources would receive preferential treatment. Herein lies the paradox: certain consumers unable to pay a market-clearing price would not be able to afford either to pollute or to treat wastes in competition with other consumers. Yet these very consumers might be important and useful for the good of the people. We are therefore faced with the rather common situation that what may be good for a firm, or an industry, or even the nation as measured by the gross national product may not be good for society as a whole. In fact, Galbraith [3] has recently written that our society should consider "quality of life as well as the Gross National Product" when making technical decisions. Since both regulation and marketing solutions have certain strong points in their favor, the obvious answer would be to devise a new system including these advantages and at the same time eliminating the major objections.

Establishment of the firm. In order to preserve our watercourses it is recommended that, where needed and desired, River Resource Allocation Boards be established. For example, a board might consist of a municipal official, an area farmer, a local conservationist, and a leading local manufacturer whose firm uses water. The board shall set out to determine the identity of all stream users and obtain from them (or determine for them) the separate costs of using the water and the measurable additive benefits of its use to the consumer and society.

The board—by a procedure described later—shall set the price for each unit of pollution-carrying resource used by the consumer. The product of the number of units used and the unit price will represent the actual cost to each consumer, to be paid as revenue to the board. Any revenue so collected will be used by the board in carrying out its stated objectives, which are more fully specified later. The board, in effect, shall act as a firm (and is hereafter referred to as the "firm") in its accounting and operational procedures. However, any profits which accrue will be "plowed back" into the firm to meet its stated objective of preserving the pollution-carrying resources.

Unit charges to consumers will be based on the financial condition of the firm. For example, when the stated objectives are being met readily and the firm is financially "flush," only a token unit charge will be necessary. On the other hand, when stream improvements are urgently required and/or the finances of the firm are low, relatively high unit resource charges will be called for. Submarginal consumers—such as recreational and farming users—may be partially subsidized by the firm if the latter has (a) the finances necessary and (b) the majority approval of the board.

The firm will also have the power to deny use of the pollution-carrying resource of the stream by (a) direct edict, (b) pricing the unit charges too high, or (c) denying subsidization to the user. This power symbolizes the will of the local firm and represents the best interests of the local society. For example, a slaughterhouse in the area may not be desired by the firm and it would be denied use of the stream resource by one of the three methods described above. In another case, the firm might recognize the benefits of a slaughterhouse to its community and even wish to subsidize part of its unit charge in order to enhance its operation.

In either case the will of the local people has been the deciding factor in the decision. This, the democratic way, is a most important aspect in the success of the proposed system. Nevertheless, the firm will not conflict with the objectives or the operations of any state, regional, or federal agency which exists to administer a law preserving minimum stream-quality criteria.

The firm would retain the services of legal, economic, and engineering experts to assist in its studies and decisions. This system is based on the theory that the local or regional river basin administration is the proper one to determine the highest level of water quality and the price of the resource. This is valid only where the regional consumers are willing to assume the major portion of the costs of using these resources. Likewise, state agencies should administer the minimum stream standards and approve abatement plans. However, the local firm can best determine the optimum use for its own resources that are available locally. Its case may have to be presented and defended to the state and federal governments in order to obtain extra monies for local use. The major argument for local control involves subsidies: only local representatives can carry out valid and equitable subsidies, since the firm members live and work in the area and represent the various interests of the region. Another strong point in favor of local control pertains to the direct effects of the wastes: wastes discharged into a stream affect primarily the people who reside, work, and travel in the vicinity of this stream.

Objectives of the firm.

1. To sell pollution-carrying resources in such a manner as to encourage consumers to utilize resources carefully.
2. To protect the low-level (minimum) resources in the stream by a system of charges that discourages the use of the last units of pollution-carrying resource.

3. To make certain that all potential consumers have an equal opportunity to obtain a portion of the resources at a reasonable price, by increasing the unit cost according to the number of units purchased or as fewer units remain.
4. To exert local influence over the use of the pollution-carrying resources in the form of prices and subsidies.
5. To minimize the cost of these resources by taking advantage of economies of scale to treat combined wastes where desirable or, by introducing other engineering systems, to make maximum quantities of resource available for sale.

Activities and procedures of the firm. The river resources allocation firm must conduct a market survey similar to that of any other industrial firm. The survey should reveal the following:

1. A listing of all of its potential customers.
2. The types and relative amounts of stream resource (product) desired by all consumers.
3. The amount of each type of stream resource which is available for sale by the firm. This will depend basically upon the decision of the firm to maintain a specific pollution index of the receiving stream.
4. An approximate value (reasonable market price) of each type of stream resource on a unit basis. This can be computed as a first approximation by ascertaining the total measurable benefits, ΣB,* of the use of each type of water resource in dollars and dividing this value by the number of units available for sale by the firm.

As an example, if the total measurable benefits of using the oxygen resource of a stream were $100,000 per year and one million pounds of oxygen resources were available for sale each year, the unit charge for

the first quantity of oxygen would be $0.10 per pound. Since most consumers cannot treat their wastes to reduce oxygen demand for this price, they would become interested purchasers.

A plan of increasing the price of the resource as it becomes used up seems ideal and in line with the economic principle of scarcity value. When the price of the resource reaches a level at which it is uneconomical for the consumer to buy any more, he will either curtail his production at that level or seek his own solution—such as waste treatment—at a cost somewhat lower than that charged by the firm. In either case, the firm's underlying objective of preserving a minimum stream quality will be achieved.

It is suggested that the firm hold an auction each year to assign as much of its pollution-carrying resource as possible. All potential consumers will be notified of the auction and told what resources will be sold and the lowest unit price anticipated for each resource. Let us assume that the meeting is attended by ten potential consumers interested in purchasing the firm's yearly one million pounds of oxygen resources, which were announced beforehand to be sold at the lowest price of $0.10 per pound. Since the unit price is so low, let us continue to assume that all ten consumers wish to purchase as many of these units as they can (up to their needs) at this price. The firm makes a decision that only the first 100,000 pounds may be bought at this price and that each consumer will be entitled to purchase an equal amount, i.e. 100,000/10 or 10,000 units of oxygen. Now, only 900,000 units are left for sale at a higher price of $150,00/900,000 or about $0.17 per pound. Not only did the pounds of product for sale decrease for the second allocation, but the total benefits to all users also increased, since damages would increase at the lower level of oxygen. Since both numerator and denominator reflect an increased unit price for the resource, the net effect is a much higher price.

Let us continue our assumptions, saying that one customer has sufficient units to satisfy his needs and that another cannot afford to purchase any more units at the next price. Each of the eight remaining consumers would be offered 100,000/8 or 12,500 units of oxygen at this price. Eventually, the last 100,000 units might cost $1,000,000/100,000 or $10 per pound. This last unit price would automatically be high enough to discourage any consumer from purchasing all of

---

*ΣB = a summation of the dollar value of a unit of resource for the following:
  a) monies paid by consumers to purchase water service,
  b) damages done to this resource by existing pollution,
  c) losses incurred when opportunities of resource usage are foregone because of existing pollution.

For example, ΣBO_2 = revenue paid + damages + opportunities foregone; revenue paid = \$/gal/lb O_2/gal = \$/lb O_2.

the remaining resources. It is suggested that these sales transactions be made final only after the firm has had a chance to review its sales and purchasers.

Difficulty arises in situations where two firms operate on the same basin, Firm A on the upper and Firm B on the lower section, and where Firm A sells all the stream resources available, leaving none for Firm B. One obvious remedy for this problem would be to merge the two firms into one whenever possible.

The firm will be faced with an important decision at the outset: what water-quality level should be maintained? In order to make this decision the firm will need to know the benefits of maintaining the stream at various water-quality levels. Total benefits (assumed to be equal to the expenditure for recreational use, the cost of damages for all uses and the opportunities foregone, in dollars), as mentioned earlier in the discussion, can be obtained by detailed surveys of the possible uses and the damages and wasted opportunities caused by the existing water quality. These total benefits can be obtained for various levels of water quality and expressed as pollution indices, determined from weighted averages of ratios of existing contaminant levels to allowable contaminant levels for each water use. At Syracuse University we are currently investigating this approach, which is designed to yield a plot of total benefits versus pollution index (Fig. 5.1).

An increasing pollution index (PI) indicates increasing contamination. The firm may then decide to main-

tain the water quality at a pollution index of A, from which the total benefit (B) can be calculated as B_A. Resources (R) available for sale at a water quality comparable to PI_A can be measured at a first price of B_A/R_A obtained for marketing by the firm as shown in Fig. 5.2.

The higher the pollution index (relative contamination) the firm is willing to tolerate, the more stream resources, such as dissolved oxygen, it will have available for sale to polluters at a lower unit cost.

Summary of benefit analyses. Our society desperately needs dollar values for the true benefits of abating pollution of a given water resource to various levels of water quality. All benefits accruing to the river basin community must be evaluated. The total benefits must be related to water quality, which in turn determines the stream resources available for marketing among the many polluters of the stream. With this information a realistic price can be placed upon each unit of water resource. The firm or water resources board is provided with true benefit information for its natural resources maintained at various quality levels. This procedure enhances the job of state, regional, and federal governments and prevents pollution by giving polluters an opportunity to purchase excess natural resources or to treat wastes, whichever is more economical for them.

Initial estimates indicate that the cost of treating industrial wastes on a level comparable to secondary

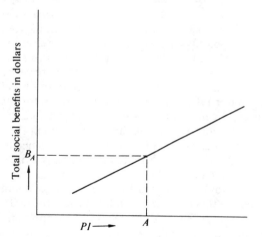

Fig. 5.1   Pollution benefits as related to water quality.

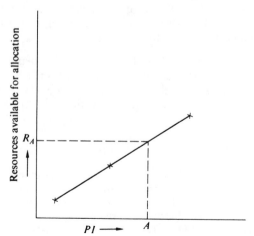

Fig. 5.2   Pollution benefits.

treatment of municipal wastes would be in the range of $2.6 to $4.6 billion during the 1969–73 federal fiscal period [7]. This includes $1.8 to $3.6 billion for new industrial treatment works and between $0.8 and $10.0 billion for replacing equipment. However, these estimates are based upon the minimal levels of control considered necessary to comply with water-quality standards. Should implementation of the standards call for higher levels of industrial waste reduction, these cost estimates could rise sharply. Meanwhile, industrial waste-abatement requirements could be met more efficiently through better in-plant controls and process changes and joint municipal and industrial treatment systems.

The estimated costs of operating and maintaining industrial waste-treatment facilities will range from $3.0 to $3.4 billion during 1969–73. As in the case of sewage-treatment works, these costs will continue to rise with increases in new treatment plants.

Manufacturing is the principal source of controllable water-borne wastes. In terms of the measurements of strength and volume usually quoted, wastes of manufacturing establishments are about *three times as great as those of the nation's sewered population* (Table 5.2). Moreover, the volume of industrial production, which gives rise to industrial wastes, is increasing by about 4.5 per cent a year or three times as fast as the population. Within industry as a whole, waste-load estimates, based on an estimate of the "average" quantity of pollutant per product unit, indicate that the chemical, paper, and kindred food industries generated about 90 per cent of the 5-day BOD (BOD_5) in untreated industrial waste-water.

Many industrial wastes differ markedly in chemical composition and toxicity from the wastes found in normal domestic sewage. Thus, the BOD_5 often is not an adequate indicator of the nature of industrial effluents. For example, industrial wastes frequently contain persistent organics which resist the secondary treatment procedures normally applied to domestic sewage and, in order to treat some industrial effluents, specific organic compounds must be stabilized or trace elements must be removed as part of the process. Obviously, these extra treatments increase the cost of joint municipal and industrial waste treatment. (Table 5.3)

An editorial in *Forbes Magazine* [1] reports the conclusions of a recent conference of 100 leaders of business, education, and government. Their main concern was why pollution abatement through industrial waste treatment remains hard to achieve. According to the editorial, seven difficulties stand in the way of any solution.

The problem of gathering adequate water-pollution data with regard to the movement and final disposition of wastes, the exact environmental effects, and the measurements of frequency and duration of specific pollutants

The difficulty of measuring costs against benefits

The difficulties in defining "purity," and in determining the degree of "acceptable" impurity that might be allowed

The difficulties in sorting out the relationships among separate chemical and organic pollutants, which may cancel one another out, add to each other's strength, or combine to produce even stronger new pollutants

Table 5.2 Comparison of annual totals of industrial and domestic wastes in the United States.

Source	Waste-water flow, billion gal	BOD, million lb	Settleable and suspended solids, million lb
All manufacturing	13,100	22,000	18,000
Sewered population*	5,300	7,300	8,800
	(120 gal)	(1/6 lb)	(1/5 lb)

*Annual totals calculated by multiplying individual daily figures (in parentheses) by 120 million persons × 365 days.

The difficulties of informing the public of the costs . . . and of their need to cooperate

The difficulties in offering incentives for industry to take pollution abatement measures as long as states have different standards and enforcement measures

The difficulties of urging industries or governments to invest heavily in abatement measures before scientifically proven standards are determined.

It is readily apparent, from these difficulties and the previous discussion on resource allocation, that there should be a greater emphasis upon economics rather than on regulation alone to solve the problem of industrial pollution. One important aspect we might mention here is that of tax allowances.

New guidelines for tax depreciation of waste-treatment facilities, which provide broader classification of facilities and allow for more objective consideration by the tax examiner, were made available to industry in 1964; however, they still provide for rather long depreciation allowances or schedules. This makes waste treatment a somewhat costly invest-ment, especially as no profit is realized from such expenditure. The size and nature of the industry will dictate in large measure not only the depreciation scheduling but also whether separate waste treatment can be economically feasible. Small companies may be induced for economic reasons to combine their wastes with municipal sewage (when the latter is adequately treated) for combined treatment, but if municipal treatment services are unavailable or incompatible with that of the industry, an insurmountable economic hardship to the industry can result. Therefore, certain tax improvements for waste treatment capital assets seem in order at both the federal and state levels. At present (1968) little positive action is being taken at the federal level to improve the tax situation, but some states are making an attempt. For example, New York State recently passed a bill which provides for (1) local tax forgiveness and (2) a 1-year depreciation write-off on state income taxes for the full value of the waste-treatment facility. Low-interest loans backed by the state might be an added inducement, especially to assist smaller industries in constructing pollution-abatement facilities.

Table 5.3 Costs of waste treatment.

The following is a table of capital and operating costs of treatment for a number of industrial wastes. All cost figures are for December 1975 and represent a selected sampling of specific projects. These costs are not intended to be fixed for all treatment of given wastes and must be updated to 1991 using ENR price adjustment rates.

Industry	Type of treatment reported	*Costs
Textile: Chapter 25.1–25.3		
Synthetic weaving mill	Secondary	$4761/MG
Cotton finishing plant	Secondary	$65–170/MG
Woven carpets and rugs	Chemicals and settling basin	$1690/MG
Tufted carpets and rugs	Settling basin and lagooning	$33/MG
Tannery: Chapters 25.4–25.5	Secondary	$495/MG
Cannery: Chapters 26.1–26.4		
General	Primary	$54/MG
General	Secondary	$182/MG
General	Screening, aeration and biological treatment	$243/MG
General	Lagoons and oxidation ditch	$450/MG
Corn and Pea	Biological treatment	$1150/MG
Dairy: Chapters 26.5–26.6		
Whey	Lagooning	$5074/MG
	Methane fermentation	$12,014/MG
Cottage cheese whey	Reverse osmosis	$1037/MG
Cheese whey	Roller drying	$38,640/MG
	Spray drying	$23,000/MG

Table 5.3 (*continued*)

Industry	Type of treatment reported	*Costs
Distillery: Chapters 26.7–26.9	Activated sludge	$297/MG
	Bio disc	$1100/MG
Meat Packing: Chapters 26.10–26.11	Screening, settling and biological treatment	$797–6701/MG
	PVC trickling filter	$50/MG (1)
	Oxidation ditch	$122/MG (2)
Celulose meat casings	Secondary	$440/MG
Feedlot wastes: Chapter 26.12	Mechanical aeration and lagooning	$9100/MG
Beet sugar: Chapters 26.13–26.14	Lagoon for lime sludge, aerated activated sludge, clarifier	$2900/MG
Miscellaneous food processing:		
Potato starch plant: Chapters 26	Evaporation	$6900/MG $8696/MG (3)
	Biological treatment	$4225/MG —
	Protein recovery and biological treatment	$9227/MG $2049/MG (3)
	Protein recovery and evaporation	$13,112/MG $7685/MG (3)
	Protein recovery, ion exchange, and biological treatment	$35,202/MG $20,681/MG (3)
Potato processing waste	Primary and secondary clarifiers, aerated lagoons and vacuum filter	$495/MG
Pickle wastes: Chapter 26.19	Neutralization and settling	$1769/1000 cucumbers
Water treatment plant: Chapter 26.22		
Lime sludge	Thickening and vacuum filter	$32–43/MG
	Thickening, centrifuge and recalcination	$4–13/MG
Alum sludge	Lagooning	$.18–9.80/MG
Pulp and paper mill: Chapter 27.1	Biological treatment	$137/MG
Metal finishing: Chapter 27.6	Chemical precipitation	$568/MG
Metal plating: Chapter 27.6	Gas chlorination	$4252/MG
	NaOCl	$7506/MG
	Neutralization, settling, and sludge beds	$1398/MG
Motor vehicle plant: Chapter 27.7	Activated sludge and chrome reduction	$315–449/MG
	Screening and settling	$525/MG
Oil refinery: Chapter 27.9	Air flotation	$45/MG
	Activated sludge	$106/MG
	Trickling filter	$126/MG
	Evaporation pond	$7/MG
Fuel oil wastes: Chapter 27.11	Ultrafiltration	$12–610/MG
	Reverse osmosis	$14–610/MG
	Centrifugation	$37–1220/MG
	Electrodialysis	$24–610/MG
Chemical industry: Chapter 28.15		
Nitrogen and phosphorous Removal:		
Nitrogen	Biological treatment	$40–200/MG
	Ammonia stripping	$18–50/MG
	Other	$200–2000/MG
Phosphorous	Biological treatment	$60–200/MG
	Chemical precipitation	$20–180/MG
	Other	$200–2000/MG

Table 5.3   (*continued*)

Industry	Type of treatment reported	*Costs
Fertilizer: Chapter 28.9	Oil separation and settling pond	$.20–1.09/ton
Organic chemicals: Chapter 28.15	Secondary	$737/MG
Explosives: Chapter 28.5		
Nitroglycerine wastes	Neutralization	$256/MG
	Ozone oxidation	$400/MG
Nitrate removal	Biodenitrification	$3.45–30.00/MG
	Algae harvesting	$20–35/MG
	Ion exchange	$170–300/MG
	Electrodialysis	$100–250/MG
	Reverse osmosis	$100–600/MG
	Distillation	$400–1000/MG
Pesticides: Chapter 28.7	Open dump or pit	$1.70/ton of product
	Ocean dumping	$2.80–33.60/ton of product
	Photochemical degredation	$102/MG
	Activated carbon	$35/MG
	Biological treatment	$.54–12.19/MG
	Landfill	$2–7/ton
Plastics and resins: Chapter 28.8		
Cellulosics	Secondary	$36/MG
Acrylics	Secondary	$735/MG
Urea	Lagoons with no overflow	$193/MG
Melamine resins	Lagoons with no overflow	$193/MG
Phenolics	Incineration	$12,495/MG
Energy industry: Chapter 29.1		
Cooling tower blowdown		
Inhibiter type:		
Chromate	Chemical treatment	$.19/lb removed
	Ion exchange	$.24/lb removed
Phosphates	Adsorption	$.48/lb removed
	Chemical precipitation	$.14–.16/lb removed
Organic	Carbon adsorption	$29–120/MG
Coal industry: Chapter 29.4		
Acid mine drainage	Equalization, aeration, settling and sludge lagooning	$136–638/MG (.1MGD Plant) $49–526/MG (1MGD Plant) $37–514/MG (7MGD Plant)
Radioactive wastes: Chapters 30.1 to 30.5	Burial	$2/CuFt (low-level waste) $10/CuFt (Alpha waste) $25/Gal (high-level waste)

*Data obtained from selected references given at the end of Section 5.4.
(1)   Costs for trickling filter only.
(2)   Capital cost only.
(3)   First figure is cost of treatment. Second figure is sales of reclaimed waste.
(4)   Capital cost only (in millions of dollars).

5.4 Costs of Waste Treatment

Some industrial plants, especially the smaller ones, are able to pay for waste-treatment facilities out of cash reserves or current profit. However, more industries are resorting to deferred payments for these facilities by borrowing funds from a variety of investors including banks, insurance companies, and even from their own stockholders through stock issuances. Another interesting possibility for obtaining abatement capital is that of selling industrial revenue bonds, which have been deemed tax exempt by the 1968 Revenue and Expenditure Control Act. These bonds are payable solely from the lease payments of a public or private corporation; they are sold on the basis of the credit rating of the corporation (which must be high in all categories as determined by Moody and Standard and Poor). This type of financing usually includes the following:

(1) A 1 to 2% lower borrowing interest rate than for other types of financing is demanded.
(2) A corporation can depreciate the equipment at a very rapid rate (usually over a five-year period.
(3) Congress allows these expenditures an "investment tax credit."
(4) A corporation can deduct lease payments as an interest expense.
(5) In some states equipment purchases can be made without a sales charge.

Hewson and Cook (1970) present a table (Table 5.4) derived from information privately provided by the Confederation of British Industries, which gives an indication of typical ranges of charge per 1000 cubic meters for the acceptance of trade effluents from a variety of process industries discharged to public sewers in conformity with local authorities' consent requirements. Hewson and Cook also believe that (for the majority of pretreatments required by industry for neutralization of acidity, removal of oil, and elimination of specifically toxic compounds and any substances known to be inhibitory to sewage works processes) the cost would be in the range of 10 to 30 pounds currency per 1000 m^3. He gives 5 pounds per 1000 m^3 for the cost of essentially untreated discharges from industry through a long pipeline to deep water and up to 20 to 30

Table 5.4 After Hewson and Cook (1970).

Industry	Typical range of trade effluent charges	
	(pounds currency per 1000 m^3)	(pence per 1000 gallons)
Metal finishing	11 – 35	12– 60
Textile finishing	22 – 33	24– 36
Tanning	22 –110	24–120
Cotton processing	27.5–110	30–120
Gas liquors	up to 440	up to 480

pounds per 1000 m^3 for effluents requiring presettlement, neutralization, and other chemical treatment before controlled-rate discharge to an estuary.

Although Smith (1968) gives the following two figures (5.3 and 5.4) as construction and total construction and operating costs for domestic sewage wastes, these figures offer some basis for comparison with industrial wastes. Generally these costs can be projected to other industrial wastes by relating pollutional loads such as BOD and/or suspended solids.

Can We Afford Pollution Control?

No modern society can afford not to protect its environment. For if we have an impure environment, life will be no life at all; at least one not worth living in. The ardent unilateral industrialist and even many politicians will not argue that life without industrial jobs can get pretty trying as well. What good is breathing clear and clean air or drinking and recreating in sparkling pure water if one has not a job from which to obtain the money to buy food, clothing, and shelter? On the surface that sounds like good reasoning. But it should be exposed for what it really is. It is only a smokescreen—a defense mechanism widely perpetuated by single-minded and motivated industrialists of all countries of the world. Unfortunately, the politically minded persons in governments have accepted this subterfuge as the gospel. It is not difficult to guess why. They conclude that a booming economy—without constraining pollution control costs—is vital assurance of their political future. This conclusion as well as industry's smokescreen is nonsense and should be exposed once and for all time as simply untrue and invalid (Nemerow, 1979).

Let's first clear the air about costs and then proceed to other matters which determine whether any price at all is worth paying to protect environmental quality. Water pollution control in many "wet" industries costs about one to two percent of their production costs. This is not a ficticious figure but one we have derived from numerous industrial plants in the United States. The percentage would be even less for developing countries. Air pollution control costs may run two to five percent of production costs. Innovative practices can lower these percentages even further. We are not convinced that industry's argument that $1.01 or even $1.05 to produce important goods for society rather than $1.00 would represent an "unbearable" burdensome cost to industry or the consumer. We are assured that this meager increase in production costs is not only well within economic limits, but also is a small price to pay for a clean environment. Industrialists are prone to quote costs not in percentages but in dollars, which may

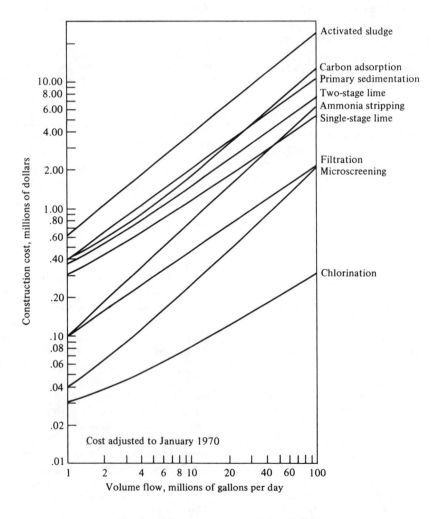

Fig. 5.3 Construction cost (after Smith).

seem exhorbitant to most sideline observers. The environmentalists are still shying away from quantifying the benefits.

And with good reason, for it is not an easy task, but can be done by expending sufficient time and effort. One attempt has been made by the American E.P.A. to include new jobs in environmental construction as one of the benefits of abatement. But the price tag still appears excessive and unrealistic. In fact, Management Informational Services (1986) reported that spending on pollution control equipment created nearly 167,000 jobs in 1985. Because the firms providing the equipment have to pay suppliers and workers, and those firms and employers spend the money they get, each dollar originally spent for control equipment generated

$2.27 in money for the marketplace. It may be helpful to provide an answer to the question of who can pay these costs.

We do not propose that industry absorb the costs. On the contrary, these costs should be passed on to the consumer, whoever he is and wherever he resides. An industrialist within any given western country may reply that if he raises the price, his sales will be reduced. The market will not "stand still" for any more price rises—no matter how small, he explains. The same industrialist in an eastern European country may plead that he is not permitted by governmental regulations to raise the price of his product to include these costs. Both industrialists have valid arguments for not raising product costs. Perhaps it is governments and consumers who

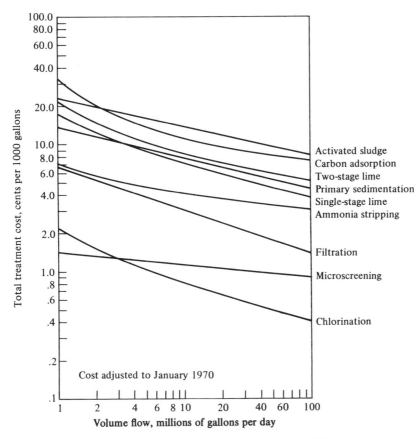

Fig. 5.4 Total treatment cost (after Smith).

are to blame for not accepting these costs. Once they understand that, first, the true costs of environmental protection are relatively small and, second, the costs will ensure that all people will be able to live—not just exist—and enjoy clean air and water and all of its amenities, I am convinced they will accept them.

But, in case they do not, they can show their refusal to accept the costs by abstaining from the purchase of the products. If products remain unsold, they may not be as vital to society as their manufacturers had assumed. If, on the other hand, they were truly vital but too high priced, governments may decide to defray part of the production costs as a societal benefit. In all cases, moreover, industry should not be expected—and rightfully—to absorb pollution abatement costs entirely.

Another question often arises during industrial production decisions. Should industry be located in developed nations where superior technology exists or in developing nations where economic stimuli and balancing of world industrial production is needed? Worldwide agencies such as the United Nations are inclined to favour the latter, while strong nationalists and ardent industrialists generally favour the former. Unfortunately, neither viewpoint takes proper consideration of the matter of pollution control. Some consider the environmental problem as insignificant when related to other production concerns, while others treat it as a costly burden either to be avoided completely or as weighing excessively in its decision of where to locate new production facilities. What is required is a progressive viewpoint that pollution control costs be treated as a normal part of production costs. When that is an accepted procedure, industrial location becomes a production decision. In effect, it is my opinion that production should take place wherever and whenever it is profitable for industry to do so. If waste treatment costs at a particular country site are excessive so as to make production costs sufficiently high and worldwide competition impossible, another site in that country or another country should be selected. The world consumer is entitled to the lowest possible product cost. However, this should not be achieved at the expense of a contaminated environment.

Another ploy used by industry is that they lack the technology suitable for solving pollution abatement requirements. Are they justified in making this claim? They advance this defense based primarily on two facts: (1) new abatement equipment is untried and costly and (2) the environmental quality levels are much too stringent to allow use of more conventional, tried, and less expensive

equipment. These arguments also appear reasonable on the surface. Data and efficiencies are available for all new treatment equipment. Otherwise it could not be sold at all. Reliable equipment manufacturers invest considerable sums of money in pilot plant testing in order to facilitate their sales and to guarantee operational efficiencies. Some large equipment units such as scrubbers, filters, and coolers are expensive—but hardly so when related once again to total plant costs. Industry still needs to think of this equipment as necessary for production, not as a burdensome and wasteful added cost to satisfy some bureaucrat. If an industrial plant has been gradually improving its production efficiency related to pollution control, the new effluent guidelines need not be considered too stringent. These guidelines should represent only another step in the direction of a clean environment. Unfortunately, some industries have neglected pollution control in the past, and these guidelines do represent to them a gigantic leap towards bankruptcy. It is useless to chastise them for their past ommissions, but it is useful to guide them firmly and patiently into making proper amends for the past. The environmental quality requirements are not too stringent, but the gap in remedial action is too wide for them now. It must be bridged nevertheless and at a faster pace than in the past.

References: Economics

1. "Cease pollution" (Editorial), *Forbes Magazine*, 1 June 1968.
2. Fair, G. M., "Pollution abatement in the Ruhr district," *J. Water Pollution Control Federation* **34**, 749 (1962).
3. Galbraith, J., *The New Industrial State*, Houghton Mifflin Co., New York (1967).
4. Kneese, A. V., *The Economics of Regional Water Quality Management*, The Johns Hopkins Press, Baltimore (1964).
5. Moore, J. G., "Water quality management in transition." *Civil Eng.* **38**, 30 (1968).
6. Nemerow, N. L., "Economics of industrial waste treatment," *Water Sewage Works J.* 238 (1968).
7. U.S. Department of the Interior, *The Cost of Clean Water*, Vol. 1, Summary Report, Jan. 10, 1968, Washington, D.C.

References: Costs of Waste Treatment

1. Anonymous, "Reuse comes out ahead," *Water and Wastewater Engineering* 9, 10, (1972).

2. Anonymous, "Projected wastewater treatment costs in the organic chemical industry," *Water Poll. Control Res. Ser.*, 12020 GND-07/71, 71, (1971).

3. Anonymous, "Disposal of wastes from water treatment plants," *A.W.W.A. Research Foundation Report*, 148, (1969).

4. Anonymous, "State-of-the-art, sugar beet processing waste treatment," *EPA Water Poll. Control Res. Ser.*, 1260 DSI 07/71, (July 1971).

5. Anonymous, "Aerobic secondary treatment of potato processing wastes," *EPA Water Poll. Control Res. Ser.*, 12060 EHV 12/70, 74, (1970).

6. Atkins, P. R., "The pesticide manufacturing industry-current waste treatment and disposal practices," *EPA Water Poll. Control Res. Ser.*, 12020 FYE 01/72 (January 1972).

7. Baker, D. A. and J. White, "Treatment of meat packing plant wastes using PVC trickling filters," Proc. 2nd Nat. Symp. on food processing wastes, *EPA Water Poll. Control Res. Ser.*, 12060-03/71, 289, (1971).

8. Barnhardt. E. L., "Biological treatment of pulp and paper mill wastes," *Proc. 10th Ontario Industrial Waste Conf.*, 43 (1963).

9. Besselievre, E. B., "Realistic approach to plating waste treatment," *Proc. 5th Ontario Ind. Waste Conf.* 92 (1958).

10. Bituminous Coal Research, Inc., "Studies of limestone treatment of acid mine drainage," *EPA Water Poll. Control Res. Ser.*, 14010 EIZ 12/71, (December 1971).

11. Brink, R. J., "Operating costs of waste treatment in General Motors," *Proc. 19th Ind. Waste Conf. Purdue Univ.*, 12 (1964).

12. Canin, K. Q., "Cost of waste treatment in meat packing industry," *Proc. 25th Ind. Waste Conf. Purdue Univ.*, 193 (1970).

13. Church, B. D., *et al.*, "Continuous treatment of corn and pea processing wastes water with Fungi Imperfecti," Symposium Proc. 2nd Nat. Food Proc. Wastes, *EPA Water Poll. Control Res. Ser.*, 12060-03/71, 203, (1971).

14. "Cost of clean water," III, Motor Vehicles and Parts, U.S. Dept. Interior (1967).

15. "Cost of clean water," III, Paper Mills, U.S. Dept Interior (1967).

16. "Cost of clean water," III, Oil Refinery, U.S. Dept Interior (1967).

17. "Cost of clean water," III, Plastics Materials and Resins, U.S. Dept. of Interior (October 1967).

18. Desai, S. V., *et al.*, "Pervaporation for removal of dilute contaminants from petrochemical effluent streams," *Amer. Inst. Chem. Engrs. Symposium Ser.* 69 (1973).

19. Ellis, M. M. and P. W. Fisher, "Clarifying oil field and refinery waste water by gas flotation," *J. Petroleum Tech.*, 426 (April 1973).

20. Eye, J. P. and L. Liu, "Treatment of wastes from a sole leather tannery," *J. Water Poll. Control Fed.* 43 2291 (1971).

21. Glover, G. E., "Cooling tower blowdown treatment costs," *Ind. Process Design for Water Pollution Control* 2, 74 (1974).

22. Goldsmith, R. L., *et al.*, "Membrane processing of cottage cheese whey for pollution abatement," Proc. 2nd Nat. Symp. Food Proc. Wastes, *EPA Water Poll. Control Res. Ser.*, 12060-03/71, 413, (1971).

23. Hewson, J. L. and W. J. M. Cook, "Water in the chemical and allied industries," S.C.I. Monograph No. 34, Soc. Chemical Industry, 114, (1970) London, England.

24. Jones, H. R., "Pollution control in the textile industry," *Pollution Technology Review*, II, Noyes Data Corp., 205, (1973).

25. Kramer, A. E. and H. Nierstrasz, "Design and operation problems of a treatment plant for metal finishing wastes," *Proc. 12th Ontario Industrial Waste Conf.* 119 (1965).

26. Martin, J. J., Jr., "Chemical treatment of plating waste for elimination of chromium, nickel and metal ions," *J. of New England Water Poll. Control Assoc.* (1972).

27. Park, W. P., "A waste treatment system for confined hog operations," EPA Project No. 13040 EVM, 660/2-74-047, May (1974).

28. Parker, C. D., "Methane fermentation of whey," Proc. 2nd Nat. Symp. on Food Processing Wastes, *EPA Water Poll. Control Res. Ser.*, 12060-03/71, 501 (1971).

29. Parker, C. D. and G. P. Skerry, "Cannery waste treatment by lagoons and oxidation ditch at Shepparton, Victoria, Australia," Proc. 2nd Nat. Symposium on Food Proc. Wastes, *EPA Water Poll. Control Res. Ser.*, 12060-03/71, 251 (1971).

30. Patterson, J. W. and R. A. Minear, "State of the art for the inorganic chemicals industry: commercial

explosives," *EPA Technology Ser.*, EPA 600/2-74-0096, March (1975).

31. Paulson, W. C., *et al.*, "Oxidation ditch treatment of meat packing wastes," Proc. of 2nd Nat. Symp. on Food Proc. Wastes, *EPA Water Poll. Control Res. Ser.*, 12060-03/71, 617 (1971).

32. Pfaff, J. W., "Treatment of waste liquids from the manufacture of celulose meat casing," *Proc. 10th Ontario Ind. Waste Conf.*, 147 (1963).

33. "Processes, procedures, and methods to control pollution from mining activities," *EPA Report*, 430/9-73-011, 282 (1973).

34. Quirk, T. P. and J. Hellman, "Activated sludge and trickling filtration treatment of whey effluents," Proc. 2nd Nat. Symp. on Food Proc. Wastes, *EPA Water Poll. Control Res. Ser.*, 12060-03/71, 447 (1971).

35. Radelmeir, W. R., "The economics of alternative methods of whey disposal at Southern Ontario cheese factories," *15th Ontario Industrial Waste Conf. Proc.*, June (1968).

36. Rose, W. W., *et al.*, "Protection and disposal for liquid wastes from cannery and freezing fruits and vegetables," Proc. 2nd Nat. Symp. on Food Proc. Wastes, *EPA Water Poll. Control Res. Ser.*, 12060-03/71, 109 (1971).

37. Rowe, W. D. and W. F. Holcomb, "The hidden commitment of nuclear wastes," *Nuclear Technology* **12** 286 (1974).

38. Smith, R. L., "Cost of Conventional and Advanced Treatment of Wastewater," *Journ. Water Poll. Control Fed.* **40**, 1546 (1968).

39. Stabile, R. L., *et al.*, "Economic analyses of alternative methods for processing potato starch plant effluents," Proc. 2nd Nat. Symp. on Food Proc. Wastes, *EPA Water Poll. Control Res. Ser.*, 12060-03/71, 185 (1971).

40. Streebin, L. E., *et al.*, "Demonstration of a full scale waste treatment system for a cannery," *U.S.E.P.A. Pub.*, 12060DSB 09/71 (1971).

41. Thomas, J. L. and L. G. Koehrsen, "Activated sludge-biodisc treatment of distillery wastes," *Environmental Protection Technology Series*, EPA-660/2-74-014 (1974).

42. Wenne, R. E. and J. R. Gaissman, "Recycling spent cucumber pickling brines," Proc. 4th Nat. Symp. on Food Proc. Wastes, EPA 660/2-73-031, December (1973).

43. Nemerow, N.L. "Can We Afford Pollution Control," *Environment Protection Engineering,* **5**, No. 4, 447, Bogazici University, Istanbul, Turkey 1979.

44. Management informational Services, "Pollution control has created thousands of jobs, report says" *The Miami Herald*, 10 A, (Jan. 29th, 1986).

Questions

1. Who should administer the economics aspect of pollution control?
2. What problem or conflict would exist when both the EPA and "firm" systems are in operation simultaneously?
3. What part do taxes play in industrial-waste treatment?
4. Why is economics of wastewater treatment important to industry and society?
5. What should be the industrial goal as far as pollution abatement is concerned?
6. What are the categories of benefits derived from industrial waste treatment?
7. What are the potential methods of relating measurable benefits to costs in order to make policy decisions?
8. What are the most reasonable geographical boundary limits of benefits of pollution control?
9. When does EPA feel it economical to build wastewater treatment plants?
10. What does the EPA feel the economic limits are?
11. What is the method of equitably rationing stream assets proposed by your authors?

Part 2 | THEORIES

CHAPTER 6

VOLUME REDUCTION

In general, the first step in minimizing the effects of industrial wastes on receiving streams and treatment plants is to reduce the volume of such wastes. This may be accomplished by: (1) classification of wastes; (2) conservation of waste water; (3) changing production to decrease wastes; (4) reusing both industrial and municipal effluents as raw water supplies; or (5) elimination of batch or slug discharges of process wastes.

6.1 Classification of Wastes

If wastes are classified, so that manufacturing-process waters are separated from cooling waters, the volume of water requiring intensive treatment may be reduced considerably. Sometimes it is possible to classify and separate the process waters themselves, so that only the most polluted ones are treated and the relatively uncontaminated are discharged without treatment. The three main classes of waste are:

a) Wastes from manufacturing processes. These include waters used in forming paper on traveling wire machines, expended from plating solutions in metal fabrication, discharged from washing of milk cans in dairy plants, and so forth.

b) Waters used as cooling agents in industrial processes. The volume of these wastes varies from one industry to another, depending on the total Btu's to be removed from the process waters. One large refinery discharges a total of 150 million gallons per day (mgd), of which only 5 mgd is process waste; the remainder is only slightly contaminated cooling-water waste. Although cooling water can become contaminated by small leaks, corrosion products, or the effect of heat, these wastes contain little, if any, organic matter and are classed as nonpollutional from that standpoint.

c) Wastes from sanitary uses. These will normally range from 25 to 50 gallons per employee per day. The

volume depends on many factors, including size of the plant, amount of waste-product materials washed from floors and the degree of cleanliness required of workers in the process operation.

Unfortunately, in most older plants, process, cooling, and sanitary waste waters are mixed in one pipeline; before 1930, industry paid little attention to segregating wastes to avoid stream pollution.

6.2 Conservation of Wastewater

Water conserved is waste saved. Conservation begins when an industry changes from an "open" to a "closed" system. For example, a paper mill which recycles white water (water passing through a wire screen upon which paper is formed) and thus reduces the volume of wash waters it uses is practicing water conservation. Concentrated recycled waste waters are often treated at the end of their period of usefulness, since usually it is impractical and uneconomical to treat the wastewaters as they complete each cycle. The savings are twofold: both water costs and waste-treatment costs are lower. However, many changes to effect conservation are quite costly and their benefits must be balanced against the costs. If the net result is deemed economical, then new conservation practices can be installed with assurance.

A paperboard mill may discharge 10,000 gallons of waste water per ton of product, although there are many variations from one mill to the next. Paper mills may release as much as 100,000 gallons or as little as 1,000 gallons of waste water per ton of product. The latter figure is usually the result of a scarcity of water and/or an awareness of the stream-pollution problem and demonstrates what can be accomplished by effective waste elimination and conservation of water. One large textile mill reduced its water consumption by 50 per cent during a municipal water shortage, without any drop in production. The author observed that, despite the savings to the mill, water usage returned to its original level once the shortage was over. This incident further illustrates the "cheapness" of water in the public's mind.

Steel mills reuse cooling waters to quench ingots, and coal processors reuse water to remove dirt and other noncombustible materials from coal. Many industries have installed countercurrent washing to reduce water consumption. By the use of multiple vats, the plating industry utilizes make-up water, so that only the most exhausted waters are released as waste. Automation, in such forms as water-regulating devices, also aids in conservation of water. Introduction of conservation practices requires a complete engineering survey of existing water use and an inventory of all plant operations using water and producing wastes, so as to develop an accurate balance for peak and average operating conditions.

6.3 Changing Production to Decrease Wastes

This is an effective method of controlling the volume of wastes but is difficult to put into practice. It is hard to persuade production men to change their operations just to eliminate wastes. Normally, the operational phase of engineering is planned by the chemical, mechanical, or industrial engineer, whose primary objective is cost savings. The sanitary engineer, on the other hand, has the protection of public health and the conservation of a natural resource as his main considerations. Yet there is no reason why both objectives cannot be achieved.

Waste treatment at the source should be considered an integral part of production. If the chemical engineer argues that it would cost the company money to change its methods of manufacture in order to reduce pollution at the source, the sanitary engineer can do more than simply enter a plea for the improvement of mankind's environment. He can point out, for instance, that reduction in the amount of sodium sulfite used in dyeing, of sodium cyanide used in plating, and of other chemicals used directly in production has resulted in both lessening of wastes and saving of money. He can also mention the fact that balancing the quantities of acids and alkalis used in a plant often results in a neutral waste, with a saving of chemicals, money, and time spent in waste treatment. Rocheleau and Taylor [15] point out several measures that can be

used to reduce wastes: improved process control, improved equipment design, use of different or better quality raw materials, good housekeeping, and preventative maintenance.

6.4 Reusing Both Industrial and Municipal Effluents for Raw Water Supplies

Practiced mainly in areas where water is scarce and/or expensive, this is proving a popular and economical method of conservation; of all the sources of water available to industry, sewage plant effluent is the most reliable at all seasons of the year and the only one that is actually increasing in quantity and improving in quality. Though there are many problems involved in the reuse of effluents for raw water supply, it must be remembered that *any* water supply poses problems to cities and industries. Since the problems of reusing sewage effluents are similar to those of reusing industrial effluents, they will be discussed here jointly.

Many industries and cities hesitate to reuse effluents for raw water supply. The reasons given [7] include lack of adequate information on the part of industrial managers, difficulty of negotiating contracts satisfactory to both municipalities and industrial users, certain technical problems such as hardness, color, and so forth, and an esthetic reluctance to accept effluents as a potential source of water for any purpose. Also, treatment plants are subject to shutdown and slug (sudden) discharges, both of which may make the supply undependable or of variable quality. In either case, industry may need an alternate source of water supply for these emergency situations. In addition, the "resistance to change in practice" factor cannot be overlooked as a major obstacle. However, as the cost of importing a raw water supply increases, it would seem logical to reuse waste-treatment plant effluents to increase the present water supply by replenishing the ground water. It cannot be denied that the ever-available treatment-plant effluent can produce a low-cost, steady water source through ground-water recharge. If any portion of a final industrial effluent can be reused, there will be less waste to treat and dispose of. Similarly, reuse of sewage effluent will reduce the quantity of pollution discharged by the municipality.

The greatest manufacturing use of water is for cooling purposes. Since the volume of this water requirement is usually great, industries located in areas where water is expensive should consider reuse of effluents. Even if the industry is fortunate enough to have a treated municipal water supply available, the cost will usually be excessive in comparison, which may have a generally beneficial effect. The reuse of municipal and industrial effluents saves water and brings revenue into the city. The design of waste-water treatment plants will be greatly influenced because the effluent must satisfy not only conventional stream requirements but those of industry as well.

Many cases are cited in the literature of industrial reuse of intermediate, untreated effluents, such as white waters from paper machines as spray and wash waters. The practice of reusing treated industrial effluents, however, is still in its infancy; there are more instances of industrial reuse of municipal effluents. For example, Wolman [17] has described the design and performance of a sewage-effluent treatment plant producing treated water at a rate of about 65 mgd for use in steel-mill processing operations. The plant employed a conventional coagulation treatment, using alum combined with chlorination; final water averaged 5 to 10 ppm turbidity, with little or no coliform bacterial contamination. The most serious problem encountered was the presence of a high concentration of chlorides. Operating costs, exclusive of interest and amortization but including pumping costs, were $1.75 per million gallons (though this figure does not include the cost of raw-sewage treatment). It is interesting to compare this with the usual municipal cost of collecting, treating, and distributing raw water of $50 to $250 per million gallons, excluding fixed charges. Even when one adds $15 to $50 per million gallons for treating the raw sewage, the reusable effluent is much more economical than water obtained by developing a separate source of raw water. Treatment-plant reuse facilities at the Sun Oil Toledo refinery have been evaluated [9] for use as make-up water in the cooling towers. The cost savings resulting from elimination of municipal fresh-water make-up were found to be $100,000 per year.

Keating and Calise [7] list five main differences between most sewage-plant effluents and typical surface- or well-water supplies: (1) higher color; (2) higher nitrogenous content; (3) higher BOD content; (4) higher total dissolved solids; and (5) the presence of phosphates, due to detergents. Industrial effluents may also possess these characteristic differences, as well as others such as higher temperature. Despite these contaminants, in many parts of the United States, the effluent from properly operated secondary sewage plants is actually superior to available surface- or well-water supplies.

The number and variety of return-flow and on-site reuse systems are increasing. Information from the latest national census shows that the overall reuse rate increased from 106 per cent (of water reused) to 136 per cent between 1954 and 1959 alone. Reuse in all industries other than steam-electric generation increased from 82 to 139 per cent during this same period. In 1959 the primary metal, chemical, paper, oil, and food industries were especially large reusers of water.

In 1957, El Paso Products Company founded a petrochemical complex near Odessa, Texas, designed to use sewage-plant effluent for cooling and boiler water. After pretreatment the only problem encountered was foaming (largely eliminated by the current changeover to "soft" detergents in domestic use). Reuse of sewage effluents often frees municipal or surface water for other valuable purposes. For example, reutilization of sewage for agricultural purposes in Israel will add a potential of 10 per cent to its total water supply. It was found that the soil structure is improved by the organics in sewage, but where industrial wastes, particularly heavy metals, are present further treatment beyond oxidation ponds is needed.

"Dry" cleaning of processing equipment instead of washing with water can greatly reduce the volume of waste water. However, this will still leave a solid waste for disposal rather than a liquid one. Hoak [6] presents a set of conservation techniques largely adapted from his experiences in steel mills:

1. Install meters in each department to make operators cost- and quantity-conscious
2. Regulate pressure to prevent needless waste
3. Use thermostatic controls to save water and increase efficiency
4. Install automatic valves to prevent loss through failure to close valves when water is no longer needed
5. Use spring-closing sanitary fixtures to prevent constant or intermittent flow of unused water

6. Descale heat exchangers to prevent loss of heat transfer and subsequent inefficient and excessive use of cooling water
7. Insulate pipes so that water is not left running to get it either cold or hot
8. Instigate leak surveys as a routine measure
9. Use centralized control to prevent wastages from improper connections
10. Recirculate cooling water, thereby saving up to 95 per cent of the water used in this process
11. Reuse, for example, blast-furnace cooling water for gas washing and clarified scale-pit water on blooming mills
12. Use high-pressure, low-volume rinse sprays for more efficiency and use a slight amount of detergent, wetting agent, or acid to improve the rinsing operation
13. Recondition waste water (often some slight inplant treatment will provide water suitable for process use)

Eden and Truesdale [3] give typical analyses of effluents from three towns in the south of England (Table 6.1). Eden found that the total-solids content appears to increase by about 340 mg/liter between the water supply and the sewage effluent derived

Table 6.1 Typical analyses of sewage effluents after conventional primary and secondary treatment (after Eden and Truesdale [3]).

Constituent*	Source		
	Stevenage	Letchworth	Redbridge
Total solids	728	640	931
Suspended solids	15		51
Permanganate value	13	8.6	16
BOD	9	2	21
COD (chemical oxygen demand)	63	31	78
Organic carbon	20	13	
Surface-active matter			
Anionic (as Manoxol OT)	2.5	0.75	1.4
Nonionic (as Lissapol NX)			0.4
Ammonia (as N)	4.1	1.9	7.1
Nitrate (as N)	38	21	26
Nitrite (as N)	1.8	0.2	0.4
Chloride	69	69	98
Sulfate	85	61	212
Total phosphate (as P)	9.6	6.2	8.2
Total phenol			3.4
Sodium	144	124	
Potassium	26	21	
Total hardness	249	295	468
pH value	7.6	7.2	7.4
Turbidity (A.T.U.)†			66
Color (Hazen units)	50	43	36
Coliform bacteria (no./ml)	1300		3500

*Results are given in milligrams per liter, unless otherwise indicated.
†Absorptiometric turbidity units.

from it. The total-solids concentration is one of the chief limiting factors in reuse of any waste water; the number of times sewage can be reused for industrial water supply is controlled by the pickup of dissolved solids, which can be removed only by expensive treatment methods. Some discussion of the contaminants listed in Table 6.1 is relevant to potential reuse of sewage effluents for industrial water. Many industrial purposes would demand concentrations of suspended solids less than 2 mg/liter, but sewage effluents contain considerably more than this and even after tertiary treatment often contain at least 7 mg/liter. The organic constituents of sewage effluents are still largely unknown. Absorption has been suggested as a method for reducing most of the organic matter. At Lake Tahoe, for example, it has been possible to reduce the organic matter (as measured by COD) to less than 16 mg/liter by a combination of coagulation, filtration, and absorption. Detergents can also be removed in this manner to a theoretical minimum level of about 0.2 mg/liter. Additional removal of ammonia, nitrite, and nitrate is relatively expensive and difficult. Ammonia, which can be air-stripped at high pH values, is objectionable in concentrations of more than 0.1 mg/liter for drinking-water supplies which are to be chlorinated. Removal of phosphates is important whenever the water used by industry will be subjected to algae growth conditions. The Tahoe method will reduce the phosphate to less than 1.0 mg/liter; controlled activated-sludge and lime-precipitation methods are also effective. At high chlorine levels it has been found possible even to remove many viruses. Since sewage effluents contain many types of microorganisms, they should be sterilized even for industrial process use. In addition, color and hardness in sewage effluents may be harmful to certain industries.

6.5 Elimination of Batch or Slug Discharges of Process Wastes

In "wet" manufacturing of a product, one or more steps are sometimes repeated, which results in production of a significantly higher volume and strength of waste during that period. If this waste is discharged in a short period of time, it is usually referred to as a slug discharge. This type of waste, because of its concentrated contaminants and/or surge in volume,

can be troublesome to both treatment plants and receiving streams. There are at least two methods of reducing the effects of these discharges: (1) the manufacturing firm alters its practice so as to increase the frequency and lessen the magnitude of batch discharges; (2) slug wastes are retained in holding basins from which they are allowed to flow continuously and uniformly over an extended (usually 24-hour) period.

References

1. Applebaum, S. B., "Industry does benefit from pollution control," *Water Wastes Eng.* **3**, 46 (1966).
2. Clarke, F. E., "Industrial re-use of water," *Ind. Eng. Chem.* **54**, 18 (1962).
3. Eden, G. E., and G. A. Truesdale, "Reclamation of water from sewage effluents," Paper read at Symposium on Conservation and Reclamation of Water, 28 November 1967, London, Reprint no. 519, Water Pollution Research Laboratory, London (1968).
4. "Flourishing on sewage-plant effluent," *Chem. Process.* **29**, 30 (1966).
5. Hershkovitz, S. Z., and F. Feinmesser, "Utilization of sewage for agricultural purposes," *Water Sewage Works* **113**, 181 (1967).
6. Hoak, R. D., "Water resources and the steel industry," *Iron Steel Engr.*, May 1964, p. 87.
7. Keating, R. J., and V. J. Calise, *The Treatment of Sewage Plant Effluent for Water Reuse in Process and Boiler Feed*, Technical Reprint T-129, Graver Water Conditioning Company, Union, N.J. (1954).
8. Marks, R. H., "Waste water treatment," *Power*, **111**, S32 (1967).
9. Mohler, E. F., Jr., H. F. Elkin, and L. R. Kumnick, "Experience with reuse and biooxidation of refinery wastewater in cooling tower systems," *J. Water Pollution Control Federation* **36**, 1380 (1964).
10. Morris, A. L., "Water renovation," *Ind. Water Eng.* **4**, 18 (1967).
11. National Association of Manufacturers and Chamber of Commerce of the United States, in cooperation with National Task Committee on Industrial Wastes, *Water in Industry*, Washington, D.C. (1967).
12. Rawn, A. M., and F. R. Bowerman, "Planned water reclamation," *J. Sewage Ind. Wastes* **29**, 1134 (1957).
13. Renn, C. E., "Serendipity at Hempstead—A study

in water management," *Ind. Water Eng.* **4**, 25 (1967).

14. Rice, J. K., "Water management to reduce wastes and recover water in plant effluents," *Chem. Eng.* **73**, 125 (1966).

15. Rocheleau, R. F., and E. F. Taylor, "An industry approach to pollution abatement," *J. Water Pollution Control Federation* **36**, 1185 (1964).

16. Unwin, H. D., "In plant wastewater management," *Ind. Water Eng.* **4**, 18 (1967).

17. Wolman, A., "Industrial water supply from processed sewage treatment plant effluent at Baltimore, Maryland," *Sewage Works J.* **20**, 15 (1948).

Suggested Additional Reading

Alexander, D. E., "Wastewater transformation at Amarillo. II. Industrial phase," *Sewage Ind. Wastes* **31**, 1107 (1959).

Berg, E. J., "Considerations in promoting the sale of sewage treatment plant effluent," *Sewage Ind. Wastes* **30**, 96 (1959).

Besselievre, E. B., "Industries recover valuable water and by-products from their wastes," *Wastes Eng.* **30**, 760, (1959).

Besselievre, E. B., "Industry must reuse effluents," *Wastes Eng.* **31**, 734 (1960).

Black, A. P., "Statement by Dr. A. P. Black," *J. Sanit. Eng. Div. Am. Soc. Civil Engrs.* **90** (**SA4**), 11 (1964).

Burrell, R., "Uses of effluent water in sewage treatment plants," *Sewage Works J.* **18**, 104 (1946).

California State Water Pollution Control Board, *Direct Utilization of Waste Waters*, Sacramento, Calif. (1955).

California State Water Pollution Control Board, "Industry utilizes sewage and waste effluent for processing operations," *Waste Eng.* **28**, 444 (1957).

Cecil, L. K., "Sewage treatment plant effluent for water reuse," *Water Sewage Works* **111**, 421 (1964).

Clarke, F. E., "Industrial re-use of water," *Ind. Eng. Chem.* **54**, 18 (1962).

Cohn, M. M., "A million tons of steel with sewage," *Wastes Eng.* **27**, 309 (1956).

Connell, C. H., "Utilization of waste waters," *Ind. Wastes* **2**, 148 (1957).

Connell, C. H., and E. J. Berg, "Industrial utilization of municipal wastewater," *Sewage Ind. Wastes* **31**, 212 (1959).

Connell, C. H., and M. C. Forbes, "Once used municipal water as industrial supply," *Water Sewage Works*, **111**, 397 (1964).

"Copper mining plant squeezes water dry," *Public Works*, **88**, 125 (1957).

Eliezer, R., R. Everett, and J. Weinstock, *Contaminant Removal from Sewage Plant Effluents by Foaming*, Advanced Waste Treatment Research Publication no. 5, U.S. Public Health Service, Cincinnati (December 1963).

Elkin, H. F., "Successful initial operation of water re-use at refinery," *Ind. Wastes* **1**, 75 (1955).

Gerster, J. A., *Cost of Purifying Municipal Waste Waters by Distillation*, Advanced Waste Treatment Research Publication no. 6, U.S. Public Health Service, Cincinnati (November 1963).

Geyer, J. C., "Reuse of sewage effluents for industrial water supply," in Proceedings of the Sixth Southern Municipal and Industrial Waste Conference, April 1957, at North Carolina State College, Raleigh.

Gloyna, E., J. Wolff, J. Geyer, and A. Wolman, "A report upon present and prospective means for improved re-use of water," Unpublished observations.

Hoak, R. D., "Industrial water conservation and re-use," *Tappi*, **44**, 40 (1961).

Hoak, R. D., "Water resources and the steel industry," *Iron Steel Eng.*, **41**, 1 (1964).

Hoppe, T. C., "Industry will reuse effluent in future waste economy drive," *Wastes Eng.* **31**, 596 (1960).

Hoot, R. A., "Plant effluent use at Fort Wayne," *Sewage Works J.* **20**, 908 (1948).

Howell, G. A., "Re-use of water in the steel industry," *Public Works*, **94**, 114 (1963).

Jenkins, S. H., "Re-use of water in industry. II. The composition of sewage and its potential use as a source of industrial water," *Water Sewage Works* **111**, 411 (1964).

Kabler, P. W., "Bacteria can be a nuisance," *Chem. Eng. Progr.* **59**, 23 (1963).

Keating, R. J., and V. J. Calise, "Treatment of sewage plant effluent for industrial re-use," *Sewage Ind. Wastes* **27**, 773 (1955).

Keefer, C. E., "Bethlehem makes steel with sewage," *Wastes Eng.* **27**, 310 (1956).

Middleton, F. M., "Advance treatment of waste waters for re-use," *Water Sewage Works* **111**, 401 (1964).

Morris, J. C., and W. J. Weber, *Preliminary Appraisal of Advanced Wastes Treatment Process*, Advanced Waste Treatment Research Publication no. 2, U.S. Department of Health, Education and Welfare, Washington, D.C. (1964).

Morris, J. C., and W. J. Weber, *Adsorption of Biochemically Resistant Materials from Solution*, Advanced Waste Treatment Research Publication no. 9, U.S. Department of Health, Education and Welfare, Washington, D.C. (1964).

Powell, S. T., "Some aspects of the requirements for the quality of water for industrial uses," *Sewage Works J.* **20**, 36 (1948).

Powell, S. T., "Adaptation of treated sewage for industrial use," *Ind. Eng. Chem.* **48**, 2168 (1956).

Randall, D. J., "Reclamation of process water," *Water Sewage Works* **111**, 414 (1964).

Scherer, C. H., "Sewage plant effluent is cheaper than city water," *Wastes Eng.* **30**, 124 (1959).

Scherer, C. H., "Wastewater transformation at Amarillo," *Sewage Ind. Wastes* **31**, 1103 (1959).

Sessler, R. E., "Waste water use in a soap and edible-oil plant," *Sewage Ind. Wastes* **27**, 1178 (1955).

Silman, H., "Re-use of water in industry. I. The re-use of water in the electroplating industry," *Chem. Ind.* **49**, 2046 (1962).

Stanbridge, H. H., "From pollution prevention to effluent re-use. Part I," *Water Sewage Works* **111**, 446 (1964).

Stanbridge, H. H., "From pollution prevention to effluent re-use. Part II," *Water Sewage Works* **111**, 494 (1964).

Stephan, D. G., "Water renovation, what it means to you," *Chem. Eng. Progr.* **59**, 19 (1963).

Stone, R., and J. C. Merrell, Jr., "Significance of minerals in waste-water," *Sewage Ind. Wastes* **30**, 928 (1958).

Tolman, S. L., "Reclaiming valuable water and bark," *Wastes Eng.* **30**, 21 (1959).

Veatch, N. T., "Industrial uses of reclaimed sewage effluents," *Sewage Works J.* **20**, 3, (1948).

Williamson, J. N., A. M. Heit, and C. Calmon, *Evaluation of Various Adsorbents and Coagulants for Waste Water Renovation*, Advanced Waste Treatment Research Program no. 12, U.S. Department of Health, Education and Welfare, Washington, D.C. (1964).

Wolman, A., "Industrial water supply from processed sewage treatment plant effluent at Baltimore, Maryland," *Sewage Works J.* **20**, 15 (1948).

Questions

1. What are the three major classifications of industrial wastes at an industrial plant?
2. What are the implications of these three types of wastes?
3. What do we mean by industrial water conservation?
4. What is another method of reducing volume? Give examples.
5. What advantage do we get by reducing the waste volume?
6. What is another method of reducing volume?
7. What is usually the greatest factor influencing an industry to reuse its wastewater?
8. What is usually the greatest deterrent to industrial reuse of wastewaters?
9. How can we encourage water conservation in an industrial plant?
10. What is another method of reducing the volume of wastewaters?

CHAPTER 7

STRENGTH REDUCTION

Waste strength reduction is the second major objective for an industrial plant concerned with waste treatment. Any effort to find means of reducing the total pounds of polluting matter in industrial wastes will be well rewarded by the savings due to the reduced requirements for waste treatment. The strength of wastes may be reduced by: (1) process changes; (2) equipment modifications; (3) segregation of wastes; (4) equalization of wastes; (5) by-product recovery; (6) proportioning wastes; and (7) monitoring waste streams.

7.1 Process Changes

In reducing the strength of wastes through process changes, the sanitary engineer is concerned with wastes that are most troublesome from a pollutional standpoint. His problems and therefore his approach differ from those of the plant engineer or superintendent. Sometimes tremendous resistance by a plant superintendent must be overcome in order to effect a change in process. The superintendent possesses considerable security because he can do a familiar job well; why should he jeopardize his position merely to prevent stream pollution? The answer is obvious. Industry dies when its progress stops. No manufacturer can meet present-day market competition without continually, and critically, reviewing and analyzing his production techniques. In addition, pollution abatement can no longer be considered by industry as a "optional" act; on the contrary, it must be regarded as a vital step in preserving water resources for all users. Many industries have resolved waste problems through process changes. Two such examples of progressive management are the textile and metal-fabricating industries. On the other hand, the leather industry still generally uses lime and sulfides (major contaminants of tannery wastes), although it is known that amines and enzymes could be substituted. The lag between research and actual application is often extensive and is caused by many operational difficulties.

Textile-finishing mills were faced with the disposal of highly pollutional wastes from sizing, kiering, desizing, and dyeing processes. Starch had been traditionally used as a sizing agent before weaving and this starch, after hydrolysis and removal from the finished cloth, was the source of 30 to 50 per cent of the mill's total oxygen-demanding matter. The industry began to express an interest in cellulosic sizing agents, which would exhibit little or no BOD or toxic effect in streams. Several highly substituted cellulosic compounds, such as carboxymethyl cellulose, were developed and used in certain mills, with the result that the BOD contributed by desizing wastes was reduced almost in direct relation to the amount of cellulosic sizing compound used.

In the metal-plating industries [1], seven changes of process or materials have been suggested. Thus, to eliminate or reduce cyanide strengths: (1) change from copper–cyanide plating solutions to acid–copper solutions; (2) replace the $CuCN_2$ strike before the copper-plating bath with a nickel strike; (3) substitute a carbo-nitriding furnace, which uses a carburizing atmosphere and ammonia gas, for the usual molten cyanide bath. For other purposes: (4) use "shot blast" or other abrasive treatment on non-intricate parts instead of H_2SO_4, in pickling of steel; (5) substitute H_3PO_4 for H_2SO_4 in pickling; (6) use alkaline derusters instead of acid solutions to remove light rust which occurs during storage (the overall pH will be raised nearer to neutrality by this procedure, which will also alleviate corrosive effects on piping and sewer lines); (7) replace soluble oils, and other short-term rust-preventive oils applied to parts after cleaning, with "cold" cleaners. These cleaners can be used in both the wash solution and the rinse solution. They inhibit rust chemically rather than by a film of oil or grease. These process changes will become more understandable to the reader after the discussion of metal-plating wastes in Chapter 24.

A Pennsylvania coal-mining company modified its process to wash raw coal with acid mine waste rather than a public or private water supply. In this way, the mine drainage waste is neutralized while the coal is washed free from impurities. In one analysis, for example, the initial mine water had a pH of 3, an acidity of 4340 ppm as $CaCO_3$, and an iron content of 551 ppm; the waste water finally discharged from the process had a pH of 6.7 to 7.1 and an iron content of less than 1 ppm.

7.2 Equipment Modifications

Changes in equipment can effect a reduction in the strength of the waste, usually by reducing the amounts of contaminants entering the waste stream. Often quite slight changes can be made in present equipment to reduce waste. For instance, in pickle factories, screens placed over drain lines in cucumber tanks prevent the escape of seeds and pieces of cucumber which add to the strength and density of the waste. Similarly, traps on the discharge pipeline in poultry plants prevent emission of feathers and pieces of fat.

An outstanding example of waste strength reduction (with a more extensive modification of equipment) occurred in the dairy industry. Trebler [8] redesigned the large milk-cans used to collect farmers' milk. The new cans were constructed with smooth necks so that they could be drained faster and more completely. This prevented a large amount of milk waste from entering streams and sewage plants. Dairymen have also installed drip pans in assembly lines to collect milk which drains from the cans after they have been emptied into the sterilizers. The drip-pan contents are returned to the milk tanks daily.

In the chemical industry Hyde [4] described a chemical plant which effected a 23 per cent decrease in average BOD, through the installation of calandrias on open-bottom steam stills and by using refrigerated condensers ahead of vacuum jets, among other process modifications.

7.3 Segregation of Wastes

Segregation of wastes reduces the strength and/or the difficulty of treating the final waste from an industrial plant. It usually results in two wastes: one strong and small in volume and the other weaker, with almost the same volume as the original unsegregated waste. The small-volume strong waste can then be handled with methods specific to the problem it presents. In terms of volume reduction alone, segregation of cooling waters and storm waters from process waste will mean a saving in the size of the final treatment plant. Many dye wastes, for example, can be more economically and effectively treated in concentrated solutions. Although this type of segregation may increase the strength of the waste being treated, it will normally produce a final effluent containing less polluting matter.

Another type of segregation is the removal of one particular process waste from the other process wastes of an industrial plant, which renders the major part of the waste more amenable to treatment, as illustrated in the following examples.

A textile mill manufacturing finished cloth produced the wastes listed in Table 7.1. The combined waste was quite strong, difficult and expensive to treat, and very similar to laundry waste. However, when the liquid kiering waste was segregated from the other wastes, chemically neutralized, precipitated, and settled, the supernatant (that part which remained on the surface) could be treated chemically and biologically along with the other three wastes, because the strength of the resulting mixture was considerably less than that of the original combined waste. This type of segregation is also practiced in metal-finishing plants, which produce wastes containing both chromium and cyanide, as well as other metals. In almost all cases, it is necessary to segregate the cyanide-bearing wastes, make them alkaline, and oxidize them. The chromium wastes, on the other hand, have to be acidified and reduced. The two effluents can then be combined and precipitated in an alkaline solution to remove the metals. Without segregation, poisonous hydrogen cyanide gas would develop as a result of acidification. A recent patent [5] allows the separation of paint from waste water by precipitation with ferric chloride and/or ferric sulfate along with calcium hydroxide.

Segregation of certain wastes is of great advantage in all industries. It is dangerous, however, to arrive at a blanket conclusion that segregation of strong or dangerous wastes is always desirable. Just the reverse

Table 7.1 Wastes from a textile mill.

	Grey water*	White water*	Dye waste*	Kier waste*	Combined waste
pH	4.0	7.3	11.0	11.8	9.4
Total solids, ppm	2680	420	2880	18,880	1560
Suspended solids, ppm	224	67	148	218	156
Oxygen consumed, ppm	1560	31	556	4,900	460

* Defined in Chapter 22.

technique—complete equalization—may be necessary in certain circumstances.

7.4 Equalization of Wastes

Plants which have many products, from a diversity of processes, prefer to equalize their wastes. This requires holding wastes for a certain period of time, depending on the time taken for the repetitive processes in the plant. For example, if a manufactured item requires a series of operations that take eight hours, the plant needs an equalization basin designed to hold the wastes for that eight-hour period. The effluent from an equalization basin is much more consistent in its characteristics than is each separate influent to that same basin. Stabilization of pH and BOD and settling of solids and heavy metals are among the objectives of equalization. Stable effluents are treated more easily and efficiently than unstable ones by industrial and municipal treatment plants. Sometimes equalization may produce an effluent which warrants no further treatment. The graph in Fig. 7.1 illustrates one of the beneficial effects of equalization.

A large chemical corporation producing a predominantly acid waste has found it an advantage to equalize its wastes for a 24-hour period in an earthen holding basin. Following this equalization, a nearby plant, producing a highly alkaline waste, pumps *its* waste into the acid-waste effluent for neutralization. Considerably greater neutralizing power would be required for the acid waste were it not equalized, to iron out the peaks before neutralization.

A textile-finishing mill, which discharged its waste into a domestic secondary sewage-treatment plant, upset the efficiency of the plant. Although this waste represented only about 10 per cent of the total being

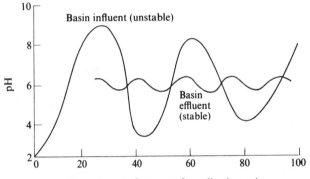

Figure 7.1

treated, it caused fluctuations, primarily in pH and BOD, which were responsible for the plant's difficulties. The solution was to build an equalization basin capable of detaining the wastes long enough to reduce the fluctuations in pH and BOD. In addition, the mill decided to deliver the equalized waste to the city treatment plant at three different rates of flow: the highest flow rate corresponded to the time when the greatest amount of sewage was reaching the plant, and vice versa. This gave a more constant dilution of the mill's waste with domestic sewage.

7.5 By-Product Recovery

This is the utopian aspect of industrial-waste treatment, the one phase of the entire problem which may lead to economic gain. Yet, many consultants deprecate this approach to the solution of waste problems. Their attitude is based mainly on statistics concerning the low percentage of successful by-products developed from waste salvage. However, any use of waste materials obviously eliminates at least some of the waste which eventually must be disposed of and the search for by-products should be encouraged, if only because it provides management with a clearer insight into processing and waste problems. All wastes contain by-products, the exhausted materials used in the process. Since some wastes are very difficult to treat at low cost, it is advisable for the industrial management concerned to consider the possibility of building a recovery plant which will produce a marketable by-product and at the same time solve a troublesome waste problem. There are many examples of positive results from adapting waste-treatment procedures to by-product recovery.

Metal-plating industries use ion exchangers to recover phosphoric acid, copper, nickel, and chromium from plating solutions. The de-ionized water, without any further treatment, is ideal for boiler-feed requirements. For final recovery of valuable chromium, copper, and nickel, companies use vacuum evaporation of the concentrated plating solutions. A nickel-wire plating plant, faced with a nickel shortage, made the plating waste alkaline with soda ash and precipitated nickel as the carbonate, then dried the sludge and treated it to recover the nickel. A silver-plating plant spends about $120,000 a year on waste treatment, of which $60,000 is returned as credit for silver recovered from the waste. The electrical industry recovers silver, gold, and (as by-products) water, valuable metals, and acids. Plants such as Scotscraft, Inc., report the recovery and reuse of by-product cyanide from plating wastes. A system of evaporation is used here to effect an overall plant saving in cost.

Specialty paper mills, with the aid of multiple-effect evaporators, recover caustic soda from cooking liquors. Chemical plants spray dilute waste acids into hot, lead-lined, brick-faced towers to concentrate the acids for reuse. Pharmaceutical houses recover the mold by drying the cake from vacuum filters or evaporating spent broth in multiple-effect evaporators. Distilleries screen the "slop" and thicken it for by-product use. Yeast factories evaporate a portion of their waste and sell the residue for cattle feed.

Even sewage plants have entered the by-product business. Methane gas from sewage digesters is com-

monly utilized for heat and power, and some cities make fertilizers and vitamin constituents from digested and dried sewage sludges. The sewage plant in Bradford, England, recovers grease by cracking with sulfuric acid and precipitating with alum and iron salts.

Classic examples of multiple usages of waste are the sulfite waste-liquor by-products from paper mills. They are used in fuel, road binder, cattle fodder, fertilizer, insulating compounds, as boiler-water additives and flotation agents, and in the production of alcohol and artificial vanillin. There are some 2000 U.S. patents for products made from waste sulfite liquor.

Packing houses and slaughterhouses recover waste blood, which is used as a binder in laminated wood products and in the manufacture of glue; they also sell waste greases to rendering plants.

The dairy industry treats skim milk with dilute acid to manufacture casein. Casein manufacturers in turn utilize their waste to precipitate albumin. The resulting albumin waste is used in the crystallization of milk sugar, and the residue from *this* process is utilized as poultry feed. Calcium and sodium lactate are also produced from skim milk, and dried and evaporated buttermilk is used for chicken feed. It is even rumored that chocolate ice cream originated as a by-product of the dairy industry.

Some companies, such as rendering plants, are in business primarily to develop by-products from other plants' waste products. Many rendering plants make feeds and fertilizers from chicken feet and feathers and recover grease, which is used to make soap.

Once a by-product is developed and put into production it is difficult to identify the new product with a waste-treatment process. For example, when sugar is extracted from sugar cane, a thick syrupy liquid known as blackstrap molasses is left. This molasses used to be so cheap that it was almost given away. Today it has many uses, one of the best-known being in the production of commercial alcohol. People have even found a use for the cane stalks: an insulating wallboard, called Celotex, is made from it.

These are only a few of the many ways in which industry can turn wastes into usable products. Although the problem of waste disposal usually persists, it is greatly lessened by the utilization of waste for by-products. In the final analysis both economic considerations and compliance with the requirements of pollution abatement play the major roles in any decisions involving by-product recovery.

7.6 Proportioning Wastes

By proportioning its discharge of concentrated wastes into the main sewer a plant can often reduce the strength of its total waste to the point where it will need a minimum of final treatment or will cause the least damage to the stream or treatment plant. It may prove less costly to proportion one small but concentrated waste into the main flow, according to the rate of the main flow, than to equalize the entire waste of the plant in order to reduce the strength.

7.7 Monitoring Waste Streams

Sophistication in plant control should include that of waste-water controls. Remote sensing devices that enable the operator to stop, reduce, or redirect the flow from any process when its concentration of contaminants exceeds certain limits are an excellent method of reducing waste strengths. In fact, accidental spills are often the sole cause of stream pollution or malfunctioning of treatment plants and these can be controlled, and often eliminated completely, if all significant sources of wastes are monitored.

7.8 Accidental Spills

Accidental discharges of significant process solutions represent one of the most severe pollution hazards. Since many accidental discharges go unobserved and are usually small in volume, they should be given special attention by the waste engineer. However, it is almost impossible to prevent every potential accident from occurring. There are some measures that can be taken to reduce the likelihood of accidents and severity when and if they occur. Some suggestions for general use include the following:

(a) Make certain that all pipelines and valves in the plant are clearly identified.
(b) Allow only certain designated and knowledgable persons to operate these valves.
(c) Install indicators and warning systems for leaks and spills.

(d) Provide for detention of spilled wastewater in holding basins or lagoons until proper waste treatment can be accomplished.
(e) Monitor all effluents—quantity and quality— to provide a positive public record, if necessary.
(f) Establish a regular maintenance program of all pollution-abatement equipment and all production equipment which may result in a liquid discharge to the sewer.

References

1. Davis, L., "Industrial wastes control in the General Motors Corporation," *Sewage Ind. Wastes* **29**, 1024 (1957).
2. Dillon, K. E., "Waste disposal made profitable," *Chem. Eng.* **74**, 146 (1967).
3. "Factory recovers cyanides from plating wastes," *Water Works Wastes Eng.* **2**, 65 (1965).
4. Hyde, A. C., "Chemical plant waste treatment by ten methods," *J. Water Pollution Control Federation* **37**, 1486 (1965).
5. Koelsh-Folzer-Werke, A. G., "Separating paint from waste or circulating water containing paint," British Patent 1,016,673 (1966); *Chem. Abs.* **64**, 649423 (1966).
6. Rosengarten, G. M., "Union Carbide Corporation's water pollution control program," *Water Sewage Works* **114**, R181 (1967).
7. Sanders, M. E. "Implementation to meet the new water quality criteria," *Water Sewage Works* **114**, R-5 (1967).
8. Trebler, H. A., "Waste saving by improvements in milk plant equipment," in Proceedings of 1st Industrial Waste Conference, Purdue University, November 1944, pp. 6–21.

Questions

1. What do we mean by strength reduction? Why should we use it?
2. Give an example of how a process change can reduce the strength of wastewater.
3. Give a classic example of an equipment modification to reduce the strength of wastewater.
4. How does segregation reduce the strength of the wastewater? Give an example.
5. What do we mean by equalization to reduce the strength of wastewater?

6. When is it profitable to install by-product recovery? How does this help in strength reduction?
7. Give two examples of by-product recovery by industry.
8. How can proportioning industrial wastes reduce their strengths?
9. What advantage is gained by installing modern methods of monitoring waste contaminants as far as strength reduction is concerned?

CHAPTER 8

NEUTRALIZATION

Excessively acid or alkaline wastes should not be discharged without treatment into a receiving stream. A stream even in the lowest classification—that is, one classified for waste disposal and/or navigation—is adversely affected by low or high pH values. This adverse condition is even more critical when *sudden* slugs of acids or alkalis are imposed upon the stream.

At a pH of only 6.5, trout of three species (brook, brown, and rainbow) have shown significantly lessened hatching of eggs and growth. When the pH is lowered to 5.5, bass, walleyed pike, and rainbow trout have been reported to be eliminated (22, 1981), as well as declines in trout and salmon. Below pH 5, most fish are unable to survive. This low pH causes female fish to deter laying of their eggs and, if laid, the fish are very sensitive in the egg, larval, and fish frog stages. Low pH can interfere with the salt balance freshwater species of fish need to maintain in their body tissues and blood plasma. Acid ioniges or other entities activate many metals already present, such as aluminum, which can be toxic to the fish even at pH values normally considered safe.

There are many acceptable methods for neutralizing overacidity or overalkalinity of waste waters, such as: (1) mixing wastes so that the net effect is a near-neutral pH; (2) passing acid wastes through beds of limestone; (3) mixing acid wastes with lime slurries or dolomitic lime slurries; (4) adding the proper proportions of concentrated solutions of caustic soda (NaOH) or soda ash (Na_2CO_3) to acid wastes; (5) blowing waste boiler-flue gas through alkaline wastes; (6) adding compressed CO_2 to alkaline wastes; (7) producing CO_2 in alkaline wastes; (8) adding sulfuric acid to alkaline wastes.

The material and method used should be selected on the basis of the overall cost, since material costs vary widely and equipment for utilizing various agents will differ with the method selected. The volume, kind, and quantity of acid or alkali to be neutralized are also factors in deciding which neutralizing agent to use.

In any lime neutralization treatment, the waste engineer should establish a minimum acceptable effluent pH and allow adequate reaction time for an acid effluent to reach this minimum pH. This will usually save considerable unnecessary expense [13]. In many cases, a mill can cut down on neutralization costs by providing sufficient detention time and sacrificing some efficiency in subsequent biological treatment (if used). During storage of alkaline wastes in contact with air, CO_2 will slowly dissolve in the waste and lower the pH. However, detention time alone, within feasible limits, will not effect as low a final pH as can be obtained by the use of neutralizing chemicals. Since biological treatment is more efficient at pH values nearer neutrality, prior neutralization by chemicals renders such treatment more effective.

8.1 Mixing Wastes

Mixing of wastes can be accomplished within a single plant operation or between neighboring industrial plants. Acid and alkaline wastes may be produced individually within one plant and proper mixing of these wastes at appropriate times can accomplish neutralization (Fig. 8.1), although this usually requires some storage of each waste to avoid slugs of either acid or alkali.

If one plant produces an alkaline waste which can be pumped conveniently to an area adjacent to a plant discharging an acid waste, an economical and feasible system of neutralization results for each plant. For example, a building-materials plant producing an alkaline (lime and magnesia) waste pumps the slurry, after some equalization, about one-half mile to mix with the effluent from a chemical plant producing an acid waste. The neutralized waste resulting from this combination is more readily treat-

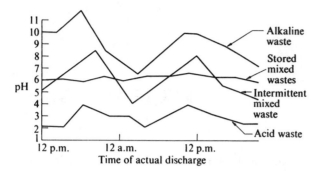

Figure 8.1

able for final disposal and both plants thus solve problems in economics, politics, and engineering. In another instance, Hyde [8] reports the use of a 500,000-gallon reservoir ahead of an anaerobic digestion pond to mix various plant wastes prior to treatment. The resulting pH of the reservoir effluent ranges from 6.5 to 8.5.

8.2 Limestone Treatment for Acid Wastes

Passing acid wastes through beds of limestone was one of the original methods of neutralizing them [4, 15]. The wastes can be pumped up or down through the bed, depending on the head available and the cost involved, at a rate of about 1 gallon per minute (gpm) per square foot or less. Neutralization proceeds chemically according to the following typical reaction:

$$CaCO_3 + H_2SO_4 \rightarrow CaSO_4 + H_2CO_3.$$

The reaction will continue as long as excess limestone is available and in an active state. The first condition can be met simply by providing a sufficient quantity of limestone; the second condition is sometimes more difficult to maintain. A sulfuric acid solution must be diluted to an upper limit of about 5 per cent and applied at a rate less than 5 gpm/ft² to avoid fouling the bed. According to Jacobs [10], no attempt should be made to neutralize sulfuric acid above 0.3 per cent concentration or at a rate of feed less than 1 gpm/ft² because of the low solubility of calcium sulfate. Excessive acid will precipitate the calcium sulfate and cause subsequent coating and inactivation of the limestone.

Disposing of the used limestone beds can be a serious drawback to this method of neutralization, since the used limestone must be replaced by fresh at periodic intervals, the frequency of replacement depending on the quantity and quality of acid wastes being passed through a bed. When there are extremely high acid loads, foaming may occur, especially when organic matter is also present in the waste.

8.3 Lime-Slurry Treatment for Acid Wastes

Mixing acid wastes with lime slurries is an effective procedure for neutralization [5, 17–19]. The reaction is similar to that obtained with limestone beds. In this case, however, lime is used up continuously because it is converted to calcium sulfate and carried out in the waste. Though slow acting, lime possesses a high neutralizing power and its action can be hastened by heating or by oxygenating the mixture. It is relatively inexpensive, but in large quantities the cost can be an important item.

Hydrated lime is sometimes difficult to handle, since it has a tendency to arch, or bridge, over the outlet in storage bins and possesses poor flow properties, but it is particularly adaptable to neutralization problems involving small quantities of acid waste, as it can be stored in bags without the erection of special storage facilities.

In an actual case [2], neutralization of nitric and sulfuric acid wastes in concentrations up to about 1.5 per cent (in the case of sulfuric acid) was accomplished satisfactorily by using a burned dolomitic stone containing 47.5 per cent CaO, 34.3 per cent MgO, and 1.8 per cent CaCO₃. The concentration of acid was limited to the stated 1.5 per cent, at least in part, because of the absence of dilution water to vary the percentage. This stone provided the additional advantage of holding residual sulfation to a minimum, an impossibility with any of the high-calcium limes [9].

8.4 Caustic-Soda Treatment for Acid Wastes

Adding concentrated solutions of caustic soda or sodium carbonate to acid wastes in the proper proportions results in faster, but more costly, neutralization. Smaller volumes of the agent are required, since these neutralizers are more powerful than lime

Chemical	Cost, $/ton (approx.)	Basicity factor†	Cost, $/ton of basicity
NaOH (78% Na_2O)	106	0.687	154
Na_2CO_3 (58% Na_2O)	57	0.507	112
MgO	83	1.306	64
High-calcium hydrated lime	14	0.710	20
Dolomitic hydrated lime	14	0.912	15
High-calcium quicklime	11	0.941	12
Dolomitic quicklime	11	1.110	10
High-calcium limestone	4	0.489	8
Dolomitic limestone	4	0.564	7

Table 8.1 Cost comparison of various alkaline agents.* (After Hoak [6].)

*Based on 1954 cost quotations.
†A measure of the alkali available for neutralization (grams of equivalent CaO per gram).

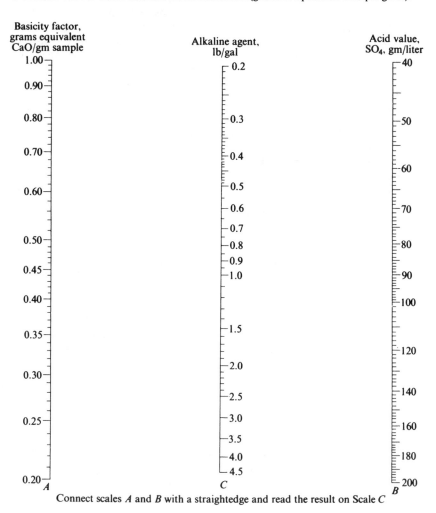

Fig. 8.2 Nomograph for treatment of acid wastes: a chart for determining the amount of alkaline agent needed. (After Hoak [7].)

Connect scales *A* and *B* with a straightedge and read the result on Scale *C*

or limestone. Another advantage is that the reaction products are soluble and do not increase the hardness of receiving waters. Caustic soda is normally bled into the suction side of a pump discharging acid wastes. This method is suitable for small volumes, but for neutralizing large volumes of acid waste water, special proportioning equipment (see Chapter 9) should be provided, as well as a suitably sized storage tank for the caustic soda, with a multiple-speed pump for direct addition of the alkali to the flow of acid wastes.

We have now discussed four methods of neutralizing acid wastes. Before we move on to alkaline wastes, let us compare the basicity and costs of the acid-neutralizing methods and agents we have considered (Table 8.1).

Since the basicity factor, as shown in Table 8.1, is one of the vital factors in selecting a neutralizing agent, Hoak [7] provides not only a method for computing this factor but also a nomograph for calculating the pounds of neutralizing agent required per gallon of waste (Fig. 8.2). He determines the acid value by titrating a 5-ml sample of sulfuric-acid waste with an excess amount of 0.5 N NaOH and back-titrating with 0.5 N HCl to a phenolphthalein endpoint. The basicity factor of the lime (or neutralizing agent) is determined by titrating a 1-gm sample of alkaline agent with an excess of 0.5 N HCl, boiling the sample for 15 minutes, and back-titrating with 0.5 N NaOH to the phenolphthalein endpoint. The acid value (line B) and basicity factor (line A) are then connected in Hoak's nomograph to find the pounds of alkaline agent required per gallon of acid waste (line C).

When sodium hydroxide is used as a neutralizing agent for carbonic and sulfuric acid wastes, the following reactions take place:

$$Na_2CO_3 + \underbrace{CO_2 + H_2O}_{\text{carbonic acid waste}} \rightarrow 2NaHCO_3,$$

$$2NaOH + CO_2 \rightarrow Na_2CO_3 + H_2O;$$

$$NaOH + \underbrace{H_2SO_4}_{\text{sulfuric acid waste}} \rightarrow NaHSO_4 + HOH,$$

$$NaHSO_4 + NaOH \rightarrow Na_2SO_4 + HOH.$$

Both these neutralizations take place in two steps and the end-products depend on the final pH desired.

For example, one treatment may require a final pH of only 6, and thus $NaHSO_4$ would make up the greater part of the products; another treatment may require a pH of 8, with most of the product being Na_2SO_4.

We shall now take up the subject of neutralization of alkaline wastes.

8.5 Using Waste Boiler-Flue Gas

Blowing waste boiler-flue gas through alkaline wastes is a relatively new and economical method for neutralizing them. Most of the experimental work has been carried out on textile wastes [1, 14, 18, 20, 21]. Well-burned stack gases contain approximately 14 per cent carbon dioxide. CO_2, dissolved in waste water, will form carbonic acid (a weak acid), which in turn reacts with caustic wastes to neutralize the excess alkalinity as follows:

$$\underset{\substack{\text{flue} \\ \text{gas}}}{CO_2} + \underset{\substack{\text{waste} \\ \text{water}}}{H_2O} \longrightarrow \underset{\substack{\text{carbonic} \\ \text{acid}}}{H_2CO_3},$$

$$\underset{\substack{\text{carbonic} \\ \text{acid}}}{H_2CO_3} + \underset{\substack{\text{caustic} \\ \text{soda in} \\ \text{waste water}}}{2NaOH} \longrightarrow \underset{\substack{\text{soda ash}}}{Na_2CO_3} + 2H_2O,$$

$$\underset{\substack{\text{excess} \\ \text{carbonic} \\ \text{acid}}}{H_2CO_3} + \underset{\substack{\text{soda ash} \\ \text{in waste}}}{Na_2CO_3} \overset{H_2O}{\longrightarrow} \underset{\substack{\text{sodium bi-} \\ \text{carbonate} \\ \text{in waste}}}{2NaHCO_3} + H_2O.$$

The equipment required usually consists of a blower placed in the stack, a gas pipeline to carry the gases to the waste-treatment site, a filter to remove sulfur and unburned carbon particles from gases, and a gas diffuser to disperse the stack gases in the waste water. Stack gases evolve hydrogen sulfide from waste waters which contain any appreciable quantity of sulfur, and this H_2S must be burned, absorbed, or vented positively to the upper atmosphere to prevent nuisance conditions.

8.6 Carbon-Dioxide Treatment for Alkaline Wastes

Bottled CO_2 is applied to waste waters in much the same way as compressed air is applied to activated-sludge basins. It neutralizes alkaline wastes on the same principle as boiler-feed gases (i.e., it forms

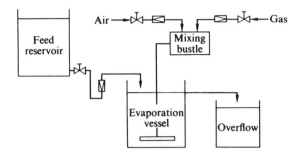

Fig. 8.3 Submerged combustion pilot unit as used by Remy and Lauria [16].

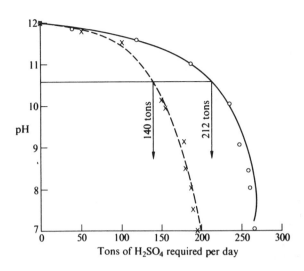

Fig. 8.4 Acid required to neutralize industrial wastes in sewer.

a weak acid (carbonic acid) when dissolved in water) but with much less operating difficulty. The cost may be prohibitive, however, when the quantity of alkaline wastes is large. A textile mill [14] producing about 6 mgd of alkaline waste studied the practical aspects of this method and found that installation of the equipment necessary to provide bottled CO_2 would cost about $150,000 and the power and fuel to generate it about $275 per day: a considerable expense, even for so large a plant.

8.7 Producing Carbon Dioxide in Alkaline Wastes

Another way to produce carbon dioxide is to burn gas under water. This process is called submerged combustion and has been used in the disposal of nylon wastes [16] to neutralize the waste prior to biological treatment. In pilot-plant studies, the researchers [16] investigated submerged combustion on a continuous basis, using an evaporation vessel, a burner with flame jets submerged below the waste surface in the vessel, a bustle where air and natural gas were mixed to form a combustible mixture, and other equipment to measure air, gas, and waste flows and the weight of waste volatilized during each run. A schematic drawing of the submerged-combustion plant used is shown in Fig. 8.3. They concluded that submerged combustion, rather than aeration, should be used to treat part of the plant waste, for economic reasons. (The researchers in this case, however, were primarily concerned with stripping toxic materials from the waste, rather than with neutralizing it.) Krofchak [11] describes this method of neutralization and suggests its use for spent pickle

liquors and spent electrolytes from nickel refining. CO_2 may also be produced by fermentation of an alkaline, organic waste; the resulting pH is thus lowered. Ebara-Infilco, Ltd., in 1965 patented such a process for fermenting alkaline beet-sugar wastes with yeast; the CO_2 produced can be used for neutralization and the excess yeast as forage.

8.8 Sulfuric-Acid Treatment for Alkaline Wastes

The addition of sulfuric acid to alkaline wastes is a fairly common, but rather expensive, means of neutralization. Sulfuric acid can cost as much as two or three cents per pound, although it may be as low as one cent per pound in large quantities. Storage and feeding equipment requirements are low as a result of its great acidity but it is difficult to handle because of its corrosiveness. The neutralization reaction which occurs when it is added to waste water is as follows:

$$2NaOH + H_2SO_4 \rightarrow Na_2SO_4 + 2H_2O.$$

waste	sulfuric	resulting
water	acid as	neutral
	neutralizer	salt

A titration curve of the alkaline waste neutralized with various amounts of H_2SO_4 is helpful to ascertain the quantities of acid required for neutralization to definite pH values and the relevant costs. Figure 8.4 represents the titration curve of an actual mixed alkaline waste in Niagara Falls, N.Y.

8.9 Acid-Waste Utilization in Industrial Processes

In some situations it may be possible to use acid wastes to effect a desired result in industrial processing—to wash, cool, or neutralize products. For example, Dillon [3] reports the use of acid mine drainage water for cleaning raw coal. Mine waste water occurs in large quantities in the coal industry. These waters are usually acid and contain sulfates of iron and aluminum; if they are used to wash raw coal, neutralization results, since coal contains calcium and magnesium carbonates. Dillon describes the treatment of 600 tons of raw coal per hour with an average of 225 gpm of mine waste water. The pH of the mine water is thereby raised from 3.0 to neutrality.

References

1. Beach, C. J., and M. G. Beach, "Treatment of alkaline dye wastes with flue gas," in Proceedings of 5th Southern Municipal and Industrial Waste Conference, April 1956, p. 162.
2. Dickerson, B. W., and R. M. Brooks, "Neutralization of acid wastes," *Ind. Eng. Chem.*, **42**, 599 (1950).
3. Dillon, K. E., "Waste disposal made profitable," *Chem. Eng.*, p. 146, 13 March 1967.
4. Gehm, H. W., "Neutralization with up-flow expanded limestone bed," *Sewage Works J.*, **16**, 104 (1944).
5. Hoak, R. D., "Neutralization studies on basicity of limestone and lime," *Sewage Works J.*, **16**, 855 (1944).
6. Hoak, R. D., "Acid iron wastes neutralization," *Sewage Ind. Wastes* **22**, 212 (1950).
7. Hoak, R. D., "A neutralization nomograph," *Ind. Wastes*, **3**, D-48 (1958).
8. Hyde, A. C., "Chemical plant waste treatment by ten methods," *J. Water Pollution Control Federation* **37**, 1486 (1965).
9. Jacobs, H. L., "Acid neutralization," *Chem. Eng. Progr.* **43**, 247 (1947).
10. Jacobs, H. L., "Neutralization of acid wastes," *Sewage Ind. Wastes* **23**, 900 (1951).
11. Krofchak, O., "Submerged combustion evaporation of acid wastes," *Ind. Water and Wastes*, **7**, 63 (1962).
12. Leidner, R. N., "Burns Harbor—Waste treatment planning for a new steel plant," *J. Water Pollution Control Federation* **38**, 1767 (1966).
13. Lewis, C. J., and L. J. Yost, "Lime in waste acid treatment," *Sewage Ind. Wastes* **22**, 893 (1950).
14. Nemerow, N. L. "Holding and aeration of cotton mill finishing wastes," in Proceedings of 5th Southern Municipal and Industrial Waste Conference, April 1956, p. 149.
15. Reidl, A. L., "Neutralization with up-flow limestone bed," *Sewage Works J.*, **19**, 1093 (1947).
16. Remy, E. D., and D. T. Lauria, "Disposal of nylon wastes," in Proceedings of 13th Industrial Waste Conference, May 1958, Purdue University Engineering Extension Series no. 96, p. 596.
17. Rudolfs, W., "Pretreatment of acid wastes," *Sewage Works J.*, **15**, 48 (1943).
18. Rudolfs, W., "Neutralization with lime," *Sewage Works J.*, **15**, 590 (1943).
19. Smith, F., "Neutralization of pickle liquor," *Sewage Works J.*, **15**, 157 (1943).
20. Steele, W. R., "Application of flue gas to the disposal of caustic textile wastes," in Proceedings of 3rd Southern Municipal and Industrial Waste Conference, March 1954, p. 190.
21. "Treatment of alkaline sulfur dye waste with flue gas," Research Report no. 8, *Proc. Am. Soc. Civil Engrs.* **82** (SA-5), 1078 (1956).
22. Boyle, R.H. "An american Tragedy," *Sports Illustrated* **55**, No. 13, 75 (September 21, 1981).

Questions

1. Name seven major methods of neutralizating both acid and alkaline wastes.
2. What are the four major factors to be considered when neutralizing wastes?
3. What effect can the storage of alkaline wastes have on pH and resulting neutralization?
4. Under what conditions would you select mixing of wastes as a solution to neutralization?
5. When would you use limestone beds or filters for acid wastes?
6. What is the advantage of using lime slurry rather than limestone beds for neutralization?
7. When would you use NaOH or Na_2CO_3 for neutralization?

CHAPTER 9

EQUALIZATION AND PROPORTIONING

9.1 Equalization

Equalization is a method of retaining wastes in a basin so that the effluent discharged is fairly uniform in its sanitary characteristics (pH, color, turbidity, alkalinity, BOD, and so forth). A secondary but significant effect is that of lowering the concentration of effluent contaminants. This is accomplished not only by ironing out the slugs of high concentration of contaminants but also by physical, chemical, and biological reactions which may occur during retention in equalization basins. For example, the recent increases in industrial wastes reported by Fall [1] at Peoria have greatly varied the organic loading at the treatment plant. A retention pond serves to level out the effects of peak loadings on the plant while substantially lowering the BOD and suspended-solids load to the aeration unit. Air is sometimes injected into these basins to provide: (1) better mixing; (2) chemical oxidation of reduced compounds; (3) some degree of biological oxidation; and (4) agitation to prevent suspended solids from settling.

The size and shape of the basins vary with the quantity of waste and the pattern of its discharge from the factory. Most basins are rectangular or square, although Metzger [5] has recently found that triangular tanks produce satisfactory flow distribution. The capacity should be adequate to hold, and render homogeneous, all the wastes from the plant. Almost all industrial plants operate on a cycle basis; thus, if the cycle of operations is repeated every two hours, an equalization tank which can hold a two-hour flow will usually be sufficient. If the cycle is repeated only each 24 hours, the equalization basin must be big enough to hold a 24-hour flow of waste. Herion and Roughhead [3] report the use of 72-hour equalization for a pharmaceutical waste to ensure ample mixing. This period (three times the 24-hour cycle of operations) was selected as the proper detention time in order not to disrupt the biota of the activated-sludge

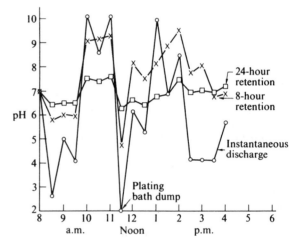

Fig. 9.1 Effect of equalization.

units. In a wool-finishing-mill waste containing dieldrin (a mothproofing insecticide) an equalization period of 44 days was necessary to yield a receiving stream concentration of less than 0.0005 mg/liter. Figure 9.1 compares the effects of 8-hour and 24-hour detention periods on the final pH of metal-plating wastes.

The mere holding of waste, however, is not sufficient to equalize it. Each unit volume of waste discharged must be adequately mixed with other unit volumes of waste discharged many hours previously. This mixing may be brought about in the following ways: (1) proper distribution and baffling; (2) mechanical agitation; (3) aeration; and (4) combinations of all three.

Proper distribution and baffling is the most economical, though usually the least efficient, method of mixing. Still, this method may suffice for many plants. Horizontal distribution of the waste is achieved by using either several inlet pipes, spaced at regular intervals across the width of the tank, or a perforated pipe

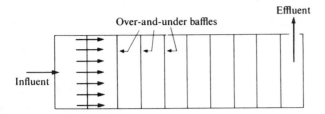

Fig. 9.2 Top view of an equalizing basin, with perforated inlet pipe and over-and-under baffles.

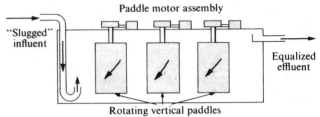

Fig. 9.3 Side view of an equalizing basin, with mechanical agitators instead of baffles.

across the entire width. Over-and-under baffles are advisable when the tank is wide, since they provide more efficient horizontal and vertical distribution (Fig. 9.2). Baffling is especially important when several different types of waste enter the basin at various locations across the width. The influent should be forced to the bottom of the basin so that the entrance velocity prevents suspended particles from sinking and remaining on the bottom.

Mechanical agitation eliminates most of the need for baffles and generally provides better mixing than baffles alone. One typical arrangement [6], shown in Fig. 9.3, utilizes three wooden gate-type agitators spaced equidistantly along the center line of the length of the tank. Agitators operated at a speed of 15 rpm by a 3-hp motor are usually adequate.

The design in Fig. 9.3 approximates the theoretically ideal tank, because of its relatively high efficiency at similar detention times, as a result of mechanical mixing, and also because it prepares varied chemical wastes for direct disposal or final treatment. If subsequent treatment is necessary, the process is made easier because the problem of wastes with rapidly changing characteristics varying from one extreme to the other is eliminated. Rudolfs and Millar [6] recommended this method of equalization when: (1) limited space is available; (2) removal of suspended solids is not desired; (3) there are rapid fluctuations in the characteristics of the wastes; and (4) facility of subsequent treatment is a goal.

This type of equipment is good not only for equalization but also for dilution, oxidation, reduction, or any other function in which one wants chemical compounds discharged at one time to react with compounds discharged before or after them, to produce a desired effect.

Aeration of equalizing basins is the most efficient way to mix wastes, but also the most expensive. To aerate an equalizing basin takes about half a cubic foot of air per gallon of waste. Aeration facilitates mixing and equalization of wastes, prevents or decreases accumulation of settled material in the tank, and provides preliminary chemical oxidation of reduced compounds, such as sulfur compounds. It is of special benefit in situations where wastes have varying character and quantity, excess of reduced compounds, and some settleable suspended solids.

9.2 Proportioning

Proportioning means the discharge of industrial wastes in proportion to the flow of municipal sewage in the sewers or to the stream flow in the receiving river. In most cases it is possible to combine equalization and proportioning in the same basin. The effluent from the equalization basin is metered into the sewer or stream according to a predetermined schedule. The objective of proportioning in sewers is to keep constant the percentage of industrial wastes to domestic-sewage flow entering the municipal sewage plant. This procedure has several purposes: (1) to protect municipal sewage treatment using chemicals from being impaired by a sudden overdose of chemicals contained in the industrial waste; (2) to protect biological-treatment devices from shock loads of industrial wastes, which may inactivate the bacteria; (3) to minimize fluctuations of sanitary standards in the treated effluent.

The rate of flow of industrial waste varies from instant to instant, as does the flow of domestic sewage, and both empty into the same sewage system. Therefore, the industrial waste must be equalized and retained, then proportioned to the sewer or stream according to the volume of domestic sewage or stream flow. To facilitate proportioning, an industry should construct a holding tank with a variable-speed pump to control the effluent discharge. Because the domestic-sewage treatment plant is usually located some distance from an industry, signalling the time and amount of flow is difficult and sometimes quite expensive. For this reason, many industries have separate pipelines through which they pump their wastes to the municipal treatment plant. The wastes are equalized separately at the site of the municipal plant and proportioned to the flow of incoming municipal waste water. Separate lines are not, of course, always possible or even necessary. One textile mill found that it could effectively proportion its waste to the variable domestic-sewage flow by adjusting the valve on the holding-tank effluent pump three times a day: 8 a.m., 12 noon, and 7 p.m.

There are two general methods of discharging industrial waste in proportion to the flow of domestic sewage at the municipal plant: manual control, related to a well-defined domestic-sewage flow pattern, and automatic control by electronics.

Manual control is lower in initial cost but less accurate. It involves determining the flow pattern of domestic sewage for each day in the week, over a period of months. Usually one does this by examining the flow records of the sewage plant or by studying the hourly water-consumption figures for the city. It is better to spend time on a careful investigation of the actual sewage flow than to make predictions based on miscellaneous, nonpertinent records. Actual investigative data should be used to support those records which are applicable to the case.

Automatic control of waste discharge according to sewage flow involves placing a metering device, which registers the amount of flow, at the most convenient main sewer connection. This device translates the rate of flow in the sewer to a recorder which is located near the industrial plant's holding tank. The pen on the recorder actuates either a mechanical (gear) or a pneumatic (air) control system for opening or closing the diaphragm of the pump. There are, of course, many variations of automatic flow-control systems. Although their initial cost is higher than that of manual control, they will usually return the investment many times by the savings in labor costs.

Some industrial and municipal sewage-plant superintendents think that the best time to release a high proportion of industrial waste to the sewer is at night, when the domestic-sewage flow is low. Whether night release is a good idea depends on the type of treatment used and the character of the industrial waste. If the treatment is primarily biological and the industrial wastes contain readily decomposable organic matter and no toxic elements, discharging the largest part of the industrial waste to the treatment plant at night is indeed advisable, since this ensures a relatively constant organic load delivered to the plant day and night.

One equipment-manufacturing company recommends a three-component system for automatic proportioning of wastes into sewers (Fig. 9.4). These three components are: (1) a kinematic manometer with integral pneumatic transmitter; (2) a remotely

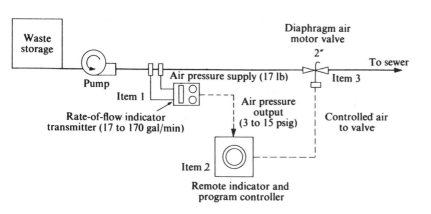

Fig. 9.4 Waste-metering system. (Courtesy Fischer and Porter Company.)

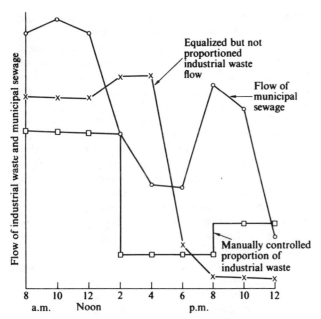

Fig. 9.5 Effect of proportioning.

Another arrangement for proportioning industrial wastes, in a situation where pipelines are flowing only partly full or waste flows in open channels, is the use of a weir, flume, or Kennison nozzle in the main flow line to measure the flow. A float-operated transmitter (either electrical or pneumatic) is connected to this measuring device and the electrical or pneumatic signals are used to actuate a flow splitter in a proportioning weir tank (such as is provided by Proportioneers, Inc.).

The Belle, West Virginia, works of the Du Pont Nemours Company has been impounding its waste in two 2.5-million-gallon tanks and releasing it to the Kanawha River in proportion to the river flow for over 10 years [3a]. This has been necessary owing to the flashiness of the river flows. Figure 9.5 compares the effects on the flow at a municipal treatment plant of both equalization and proportioning.

located indicator–program controller which receives air signals and has a precut time-pattern cam for continuously adjusting the set point of the pneumatic controller to give a waste-flow rate in accordance with the desired pattern; (3) a diaphragm-actuated, motor-controlled valve which is actuated by the air signal from the program controller. Practically speaking, the length of the pneumatic capillary tubing limits the physical separation between the sensing components, but this difficulty can be overcome with an electrical system.

The typical waste-flow proportioning system (Bubbler System) shown in Fig. 9.4, as supplied to me by Fischer and Porter Company, consists essentially of the three separate devices described above. Item 1, with a linear air-pressure output of 3 to 15 pounds per square inch, has a flow range of 17 to 170 gpm of an industrial waste (specific gravity assumed, 1.1). Item 2 is a remotely located indicator for receiving air signals from item 1, as explained in the text. Item 3 is an automatic valve capable of operating at a maximum pressure drop of 10 pounds at maximum flow rate. This valve is actuated by air signals from the program controller, item 2.

References

1. Fall, E. B., "Retention pond improves activated sludge effluent quality," *J. Water Pollution Control Federation*, **37**, 1194 (1965).
2. Gibbs, C. V., and R. H. Bothel, "Potential of large metropolitan sewers for disposal of industrial wastes," *J. Water Pollution Control Federation* **37**, 1417 (1965).
3. Herion, R. W., and H. O. Roughhead, "Two treatment installations for pharmaceutical wastes," in Proceedings of 18th Industrial Waste Conference, 1964, Purdue University Engineering Extension Series, Bulletin no. 115, p. 218.
3a. Hyde, A. C., "Chemical plant waste treatment by ten methods," *J. Water Pollution Control Federation*, **37**, 1486 (1965).
4. *Manual for Sewage Plant Operators*, Texas Water and Sewage Works Association, Austin (1955), pp. 342–345.
5. Metzger, I., "Triangular tank for equalizing liquid wastes," *Water Sewage Works*, **114**, 9 (1967).
6. Rudolfs, W., and J. N. Millar, "A method for accelerated equalization of industrial wastes," *Sewage Works J.*, **18**, 686 (1946).
7. Wilroy, R. D., "Industrial wastes from scouring rug wool and the removal of dieldrin," in Proceedings of 18th Industrial Wastes Conference, 1964, Purdue

University Engineering Extension Series, Bulletin no. 115, p. 413.

Questions

1. Define equalization including its purpose.

2. What are four methods of mixing to effect equalization?

3. What are the objectives of proportioning of industrial wastes?

4. What are the problems associated with proportioning industrial wastes into municipal sewers?

CHAPTER 10

REMOVAL OF SUSPENDED SOLIDS

10.1 Sedimentation Theory

Although sedimentation is a method of treatment utilized in almost all domestic-sewage treatment plants, it should be considered for industrial-waste treatment only when the industrial waste is combined with domestic sewage or contains a high percentage of settleable suspended solids, such as are found in cannery, paper, sand-and-gravel, coal-washery, and certain other wastes. The efficiency of sedimentation tanks depends, in general, on the following factors:

Detention period	Velocity of particles
Waste-water character- istics	Density of particles Container-wall effect
Tank depth	Number of basins
Floor surface area and overflow rate	(baffles) Sludge removal
Operation (cleanliness)	Pretreatment (grit
Temperature	removal)
Particle size	Flow fluctuations
Inlet and outlet design	Wind velocity

Although settling tanks have been used for other purposes, such as grease flotation, equalization, and BOD reduction, they are primarily used for removing settleable suspended matter. Theoretically, a suspended particle in a waste-water solution will continue to settle at a fixed velocity relative to the solution, as long as the particle remains discrete; when it coalesces with other particles, its size, shape, and resulting density will change, as will its settling velocity. Coagulation, or self-flocculation, of particles causes an increase in velocity. In liquid wastes containing high percentages of suspended solids, greater reductions in the suspended solids will occur primarily because of increased flocculation. The fixed settling velocity will also be altered by changes in the temperature and density of the liquid solvent through which the particle is moving. Rising layers of warmer liquid can cause eddying and a disturbance in the settling of particles; an increased density in the lower layers of liquid can deter the particle from settling to the bottom. These factors can interfere with settling to such an extent that particles may be carried out of the tank with the effluent.

Depth of tank is also of great importance. The deeper the tank (all other factors being equal) the better the chance of preventing the deposited solids from being resuspended—e.g., by sudden scouring due to turbulence caused by unequal flow distribution or by exposure to wind or temperature effects—and thus being carried out with the effluent. This is especially important when sludge is stored in sedimentation basins for lengthy periods before pumping. If the solids are continuously removed from the bottom of settling tanks as soon as they land, shallower tanks can be built.

Surface area is another factor affecting tank efficiency, and engineers agree that floor area must be adequate to receive all the particles to be removed from the waste waters. However, many state health departments, when establishing acceptable dimensions for settling basins, do so on the basis of standard detention periods. In certain designs this method may not provide adequate floor area and complete settling is not achieved.

Figure 10.1 illustrates the effect of doubling the floor area and halving the depth of a settling basin, with volume and detention time remaining constant. Theoretically, the basin in Fig. 10.1 (b) will remove twice as many discrete particles as the basin in (a). Therefore, the engineer should strive to design settling basins which are as shallow as possible and contain ample floor area. However, tanks less than six feet deep have been found impractical from an operational standpoint, because they are subject to upsetting by scouring or velocity of currents. The floor area is increased most satisfactorily by extending the length of the basin.

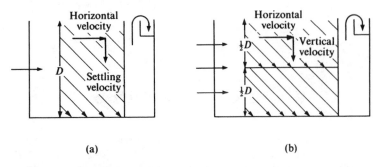

(a)   (b)

Fig. 10.1 Effect of doubling the floor area and halving the depth of a settling basin.

Since the percentage of particles reaching the bottom of the settling basin also depends on the rate of waste flow, an expression correlating horizontal flow with the floor or surface area has been devised. It is commonly referred to as the *overflow rate* and is expressed as gallons per square foot per day. Typical overflow rates vary from 200 to 800 gallons per square foot per day for primary sedimentation basins and from 1000 to 3000 gallons per square foot per day for final tanks, since particles in the final tanks usually settle more rapidly than those in primary basins. Exceptions include grit particles, which settle faster than the average particle in primary basins, and activated sludge floc, which tends to slow down the settling rate in secondary basins. Because of these discrepancies, both primary and secondary settling basins are often designed for the same overflow rates. Lower overflow rates for domestic-type wastes generally result in the removal of more suspended solids and BOD, as shown in Fig. 10.2 [10]. For further reading on theories of sedimentation the reader is referred to Eckenfelder [7] and O'Conner and Eckenfelder [13].

Unfortunately, actual settling velocities may vary from theoretical formulations. Turbulence and flocculation are the main causes of variation. Another factor is that velocities do not remain constant throughout a cross-sectional area of a tank. The settling velocity of discrete particles of diameter d in a quiescent viscous fluid is given by

$$V = \frac{4}{3} \cdot \frac{gd}{C_d} \left(\frac{\rho_s - 1}{\rho} \right)$$

where C_d is the drag coefficient between the fluid and the particle, g is the acceleration due to gravity, and ρ_s and ρ are the densities of the particle and the fluid. The drag coefficient does not remain constant but varies with the Reynolds number, R, which equals $\rho dV/\mu$. The correlation between C_d and R has been plotted in various textbooks, but a trial-and-error procedure is still required to obtain V.

Turbulence in sedimentation basins has both a positive and a negative effect on the settling velocity of a particle. It causes eddies, which carry some particles down and some up (as shown in Fig. 10.3), and thus it both helps flocculation and hinders sedimentation. Settling or rising velocities can be unequal, depending on the local circumstances causing the turbulence, such as increased horizontal velocity of water at the inlet.

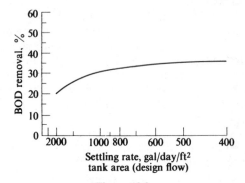

Figure 10.2

Fig. 10.3 Effect of turbulence on particle path.

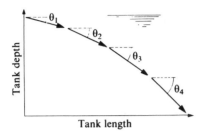

Fig. 10.4 Flocculation increases settling rate.

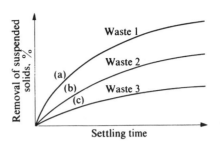

Fig. 10.5 (a) Fast and good settling characteristics typical of heavy suspended solids. (b) Medium and normal settling characteristics typical of homogeneous mixture of solids. (c) Slow and poor settling characteristics typical of highly colloidal and finely divided solids.

Other factors that induce eddying include wind, unequal distribution of flow, changes in temperature, and changes in density of the liquid at various depths. Eddying generally decreases the settling velocity and efficiency of operation, while flocculation generally increases the overall total of solids removed. The influence of flocculation is illustrated in Fig. 10.4, where θ is the angle of vertical settling of the average particle. Since shallow tanks appear to induce more flocculation, for this and other reasons they are preferred to deep tanks, provided scouring of settled particles is prevented. The average settling velocities of particles in industrial wastes vary appreciably (Fig. 10.5).

The percentage of suspended solids removed depends on the tank design, which in turn depends on the demands of the particular situation. In recent years, design engineers have been using either circular or square tanks instead of the conventional rectangular basins, for reasons of space and/or economics. Circular tanks require less form work, materials, and land space than rectangular basins for large flows and for any size of tank. However, they are less efficient, owing to (1) reduced length of effective settling zone and (2) short circuiting (waste water leaving the tank prior to theoretical detention time). The efficiency of circular tanks has been increased somewhat by the introduction of peripheral feed with center draw-off. This system eliminates the turbulence at the inlet.

The relative percentages of total transverse distance occupied by the inlet zones of circular and rectangular talks are shown in Fig. 10.6. Since the inlet zone of a circular tank occupies such a large portion of the horizontal particle path, special care must be used in designing inlet and outlet devices. The slightest disturbance in flow conditions will tend to disrupt the operation of a circular tank, but with long, narrow, rectangular tanks, the design of the inlet and outlet zones becomes less important.

Short circuiting means that effective sedimentation is not taking place in the entire volume of the settling tank; that is, a given entering volume of waste is hindered from spreading uniformly throughout the tank in a quiescent manner, so that it reaches the effluent weir before the theoretical detention time has been utilized. This is essentially true in all tanks, regardless of shape, but it seems to occur most readily in circular and square tanks, as illustrated in Fig. 10.7. To avoid short circuiting, some state regulatory agencies specify a minimum distance between the inlet and exit of the tank. It has also been demonstrated graphically by Camp [4] (see Fig. 10.8)

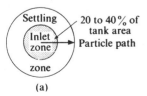

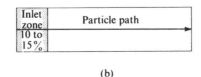

Fig. 10.6 Inlet zone of a circular tank (a) occupies 20 to 40 per cent of tank area. Inlet zone of a rectangular tank (b) occupies only 10 to 15 per cent of tank area.

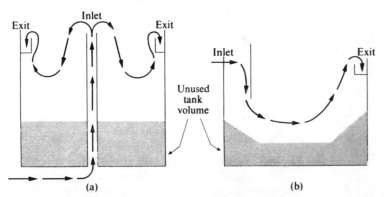

Fig. 10.7 (a) Circular tank. (b) Square tank.

that different shapes of sedimentation tanks cause different degrees of short circuiting. Villemonte *et al.* [17] recently showed that hydraulic efficiencies, predicted by basin dispersion curves, are related directly to the basin performance, measured by suspended-solids reduction.

In Fig. 10.8, the higher peaks occurring over shorter ranges of t/T indicate the absence of short circuiting. Curve A is a theoretical one for an ideal, instantaneous dispersion of a slug with entire tank contents. (Short circuiting approximates this.) Curve B is for a circular tank and indicates that some suspended contaminant reaches the outlet after about 15 per cent of the detention period, and the greatest concentration of matter reaches the outlet after about 50 per cent of the detention period. Curve C shows the situation in a wide rectangular tank, which approximates a square one. Curve D refers to a long, narrow, rectangular tank and indicates that no contaminant reaches the end of the tank until after

50 per cent of the detention period and most reaches the outlet after about 80 per cent of the detention period. Curve E is the dispersion curve of a round-the-end, long, baffled, rectangular chamber, with great length compared to width and depth. This type of tank gives a theoretical maximum contaminant content in the effluent after 100 per cent of the detention period, but little or none before this time. The student can readily appreciate, from a study of Fig. 10.8, the importance of proper design of sedimentation tanks. Preference should be given, wherever possible, to long, rectangular tanks with proper baffling.

A major objective of sedimentation is to produce sludge with the highest possible solids concentration. As the reader will see in Chapter 14, the volume and weight of sludge requiring final disposal is a major factor in waste treatment. A relatively new piece of equipment to achieve this objective is the Clarithickener, which combines sludge separation in circular settling tanks and thickening by means of slowly

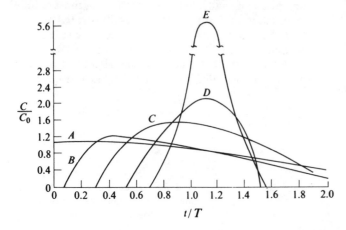

Fig. 10.8 Typical dispersion curves for various tanks (see text for explanation). The vertical axis shows the ratio of the actual concentration of contaminant (C) to the concentration of contaminant mixed with the entire tank volume (C_0); the horizontal axis shows the ratio of the actual time (t) a concentration takes to reach the end of the tank to (T), the total detention period (vol/rate). (After Camp [4].)

rotating picket-fence arms. Other methods of decreasing short circuiting include effective inlet and outlet design, properly located baffling, inboard weirs, and modification of existing sedimentation tanks to obtain better flow distribution.

Although the differences between domestic sewage and industrial wastes are often quite significant, some general statements made for domestic sewages hold true for all wastes. Normally, with detention periods of two hours, primary sedimentation basins remove 50 to 70 per cent of the suspended solids in the influent. Data collected from waste-water treatment-plant superintendents [3] are presented in Tables 10.1 and 10.2, to show design criteria and efficiencies of removal for rectangular and circular tanks.

In an unpublished study carried out by Nemerow and McWeeny (1972), the settling basin was separated into two zones by placing a series of vertical baffles perpendicular to the flow, with the top of the baffles below the surface of the sewage. The waste flows across the tops of the baffles following the path of least resistance, while the particles settle through the baffles into the sludge collection zone. The baffles dampen out the turbulence caused by sludge removal along the bottom of the basin. A removal efficiency of 85 per cent of suspended solids at 691 gallons per day per square foot of suspended solids was obtained, a 20 per cent increase over the current design expectation. A removal efficiency of 65 per cent of suspended solids was obtained at a loading rate of 4600 gallons per square foot with a baffle spacing of 0.25 feet.

10.2 Design of Sedimentation Process Units

Sedimentation processes are very effective in removing suspended solids in industrial wastewater. Clarifiers, either rectangular or circular in shape, are most commonly used in the application of sedimentation in wastewater treatment facilities. The design of the clarifers are based on several factors as follows:

- Influent TSS Concentration.
- Effluent TSS Concentration.
- Surface Loading.
- Detention Time.
- Sludge Generation.

Clarification is used as a process to remove suspended solids at different stages of industrial wastewater treatment. It is often used in the primary treatment stage to remove TSS or colloidal solids prior to treatment for removal of dissolved inorganics or organics. A typical example of this application is in the treatment of metal finishing industry wastewater where the suspended solids are removed by primary clarification before other physico-chemical processes are used to remove the dissolved heavy metals. Sedimentation is also commonly used in the secondary treatment stage usually following biological treatment. An example of this application is the treatment of pulp and paper mill wastewater where primary clarification is followed by biological treatment, and then the biological solids are removed by secondary clarification. In this case the clarifier is used not only to remove the TSS, but also to act as a thickener for the sludge generated. The design of the clarifier is based on different considerations depending upon the stage of the treatment in which clarification is used.

Another consideration in the design of sedimentation processes is the characteristics of the suspended solids. In some cases the suspended solids could be discrete particles as in the case of grit, sand or suspended metal scales or particles. These type of solids settle easily following the principles of discrete settling (Figure 10-9). In other cases, the TSS could be composed of floc-type particles, and the settling characteristics of these suspended solids are different from the discrete type solids (Figure 10-9). Examples of the floc-type particles are biological flocs or chemically coagulated and flocculated particles. Due to the different settling characteristics of the suspended solids, it is important that batch settling tests be conducted using settling columns prior to design of sedimentation unit processes. In the settling test (Figure 10-9), the height of the interface between the clear supernatant liquid and the layer of suspended solids is noted with time as settling occurs in the column. (The data is plotted as shown in Figure 10-9). The settling of the solids takes place in essentially two phases. The initial rate of settling (AB) known as hindered settling is used to compute the area required for clarification. The second phase of settling (CD) represents the thickening of the sludge. A graphical method of combination of the rates of settling in the AB and CD portions of the curve is used to compute the area required for thickening. The larger of the two areas for

Table 10.1 Rectangular primary settling-tank data.*

Plant location	No. of tanks	Length, ft	Width, ft	Depth, ft	Length/width	Length/depth	Flow, mgd	Detention, hr	Overflow, gpd/ft²	Weir rate, gpd/ft	Raw suspended solids, mg/liter	Removal suspended solids, %	Raw BOD, mg/liter	Removal BOD, %
Hartford, Conn.	8	100	68	8.8	1.5	11.4	24.30	3.53	450	56,800	173	61	240	42
Detroit, Mich.	8	270	117	13	2.3	20.8	418.00	1.41	1650	408,000	184	44	153	39
Racine, Wis.	4	140	40	10.5	3.5	13.3	17.03	2.48	760	106,500	149	67	133	48
New York City, Bowery Bay	3	124	50	12	2.5	10.3	41.00	0.98	2210	284,000	152	39	169	22
New York City, Tallmans Island	3	124	50	11.6	2.5	10.7	31.00	1.25	1670	215,000	137	55	128	39
Fort Wayne, Ind.	3	100	33	13	3.3	7.7	18.70	1.25	1890	94,500	409	61	231	34
Rochester, N.Y.	2	37	12	8	3.1	4.6	0.81	1.56	914	41,000	233	21	260	21
Marshalltown, Iowa	3	80	16	8	5.0	10.0	1.22	1.51	950	13,550	436	58	414	42
Kenosha, Wis.	4	132	32	10.4	4.1	12.7	12.77	2.49	755	100,000	138	48	102	48
Jackson, Mich.	3	67.3	31	10.4	2.2	6.7	0.17	1.22	1470	118,000	193	16.1	134	22
Hammond, Ind.	6	120	16	13.25	7.5	9.0	20.70	1.32	1800	24,000	273	30	206	25
New York City, 26th Ward	4	162	67	12	2.4	13.5	41.00	2.16	930	35,500	139	31	127	28
New York City, Hunts Point	4	168	108.9	12	1.5	14.0	95.00	1.70	1300	97,000	140	48	113	30
Abington, Pa.	2	50	14	10	3.6	5.0	1.24	2.02	855	44,400	237	39	198	29
Portsmouth, Va.	4	100	15.25	10	6.5	10.0	7.36	1.49	1200	46,000	153	63	185	45
Canton, Ohio	3	124	32	10.6	3.9	11.7	17.00	1.33	1430	214,000	577	40	253	33
Niles, Mich.	6	75	14	9	5.4	8.3	2.30	1.86	362	27,200	250	69.2	106	57
Dallas, Tex.	2	180	50	12	3.6	15.0	19.40	2.00	1080	24,000	358	66	256	41
Richmond, Ind.	4	95	16	14.5	5.9	6.5	6.10	2.64	990	25,000	159	40	133	23
Lansing, Mich.	16	87.5	16	10	5.5	8.7	16.45	2.45	735	23,700	445	76	201	68
Winsted, Conn.	2	65	12	9	5.5	7.2	0.50	5.00	320	20,800	130	75	170	51
Waterbury, Conn.	3	212.5	33	10	6.4	21.2	13.94	2.71	660	14,500	144	54	166	33
Oklahoma City, Okla.	3	85	33	10	2.5	8.5	5.19	2.91	619	20,400	242	50	228	31
Tampa, Fla.	4	170	40	13	4.2	13.1	12.30	5.12	455	17,300	215	69	183	37
Roanoke, Va.	2	120	32	10.5	3.8	11.4	7.76	1.87	1010	120,000	230	67	190	51
Blackstone Valley, R.I.	2	230	68	10.8	3.4	21.1	12.21	4.97	390	62,000	212	62	333	12
East Hartford, Conn.	2	125	32	7.5	3.9	16.7	1.50	7.18	187	12,500	212	54	242	50
Milford, Conn.	2	55	16	9.75	3.5	5.1	9.70	4.40	400	21,800	150	79	130	72
Springfield, Mass.	4	115	50	14.5	2.3	7.9	17.5	3.36	761		160	49	145	26
Orrville, Ohio	2	43.8	16	10.4	2.7	4.2	0.73	3.65	515		342	64	415	18
New Haven, Conn.	3	145	31	11.5	4.7	12.6	14.7	1.90	1090		176	49		
Cleveland, Ohio (Easterly)	8	115	50	15	2.3	7.7	97.7	1.27	2120		240	37	149	35

*Data from plant superintendents. See reference 8, pp. 90–91.

Table 10.2 Circular primary tanks: long-term performance data.*

Location	Data period Years	Data period No.	Average flow, mgd	No. of tanks	Diameter, ft	Side-water depth, ft	Detention, hr	Over-flow, gpd/ft²	Suspended solids Raw, mg/liter	Suspended solids Efflu-ent, mg/liter	Suspended solids Re-moval, %	BOD Raw, mg/liter	BOD Efflu-ent, mg/liter	BOD Re-moval, %	Sludge Solids, %	Sludge Volatile matter, %
Washington, D.C.	1944–45	2	136.3	12	106	14	1.88	1350	163	83	49	173	120	30.5	8.05	67.5
Winnipeg, Man.	1943–44	2	22.8	2	115	12	1.98	1100	348	159	55	310	231	25.5	9.0	70.5
Battle Creek, Mich.	1938–42	5	4.92	2	80	10	3.66	490	282	85	70	264	174	34.1	5.5	82.5
Buffalo, N.Y.	1939–41	3	135	4	160	15	1.6	1690	209	114	46	138	107	22.5	5.8	59
Albuquerque, N. Mex.	1939–46	7	5	1	80	12.2	2.21	995	254	91	61	282	150	44.5	3.9	81
Yakima, Wash.	1942	1	9.5	4	90	9	4.32	373	110	23	74	175	92	50	7.0	74.4
Appleton, Wis.	1938–45	7	4.8	2	70	10	2.90	623	276	63	77	284	141	50	5.6	58
Baltimore, Md.	1939–44	4	89.5	3	170	12	1.64	1360	214	83	61	281	204	27.5	3.9	82.7
Springfield, Ohio	1937–40	4	14.8	2	90	10	1.55	1160	166	63	62	90	43	52		
Mansfield, Ohio	1944–45	2	3	1	65	12	2.38	905	208	87	58	227	139	38.8	4.2	76
Cedar Rapids, Iowa	1936–44	9	4.21	1	70	11.5	1.95	1060	354	132	63	383	291	24	5.5	81.2
Austin, Tex.	1944–45	2	5.64	1	75	12	1.69	1275	263	95	64	285	152	46.3	4.0	83
Denver, Colo.	1939–43	5	46	4	140	9.7	2.34	750	187	44	77	212	108	49	5.4	76
Ypsilanti, Mich.	1943–45	3	1.66	2	40	9	2.5	660	226	87	62	141	95	33	8.2	71.4
Monroe, Mich.	1938–46	8	4.3	2	85	7.5	3.55	378	329	75	77	135	73	46	5.2	67.7

*Data from plant superintendents and/or annual reports. See reference 8, pp. 90–91.

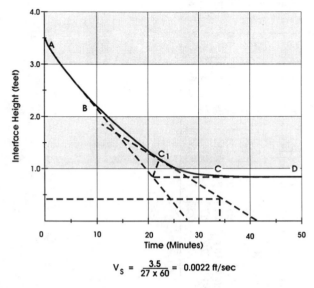

$$V_s = \frac{3.5}{27 \times 60} = 0.0022 \text{ ft/sec}$$

Fig. 10.9   Graphical Analysis of Settling Test Results

clarification and thickening is used for sizing the clarifier.

$$V_s = \frac{Q}{A_c}$$

$$\text{or } A_c = \frac{Q}{V_s}$$

where V_s = settling velocity in the hindered settling zone

Q = hydraulic flow

A_c = surface area required for clarification

V_s is computed as the slope of the line AB

The area required for thickening is computed from the following equation:

$$A_t = \frac{Qt_u}{H_o}$$

where A_t = surface area required for thickening to desired solids concentration in sludge

Q = hydraulic flow into the clarifier

H_o = initial interface height of the settling column

t_u = time required to reach desired solids concentration in sludge.

t_u is computed by the following steps: (1) draw tangents to each of the portions AB and CD of the curve; (2) draw the bisector of the angle formed by the intersection of the two tangents; C_1 represents the critical concentration in the transition between the hindered and compression settling phases; (3) draw a tangent at C_1; and (4) draw a line parallel to the time axis at the interface height (H_u) corresponding to the desired solids concentration in the sludge (C_u). The time-scale intercept of the tangent with the sludge concentration line is the required time t_u. H_u is computed as follows:

$$H_u = \frac{C_o H_o}{C_u}$$

where C_o = solids concentration in the influent

C_u = desired solids concentration in the sludge

H_u = interface height at desired solids concentration in sludge

EXAMPLE

The results of a batch settling column test for an industrial wastewater is given below:

Time (Min)	Interface Height (Ft.)
0	3.5
5	2.8
10	2.2
15	1.5
20	1.2
25	0.9
30	0.7
35	0.6
40	0.5
45	0.5

Using the test data, determine the size of the clarifier. Flow = 0.8 mgd, influent solids concentration = 2,000 mg/L, sludge solids concentration = 1.5%.

Solution:

The data from the settling test is plotted as shown in Figure 10-9. From the figure, V_s is computed as follows:

$$V_s = \frac{3.5}{27 \times 60} = 0.0022 \text{ ft/sec}$$

Area required for clarification $A_c = \dfrac{Q}{\bar{V}_s}$

$$= \frac{0.8 \times 1.547 \text{ ft}^3/\text{sec}}{0.0022 \text{ ft/sec}}$$

$$= 563 \text{ ft}^2$$

H_u is computed as follows:

$$H_u = \frac{C_o H_o}{C_u}$$

$$C_o = 2000 \text{ mg/L}$$

$$= \frac{2000}{16020} = 0.125 \text{ lb/ft}^3$$

$$C_u = \frac{1.5 \times 10,000}{16020} = 0.94 \text{ lb/ft}^3$$

$$H_u = \frac{C_o H_o}{C_u}$$

$$= \frac{0.125 \times 3.5}{0.94} = 0.47 \text{ ft}$$

From the figure $t_u = 33$ min for $H_u = 0.47$ ft.

Area required for thickening A_t

$$= \frac{Q t_u}{H_o}$$

$$= \frac{0.8 \times 1.547 \times 33 \times 60}{3.5}$$

$$= 700 \text{ ft}^2$$

Therefore the size of the clarifier selected in 700 ft^2.

10.3 Flotation

Flotation is the process of converting suspended substances and some colloidal, emulsified, and dissolved substances to floating matter [11]. The term "flotation" includes both violently agitated froth flotation, as used in the separation of ores in the mining industry, and quiescent flotation, which is now becoming popular as an efficient method for the removal of most suspensions from waste waters.

Small and difficult-to-settle particles in suspension can be flocculated and buoyed to the liquid surface by the lifting power of the many minute air bubbles which attach themselves to the suspended particles. Floated agglomerated sludges can be readily and continuously removed from the surface of the liquid by skimming. These skimmings are usually collected as a concentrated sludge and normally drain quite readily. A convenient practice is to detain the sludge float in a receiving tank for a few hours before draining the subnatant liquor from the bottom. The solids content of the float can be more than doubled by this concentration method; water is actually squeezed out of the float while the particles compact. Such a sludge float is usually quite stable and free from odors. Since the flotation process brings partially reduced chemical compounds into contact with oxygen in the form of tiny air bubbles, satisfaction of any immediate oxygen demand of the waste water is thereby aided.

Typical vacuum flotation units first aerate the waste with air diffusers or mechanical beaters. Aeration periods are brief, some as short as 30 seconds, and require only about 0.025 to 0.05 cubic feet of air per gallon of waste water. A brief de-aeration period is then provided at atmospheric pressure, to remove large bubbles. The waste, at this point nearly saturated with dissolved air, passes to an evacuation tank which is enclosed and maintained under a vacuum of about nine inches of mercury. This vacuum gives rise to bubbles, which cause flotation.

Pressure flotation differs from vacuum flotation in that air is injected into the waste under pressure, and bubbles of air are then formed when the waste is exposed to atmospheric pressure. Wastes are normally pressurized to about 30 to 40 pounds per square inch and retained at this pressure for approximately a minute. Some coagulant aids (alum and/or silica) and a small volume of air can be bled into the system at the suction end of the pump, where waste water enters the tank. Passage through the pump usually suffices to provide good mixing of the chemicals and air with the waste. When released to the atmosphere in the flotation tank, the tiny, rising bubbles trap suspended, colloidal, and (some) emulsified particles. The floated sludge is usually continuously skimmed and removed from the tank by sludge pumps.

Vrablik [19] makes a distinction between two methods of flotation: dissolved-air and dispersed-air. Dispersed-air flotation generates gas bubbles by the mechanical shear of propellers, diffusion of gas through porous media, or by homogenizing a gas and liquid stream. Dissolved-air flotation generates gas bubbles by precipitation from a solution supersaturated with the gas. These bubbles are much smaller than dispersed-air bubbles, generally not exceeding 80 microns* in diameter, while dispersed-air bubbles often reach 1000 microns in diameter.

To understand the theory of dissolved-air flotation, the student must investigate the gas, liquid, and solid phases as they are brought into intimate contact with each other. Henry's law indicates the relationship between the solubility of gas (in this case, dissolved air) and the total pressure:

$$C = kp,$$

where C is the concentration of gas in solution, k is Henry's law constant, and p is the absolute pressure above the solution at equilibrium.

By attachment to, or inclusion in, a suspended-solids structure or liquid phase, the bulk density of the paired system may be less than the density of the parent system, causing the agglomeration to be floated to the top. The gas bubbles therefore render a buoyancy to the original suspended particle in accordance with Archimedes' principle: the resultant pressure of a fluid on an immersed body acts vertically upward through the center of gravity of the displaced fluid and is equal to the weight of the fluid displaced. The resultant upward force exerted by the fluid on the body is called *buoyancy* and this force is responsible for the floating of solids which were originally somewhat heavier than the surrounding fluid.

Since we are usually dealing with large volumes of water in waste treatment, detention time in flotation chambers becomes a critical factor. Detention time, in turn, is dependent primarily on the rate of rise of air bubbles in the water. This can best be expressed by Stokes's law, which holds true for particles with a diameter of less than 130 microns:

$$V = kD,^2$$

where V is the rate of bubble rise (ft/min), k is Stokes's conversion factor (this includes all the factors which affect the rise or fall of bubbles, such as density or viscosity of the liquid, excluding the density of the bubble), and D is the diameter of the air bubble. The Stokes relationship is shown quantitatively in Fig. 10.10.

Typical results obtained from samples of several

---

*1 micron = 0.0001 cm = 0.0000394 in.; 1 in. = 2.54 cm.

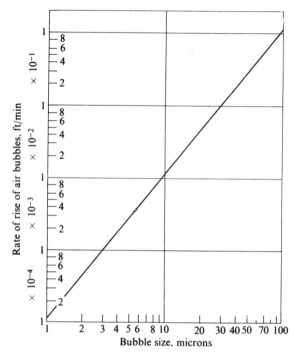

Fig. 10.10   Rate of rise of air bubbles in tap water (calculated by means of Stoke's law) as a function of bubble size. (After Vrablik [18].)

industrial wastes [11] treated by dissolved-air flotation show suspended solids and BOD reductions of 69 to 97.5 and 60 to 91.8 per cent, respectively (Table 10.3).

Since almost twice as much air can be dissolved in water, all other factors being equal, at 0°C than at 30°C, the temperature of waste water is a significant factor in the effectiveness of the flotation process. This relationship is shown in Fig. 10.11.

Generally, air bubbles are negatively charged, the anions collecting mainly on the gas side of the interface, while the cations spread themselves out thinly on the water side of the interface. Since suspended particles or colloids may have a significant electrical charge, either attraction or repulsion will occur between these and the air bubbles.

Vrablik [18] made an extensive study of the three different processes by which flotation may be caused: (1) adhesion of a gas bubble to a suspended liquor or solid phase; (2) the trapping of gas bubbles in a

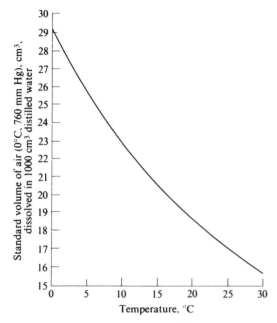

Fig. 10.11 Solubility of air in distilled water at various temperatures. (From *Handbook of Chemistry and Physics*, 36th edition, Chemical Rubber Publishing Company (1955), p. 1609.)

floc structure as the gas bubble rises; (3) the absorption of a gas bubble in a floc structure as the floc structure is formed. These three phenomena are illustrated in Fig. 10.12.

An illustration of pressure flotation is shown in Fig. 10.13.

Finally, the engineer should be aware of both the advantages and disadvantages of flotation as a waste-treatment process [8]. The advantages are as follows:

1) Grease and light solids rising to the top and grit and heavy solids settling to the bottom are all removed in one unit

2) High overflow rates and short detention periods mean smaller tank sizes, resulting in decreased space requirements and possible savings in construction costs

3) Odor nuisances are minimized because of the short detention periods and, in pressure and aeration-type units, because of the presence of dissolved oxygen in the effluent

4) Thicker scum and sludge are obtained, in many cases, from a flotation unit than from gravity settling and skimming.

The disadvantages are as follows:

1) The additional equipment required results in higher operating costs

2) Flotation units generally do not give as effective treatment as gravity-settling units, although the efficiency varies with the waste

3) The pressure type has high power requirements, which increase operating cost

4) The vacuum type requires a relatively expensive airtight structure capable of withstanding a pressure

Table 10.3 Typical efficiencies of dissolved-gas flotation treatment of wastes [14].

Waste source	Suspended solids in influent, ppm	Reduction obtained, %	BOD in influent, ppm	Reduction obtained, %
Petroleum production	441	95.0		
Railroad maintenance	500	95.0		
Meat packing	1400	85.6	1225	67.3
Paper manufacturing	1180	97.5	210	62.6
Vegetable-oil processing	890	94.8	3048	91.6
Fruit-and-vegetable canning	1350	80.0	790	60.0
Soap manufacture	392	91.5	309	91.6
Cesspool pumpings	6448	96.2	3399	87.0
Primary sewage treatment	252	69.0	325	49.2
Glue manufacture	542	94.3	1822	91.8

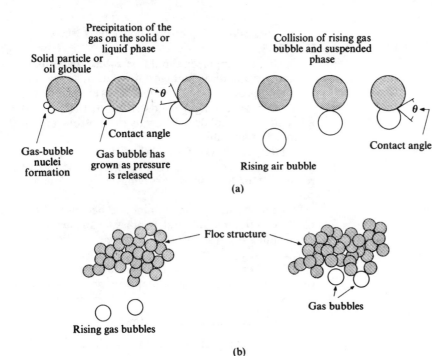

Solid particle or oil globule

Precipitation of the gas on the solid or liquid phase

Collision of rising gas bubble and suspended phase

θ

Contact angle

Gas-bubble nuclei formation

Gas bubble has grown as pressure is released

Rising air bubble

θ

Contact angle

(a)

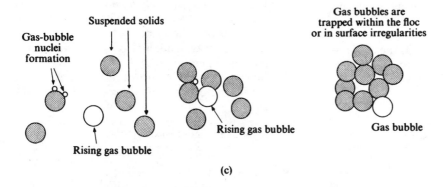

Floc structure

Rising gas bubbles

Gas bubbles

(b)

Fig. 10.12 Three methods of dissolved-air flotation. (a) Adhesion of a gas bubble to a suspended liquid or solid phase. (b) The trapping of gas bubbles in a floc structure as the gas bubbles rise. (c) The absorption and adsorption of gas bubbles in a floc structure as the floc structure is formed. (After Vrablik [18].)

Gas-bubble nuclei formation

Suspended solids

Gas bubbles are trapped within the floc or in surface irregularities

Rising gas bubble

Gas bubble

Rising gas bubble

(c)

of nine inches of mercury; any leakage to the atmosphere will adversely affect performance

5) More skilled maintenance is required for a flotation unit than for a gravity-settling unit.

Quigley and Hoffman [14] deserve credit for daring to refer to flocculation and dissolved-air flotation as "secondary treatment" when it follows sedimentation. They describe an effective dissolved-air flotation system for treating oil-refinery wastes. By recycling

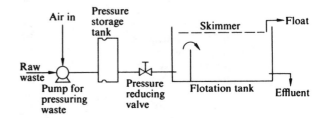

Air in

Pressure storage tank

Skimmer

Float

Raw waste

Pump for pressuring waste

Pressure reducing valve

Flotation tank

Effluent

Fig. 10.13 Schematic drawing of pressure flotation system.

treated effluent and using lime as a coagulant they were able to obtain oil removals of 68 to 96 per cent.

10.4 Screening

Screening of industrial wastes is generally practiced on wastes containing larger suspended solids of variable sizes, e.g., from canneries, pulp and paper mills, or poultry processing plants. It is an economical and effective means of rapid separation of these larger suspended solids from the remaining waste material. In many cases screening alone will reduce the suspended solids to a low enough concentration to be acceptable for discharge into a municipal sewer or a nearby stream. Often considerable BOD is also removed by the screening process, the percentage removed varying almost directly with the size of the screen and the amount of BOD associated with the screenable solids. Screens are available in sizes ranging from coarse (10 or 20 mesh) to fine (120 to 320 mesh).

The North and Sweco screens are typical of two major types used in industry today. The former are generally rotary, self-cleaning, gravity-type units (Fig. 10.14). The latter are mostly circular, overhead-fed, vibratory units (Fig. 10.15).

The rotary, gravity-type, waste-disposal screens are manufactured in several sizes to handle almost any volume of waste liquid. In general, they vary from 3 to 5 feet in diameter and from 4 to 12 feet in length and weigh between one and five tons. These screens separate solid and liquid constituents from waste materials at a location where they gravitationally flow or can be pumped into the screened cylinder. The machine's large drum rotates at 4 rpm. The lift paddles within the drum pick up the solid material from the water and deposit it into a stationary, perforated hopper within the cylinder. The hopper holds a spiral screw conveyor that moves the solids to the rear end of the machine and out through the discharge spout. In the process, it compresses the wastes and squeezes out more liquid, which drains through the perforated hopper back into the cylinder. The water in the cylinder drains through the wire mesh and collects in a steel or wooden tank which is part of the machine. The fine wire mesh is at all times kept

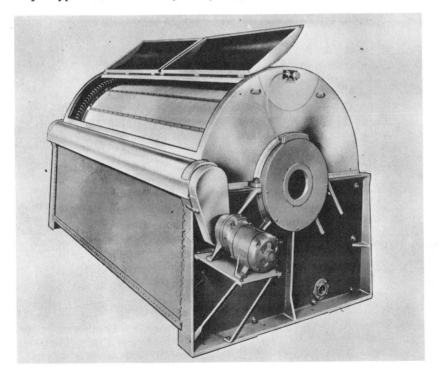

Fig. 10.14 North water filter. (Photograph courtesy Green Bay Foundry and Machine Works, Green Bay, Wisconsin.)

Fig. 10.15 The 48-inch-diameter Sweco separator shown is screening lint from waste water at the Eastern Overall Company, in Baltimore, Maryland. The waste water is fed onto the 60-mesh market-grade screen at a rate of 300 gpm. The screened waste water is discharged to the sewer. (Photograph courtesy Sweco Inc., Los Angeles, California.)

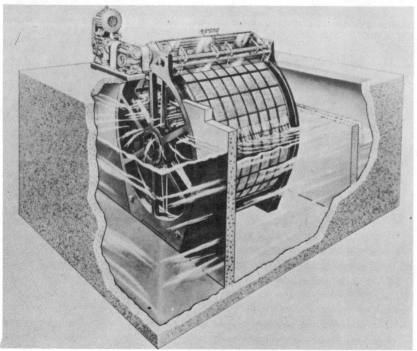

Fig. 10.16 Cutaway view of a 7½-foot-diameter microstrainer. (Photograph courtesy Crane Company, King of Prussia, Pennsylvania.)

free from clogging by a continuous spray pipe with jet nozzles, located above the rotating cylinder. They have been used successfully in treating wastes from meat-packing, canning, grain-washing, tanning, malting, woolen, and sea-food plants.

The circular, vibratory screens have been quite effective in screening wastes from food-packing processes such as meat and poultry packing or fruit and vegetable canning. Vibration is designed usually to remove solids at the periphery of the screen, although Swallow [16] reports the use of a new center-discharge separator.

Microstraining, a particular screening device, was first introduced by Dr. P. L. Boucher in England in 1945 for water clarification and there were (1965) about 70 water-treatment plants in the United States utilizing this process [2]. It involves the use of high-speed, continuously backwashed, rotating drum filters work-

Table 10.4 Results obtained on humus tank effluent at Eastern Sewage Works, London, England, December 30, 1966–January 13, 1967.

Characteristic	Effluent from			
	Humus tank	Micro-strainer	Ozonizer	Sand filter
Suspended solids	51†	19	15	10
Total solids	931			928
BOD	21	13	11	9
COD	78	54	44	39
Permanganate value	16	10	6	5
Organic carbon		23	19	10
Surface-active matter				
Anionic (as Manoxol OT)	1.4	1.4	0.6	0.6
Nonionic (as Lissapol NX)	0.37		0.07	0.07
Ammonia (as N)	7.1	7.5	7.4	7.6
Nitrite (as N)	0.4	0.4	0.02	0.01
Oxidized nitrogen (as N)	26	26	26	27
Total phosphorus (as P)	8.2			7.4
Orthophosphate (as P)	6.6			7.0
Total hardness (as $CaCO_3$)				468
Chloride				98
Sulfate	212			213
Color (Hazen units)	36		4	7
Turbidity (ATU)‡	66		27	13
Total phenol	3.4			0.9
Temperature (°C)	8.1	8.0	7.9	7.7
Dissolved oxygen (% saturation)	52	52	99	94
Conductivity (μmho/cm³)	1173	1175	1170	1150
Langlier index	−0.08			+0.12
pH	7.4	7.4	7.4	7.5
Pesticides (μg/l)				
α BHC		0.025	0.007	
γ BHC		0.035	0.030	
Aldrin		0.004	0.000	
Dieldrin		0.193	0.032	
pp DDT		0.031	0.030	

*After E.W.J. Diaper, *Water Wastes Eng.* **5**, 56 (1968).
†All results are given in milligrams per liter, unless otherwise specified.

ing in open gravity-flow conditions (see cutaway picture in Fig. 10.16). The principal filtering fabrics employed have apertures of 35 or 25 microns and are fitted on the drum periphery. Head loss is between 4 and 6 inches. Results in London, England, showed that microstraining removes most of the suspended solids remaining after biological treatment (Table 10.4).

The Bauer Company manufactures a perforated plate screen (referred to as a Hydrasieve), which is installed at a slight angle to the vertical. Wastewater is passed down the screen from the top with water going through the screen and solids collecting at the bottom. The efficiency of removal depends primarily upon the size of the screen opening and the wastewater application rate.

References

1. Bewtra, J: K., "Diagram for the settling of discrete particles in viscous fluids," *Water Sewage Works*, **114,** 60 (1967).
2. Boucher, P. L., "Micro-straining, microzon, and demicellization applied to public and industrial water supply," in Proceedings of Water Treatment Symposium, May 1965, Adelaide, S. Australia.
3. Bramer, H. C., and R. D. Hoak, "Measuring sedimentation-flocculation efficiencies," *Ind. Eng. Chem. Process Design Develop.* **5,** 316 (1966).
4. Camp, T. R., "Studies of sedimentation basin design," *Sewage Ind. Wastes* **25,** 1 (1953).
5. Clark, J. W., and W. Viessman, Jr., *Water Supply and Pollution Control*, International Textbook Company, Scranton, Pa. (1965), pp. 274–294.
6. Dobbins, W. E., "Advances in sewage treatment design," Paper read to the Sanitary Engineering Division of the A.S.C.E. (Metropolitan Section) Conference at Manhattan College, New York City (May 1961).
7. Eckenfelder, W. W., *Industrial Water Pollution Control*, McGraw-Hill Book Co., New York (1966).
8. Federation of Sewage and Industrial Wastes Association, *Sewage Treatment Design*, Manual of Practice no. 8 (American Society of Civil Engineers Manual of Engineering Practice no. 36) (1959), p. 78.
9. Fitch, B., "Current theory and thickener design," *Ind. Eng. Chem.* **10,** 18 (1966).
10. Great Lakes–Upper Mississippi River Board of State Sanitary Engineers, *Recommended Standards for Sewage Works*, Harrisburg, Pa., May 10, 1960.
11. Hess, R. W., *et al.*, "1952 Industrial wastes forum," *Sewage Ind. Wastes* **25,** 709 (1953).
12. Katz, W. J., "Adsorption—Secret of success in separating solids by air flotation," *Ind. Wastes* **30,** 11 (1959).
13. Eckenfelder, W. W., *Industrial Water Pollution Control*, McGraw-Hill Book Co., New York (1966), Chap. 2, p. 28.
14. Quigley, R. E., and E. L. Hoffman, "Flotation of oily wastes," in Proceedings of 21st Industrial Wastes Conference, Purdue University, May 1966, p. 527.
15. Swallow, D. M., "Design and operation of the center-discharge separator," in Proceedings of the Seminar on Water Pollution Control, during 30th Exposition of Chemical Industries, New York, Nov. 30, 1965, p. 20.
16. Villemonte, J. R., "Hydraulic characteristics of circular sedimentation basins," in Proceedings of 17th Industrial Waste Conference, at Purdue University, 1962, p. 682.
17. Villemonte, J. R., *et al.*, "Hydraulic and removal efficiencies in sedimentation basins," *J. Water Pollution Control Federation* **38,** 371 (1966).
18. Vrablik, E. R., "Fundamental principles of dissolved-air flotation of industrial wastes," in Proceedings of 14th Industrial Waste Conference, Purdue University Engineering Extension Series, Bulletin no. 104, May 1960, p. 743.

Suggested Additional Reading

Sedimentation

Camp, T. R., "Sedimentation and the design of settling tanks," *Trans. Am. Soc. Civil Engrs.*, **111,** 895 (1946).

Dobbins, W. E., "Effect of turbulence on sedimentation," *Trans. Am. Soc. Civil Engrs,* **109,** 629 (1944).

Federation of Sewage and Industrial Wastes Association, *Sewage Treatment Design*, Manual of Practice no. 8 (American Society of Civil Engineers Manual of Engineering Practice no. 36), (1959), pp. 90–91.

Hazen, A., "On sedimentation," *Trans. Am. Soc. Civil Engrs,* **53,** 45 (1904).

Rich, L. G., *Unit Operations in Sanitary Engineering*, John Wiley & Sons, New York (1961), Chapter 4, pp. 81–109.

Flotation

Beebe, A. H., "Soluble oil wastes treatment by pressure flotation," *Sewage Ind. Wastes* **25**, 1314 (1953).

D'Arcy, N. A., Jr., "Dissolved air flotation separates oil from waste water," *Oil Gas J.* **50**, 319 (1951).

Rich, L. G., *Unit Operations in Sanitary Engineering*, John Wiley & Sons, New York (1961), Chapter 5, pp. 110–35.

Screening and Microstraining

For full bibliography see: Boucher, P. L., *J. Inst. Public Health Engrs.* **60**, 294 (1961); and Reference 2 above.

Campbell, R. M., and M. B. Prescod, *J. Inst. Water Engrs* **19**, 101 (1965).

Boucher, P. L., "Micro-straining and ozonisation of water and waste water," in Proceedings of 22nd Industrial Waste Conference, Purdue University Engineering Extension Series, Bulletin no. 129, May 1967.

Diaper, E. W. J., "Micro-straining and ozonisation of water and waste water," *Water Wastes Eng.* **5**, 56 (1968).

Hazen, R. "Application of the microstrainer to water treatment in Great Britain," *J. Am. Water Works Assoc.* **45**, 723 (1953).

Questions

1. What are three major methods of removing suspended solids?
2. When would you use sedimentation for removal of suspended solids?
3. When should you use flotation for removal of suspended solids?
4. Would you even use both sedimentation and flotation together?
5. Why and when would you use screening for suspended-solids removal?
6. What are the most important factors affecting industrial wastewater sedimentation?
7. In dissolved air flotation what is the apparent anomaly which exists because of the size of the air bubble?
8. What are the three different methods by which suspended matter can be removed by dissolved air flotation with chemical coagulant addition?
9. What are the two major types of screening devices mentioned in this chapter? What is a third type of screening device type not pictured in the chapter? What are the major advantages of each type?

CHAPTER 11

REMOVAL OF COLLOIDAL SOLIDS

11.1 Characteristics of Colloids

A colloid may be defined as a particle held in suspension by its extremely small size (1 to 200 millimicrons), its state of hydration, and its surface electrical charge. There are two types of colloids: *lyophobic* and *lyophilic*. Because of the difference in their characteristics, they react differently to alterations in their environment. Table 11.1 will assist the student in understanding their properties. Colloids are often responsible for a relatively high percentage of the color, turbidity, and BOD of certain industrial wastes. Since it is important to remove colloids from waste waters before they can get into streams, one must understand their physical and chemical characteristics.

Colloids exhibit Brownian movement, a bombardment of the particles of the disperse phase by molecules of the dispersion medium. They are essentially nonsettleable because of their charge, small size, and low particle weight. They are dialyzable; that is, they can be separated from their crystalloid (low molecular weight) counterparts by straining through a semipermeable membrane. The colloids diffuse very slowly compared to soluble ions. Colloidal particles, in general, exhibit very low (if any) osmotic pressure because of their large size relative to the size of soluble ions. They also possess the characteristic of imbibition (the taking in of water by gels). In fact, it is by this very process that bacteria spores (often considered colloidal) take up water and germinate. Colloidal gels are very often used as ultrafilters, having pores sufficiently small to retain the dispersed phase of a colloidal system but large enough to allow the dispersion medium and its crystalloid solutes to pass through. For example, Perona *et al.* [10] found that the formed membranes may be used to remove up to 90 percent of the colored material and somewhat less of the COD and total dissolved solids of pulp-mill sulfite wastes. Colloidal systems show a wide range in viscosity or plasticity. Usually the lyophobic colloidal suspensions exhibit a viscosity only slightly higher than that of the pure dispersing medium (Fig. 11.1) and this concentration increases only very slightly when the concentration of the dispersed material is increased. On the other hand, lyophilic systems may reach very high values of viscosity. With these types of colloids, a parabolic, rather than a linear, relationship exists between viscosity and the concentration of dispersed phase, as shown in Fig. 11.1. Woodard and Etzel [21] have shown that, under certain conditions, one may change a lyophilic colloid in an industrial waste to a lyophobic one. In this case, lignin was altered by the addition of acetone and sodium hydroxide to render the colloid less stable and to enhance color removal.

Many colloidal systems, especially lyophilic (gel) systems, possess the property of elasticity ("springiness" or resistance). This property enables the gels to resist deformation and thereby recover their original shape and size once they have been deformed. If a concentrated beam of light is passed through a colloidal solution in which the dispersed phase has a different refractive index from that of the dispersion medium, its path is plainly visible as a milky turbidity when viewed perpendicularly. This is known as the Tyndall effect (see Table 11.1).

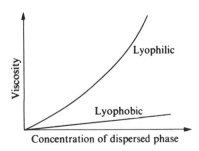

Fig. 11.1 Effect of colloidal type on viscosity.

Table 11.1 Types and characteristics of colloidal sols.

Characteristic	Lyophobic (hydrophobic)	Lyophylic (hydrophylic)
Physical state	Suspensoid	Emulsoid
Surface tension	The colloid is very similar to the medium	The colloid is of considerably less surface tension than the medium
Viscosity	The colloid suspension is very similar to the dispersing phase alone	Viscosity of colloid suspension alone is greatly increased
Tyndall effect	Very pronounced (ferric hydroxide is an exception)	Small or entirely absent
Ease of reconstitution	Not easily reconstituted after freezing or drying	Easily reconstituted
Reaction to electrolytes	Coagulated easily by electrolytes	Much less sensitive to the action of electrolytes, thus more is required for coagulation
Examples	Metal oxides, sulfides, silver halides, metals, silicon dioxide	Proteins, starches, gums, mucilages, and soaps

An important property of colloidal particles is the fact that they are generally electrically charged with respect to their surroundings. An electric current passing through a colloidal system causes the positive particles to migrate to the cathode and the negative ones to the anode.

11.2 Chemical Coagulation

The removal of oxygen-demanding and turbidity-producing colloidal solids from waste waters is often called intermediate treatment, since colloids are intermediate in size between suspended and dissolved solids. The most common and practical method of removing these solids is by chemical coagulation. This is a process of destabilizing colloids, aggregating them, and binding them together for ease of sedimentation. It involves the formation of chemical flocs that absorb, entrap, or otherwise bring together suspended matter, more particularly suspended matter that is so finely divided as to be colloidal.

The chemicals most commonly used are: alum, $Al_2(SO_4)_3 \cdot 18H_2O$; copperas, $FeSO_4 \cdot 7H_2O$; ferric sulfate, $Fe_2(SO_4)_3$; ferric chloride, $FeCl_3$; and chlorinated copperas, a mixture of ferric sulfate and chloride. Aluminum sulfate appears to be more effective in coagulating carbonaceous wastes, while iron sulfates are more effective when a considerable quantity of proteins is present in the waste. The use of organic polymers, which have the ability to act as either negatively or positively charged ions has made a significant impact on the efficiency of removal of colloids by chemical coagulation. These polymers, acting as a coagulant aid, and applied in conjunction with the coagulant, enhance the formation of flocs and result in improved settling characteristics. Smaller dosages and the elimination of many storage problems are among the major advantages of these polymers. Dey [2] presents results obtained in various industries where water-soluble polymeric coagulation chemicals are used to achieve improved waste solids settling. Schaffer [15] found that these polymers were useful in maintaining higher solids concentrations in an anaerobic contact treatment process for meat-packing wastes.

The process of chemical coagulation involves complex equilibria among a number of variables including colloids of dispersed matter, water or another dispersing medium, and coagulating chemicals. Driving forces—such as the electrical phenomenon, surface effects, and viscous shear—cause the interaction of these three variables.

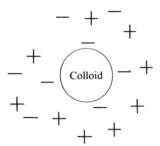

Fig. 11.2 Stable colloid.

11.3 Coagulation by Neutralization of the Electrical Charges

This can be accomplished by:

1) Lowering the zeta potential of the colloids (Fig. 11.2). Zeta potential is the difference in electrical charge existing between the stable colloid and the dispersing medium.

2) Neutralizing the colloidal charge by flooding the medium with an excess of oppositely charged ions, usually hydrous oxide colloids formed by reaction of the coagulant with ions in the water. The coagulant colloids also become destabilized by the reaction with foreign, oppositely charged, colloids and produce hydrous oxide, which is a floc-forming material.

From the standpoint of electrical charges, there are two predominant types of colloid in waste waters:

1) Colloids naturally present, including several proteins, starches, hemicelluloses, polypeptides, and other substances, all possess negative charges (mostly lyophilic in nature).

2) Colloids artificially produced by coagulants, usually the hydroxides of iron and aluminum (mostly lyophobic in nature), are mainly positively charged ions.

In most scientific circles it is believed that the charge on colloidal particles is due mainly to the preferential adsorption of ions (H^+ or OH^-), from the dispersing medium. The charge may also be due, in part, to the direct ionization of a portion of its structural groups, such as NH_2^+ and COO^-.

Hydrous aluminum and iron oxides, as well as other metal sols, can acquire both positive and negative charges. Excess Fe^{+++} makes colloids positively

charged. The following expression depicts a resultant positively charged colloid:

$$\begin{matrix} FeO \\ FeO \cdot x \ HOH \quad Fe^{+++} \leftarrow \end{matrix} \begin{cases} OH^- \\ OH^- \\ OH^- \end{cases}$$
$$O$$

Excess OH^- makes colloids negatively charged. The following expression depicts a resultant negatively charged colloid:

$$\begin{matrix} FeO \\ FeO \cdot x \, H_2O \quad OH^- \leftarrow H^+ \\ O \end{matrix}$$

However, a colloid can acquire a charge by means other than adsorption. A protein dissolved in solution can be schematically illustrated as follows:

$$COO^- \text{——— [protein base molecule] ——— } NH_2^+.$$

It may become necessary to add up all the positive NH_2^+ groups and the negative COO^- groups to ascertain the final ionic charge of the solution, because of the inherent charge brought about by direct ionization of the particle. The sol is thereby stabilized by inherent ionization of groups within the molecule itself.

Any alteration of the type and number of double-layer ions should reduce the zeta potential to such a point that the colloid will lose its stability. Stability is defined as the ability to resist precipitation and/or coagulation into a relatively large particle. A colloid is most stable when it possesses the greatest electrical charge and smallest size. The coagulating power of ions rises rapidly as the electrical charge increases, as is stated by the Schulze-Hardy rule. Table 11.2 illustrates the minimum concentration of various chemical coagulants required for anions and cations to complete the reaction. Ratios of concentrations of electrolytes required for valences of 1, 2, or 3 are in the order of 729:11.4:1.

Electrolytes and colloids react readily to changes in the pH of the waste water. Most negatively charged particles, including the majority ·of contaminating colloids present in waste waters, coagulate at an optimum pH value of less than 7.0. Flocculent hydroxide colloids, on the other hand, are insoluble only at pH values above 7.0 and usually over 9.0. Lime is normally added to raise the pH, as well as to aid in precipitation of colloids.

Table 11.2 Valence and coagulant dosage.

Electrolyte	Anion or cation valence	Minimum concentration required, mmols/liter
	Anion	
KCl	1	103
KBr	1	138
KNO_3	1	131
K_2CrO_3	2	0.325
K_2SO_4	2	0.219
$K_3Fe(Cn)_6$	3	0.096
	Cation	
NaCl	1	51
KNO_3	1	50
K_2SO_4	1	63
$MgSO_4$	2	0.81
$ZnCl_2$	2	0.68
$BaCl_2$	2	0.69
$AlCl_3$	3	0.09

Alum has a pH range of maximum insolubility between 5 and 7; the ferric ion coagulates only at pH values above 4; and the ferrous ion only above 9.5. Copperas ($FeSO_4 \cdot 7H_2O$) is a useful coagulant only in highly alkaline wastes. Lime, a coagulant in itself, is often added with iron salts to raise the pH to the isoelectric point of the coagulant. At this point, the colloid has its minimum electrical charge and is least stable. Since lime is quite insoluble at pH values of 9 and over, coagulation with lime and copperas together increases the pH range. Aeration of waste waters before addition of lime enhances coagulation by evolving lime (thus consuming carbon dioxide and supplying oxygen for converting iron to the oxide and hydroxide states).

Since the ferrous ion when oxidized to the ferric ion can also be used as a coagulant at low pH values, oxidation may be carried out by chlorination, as follows:

$$6Fe^{++}SO_4 \cdot 7H_2O + 3Cl_2 \rightleftarrows 6Fe^{+++} + 6SO_4^{=} +$$
Copperas
$$+ 6Cl^- + 42H_2O.$$

Negative ions already present in waste waters extend the useful range of pH in the acid category

and positive ions extend the useful pH range in the basic category. Thus, in soft waters, the negatively charged color colloids coagulate best in the acid pH range, and positively charged iron and aluminum ions are good precipitating chemicals in alkaline waters. Prechlorination of alum-treated wastes sometimes increases color removal. Finely divided clay, activated silica, bentonite, or other coagulant aids are often used for relatively clear waters. The addition of any of these produces an effect similar to that of seeding clouds with silver-iodide crystals: they provide nuclei about which the precipitate can gather, agglomerate, and flocculate, with a resultant increase in density and settling rate.

Sometimes the presence of iron and manganese in waste waters will add to the effect of the cationic coagulants. An increase in the concentration of the coagulant shortens the time of the coagulation reaction considerably. Gentle agitation of the waste water also enhances coagulation, by increasing the number of collisions and thus bringing about more rapid floc formation.

11.4 Removal of Colloids by Adsorption

A large number of compounds which are not amenable to other types of treatment may be removed from wastes by adsorption. For example, pesticides, such as 2,4-D herbicides and carbamate insecticides may be removed by adsorption onto powdered activated carbon but not onto clay materials such as illite, kaolinite, and montmorillonite [16]. In addition, colloidal suspensions of DDT, chlorobenzene, and p-chlorobenzenesulfonic acid resulting in DDT production may be removed by using activated carbon [8]. Cooper and Hager [1] also suggest activated carbon for advanced waste treatment where reclamation is of paramount importance. They present three typical activated-carbon treatment systems and a granular-carbon reactivation system (Fig. 11.3). The granular carbon used in most reactivation systems in the world is made from bituminous coal. Cooper and Hager also present a summary of properties for two types of this coal (see Table 11.3) and claim this treatment is especially effective in removing biologically resistant (refractory) compounds.

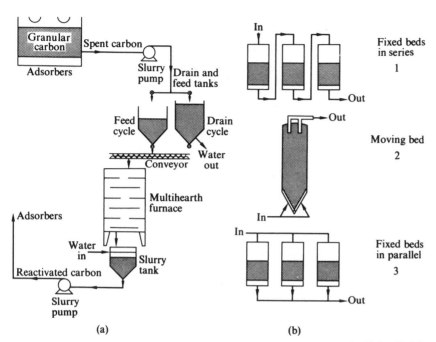

Fig. 11.3 (a) Granular carbon reactivation cycle. (b) Adsorber configuration for granular carbon waste treatment.

Table 11.3 Properties of coal-derived granular carbon for waste treatment.

Characteristic	Type SGL	Type CAL
Mesh size, U.S. Sieve Series	8 × 30	12 × 40
Effective size, mm	0.8–0.9	0.50–0.60
Uniformity coefficient	1.9 or less	1.7 or less
Mean particle diameter, mm	1.5	0.9
Real density, g/cm³	2.1	2.1
Apparent density		
g/cm³	0.48	0.44
lb/ft³	30.0	27.5
Particle density wetted with water, g/cm³	1.4–1.5	1.3–1.4
Total surface area (N_2 BET method), m²/g	950–1050	1000–1100
Pore volume, cm³/g	0.85	0.94

References

1. Cooper, J. C., and D. G. Hager, "Water reclamation with activated carbon," *Chem. Eng. Progr.* **62**, 85 (1966).

2. Dey, R. F., "Use of organic polymers in treatment of industrial wastes," in Proceedings of 12th Ontario Industrial Waste Conference, June 1965, pp. 89–104.

3. Fair, G. M., and J. Geyer, *Elements of Water and Waste Water*, John Wiley & Sons, New York (1958), p. 616.

4. Hogg, R., T. W. Healy, and D. W. Fuerstenau, "Mutual coagulation of colloidal dispersions," *Trans. Faraday Soc.* **62**, 1638 (1966).

5. Johnson, R. L., F. J. Lowes, Jr., R. M. Smith, and T. J. Powers, *Evaluation of the Use of Activated Carbon and Chemical Regenerants in the Treatment of Waste Water*, Publication no. 999–13, U.S. Public Health Service, Washington, D.C. (1964).

6. Joyce, R. S., and V. A. Sukenik, *Feasibility of Granular Activated Carbon Adsorption for Wastewater Renovation*, Environmental Health Service Supply and Pollution Control Publication no. 999-WP-28, U.S. Public Health Service, Washington, D.C. (1965).

7. Kawamura, S., and T. Yoshitaro, "Applying colloid titration techniques to coagulant dosage control," *Water Sewage Works* **113**, 398 (1966).

8. Kul'skii, L. A., and A. G. Shabolina, "Adsorption of DDT from colloidal solutions on the iodine KAD activated carbon," *Chem. Abstr.* **67**, no. 14686y, 1967.

9. Middleton, A. E., "Activated silica solution applications," *Water Sewage Works* **100**, 251 (1963).

10. Perona, J. J., *et al.*, "Hyperfiltration—Processing of pulp mill sulfite wastes with a membrane dynamically

formed from feed constituents," *Environ. Sci. Technol.* **1**, 991 (1967).

11. Rudolfs, W., and J. L. Belmat, "A separation of sewage colloids with the aid of the electron microscope," *Sewage Ind. Wastes* **24**, 247 (1952).

12. Rudolfs, W., and H. W. Gehm, "Chemical coagulation of sewage," *Sewage Works J.* **8**, 195, 422, 537 and 547 (1936).

13. Rudolfs, W., and H. Gehm, "Colloids in sewage treatment: 1. Occurrence and role. A critical review," *Sewage Works J.* **11**, 727 (1939).

14. Sawyer, C. N., and P. E. McCarty, *Chemistry for Sanitary Engineers*, McGraw-Hill Book Co., New York (1960).

15. Schaffer, R. B., "Polyelectrolytes in industrial waste treatment," *Water Sewage Works* **111**, 300R (1964).

16. Schwartz, H. G., Jr., "Adsorption of selected pesticides on activated carbon and mineral surfaces." *Environ. Sci. Technol.* **1**, 332 (1967).

17. Sennet, P., and J. P. Oliver, "Colloidal dispersions, electrokinetic effects and concept of zeta potential," *Ind. Eng. Chem.* **57**, 32 (1965).

18. Weber, W. J., "Adsorption," in Proceedings of Summer Institute for Water Pollution Control, Manhattan College, New York, 1967.

19. Weber, W. J., and J. C. Morris, *Adsorption of Biochemically Resistant Materials from Solution*, Publication no. 999-WP 33 W62–24, U.S. Public Health Service, Washington, D.C. (1966).

20. Williamson, J. N., A. M. Heit, and C. Calmon, *Evaluation of Various Adsorbents and Coagulants for Waste Water*, Publication no. 999-WP, U.S. Public Health Service, Washington, D.C. (1964).

21. Woodard, F., and J. Etzel, "Coacervation and chemical coagulation of lignin from pulpmill black liquor," *J. Water Pollution Control Federation* **37**, 990 (1965).

Questions

1. What are colloids?
2. Why is it important to remove colloids from wastes?
3. What is the approximate size of colloids?
4. Are colloids all of one type? If of more than one type, what are the differences?
5. What importance should the industrial-waste engineer attribute to changes in density of colloids?
6. In chemical coagulation of colloids, what are the two major economic considerations?
7. What is meant by the zeta potential and why is this important to industrial-waste engineers in the chemical coagulation of colloids?
8. For what reasons do we add lime to industrial wastes for colloidal-solids removal?
9. When would you recommend the use of activated carbon for removing colloidal solids? What is the mechanism involved?

CHAPTER 12

REMOVAL OF INORGANIC DISSOLVED SOLIDS

The removal of dissolved minerals from waste waters has been given relatively little attention by waste-treatment engineers, because minerals have been considered less pollutional than other constituents, such as organic matter and suspended solids. However, as we learn more about the causes and effects of pollution, the importance of reducing the quantity of certain types of inorganic matter which sewage plants permit to enter streams is apparent. Chlorides, phosphates, nitrates, and certain metals are examples of the more common and significant inorganic dissolved solids. Among the methods employed mainly for removing inorganic matter from wastes are: (1) evaporation; (2) dialysis; (3) ion exchange; (4) algae; (5) reverse osmosis; and (6) miscellaneous methods. Other treatment methods which remove minerals incidentally but are aimed primarily at other contaminants are discussed in Chapters 10, 11, and 13. One should not overlook the minerals contributed by natural runoff from overland flow. The amount of dissolved solids which these natural flows contain often exceeds that contributed by waste waters from industry.

12.1 Evaporation

Evaporation is a process of bringing waste water to its boiling point and vaporizing pure water. The vapor is either used for power production, or condensed and used for heating, or simply wasted to the surrounding atmosphere. The mineral solids concentrate in the residue, which may be sufficiently concentrated for the solids either to be reusable in the production cycle or to be disposed of easily. This method of disposal is used for radioactive wastes, and paper mills have for a long time been evaporating their sulfate cooking liquors to a degree where they may be returned to the cookers for reuse.

Major factors in the selection of the evaporation method are: (1) *Economics:* does the value of the reusable residue outweigh the cost of fuel for evapora-

tion? (2) *Initial dissolved solids:* are there enough solids in the waste, of a variable nature, to warrant evaporation? Generally, 10,000 ppm are required. (3) *Foreign matter:* is there foreign matter present which could cause scale formation or corrosion or interfere with heat transfer in evaporation? (4) *Pollution situation:* what effect will the minerals have on the receiving stream? For example, caustic soda kills fish, ammonium salts initiate troublesome algae growths and in some cases stimulate bacterial growth upon organic matter already present [1], salt interferes with water use by industries and municipalities, and so forth.

Today many evaporators are heated by steam condensing on metallic tubes, through which flows the waste to be concentrated or evaporated. The steam is at a low pressure, usually less than 50 pounds per square inch (psi) (absolute). Most evaporators operate with a slight vacuum on the vapor side, to lower the boiling point and to increase the rate of vapor removal from the evaporator. Vacuum systems are especially preferable to atmospheric evaporators when the decomposition of organic matter is involved. Care must be exercised, however, that the vacuum is not great enough to permit priming of the waste water into the vapor.

Evaporating a waste presents many problems, which include concentration changes during evaporation, foaming, temperature sensitivity, scale formation, and the materials used in evaporator construction. In industrial-waste concentration, scale formation usually presents the major obstacle. As crust is deposited on the heating surface, the overall heat-transfer coefficient decreases, causing the efficiency to drop until it is necessary to shut down and clean the tubes—a complicated process when the scale is hard and tenacious.

Chrome, nickel, and copper acid-type plating wastes may be reclaimed from the rinse tank by evaporation in glass-lined equipment, or other suitable evaporators, and the concentrated solution returned to the plating system [16]. Initial cost of equipment is high, so that the quantity and value of chemicals to be recovered, plus the estimated cost of operation of a treatment system if evaporative recovery were not practiced, are criteria one must use to justify purchasing such equipment.

Efficiency of evaporation is directly related to heat-transfer rate—expressed in British thermal units per hour (Btu/hr)—through the heating surface (tube wall). This rate is equal to the product of three factors: the overall heat-transfer coefficient, the heating surface area, and the overall change in temperature between the waste and the steam. It is expressed mathematically as

$$q = UA(t_s - t_w) = UA\,\Delta t,$$

where q is the rate of heat transfer (Btu/hr), U is the overall coefficient (Btu/ft^2/hr/°F), A is the heating-surface area (ft^2), t_s is the temperature of steam condensate (°F), t_w is the boiling temperature of waste (°F), and $\Delta t = t_s - t_w$ is the overall temperature change between steam and waste. Typical values of U for various types of evaporators are given in Table 12.1. These figures are estimated within broad ranges, by considering the viscosity of the waste, scale formation, and operating temperatures (greater temperature differentials yield higher coefficients). Tube wall thickness also influences U; the greater the thickness, the lower the value of U.

Table 12.1 Typical overall coefficients in evaporators [4].

Type of evaporator	Overall coefficient, Btu/ft^2/hr/°F
Long-tube vertical	
Natural recirculation	200–600
Forced circulation	400–2000
Short-tube	
Horizontal tube	200–400
Calandria type	150–500
Coil	200–400
Agitated-film	
Newtonian liquid viscosity	
1 centipoise	400
100 centipoises	300
10,000 centipoises	120

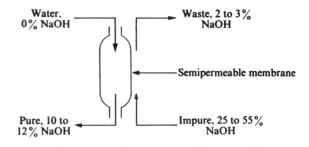

Fig. 12.1 Typical dialysis flow diagram.

12.2 Dialysis

Dialysis is the separation of solutes by means of their unequal diffusion through membranes [2, 6, 9, 12, 13, 27–30, 32, 34]. It is most useful in recovering pure solutions for reuse in manufacturing processes, for example, caustic soda in the textile industry [17]. Recovery involves separation of a crystalloid (NaOH) from a sol in which about 96 per cent of the impurities are in the form of hemicellulose and the rest include pectins, waxes, and dyes.

There are some eight to ten commercial dialyzers presently on the market. In our example, they all operate on the simple principle of passing a concentrated, impure caustic solution upward, countercurrent to a downstream water supply, from which it is separated by a semipermeable membrane (Fig. 12.1). The caustic soda permeates the membrane and goes into the water more rapidly than the other impurities contained in the waste. The concentration of caustic is always greater in the impure solution than in the water, and the water which flows through the membrane into the impure caustic solution tends to dilute it. The quantity of sodium hydroxide diffusing through the diaphragm depends upon the time, the area of the dialyzing surface, the mean concentration difference, and the temperature. These factors are expressed in the equation

$$Q = KAt\,(\Delta c),$$

where K is the overall diffusion coefficient, t is the time in minutes, A is area of dialyzing surface, and

$$\Delta c = \Delta c_{av} = \frac{(\Delta c_1 - \Delta c_2)}{2.3\log_{10}\Delta c_1/\Delta c_2},$$

where Δc_1 and Δc_2 are the differences in concentration between the two solutions at the top and bottom of the diaphragm, respectively.

When one actually computes the weight of NaOH recovered, it becomes apparent that the quality and type of diaphragm are of paramount importance. This is evident from the following equation:

$$W = UA \, \Delta c_{\text{log mean}},$$

where W is the weight of material passing through the membrane in a unit of time (gm/min), U is the overall dialysis coefficient, and $\Delta c_{\text{log mean}}$ is the logarithmic mean concentration gradient across the membrane ($=\Delta c_{\text{av}}$). Also,

$$\frac{1}{U} = \frac{1}{U_1} + \frac{1}{U_2},$$

where U_1 is the combined film resistance (cm/min) and U_2 is the membrane resistance (cm/min). Each diaphragm shows a different membrane resistance (U_2). The restrictive characteristics of porous membranes are due to both a mechanical sieve action and a physicochemical interaction between solute, solvent, and membrane. Cellulose nitrate, parchment, and cellophane are the principal membranes in use today.

Smith and Eisemann [32] present an excellent evaluation of electrodialysis. Dialysis is an operation requiring very little operator attention and, although its main role is to conserve raw materials and to reduce plant waste, at the same time it aids in waste treatment. With the introduction of acid-resistant membranes, dialysis has been used successfully in the recovery of sulfuric acid in the copper, stainless-steel, and other industries. Some operations can recover as much as 70 to 75 per cent of the acid, but a recovery of as little as 20 per cent may justify the process. In dialysis the driving force of separation is natural diffusion because of concentration gradient; in electrodialysis this natural force is enhanced by the application of electrical energy. McRae [15] found that, for a secondary effluent containing 900 ppm of dissolved solids, electrodialysis could achieve 44 per cent reduction, with costs ranging from 10 to 15 cents per 1000 gallons. He finds this process useful for treating the wastes from dairies.

Due to the resulting improvements in mechanical and thermal properties, ceramic-membrane filters now generally outperform polymer membranes, their established competitor (High Technology/August 1987). Ceramic membranes are made by blending dry mineral powders, such as alumina, spinel, cordierite, and zirconia, in various proportions with a solvent to form either a slurry, which is poured into a mold, or a dough, which is extruded. The resulting configurations, either hollow fibers, flat plates, or honeycombs, are then dried and fused together. Layers of supporting material are added to complete the filter. Because ceramic membranes tolerate acids and bases, they can be more thoroughly cleaned than other designs. Also, these membranes can be cleaned with steam up to 140° C., whereas polymeric membranes cannot be steam-cleaned. Ceramic membranes can also be heated up to 500°C. to burn out impurities trapped during filtration. A major drawback is their brittleness, as well as their relatively high cost.

12.3 Ion Exchange

Ion exchange is basically a process of exchanging certain undesirable cations and anions of the waste water for sodium, hydrogen, or other ions in a resinous material. The resins, both natural and artificial, are commonly referred to as *zeolites*. The ion-exchange process was originally developed to reduce hardness in domestic water supplies, but has recently been used to treat industrial waste waters, such as metal-plating wastes. The softening reactions may be illustrated as follows [22]:

$$\left.\begin{matrix}\text{Ca}\\ \text{Mg}\end{matrix}\right\}\begin{matrix}(\text{HCO}_3)_2\\ \text{SO}_4\\ \text{Cl}_2\end{matrix} + \text{Na}_2\text{Z} \longrightarrow \left.\begin{matrix}\text{Ca}\\ \text{Mg}\end{matrix}\right\}\text{Z} + \begin{matrix}2\text{NaHCO}_3\\ \text{or}\\ \text{Na}_2\text{SO}_4\\ \text{or}\\ 2\text{NaCl}\end{matrix}$$

where Z is the symbol for the zeolite radical. When the ability of the zeolite bed to produce soft water is exhausted, the softener is temporarily cut out of service. It is then backwashed to cleanse and hydraulically regrade the bed, regenerated with a solution of common salt, which removes the calcium and magnesium in the form of their soluble chlorides and simultaneously restores the zeolite to its original condition, rinsed free of these and the excess salt, and finally returned to service. The reaction may be indicated as follows:

$$\left.\begin{matrix} Ca \\ Mg \end{matrix}\right\}Z + 2\,NaCl \rightarrow \left.\begin{matrix} Ca \\ Mg \end{matrix}\right\}Cl_2 + Na_2Z.$$

Ion exchange as a means of waste treatment is only a new application of a traditional method of water softening. If the proper approach is used, it offers great potential for material and water conservation. For instance, in the treatment of metal-plating wastes [16], rinse water is passed through beds of cationic and anionic resins selected for the particular application and the deionized water is then recycled through the rinse tank. This method may be applied on a continuous basis to the removal of contaminating metals [16] from chromic-acid solutions, permitting the return of pure chromic-acid solution to the process tank. In the case of nickel- and copper-plating solutions, both the contaminating metals and the metal to be plated are cationic, and therefore all will be extracted. Cation-exchange resins are suggested [21] for use in the steel industry to remove the iron from spent liquor and to recover sulfuric acid and iron oxide for further use. Unless the aim of the procedure is recovery of metals, ion exchange becomes simply a concentration method, and some treatment for the regenerated solution must be devised.

Walther [36] reports the use of a continuous ion-exchange unit, consisting of a stainless-steel loop divided into sections by butterfly valves, which successfully removed over 700 mg/liter of dissolved inorganic solids. The unit contains about 15 cubic feet of ion-exchange resin which moves around the loop in about three minutes. When the resin becomes saturated with hardness, it is removed from the loop and regenerated resin is exchanged. The spent resin is then regenerated and returned to the loop on the next cycle.

Organic matter and pH have a pronounced effect on the operation and efficiency of resin beds; the leaching of organic matter from certain resins may have a detrimental effect on the metals plated. Chemicals used for regenerating resin beds may also require special treatment before disposal.

General appraisal. Demineralization (ion exchange) is most useful when water of the highest quality is required, but it involves complex chemical reactions and therefore requires careful operation and supervision at all times. Furthermore, ion-exchange processes sometimes utilize chemicals which are hazardous to personnel and equipment. These are matters to think about before selecting an ion-exchanger system in preference to an evaporator; although evaporators, too, are uneconomical in certain instances, e.g. when the flow is light. Dialysis is normally economical and can compete in efficiency with both evaporation and ion exchange when the recovery of a pure compound is considered essential. The decision whether to use evaporation or demineralization can be intelligently made only after a thorough evaluation of the heat balance of the plant and expected operating conditions [26]. These factors, as well as operating costs, must be considered in relation to the capital investment needed for either system.

12.4 Algae

Algae require nine minor essential elements (Fe, Mn, Si, Zn, Cu, Co, Mo, B, and Va) and seven major essential elements (C, N, P, S, K, Mg, and Ca) for their optimum growth. The use of algae for removing minerals from waste waters have been investigated; most investigations have been carried out on sewage effluents. One such study carried out in the author's laboratory [7] involved a suburban housing-development treatment plant and utilized primary sedimentation, trickling filtration, and stabilization ponds. Although the sedimentation and filtration did not remove any phosphorus, the algae actively growing in the ponds caused a reduction of about 42 per cent of the phosphate content. Other mineral concentrations were not measured.

If this method is used to remove minerals such as phosphate over a period, algae must also be removed from the effluent before this is released into a stream used for water supplies and recreation. Golueke and Oswald [7] observed three steps in harvesting oxidation-pond algae: (1) collection and initial concentration, (2) dewatering or secondary concentration, and (3) final drying. They found chemical precipitation and centri-

Table 12.2 Elemental composition of green algae. (After Krauss [10].)

Element	Range of dry weight, %
Chlorella	
Carbon	51.4–72.6
Hydrogen	7.0–10.9
Oxygen	11.6–28.5
Scenedesmus	
Nitrogen	2.2–7.7
Phosphorus	1.1–2.0
Sulfur	0.28–0.39
Magnesium	0.36–0.80
Potassium	0.85–1.62
Calcium	0.005–0.08
Iron	0.04–0.55
Zinc	0.0006–0.005
Copper	0.001–0.004
Cobalt	0.000003–0.0003
Manganese	0.002–0.01

fugation to be most economical. The harvested algae can be sold as animal feed supplements. Oswald [25] describes *Chlorella* and *Scenedesmus* as the most active algae in stabilization ponds, because they are extremely hardy. Krauss [10] presents the elemental composition of these two algal types to validate their fixation of minerals (Table 12.2).

Table 12.2 shows the extent to which algae take up minerals from any solution in which they grow. In fact, the continued photosynthesis of algae depends directly on the ability of the culture medium (waste water) to supply these inorganic components over a long period, at a rate sufficient to support the growth potential of the algae. There is some evidence that the uptake (and the algal growth) depends on the availability, as well as the presence, of inorganic nutrients. Thus, insolubility and colloidal characteristics of the nutrients may hamper algal growth, but hardness in waste waters can contribute to it. A statistical study of Massachusetts lakes and reservoirs carried out in 1900 showed that the hard water supplies yielded more algae than the soft (Table 12.3). Bogan [3] capitalized on the ability of algae to utilize phosphorus in providing tertiary treatment of the sewage from Seattle, Washington,

which utilized both algal activity and lime and which removed over 90 per cent of the phosphorus in the secondary sewage-plant effluent. Oxidation-pond usage has been increasing since the advent of lower-cost mechanical aeration.

12.5 Reverse Osmosis

Reverse osmosis is a membrane permeation process for separating relatively pure water or some other solvent from a less-pure solution. The solution is passed over the surface of a specific semipermeable membrane at a pressure in excess of the effective osmotic pressure of the feed solution. The permeating liquid is collected as the product and the concentrated feed solution is generally discarded. The membrane must be (1) highly permeable to water, (2) highly impermeable to solutes, (3) capable of withstanding the applied pressure without failure, (4) as thin as possible consistent with the strength requirement, (5) chemically inert, mechanically strong, and creep resistant, and (6) capable of being fabricated into configurations of high surface-to-volume ratios. A number of commercial units in practice treat brackish waters of less than 2000 to 3000 ppm of total dissolved solids.

Although several types of membranes have been developed, two types of membranes are generally used in commercial equipment. The first is a symmetric of "skinned" cellulose acetate membranes made in flat or tubular forms. Generally the membranes are approximately 100 μm thick with a surface skin of about 0.2 μm which acts as the rejecting surface. Typically they operate at 40 to 50 atmospheres pressure and produce a waterflux of 10 to 20 gallons of water per square foot per day with a salt-rejection of about 95 per cent. These membranes have generally exhibited a decrease in flux rate with time due to both compaction (creep) and fouling of the membrane. Therefore, operating pressures are kept low to avoid creep, and suspended solids in the feed solution are kept as low as possible to prevent fouling.

The second type of membrane is an aromatic polyamide, or polyamide-hydrazide. The membrane in commercial units is in the form of hollow fine fibers. The patented membrane is claimed to operate at 27 atmospheres pressure and in a water flux of 1 to 2 gallons per square foot per day with a salt (NaCl) rejection of about 95 per cent.

Some limitations of existing membranes other than the total dissolved solids concentration are described by Lonsdale and Podall [38] in the following manner: (1) The relatively high cost of operation could be reduced if the water flux could be substantially increased without loss in salt-rejection or other properties. (2) Flux decline is serious with high-flux membranes. (3) Certain species are inadequately rejected—for example, boric acid, phenol, and nitrates. (4) For certain applications, existing membranes are not sufficiently resistent to chemical or microbiological attack, or their mechanical or thermal stability is inadequate. (5) Feedwater pH should generally be on the acidic side (pH 5 to 7) for best operation and minimize membrane hydrolysis. For industrial waste treatment some prelim-

inary promising results have been reported with sulfite, kraft pulping, and textile dyeing wastes.

Substantial energy needed to constantly pressurize the incoming salty water is one of the drawbacks of the reverse osmosis system. Reliable Water Company (see Figure 12.2) system uses an energy recovery mechanism that reclaims most of the fluid pressure from the brine waste as it leaves the system. Using hydraulic oil pumps transfer barriers and special valves, the system extracts the pressure in the brine and transfers it to the incoming waste salt water, substantially reducing the energy requirements.

Osantowski and Geinopolos (39, 1979) obtained excellent rejections of dissolved solids for desalting processes after pretreating deinking paper mill and

Table 12.3 Occurrence of *Cyanophyceae* and *Chlorophyceae* in Massachusetts lakes and reservoirs [36].

Characteristic	Chemical analysis, ppm	Often above 1000/cm³		Below 100/cm³	
		Cyano-phyceae	*Chloro-phyceae*	*Cyano-phyceae*	*Chloro-phyceae*
Color	0–30	2	2	11	0
	30–60	2	2	3	1
	60–100	3	1	7	2
	> 100	0	0	1	1
Chlorides (excess above normal)	0	2	1	3	1
	0.1–0.3	1	1	10	5
	0.4–2.5	1	0	9	6
	> 2.5	3	3	0	0
Hardness	0–5	0	0	6	4
	5–10	2	1	10	5
	10–20	2	1	5	2
	> 20	3	3	1	1
Albuminoid ammonia (dissolved)	0–0.10	0	0	4	3
	0.1–0.15	0	0	6	4
	0.15–0.20	2	2	7	3
	> 0.20	5	3	5	2
Free ammonia	0–0.01	0	0	10	4
	0.01–0.03	0	0	8	5
	0.03–0.10	3	2	4	3
	> 0.1	4	3	0	0
Nitrates	0–0.05	1	0	12	6
	0.05–0.10	3	2	10	6
	0.10–0.20	1	0	0	0
	> 0.20	2	3	0	0

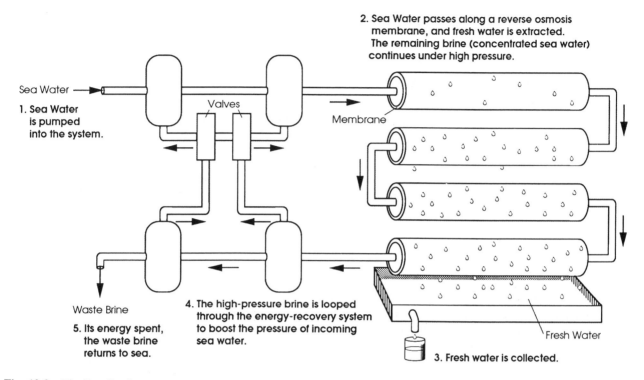

2. Sea Water passes along a reverse osmosis membrane, and fresh water is extracted. The remaining brine (concentrated sea water) continues under high pressure.

Sea Water →

1. Sea Water is pumped into the system.

Valves

Membrane

Waste Brine

5. Its energy spent, the waste brine returns to sea.

4. The high-pressure brine is looped through the energy-recovery system to boost the pressure of incoming sea water.

3. Fresh water is collected.

Fresh Water

Fig. 12.2 The Desalination Process Using Energy Recovery and Artificial Intelligence Control

slaughterhouse and meat packing wastes. They used reverse osmosis, ion exchange and electrodialysis. They found reverse osmosis was the most economical process for providing reusable quality water in both the papermill and food products plants. Reuse quality requirements could be met in most cases by blending the reverse osmosis product water with undesalted wastewater. On the other hand, electrodialysis provided the optimum performance of the three desalting technologies investigated at an organic chemicals plant.

12.6 Miscellaneous Methods

Chemical precipitation or coagulation have been used to remove some inorganic matter from waste waters. For example, elevated pH values aid in the removal of heavy metals by precipitation of the hydroxide or carbonate, and, under some conditions, treatment of waste waters with calcium hydroxide is reasonably

effective in the removal of nitrogen and phosphorus.

Oxidation-reduction chemical reactions are used in certain cases to alter inorganic matter and thus enhance its removal. For example, chromate must be reduced, usually with ferrous sulfate or sulfur dioxide under acid conditions, to the trivalent form as a preliminary to precipitation with lime and subsequent removal as a chromic-hydroxide sludge (see reactions on page 508. Likewise, cyanides must be completely oxidized, usually with chlorine under alkaline conditions, to split them up into harmless and volatile nitrogen gas and carbonate ions (see reaction on page 507).

The ultimate disposal or salts has always been and still remains a major problem to the environmental engineer. One novel suggestion for the use of concentrated salt waste has been proposed (*New York Times* page A9, August 3, 1981). Salt solutions with at least 10% salt content can be added to the warm surface waters of lakes. The increased density of these warm sur-

Table 12-4.   Refractory Containment Removal Techniques

Membrane filtration	Evaporation	Absorption
Limitations (a) Life of membrane (b) Loss in flux rate (c) Relatively small amount of effluent which can be produced (d) Relatively limited type of materials which can be removed	Requires source of relatively inexpensive energy input Wastewater should be very high in solids content Wastewater could cause scaling of pipes in evaporator thus interfering with heat transfer	Cost of new carbon is relatively high Can clog quite easily with finely-divided suspended solids Cost of regenerating equipment is high, but carbon costs are reduced to $\frac{1}{10}$ of above costs by regeneration of carbon
When to use With a relatively small volume of primarily two component contaminants; when one component is quite valuable; when recovered in relatively pure form. Each component should be as different as possible in molecular size and noncorrosive to membranes	With a high solids, noncorrosive, nonscale-forming waste; when an inexpensive source of heating is available	With a highly soluble, single-state waste which, however, has a high enough molecular weight to be easily removed by adsorption. Preferably, the contaminant should be organic and can either be recovered by solvent extraction if valuable, or destroyed by burning if not valuable

face waters tends to cause them to sink to the bottom where they stay and serve as a reservoir for hot water energy. The hot water energy is pumped out periodically to drive turbines to produce electrical energy. The hot water can also be used for space heating or for agricultural or industrial processes.

12.7 Refractories

Refractories can be defined for our purpose as solids in wastewater—generally in the dissolved stage—which are not susceptible to removal by the usual "primary" or "secondary" treatment techniques, including those of chemical coagulation. They may have to be removed because (1) the increased water reuse results in a gradual build-up in water supplies downstream and deterioration of the water for its best

usage, and (2) we are learning more about the potential hazardous effects of refractories in water. For examples see the following list:

a) fluorides causing mottled teeth;
b) nitrates causing methemoglobenemia;
c) ABS interfering with surface reaeration and adding tastes to water;
d) metals causing blood poisoning;
e) certain insecticides and benzene-structured organics causing potential nerve damage and/ or carcinogenic reactions.

At the present time it is feasible to remove these refractory materials by (1) membrane filtration, (2) evaporation, or (3) adsorption. The theories of these methods are described in Chapters 11 and 12. Some of the major limitations and potentials for use are given in Table 12.4 derived by the authors.

References

1. Amberg, H. R., "The effect of nutrients upon the rate of stabilization of spent sulfite liquor in receiving waters," *Proc. Am. Soc. Civil Engrs.* **81**, 821 (1955).
2. Bassett, H. P., "Super filtration by dialysis," *Chem. Met. Eng.* **42**, 254 (1938).
3. Bogan, R. H., *Pilot Plant Evaluation of a Tertiary Stage Treatment Process for Removing Phosphorus from Sewage*, A report prepared for the city of Seattle (December 1959).
4. Brown, G. G., D. Katz, A. S. Foust, and R. Schneidewind, *Unit Operations*, John Wiley & Sons, New York (1950), p. 484.
5. Bryson, J. C., *Control of Algae through Phosphate Control*, Unpublished report, Syracuse University, Syracuse, N.Y., September 1961.
6. Eynon, D. J., "Operation of Cerini dialysers for recovery of caustic soda solutions containing hemicellulose," *J. Soc. Chem. Ind.* **52**, 173T (1933).
7. Golueke, C. G., and W. J. Oswald, "Harvesting and processing sewage grown planktonic algae," *J. Water Pollution Control Federation*, **37**, 471 (1965).
8. Keating, R. J., and R. Dvorin, "Dialysis for acid recovery," in Proceedings of Industrial Waste Conference, Purdue University, 1960, pp. 567–76.
9. Kirk, R. E., and D. F. Othmer, *Encyclopedia of Chemical Technology*, Interscience, New York (1950) p. 5.
10. Krauss, R. W., "Photosynthesis in the algae," *Ind. Eng. Chem.* **48**, 1449 (1956).
11. Kunin, R., and F. Y. McGarvey, "Status of ion exchange technology," *Ind. Eng. Chem.* **55**, 51 (1963).
12. Lee, J. A., "Caustic soda recovery in rayon industry," *Chem. Met. Eng.* **42**, 483 (1935).
13. Lovett, L. E., "Application of osmosis to recovery of caustic soda solutions containing hemicelluloses in rayon industry," *Trans Electrochem. Soc.* **73**, 163 (1938).
14. McCabe, W. L., and J. C. Smith, *Unit Operations of Chemical Engineering*, McGraw-Hill Book Co., New York (1956) p. 530.
15. McRae, W. A., "Electrodialysis in wastewater reclamation," in Proceedings 2nd Water Quality Research Symposium, New York State Department of Health, April 14, 1965, pp. 97–119.
16. Merrill, G. R., A. R. Macommer, and H. R. Mansberger, *American Cotton Handbook*, 2nd ed., Textile Book Publishers, New York (1949).
17. Michalson, A. W., and C. W. Burhans, Jr., "Chemical waste disposal by ion exchange," *Ind. Water Wastes* **1**, 11 (1962).
18. Nemerow, N. L., and J. C. Bryson, "How efficient are oxidation ponds?" *Wastes Eng.* **34**, 133 (1963).
19. Nemerow, N. L., and W. R. Steele, "Dialysis of caustic textile wastes," in Proceedings of 10th Industrial Waste Conference, Purdue University, May 1955, pp. 74–81.
20. *Textile Wastes–A Review*, New England Interstate Water Pollution Control Commission, December 1950.
21. "New process developed to recover acid and iron from spent pickle liquor," *Iron Steel Engr* **42**, 167 (1965).
22. Nordell, E., *Water Treatment*, Reinhold, New York (1951), p. 341.
23. Ohio River Valley Water Sanitation Commission, *Methods for Treating Metal Finishing Wastes*, January 1953, p. 58.
24. Okey, R. W., and P. L. Stavenger, "Membrane technology. A process report," *Ind. Water Eng.* **4**, 36 (1967).
25. Oswald, W. J., "Fundamental factors in oxidation pond design," in Conference on Biological Waste Treatment, at Manhattan College, New York, April 20–22, 1960, Paper no. 44.
26. Paulson, C. F., "Chromate recovery by ion-exchange," in Proceedings of 7th Industrial Waste Conference, Purdue University, 1952, p. 209.
27. Powell, S. T., *Water Conditioning for Industry*, McGraw-Hill Book Co., New York (1954), p. 214.
28. "Reverse osmosis. An old concept in new hardware," *Ind. Water Eng.* **4**, 20 (1967).
29. Roetman, E. T., "Viscose rayon manufacturing wastes and their treatment," *Water Works Sewerage* **91**, 295 (1944).
30. Roetman, E. T., "Stream pollution control at Front Royal, Virginia, rayon plant," *Southern Power Ind.* **62**, 86 (1944).
31. Rudolfs, W., "A survey of recent developments in the treatment of industrial wastes," *Sewage Works J.* **9**, 998 (1937).
32. Smith, J. D., and J. L. Eisemann, "Electrodialysis in

waste water recycle," in Proceedings of 19th Industrial Waste Conference, Purdue University, 1964, pp. 738–760.

33. U.S. Department of Health, Education and Welfare, *Cost of Purifying Municipal Waste Waters by Distillation*, Publication no. AWTR-6, Washington, D.C. (1963).

34. U.S. Department of Health, Education and Welfare, *Advanced Waste Treatment Research*, AWTR-14 S Summary Report (PHS Publication NQ 999-WP-24), Washington, D.C. (1965).

35. Volbrath, H. B., "Applying dialysis to colloid-crystalloid separations," *Chem. Met. Eng.* **43**, 303 (1936).

36. Walther, A. T., "LaGrange tests ion exchange unit," *Water Sewage Works* **112**, 212 (1965).

37. Whipple, G. C., *Microscopy of Drinking Water*, John Wiley & Sons, New York (1948), pp. 214–215.

38. Lonsdale, H. K., and H. E. Podall, "Reverse osmosis membrane research in a symposium," June 1971, New York: Plenum Press (1972).

39. Osantowski, R. and Ceinopoloc, A., "An Evaluation For Water Reuse in Selected Industries Using Advanced Waste Treatment Processes," Proc. Ind. Wastes Symposium, 52nd Ann. WPCF Conf., Houston, Texas, 1-12 (Oct. 1979).

Questions

1. What percentage of industrial plants practice removal of minerals?
2. What are the significant minerals present in industrial wastes?
3. Name six major methods whereby inorganic ions can be removed?
4. Describe the principles and problems in using evaporation and its major use.
5. Describe the principles and problems in using dialysis and its major use.
6. Describe the principles and problems in using ion-exchange and its major use.
7. What are algae used for and when are they most useful in waste treatment?
8. What miscellaneous methods are also available for removing minerals?

CHAPTER 13

REMOVAL OF ORGANIC DISSOLVED SOLIDS

The removal of dissolved organic matter from waste waters is one of the most important tasks of the waste engineer, and, unfortunately, also one of the most difficult. These solids are usually oxidized rapidly by microorganisms in the receiving stream, resulting in loss of dissolved oxygen and the accompanying ill effects of deoxygenated water. They are difficult to remove because of the extensive detention time required in biological processes and the elaborate and often expensive equipment required for other methods. In general, biological methods have proved most effective for this phase of waste treatment, since bacteria are adept at devouring organic matter in wastes, and the greater the bacterial efficiency the greater the reduction of dissolved organic matter. Microorganisms, however, are quite "temperamental" and sensitive to changes in environmental conditions, such as temperature, pH, oxygen tension (level of oxygen concentration), mixing, toxic elements or compounds, and character and quantity of food (organic matter) in the surrounding medium. It is the responsibility of the engineer to provide optimum environmental conditions for the proliferation of the particular biological species desired.

There are many varieties of biological treatment, each adapted to certain types of waste waters and local environmental conditions such as temperature and soil type. Some specific processes for treating organic matter are: (1) lagooning in oxidation ponds; (2) activated-sludge treatment; (3) modified aeration; (4) dispersed-growth aeration; (5) contact stabilization; (6) high-rate aerobic treatment (total oxidation); (7) trickling filtration; (8) spray irrigation; (9) wet combustion; (10) anaerobic digestion; (11) mechanical aeration system; (12) deep well injection; (13) foam phase separation; (14) brush aeration; (15) subsurface disposal; and (16) the Bio-Disc system.

13.1 Lagooning

Lagooning in oxidation ponds is a common means of both removing and oxidizing organic matter and waste waters as well. More research is needed on this method of treatment, which originally developed as an inexpensive procedure for ridding industry of its waste problem. An area adjacent to a plant was excavated, and waste waters either flowed or were pumped into the excavation at one end and out into a receiving stream at the other end. The depth of the lagoon depended on how much land was available, the storage period desired or required, and the condition of the receiving stream. Little attention was paid originally to the effect of depth on bacterial efficiency. In fact, reduction of dissolved organic matter was usually not anticipated or even desired, since it was presumed, and with good reason, that biological degradation of organic matter would lead to oxygen depletion and accompanying nuisances from odors. Thus, the lagoons served solely to settle sludge and equalize the flow. Now, modern techniques have led to new theories about the stabilization of organic matter in lagoons.

We now know that stabilization or oxidation of waste in ponds is the result of several natural self-purification phenomena. The first phase is sedimentation. Settleable solids are deposited in an area around the inlets to the ponds, the size of the area depending on the manner of feeding in the waste and location of the inlet. Some suspended and colloidal matter is precipitated by the action of soluble salts; decomposition of the resulting sediment by microorganisms changes the sludge into inert residues and soluble organic substances, which in turn are required by other microorganisms and algae for their metabolic processes.

Decomposition of organic material is the work of microorganisms, either aerobic (living in the presence of free oxygen) or anaerobic (living in absence of free oxygen). In a pond in which the pollution load is exceedingly high or which is deep enough to be void of oxygen near the bottom, both types of micro-organism may be actively decomposing organic material at the same time. A third type of micro-organism, the facultative anaerobic, is capable of growth under either aerobic or anaerobic conditions and aids in decomposing waste in the transition zone between aerobic and anaerobic conditions. It is desirable to maintain aerobic conditions, since aerobic microorganisms cause the most complete oxidation of organic matter. Anaerobic fermentation has proved effective for treating citrus, slaughterhouse, and certain paper-mill wastes, while aerobic bacteria have been most effective in oxidizing dairy, textile, and other highly-soluble organic wastes.

Table 13.1 gives a general scheme of the microbial degradation of the organic constituents in sewage. It also points out the difference between aerobic and anaerobic decomposition.

Algae are significant in stabilization ponds in that they complete nature's balanced plant-animal cycle. Whether seasonal or perennial, algae utilize CO_2, sulfates, nitrates, phosphates, water, and sunlight to synthesize their own organic cellular material and give off free oxygen as a waste product. This oxygen, dissolved in pond water, is available to bacteria and other microbes for their metabolic processes, which include respiration and degradation of organic material in the pond. Thus, we have a completed cycle in which: (a) microorganisms use oxygen dissolved in the water and (b) break down organic waste materials to produce (c) waste products such as CO_2, H_2O, nitrates, sulfates, and phosphates, which (d) algae use as raw materials in photosynthesis, thereby (e) replenishing the depleted oxygen supply and keeping conditions aerobic, so that the microorganisms can function at top efficiency (see Fig. 13.1). However, one drawback of algae should be mentioned, namely, when they die, they impose a secondary organic loading on the pond. Another disadvantage is

Table 13.1   Biological degradation of organic constituents in sewage.

Substance decomposing	Class of microbial enzymes	End-products	
		Anaerobic decomposition	Aerobic decomposition
Proteins	Proteinase*	Amino acids Ammonia Hydrogen sulfide Methane Carbon dioxide Hydrogen Alcohols Organic acids Phenols Indols	Ammonia, nitrites, nitrates Hydrogen sulfide, sulfuric acid Alcohols Organic acids Carbon dioxide Water
Carbohydrates	Carbohydrase*	Carbon dioxide Hydrogen Alcohols Fatty acids	Alcohols Fatty acids Carbon dioxide Water
Lipids (fats)	Lipase*	Fatty acids Carbon dioxide Hydrogen Alcohols	Fatty acids and glycerol Alcohols Carbon dioxide Water

*Class of enzymes only. Dozens of enzymes may be utilized in this degradation.

a seasonal one: algae are less effective in winter.

Ice and snow cover during winter months interferes with the stabilization process in the following manner:

1) It prevents sunlight from penetrating the pond, causing a reduction in the size and number of algae present. Algae are not necessarily killed by the absence of sunlight (those known as facultative chemo-organotrophs can carry on metabolic processes despite darkness), but they release little or no oxygen without sunlight.

2) It prevents mixing and reaeration by wind action.

3) It prevents reaeration by atmosphere–water dynamic equilibrium phenomena.

4) It usually results in anaerobic conditions if it continues over an extended period of time.

These factors tend to result in a lowered pond or lagoon efficiency during the winter.

Hermann and Gloyna [10] (disregarding the part played by minerals) describe the reaction in high-rate ponds in which sewage is oxidizing as:

$$C_{11}H_{29}O_7N + 14O_2 + H^+ \rightarrow 11CO_2 + 13H_2O + NH_4^+.$$

The canning industry, one of the first to attempt lagooning, soon found it difficult to maintain aerobic conditions in basins; other industries experienced similar situations. As industries became aware that biological degradation occurs in lagoons, they made attempts to encourage and control the oxidation and began to refer to such lagoons as waste-oxidation basins.

Most modern oxidation basins have a maximum water depth of four feet and operate on a continuous-flow basis. Engineers try to maintain in the basin near-neutral pH, adequate oxygen concentration, and sufficient nutrient minerals for biological oxidation. Chemical neutralizers are used to alter pH values, oxygen concentrations are maintained by reducing detention times and using shallow basins, and mineral-salts nutrients may be added as needed, to accelerate biological activity. BOD removals range from as low as 10 per cent to as high as 60 to 90 per cent.

In an interesting full-scale study [26], the author treated an air-base oxidation pond, at 43° north latitude with ice cover during the winter, with an elevated loading of 130 pounds of BOD in the waste water per acre of pond area. The BOD reductions at these relatively high loadings ranged from 87.7 per cent in August to 53 per cent in January, with a yearly average of 69.3 per cent. In another pilot-plant study [25], the writer achieved BOD removals in excess of 80 per cent, using close-baffled four-foot-deep, or unbaffled eight-foot-deep, basins, during the critical summer period in central New York State, at elevated loadings of 312 to 467 pounds per acre per day. A photograph of the five parallel pilot-plant basins appears as Fig. 13.2.

Oswald [30] believes that in such heavily loaded ponds, particularly during periods when methane fermentation is either nonexistent or limited by temperature and when algal photosynthesis is not taking place in the surface layers, a buildup of organic acid occurs, with a subsequent lowering of the pH level and emission of hydrogen sulfide from the pond. The writer, however, did not experience these odors, even at the high loadings described above. Oswald offers the explanation that, if methane fermentation becomes established in the bottom deposits, high rates of BOD removal may be attained without

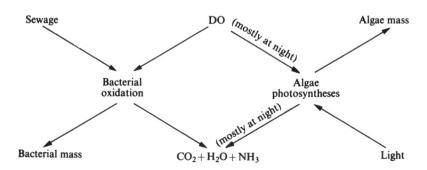

Fig. 13.1 The role of algae in stabilization ponds [37].

Fig.   13.2 Accelerated-oxidation pilot-plant basins [25].

appreciable odors. He also believes that ponds in which both photosynthetic oxygenation and methane fermentation occur (facultative ponds) must be restricted to about 50 pounds of BOD per acre per day, because conditions are at times unfavorable for either process. The author, at this point, does not necessarily agree with these findings. Furthermore, Oswald's later high-rate oxidation ponds for treating sewage in warmer climates have been loaded to over 600 pounds of BOD per acre per day or over, being aerated for an hour each midnight.

The reader is referred to the discussion in Section 12.4 of the necessity of preventing algae growth in bodies of water that are used for water supplies and recreational activities.

13.2 Activated-Sludge Treatment

The activated-sludge process has proved quite effective in the treatment of domestic sewage, as well as a few industrial wastes from large plants. In this process, biologically active growths are created, which are able to adsorb organic matter from the wastes and convert it by oxidation–enzyme systems to simple end-products like CO_2, H_2O, NO_3, and SO_4. Biological slimes develop naturally in aerated organic wastes which contain a considerable portion of matter in the colloidal and suspended state, but for the efficient removal of organic dissolved solids there must be high floc concentrations, to provide ample contact surface for accelerated biological activities. The flocs (zoogleal masses) are living masses of organisms, food, and slime material and are highly active centers of biological life—hence the term "activated sludge." They require food, oxygen, and living organisms in a delicately controlled environment.

Various degrees of efficiency are obtained by controlling the contact period and/or the concentration of active floc. The contact period can be regulated by careful design of the hydraulic systems of aeration basins, the average time of aeration being 6 hours for domestic sewage and 6 to 24 hours for various industrial wastes. The desired concentration of active floc is maintained by recirculating a specific volume of secondary settled sludge, normally about 20 per cent.

Higher sludge quantities lead to greater BOD removal and create a need for more air and food (organic matter) for proper balance. Also, "old," heavy sludge tends to become mineralized and devoid of oxygen, which results in a less-active floc. The reverse is true of a "young," light, sludge floc. The "age" of the growths, therefore, becomes an important consideration.

Busch [4] summarizes the situation by saying that for optimum activity the kinetics of activated sludge require: a young, flocculent sludge in the logarithmic stage of growth; maintenance of the logarithmic growth state by controlled sludge wastage; continuous loading of the organisms; and elimination of anaerobic conditions at any point in the oxidative treatment.

Hazeltine [9] has said of the present status of domestic-sewage activated-sludge treatment that BOD removals are usually above 90 per cent when the loadings are below 0.3 pound of BOD per pound of suspended solids in the waste under aeration. Efficiencies are difficult to predict when these loadings are increased to 0.5 pound per pound. Normally, the BOD loading is related to the aeration-tank capacity; about 30 to 35 pounds of BOD per 1000 cubic feet can be treated in plants with about 2000 ppm of suspended solids under aeration.

Sawyer [34] lists the limitations of the domestic-sewage activated-sludge process as follows: BOD loadings are limited to about 35 pounds per 1000 cubic feet of tank capacity, thus requiring relatively long detention time and resulting high capital investment; there is a high initial oxygen demand by the mixed liquors; there is a tendency to produce bulking sludge; the process cannot produce an intermediate quality of effluent; high sludge-recirculation ratios are required for high-BOD wastes; there are high solids loadings on final clarifiers; and large air requirements accompany the process.

The Kraus process [17] attempts to overcome some of the sludge-bulking problems of conventional activated-sludge plants by controlling the sludge volume index (a measure of the volume occupied by one gram of suspended solid). The process is similar to that of conventional activated-sludge treatment, employing separate reaeration for sludge, except that some digester sludge, digester supernatant, and activated sludge are aerated together for as much as 24 hours, in what he terms a nitrifying aeration tank.

BOD loadings as high as 170 pounds per 1000 cubic feet per day have recently been used, with removals near 90 per cent [18].

Von der Emde [7] notes that ciliated and flagellated protozoa, as well as bacteria, are normally prevalent in activated sludge. When the BOD loading is high or very low, flagellates replace the ciliates, regardless of the level of oxygen present. Where there are short aeration periods or when only traces of oxygen are maintained, bacteria only are observed in the sludge.

Many characteristics of industrial wastes—e.g. toxic metals, lack of nutrients required for biological oxidation, organic nondegradable matter, high temperature, and high or low pH values—give rise to problems requiring careful analysis. When the suitability of this process for a particular industrial waste is in question, laboratory and/or field pilot plant will yield the results necessary for decision making.

Heukelekian (1941) believed that bulking of activated sludge should result when conditions are not so unfavorable as to destroy the purification mechanism and yet sufficiently unfavorable as to bring a shift in the delicate biological balance. One of these unfavorable conditions should be an inadequate oxygen supply. If the oxygen supply in relation to the demand becomes inadequate, Sphaerotilus and other filamentous organisms attain the ascendency and the sludge becomes bulking. Biochemical activities of these organisms would bring about the purification in a way similar to the desirable sludge organisms except with lower and more efficient oxygen utilization at the lower tensions. In other words when the sludge is diffuse and filamentous, it exposes more surface which might enable the sludge to obtain the limited amount of oxygen present in the medium immediately surrounding it. This type of sludge, however, usually produces a sparkling effluent. Dairy wastes exhibit this problem, which has been overcome by the addition of ammonium chloride to provide a more favorable carbon-to-nitrogen ratio.

13.3 Modified Aeration

Modified, tapered, and step aeration are variations of the activated-sludge treatment. The objective is to supply the maximum of air to the sludge when it

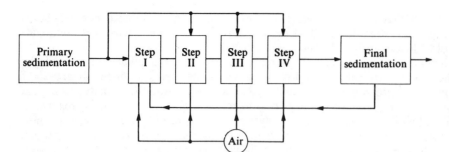

Fig. 13.3 Schematic diagram of step-aeration treatment. Step I, high sludge seed (4000 ppm); step II, 2000 ppm; step III, 1000 ppm; step IV, 800 ppm.

is in the optimum condition (sludge age) to oxidize adsorbed organic matter. The location of the aerator and the quantity of air supplied is varied, depending on sludge solids and organic matter to be oxidized. Lower volumes of air and shorter detention times are claimed for these processes, while the mechanisms and theories of operation are similar to those of activated sludge.

Step aeration attempts to eliminate the problems encountered with plain aeration by providing a two- to three-hour aeration only. Highly activated and concentrated sludge floc is returned to the aeration tank at the proper location (usually the inlet); this reduces bacterial lag, accelerates logarithmic bacterial growth, and provides abundant surfaces for adsorption of new cells. The chief advantage of this process is the flexibility it offers the operator. Figure 13.3 shows that one can obtain almost any desired ratio of primary effluent to sludge seed returned.

13.4 Dispersed-Growth Aeration

Dispersed-growth aeration is a process for oxidizing dissolved organic matter in the absence of flocculent growths [11]. The bacteria (seed) for oxidizing are present in the supernatant liquor after wastes have been aerated and settled. A portion of this supernatant liquor is retained for seeding incoming wastes, while the settled sludge from the secondary settling tank is digested or treated by other sludge-treatment methods. This process has been successfully used to treat many types of dissolved organic wastes [13, 21–24]. Its advantage is that it eliminates certain problems associated with sludge seeding. With many industrial wastes, it is difficult to build up any significant sludge concentration; in such cases, dispersed-growth aeration (which is not dependent on sludge) finds ready acceptance. Dispersed-growth aeration

does require more air to achieve the same BOD reduction as the activated-sludge process. However, when one considers that the initial BOD in dispersed-growth aeration is usually quite high, the amount of air required per pound of BOD removed is about the same as that used in the activated-sludge process, even though aeration periods to reach the same BOD reduction are normally quite lengthy (24 hours as compared to 6). Treatment by dispersed-growth aeration involves complete removal by oxidation, rather than by adsorption and partial oxidation, as in activated-sludge treatment.

Heukelekian [11] originally conceived this idea of seeding concentrated, soluble organic wastes with dispersed, instead of flocculent, growths, when he discovered that bacteria in culture mediums normally grow in the dispersed state or in small groups and that seeding is essential for high-rate biological activity. If a waste contains only soluble material, no flocculent growth should form. In his early work on penicillin and streptomycin wastes [12, 13], he made the following claims for the dispersed-growth aeration process:

1) It is better adapted than activated-sludge methods for the treatment of concentrated, soluble organic wastes because: (a) activated sludge has a tendency to bulk with concentrated organic wastes; (b) it is difficult to develop an activated sludge from a soluble waste.

2) Little sludge is formed with dispersed growths when soluble substrates are decomposed.

3) The percentage of BOD reduction decreases as the strengths of penicillin and streptomycin wastes are increased, but 80 per cent reduction may be expected with wastes up to 3000 ppm of BOD. Greater BOD reductions are possible when the BOD is less than 1000 ppm and the waste is aerated for 24 hours.

4) The effluent has a higher turbidity than the raw waste, and color is not removed.

5) The process may be used as a pretreatment unit for conventional biological-treatment processes.

6) The seed material can readily be developed and adapted from soil or sewage within a few days.

7) Optimum results are obtained with air rates of 2 to 3 cubic feet per gallon per hour; stronger wastes require higher air rates.

8) The initial pH of the waste does not seem a critical factor, since the pH increases during aeration. The BOD of raw waste with a pH of 6.4 is reduced as much as the BOD of the same waste adjusted to 7.2.

The author [28] also found this method of treatment suitable for rag and jute paper-mill wastes. In a basic study of the oxidation of glucose by dispersed-growth aeration [27], the authors found 5 million bacteria per milliliter when using nutrient broth as a medium. The two major types of bacteria found during the 24-hour aeration period were:

1) Dispersed, short, thick, round-ended rods; approximate size, 2 to 2.5 microns × 1 micron. Some of these organisms appeared as fingerlike capsules. As the aeration period progressed, there was an apparent increase in the number of slime-enmeshed bacteria (Fig. 13.4).

2) Sphaerotilus-like organisms, often as unsheathed forms (Fig. 13.5); these were more abundant after 6 hours of aeration and reached an apparent maximum after 24 hours.

In studying the suitability of dispersed-growth aeration for industrial wastes containing both proteins and carbohydrates, Struzeski and Nemerow [38] found such wastes amenable to oxidation by this process. Biological oxidation was enhanced by an increase in temperature, as shown in Fig. 13.6, and initial pH values up to 9.5 did not hamper it. It was also found that, when soluble protein–carbohydrate wastes are to be treated by dispersed-growth aeration, units must be designed to allow ample detention time, since air rates above the critical level (1050 cubic feet of air per pound of BOD per day) do not increase the reduction of BOD.

13.5 Contact Stabilization

Biosorption is the commercial name of one equipment manufacturer's high-rate biological-oxidation process, used mainly for domestic sewage. It was originally developed at Austin, Texas, by Ullrich and

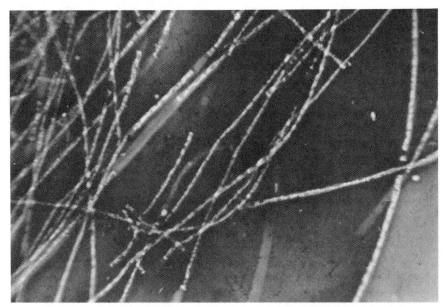

Fig. 13.4 *Sphaerotilus*-like organism, sheathed and unsheathed (× 620).

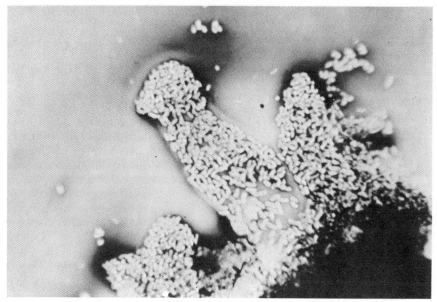

Fig. 13.5   Round-ended rods in a capsule of slime (× 620).

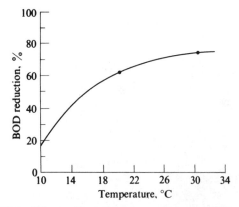

Fig. 13.6   Effect of temperature on average BOD reduction of a synthetic protein-glucose waste, using a dispersed-growth-aeration system, after 24 hours of aeration and no settling.

activated sludge from a stabilization–oxidation tank, or aerobic digester, for a short period of time (15 to 20 minutes). This activated-sludge—raw-waste mixture is then clarified by settling for about two hours, after which the settled sludge (consisting of activated-sludge floc with adsorbed impurities from the raw waste) goes through intense biological oxidation in the stabilization—oxidation basin for an aeration period of one to two hours. It then returns to the mixing tank and is again mixed with raw waste, so that it can absorb and adsorb added organic matter, and so on, in a continuous process. Excess or waste sludge can be taken from the system after either the clarifying or stabilizing steps, for anaerobic digestion or for dewatering on vacuum filters (see Fig. 13.7).

Ullrich and Smith claimed that this process requires less aeration-tank capacity than other processes, since the real aeration or reactivation takes place in the settled and concentrated sludge, not in the mixed liquor. Because the sewage and returned sludge is given only a brief mix, a small mixing compartment is needed. Pertinent pilot-plant and full-scale operating results are given in Table 13.2 [40, 41].

Smith [40]. It is essentially a modification of the activated-sludge process and is similar in some respects to the step-aeration process, but generally requires less air and plant space than these other two methods. In the contact-stabilization process, raw waste is mixed by aeration with previously formed

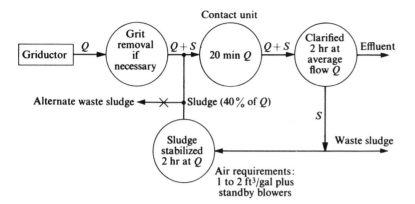

Fig. 13.7 Schematic arrangement of contact stabilization process.

13.6 High-Rate Aerobic Treatment

High-rate aerobic treatment (total oxidation) has developed in the last ten years as a means of oxidizing organic wastes [39]. This process consists of comminution of the waste, long-period aeration (one to three days), final settling of the sludge, and return of the settled sludge to the aeration tank. There is no need for primary settling or sludge digestion, but the aeration system must be large, to provide the required aeration period. The total-oxidation process is particularly useful in small installations, since it does not require a great deal of supervision. Little difficulty occurs with bulking on the sludge, even though the settling period is relatively short at times. In fact, since the solids resulting from this process are mostly of low volatility and therefore high in ash, the settling rate is quite fast. Return of the sludge is continuous and very rapid in comparison with normal activated-

sludge practice. By returning sludge at a high rate (100 to 300 per cent of flow), the system is kept completely aerobic at all times. The concentration of solids in the mixed liquor after a long period reaches a high level, and a portion of the sludge can then be wasted to reduce the concentration to 3000 to 5000 ppm. Lesperance [19] suggests that a waste sludge can be expected equal to 0.15 pound per pound of BOD removed. The small volume of wasted sludge is then stored and further concentrated until removed by tank car or other means to an area away from the plant.

The high-rate aerobic treatment, though it produces little waste sludge, has the disadvantages of requiring about three times as much air as conventional activated-sludge plants and of releasing some floc in the effluent. On the other hand, it needs very little operational maintenance and is well-suited for shock loadings from industrial operations.

Table 13.2 Biosorption operating data. (After Ullrich and Smith [40, 41].)

Item	1951 (Pilot plant)		1955 (Full scale)	
	Mean	% removal	Mean	% removal
Influent BOD, ppm	264		307	
Effluent BOD, ppm	19.5	92.5	20	93.4
BOD loading, lb/1000 ft³			144	
Detention time, hr	2.83		3	
Air required, ft³/lb BOD removed			665	
Suspended-solids influent, ppm	226		226	
Suspended-solids effluent, ppm	13.9	93.8	18	92.1

Another version of this treatment is referred to as completely mixed systems [20]. It operates on the assumption that, if microorganisms are kept in a constant state of growth, they operate at maximum efficiency and are adapted to the particular character and concentration of the waste. This constant-growth state can be maintained only if: (1) the microorganisms and raw wastes are thoroughly and continuously mixed; (2) the organic concentration is held constant; and (3) the effluent is separated from the microorganisms at a constant rate that is equal to the waste-feed rate. Figure 13.8 depicts a typical complete-mixing activated-sludge system [20]. A loading of 60 pounds of oxidizable organics per 1000 cubic feet of aeration tank is possible with this type of treatment.

13.7 Trickling Filtration

Trickling filtration is a process by which biological units are coated with slime growths (zoogleal forms) from the bacteria in the wastes. These growths adsorb and oxidize dissolved and colloidal organic matter from the wastes applied to them. When the rate of application is excessive—10 to 30 million gallons per acre per day (mgad)—and continuous, the humus collected on the filter-bed surfaces is sloughed off continuously. Crushed stone, such as traprock, granite, and limestone, usually forms the surface material in the filter, although other materials, such as plastic rings, have proved very effective. The main advantages of plastic media are light weight, chemical resistance and high specific surface, i.e., square feet per cubic feet of bed volume. Since smaller stones provide more surface per unit of volume, the contact material must be small in order to support a large surface of active film, but not so small that its pores become filled by the growths or clogged by accumulated suspended matter or sloughed film. Crushed stone, 3 to 5 inches in diameter, is used, with the smallest stone at the top. The integral parts of a trickling-filter system are the distribution nozzles, contact surface, and underdrain units (Fig. 13.9). The process may be summarized as follows:

1) An active surface film grows on the stone or contact surface.
2) Concentration of colloidal material and gelatinous matter occurs.
3) These adsorbed substances are attacked by bacteria and enzymes and reduced to simpler compounds, so that NH_3 is liberated and oxidized by chemical and bacterial means, giving a gradual reduction of NH_3 and an increase of NO_2 and NO_3 (Fig. 13.9).
4) A flocculent, humuslike residue or sludge, containing many protozoa and fungi, accumulates on the surface. When it gets too heavy it will slough off and resettle (a continuous process with biofilters). Part of the oxygen is supplied by spraying waste, blowing air into the filter, or allowing waste to drip into the filter. Another portion is supplied by convection due to the temperature difference between the incoming waste and the bed. The larger the surface, the greater the number of bacterial organisms that come into contact with the liquid to be purified; the greater the number of organisms, the higher the purification of the liquid. The smaller the pieces of rock in the surface media, the greater the purification; too-small particles, however, promote clogging. We can summarize by saying that trickling filters act as both

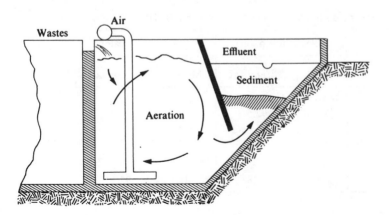

Fig. 13.8 Complete-mixing activated-sludge system. (After McKinney et al. [20].)

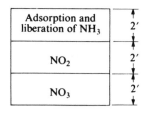

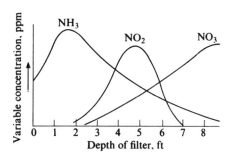

Fig. 13.9 Changes in nitrogen occurring in filter.

strainers and oxidizers.

Zobell [44] pointed out the importance of providing solid contact surfaces to further the physiological activities of bacteria growing in dilute nutrient solutions, such as most industrial wastes are. The following phenomena, according to Zobell, cause this increase in biological oxidation:

1) Solid surfaces make possible the concentration of nutrients and enzymes by adsorption to the surface.

2) The interstices between bacterial cells and surfaces act as concentration points; they retard the diffusion of exoenzymes and metabolites away from the cell, thereby favoring both digestion and adsorption of foodstuffs.

3) The interstices between surfaces and cells produce optimum conditions for oxidation–reduction and other physicochemical reactions.

4) Surfaces function as attachment points for microorganisms which are obligatory periphytes.

A typical, standard-rate, stone-bed trickling filter provides about 100 square feet of surface material per square foot of ground on which the filter is constructed. Velz [42] proposed the performance equation for trickling filters as

$$\frac{L_D}{L} = 10^{-kD},$$

where L_D is the removable fraction of BOD remaining at depth D, L is total removal, k is the logarithmic extraction rate, and D is depth of the bed. The reader will note the similarity between this equation and the monomolecular rate of decomposition of organic matter in streams:

$$\frac{L_t}{L} = 10^{-kt}.$$

The student should realize that the contact time in a filter is relatively short, compared with an activated-sludge process. However, the organic matter (bacterial food) resides in the bed longer than computed from the detention time. Howland [14] has contributed to our knowledge of contact time in filters. Assuming that a sheet of water is flowing steadily down an inclined plane under laminar flow conditions, he expresses the contact time as

$$T = \left(\frac{3v}{gs}\right)^{1/3} \frac{l}{q^{2/3}},$$

where T is the time of flow down the inclined plane, l is the length of the plane, s is the sine of the angle which the plane makes with the horizontal, g is the acceleration due to gravity, v is the kinematic viscosity of water (μ/ρ), and q is the rate of flow per unit width of the plane.

Howland indicates that the amount of oxidizable organic matter removed in a filter depends directly on the length of time of the flow. He recommends a deep filter containing the smallest practical media, to achieve an optimum contact time and maximum efficiency. Although some researchers recommend shallower filters and larger stone to reduce both initial and operating costs, the tendency today appears to be to follow Howland's recommendations, because engineers want an increasing degree of removal of BOD. Still, because of clogging and head-loss difficulties, there is a limit to how deep a bed and how small a stone size one can utilize.

Ingram [15] suggests the following drawbacks of trickling filters: they occupy too much space; they exhibit seasonal variation in efficiency; clogging and pooling present problems; there are limitations on hydraulic and organic loading; and there are limitations

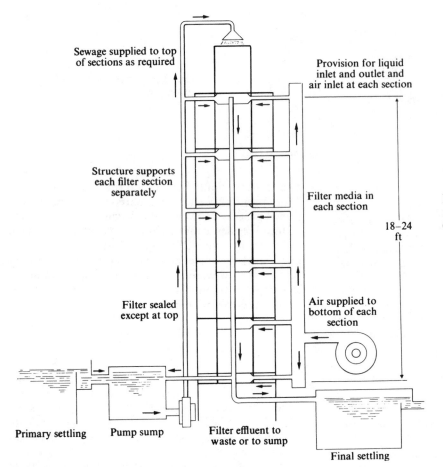

Fig. 13.10 Diagram of experimental controlled-filtration system. (After Ingram [15].)

on the strength of sewage applied. He proposes a trickling-filtration process called controlled filtration, which utilizes deep filters (18 to 24 feet). He was able to achieve greater than 70 per cent BOD removal (the removal expected in high-rate filters loaded at a normal rate of 20 mgad and 1300 pounds of BOD per acre-foot per day) with a minimum hydraulic loading of twice—and an organic loading of $1\frac{1}{2}$ to $10\frac{1}{2}$ times—these normal standards. His experimental filter is shown in Fig. 13.10.

minimum hydraulic loading of twice—and an organic loading of $1\frac{1}{2}$ to $10\frac{1}{2}$ times—these normal standards. His experimental filter is shown in Fig. 13.10.

Behn [2] points out that deviations from the usual reaction rates sometimes occur because of temperature of the waste and degree of filter saturation. Rankin [33] has been concerned with recirculation of filter effluent and concludes, from a study of a number of treatment plants, that performance

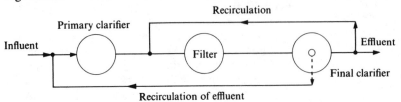

Fig. 13.11 Single-stage trickling filter (with recirculation).

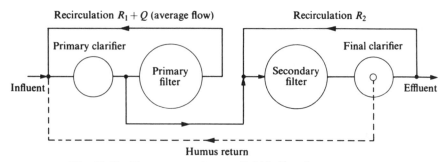

Recirculation $R_1 + Q$ (average flow)　　　　Recirculation R_2

Fig. 13.12 Two-stage series-parallel biofiltration process.

appears to depend primarily on the ratio of recirculation to raw waste-water flow, rather than on dosing rate, loading of the filter, or depth of filter (within the ranges studied). In smaller plants, with only one filter, single-stage filters appear to be most feasible, while for larger plants, where multiple filters are necessary or where stronger wastes are being treated, a two-stage series-parallel arrangement of filters yields a better effluent than single-stage filters with the same tank and filter capacity and the same volume of re-circulated liquor. Diagrams of each of these systems are shown in Figs. 13.11 and 13.12.

13.8 Spray Irrigation

Spray irrigation is an adaptation of the familiar method of watering agricultural crops by portable sprinkling-irrigation systems; wastes are pumped through portable pipes to self-actuated sprinkler heads. Lightweight aluminum or galvanized piping, equipped with quick-assembly pipe joints, can be easily moved to areas to be irrigated and quickly assembled. Wastes are applied as a rain to the surface of the soil, with the objective of applying the maximum amount that can be absorbed without surface run-off or damage to the cover crops. A spray-irrigation system is composed of the following units: (1) the land on which to spray; (2) a vegetative cover crop to aid absorption and prevent erosion; (3) a mechanically operated screening unit; (4) a surge tank or pit; (5) auxiliary stationary screens; (6) a pump which develops the required sprinkler-nozzle pressure; (7) a main line; (8) lateral lines; and (9) self-actuated revolving sprinklers operating under 35 to 100 psi nozzle pressure.

With good cover crops (dense, low-growing

grasses) and fairly level areas, waste to a depth of 3 to 4 inches can be applied at a rate of 0.4 to 0.6 inch per hour. The process is generally limited to spring, summer, and autumn. Anderson *et al.* [1], in a recent study with citrus wastes, found that aerobic conditions are maintained without odors to a depth of at least three feet.

13.9 Wet Combustion

Wet combustion is a process [43] of pumping organics-laden waste water and air into a reactor vessel at elevated pressure (1200 psi) (Fig. 13.13). The organic fractions undergo rapid oxidation, even though they are dissolved or suspended in the waste. This rapid oxidation gives off heat to the water by direct convection, and the water flashes into steam. Inorganic chemicals, which are present in many industrial wastes, can be recovered from the steam in a separate chamber. Heat from an external source is applied just to start the process; thereafter, it requires only 12 to 20 per cent of its own heat to maintain itself. The remaining 80 to 88 per cent can be utilized as process steam or to drive turbines for electrical or mechanical power. This process has a good potential where steam is essential and inexpensive enough to justify the cost of the equipment and where the inorganic chemicals in the waste are worth recovering and reusing. The wet combustion process can maintain itself only when the waste has a high percentage of organic material (usually about 5 per cent solids and 70 per cent organic).

13.10 Anaerobic Digestion

Anaerobic digestion is a process for oxidizing organic

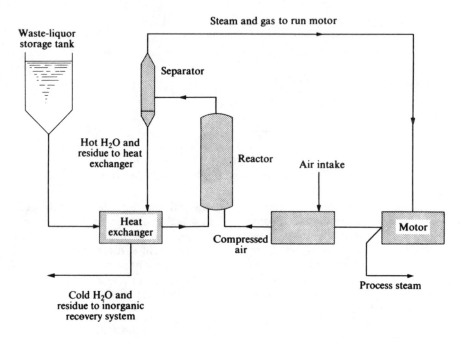

Fig. 13.13 Schematic arrangement of wet-combustion-process units.

matter in closed vessels in the absence of air. The process has been highly successful in conditioning sewage sludge for final disposal. (Since digestion is primarily used for the treatment of sludge, rather than liquid wastes, the theory of its operation is described in more detail in Chapter 14.) It is also effective in reducing the BOD of soluble organic liquid wastes, such as yeast, cotton-kiering, slaughterhouse, dairy, and white-water (paper-mill) wastes. Generally, anaerobic processes are less effective than aerobic processes, mainly because of the small amount of energy that results when anaerobic bacteria oxidize organic matter. Anaerobic processes are therefore slow and require low daily loadings and/or long detention periods. However, since little or no power need be added, operating costs are very low. Where liquid waste volumes are small and contain no toxic matter and there are high percentages of readily oxidized dissolved organic matter, this process has definite advantages over aerobic systems. The pH in digesters must be controlled to near the neutral point.

Buswell [5] proposed the following general equation for conversion of organic matter in industrial wastes to carbon dioxide and methane:

$$C_nH_aO_b + \left(n - \frac{a}{4} - \frac{b}{2}\right)H_2O$$

$$= \left(\frac{n}{2} - \frac{a}{8} + \frac{b}{4}\right)CO_2 + \left(\frac{n}{2} + \frac{a}{8} - \frac{b}{4}\right)CH_4.$$

In the United States, anaerobic treatment plants have been built to treat yeast, butanol–acetone, brewery, chewing-gum, and meat-packing wastes. Pettet *et al.* [32], in a review of British practices, found that slaughterhouse waste appears to respond extremely well to anaerobic digestion, although up to 1959 there were no full-scale anaerobic-digestion plants in Great Britain. In the United States, BOD reductions of 60 to 92 per cent have been attained with all these wastes, at loadings of 0.003 to 0.191 pound of BOD per cubic foot of digester per day. Concentrations of organic matter ranged from 1565 to 17,000 ppm BOD.

13.11 Mechanical Aeration System

Cavitation is a typical process for mechanical aeration of wastes. The complete Cavitator assembly con-

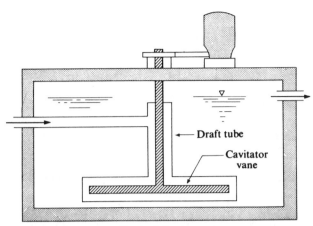

Fig. 13.14 Typical Cavitator system.

sists of a vertical-draft tube with openings for connection to the influent pipe and a rotor assembly of the multiblade type, supported by an adjustable ball thrust bearing mounted at the motor level. The rotor is mounted on a stainless-steel shaft and the entire unit, including the draft tube, is supported by a structural-steel bridge. A cross-section is shown in Fig. 13.14 [36]. As soon as the rotor exceeds a certain critical speed, air is drawn in from the atmosphere through the vertical hollow tube and dispersed organic solutions may be disposed of in this manner. To be effective, the wastes must be placed in a geological formation which prevents the migration of the wastes to the surface or to ground-water supplies. The rock types most frequently used are the more porous ones such as limestones, sandstones, and dolomites, since the porosity may help develop a filter cake which plugs the well. Other factors, in addition to geology, to be considered are depth and diameter of well, injection pressures, and the volume and characteristics of the wastes. At the end of 1966, there were 78 industrial disposal wells in the United States, most of which are used for chemical and refinery wastes (86 per cent); most are less than 4000 feet deep (74 per cent), dispose of less than 400 gpm per well (87 per cent), and operate at less than 300 psi (57 per cent). Costs of injection disposal installations vary from $30,000 for a shallow (1800-foot) well not requiring pretreatment to over $1,400,000 for a very deep (12,000-foot) well with intricate pretreatment. Actual costs vary depending on depth, surface equipment, pretreatment, diameter of well, injection pres-

sure, variability of composition of waste water, and availability of drilling equipment.

Donaldson [6] reports on a wide variety of industrial wastes being injected into formations ranging in age from Precambrian to modern-day. In the United States, up to 1964, more than 30 wells, ranging in depth from 300 to 12,000 feet, were being used for waste disposal into subsurface formations which include unconsolidated sand, sandstone, regular limestone, and fractured gneiss. Although subsurface into the waste. The rotor creates a zone of cavitation in its turbulent trail and air moves in to fill the areas of rarefied underpressure. The amount of air which is being entrained depends on the size and shape of the rotor, the rpm, and the water depth. The manufacturers claim that their system utilizes at least 25 per cent of the available oxygen in the air, in contrast with conventional aeration equipment, which utilizes only 5 per cent. At least one waste-treatment plant (dealing with canning wastes and sewage) attained over 90 per cent BOD removal with an air supply of 110 cubic feet per pound of BOD per day. Operational costs [36] were $12.80 per day for an equivalent population of 12,000 persons. A recent modification of this system employs mechanical mixing by a rotor submerged (but near the surface of) the waste water. Power costs are thus reduced, with no apparent loss of aeration or mixing efficiency. This system promises to be the most economical one for secondary treatment of wastes with a highly dissolved organic content.

13.12 Well Injection

Disposal of wastes containing dissolved organic matter by injecting them into deep wells has been successful in areas of low or nonexistent stream flow, especially when wastes are malodorous or toxic and contain little or no suspended matter. Deep well injection has been used successfully to dispose of organic solutions from chemical, pharmaceutical, petrochemical, paper, and refinery wastes; in addition, many injection offers an economical method of final disposal where receiving surface water is inadequate to carry the waste water away safely, circumstances can limit its effectiveness, e.g. the area lacks suitable underground formations for waste injection, the initial capital expense is excessive, or the pretreatment

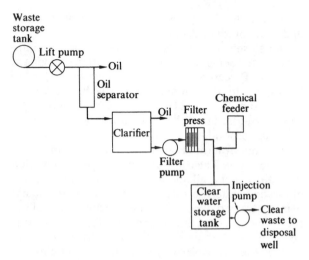

Figure 13.15 Typical surface equipment for deep well waste injection from surface waste storage.

required may be too extensive and expensive.

The industrial waste engineer must work closely with a geologist familiar with the subsurface formations in the area in order to select the proper waste-disposal zone. A well is drilled and core samples are analyzed for specific characteristics such as permeability and reactivity with the waste. Tests are carried out to determine the injection pressure required at various waste-water flows. Certain procedures, such as fracturing and acidizing, may be used to improve the soil permeability and thus reduce the injection pressure required at various flows.

Schematic drawings of typical complete subsurface waste-disposal systems are shown in Figs. 13.15 and

13.16. Although cement tanks up to 50,000-gallon capacity are commonly employed within the basement of a factory, large, shallow, open ponds may be used where land is available and where some settling and oxidation is required as a pretreatment. The oil separator is required for petroleum refinery wastes, since oil tends to plug disposal formation and the oil can be recovered and reused. The usual separator consists of a tank with many internal baffles to cause the oil to separate and rise. If a clarifier is then used, heavier material such as dirt, resin flocs, and suspended grease can settle out. Mechanical equipment such as sludge rakes and surface skimmers can also be used with this equipment. Since not all solids are completely removed by the treatment so far described, filters are then used to protect sand or sandstone formations from plugging. The screens are usually metal and coated with diatomaceous earth, but in some situations sand filters are preferred. If wastes contain slime that will form bacteria, algae, iron bacteria, sulfate-reducing bacteria, or fungi, a suitable bactericide (such as quaternary amines, formaldehyde, chlorinated hydrocarbons, chlorine, or copper sulfate) is used to control their detrimental effects. The clear-water storage tank is normally equipped with a float switch designed to operate the injection pump at certain liquid levels. The size and type of injection pump is controlled by wellhead pressure, waste-water flow, and waste-water characteristics such as pH and corrosiveness. The multiplex piston pump is most commonly used when wellhead pressures of greater than 150 psi are required, whereas single-stage centrifugal pumps are used at lower pressures.

To construct the well, first a 15-inch-diameter hole

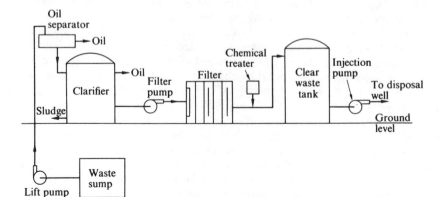

Fig. 13.16 Typical surface equipment for deep well waste injection from waste sump underground. (From Bureau of Mines Information Circular no. 8212.)

is drilled to 200 feet below the deepest fresh-wat aquifer and a 10½-inch (O.D.) casing is set and cemented to the surface. Next, a 9-inch-diameter hole is drilled to the bottom of the potential disposal formation, a 7-inch (O.D.) casing is set at the total depth of the hole, and cement is circulated in the annulus between the injection casing and the 9-inch hole to the surface (Fig. 13.17). This method has been proved to seal off water aquifers from the well and to protect other water resources. Table 13.3 summarizes Donaldson's findings for 20 separate installations, presenting much valuable information such as costs, associated problems, well depth, injection pressures, and formation type.

For the disposal of acid wastes, there are five requirements for deep well disposal: (1) a satisfactory disposal horizon; (2) an horizon filled with salt water; (3) an horizon located at a sufficient depth; (4) a suitable cap rock; (5) a waste compatible with the natural water in the disposal horizon. The possible dangers of deep well disposal include: (1) contamination of potable water supplies either by lateral migration to

existing unplugged dry holes or producing wells, or by vertical migration through the subsurface, or by vertical migration due to mechanical failure; and (2) possible movements along old fault planes. The representative cost for a 4000-foot well is $450,000, which includes the cost of well and equipment, aboveground pumping and equipment, and holding-tank and collection equipment.

The reader may use the following checklist in the design of deep-well disposal systems:

Factors to consider in subsurface disposal of industrial wastes by deep well injection
A. State laws and legal aspects
   1) State recognition of this method
   2) Subsurface trespass
B. Geology
   1) Employment of a geologist or well contractor
   2) Disposal formation
      a) Porosity
      b) Permeability
      c) Composition (sandstone or limestone)
C. Waste characteristics
   1) Volume reduction
   2) Injection flow rate
   3) Injection pressure
   4) Corrosiveness
   5) Biological effects
D. Surface equipment needs
E. Wells
   1) Number
   2) Size
   3) Monitoring
F. Economics

13.13 Foam Phase Separation

Figure 13.18 illustrates the equipment used for foam phase separation. A sparger producing small gas bubbles (usually air) causes these bubbles to rise through the liquid and adsorb surface-active solutes and suspended matter. When the bubbles reach the surface, a foam forms, which is forced out of the foamer, collapsed, and discharged as a concentrated waste.

If the following assumptions are made: (1) complete mixing in the foamer; (2) sufficient depth of liquid to reach maximum solute adsorption of the gas–liquid

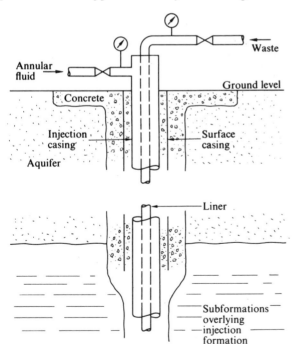

Fig. 13.17 Typical injection well. (From Bureau of Mines Information Circular no. 8212.)

Table 13.3 Summary of disposal systems.

Company	Type of waste	Injection rate, gpm	Injection pressure, psi	Subsurface depth of wells, feet
A	Brine; chlorinated hydrocarbons	200	500	12,045
B	Clear 4 % solution Na$_2$SO$_4$	300	45	295
C	Masic waste, pH ±10	70	1000	6,160
D	Magnesium; calcium hydroxides	200	Vacuum	400
	Manufacturing waste, pH may change from 1 to 9 in 8 hr	400	Vacuum	4,150
E	Lachrymator waste from acrolein and glycerine units	700	150–170	1,960
F	Aqueous solution—phenols, mercaptans, and sulfides	215	30–90	1,795
G	Phenols; mercaptans; sulfides; brine	100	40–100	1,980
H	Phenols; chlorinated hydrocarbons	200	450	4,000
	Brine	200	150	4,000
	Phenols; mercaptans; sulfides	50	—*	4,000
I	Coke oven phenols; quench water	50	300	563
J	Organic wastes	60	500	1,472
K	Sulfuric acid waste	400	Vacuum	1,830
L	Detergents; solvents; salts	254	280	1,807
M	38% HCl solution	14		1,110
N	Stripping steam condensate; cooling tower blowdown	50	10–20	1,110
	Aqueous petroleum refinery effluent	400	50–70	1,110
O	Phenols; brine	75	400	7,650

* Information not available.

Table 13.3 (*continued*)

Formation age, type, and name	Total cost of system ($)	Date started	Problems	Solutions and remarks
Precambrian fractured gneiss (unnamed)	1,419,000	March 1962	Microorganisms in waste	
Sandstone	—*	June 1951	None	
Cambrian sandstone	250,000	Nov. 1960	Inadequate filtration	Larger filter planned
Permian salt bed (Hutchinson)	—*	—*	None	
Ordovician vugular limestone (Arbuckle)	500,000	Dec. 1957	Corrosion and water hammer	Heavier tubing planned
Pleistocene	135,000	1956	Sand incursion increased injection pressure	Back-washing every 4 months
Pleistocene	30,000	Sept. 1959	Sand incursion	Periodic back-washing
Pleistocene	—*	March 1960	Sand incursion	Periodic back-washing
Devonian vugular limestone (Dundee)	—*	1950	None	
Devonian vugular limestone (Dundee)	—*	1931	None	
Devonian vugular limestone (Dundee)	—*	1950	None	
Silurian sandstone (Sylvania)	25,000	Aug. 1956	High wellhead pressure	Acidizing and fracturing
Devonian sandy limestone (Dundee); Traverse, Dundee & Monroe	400,000	1954	None	
Permian sandstone	562,000	Jan. 1960	Microorganisms decreased injectivity	Formaldehyde
Ordovician vugular limestone (Arbuckle)	300,000	Feb. 1960	Mechanical failure of surface equipment	
Unconsolidated sand (Glorieta)	—*	April 1962	None	
Unconsolidated sand (Glorieta)	—*	1959	None	
Unconsolidated sand (Glorieta)	—*	1958	None	
Eocene sand and clay (Frio)	—*	1958	High injection pressure	Periodic acidizing

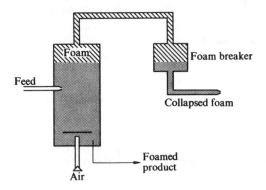

Fig. 13.18   A column foam fractionator.

interface; (3) constant liquid density; (4) no bubble rupture in the foam phase; and (5) negligible volume of the liquid layer containing the surface excess of solute—one can arrive at a material-balance equation

$$C_F - C_B = 1000 \frac{G}{F} \Gamma_B S,$$

where C_F and C_B are feed and bottom product concentrations in mg/liter, G is the volumetric gas rate in liters/minute, Γ_B is solute surface excess corresponding to C_B in mg/cm², and S is the specific surface of bubbles in foam phase in cm²/cc.

At flows of air to liquid feed of $G/F \geq 3$, it is reported that the COD is reduced by 25 per cent and alkyl benzene sulfonate (ABS) concentrations are reduced by 50 to 75 per cent. At air rates of 1.5 liters/mg of ABS in secondary effluents, removals of 0.4 ppm ABS have been reported. The success of the process, in general, depends upon the foamability of the liquid waste, which is said to be of low order of magnitude. Bruner and Stephen [3] have calculated foam separation costs (not including the foamate disposal) as follows:

mgd	cents/1000 gal.
1	3.6
10	1.9
100	1.4

Schoen *et al.* [35] used this treatment successfully to separate radium from uranium-mill waste water. They found the pH of the waste water very important

in selecting foaming agents. An increase in foaming agent will generally produce a similar increase in foam during treatment. Grieves and Crandall [8] also experimented with both iron and alum as coagulants, using bentonite as an aid, in foaming low-quality waters.

13.14 Brush Aeration

According to Pasveer [31] the brush aeration system was evolved between 1925 and 1930 by Dr. Kessener for use in the activated-sludge process. It is essentially an extended aeration process providing over 24 hours of aeration. Since the 1930s, it has found application, particularly in the Netherlands and in a few plants in Britain, and about a dozen plants were constructed in Canada and the United States prior to 1964 [29] (see Fig. 13.19). Most of these aeration systems are installed in "oxidation ditches." The design of the oxidation ditch combines an aeration tank and a holding tank in a single unit; the aeration rotor circulates the mixed waste through the whole ditch by means of the rotating cage, but aeration occurs only in the vicinity of the rotor. The rotor is fixed at both ends and set transversely across the aeration ditch and rotates in the direction of waste flow. Aeration is obtained by means of long, rectangular, angle irons welded to the rotating cage. Although the results have been obtained mostly with domestic sewage, it is apparently adaptable to any organic industrial waste.

13.15 Subsurface Disposal

Three other methods of disposing of dissolved organic wastes below the ground surface are injection, placement in underground cavities, and spreading. Since injection is discussed in some detail in Section 13.12 and placement in underground cavities is limited to either small volumes of wastes or particular situations of subsurface formation, they will only be mentioned here as possibilities. Koenig [16] reports, however, that, in 1956, 244 cavities were used for storage, mostly for hydrocarbons and mostly in salt mines.

Spreading may be defined as the dispersal of liquid wastes on the ground in order to enhance their infiltration into it. Reclamation of waste waters by spreading

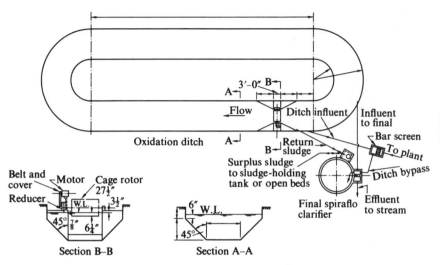

Fig. 13.19 Typical layout of an oxidation ditch treatment plant. (Courtesy Lakeside Engineering Corp.)

on land, with subsequent withdrawal of ground water, has been extensively practiced, mostly for secondary sewage effluents. Infiltration rates govern the use of this method, while ultimate effects on underground water supplies govern its acceptability. Because of numerous physical limitations this method should be considered mainly for small volumes of concentrated, organic wastes in particularly suited soils. Koenig [16] gives comparative costs for this method of disposal and for injection, wet oxidation, and incineration (Fig. 13.20).

13.16 The Bio-Disc System

The Bio-Disc system was developed independently in West Germany (by Hartmann and Pöpel) and the United States (by Welch and Antonie).* It consists of a series of flat, parallel discs which are rotated while partially immersed in the waste being treated. Biological slime covers the surface of the discs and adsorbs and absorbs colloidal and dissolved organic matter present in the waste water. Excess slime generated by synthesis of the waste materials is sloughed off gradually into the mixed liquor and subsequently separated by settling

(Fig. 13.21). The rotating discs carry a film of the waste water into the air where it absorbs the oxygen necessary for aerobic biological activity of the slime. Disc rotation also provides contact between the slime

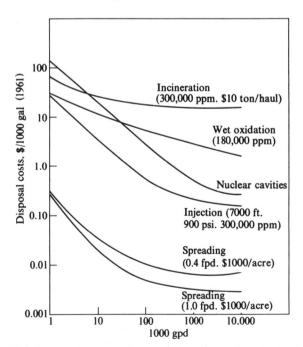

Fig. 13.20 A comparison of unit disposal costs (1961 figures). (After Koeing [16].)

*For references see the section on Bio-Disc in **Suggested Additional Reading.**

Fig. 13.21 RBC disc system. (Courtesy Allis Chalmers Company.)

and the waste water. Thus, the rotating discs provide: (1) mechanical support for a captive, microbial population; (2) a mechanism of aeration, the rate of which can be adjusted by changing the rotational speed; and (3) contact between the biological slime and the waste water, the intensity of which can be varied by changing the rotational speed.

Use of closely spaced parallel discs achieves a high concentration of active biological surface area. This high concentration of active organisms and the ability to achieve the required aeration rate by adjusting the rotational speed of the discs enables this process to give effective treatment to highly concentrated wastes. At a loading of 11 lb BOD/day/1000 ft² of surface area, 90 per cent BOD removal is obtained in 2000 ppm BOD dairy waste. Secondary treatment of domestic sewage is accomplished with a retention time of 1 hour or less, and 90 per cent BOD reduction is obtained at a loading of 5 lb BOD/day/1000 ft². Because a buoyant plastic material is used for the discs and negligible head loss is encountered through the RBC itself the power requirement for this process is very low. Its simplicity of construction and operation has demonstrated that minimal unskilled maintenance is all that is required for efficient operation.

The RBC process has gained wide acceptance in Europe. In the past ten years over 400 Bio-Disc plants have been constructed there ranging in size from 24,000 to 55,000 population equivalents for treatment of domestic and industrial wastes. This process is now being introduced commercially to the United States.

13.16A Biological Treatment Design

EXAMPLE

An integrated pulp and paper mill generates wastewater from the diffuent operations with the following characteristics:

	Process A	Process B
Flow	3.5 mgd	6.0 mgd
BOD	2500 mg/L	1100 mg/L
TSS	1000″	450 mg/L
pH	10	6.9
Temperature-Summer	25°C	25°C
Winter	15°C	15°C

The treatment scheme proposed to be used consists of flow equalization, primary clarification, activated sludge, secondary clarification and final treatment in aerated stabilization basins. Design the treatment units given the following design criteria:

Activated Sludge Effluent BOD	= 100 mg/L
TSS	= 100 mg/L
pH	= 6.5–9.0
K	= 0.0015 L/mg-d at 20°C
Y	= 0.6
b	= 0.01/day
O	= 1.05
MLSS	= 3000 mg/L
MLVSS	= 2500 mg/L
F/M	= 0.15 to 0.4
Minimum Solids Retention Time	= 10 days
Sludge Recycle Rate	= 40%

SOLUTION

Equalization

Combined flow Q = 9.5 mgd

Combined BOD $= \dfrac{3.5 \times 2500 + 6.0 \times 1100}{9.5}$

$= 1616$ mg/L

Combined TSS $= \dfrac{2.5 \times 1000 + 6.0 \times 450}{9.5}$

$= 547$ mg/L

Dentention time = 15 minutes
Volume of equalization tank = Qt $= \dfrac{9.5 \times 15 \times 10^6}{60 \times 24}$

$= 99,000$ gallons

Mixing requirements: @ 0.05 hp/10^3 gal
$= 0.05 \times 99 = 4.95$ hp

Aeration requirements: @ 1.5 ft^3 air/10^3 gal$_3$
$= 1.5 \times 99 = 148.5$ ft^3

Primary Clarifications

Surface loading or Overflow rate ranges generally from 500 to 1000 gal/day/ft^2. Recommend select the loading based on settling tests.

Assume loading of 750 gal/day/ft^2

Surface loading (u) $= \dfrac{Q}{A}$

Area A $= \dfrac{Q}{u} = \dfrac{9.5 \times 10^6}{7540} = 12,667$ ft^2

Diameter of Clarifer $= \sqrt{\dfrac{12,667 \times 4}{}}$

$= 127$ ft.

Detention time ranges from 2 to 4 hours.
Assume detention time of 3 hours.

Volume of clarifer = Qt
(V) $= 9.5 \times 10^6 \times \dfrac{3 \text{ hrs}}{24 \text{ hrs/d}}$

$= 1,187,500$ gal.

@ 7.48 gal/ft^3
V $= \dfrac{1,187,500}{7.48} = 158,757$ ft^3

Depth of Clarifer $= \dfrac{V}{A}$

$= \dfrac{158,757}{12,667}$

$= 12.5$ ft.

Depth generally ranges from 10 to 15 ft. Therefore, calculated depth is within range.

Circumference of clarifer = D
$= \times 127 = 399$ ft.

Check effluent weir loading (ranges from 10,000 gpd/ft. to 25,000 gpd/ft.)

Weir loading $= \dfrac{9.5 \times 10^6}{399} = 23,810$ gpd/ft. is within range

Activated Sludge

Assume 50% BOD removal in primary clarifier. Therefore, BOD in influent to activated sludge unit is 808 mg/L.

Calculate reaction range (k) for summer and winter temperatures.

$K_T = k_{20} \theta^{T-20}$ $\theta = 1.04$

Summer: $k_{25} = (0.0015)1.04^{25-20} = 0.00183$ L/mg-day

Winter: $k_{15} = (0.0015)1.04^{15-20} = 0.00123$ L/mg-day

Calculate volume of aeration tank:

Summer Condition

$\dfrac{F}{M} = k\, s_e$

$= 0.00183(100) = 0.183$ days^{-1}

$\dfrac{1}{SRT} = Y\left(\dfrac{F}{M}\right) - b$ Use this value since greater than minimum design criteria.

$= 0.64(0.183)^{-1} - 0.01$

$= 0.017$ days

Therefore, SRT = 9.4 days

Use SRT = 10 day minimum per design criteria.

$$XV = \frac{Y\,(S_o - S_e)\,SRT}{1 + b\,(SRT)}\,Q$$

$$= \frac{0.64\,(808 - 100)\,10\,(9.5)}{1 + 0.01\,(10)}$$

$$= 39133$$

@ MLVSS = X = 2500 mg/L

Therefore, volume of aeration tank $= \dfrac{39133}{2500} =$ 15.6 million gallons

Winter Conditions

$$\frac{F}{M} = k\,S_e$$

$$= 0.00123(100) = 0.123\ \text{days}^{-1}$$

Use $\dfrac{F}{M} = 0.15$ per minimum design criteria.

$$\frac{1}{SRT} = Y\left(\frac{F}{M}\right) - b$$

$$= 0.64\,(0.15) - 0.01$$

$$= 0.086\ \text{days}^{-1}$$

Therefore, SRT = 11.6 days

Use this value for SRT since greater than minimum design criteria.

$$XV = \frac{Y\,(S_o - S_e)\,SRT}{1 + b\,(SRT)}\,Q$$

$$= \frac{0.64\,(808 - 100)\,11.6\,(9.5)}{1 + 0.01\,(11.6)}$$

$$= 44584$$

Therefore, volume of aeration tank $= \dfrac{44584}{2500}$

$$= 17.8\ \text{million gallons}$$

Therefore, the larger volume required for winter conditions determines the design volume of the aeration tank.

Detention time $= \dfrac{V}{Q} = \dfrac{17.8}{9.5} = 1.87$ days

Check $\dfrac{F}{M}$ ratio

$$\frac{F}{M} = \frac{9.5\ \text{mgd} \times 808\ \text{mg/L} \times 8.34}{17.8\ \text{mg} \times 2500\ \text{mg/L} \times 8.34} = 0.17\ \text{days}^{-1}$$

Check at maximum flow condition.

$$\frac{F}{M} = \frac{9.5 \times 808 \times 8.34}{13.3 \times 2500 \times 8.34}$$

$$= 0.23$$

Both $\dfrac{F}{M}$ are within design criteria range.

(b) Waste sludge production

$$SRT = \frac{\text{lbs MLSS}}{\text{lbs solids wasted/d} + \text{lbs solids in effluent}}$$

Calculate waste sludge production for minimum SRT = 10 days.

$$SRT = \frac{17.8\ \text{mgal} \times 8.34 \times 3000\ \text{mg/L}}{S_W + (9.5\ \text{mgd} \times 8.34 \times 100\ \text{mg/L})}$$

$$10 = \frac{445356}{S_W + 7923}$$

Therefore, sludge wasted $S_w = 36{,}613$ lbs/day.

Calculate oxygen requirements:

lbs oxygen requires/day = y′ (lbs BOD removed/day) + b′ (lbs MLVSS)

lbs BOD removed/day = 9.5 mgd × (808 - 100) × 8.34
= 56,095 lbs/day

lbs MLVSS = 2,500 mg/L × 17.8 × 8.34
× 371,130 lbs/day

lbs Oxygen = 0.6 (56,095) + 0.1 (371,130)
= 70,770 lbs/day
= 2,950 lbs/hr

Aeration horsepower requirements:

Oxygen transfer rate for design conditions is calculated as follows:

$$N = N_o\ \frac{B \times {}^{C}S\ \text{at}\ {}^{-C}L}{{}^{C}Sc}\ 1.024\ T\text{-}20$$

Calculate transfer rate based on summer conditions.

$N = 3.0 \dfrac{0.9 \times (8.5 - 2.0)}{9.2} 1.024^{25-20} (0.8)$

$= 1.7$ lbs oxygen/hp-hr

Horsepower $= \dfrac{2950 \text{ lbs/hr}}{1.7 \text{ lbs/hp-hr}}$

$= 1735$ hp

Check power requirements to ensure adequate mixing in aeration tank. Assume approximately 100 hp/million gallons required for mixing.

$\dfrac{1735 \text{ hp}}{17.8 \text{ mg}} = 97.5$ hp/million gallons

Therefore, horsepower provided for aeration would be adequate for mixing in the aeration tank.

Final Clarifier

Solids loading to final clarifier is calculated as follows:

Sludge recirculation $= 40\%$

Recirculation flow(R) $= 0.4 \times 9.5 = 3.8$ mgd

Solids loading to clarifier $= (Q + R) \text{ MLSS} \times 8.34$
$= (9.5 + 3.8) \, 3000 \times 8.34$
$= 332,766$ lbs/d

Final clarifier area is calculated based on requirement for thickening area and clarification area.

Thickening area is calculated assuming solids loading rate of 15 lbs/d/ft^2 (range 10 - 20 lbs/d/ft^2).

Thickening area $= \dfrac{332,766}{15}$

$= 22,184$ ft^2

Clarification area is calcualted assuming an overflow rate (OFR) of 600 gpd/ft^2 (range 400 to 1,000 gpd/ft^2).

Clarification area $= \dfrac{Q}{\text{OFR}}$

$= \dfrac{9.5 \times 10^6}{600}$

$= 15,833$ ft^2

Therefore, the thickening area governs.

Area $= 22,184$ ft^2

Diameter of final clarifier $= 168$ ft

Assume a detention time of 3 hours.

Volume of clarifier $= Qt$

$= 9.5 \times 10^6 \times 3$
$= 1,187,500$ gals

$= \dfrac{1,187,500}{7.48}$ ft^3

$= 158,757$ ft^3

$\therefore$ Depth of clarifier $= \dfrac{158757}{22184} = 7$ ft

Depth of clarifier $= \dfrac{158757}{22184}$

$= 7$ ft

O.K. (range 5 to 15 ft)

Check effluent weir loading $= \dfrac{Q}{d}$

$= \dfrac{9.5 \times 10^6}{\times 168}$

$= 18,008$ gpd/ft

O.K. (range 10,000 to 20,000 gpd/ft)

13.17 Collection and Reclamation (scavenging)

The scavenger hauls, treats, reclaims, and disposes of a variety of industrial wastes acquired through contract or purchase from firms that do not choose to treat their own wastes.

Scavenging firms may be as varied as the customers they serve. Some small firms may only provide hauling and land disposal services; others specialize in a single line of wastes, such as solvents, which may be profitably reclaimed. Large firms may provide a full range of treatment and consulting services.

Recycling Laboratories, for example, located in Canastota, New York, specializes in the reclamation of solvents, particularly the chlorinated hydrocarbons such as trichlorethylene. Typically, they purchase used solvents, redistill them, and then resell them at a profit. They also handle some alcohols and thinners and, for a fee, will accept petroleum products, which they use in their steam plant. Recycling Laboratories is a small, fairly new firm and most of their business comes from the immediate central New York area. They envision expanding one line at a time as they acquire experience in dealing with different sorts of wastes.

Chem-Trol Pollution Services, Inc. of Model City, New York (near Buffalo) represents the other end of the spectrum. It is a large firm, which operates on a 240-acre site with 7 separate lagoons. Lagoon storage totals six million gallons, while closed-tank storage totals an additional two million gallons. It processes wastes using the following techniques: (1) filtration; (2) thermal oxidation (incineration); (3) neutralization; (4) distillation; (5) chemical fixation; and (6) physical separation (such as centrifuging). Residues are land-filled. Many of its operating practices are subject to proprietary considerations but may be considered to generally follow the usual treatment techniques.

Advantages of scavenging treatment:
(1) Economy of scale exists in larger plants.
(2) Operating efficiencies other than scale—i.e., specialization, neutralization, and equalization—increase as the size of the operation increases.
(3) Treatment expense only is involved for the small industrial plant, rather than capital investment.
(4) Resource-recovery potential exists in the larger reclamation plant.

Disadvantages of scavenging treatment:
(1) Newness of the field: There is not now a trade association nor a real trade publication. Relatively little information exchange occurs, particularly with regard to proprietary practices. Regulation and standards vary with locality.
(2) Transportation expense, typically by truck, may be high.
(3) Batch processes are usually necessary with attendant high labor and overhead costs.
(4) Dependence on secondary markets is risky for the scavenger.
(5) Poor public image may exist unless a proper public-relations program is used.

Nevertheless, the scavenger fills a very real need in this time of increasingly strict environmental controls and is especially useful for the treatment of low volumes of highly concentrated organic wastes.

13.18 Miscellaneous (ozonization, photolysis, pure oxygen, etc.)

Organic matter in industrial wastewaters can be completely oxidized through contact and reaction with ozone (O_3). Ozone is a gas that is produced by passing oxygen through an electrical field. Approximately 11 kwh of electrical power are required to produce 1 lb of ozone from air. From 1.5 to 2.5 lb of ozone are used for each pound of dissolved organic matter oxidized. If pure oxygen is used instead of air as a source of ozone, the power requirements are reduced by about 50 per cent. Small size ozone-generating units, producing up to 75 lb of ozone per day, are available commercially for use by industries with small volumes of organic wastes. The ozone produced is usually dissolved in wastewater by injection nozzles or cavitation. This form of treatment has been found especially suited for oxidizing phenolic wastes. Ozone contact time can usually be kept below 30 minutes, especially for final or tertiary-type treatment, where organic matter concentration is relatively low and allowable effluent limits are also low.

Photolysis

The interaction of photocatalysts with radiation below about 4200 °A produces active oxygen species, which destroy organic matter by complete oxidation to CO_2 and H_2O. Certain oxides, notably zinc oxide and titanium dioxide are known to be photosensitizers or photocatalysts. Kinney and Ivanuski [45] recently reported that photcatalytic oxidation of dissolved organic matter by irradiation of slurries of zinc titanate (Zn_2TiO_2), zinc oxide (ZnO), titanium dioxide (TiO_2), and beach sand by sunlamps was effective. Dissolved organic matter in a sample of domestic sewage was reduced 50 per cent in 24 hours and 75 per cent in 70 hours. The reaction appears to follow first-order kinetics in most cases. Zinc oxide appears to be superior for this purpose. At concentrations of 100 to 200 mg/l organic carbon, 80 per cent of phenol, 67 per cent benzoic acid, 44 per cent acetic acid, 40 per cent sodium stearate, and 16 per cent sucrose were oxidized in 24 hours with 10 gr/l zinc oxide catalyst. Continued illumination reduced organic carbon to a few mg/l in most cases. The photocatalytic properties of illuminated beach sand, which oxidized 87 per cent of phenol in 72 hours, strongly suggest that photocatalysts are widely distributed in nature. Further it suggests that photocatalytic oxidation is a mechanism whereby dissolved organic matter is oxidized in the natural environment of streams and lakes. The researchers [45] suggested that three conditions must be satisfied to achieve oxidation of an organic molecule: (1) the molecule must be adsorbed at, or be in the vicinity of, the active site on the catalyst; (2) light energy of suitable wavelength (below about 4200 °A) must impinge on the active site; (3) dissolved oxygen must be present to replace the active oxygen species displaced by the radiation. From these considerations and with everything else being equal, the higher the concentration of contaminant, the faster the rate of oxidation. Thus it would appear that photocatalytic oxidation will find its greatest utility in problems of industrial-waste treatment, where massive contamination is involved. The researchers also believe that vigorous agitation at elevated temperature would favor faster kinetics, despite lower dissolved-oxygen levels.

Pure Oxygen Treatment

The Union Carbide Corporation has refined and improved upon a biological aeration system using pure oxygen in place of air. In this system the aeration tanks are completely covered to provide a gas-tight enclosure above the mixed-liquor. Both the liquid and gas phases are staged with a concurrent flow of the gas and liquid through the multistage system. The feed wastewater along with the recycled sludge is introduced into the first stage, together with the oxygen feed gas. The oxygen gas is fed into the enclosure above the mixed-liquor. Mechanical agitation provides the required bulk fluid motion to maintain the sludge in suspension and to ensure a uniform liquid composition. Blowers in each stage recirculate the gas in the enclosure through the mixed-liquor. The gas is piped through a hollow agitator shaft and dispersed into the mixed-liquor through a rotating-sparger device. The pressure of the gas enclosures above the mixed-liquor is automatically controlled at a few inches of water above atmospheric pressure.

The oxygen gas is fed in direct proportion to the demand of the mixed-liquor. The aeration gas flows freely from stage to stage with only a slight pressure differential to prevent back-mixing of the gas between adjacent stages. The mixed-liquor exiting the final stage is passed into a conventional settler for clarification. The vent gas will usually comprise only about 10 per cent of the volumetric flow rate of the oxygen feed gas and will be about 50 per cent oxygen. Thus it is claimed to be 95 per cent efficient in oxygen transfer. The manufacturer claims high treatment-rate performance capability of the oxygenation system. They report greater than 90 per cent reductions in both BOD and suspended solids after shorter detention times and higher loadings than the conventional aeration-activated sludge-treatment systems.

References

1. Anderson, D. R., W. D. Bishop, and H. L. Ludwig, "Percolation of citrus wastes through soil," in Proceedings of 21st Industrial Waste Conference, Purdue University Engineering Extension Series, Bulletin no. 121, 1966, p. 892.
2. Behn, V. C., "Trickling filter formulations," in Conference on Biological Treatment, at Manhattan College, New York, April 20–22, 1960, Paper no. 26.
3. Bruner, C. A., and B. G. Stephen, "Foam fractionation," *Ind. Eng. Chem.* **57**, 40 (1965).
4. Busch, A. W., and A. A. Kalinske, "The utilization of the kinetics of activated sludge in process and equipment of design," in J. McCabe and W. W. Eckenfelder (eds.), *Biological Treatment of Sewage and Industrial Wastes*, Reinhold, New York (1956), p. 277.
5. Buswell, A. M., and W. D. Hatfield, "Anaerobic fermentations," Bulletin no. 32, State of Illinois, Division of State Water Survey, Urbana, Ill. (1939).
6. Donaldson, E. C., *Subsurface Disposal of Industrial Wastes in the United States*, Information Circular 8212, Bureau of Mines, U.S. Department of the Interior, Washington, D.C. (1964).
7. Emde, W. von der, "Aspects of high rate activated sludge process," in Conference on Biological Waste Treatment at Manhattan College, New York, April 20–22, 1960, Paper no. 35.
8. Grieves, R., and C. Crandall, "Water clarification by foam separation: Bentonite as a flotation aid," *Water Sewage Works* **113**, 432 (1966).

9. Haseltine, T. R., "A rational approach to the design of activated sludge plants," in J. McCabe and W. W. Eckenfelder (eds.), *Biological Treatment of Sewage and Industrial Wastes*, Reinhold, New York (1956), p. 257.

10. Hermann, E. R., and E. F. Gloyna, "Waste stabilization ponds," *Sewage Ind. Wastes* 30, 511 and 646 (1958).

11. Heukelekian, H., "Aeration of soluble organic wastes with non-flocculent growths," *Ind. Eng. Chem.* 41, 1412 (1949).

12. Heukelekian, H., "Treatment of streptomycin wastes," *Ind. Eng. Chem.* 41, 1412 (1949).

13. Heukelekian, H., "Characteristics and treatment of penicillin wastes," *Ind. Eng. Chem.* 41, 1535 (1949).

14. Howland, W. E., "Flow over porous media as in a trickling filter," in Proceedings of 12th Industrial Waste Conference, Purdue University, 1957, p. 435.

15. Ingram, W. T., "A new approach to trickling filter design," *Proc. Am. Soc. Civil Engrs*, 82, Paper no. 999, (1956).

16. Koenig, L., *Ultimate Disposal of Advanced-Treatment Waste*, Environmental Health Series AWTR-8, U.S. Department of Health, Education and Welfare, Washington, D.C. (1964).

17. Kraus, L. S., "The use of digested sludge and digester overflow to control bulking of activated sludge," *Sewage Works J.* 17, 1177 (1945).

18. Kraus, L. S., "Dual aeration as a rugged activated sludge process," *Sewage Ind. Wastes* 27, 1347 (1955).

19. Lesperance, T. W., "Extended aeration and high rate treatment," *Water Works Wastes Eng.* 2, 40 (1965).

20. McKinney, R. E., J. M. Symons, W. G. Shifrin, and M. Vezina, "Design and operation of a complete mixing activated sludge system," *Sewage Ind. Wastes* 30, 287 (1958).

21. Nemerow, N. L., "Oxidation of enzyme desize and starch rinse textile wastes," *Sewage Ind. Wastes* 26, 1231 (1954).

22. Nemerow, N. L., "Oxidation of cotton kier wastes," *Sewage Ind. Wastes* 25, 1060 (1955).

23. Nemerow, N. L., "Holding and aeration of cotton mill finishing wastes," in Proceedings of 5th Southern Municipal and Industrial Waste Conference, Chapel Hill, N. C., April 1956, p. 149.

24. Nemerow, N. L., "Dispersed growth aeration of cotton finishing wastes: II. Effect of high pH and lowered air rate," *Am. Dyestuff Reptr.* 46, 575 (1957).

25. Nemerow, N. L., "Accelerated waste oxidation pond studies," in Proceedings of Third Conference on Biological Waste Treatment, Manhattan College, New York, April 20–22, 1960.

26. Nemerow, N. L., and J. C. Bryson, *Hancock Air Force Base Waste Stabilization Research Report*, Syracuse University Reports to U.S. Air Force, March 1960.

27. Nemerow, N. L., and J. Ray, "Biochemical oxidation of glucose by dispersed growth aeration," Chapters 1–7 of *Biological Treatment of Sewage and Industrial Wastes*, Reinhold, New York (1956).

28. Nemerow, N. L., and W. Rudolfs, "Rag, rope, and jute wastes from specialty paper mills: V. Treatment by aeration," *Sewage Ind. Wastes* 24, 1005 (1952).

29. Ontario Water Resources Commission, *Evaluation of the Oxidation Ditch as a Means of Wastewater Treatment in Ontario*, Research Publication No. 6, Ottawa (July 1964).

30. Oswald, W. J., "Fundamental factors in oxidation pond design," in Proceedings of Third Conference on Biological Waste Treatment, at Manhattan College, New York, April 20–22, 1960, Paper no. 44.

31. Pasveer, A., "New developments in the application of Kessener brushes in the activated-sludge treatment of trade-waste waters," in *Waste Treatment* (ed. P. Isaac), Pergamon Press, New York (1959), pp. 126–155.

32. Pettet, A. E. J., T. G. Tomlinson, and J. Hemens, "The treatment of strong organic wastes by anaerobic digestion," *J. Inst. Public Health Engrs* 170 (1959).

33. Rankin, R. S., "Performance of biofiltration plants by three methods," *Proc. Am. Soc. Civil Engrs* 79, Separate No. 336 (1953).

34. Sawyer, C. N., "Activated sludge modifications," *J. Water Pollution Control Federation* 32, 233 (1960).

35. Shoen, H. M., E. Rubin, and D. Ghosh, "Radium removal from uranium mill wastewater," *J. Water Pollution Control Federation* 34, 1026 (1962).

36. Schulze, K. L., and H. S. Foth, "New low cost secondary treatment by new cavitation system," *Water Sewage Works* 102, 74 (1955).

37. *Sewage Stabilization Ponds in the Dakotas*, Joint report by North and South Dakota Departments of Health and the United States Department of Health Education and Welfare, (1957).

38. Struzeski, E. J., and N. L. Nemerow, "Dispersed growth aeration of protein—glucose mixtures," in Proceedings of 12th Industrial Waste Conference,

Purdue University, May 1957, p. 145.

39. Tapleshay, J. A., "Total oxidation treatment of organic wastes," *Sewage Ind. Wastes* **30**, 652 (1958).
40. Ullrich, A. H., and M. W. Smith, "The Biosorption process of sewage and waste treatment," *Sewage Ind. Wastes* **23**, 1248 (1951).
41. Ullrich, R. A., and M. W. Smith, "Operation experience with activated sludge—Biosorption at Austin, Texas," *Sewage Ind. Wastes* **29**, 400 (1957).
42. Velz, C. J., "A basic law for the performance of biological filters," *Sewage Works J.* **20**, 607 (1948).
43. "Wet combustion of wastes," *Power Eng.* **59**, 63 (1955).
44. Zobell, C. E., "The influence of solid surface upon the physiological activities of bacteria in sea water," *J. Bacteriol.* **33**, 86 (1937).
45. Kinny, L. C., and Ivanuski, V. R., "Photolysis Mechanisms for Pollution Abatement," Taft Water Research Center Report No. TWRC-13, Cincinnati, Ohio, October 1969.

Suggested Additional Reading

Flower, W. A., "Spray irrigation—A positive approach to a perplexing problem," in Proceedings of 20th Industrial Waste Conference, Purdue University, 1965, p. 679.

Ling, J. T., "Pilot study of treating chemical wastes with an aerated lagoon," *J. Water Pollution Control Federation* **35**, 963 (1963).

Luley, H. G., "Spray irrigation of vegetable and fruit processing wastes," *J. Water Pollution Control Federation* **35**, 1252, (1963).

Oswald, W. J., and H. B. Gotaas, "Photosynthesis in sewage treatment," *Trans. Am. Soc. Civil Engrs.* **122**, 73 (1957).

Parker, C. D., "Food treatment waste treatment by lagoons and ditches at Shepparton, Victoria, Australia," in Proceedings of 21st Industrial Waste Conference, Purdue University, 1966, p. 284.

Deep Well Injection

Barraclough, J. T., "Waste injection into deep limestone in northwestern Florida," *Groundwater* **4**, 22 (1966).

Hundley, C. L., and J. T. Matulis, "Deep well disposal," in Proceedings of 17th Industrial Waste Conference, Purdue University, 1962, p. 175.

Koenig, L., "Advanced waste treatment," *Chem. Eng.* **70**, 210 (1963).

Powers, T. J., and G. W. Querio, "Check on deep-well disposal for specially troublesome waste," *Power* **105**, 94 (1961).

"Production waste goes underground at Holland-Suco. Mich.," *Water Sewage Works* **113**, 329 (1966).

Querio, C. W., and T. J. Powers, "Deep well disposal of industrial waste water," *J. Water Pollution Control Federation* **34**, 136 (1962).

Selm, R. P., "Deep well disposal of industrial wastes," in Proceedings of 14th Industrial Waste Conference, Purdue University, 1959.

Talbot, J. S., "Deep well method of industrial waste disposal," *Chem. Eng. Progr.* **60**, 1 (1964).

Warner, D. L., "Deep well waste injection—Reaction with aquifer water," *J. Sanit. Eng. Div. Am. Soc. Civil Engrs.* **92**, no. SA4, 95 (1966).

"Waste well goes down over two miles," *Eng. News-Record* **165**, 32 (1960).

Winar, R. M., "The disposal of wastewater underground," *Ind. Water Eng.* **4**, 21 (1967).

Foam Phase Separation

Advanced Waste Treatment Research, Publication no. AWTR-14, U.S. Public Health Service, Cincinnati, Ohio (1955).

Brown, D. J., "A photographic study of froth flotation," *Fuel Soc. J. Univ. Sheffield* **16**, 22 (1965).

Eldib, I. A., "Foam fractionation for removal of soluble organics from wastewater," *J. Water Pollution Control Federation* **33**, 914 (1961).

Gassett, R. B., O. J. Sproul, and P. F. Atkin, Jr., "Foam separation of ABS and other surfactants," *J. Water Pollution Control Federation* **37**, 460 (1965).

Grieves, R. B., C. J. Crondall, and R. K. Wood, *Air Water Pollution* **8**, 501 (1964).

Rubin, E., R. Everett, Jr., J. J. Weinstock, and H. M. Shoen, *Contaminant Removal from Sewage Effluents by Foaming*, Publication no. 999-WP-5, U.S. Public Health Service, Cincinnati, Ohio (1963).

Bio-Disc

Antonie, R. L., and F. M. Welch, "Preliminary results of a novel biological process for treating dairy wastes," in Proceedings of 24th Industrial Waste Conference, Purdue University, 1969.

Hartmann, H., "Investigation of the biological purification of sewage using the Bio-Disc filter," Stuttgarter Berichte zur Siedlungswasserwirtschaft no. 9, R. Oldenbourg, Munich (1960).

Pöpel, F., "Estimating construction and output of Bio-Disc

filter plants," Stuttgarter Berichte zur Siedlungs-wasserwirtschaft no. 11, R. Oldenbourg, Munich (1964).

Welch, F. M., "Preliminary results of a new approach in the aerobic biological treatment of highly concentrated wastes," in Proceedings of 23rd Industrial Waste Conference, Purdue University, 1968.

Questions

1. Why has removal of organic dissolved solids long been the most important and most difficult phase of industrial-waste treatment?
2. What is the basis for most treatment processes for dissolved organic removal?
3. What is necessary for optimum biological treatment?
4. What are the advantages and disadvantages of lagooning? Of activated sludge?
5. Name and describe some modifications of the activated-sludge process.
6. What is the principle of dispersed-growth aeration? What are its advantages and disadvantages?
7. What is meant by trickling filtration and when would you use it in preference to biological aeration?
8. Is spray irrigation a suitable method of disposing of dissolved organic matter?
9. Can you use subsurface disposal for dissolved organic solids? When?
10. What is the theory of the Bio-Disc treatment system? Does it resemble trickling filtration or activated-sludge treatment?
11. What methods are used for nonbiological oxidation of dissolved organic wastes? Explain the principles of ozonation, chlorination, pure oxygen, and photolysis treatments.
12. What is wastewater scavenging and what are its limitations?

CHAPTER 14

TREATMENT AND DISPOSAL OF SLUDGE SOLIDS

Of prime importance in the treatment of all liquid wastes is the removal of solids, both suspended and dissolved. Once these solids are removed from the liquids, however, their disposal becomes a major problem. Unfortunately, waste engineers spend more time and money removing the solids than finally treating and disposing of them, so that often a poor solids-disposal program will cause trouble in an otherwise properly designed and operated waste-treatment plant. When the solids-disposal system is poor, the solids tend to build up in the flow-through treatment units and overall removal efficiencies then begin to decrease. Therefore, proper sludge handling enhances the overall treatment of all wastes. The following list contains most of the methods commonly used to deal with sludge solids: (1) anaerobic and aerobic digestion; (2) vacuum filtration; (3) elutriation; (4) drying beds; (5) sludge lagooning; (6) wet combustion; (7) atomized suspension; (8) drying and incineration; (9) centrifuging; (10) sludge barging; (11) landfill; and (12) miscellaneous methods.

14.1 Anaerobic and Aerobic Digestion

Anaerobic digestion is a common method of readying sludge solids for final disposal. All solids settled out in primary, secondary, or other basins are pumped to an enclosed airtight digester, where they decompose in an anaerobic environment. The rate of their decomposition depends primarily on proper seeding, pH, character of the solids, temperature, and degree of mixing of raw solids with actively digesting seed material. Digestion serves the dual purpose of rendering the sludge solids readily drainable and converting a portion of the organic matter to gaseous end-products. It may reduce the volume of sludge by as much as 50 per cent organic matter reduction. After digestion, the sludge is dried and/or burned, or used for fertilizer or landfill.

Two main groups of microorganisms, hydrolytic and methane, carry out digestion. Hydrolytic bacteria exist in great numbers in sewage and waste sludges and are capable of rapid rates of reproduction; they are saprophytic microorganisms that attack complex organic substances and convert them to simple organic compounds. Among these saprophytes are many acid-forming bacteria which produce fatty acids of low molecular weight, such as acetic and butyric, during degradation processes. In some cases, such acids are produced in quantities sufficient to lower the pH to a level where all biological activity is arrested.

Fortunately, methane bacteria, the other group of microorganisms, are capable of utilizing the acid and other end-products formed by the hydrolytic bacteria. Methane producers, however, are sensitive to pH changes and proliferate only within a narrow pH range of 6.5 to 8.0, with an optimum of 7.2 to 7.4; furthermore, they are few in number and reproduce slowly. Consequently, organic acids may form faster than they can be assimilated by the limited population of methane bacteria. As a result, the pH may be lowered and conditions made even more unfavorable for methane bacteria. When this happens, lime is usually added and the digestion process stopped until normal conditions return.

The proper environment for both types of bacteria requires a balance between population of organisms, food supply, temperature, pH, and food accessibility. The following factors are measures of the effectiveness of digestive action: gas production (both quantity and quality), solids balance (total, volatile, and fixed), BOD, acidity and pH, volatile acids, grease, sludge characteristics, and odor.

As mentioned before, fermentation (digestion) of organic matter proceeds in two stages: (1) hydrolytic action, converting organic matter to low-molecular-weight, organic acids and alcohols, and (2) evolution of carbon dioxide and the simultaneous reduction to methane (carbon dioxide is actually consumed). The following general equations represent the digestion of carbohydrates, fats, and proteins:

Carbohydrates:

$$(C_6H_{10}O_5)_x + x\ H_2O \rightarrow x(C_6H_{12}O_6),$$
$$C_6H_{12}O_6 \rightarrow 2C_2H_5OH + 2CO_2,$$
$$2CH_3CH_2OH \xrightarrow{+CO_2} 2CH_3COOH + CH_4,$$
$$CH_3COOH \rightarrow CH_4 + CO_2.$$

Fats:

$$
\begin{array}{l}
H_2C\!-\!O\!-\!\overset{\overset{\displaystyle O}{\|}}{C}\!-\!R_1 \\[2mm]
\\
HC\!-\!O\!-\!\underset{\displaystyle O}{\overset{\|}{C}}\!-\!R_2 \ +\ 3HOH\ \rightarrow \\[2mm]
\\
H_2C\!-\!O\!-\!\underset{\displaystyle O}{\overset{\|}{C}}\!-\!R_3
\end{array}
$$

Glycerol and Acid:

$$
\begin{array}{l}
H \\
H\!-\!C\!-\!OH \\
H\!-\!C\!-\!OH \\
H\!-\!C\!-\!OH \\
H \\
\text{Glycerol}
\end{array}
\qquad
\begin{array}{l}
HO\!-\!\overset{\overset{\displaystyle}{}}{C}\!-\!R_1 \\
\| \\
O \\
HO\!-\!C\!-\!R_2 \\
\| \\
O \\
HO\!-\!C\!-\!R_3 \\
\| \\
O \\
\text{Acid}
\end{array}
$$

Alpha oxidation of acids:
$$4RCH_2COOH + 2HOH \rightarrow 4RCOOH + CO_2 + 3CH_4.$$

Beta oxidation of acids:
$$2RCH_2CH_2COOH + CO_2 + 2HOH \rightarrow$$
$$2RCOOH + 2CH_3COOH + CH_4,$$
$$CH_3COOH \rightarrow CH_4 + CO_2.$$

Proteins:

$$
\begin{array}{l}
H \\
R\!-\!C\!-\!COOH \xrightarrow[\text{Deaminase}]{HOH} \\
NH_2
\end{array}
NH_3 +
\begin{array}{l}
H \\
R\!-\!C\!-\!COOH, \\
OH
\end{array}
$$

$$
\begin{array}{l}
H \\
R\!-\!C\!-\!COOH \xrightarrow[\text{Decarboxylase}]{} \\
OH
\end{array}
\begin{array}{l}
H \\
R\!-\!C\!-\!H + CO_2, \\
OH
\end{array}
$$

$$2RCH_2OH + CO_2 \rightarrow 2RCOOH + CH_4,$$
$$RCOOH \rightarrow CH_4 + CO_2.$$

One hypothesis is that each molecule of methane arises from a reduction of one molecule of carbon dioxide. In other words, carbon dioxide acts as the hydrogen acceptor, while the alcohol acts as the hydrogen donor, as in the following equation:

ethyl alcohol   carbon dioxide   acetic acid
$$2C_2H_5OH + CO_2 + H_2O \rightarrow 2CH_3COOH + CH_4 + 2H_2O$$
hydrogen donor hydrogen acceptor   methane

One can readily see that carbon dioxide is an important food constituent. In mixed cultures, carbon dioxide is produced by other organisms and therefore becomes more available than sulfates or nitrates. Buswell [11] describes fermentation as a chain of reactions involving the transfer of hydrogen.

The slowest reaction in the degradation process, production of methane, is therefore the rate-controlling reaction. The essential physiological characteristics of methane bacteria are: (1) they are obligate anaerobes; (2) they require carbon dioxide as a hydrogen acceptor; (3) as hydrogen donors, they use simple organic substrates, such as calcium acetate, butyrate, and ethyl and butyl alcohols; (4) their nitrogen source is ammonia; (5) they develop at a slow rate owing to low energy yields; (6) they do not form spores; (7) they are very sensitive to changes in pH.

Buswell [11] concluded that the higher the percentage of carbon atoms in the fatty-acid substrate, the higher the percentage of methane in the gas. Barker [4] established the following unique features of methane fermentation:

1) It takes place in mixed or enriched cultures and hence may be maintained continuously on a large scale.

2) It is applicable to any type of substrate except lignin and mineral oil.

3) The reaction is quantitative and converts the entire substrate to carbon dioxide and methane.

4) There is no specific temperature limitation in the range of 0 to 55°C, but once the culture has been acclimated to a certain temperature, a drop of two degrees may completely interrupt methane fermentation and render obstructive the accumulated acids.

5) The presence of inert solid matter is important, and hence addition of straw or sawdust to industrial wastes may be required.

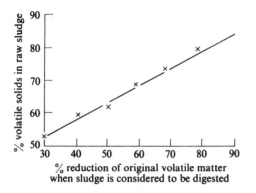

Fig. 14.1 Reduction of volatile matter in raw sludge by digestion [29].

6) If the substrate concentration is too great, volatile acids build up and inhibit the fermentation, especially when their build-up occurs faster than their subsequent conversion to methane. Keeping the volatile acid level below 3000 ppm, and closer to 2000, helps the situation, but alkali addition will not alleviate it, since it is not a pH effect. Mineral salts begin to inhibit the fermentation at 4000 ppm, and 50 ppm of nitrate nitrogen inhibit it completely.

The extent of reduction of volatile solids by digestion depends in part on the amount of volatile matter in the raw sludge. Schlenz [29] found that when volatile solids in raw sludge increased from 55 to 80 per cent, the reduction in volatile matter increased from 35 to 85 per cent. This is shown graphically in Fig. 14.1.

The usual unit-capacity requirements may be reduced, provided the operations are controlled and carried out as follows [31]: (1) tank contents must be agitated to maintain an even mixture of raw and digesting solids; (2) raw sludge must be added continuously to the digestion unit; (3) raw sludge must be concentrated or prethickened before being added to the digester. Two-stage digestion, with the first stage used primarily for active digestion and the second stage for storage and sludge consolidation, is often carried out in two separate tanks. It is usually more economical in large plants with continuous operation.

Aerobic digestion is now playing an important role in small plants. It is claimed that less-skilled operators are required; also, air is normally available in these plants, since secondary treatment of the liquid-waste fraction is becoming rather commonplace [13].

14.2 Vacuum Filtration

Vacuum filtration is a means of dewatering sludge solids which has become popular because the volume of solids for ultimate disposal is reduced and the sludge is drier than it would otherwise be, so that "handleability" is improved. Large plants are increasing their use of vacuum filtration. Some plants filter chemically precipitated and/or plain settled sludge, while others filter digested sludge. In a typical vacuum-filtration unit, a porous cylinder overlying a series of cells revolves about its axis with a peripheral speed somewhat less than one foot per minute, its lower portion passing through a trough containing the sludge to be dried. A vacuum inside the cylinder picks up a layer of sludge as the filter surface passes through the trough, and this increases the vacuum. When the cylinder has completed three-quarters of a revolution, a slight air pressure is produced on the appropriate cells, which aids the scraper, or strings, to dislodge the sludge in a thin layer. Sometimes it is necessary to add chemicals, such as lime and ferric chloride, as sludge conditioners prior to filtration. Filtering rates should be from 2 to 10 pounds of dry solids per square foot per hour. Vacuum filters are available in diameters up to about 20 feet and in many different lengths.

The quality of the filter medium (the material covering the cylinder) is important in the performance and life of the filter. In the past, woven-fabric filter media have been widely used. The physical process of solids retention on woven filters is a combination of at least three actions: (1) straining action, in which particles *larger* than the filter-medium openings cling to the filter; (2) adsorption, or attraction, to the filter of particles smaller than the openings in the filter medium; (3) filtration of particles of different sizes, which cling to already filtered, caked material. The first two actions prevail at the onset of filtration but, as the "cake" builds up, the third is responsible for the greatest amount of solids removal. Thus the problem arises that, unless the cake is removed completely and the fiber filter medium kept clean continually, the filter will clog or "blind."

(a)

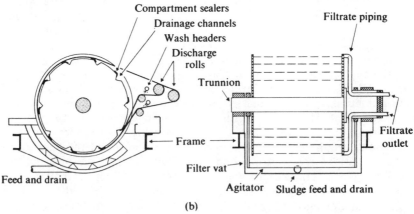

(b)

Fig. 14.2 (a) The Coilfilter, a patented machine for the vacuum filtration of sludge. This particular machine, in use since 1953 at the sewage-treatment plant at St. Charles, Illinois, has filtering media made up of two layers of alloy steel coiled springs, each spring made endless by joining its two ends with a threaded plug. These springs discharge the filter cake after each revolution of the cylinder, and are then washed before they re-enter the vat for another cycle. The material at the left which looks like a length of corduroy is actually a layer of sludge [10]. (b) Schematic drawing of the Coilfilter shown in (a). (Courtesy Komline Sanderson Co.)

Tiller and Huang [35] report that there is a paucity of theory and research on filtration through porous media. Three reasons for this deficiency are: (1) complexity of vacuum-filtration machinery, (2) difficulty of experimentally reproducing the precipitates found in filter beds, and (3) insufficient interest on the part of researchers. They also report that, although flow through filter beds is almost always viscous, no reliable theory has been developed as to the relation between permeability and porosity of the filter medium, as affected by compressive pressure.

A major step toward lengthening the life and decreasing the operational problems of vacuum-filtration systems is the use of stainless-steel, coil-spring filter media. A representative unit of this type, the Coilfilter, is shown in Fig. 14.2 [10].

14.3 Elutriation

Elutriation is a process of improving filtration by washing the sludge. It reduces the alkalinity—and therefore the lime coagulant demand—of sludge by upgrading the biochemical quality of the sludge water before chemicals are added [16]. There are three practical methods of washing sludge solids; the equipment used in all cases is relatively simple, with upward-flow tanks frequently used.

1) Single-stage elutriation, which involves one batch at a time, is a fill-and-draw procedure: sedimentation and decantation are performed in a single step.

2) Two-stage elutriation involves repeating the single-stage steps on the elutriated sludge, using fresh water on the second wash. In small plants, the same settling tank may be used for both stages.

3) In larger plants (6000 to 24,000 pounds of solids per day), a second tank, connected in series with the first, is usually employed. Such a two-tank system can also be used for countercurrent washing. With this system, the fresh water is added only to the second-stage washing, and the decanted elutriate (or top water) from this tank flows by gravity to mix with the sludge entering the first tank.

Since the degree of chemical fouling [15] resulting from digestion can be conveniently measured in terms of alkalinity, an elutriated sludge can be defined as one that has had the alkalinity of its water reduced by dilution with water of lower alkalinity, sedimentation, and decantation. Advantages of elutriation as a preliminary to sludge dewatering on vacuum filters include elimination of ammonia odors and of the need

to use lime in sludge conditioning. Elutriation may also reduce the capacity requirements of secondary digesters (used for storage and additional digestion to ensure optimum filtration), and it is particularly helpful in that it permits small plants to use vacuum filters to advantage. Genter claims that elutriation reduces the ratio of sludge water to the mineralized sludge solids; thus there is a marked decrease in the chemicals required for conditioning. The savings in ferric chloride are illustrated in Fig. 14.3, which is based on Genter's data [15].

Genter [16] also discusses a method of predicting the final alkalinity of elutriated sludge by a formula. Assuming that a equals the volumes of pure water added to one volume of fouled sludge mixture, he obtains the following relationships:

$a + 1 =$ total volume of mixed sludge and clean water;

$1/(a + 1) =$ fraction of original concentration of fouling agent left if solids are allowed to settle back to a washed sludge equivalent to the original volume and the added volume of water is siphoned off;

$1/(a + 1)^2 =$ fraction of original concentration of fouling agent if this same dilution, sedimentation, and decantation technique is repeated.

Therefore, the fraction of original fouling agent left in the final sludge is $1/(a^2 + 2a + 1)$ if the second wash water is decanted for a new first wash and the two elutriation tanks are placed on countercurrent series. For example, if four volumes of pure water are used to wash a digested sludge of 3000 ppm

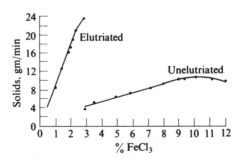

Fig. 14.3 Effect of elutriation on $FeCl_3$ required for conditioning of sludge [15].

alkalinity, the alkalinity left in the elutriated sludge after countercurrent washing in two tanks is

$$\frac{3000}{(4)^2 + (2 \times 4) + 1} = 120 \text{ ppm.}$$

14.4 Drying Beds

Sludge-drying beds remove moisture from sludge, thereby decreasing its volume and changing its physicochemical characteristics, so that sludge containing 25 per cent solids can be moved with a shovel or garden fork and transported in watertight containers.

Sludge filter beds are made up of 12 to 24 inches of coarse sand, well-seasoned cinders, or even washed grit from nearby grit chambers and about 12 inches of coarse gravel beneath the sand. The upper 3 inches of gravel particles are $\frac{1}{8}$ to $\frac{1}{4}$ inch in diameter. Below the gravel, the earth floor of the bed is pitched to a slight grade into open-joint tile underdrains 6 or 8 inches in diameter. These tiles may be laid from 4 to 20 feet apart on centers, depending on the porosity of the coarse gravel. Disposing of the underdrain liquor sometimes poses a problem; this should never be discharged without an analyses of its constituents and usually some form of treatment. Several smaller, rectangular beds serve the purpose better than one large filter bed. These beds may be covered with glass or plexiglass when weather conditions demand, in which case ventilation must be provided to dissipate the hot, wet air above the beds.

Generally speaking, raw settled sludge does not drain well on sand drying beds. Some form of pretreatment—digestion, elutriation and/or chemical treatment—is usually required. Well-digested sewage sludge will dewater more readily than partly digested sludge [30]. However, prolonged storage of digested sludges decreases drainability, since the gases present initially permit more drainage of moisture through the filtering medium, thus reducing the evaporation cycle. A high total-solids content in digested sludges naturally permits greater removal of dry solids per year from sludge beds.

Drying time is dependent on dosing depth, 8 inches being generally accepted as most desirable for rapid drying, and on climate. It is, naturally, short in regions of plentiful sunshine, scant rainfall, and low relative humidity, such as certain arid areas of the South where summers are long. Wind velocity also affects speed of sludge-drying on the beds. In fact, all the factors enhancing evaporation will also aid in drying sludge. Cox [12] derived the following equation for calculating the rate of evaporation of water, which may also apply to sludge drying although exact values of constants may vary from water to sludge water:

$$E = \frac{(e_a - e_d + 0.0016 \,\Delta T)}{(0.564 + 0.051 \,\Delta T + W/300)},$$

where

E = evaporation (inches/day)
e_a = saturated vapor pressure at air temperature
e_d = actual vapor pressure
ΔT = difference between mean temperature of the air and that of the water
W = velocity of the wind (miles/day)

Meyer's formulation is also widely used:

$$E = C(V - v)\left(1 + \frac{W}{10}\right),$$

where

E = evaporation (inches) for a given unit of time
V = saturation vapor pressure at the water temperature (inches of mercury)
v = actual vapor pressure of the air, 25 feet above ground
W = wind velocity (mph), 25 feet above ground
C = coefficient, varying with unit of time used and depth of water (varies from 0.36 to 0.50).

In addition to evaporation, the drying rate is also influenced by capillary action, which causes water to rise from the depths of the sludge to the evaporative surface.

In the case of domestic-sewage sludge, engineers estimate that approximately 20 to 25 pounds of dry solids can be loaded onto one square foot of properly designed sand-base drying bed each year. Haseltine [18] takes exception to this unit-of-loading estimate and suggests a "gross bed loading," which takes into account the number of pounds of solids applied per square foot per 30 days of actual bed use. For example, if sludge which has a density of 62.5 lb/ft³ and contains 10 per cent solids is applied 12 inches deep and removed after 40 days, the gross bed loading is

$$\frac{62.5 \times 0.10 \times 30}{40} = 4.69 \text{ lb/ft}^2/30 \text{ days.*}$$

Haseltine also develops the following straight-line relationship between the gross bed loading (Y) and the percentage of solids in applied sludge (X) from data supplied by 14 different plants for periods of operation up to 14 years:

$$Y = 0.96X - 1.75.$$

The gross bed loading Y varied from 0 to 10 and X varied from 0 to 14. He concluded that, next to temperature, the solids content of the sludge in drying beds is the most important factor influencing bed performance. The amount of moisture to be removed from the sludge is the third most important factor.

14.5 Sludge Lagooning

Lagoons may be defined as natural or artificial earth basins used to receive sludge. Lagooning is practiced when the economics of the situation (money and land) indicate its use, since it is a relatively inexpensive method of treating waste sludges. However, there are many other factors to be considered: (1) nature and topography of the disposal area; (2) proximity of the site to populated areas; (3) meteorological conditions, especially whether prevailing winds blow toward or away from populated areas; (4) soil conditions; (5) chemical composition of sludges, with special consideration given to toxicity and odor-producing constituents; (6) proximity to surface- or ground-water supplies; (7) effect of waste materials on the porosity of the soil; (8) means of draining off the supernatant liquor to provide more space in the lagoon; (9) fencing, and other safety measures, when lagoons are deeper than five feet; (10) nuisances, such as weed growth, odors, and insect breeding.

Lagooning of wastes in limestone areas is particularly hazardous because of the channels and cavities found underground in these formations [25]. Ordinarily groundwater moves slowly, sometimes less than a foot a day, depending on the fineness of the aquiferous sand through which it per-

colates and the degree of saturation of the sand. In limestone country, water may travel vertically and laterally at much higher velocities, so that sludge lagooned on high ground may quickly contaminate large portions of valuable ground-water supplies. Quite often, manufacturing plants bulldoze out a sludge lagoon every year or two, the frequency depending on sludge build-up and soil conditions.

Bloodgood [5] states that at least one pound of raw sewage solids can be digested per year per 0.17 cubic foot of lagoon capacity. However, if lagoons are to be used for both digestion and dewatering, one pound of raw sludge solids requires about 0.4 cubic foot per year of lagoon capacity, provided air-dried sludge is removed as soon as it becomes ready for hauling.

14.6 The Wet Combustion Process

The Zimpro process is a relatively innovative treatment for sludge. It operates on the basic principles that (1) organic matter contained in an aqueous solution can be oxidized and whatever heat value it contains released and (2) oxidation at this stage is more effective than if the water were first evaporated and the residue used as fuel in a conventional boiler. Since heat is liberated by a fuel only when it is subjected to combustion in the presence of air, the Zimpro process depends on air being forced into a reactor vessel. One objective of this process is the production of the maximum number of Btu's from the organic matter in a waste effluent per pound of compressed air fed into the reactor.

Since the Zimpro process eliminates conventional filters, chemicals, sludge-digestion units, incinerators, and auxiliary equipment, it reduces space and land requirements. The end-products are steam, nitrogen, CO_2, and ash. The effluent gases from the reactor, having been "scrubbed" with water, contain no fly ash and are practically odorless.

In the treatment of sewage sludge, oxidation is brought about by continuously pumping the sludge and a proportionate amount of air (both sludge and air at elevated temperatures and pressures) into a reactor vessel. Combustion occurs as the oxygen in the compressed air combines with the organic matter in the sludge to form CO_2, N_2, and steam, while the ash remains in the residual water. The reactor, and

---

*Specific gravity of wet sludge assumed, 1.0.

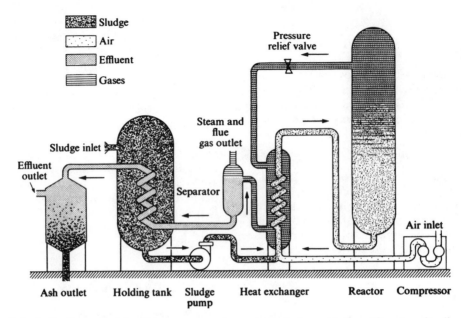

Fig. 14.4 Schematic diagram of the Zimpro process for sewage-sludge oxidation. (Courtesy Sterling Drug Co.)

the whole process system as well, is automatically maintained at a constant pressure and the products of the combustion are continuously removed from the reactor. If the concentration of volatile matter is high and the sewage sludge concentration is great enough (> 5 per cent), the steam, plus the gases (CO_2 and N_2) which are products of combustion, will contain more than enough energy to run the air compressors and pumps used in the process. The residual hot water from the reactors is utilized in heat exchangers that raise the temperature of the incoming sludge and air sufficiently to cause oxidation to begin as soon as they come together in the reactor. In this way, once the process is started, no external heat or power is required to sustain the combustion.

Equipment required for the Zimpro process includes: a compressor, an air receiver, a high-pressure sludge pump, a sludge-storage tank with agitators, heat exchangers, a reactor, a separator, and a cooler. A schematic drawing of the process is presented in Fig. 14.4. The manufacturer claims that: "Units achieve 80 to 90 per cent reduction of insoluble organic content of sewage sludge by oxidation with-

out flame. Sludge is burned without dewatering or pretreating. The unit operates continuously at pressures of 500 to 600 psig and temperatures of 420°F. End products are substantially inorganic, inert, biologically stable ash; residual water; and odor-free gaseous products of combustion (carbon dioxide, nitrogen, and steam). The plant is designed for automatic operation with minimal maintenance. An air compressor and sludge pump are the only equipment components with moving parts. Power requirement is approximately 50 hp for a one-ton unit (dry weight). Building and land-space requirements are nominal."* Teletzke [34] describes the low-pressure Zimpro treatment, which operates in the range of 150 to 300 psi and at about 300°F. He portrays low-pressure, wet-air oxidation as an economical and flexible method of producing a sterile, drainable, and completely acceptable end-product for ultimate disposal.

---

*New Zimpro Sludge Oxidation Units for Smaller Communities, Sterling Drug Co., Rothschild, Wisconsin.

14.7 Atomized Suspension

The atomized-suspension technique consists of atomizing the waste liquor or slurry in the top of a tower, the walls of which are maintained at an elevated temperature by hot gases circulating through a jacket, a method described by Gauvin [14, 26]. No air, or other foreign gas, is introduced into the equipment, which sharply distinguishes this technique from spray drying. The developers claim that in the immediate range of the nozzle the finely divided droplets (20 to 25 mm in diameter) quickly decelerate from the high initial velocity imparted by the atomizer to their slower terminal velocity and then become dispersed in the vapor produced by their own evaporation. The suspension thus created flows down the reactor in nearly streamline motion. Evaporation, quickly completed, is followed by drying. At the end of the drying zone, dried particles can be subjected to a sequence of chemical reactions, such as oxidation, reduction, nitration, sulfonation, and so forth, through the injection of the proper internal gaseous reactants (in the presence of a powdered catalyst, if necessary). When it leaves the reactor at the bottom, the suspension consists of a solid residue (which is recovered in cyclone collectors), large amounts of steam (which is condensed and utilized), and byproduct gases (which can be further processed for recovery or piped away for disposal).

Advocates of the atomized-suspension process claim that the only outside energy required is that used for pumping of the liquid—an almost negligible amount. A striking feature of the recovery flow sheet is the complete absence of blowers or compressors, although large volumes of gases and vapors are continuously flowing through the system. Need for them is eliminated by the efficient utilization of the pressure generated in the reactor during evaporation. A typical flow sheet for Gauvin's process [14, 26] is shown in Fig. 14.5.

14.8 Drying and Incineration

A large volume of sludge can be reduced to a small volume of ash, which is free from organic matter and therefore easily disposable, by a combination of heat drying and incineration [32]. Flash drying involves drying sludge particles in suspension in a stream of hot gases, which ensures practically instantaneous

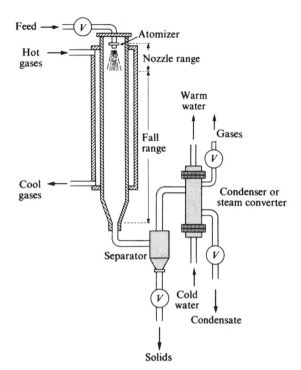

Fig. 14.5 Apparatus for the atomized-suspension technique [14, 26].

removal of moisture. When hot gases created by the drying and oxidation of the sludge itself are used directly for drying, there are no conversion losses. After the flash drying, the gas containing sludge particles usually passes to cyclone separators, where the dried sludge is separated from the moisture-carrying cooler gases.

Flash-dried sludge is utilized as fertilizer, soil conditioner, or for other valuable purposes. Unused dried sludge can be incinerated by blowing it through a duct to a burner in the combustion chamber of a furnace. The sludge blower, in addition to conveying the sludge to the furnace, also supplies the major portion of the air required for combustion. To eliminate odors, preheated gases after combustion are returned to the combusting sludge. To eliminate fly ash, the cooled gas after combustion is drawn through an ash collector by induced-draft fans, and the fly ash settles out by centrifugal action and is discharged automatically into the furnace bottom. This ash can be removed from time to time, either by shoveling or by mixing it with water and pumping it out to be used as landfill.

Whether the ultimate aim is to dry the sludge for use as a soil additive or to incinerate it to a sterile ash, it is necessary first to evaporate the free moisture from the solids, remove it in the form of a gas, and discharge it to the atmosphere. This gas is referred to as the evaporator load. Only high-temperature (1200 to 1400°F) deodorization is effective in controlling odors from sludge incinerators.

When sludge is to be incinerated, the heat released in the furnace is also of importance: the furnace volume should be ample to allow a heat release of X Btu's per cubic foot of furnace (generally held at 12,000 Btu's per cubic foot of furnace volume per hour to ensure long life of walls and furnace). The heat input is determined by multiplying the pounds of dry solids to be incinerated per hour by the gaseous products of the volatile-solids content and their heat value. The furnace volume required can therefore be computed by dividing this heat input by 12,000. Thermal efficiencies of 30 to 60 per cent can be expected from incinerators. The lower the stack temperature, the higher the thermal efficiency [20]. This relationship is shown in Fig. 14.6; in Fig. 14.7 a flow diagram is presented [20] to show heat balance for a flash-drying and incineration system.

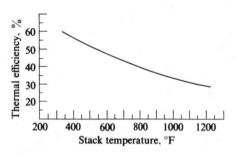

Fig. 14.6 Effect of stack temperature on thermal efficiency [20].

To calculate the rate of drying during the constant-rate period (after the temperature of the material adjusts itself to the drying conditions), either the mass-transfer or the heat-transfer equation may be used [21]:

(mass transfer) $W = k_y'(H_i - H)A;$

(heat transfer) $W = \dfrac{h_y(t - t_i)A}{\lambda_i},$

where

W = evaporation rate (lb/hr)
A = drying area (ft²)
h_y = heat-transfer coefficient (BTU/ft²/hr/°F)
k_y' = mass-transfer coefficient (lb/ft²/hr for a unit of humidity difference)
H_i = humidity of air at interface (lb water/lb dry air)
H = humidity of air (lb water/lb dry air)
t = temperature of air (°F)
t_i = temperature at interface (°F) and
λ_i = latent heat at temperature t_i (BTU/lb).

The heat-transfer coefficient, h_y, is estimated to be about $0.128\,G^{0.8}$ when air flows parallel to the sludge surface and about $0.37\,G^{0.37}$ when air flows perpendicular to the sludge surface (G = the mass velocity in lb/ft²/hr).

Pit incineration has been used to dispose of certain solid and semi-solid wastes. The incinerator consists of a rectangular pit lined with firebrick, to which air is supplied so as to retain particulates and to allow complete combustion. This disposal method is simple in concept and operation and is especially adaptable to situations where the waste requires batch incineration. It has been used for disposal of synthetic organics and is currently being studied for disposal of paint sludges in the automotive industry [3].

14.9 Centrifuging

Centrifugation is a method of concentrating sludge to enhance final disposal. One of the factors which made centrifugal concentration unacceptable in the earlier installations was its low efficiency—large amounts of fine particles were returned to the system with the supposedly clarified effluent. Newer installations [7], using 20-hp built-in drive motors, can handle 3000 to 4000 gallons per hour of waste sludge, containing 0.5 to 0.75 per cent solids on a dry basis. Only 11 hp is required once the centrifuge reaches operating speed (6100 rpm). The resulting sludge is concentrated to about 5 per cent solids and the effluent contains about 300 ppm solids. The centrifugal force throws the denser solid material to the wall of the centrifuge bowl, where it is discharged

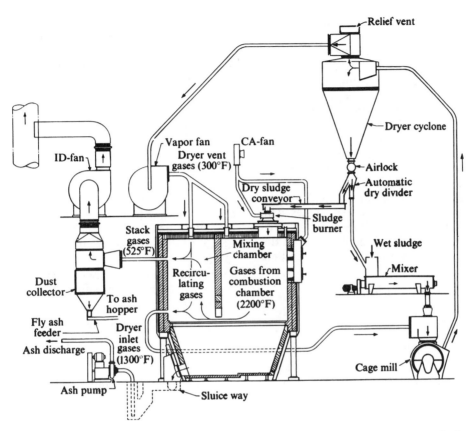

Fig. 14.7   Flow diagram for heat balance for a flash-drying and incineration system [20].

through nozzles located in the periphery. One bowl [7] is equipped with 12 nozzle openings, such that various numbers of discharge nozzles can be utilized, depending on the amount of solids in the feed liquor and the results desired. Use of the centrifuge for higher concentrations is limited by the capability of the pumps which discharge concentrated sludge from the centrifuges. The effluent from which the solids are separated travels toward the center of the centrifuge bowl through intermediate discs; as it discharges from the upper cover, it is claimed to average approximately 300 ppm solids. Centrifuged sludge is discharged from the lower cover of the centrifuge into a sump, from which it can be pumped to digesters or other final sludge-treatment units. Figure 14.8 is a schematic diagram of a centrifuge bowl. Blosser and Caron [6] expect the costs of centrifuging paper-

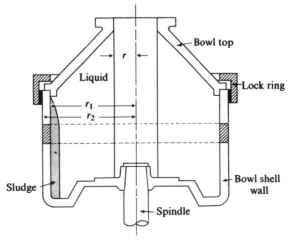

Fig. 14.8   Schematic sketch of centrifuge bowl.

mill sludges to vary from $4 to $20 per ton of dry solids, including the hauling of the cake.

Ambler [1] recently reviewed the theory of centrifugation. When a force is applied to a particle, the particle is accelerated ($F = ma$) until it reaches a velocity along the line of the force at which the resistance to its motion equals the applied force. In a settling tank, this is the force of gravity. In a centrifuge, it is the centrifugal field, $\omega^2 r$. The two differ only in direction and order of magnitude. The gravitational field is along a radius of the earth; the centrifugal field is along a radius normal to the axis of rotation and may be upward of 60,000 times that of gravity for continuous-flow centrifuges. The velocities of particle movement are generally proportional to the square root of the diameter of the particle. The effective force acting on the particle is

$$F = (m - m_1)\omega^2 r$$

and for a sphere it is

$$F = \frac{\pi}{6}(d^3)\,\Delta\rho\omega^2 r.$$

The force opposing sedimentation, according to Newton's drag law in laminar flow, is

$$F = 3\pi\mu\, dv_s.$$

At equilibrium (Stokes's law),

$$v_s = \frac{\Delta\rho\, d^2\,\omega^2 r}{18\mu}.$$

In the simplest form of a continuous centrifuge, v_s is the velocity with which the particle, if it is heavier than the fluid, approaches the bowl wall. If X is the distance the particle will travel,

$$X = v_s t - \frac{\Delta\rho\, d^2\,\omega^2 r}{18\mu}\frac{V}{Q}.$$

If X is greater than the initial distance the given particle is from the wall of the rotor, the particle will be deposited against the wall and be removed from the system. In an ideal system ($X = s/2$), half the particles of diameter d will be removed. This may be considered the cutoff point at which

$$Q = \frac{\Delta\rho\, d^2}{9\mu}\cdot\frac{V\,\omega^2 r}{s},$$

where Q is volume of flow per unit of time.

Since the term $\Delta\rho d^2/9\mu$ is concerned only with the parameters of the system that follow Stokes's law and the term $V\omega^2 r/s$ with the parameters of the rotor, the above equation may be written as

$$Q = 2V_g\Sigma,$$

in which

$$V_g = \frac{\Delta\rho\, d^2 g}{18\mu} \quad\text{and}\quad \Sigma = \frac{V\,\omega^2 r_\theta}{g s_\theta},$$

where r_θ and s_θ are the effective radius and settling distance, respectively, of the centrifuge, and Σ is an index of centrifuge size that has the dimension of (length)^2 and is the equivalent area of a settling tank theoretically capable of doing the same amount of useful work as the centrifuge.

Ambler [1] uses the above theory to formulate an index of centrifuge sizes for various centrifuge types as follows:

1) For the laboratory test-tube or bottle centrifuge,

$$\Sigma = \frac{\omega^2 V}{4.6\log\left[2r^2/(r_1 - r_2)\right]}.$$

2) For the tubular-bowl centrifuge,

$$\Sigma = \frac{\pi l\omega^2}{g}\frac{(r_2^2 - r_1^2)}{\ln\left[2r_2^2/(r_2^2 - r_1^2)\right]}.$$

3) For the disc-type centrifuge,

$$\Sigma = \frac{2\pi n\omega^2\,(r_2^3 - r_1^3)}{3gC\tan\theta},$$

where

Σ = equivalent area of the centrifuge
ω = angular velocity (rad/sec)
V = volume
r = radius from axis of rotation
r_1 = radius to inner surface
r_2 = radius to outer surface
l = light-phase discharge radius
g = gravitational constant
n = number of spaces between discs
C = concentration of solute
θ = half-included angle of the disc.

In each of these cases, Ambler bases his calculations on the behavior of a single particle under conditions of unhindered settling and on the assumption that this particle is always in equilibrium with the force field of the centrifuge under the conditions defined by Stokes's law.

14.10 Sludge Barging

Sludge barging or ocean disposal is one of the means of final disposal of sludge that has been practiced by some cities. There is little theory involved in this method of treatment. Raw, precipitated, digested, or filtered sludge solids are pumped into a waiting barge and transported to a suitable site from the shore, where it is discharged, usually by pumping out deep under the water surface. There are some advantages of this method of disposal, such as relatively lower operating costs and reduced land demands. However, experience has shown that this method of disposal results in several environmental concerns as follows: (1) long-term adverse effects on ecology of the receiving water, (2) sludge floating matter rising to the surface, (3) public objection, and (4) potential for sludge residues carried to the shore during tidal cycles and causing public health impacts. Based on these concerns, this method of disposal has been discontinued and is not a recommended practice.

14.11 Sanitary Landfill

Sanitary landfill is used to bury garbage, refuse, and sludge in a planned and methodical manner [28]. It is a relatively simple, effective, and inexpensive method for disposing of dry matter such as refuse, but sludge is usually too liquid for this procedure. However, mechanically dewatered or sand-bed-dried sludge can be disposed of in this matter.

The area proposed for the sanitary fill [28] should be easily accessible, yet remote from sources of water supply and recreational areas, and at the same time on land which is not too costly. The suitability of the soil and possible future use of the property are also important considerations.

For municipal refuse, the land area required is estimated at about one acre per year for 10,000 persons, when using a six-foot-deep compaction. Sanitary landfills should be located above the groundwater level and no closer than 500 feet to any sources of water supply, particularly when the soil is sandy, gravelly, or of limestone derivation. The area should be staked out for trenches and bench marks estab-

lished, giving the elevation to which the finished fill is to be carried and the depth to which excavations are to be dug. Normally a trench is about 15 feet wide and about 4 feet deep. At the end of each day's dumping, the sludge should be covered and compacted by a bulldozer or tractor. Bacon [2] suggests using sludge to reclaim land as an economic method of disposal especially in marginal lands and coal strip-mining areas.

One of the major concerns for sludge disposal in a landfill is the presence of hazardous or toxic constituents in the sludge. Depending on the type of constituents, the soil type underlying the landfill, and the depth to ground water, these constituents could leach out of the landfill and migrate and contaminate the groundwater. Heavy metals, precipitated during the industrial wastewater treatment, are some of the commonly found toxic constituents in sludge. Other toxic organics including petroleum based and other chlorinated solvents could also be present in the sludge. Recently promulgated hazardous waste regulations by EPA on land ban of hazardous chemicals have restricted the disposal of industrial wastes in landfills. In the context of these regulations, it is important to determine whether the constituents present in the sludge are banned from landfill disposal and whether the sludge is considered a hazardous waste based on special characteristic criteria specified by EPA such as ignitability, corrosivity, reactivity and leaching tests (EP Toxicity or TCLP) before deciding on the landfill disposal option. Also due to long term potential liabilities of contaminating the groundwater, it is difficult to find landfills offsite that will accept the industrial sludge. On site landfills may be constructed after obtaining permits from local regulatory agencies. Generally, it is required and is also recommended that landfills be constructed with appropriate liners to prevent the migration of the leachate containing contaminants to the subsurface and groundwater. These issues are discussed in more detail in Chapter 32.

14.12 Miscellaneous Methods

Other methods for disposing of sludge solids include sludge concentration, flotation and thickening. Biological means, aided only by temperature and time controls, are used to induce flotation of sludges [19]. The resultant solids, in concentrations of

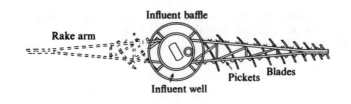

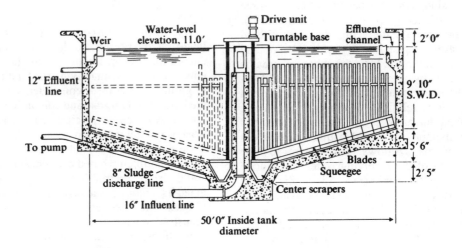

Fig. 14.9 Schematic plan of thickener mechanism, and section of tank [36].

20 per cent, do not require the addition of chemicals when they are subsequently dewatered on a vacuum filter. Optimum results with this method of concentration were found to exist at 35°C after a detention period of 120 hours [19]. However, certain types of sludge (for example, activated sludges) are not amenable to this treatment. Aside from time and temperature controls, the chief factors in the flotation method for concentration of raw sludges appear to be volatile content and pH.

In 1953 a method was developed by Torpey [36] for thickening sludge on a continuous basis without the addition of chemicals. Generally, the flow pattern permits dilute sludge—from the primary clarifiers alone or combined with secondary sludge—to be fed to the center feedwell of a thickener. A schematic drawing of one typical thickener is shown in Fig. 14.9. The solids settle, thicken in a definite "blanket" zone, and are drawn away from the bottom of the

tank. The excess liquid is decanted by a peripheral weir. The thickeners also contain a mechanism with vertical pickets attached to the rake arms. The pickets are V-shaped, and their channeling action allows entrapped water (water which is caught in sludge) and gases to escape to the surface. The degree to which the sludges can be thickened depends on several factors, the chief one being the source of the sludge [8]. The nature of the sludge is also most important. Some sludges are of a gelatinous and voluminous nature which impedes thickening beyond a certain limit, regardless of detention time. Others are more granular and release entrapped water when subjected to physical action, such as the slow mechanical mixing provided by the rotating pickets and rake arms.

Composting, a method of steeping solid wastes that contain 30 to 70 per cent water in large piles and allowing microorganisms to decompose the organic fractions, has been utilized to some degree for solid wastes from industry. The process is accelerated when the piles are turned regularly by mechanical means. Mercer *et al.* [23] found that the solid wastes of apricots and clingstone peaches were amenable to this form of treatment and that aerobic conditions were maintained by an initial daily turning for 5 to 6 days followed by turning on alternate days until the process was complete.

Your authors have listed six types of industrial sludges along with their possible origin and more specific character

Industrial types	Sources	Character
(1) metal hydroxides	plating wastes	Cr (OH)$_3$ Ni (OH)$_2$ Zn (OH)$_2$
(2) organic residues	paper mill	fines
	tannery	hair, skins, lime
	cannery	pulp, seeds, skins fruits & vegetables
	textile	fibres
	winery	dregs
	sugar ref.	lees
	poultry & meats	feather, innards, fat
(3) precipitated colloids	steel mill	Al (OH)$_3$
	pickle liquor	Fe (OH)$_3$
(4) inorganic	cement mill	sand
	steel mill	iron
(5) alkaline or neutral residues	fertilizer	sypsum + impurities
	sugar	steffans sludge
(6) organic residues from land use	agriculture	bagasse, corn
	crop debris	stalks, peanut hulls
	animal dung	cow, pig, sheep manure duck & chicken droppings

References

1. Ambler, C. H., "Theory, centrifugation equipment," *Ind. Eng. Chem.* **53**, 430 (1961).
2. Bacon, V. W., "Sludge disposal," *Ind. Water Eng.* **4**, 27 (1967).
3. Balden, A. R., "The disposal of solid wastes," *Ind. Water Eng.* **4**, 25 (1967).
4. (a) Barker, H. A., "On the biochemistry of methane formation," *Arch. Microbiol.* **7**, 404 (1936).
   (b) Barker, H. A., "Studies on the methane producing bacteria," *Arch. Microbiol.* **7**, 720 (1936).
   (c) Barker, H. A., "The production of caproic and butyric acids by the methane fermentation of ethyl alcohol," *Arch. Microbiol.* **8**, 415 (1937).
5. Bloodgood, D. E., "Sludge lagooning," *Water Sewage Works* **93**, 344 (1946).
6. Blosser, R. O., and A. L. Caroń, "Centrifugal dewatering of primary paper industry sludges," in Proceedings of 20th Industrial Waste Conference, Purdue University, May 4, 1965, p. 450.
7. Bradney, L., and R. E. Bragstad, "Concentration of activated sludge by centrifuge," *Sewage Ind. Wastes* **27**, 404 (1955).
8. Brisbin, S. G., "Sewage sludge thickening tests," *Sewage Ind. Wastes* **28**, 158 (1956).
9. Bruemmer, J. H., "Use of oxygen in sludge stabilization," in Proceedings of 20th Industrial Waste Conference, Purdue University, May 1965, p. 544.
10. Bulletin no. 102, 5–54, Komline-Sanderson Engineering Corp., Peapack, N.J.
11. Buswell, A. M., and W. D. Hatfield, *Anaerobic Fermentations*, Bulletin no. 32, Illinois State Water Survey, Urbana, Ill. (1939).
12. Cox, G. N., *A Summary of Hydrologic Data; Bayou Duplantier Watershed, 1933–1939*, University Bulletin, Louisiana State University, Baton Rouge (1940).
13. Eckenfelder, W. W., "Studies on the oxidation kinetics," *Sewage Ind. Wastes* **28**, 983 (1956).
14. Gauvin, W. H., "The atomized suspension technique," *TAPPI*, **40**, 866 (1957).
15. Genter, A. L., "Computing coagulant requirements in sludge conditioning," *Trans. Am. Soc. Civil Eng.* **111**, 635 (1946).
16. Genter, A. L., "Conditioning and vacuum filtration of sludge," *Sewage Ind. Wastes* **28**, 829 (1956).
17. Harding, J. C., and G. E. Griffin, "Sludge disposal by wet air oxidation at a five mgd plant," *J. Water Pollution Control Federation* **37**, 1134 (1965).

18. Haseltine, T. R., "Measurement of sludge drying bed performance." *Sewage Ind. Wastes* **23**, 1065 (1951).

19. Laboon, J. F., "Experimental studies on the concentration of raw sludge," *Sewage Ind. Wastes* **24**, 423 (1952).

20. Leet, C. A., C. W. Gordon, and R. G. Tucker, *Thermal Principles of Drying and/or Incineration of Sewage Sludge*, Combustion Engineering, Inc., New York (1959).

21. McCabe, W. L., and J. C. Smith, *Unit Operations of Chemical Engineering*, McGraw-Hill Book Co., New York (1956), p. 891.

22. Malina, F. H., Jr., and H. N. Burton, "Aerobic stabilization of primary wastewater sludge." in Proceedings of 19th Industrial Waste Conference, Purdue University, 1964, p. 716.

23. Mercer, W. A., W. W. Rose, J. E. Chapman, A. Katsuyama, and F. Dwinnell, Jr., "Aerobic composting of vegetable and fruit wastes," *Compost Sci.* **3**, 3 (1962).

24. Miller, D. R., "World's deepest submarine pipeline," *Sewage Ind. Wastes* **30**, 1426 (1958).

25. Powell, S. T., "Industrial wastes," *Ind. Eng. Chem.* **46**, 95A (1954).

26. Rabinovitch, W., P. Luner, R. James, and W. H. Gauvin, "The automized suspension technique. Part III," *Pulp Paper Mag. Can.* **57**, 123 (1956).

27. Rawn, A. M., and F. R. Bowerman, "Disposal of digested sludge by dilution," *Sewage Ind. Wastes* **26**, 1309 (1954).

28. Salvato, J. A., *Environmental Sanitation*, John Wiley & Sons, New York (1958), p. 288.

29. Schlenz, H. E., "Standard practice in separate sludge digestion," *Proc. Am. Soc. Civil Engrs* **63**, 1114 (1937).

30. *Sewage Treatment Plant Design*, Manual of Engineering Practice no. 36, American Society of Civil Engineers, New York (1959), p. 265.

31. *Sewage Treatment Plant Design*, Manual of Practice no. 8, Federation of Sewage and Industrial Waste Association, Washington, D.C. (1959), p. 214.

32. *Sludge Drying and Incineration*, Bulletin no. 6791, Dorr Co., Stamford, Conn. (1941).

33. Sylvester, R. O., "Sludge disposal by dilution in Puget Sound," *J. Water Poll. Control Federation* **34**, 891 (1962).

34. Teletzke, G. H., "Low pressure wet air oxidation of sewage sludge," in Proceedings of 20th Industrial Waste Conference, Purdue University, May 4–6, 1965, p. 40.

35. Tiller, F. H., and C. J. Huang, "Theory of filtration equipment," *Ind. Eng. Chem.* **53**, 529 (1951).

36. Torpey, W. N., "Concentration of combined primary and activated sludges in separate thickening tanks," *Proc. Am. Soc. Civil Engrs* **80**, Separate no. 443 (1954).

37. West, L., "Sludge disposal experiences at Elizabeth, N.J.," *Sewage Ind. Wastes* **24**, 785 (1952).

Questions

1. Why is sludge treatment the most vital part of industrial-waste treatment?

2. How does anaerobic sludge digestion solve the problem of sludge treatment of industrial wastes?

3. How does it differ from aerobic sludge digestion? When would you use the latter in preference to the former?

4. What factors influence the vacuum filterability of sludge?

5. How does elutriation aid in sludge treatment?

6. What are the principles and limitations of drying beds?

7. When would you recommend lagooning of waste sludge?

8. Under what conditions could a wet-combustion treatment be used for industrial-waste sludges?

9. What are the essential differences between Zimpro and Atomized-Suspension Techniques for sludge treatment?

10. Discuss centrifuging as a sludge-concentrating process.

11. What are the advantages and disadvantages of sludge barging for ultimate disposal? How does this compare to sludge pumping into oceans as an ultimate disposal system?

12. Under what circumstances can you place waste industrial sludges in a sanitary landfill?

13. What is the principle of sludge thickening and when is it used?

CHAPTER 15

INDUSTRIAL COMPLEXING FOR ZERO POLLUTION ATTAINMENT

The authors present a novel method of potentially reducing industrial pollution to zero without utilizing extensive and costly waste treatment.

15.1 Definition

Environmentally Balanced Industrial Complexes (EBIC's) are simply a selective collection of compatible industrial plants located together in one area (complex) to minimize both environmental impact and industrial production costs.

15.2 Objectives

These objectives (impact and cost reduction) are met by utilizing the waste materials of one plant as the raw material for another with a minimum of transportation, storage, and raw material preparation. When a manufacturing plant neither treats its wastes, nor stores or pretreats certain of its raw materials, its overall production costs must be reduced significantly.

15.3 Industrial Complex Principle and Concerns

Although the real measurable cost of industrial environmental pollution control remains relatively small when compared to total production or value-added costs, it can be a significant amount when considered by itself. In fact, the amount may be enough to influence industry management to consider whether to produce or discontinue the manufacture of specific goods. Though environmental engineers are usually not involved in that decision, the goal should be to reduce treatment costs to a minimum while protecting the environment to a maximum.

In conventional industrial solutions to waste problems, industry uses separate treatment plant units, such as physical, chemical, and biological systems. These separate systems add production costs to already high manufacturing costs. These costs are also easily identified and, even if relatively small when compared to other production costs, are opposed by industry. On the other hand, reuse costs, if any, in an EBIC will be difficult to identify and more easily absorbed into reasonable production costs.

Large, water-consuming and waste-producing industrial plants are ideally suited for location in such industrial complexes. Even though their wastes—if released to environment—might cause pollution, such wastes may be amenable to reuse by close association with satellite industrial plants using wastes and producing raw materials for others within the complex. Examples of such major industries are steel mills, fertilizer plants, sugarcane refineries, pulp and paper mills, and tanneries. Cement plants may also produce the ideal product to allow a perfect match for the phosphate fertilizer plants in a balanced industrial complex.

One needs to choose the proper mix of industries of the appropriate size and locate them in a specific area isolated from other municipal, industrial or commercial establishments. These choices will be highly influenced by marketing and socioeconomic factors.

Since 1977, Nemerow has proposed several typical complexes for tannery, pulp and paper, fertilizer, steel mill, sugarcane and textile industries [1–11]. Such complexes have the presumed advantages of minimizing production costs and adverse environmental impacts. Optimization of these advantages will meet the objectives of both industries and environmentalists.

Although the advantages of this type of complex are obvious, there are certain difficulties to overcome. One involves compatibility. There is no evidence, to date, that waste and product compatibility necessarily means industrial working compatibility. Other plant operating requirements such as labor availability, marketing of products, and taxes may not mesh as easily. Another involves optimal mass balances. Again there is no evidence, to date, to show that all plants within such a complex can operate at or near their optimum production required for economic purposes. However, lack of evidence is no reason to discard the principle, but rather reason for more complete investigation and trials.

15.4 Some Examples of Complexes

There are many possible combinations of industrial plants within these complexes. Compatibility in all aspects of the project solution is the key to the instigation and use of EBIC's. Your authors present in this discussion several such complex plans for which some data has been collected: Tannery (15.5), Sugarcane (15.6), Pulp and Paper (15.7), Textile (15.8), Fertilizer-Cement (15.9), and Steel,-Fertilizer,-Cement (15.10).

15.5 Tannery Complex

Tannery wastes from upper sole, chrome tanning mills contribute to a significant pollution problem in the United States. The wastes are hot, highly alkaline, odourous, highly colored, and contain elevated quantities of dissolved organic matter, B.O.D., total suspended solids, lime, sulfides and chromium. The treatment of such wastes has been difficult because of the conflicting pollutional parameters of pH, organic matter, and potential toxic compounds. Most successful treatment plants utilize some form of biological treatment to reduce the oxygen demand on receiving waters. This necessitates the use of well-designed and operated preliminary treatments to ensure safe and efficient biodegradation. High sludge quantities result from these treatments. Therefore, properly designed and operated tannery waste treatment systems may be costly to build and operate; while the lack of these facilities will cause excessive stream pollution. Placing the tannery in an environmentally optimized industrial complex eliminates both of these negatives.

The Slaughterhouse-Tannery-Rendering Complex

The author has presented two formal papers at technical meetings on the subject (10, 1977 and 7, 1981) and a Report (8, 1980) representing a first attempt at providing a complete mass balance of reference-validated inputs and outputs of plants within an industrial complex. The fulcrum industrial plant of this complex is a tannery. Supporting industries include slaughterhouse and rendering plants. The three-industry complex is also expanded to consist of an animal grazing and feedlot facility as well as a residential area for homes of all personnel working in the complex. As the complex is expanded to include the feedlot and residences and biogas and power plant services, the complex becomes more self sustaining. Outside service requirements are mini-

mized by the expansion. All power is generated within the complex—in the expanded third stage version. Excess products of leather, meat, meal, soap, and even electricity are sold to consumers outside the complex. Chemicals, water, cattle, and animal feed are imported to the complex. Wastewater, blood and bone meal, hide and leather trimmings, cattledung, and residential solid wastes are recovered and reused within (internally) the expanded complex. The complex can be constructed as shown in the first stage, second stage, or fully expanded to the third stage. Criteria for decision will be based upon area requirements and individual, local objectives.

Stage 1

This is the first of the three stage Industrial Complex which is balanced internally so that little or no adverse environmental impact results from any of the industrial plant's productive activities. Each stage represents a totally balanced and individual industrial complex.

The first stage consists of a three industry plant complex comprising: (1) a slaughterhouse, (2) a tannery and (3) a rendering plant (Report 8, 1980).

Stage 2

The second of the three stage industrial complex is also balanced internally so that little or no adverse environmental impact results from any of the industrial plants' productive activities. It differs from the first stage in that it provides a more complete and self sufficient complex. It also provides more reuse potential for the three industrial effluents than the first stage. In addition, it provides living space in the complex for employees of the industrial plants and feedlot and grazing area for raising the animals to the required weight. Whenever feasible, the second stage complex is recommended in preference to the first stage (Report 8, 1980).

Stage 3

This is the third of three stage industrial complex. It enlarges the smaller complex and is more balanced internally so that little or no adverse environmental impact results from any of the industrial plants' productive activities. Agriculture and municipal residence services are provided in this complex. Residential solid wastes from both industrial and municipal facilities are fermented to methane gas which is used subsequently to produce electrical energy for use in the complex. Waste sludge from the fermenter is incinerated to produce additional electrical energy for use in the complex. The schematic arrangement of the third phase of the complex is shown along with the mass balances of each unit in Fig. 15.1. External raw materials and manufactured products for external sale are given in Table 15.1.

General Discussion

As we proceed with the Three Industry Complex by adding stages, some apparent potential problems arise. For example, when we add stage 2 to the complex, we compute that a cattle grazing and feedlot area of 620 acres is required for the 135,000 cattle. This vast acreage may be difficult to obtain. In addition, 1350 tons per day of feed must be supplied from internal and external sources.

In the third stage of the complex, we are proposing to produce methane gas from solid waste residues. This gas will subsequently be used for power production. An excess of power within the complex results from this sequence of operations. An alternative to exporting power for sale outside the complex would be the production of other valuable intermediate products such as alcohol from the fermenters. This can be determined from market conditions at the time of establishment of the complex.

This three stage complex analysis is the deepest study of the new concept. As shown in Table 15.1, the managers of the three stage complex still must import four basic materials: water, calves, chemicals, and cattlefeed. About 3 million gallons of water, 2.6 million pounds of feed, 900 cattle, and about 6 tons of chemicals are needed each and every production day. This complex also will produce for external sale about 250 tons of meat, 36,000 square feet of leather, 40 tons of tallow, and almost 700,000 Kilowatts of energy each production day. Although complete economic analysis of such a system has not been made, it appears at least self-sustaining and probably will show a considerable net profit. The implications of such complexes are obvious. However, if the complex is able to produce a profit and protect the environment from any degradation, its major goals will have been achieved.

Table 15.1 External Raw Materials and Manufactured Products in Three Industry Complex (STAGE - 3)

Raw Material Required from Outside the Complex		Manufactured Products for Outside Sale	
Material	Amount	Material	Amount
1. Fresh Makeup water	2,927,599 gal/d	1. Meat products	513,341 #/d
1A. Well water (one time only)	12 MGD	2. Tanned leather	36,000 sq.ft./d
		3. Tallow	79,740 #/d
2. Calves	900/d (150 days) 540,00 #/d	4. Energy	694,710 KWH/d
3. Chemicals	495 #/d Na_2S 3960 #/d $Ca(OH)_2$ 1500 #/d H_2SO_4 2475 gal/d kerosene 1980 #/d oil or wax 2475 #/d $Cr_2(SO_4)_3$ 4208 #/d NaCl 7 #/d Cl_2		
4. Cattlefeed	2,625,000 #/d		

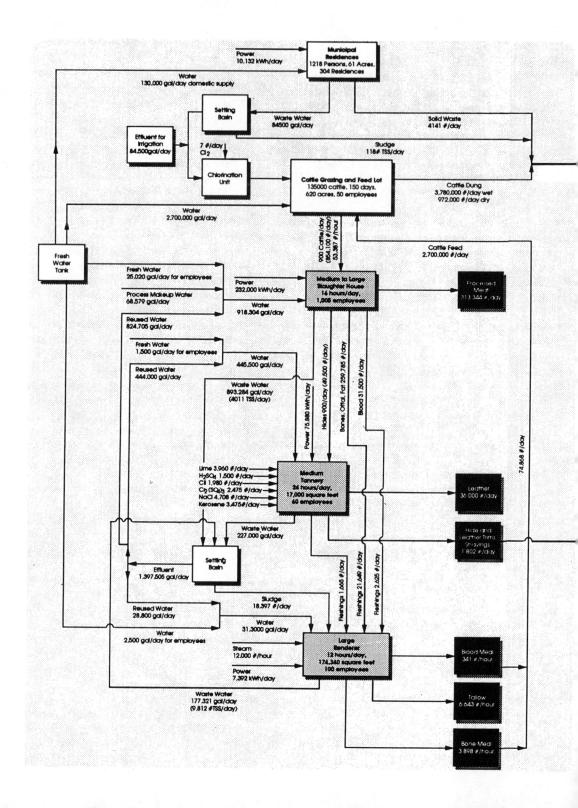

210

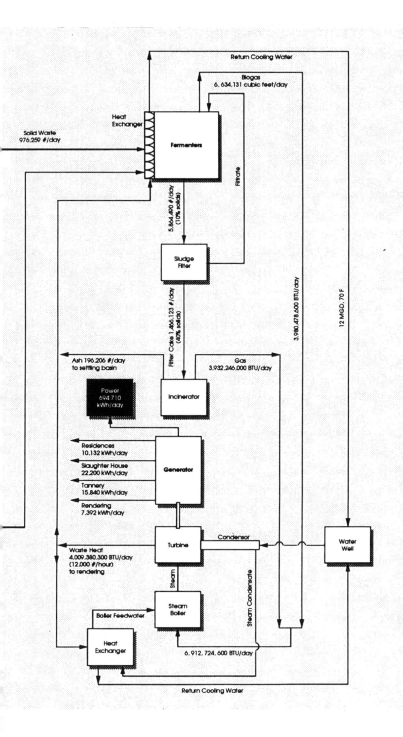

Return Cooling Water

Biogas
6,634,131 cubic feet/day

Heat
Exchanger

Solid Waste
976,259 #/day

Fermenters

Filtrate

5,864,800 #/day
(10% solids)

Sludge
Filter

Filter Cake 1,466,123 #/day
(40% solids)

Ash 196,206 #/day
to settling basin

Gas
3,932,246,000 BTU/day

3,980,478,600 BTU/day

12 MGD, 70 F

Power
694,710
kWh/day

Incinerator

Residences
10,132 kWh/day

Slaughter House
22,200 kWh/day

Generator

Tannery
15,840 kWh/day

Rendering
7,392 kWh/day

Waste Heat
4,009,380,300 BTU/day
(12,000 #/hour)
to rendering

Turbine

Condenser

Water
Well

Steam

Steam Condensate

Boiler Feedwater

Heat
Exchanger

Steam
Boiler

6,912,724,600 BTU/day

Return Cooling Water

Fig. 15.1   Three Industry Complex (Stage-3)

Conclusion

A three stage environmentally balanced complex has been designed. Mass balances of all plant inputs and outputs have been computed based upon the most recent published industrial data. From an analytical standpoint, an industrial complex consisting of a slaughterhouse, tannery, and rendering plants is technically feasible. This complex is also technically feasible when expanded to include animal grazing and feedlots as well as municipal residences (second stage). The expanded version (third stage) of the complex is more self-sustaining as far as reused products and electrical energy generation is concerned.

15.6 Sugarcane Complex

The Cane Sugar Industry

The cane sugar manufacturing industry is essential to the production of many varieties of foods. In the United States, there are about 6,400 sugarcane plantations, 94 sugar mills, and 24 sugar refineries, mostly located in Florida, Louisiana, and Hawaii. Most of the 3 million tons of cane sugar produced each year comes from Florida and Hawaii. Sugar is used in cakes, ice cream, candy, and soft drinks, as well as in other foods and beverages.

Because of recent dietary recommendations, alternative sweeteners have entered the market. Competition from the lower prices of other sweeteners has caused a reduction in refined sugar prices. This is true even though there has been a deficit in sugar produced in the United States. Florida, the largest sugar-producing state in the nation, grows about one-fifth of all U.S. sugar. It is imperative to the Florida mills, as well as sugar refineries elsewhere, that production costs be kept to a minimum to keep the industry healthy.

Brief Outline of the Sugar Manufacturing Process

In the manufacture of sugar, the sugarcane stalks are chopped into small pieces by rotary knives, and the cane juice is extracted from these pieces by crushing them through one or more roller mills. The solid residual material from this operation, consisting of fibrous residue of the cane sugar stalks, is termed "bagasse" and is a solid waste of the cane sugar industry. After the juice is extracted from the stalks, it goes to the boiler room where lime is added to precipitate insoluble sugars. The precipitate, in the form of thick surry, is vacuum-filtered to produce a filter-cake often termed "cachaza" and constitutes the second type of solid waste from sugarcane manufacturing operations. Then, the clarified juice is thickened in evaporators, and the

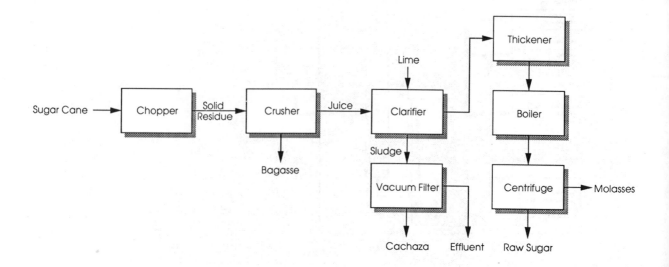

Figure 15.2 Raw Sugar Manufacture-Flow Diagram

resulting syrup containing sugar and molasses is boiled in vacuum pans to form raw sugar crystals. The sugar crystals are separated from molasses by centrifugation, and the molasses is sometimes further evaporated to recover more sugar. The final products are coarse, crystalline brown raw sugar and molasses. The raw sugar is transported for further processing in sugar refineries to produce the various forms of white refined sugar. The bulk of the molasses is used for production of various types of fermentation products and a small portion is used for animal feed. A schematic diagram of a sugar mill operation is shown in Figure 15.2.

Solid Waste Problem

The two forms of solid wastes generated in the manufacture of cane sugar are bagasse and cachaza. Every 1,000 tons of processed sugarcane generates about 270 tons of bagasse and 34 tons of cachaza.

The sugar industry is faced with the problem of proper and economical disposal of large quantities of bagasse and cachaza. The most common bagasse disposal method has involved burning as much as possible in boilers operated at sugar mills. Burning bagasse presents problems of its own. It is not a particularly clean fuel, and mills require installation and maintenance of stack scrubbers to clean the emissions. Moreover, utilization of bagasse as a boiler fuel is impaired by the high degree of moisture (45 to 60 percent). In addition, its bulkiness requires the construction of special furnaces to operate efficiently.

The other type of solid waste generated, namely, cachaza, generally is slurried for disposal by lagooning or disposed as landfill, resulting in land and water pollution. Even if a large portion (usually 70 percent) of the bagasse generated is burned directly in boilers, a considerable amount of bagasse (30 percent) remains to be disposed of with the entire quantity of cachaza.

Considering the high cellulose content of bagasse and the organic matter in cachaza, these are potential renewable sources of biomass for biochemical conversion to methane by anaerobic fermentation. In addition, the residual digested sludge can have beneficial uses as fertilizer/soil conditioner.

The Environmentally Balanced Industrial Complex Solution

Anaerobic digestion of a 2.4- to -1 mixture of bagasse-to-cachaza was demonstrated to be effective in produc-

ing methane gas and reducing organic solids (4, 1984). Despite this development, residual wastes remain to be considered.

An evaluation of the sugarcane refinery based on products and wastes after digestion suggested that a "closed loop" complex would result in the discharge of little or no final residual wastes. Figure 15.3 presents a schematic diagram of a sugarcane refinery based-environmentally balanced industrial complex. For purposes of this evaluation, the mass balances are estimated based upon the refining of 1,000 tons of sugarcane, resulting in a generation of about 270 tons of bagasse and 34 tons of cachaza (4, 1984). In many mills, these wastes are discharged to the environment with a variety of adverse impacts.

15.7 Pulp and Papermill Complex

The products of pulp and paper mills, the fifth largest in the U.S. economy, are consumed at the annual rate of about 400 pounds per person. The pulping of the wood and the formation of the paperproduct produce wastes containing considerable quantities of sulfates, fine pulp solids, bleaching chemicals, mercaptans, sodium sulfides, carbonates and hydroxides, sizing casein, clay, ink, dyes, waxes, grease, oils, and other small fibers. The overall wastes can be high or low in pH, and certain high color, suspended, colloidal, as well as dissolved solids and inorganic filters. Because of its high water consumption and wastewater discharge of 20,000 to 60,000 gallons per ton of product, the wastes contain large total quantities of organic, oxygen-demanding matter.

The high water use and wastewater production usually preclude the possibility of joint treatment with municipal sewage. These wastes also create considerable environmental impacts because of their concentrated loads of air, water, and land pollutants. The siting of new pulp and paper mills today has become a major endeavor. They must be located near vast quantities of relatively clean water, as well as receiving water resources, downwind and at a distance from residential habitation (because of common air pollutants such as SO_2 and mercaptans), usually on a rail line and near major highways for shipping, and near adequate land area for waste treatment and sludge disposal. Such sites are also difficult to find. For these and other reasons

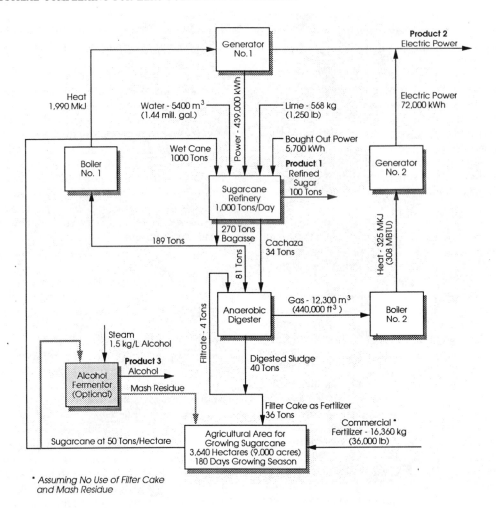

Figure 15.3   Sugarcane Refinery-Based Environmentally Balanced Industrial Complex

previously given, we recommend consideration of a pulp and paper mill complex with little or no adverse environmental effect. Figure 15.4 describes one possible complex centered about an average-sized paper mill producing 1000 tons of paper product per day.

In the first publication (11, 1977), a balanced industrial complex centered about a pulp and paper mill was presented. The preliminary mass balance was given in this paper and is produced here as Fig. 15.4 for additional clarification. Eight separate industrial plants were included as part of this complex, five of which would produce products to be used within the complex.

Timber is brought into the complex to the pulp mill

(1) where it is converted into pulp for use by the paper mill (2). Major wastes from (1) are bark, which is burned subsequently in the steam plant, and sulfate waste liquor, which is used in three internal complex plants; road binder (3), vanillan (4), and sulfate concentrating (8). Products from (3), (4) can be sold locally or internationally, while those from (8) are reused in the complex by (1) or by the hardboard manufacturing plant (7). Fine paper product from (2) can be sold in the world market. Wastes from (2) include heat, fillers, and fines which can be used internally in the groundwood pulp mill (5), which also uses a percentage of used newspaper stock.

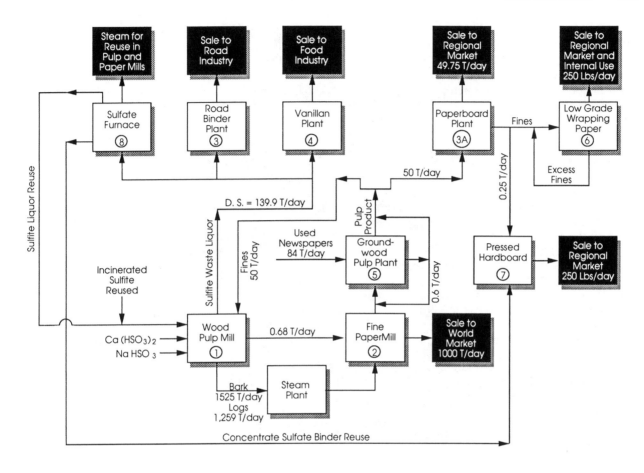

Fig. 15.4 Pulp and Paper Mill Complex

The pulp and papermill environmental problem

Pulp and paper mills are one of five major industrial water users in the United States (Nemerow). They used about 2×10^{12} gal. per day (7.5 E6 m3/day), (Gould 76). The largest percentage of this water is discharged into the environment as wastewater from pulp washing and papermaking or as steam from the drying plant. The pulp and paper industry is also the ninth largest industry in the United States, accounting for nearly 4% of the value of all manufacturing, producing 7866 millions of dollars of our Gross National Product in 1971 (API, 73). In the cooking and bleaching of the pulp, these mills may also contaminate the air surrounding the plant. Approximately one half of all weight of the wood entering the pulp mill leaves the mill as product

paper. The greatest percentage of the loss in weight ends up as solid material to be disposed of in the environment. Bark, waste pulp, and papermill fines constitute most of these solids and potentially end up on the land or in the air. Therefore, this industry—because of its great volume of water, wood and chemicals intake per mill—impacts adversely the air, water and land environments. The large quantity of wastes also represent a potential supply of valuable resources for ancillary and compatible industries.

With the above in mind, we derived from our general knowledge and from theory a pulp and papermill industrial complex for further investigation. The complex is shown in Fig. 15.4 and produces 1000 tons per day of fine paper.

Timber is brought into the complex to the *pulp mill* (1) where it is converted to pulp for use by the *paper-mill* (2). Major wastes from (1) are bark, which is burned subsequently in the steam plant, and sulfite waste liquor, which is used in three of the internal complex plants; *road binder* (3), vanillan (4), and sulfate concentrating (8). Products from (3) and (4) can be sold locally or internationally, while those from (8) are re-used in the complex by (1) or by the *hardboard manu-facturing plant* (7). Fine paper products from (2) can be sold in the world market. Wastes from (2) include heat, fillers and fines, which can be used internally in the groundwood pulp mill (5), which also uses a percentage of used newspaper stock. The pulp product from (5) will be used partially in the complex by (1) and also sold as paperboard externally. The plant (5) produces waste suspended solids which are used internally by the wrapping paper plant (6) and (7). The products of (6) and (7) can be sold regionally. In total, this pulp and papermill complex produces six products for external sale (fine paper, wrapping paper, hardboard, vanillan, paperboard and road binder) and four products for internal use (concentrated sulfate, wood pulp, wrapping paper and groundwood pulp). In addition, all major wastes of suspended solids, cooking liquor, fillers, heat and bark are reused within the complex in the manufacture of these products.

Mass balance of products

Literature review evolved typical concentrations of recoverable suspended solids in various process efflu-ents. A mass balance was prepared assuming that the total production of fine paper is 1000 tons per day. The remaining quantities are calculated based on this pro-duction.

Computation of trees required at the complex

Production of fine paper
= 1000 tons/day (907.2 kg/day $\times 10^3$).
Fiber losses from papermill
= 1.68% of production.
Therefore, suspended solids going into waste streams from papermill
= $1.68/100 \times 1000 = 16.8$ tons/day (15.24 kg/ day $\times 10^3$).

Total wood pulp produced per day
= $1000 + 16.8 = 1016.8$ tons (922.44 kg $\times 10^3$).
Quantity of sulfite liquor generated in wood pulp mill
= 300 gallon/ton (1.24×10^{-3}/kg) of pulp pro-duced.
While concentration of dissolved solids in sulfite liquor
= 11%.
Thu, dissolved solids going in sulfite liquor
= $110,000 \times 8.34 \times 300 \times 10^{-6}$
= 275.22 lb/ton (1.376×10^{-1} kg/kg) of pulp.
Total sulfite wastewater dissolved solids produced per day
= 275.22   No./tons $\times$ 1016.8   tons/day $\times$ tons/ 2000 lb
= 139.9 tons/day (1.269×10^{-5} kg/day).
On an assumption that the amount of bark produced is generally 15% (by weight) of the pulp production. There-fore, bark production
= $15/100 \times 1016.8 = 152.5$ tons/day (1.38×10^5 kg/day).
Total tonnage of trees used in the complex
= $1016.8 + 139.9 + 152.5 =$ 1309.2   tons/day (1.187×10^6 kg/day).

Groundwood pulp production

Recovery of suspended solids from papermill = 16.8 tons/day (1.52×10^4 kg/day). Assume that 100 tons (907.2×10^2 kg) of ground pulp is required for produc-tion every day. Fiber loss in the groundwood pulp plant = 0.6 tons/100 tons of the groundwood pulp (Nemerow, 78)
= $0.6/100 \times 100 = 0.6$ tons/day (544.3 kg/day).
Total groundwood pulp produced and lost per day
= $100 + 0.6 = 100.6$ tons/day (912.6×10^2 kg/ day).
Therefore, used newspaper required
= $100.6 - 16.8 = 83.6$ tons/day (758.4×10^2 kg/ day),
assuming 50% of the groundwood pulp is recycled as shown in Fig. 1, and the remaining is used in the produc-tion of paperboard.

Paperboard production

Loss of fines from groundwood pulp production is about 0.5% of the production (Nemerow, 78).
Let us say that paperboard production = X tons/ day:

$X + 15/100\ X = 50$ tons pulp/day (543.59×10^2 kg/day),

$1.005\ X = 50$,

$X = 49.75$ tons paperboard/day (451.32×10^2 kg/day).

Fines recovered from paperboard waste

$= -49.75 = 0.25$ tons/day (226.79 kg/day).

0.25 tons/day of fines can be used to produce low grade wrapping paper and pressed hardboard. With no loss of fines and with a 50-50 product production split, 250 lg (113.64 kg) of each product can be manufactured.

Sulphite recovery

The solids concentration of spent sulfite liquor drawn from the digesters may vary from 6 to 16% with an average value of 11%. These solids may contain as much as 68% lignosulfonic acid, 20% reducing sugars, and 6.7% calcium (Nemerow, 78). Complete evaporation of the sulfite waste liquor produces both a fuel which can be burned without an additional outside fuel supply and a salable by-product such as synthetic vanillan and road binder.

An overall mass balance regarding the production of different quality of papers is given in Fig. 1. No attempt is made in this paper to correlate the effects of the complete recovery of suspended solids on the reduction of final biochemical oxygen demand (BOD) of the wastewater. Similarly, no detailed information is given about the recycling of the wastewater effluent. However, it is reported in the literature that 90% of the effluent can be recycled.

wastewater, it must be presumed that a considerable portion of the dissolved and colloidal organic matter is being reincorporated into the various products. This is especially true in the case of the sulfite waste liquor—which is completely reused or recovered and contains the major portion of BOD in the complex.

Energy management

Integrated production complexes have a significant advantage over conventional plants from the energy management standpoint. Waste heat from one section of the complex can be used as process heat for another section, the concept being minimization of waste heat. The environmental problems associated with waste-heat dis-

charged to ecosystems have been well documented. It is accepted that thermal discharges may result in anomalous stratification in the receiving basin, lowering of capacity to hold oxygen, increased reaction rates and metabolism. These effects vary significantly with the chemical and meteorological conditions associated with the water body. The lethal effects of thermal pollution are sometimes obvious, whereas the sub-lethal effects on food chains and waste assimilative capacities are not easy to foresee without careful study.

The present industrial complex outlined can reduce waste heat discharged to the hydrosphere and atmosphere. The two significant areas of concern are:

(a) Utiliziation of solid wastes from the plant to achieve energy efficiency.

(b) Utilization of low grade heat from one section in another suitable section.

The first area has two possible applications. Bark from the shredding plant is used to provide heat from the steam plant. The estimated bark production for the plant is 152.5 tons/day (138.345×10^3 kg/day). Since the heating value of bark varies considerably with the type of tree and aging, an average heating value of 4000 calories/g is used (author's estimate, 77).

The heat available by combustion of bark

$= 152.5 \times 2000 \times 454 \times 4$ kcal

$= 5.5 \times 10^8$ kcal/day (2.3×10^{12} j).

Assuming incoming water temperature at 25°C, the total steam production

$= 5.5 \times 10^8/(75 + 54)$ kg $= 8.9 \times 10^5$ kg/day.

The second solid waste to energy recovery application lies in evaporation and burning of sulfite liquor. Sulfite liquor is evaporated to enough of a solid content suitable for burning. Difficulties with scaling, corrosion and fly ash may result. However, this burning procedure can also be justified since this will eliminate the need to discharge sulfite wastes to the environment.

Utilization of low grade waste heat from the proposed complex is somewhat difficult to quantify, since details of process thermodynamics is needed. Further, research on this concept will provide the needed data for such analysis. However, some conceptual comments can be made regarding the proposed complex. Low grade heat from fine papermill can be used in the groundwood pulp plant. In cooler regions of the world, waste heat from any of the effluents in the complex could be used for space heating and providing hot water for use by plant personnel.

Economy of complex or residual environmental impact

Little or no air or water pollution results from this complex. In addition, it is anticipated that no expensive wastewater treatment plant would be required for the pulp and papermill complex. This, in itself, represents a savings not only in capital equipment costs, but also in operating or production costs equal to about 1–5% of production costs. Additional operating costs will be reduced by the following practices:

(1) Reusing burned sulfite waste liquor to replace a portion of calcium or sodium bisulfite cooking liquor.
(2) Reusing groundwood pulp (50 tons/day) or 43.359×10^3 kg/day) to replace a similar weight of trees.
(3) Burning bark to generate steam for use in the fine papermill.
(4) Reusing concentrated sulfite waste liquor in making pressed hardboard.
(5) Reusing fillers, fines and heat from the papermill in making groundwood.
(6) Reusing groundwood pulp mill fines (0.6 tons/day or 544.31 kg/day) to make additional pulp or paperboard.
(7) Reusing paperboard mill fines (0.25 tons/day or 226.80 kg/day) to make both low grade wrapping paper and pressed hardboard.
(8) The sale of additional products as follows:
  (a) low grade wrapping paper—250 lb/day (113.64 kg/day);
  (b) pressed hardboard—250 lb/day (113.64 kg/day);
  (c) paperboard—49.75 tons/day (45.132×10^3 kg/day);
  (d) vanillan;
  (e) road binder.
(9) Combustion of concentrated liquor for heat recovery and pollution abatement.
(10) Use of low grade waste heat for space and water heating.

An exact and more detailed economic analysis of this complex will be made in a continuing study. It will be necessary to obtain more precise data on the production requirements of the small service industrial plants in the complex. In the meantime, we propose this analysis as a beginning to what may develop into a revolutionary new system of industrial plant design.

References

American Paper Institute, General Statistics for the U.S. Pulp and Paper Industry, Washington, DC. (1973).

M. Gould, *Water Pollution Control in the Paper and Allied Products Industry*, Industrial Wastewater Management Handbook, McGraw-Hill, New York, (1976).

N.L. Nemerow, *Industrial Water Pollution: Theories, Characteristics and Treatment*, Addison-Wesley, Reading MA. (1978).

A major producer of the present-day pulp and paper uses the kraft process. Under normal circumstances a variety of liquid, solid and air wastes are created in this process. One possibility for integrating the typical kraft mill into an EBIC is shown in Fig. 15.5.

The approach adopted for developing this complex is:

(1) Single system concept of water, power, raw materials, and wastes management. The operation of utilities system, so critical to the production process, requires efficient performance. Centralized utilities and management allows to achieve self-containment.
(2) Wastes utilization/by-product recovery: Industries based on chemicals from the other half of the tree and wastes have been included in the complex. This step, a mode of pollution control, is of significant merit since it allows two-fold benefits of minimizing production and treatment costs.
(3) Grouping of other compatible and complimentary industrial plants like chemical, forestry research, aquaculture (water hyacinth), etc.
(4) Integrated development in stages and overall management of all operations within the complex. The central starting point is pulp industry, based on which the complex develops in stages by inclusion of other industries and utilities. The gradual development in this way approaches self-sustainment.

Although complete analysis with respect to criterion for acceptability has not been made, primary techno-economic considerations make this complex extremely attractive. As resources become increasingly scarce and environmental regulations become more severe, multi-benefits will be more than apparent.

15.8 Textile Complex

Plight of Small Texile Mills

The textile industry represents one of the most competitive fields of production worldwide. Each plant attempts to reduce its cost to compete with other similar plants within its region as well as plants in other countries of the world. One answer to competition has been to increase production, sometimes by merging with other plants, and sometimes merely by expanding one plant's capacity. Lower unit costs generally result from increased production in accordance with accepted economic principles. However, some mills for one reason or another cannot increase production to reduce costs. These smaller plants are vital to local economies, but are finding it more and more difficult to compete with other larger mills.

In addition, these small textile mills are often located

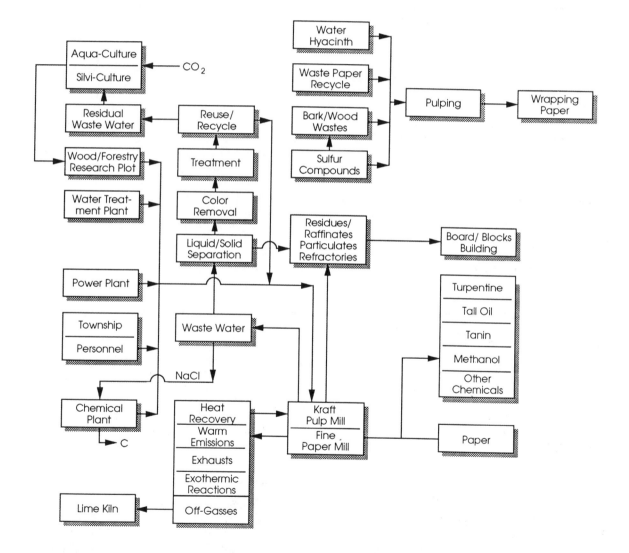

Fig. 15.5 Environmentally Balanced, Resource & Energy Optimized Self-contained Intergrated Pulp & Paper Complex

on small watercourses where their waste exert an unusually high pollutional demand on the environment. Pressure is being applied by water pollution control agencies to avoid and avert this pollution. Treatment of these wastes may also increase production costs.

When one couples the economic size and environmental pollution problems with the reality of dwindling supplies of fresh, raw process water, the smaller textile mill is currently being squeezed either out of business or to disproportionately increase its product cost or both. Larger mills are usually located where process water is more abundant and hence cheaper and where receiving streams or domestic wastewater treatment plants are more able to handle the pollutional load.

The ultimate survival of small textile mills—indeed small water industrial plants of all types—depends upon solving both the economic as well as the environmental resource problems. We offer in this section a new, innovative and potential solution to the plight of these small mills.

Cost of Raw Water

Although process water cost generally represents a minor portion of total manufacturing cost, it is significant because it is becoming an increasing percentage. Process water is also becoming a more scarce raw material. Little information is published on the actual cost of raw water. In general, municipal water utilities charge from $0.50 to $1.50 per 100 ft^3 (or 750 gallons of water). In fact, our survey of textile mills using public water supply showed they pay $0.44 to $1.43 per 1000 gallons.

For a typical small mill finishing woven fabric in a series of complex processes which uses 600,000 gallons per day, its daily cost would be $264 to $858. Even these charges may be misleading because they occurred only where this amount of water was available for sale as reported by the mills.

Conventional Wastewater Treatment

Wastewater treatment from small textile finishing mills has been either (1) separate treatment and reuse of dye wastes only or (2) complete treatment of the whole finishing mill waste (2, 1987). The first has been accomplished mainly by hyperfiltration and/or dye bath reconstitution, while the second has been done mainly by chemical coagulation and/or biological aeration.

Both methods have produced certain amounts of reusable water. However, economic considerations as well as governmental environmental regulations play major roles in the decision to produce reusable wastewater. Cost of producing acceptable quality reusable wastewater to the small mill will need further definition and reduced to a minimum before reuse becomes standard practice, regardless of receiving water quality degradation.

Costs of Conventional Wastewater Treatment

In our research of the literature thus far, we have found that capital and operating costs of small textile mill wastewater treatment depend largely upon the type and extensiveness of the treatment used. The capital costs range from as low as $31,500 for simple dye bath reconstitution to as high as $303,000 to $982,000 for chemical coagulation, filtration, and activated sludge treatment. Operational costs for similar treatments range from $40,000 to $328,000 per year.

A typical small mill produces about 25,000 pounds per day with average capital cost of $500,000 for complete treatment and annual operating cost of $150,000. This results in capital costs of $20 per pound of production per day and annual operating costs of $0.02 per pound of production per day (assuming 6 days per week and 50 weeks per year of production). These are very approximate costs for presumed average small mills. Range of true costs may vary greatly from these approximates. However, it is apparent that both capital and operating costs to these small mills represents a very significant expenditure.

Minimization of these costs by subtracting them from the benefits of wastewater reuse would constitute a real boom to the small mills.

Alternate Solutions to Dilemma

There are two potential methods for reducing waste treatment costs of the small textile plant and, at the same time, producing reusable wastewater to replace or replenish the mills costly water supply. These are (1) industrial complexing and (2) chemical coagulation. Other methods reported in the literature may reduce waste treatment costs or produce a partial supply of raw water, but will not accomplish both of our objectives. For example, dispersed growth aeration as suggested by Nemerow and Dasgupta [1, 1987] will treat the wastewater at reduced costs, but will not, by itself, produce acceptable reusable water. Also, Brandon and Porter [2, 1976] hyperfiltered dye wastes through membranes to produce both recyclable water and dyes, but failed to treat a sufficient portion of the plant's total waste at a lowered cost to result in satisfactory overall waste treatment. In order to be cost effective for the small textile manufacturer, the solution to their problems must satisfy both environmental and production concerns.

Industrial Complexing

Water-consuming and waste-producing textile finishing mills are ideally suited in these industrial complexes. Although its wastes may pollute our fragile environment, they may be amenable to reuse by close association with satellite industrial plants which are able to use them and, in turn, produce raw materials for others within the complex.

An ideal, illustrative EBIC for the small textile finishing mill is shown in Fig. 15.6. This complex contains five manufacturing plants producing 12,000 pounds of woven fabric for sale outside the complex and 13,200 pounds of cotton, 14,640 pounds of greigh

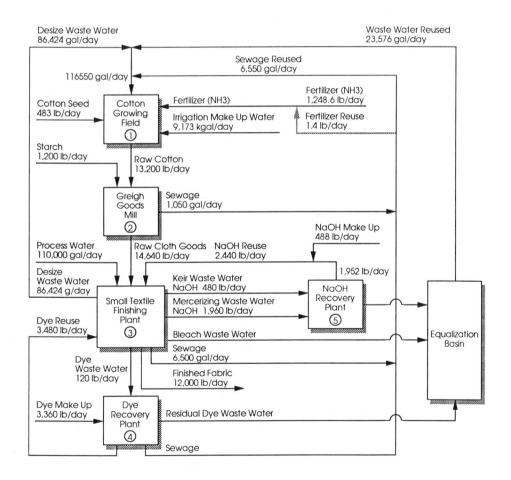

Figure 15.6   Diagram of the Integrated Five-Plant Industrial Complex

Table 15.2   Raw Material Balance - As Part of Total Production Cost

Raw Materials Needed / Amount needed	Plant Type / Amount needed	Mass Balance of Small Textile Complex* Raw Mat'l				Cost/Unit of (1982-'83)
		by the complex Quantities	Cost($)	when plants are separate Quantities	Cost($)	
	Agricultural Growing Field					
Irrigation Water		9.173mgd	4,036.00	9.29mgd	4,087.00	$0.44–1.43/1000gal
Cotton seed		483#/day	48.30	483#/day	48.30	0.10#
Fertilizer		1,248.6#/day	81.16	1,250#/day	81.25	0.065# (1987)
	Greigh Goods Manufacturing					
Starch Sizing		1200#/day	180.00	1200#/day	180.00	$0.15
	Textile Mill Finishing					
NaOH for Fabrics		488#/day	131.80	2,440#/day	658.80	$0.27#
Dyes for Fabrics		3,360#/day	1,512.00	3,480#/day	1566.00	0.45#
Process Water		0.11 mgd	104.50	0.11 mgd	104.50	0.44–1.43/1000gal
Total Mat'l Cost		$6,093.75/day		$6,726/day		

*Assume Typical Small Textile Finishing Plant Producing 12,000 #/day of woven finished cloth product.

goods, 1952 pounds of NaOH, and 120 pounds of dyes for reuse within the complex.

In addition, all sewages and wastewaters are reused without treatment within the complex. Of notable interest and importance is the reuse of 86,424 gallons of untreated finishing mill desize waste which contain 732 pounds of B.O.D.

In Table 15.2, we present a raw material balance justification which compares the raw material quantities and costs for the five separate plants manufacturing at distant locations and manufacturing within the EBIC. These data, although obtained and referenced only from authentic literature, show the cost advantage of this textile EBIC.

In addition, environmental costs would have to be considered. Within the industrial complex, we are presuming no external environmental costs are needed. As separate plants operating at distant locations from each other, environmental costs would include both domestic and industrial waste treatment charges as well as any measurable adverse environmental impact costs of the residual effluent wastes. These additional costs are currently being assessed by the authors.

From Table 15.2, it is apparent that the cost savings of the industrial complex from a material balance alone is $6,726 minus $6,093.75 or $623.25 per day which represents a savings of $52.69 per 1,000 pounds of finished cotton fabric.

From a preliminary study, the total environmental savings appear to be greater than $1248 per day, which represent the average costs of treatment for separate greigh goods and finishing mill wastes. To both savings we must add the savings from transportation of raw cotton and sized, woven goods. Presuming transportation cost is $0.026 per pound of cotton transported, we can estimate the additional cost due to transportation as $343.2 per day.

References

[1] Nemerow, N.L. and A. Dasgupta, (1985), "Zero Pollution for Textile Waste", Proc. 7th Alternative Energy Sources Conference, Miami, Florida, December 1985, Vol. 6, pg. 499, 1987, Hemisphere Pub. Co.

[2] Brandon, C.A., and J.J. Porter, (1976), "Hyperfiltration for Renovation of Textile Finishing Plant Wastewater", EPA-600-2-76-060, 157 pgs.

15.9 Fertilizer-Cement Complex

Fertilizer Plant Wastes and Production

It has been reported and generally accepted that phosphate mining in Central Florida accounts for about 75% of the U.S. needs and one-third of the world's supply. This alone makes it a vital industry not only to Florida, but also to the United States and the world.

After the rock is extracted, slurried and separated from the clay and sand by screening and flotation, it is used to produce wetprocess phosphoric acid. The rock is digested by sulfuric acid to produce a slurry of contaminated gypsum ($CASO_4.2H_2O$) and phosphoric acid. The slurry passes over a rotating disc filter where the calcium sulfate and acid are separated. The gypsum is pumped to holding ponds, whence it represents a major disposal problem for the fertilizer industry. Since 4.5 to 5 tons of gypsum are formed during the production of each ton of phosphoric acid, the industry has a formidable volume of phosphogypsum (PG) waste with which to cope.

Any recovery and reuse system for the phosphogypsum will free up reclaimable land for productive purposes by the industry or by other private or public landowners. Further benefits can be derived from the elimination of adverse environmental consequences of leachates from the gypsum heaps. Leachates carry phosphate and other mineral nutrients which could contaminate drinking water supplies and cause algal blooms (red tide) in recreational waters. Direct reuse of PG presents the potential problem of incorporating radioactivity into building or road products.

It is quite likely that the direct reuse of this PG within a closed industrial complex in making cement could eliminate all of the above problems. In addition, it anticipates a lowered production cost for both the fertilizer and cement plants for reasons already mentioned.
tioned.

One other area of waste recovery that should be mentioned is "heat" energy. A phosphate complex generates and must dissipate large amounts of energy as waste heat. The recovery and utilization of much of this energy is a very real success story. Efforts are continuing to recover even more of the "waste" energy from the more difficult to recover sources, and there is no doubt that an even greater percentage of the available "waste heat" will be put to profitable use in the future.

As early as 1968, it was reported that many firms in the U.S. had innovated processes for manufacturing useful products such as H_2SO_4 and *cement* from waste gypsum [11]. Nothing would be gained by reporting here the numerous papers which have been published describing the potential or actual use of gypsum in cement making. However, a few representative ones are in order. The British Sulfur Corp., Ltd., reported that the MASAN product transformed from PG by the Brussels based company, Ultra International SA, is a useful cement or plaster [12]. It was reported to possess a compressive strength 3 to 4 times that of Portland cement (1100 kg/cm² as compared to 300-400 for Portland cement). Moreover, the cost of cement from PG was $10/ton as compared to $30–$40/ton for Portland cement at that time.

Ellwood describes a chemical process for converting PG into *hemihydrate* powder as a cement strong enough to compete with cement in applications such as sound proof dividing walls [13].

More recently, Carmichael reported two Belgium plants which are using the Central-Prayon process for converting PG to the hemihydrate form of $CASO_4$ [14]. The gypsum is then suitable for direct use in the plaster industry or as a cement retarder.

Bhanumathidas and Kalidas reported the conversion of Anhydrite I grade of PG to calcine at 950°C to obtain a product similar to Portland cement [15]. They claim that the product "has shown remarkable cementitious behaviours in parallel to that of White Portland cement."

Clur claims that the Fedmis (South Africa) fertilizer plant disposes of about 25% of its PG production as soil conditioner, cement clinker and cement retarder [16]. "The quality of the cement compares favourably with that of local limestone-based cements, and is used in all classes of building construction and civil engineering." Clur also reports that "the technical problems of producing a good quality cement from PG have largely been solved, the future of the process would seem to depend mainly on economic and environmental factors."

Cement Plant Raw Materials and Wastes

Portland cement is made by mixing and calcining calcerous and argillaceous materials in the proper ratio [17]. The Table 15.3 summarizes the raw materials consumed in 1972.

Table 15.3   Raw Materials Consumed for Portland Cement in U.S. (1000's of short tons)

Cement rock	23,799
Limestone	90,003
Marl	2,080
Clay and shale	12,158
Blast furnace slag	759
Gypsum	4,094
Sand and sandstone	2,774
Iron materials	839
Miscellaneous	414

One can observe in Table 15.3 that limestone represents the majority mass of cement raw materials. Replacement of some or all of this calcareous material with PG would reduce the production cost of the cement as a result in savings of raw material.

Unit processes involved in cement manufacturing essentially include *storage and mixing of raw materials, drying, grinding and crushing, calcining, clinker storage, finishing additives and ball milling, and packing for delivery*. Although dry processing is practiced more than wet processing, both are shown in the following flow charts (Fig. 15.7 A & B) to provide the reviewers with a visual aid for cement production.

For each 376 barrels of finished cement by the dry process, 1,120,000 Btu of fuel are required as well as 24.1 kWh of electricity, 30 gallons of water, and 0.17 hours of direct labor. Also required are 498 pounds of limestone, 124 pounds of shale, and 16 pounds of gypsum [17].

It should be mentioned here that Nemerow, as far back as 1944, developed a wallboard for Johns Manville Corporation. This board was made of asbestos fibres and gypsum formed under high temperature and pressure.

Statement of Problems and Objectives

The problems are twofold: (1) to lower production costs, and (2) to eliminate adverse environmental impacts of industrial plants. These problems are especially severe or extensive when the particular industry is highly competitive, such as fertilizer and cement plants proposed in this research, and when these same plants produce significant wastes which pollute the environment (air, water and land).

The overall objective of this research is to determine

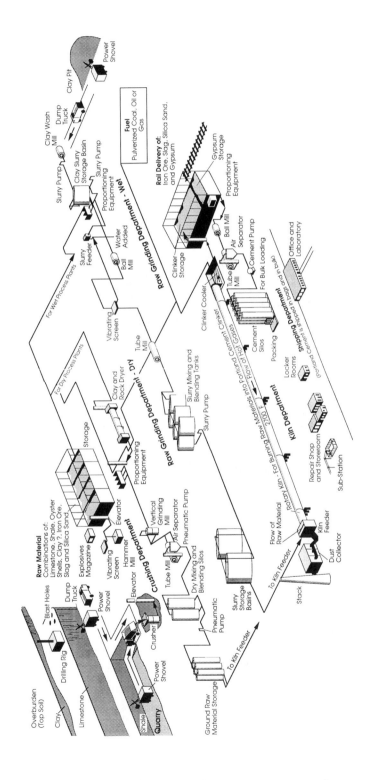

Fig. 15.7A   Isometric Flow Chart for the Manufacture of Portland Cement by Both Dry and Wet Processes.

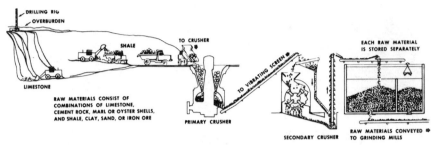

1. **Stone is first reduced to 5-in. size, then to ¾ in., and stored.**

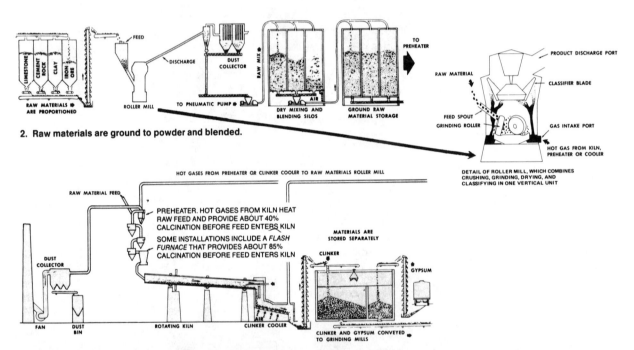

2. **Raw materials are ground to powder and blended.**

3. **Burning changes raw mix chemically into cement clinker. Note four-stage preheater, flash furnaces, and shorter kiln.**

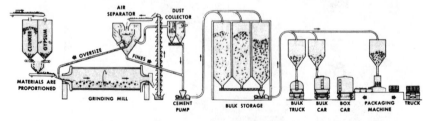

4. **Clinker with gypsum is ground into portland cement and shipped.**

Fig. 15.7B New Technology in dry-process cement manufacturing. (Reprinted from Design and Control of Concrete Mixtures, by Steven H. Kosmatka and William C. Panarese, 13th edition, 1988, with permission from the Portland Cement Association, Skokie, IL.)

the feasibility of locating, building and operating a two-industry complex, consisting of a phosphate fertilizer and a cement plant, within an Environmentally Balanced Industrial Complex (EBIC) at one site. The ultimate goal of this complex is to *lower production costs* at both plants *and eliminate all adverse environmental impacts* at the same time.

Further study should analyze and evaluate in depth the practicality of the complex briefly described above and shown schematically in Fig. 15.8A. More precisely, it is necessary to determine (a) the optimum size for each manufacturing plant included within the com-

plex; (b) the suitability of the by-products (wastes) for recovery and reuse as raw materials for ancillary adjacent plants within the complex (compatibility of plants); (c) the validity of total waste elimination from the two plants involved within the complex; and (d) the cost of production of the prime goods when manufactured at distinctly separated plants and compared to the same when manufactured within the complex as shown in Fig. 15.8B.

The main thrust of future research should be to ascertain the *extent of the economic gain* by using the Complex principle. This study will be made to *include the*

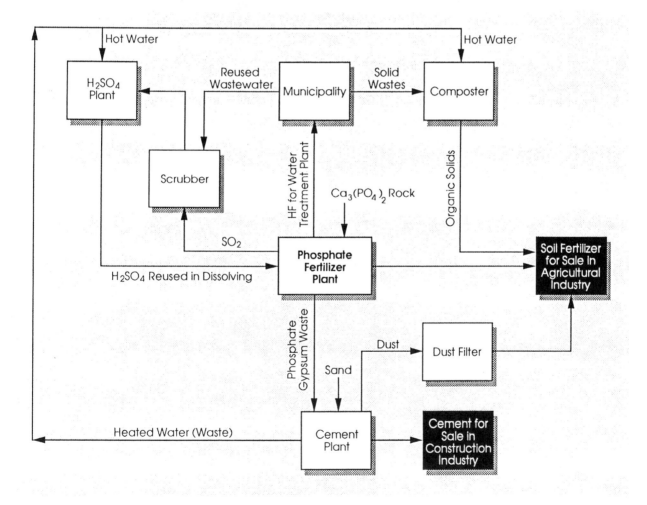

Fig. 15.8A Fertilizer Plant Complex

economic cost of environmental damage caused by wastes of all plants involved as part of the production costs.

This concept is not only a "gypsum for cement" idea, but rather a totally new balanced industrial complex plan. The question is not whether this innovation *is* economical, but rather how much reduction in cost can be obtained by this complex principle *when environmental costs* are also included and all complex plant wastes are reused including excess heat.

15.10 Steel Mill-Fertilizer-Cement Complex

Steel mills are actually five separate industrial plants in one consisting of: (1) coke plant (2) iron ore reduction plant, (3) steel production, (4) hot rolling mill, and (5) cold rolling mill. Predominant wastes originate from the coke and steel plants, although certain dusts, slag, and iron also come from the other plants.

Troublesome waste products include ammonia, cyanide, phenol, heat, and acidic ferrous sulphate or choloride pickle liquor. Steel mills also use huge volumes of water—mostly for cooling and quenching and produce like volumes of air, water and solid contaminants. They have developed a world-wide reputation as one of the most polluting industries existing in modern times. They require so much land area and employ so many people that their location in a separate industrial complex would be a natural development. Fertilizer and building material plants are likely candidates for

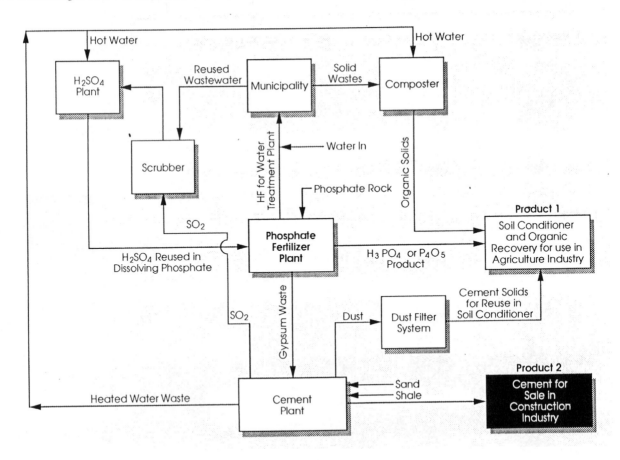

Fig. 15.8B Environmentally-Balanced Fertilizer-Cement Plant Complex-Phase 1

auxiliary industries for a steel mill complex. Such a complex is proposed and presented in Fig. 15.9.

15.11 Fossile-fueled Power Plant Complexes

Dilemma of Electric Power Plants

Electric Power Plants face the problem of producing more electricity at a lower production cost and, at the same time, minimize the external damage to the surrounding environment. This is extremely difficult to accomplish because of the problem of obtaining permits for producing nuclear power, the polluting characteristics of both oil and coal fuels, and the untried utilization of more sophisticated wind, solar and hydrogen generated power. Since fossil–fueled power plants are currently cost effective and generally acceptable to the public, it is reasonable to utilize this form of fuel and attempt to ameliorate or abate the adverse environmen-

tal consequences. The challenge is to do this effectively; that is, at a minimum production cost and with little or no adverse external environmental consequences.

Background on Coal-Fired Power Plants

Coal-fired power plants generate the major amount of electricity in the United States. In 1986 the electric utilities produced 2,487.3 billion kilowatt-hours of electricity. In the same year, coal-fired plants generated 1,385.1 billion kilowatt-hours of electricity. This was approximately 56 percent of the Nation's total production for the year 1986 [1].

The United States Department of Energy has been encouraging the use of coal as a principal fuel (in lieu of gas or oil) by the electric utility and industrial sectors. Combustion residues from coal-fired power plants—i.e., fly ash, bottom ash, boiler slag, and flue-gas-de-

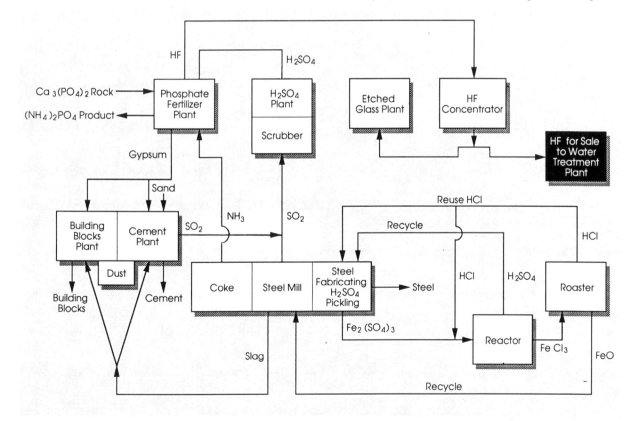

Fig. 15.9   Steel Mill-Fertilizer Cement Complex

Table 15.4 Wastewater Discharge Flowrates for Coal-Fired Power Plants [2]

Waste Streams	Mean Value (GPD/MW)[a]	Maximum Value (GPD/MW)
Cooling System		
once-through	1,140,619	55,430,000
recirculating	2,937	63,056
Wet Ash Handling		
fly ash pond overflow	3,808	16,387
bottom ash pond overflow	3,881	38,333
Flue-Gas Scrubber		
solids pond overflow	3,973	195,000
blowdown	811	8,824
Boiler Blowdown	148	3,717
Evaporator Blowdown	126	8,292
Other Streams[b]	288	6,600

[a]Million gallons per day per megawatt of installed capacity.
[b]Other streams represent the total waste discharge from ion exchange softener spent regenerant blowdown, filter backwash, clarifier blowdown, lime softener blowdown, air preheater and boiler fireside washwaters.

sulfurization (FGD) sludge—are currently exempted from the RCRA, which requires the Environmental Protection Agency (EPA) to promulgate regulations for the disposal of hazardous and non-hazardous wastes [3].

On the basis of chemical origin, EPA categorizes the wet waste streams for the steam electric power generating point source category as follows:

1. Once-through Cooling Water
2. Recirculating Cooling System Blowdown
3. Fly Ash Transport Discharge
4. Bottom Ash Transport Discharge
5. Metal Cleaning Wastes
   -air preheater
   -fireside wash, etc.
6. Low Volume Wastes
   -boiler, evaporator, softener blowdowns
   -drains, sanitary wastes, etc.
7. Ash Pile Runoff
8. Coal Pile Runoff
9. Wet Flue-Gas Cleaning Blowdown

Mean and maximum values (discharge flowrates per installed capacity) of these waste streams for coal-fired power plants are tabulated in Table 15.4. It should be noted that cooling water and wet ash handling systems, and flue-gas scrubber processes constitute the major discharge volume.

Mean discharge flowrate per installed capacity from once-through cooling water systems is approximately 900 times as much as that from recirculating cooling water systems. However, discharge from once-through systems is relatively clean, and does not need treatment prior to discharge. In Table I, the waste streams associated with the "Other Streams" represent the total wastewater discharge from several treatment processes which may not exist in every coal-fired power plant. General description of systems handling major wastewater streams are as follows:

Cooling Water Systems: In a steam electric power plant, cooling water absorbs the heat that is liberated from the steam when it is condensed to water in the condensers. Depending upon the size, location of the power plant, and availability of water body, there may be either of the cooling water systems, once-through and recirculating, described as follows:

Once-through cooling water systems: In a once-through cooling water system, the cooling water is withdrawn from the water source, passed through the system, and returned directly to the water source. Discharge flowrates from such a system in coal-fired power plants may reach up to 55.4 MGD/MW (Million gallons per day/Megawatts). In the United States, about 65 percent of the total power plants have once-through cooling water systems.

Recirculating cooling water system: In a recirculating cooling water system, the water is withdrawn from the water source and passed through the condensers several times before being discharged to the receiving water. After each pass, the heat is removed from the water by three major methods: Cooling ponds or cooling canals, mechanical draft evaporative cooling towers, and natural draft evaporative cooling towers. Discharge from such a system in coal-fired power plants may reach up to 63,057 GPD/MW (Gallons per day/Megawatts).

Ash Handling Systems: The chemical compositions of both types of bottom ash—i.e., dry or slag—are quite similar. The major species present in bottom ash are silica (20–60 weight percent as SiO_2), alumina (10–35 weight percent as Al_2O_3), ferric oxide (5–35 weight percent as Fe_2O_3), calcium oxide (1–20 weight percent as CaO), and other minor amounts of metal oxides. Fly ash generally consists of very fine particles. The major species present in fly ash are silica (30–50 weight percent as SiO_2), alumina (20–30 weight percent as Al_2O), and other species including sulfur trioxide, carbon, boron, etc. Distribution between bottom ash and fly ash varies depending on the type of boiler bottom. Typically, bottom ash to fly ash ratio is 35/65 for wet bottom boilers, and it is 15/85 for dry bottom boilers [2].

Wet ash handling (sluicing) systems produce wastewaters which are currently either discharged as blowdown from recycle systems or discharged directly to receiving streams in a once-through manner. In a coal-fired power plant, wet ash handling system discharges may change reaching up to 16,387 GPD/MW for fly ash ponds, and 38,333 GPD/MW for bottom ash ponds.

Flue-Gas-Desulfurization Processes: In the lime or limestone flue-gas-desulfurization processes, SO_2 is removed from the flue-gas by wet scrubbing with slurry of calcium oxide (lime) or calcium carbonate (limestone). The principal reactions for absorption of SO_2 by slurry are:

lime: $\quad SO_2 + CaO + \frac{1}{2}H_2O \rightarrow CaSO_3 \cdot \frac{1}{2}H_2O$

limestone: $\quad SO_2 + CaCO_3 + \frac{1}{2}H_2O \rightarrow CaSO_3 \cdot \frac{1}{2}H_2O + CO_2$

Oxygen absorbed from the flue-gas or surrounding atmosphere causes the oxidation of absorbed SO_2. The calcium sulfite formed in the principal reaction and calcium sulfate formed through oxidation are precipitated as crystals in a holding tank. The potential exits to use calcium sulfite in manufacturing cement within the proposed complex.

Table 15.5 Amount of Raw Materials Consumed for the Production of Portland Cement [27]

Raw Materials	Amount Consumed (percent)
Cement rock	23.8
Limestone	90.0
Marl	2.1
Clay and shale	12.2
Blast furnace slag	0.8
Gypsum	4.1
Sand and sandstone	2.8
Iron materials	0.8
Miscellaneous	0.4
Total:	100.0

Background on Cement Manufacturing Plants. Portland Cement is made by mixing and calcining calcerous and argillaceous materials in the proper ratio. Table 15.5 summarizes the relative amount of raw materials consumed for the production of Portland Cement. It can be observed that limestone represents the major amount of raw material consumed. Composition of regular Portland cement includes approximately 2.9 percent compound calcium sulfate ($CaSO_4$) calculated from oxide analysis.

Unit processes involved in cement manufacturing include storage and mixing of raw materials, drying, grinding and crushing, calcining, clinker storage, finishing additives and bell milling, and packaging for delivery. A schematic diagram of a typical rotary steam kiln boiler and isometric flow chart for the manufacture of Portland Cement is shown in Figure 15-10. For each 100 barrels of finished cement by the dry process, 297,872 Btu of fuel as well as 6.4 kilowatt-hour of

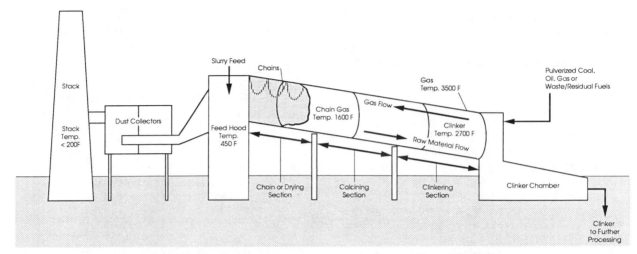

Figure 15-10.   Schematic Diagram of a Typical Rotary Steam Kiln Boiler

electricity, and 8 gallons of water are required. Also required are 132 pounds of limestone, 33 pounds of shale, and 4.3 pounds of gypsum [27].

Background on Concrete Block Manufacturing Plants. The aggregates used for mortars and concretes can be conveniently divided into dense and light weight types. The former class includes all the aggregates normally used in mass and reinforced concrete, such as sand, gravel, crushed rock and slag. The light weight class includes pumice, furnace clinker (or 'cinders' in the U.S.), foamed slag, the expanded clay, shale, and slate. Among the light weight aggregates, cinders are least expensive but variable in quality [15]. Table 15.6 summarizes some characteristics of concrete made with light weight aggregates.

Environmentally Balanced Industrial Complex. An *Environmentally Balanced Industrial Complex* has been proposed for a coal-fired power plant. Schematic flow diagram of the proposed complex is shown in Fig. 15.11. The complex consists of three industrial plants, including a coal-fired power plant, and two ancillary plants: cement and concrete block manufacturing plants. The coal-fired power plant has the following waste streams:

1. Recirculating System Cooling Water Blowdown
2. Boiler/Evaporator Blowdown
3. Fly Ash Discharge
4. Bottom Ash Discharge
5. Flue-Gas-Discharge

Cooling water blowdown, boiler blowdown, and evaporator blowdown are determined to be the major wet waste streams from a coal-fired power plant. These streams will be directed to the kiln steam boiler in the cement manufacturing plant as shown in the Fig. 15.11.

Sulfur dioxide which is released during the combustion of coal will be scrubbed with lime/limestone slurry. The calcium sulfate formed after oxidation will then be utilized in the cement manufacturing plant as a cement additive.

Waste dust from the kiln steam boiler in the cement manufacturing plant as well as fly ash and bottom ash formed during the combustion of coal in the power plant will be transported to the concrete block manufacturing plant. These solid wastes will be utilized in the production of concrete blocks.

15.12 Reduction in Production and Environmental Costs

According to the proposed theory and obvious implementation of this zero pollution solution, production costs must decrease. Likewise, by eliminating industrial waste treatment environmental costs must also decrease. The latter is especially valid when we include the benefits of eliminating adverse environmental impacts caused by the plant wastes. The sum of these two positive cost savings should be significant.

Nemerow (2, 1987) reported—as an example of these savings—that the production cost savings alone within a textile mill complex was $52.69 per 1000

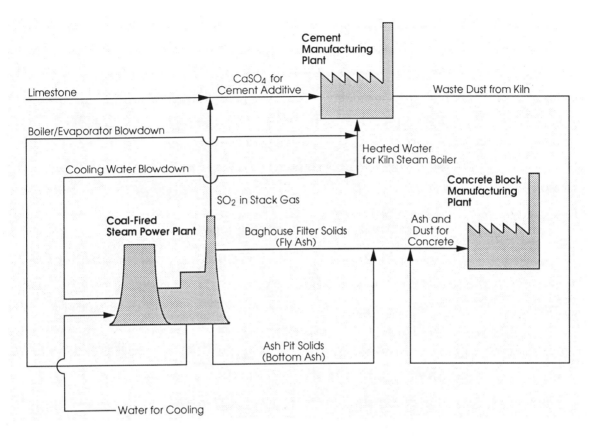

Figure 15.11 Schematic Flow Diagram of the Environmentally Balanced Industrial Complex for the Power Plant Industry

Table 15.6 Some Characteristics of Concrete Made from Light Aggregates [15]

Aggregate	Density of dry concrete (lb/ft^3)	Compressive strength (lb/in^2)	Drying shrinkage (percent)	Thermal conductivity (Btu-in/ft^{2}h° F)
Pumice	50	600	0.04–0.08	1–2
Clinker	60–95	300–1000	0.04–0.08	2.5–4
Expanded clay or shale	60–75	800–1200	0.04–0.07	2–3
Sintered pulverized fuel-ash	70–80	600–1500	0.04–0.07	2–3
Foamed slag	60–95	300–1000	0.03–0.07	1.5–3

pounds of cotton fabric from a typical small mill (12,000 pounds of cotton fabric per day). From a preliminary study, the total environmental savings (from eliminating waste treatment) appear to be greater than $104 per 1000 pounds of fabric. To both the above savings we must add the savings from transportation of raw cotton and sized, woven goods. In one instance and presuming a typical transportation cost in Southeast U.S.A. is $0.026 per pound of cotton shipped, we can add $343 per 1,000 pounds of fabric produced. The total savings of these three costs is almost $500 per 1000 pounds of cotton fabric produced. To allow the reader to grasp the significance of this, the above savings represent 77 percent of the cost of cotton required ($590 per 1000 pounds of cotton; New York Times page 48, April 12, 1988).

15.13 References

The following published references of the author's industrial complexes are included here as well as other references (12-18) for referral by readers of this book.

1. U.S.-India Joint Research on Industrial Complexing: A Solution to Phosphogypsum Fertilizer Waste Problem Nemerow, N.L. and T. N. Veziroglu, Nat. Sc. Found., Wash. D.C., March 1988.

2. Nemerow, N. L., Waite, T. D., and Tekindur, T., "Industrial Complexing and Ferrate Treatment for Reuse of Wastewater of Small Textile Mills," Proc. 8th Miami International Conference on Alternative Energy Sources Session on Environmental Problems, Miami Beach, Florida, December 15, 1987.

3. Nemerow, N. L. and A. Dasgupta, "Zero Pollution for Textile Wastes," 7th Alternative Energy Conference, Miami, Florida, December 1985.

4. Nemerow, N. L. and A. Dasgupta, "Zero Pollution: A Sugarcane refinery-Based Environmentally-Balanced Industrial Complex," 57th Annual Conference of Water Pollution Control Federation. New Orleans, Louisiana, October 1984.

5. Nemerow, N. L., "Environmentally-Balanced Industrial Complexes," The Biosphere: Problems and Solutions, Elsevier Science Publishers. B.V. Amsterdam, 1984, pp. 461-4170.

6. Tewari, R. N. and Nemerow, N. L., "Environmentally-Balanced and Resource Optimized Kraft Pulp and Papermill Complex," 37th Annual Purdue University Industrial Waste Conference, May 12, 1982, Proceedings, p. 353.

7. Nemerow, N. L. and Dasgupta, A., "Environmentally-Balanced Industrial Complexes," 36th Annual Industrial Waste Conference, 1981, Proceedings, p. 416.

8. Nemerow, N. L., "Preliminary Assessment of Environmentally-Balanced Industrial Complex: Three-Stage Evolution," Report to United States E.P.A., Contract Co. 68-02-3170, RTP, North Carolina, June 1980.

9. Nemerow, N. L., Farooq, S., and Sengupta, S., "Industrial Complexes and Their Relevance for Pulp and Paper Mills," Journal of Environmental International 3, 1, Pergamon Press. Oxford, England, 1980, p. 133.

10. Nemerow, N. L., "Environmentally Optimized Industrial Complexes," Lecture published in bound Proceedings of the National Environmental Engineering Research Institute, Nagpur, India, 1980.

11. Nemerow, N. L., Farooq, S., and Sengupta, S., "Industrial Complexes and Their Relevances for Pulp and Papermills," presented at Seminar on Industrial Wastes.

12. Chemical Week, Issue of August 3, 1968.

13. Anon., Phosphorous and Potassium, No. 85, September-October 1976.

14. Ellwood, P., "Turning By-Product Gypsum into a Valuable Asset," Chemical Engineering, March 24, 1969.

15. Carmichael, J.B., "World-wide Production and Utilization of Phosphogypsum", 2nd International Symposium on Phosphogypsum FIPR, University of Miami, Coral Gables, p. 34, December 10–12, 1986.

16. Bhanumathidas, N. and Kalidas, N., "Anhydrite: A Possible Solution for White Portland Cement", 2nd International Symposium on Phosphogypsum FIPR, p. 65, University of Miami, Coral Gables, December 10–12, 1986.

17. Clur, D.A., "Fedmis Sulphuric Acid/Cement from Phosphogypsum", 2nd International Symposium on Phosphogypsum FIPR, Univeristy of Miami, Coral Gables, pg. 141, December 10–12, 1986.

18. Shreve, R. N. and Brink, Jr., J. A., Chemical Process Industries, 4th edition, McGraw Hill Pub. Corp., pg. 153, 1977.

Part 3 | APPLICATIONS

CHAPTER 16

JOINT TREATMENT OF RAW INDUSTRIAL WASTES WITH DOMESTIC SEWAGE

Introduction

Industrial–waste problems should be solved using "systems engineering," which when applied to industrial wastes, implies combining the artful and scientific factors toward optimum solutions to treatment problems. When an engineer considers only the scientific or art factors involved in any waste–treatment problem, his solution is less than ideal. The more informed engineer utilizes as much of both art and science as possible in solving these problems. Even so, the result may be less than an optimum solution because of the interdependencies of many factors. Today's enlightened and progressive waste–treatment engineer takes both social and physical factors into consideration.

These factors can usually be applied to the major question of "which path to follow." As you can see from Fig. 16.1, industry must decide whether to treat its own waste or to contract with a municipality to accept, treat, and dispose of its wastes. Twelve potential combinations of alternatives, or side paths, are also available to an industrial plant. For example, an industrial plant may decide to partially treat its waste (alternatives 2, 4, 7, or 10) and then deliver the residual–waste volume and load to the municipal treatment plant for final treatment and disposal. It is vital then for us to point out and discuss the many factors of both types that will assist us in selecting the correct alternative path to follow and hence the overall solution to these problems.

Before examining the factors which effect joint treatment decisions, we must consider the four rules concerning objectives, conflicts, solutions and procedures as presented in the following grouping below:

		Industry	*Municipality*
(1)	Objective	To produce the best product at least cost	To provide best living conditions for people at least cost
(2)	Conflict	Production causes increased municipal costs (air, water, land environmental protection)	People depend upon industry for work and money, but destroy the clean environment at the same time
(3)	Solution	Produce goods in amounts and types so as to minimize adverse environmental impact	Only purchase goods which are produced and sold economically and which do not cause environmental degradation
(4)	Procedure	Select least cost alternative which protects the environment	Inform people of the environmental costs of production and adverse environmental impact

The following are the factors which assist us in selecting the proper sequence of treatment.

A. *Art in Industrial Wastes*
1. Precedent
2. Social relationship of industry and municipal officials
3. Political compatability of industry and municipality
4. Sewer service charge
5. Location of industry—especially in relation to the municipal plant and receiving stream
6. Competence of municipal plant operator
7. Permanence of industrial production

B. *Science in Industrial Wastes*
1. Type of municipal sewage treatment

2. Character of industrial waste
3. Required water quality of the receiving stream
4. Volume of industrial waste in relation to that of the municipality
5. Economics of the alternatives.

I suggest that we consider the art or social factors first, since they have been less publicized and are perhaps less well-known and understood by engineers.

A.1 Precedent

People are most inclined and even biased towards that which they are most familiar with or that which has been done rather uniformly and continuously in the past. There is a form of security in knowing that what

is being suggested is in keeping with past practice and that its effects have been previously experienced. If, in the past, the practice has been to include all types of industrial wastes regardless of volume or character in the municipal treatment system, this practice may be expected to be repeated in the future. On the other hand, if separate treatment of industrial wastes and municipal sewage has been encouraged, it will be difficult to change in the future. Any change in policy will require considerable effort, education, and enthusiasm of all parties concerned. The engineer must recognize that, even with proper theoretical justification, it may be impossible to overcome and alter that which has become practice by precedent.

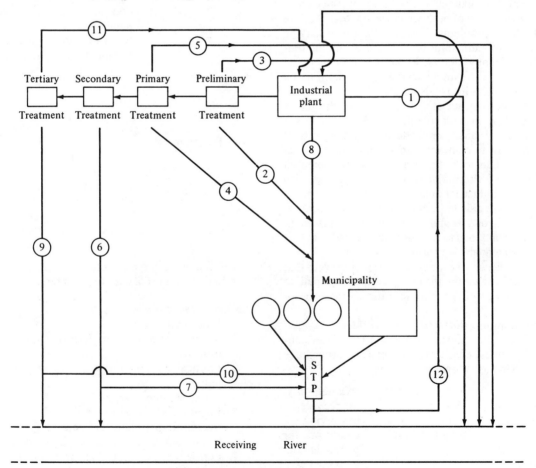

Fig. 16.1 Twelve alternatives of industrial waste treatment systems.

A.2 Social Relationship of Industry and Municipal Officials

The consulting engineer must be aware of and appraise the social state of the relationship between the appropriate members of the industrial and municipal community. For one reason or another this relationship may have been enhanced by a previous event which makes future compatibility for joint treatment possible. Likewise, some past experience, sometimes slight or seemingly insignificant, may have alienated the mutual social relationship and made future reasoning and negotiations for joint treatment rather difficult. A designing engineer must develop the art of detecting and evaluating this compatibility at the onset of his project planning. An oversight at this stage or even an improper consideration or evaluation can be disastrous in terms of the search for an overall optimum path of solution to the joint problems.

A.3 Political Compatibility of Industry and Municipality

When industry and the influential major municipal officials are of the same political leaning, cooperative solutions to waste treatment are facilitated. The engineer must determine this relationship and consider the likelihood of a change in municipal officials in an election year during the progress of his project.

A.4 Sewer Service Charge

Municipalities are composed of elected officials, each possessing his own philosophy on the subject of industrial patronage or subsidy for local industry. Some prefer to accept all wastes at a minimum charge or a flat fee or on a water-use basis. Others believe that industry should solve its own environmental problems and, should it wish to have the municipality treat its waste, it should be ready and willing to pay a sewer service charge which correctly and adequately covers capital as well as operating costs. The engineer should ascertain which type of feeling prevails in the municipality prior to the establishment of a charge. Industry usually does not object to a reasonable charge, but often objects to the philosophy of a comprehensive, complicated, or involved charge which may lead to

many unnecessary confrontations. It might prefer to treat its own waste, regardless of "on-the-surface" economics rather than face potential disagreements with public officials.

A.5 Location of Industry

The location of the industrial plant relative to both municipal sewers and treatment facilities as well as to the final receiving stream plays a vital part in influencing industry for or against combined treatment. At first consideration one might presume that economics alone is effected by plant location. True, the cost of connecting to and transmitting through existing or potential municipal sewers for a given distance directly effects costs. However, accessibility to sewers and proximity to treatment plants and receiving streams is important from a psychological viewpoint as well. Questions such as "Why have two treatment plants so close to each other?" or "Why pump and maintain a private pumping station and connecting sewer systems when we can treat our own wastes and dispose of them by gravity nearby?" are difficult to answer only in terms of dollars. Proximity enhances joint solutions regardless of the economic consideration.

A.6 Competence of Municipal Plant Operator

The experience, congeniality, and prominence of a supervisor and/or operator of a municipal treatment facility will enhance or hinder cooperative arrangements between industry and government. These qualities of the operating personnel are especially important in the initial stages of decision-making. For example, industry usually considers combined treatment feasible and even desirable when the municipal operator is held in high-esteem by his associates on a state or national level. The basis for this feeling is deeply rooted from a psychological standpoint, but most probably stems from a feeling of security in knowing that the wastes will be treated in an optimally-effective manner. An "easy-to-get-along-with" operator will enhance initial as well as continued discussions and negotiations. Combined treatment always manifests problems and difficulties of one type or another. Most can be solved with some effort, ingenuity, and cooperation. When industry knows a priori that it is dealing with a man or men possessing these characteristics, combined treatment becomes practical. An experienced operator usually knows what plant operating changes to make during critical situations in order to avoid malfunctioning. This is reassuring to both industrial and municipal officials. Conversely, the lack of those operator characteristics demands greater consideration of separate industrial-waste treatment.

A.7 Permanence of Industrial Production

Industry must show a profit for its owners in order to stay in business. The future of any given industrial facility depends primarily upon the whims of the purchasing public for its product. When whims change, a given industrial plant has three alternatives open to it: (1) *reduce production*—hopefully cutting costs so as not to allow net profits to deteriorate; (2) *change the type of production* to meet the needs of society; or (3) *cease all production* to avoid deficit financing. In all three alternatives the type and/or quantity of waste is altered considerably. Many industrial operations have found it necessary to resort to any or all of these alternatives over a 20-to-40 year period (during which time bonds for waste-treatment facilities are usually amortized). Industry has begun to recognize this as "a fact of industrial life."

Because our technology is changing so rapidly and with it the needs and demands of its customers industry hesitates to enter into long-term contractual committments, especially those concerned with waste-treatment facilities. Thus industry is now recognized as being relatively impermanent when compared to municipal operations. Certain industries are more permanent or less apt to cease production entirely (as evidenced from past records only). For example, DuPont Chemical Corporation or U.S. Steel's main production facilities could be expected to continue operation—with some modifications—with greater assurance for longer periods into the future than a small leather tannery or a peach cannery. The former types of industries would be more likely to consider joint ownership of a treatment facility than the latter. The cannery or tannery would be inclined towards private ownership of its separate facilities. There are exceptions and even variations within these latter industries mostly depending upon the size and type of ownership of the industry, as well

as specific past experiences it may have had in waste-treatment problems.

Some of the more important scientific factors associated with industrial-waste treatment include the following.

B.1 Type of Municipal Sewage Treatment

A secondary biological treatment plant, if adequately sized, can best be utilized to treat a readily decomposible, organic-laden industrial waste. Typical examples include dairies, canneries, slaughterhouses, and tanneries. However, each of these wastes, as well as other typical organic wastes, contain contaminants which can interfere with effective treatment when combined with domestic sewage. For example, dairy wastes often turn acid extremely fast and the lowered pH can affect biological oxidation; many canneries contain an extremely alkaline lye-peel waste which, when discharged in slug loads, can also hamper biological oxidation; slaughterhouse wastes contain slug loads of grease and blood which could hinder physical and biological processes; while tannery wastes contain chromium, sulfides, and lime, which are not compatible with normal sewage treatment. Proper pretreatment and plant operation, however, can remedy these problems.

Little would be gained by either party if an industrial waste (such as from a diary or textile mill) was treated in a municipal treatment plant consisting solely of primary treatment.

Industry must make an assessment of the objectives of treatment required and then examine the municipal plant available or being planned. Mutually compatible objectives and treatment facilities would be conducive to combined treatment.

B.2 Characteristics of Industrial Waste

When considering the treatment of the wastes from a tissue-paper mill, industry needs a municipal plant which concentrates its equipment units on the removal of the finely divided suspended-solids area of waste treatment. It is of little benefit to the tissue-paper mill if the municipal plant possesses only a high-rate trickling filter primarily designed for BOD removal.

Along similar lines a metal-plating waste would not be a welcome addition to the high-rate trickling filter plant because of its potentially toxic metals and acids as well as its lack of organic matter—hence the need for BOD reduction.

The waste engineer must carry out a complete analysis of the industrial waste to ascertain its compatibility for treatment by the varied possible methods. Some analyses often overlooked by the traditional sanitary engineer include the k-value, the waste deoxygenation rate; ultimate oxygen demand, toxic chemicals and metals, temperature, grease content at reduced temperatures, refractory organic matter, phosphates and nitrates, and other algae nutrients, etc. These and other important characteristics of certain industrial wastes should signal the key to its eventual successful treatment.

B.3 Receiving-stream Water Quality

It is a foregone conclusion that a stream which must be maintained in a high water-quality state requires the maximum offshore waste treatment. Generally this means a minimum of the equivalent to secondary treatment. But what is secondary treatment to an industry whose waste contains varied types of contaminants? Often the conventional biological treatment system will not adequately remove sufficient amounts of the contaminants. Sometimes specific treatment such as chemical precipitation followed by adsorption on activated carbon may remove more industrial contaminants than a secondary type trickling-filter plant. Industry has inherited the moral, if not the legal, obligation of treating its waste in a manner so as to maintain the highest possible quality water-level in the receiving stream. This cannot always be accomplished by following established state or federal rules. In many instances ingenuity must be used and sacrifices must be made in order to remove the proper amount of specific contaminants. In one specific waste problem it was necessary for management to render a decision to completely eliminate one of its three raw materials. This particular raw material was relatively inexpensive and from its use the industry was able to produce a satisfactory product at considerable profit. However, the waste resulting from processing with this particular raw material accounted for more than 80 per cent of the total plant contamination. To remove the contamination by conventional treatment proved

extremely costly and, after much deliberation, the industry reluctantly decided to discontinue the use of this particular raw material.

It is even more disconcerting for industry to be faced with a legal decision to install "so-called" complete treatment of its wastes when the receiving stream, even under critical conditions, shows little deterioration from the wasteload. In this situation, industry should endeavor to prove to regulatory agencies its dilemma so that it can operate at optimum economic efficiency. The latter is essential for the long-term benefit of society, as long as valuable river resources are not degraded in the process.

B.4 Volume Ratio of Industrial to Municipal Wastes

A relatively small volume of industrial waste can usually be assimilated in a municipal sewage-treatment system regardless of its contaminants. This fact does not always depend upon rational reasoning, but is often based on results of the very fact that an attempt to handle the wastes is made. In other words, municipal plant operators generally react optimistically towards *small volumes* of any industrial wastes, agree to try to treat them, and end up accepting them with or without certain preconditions. If the ratio of waste-volume to sewage had been greater, an attempt might never have been made to handle the waste, regardless of the potential of acceptability or treatability. Thus when ratios are high industry usually builds its own treatment plant despite the potentially favorable economics or the potential compatibility for joint treatment. There are exceptions to this generality; but they usually exist when a new facility is being contemplated by both municipal and industrial officials simultaneously.

B.5 Economics of Alternatives

Industry tends to select the least costly alternative, especially when other conditions are equal. How does industry select the least costly system? Usually industry prefers to compare alternative system costs on the basis of total capital expenditures; the least expensive capital outlay is often preferred. This method can often be misleading and even erroneous over the long-term. First, it does not take into consideration the annual cost of operation and maintenance required for effective treatment. Often the least costly alternative can be the most expensive to maintain, especially when perfection is required in efficiency of operation.

Furthermore, the cost of obtaining money for capital expense is not taken into account when considering only capital costs. This is significant during high-cost periods and when one or more alternatives entail public rather than private borrowing. It is well-known that municipalities can usually borrow money at lower interest rates than private corporations.

And last, the financial rating of the borrower is a significant and often overlooked factor in comparing capital costs only. Certain public authorities may possess such poor financial ratings that many prominent, profitable industrial firms may be able to borrow money as cheaply as or cheaper than the municipality.

Therefore economic alternatives should be based upon net annual costs which include operation, maintenance, and amortization of capital costs. The latter should be selected on the realistic basis of the financial rating of the appropriate borrower as well as on the basis of a reasonable bond amortization period, depending upon the borrower's ability to repay and the expected life of the specific equipment required in each alternative.

The question of whether the optimum alternative selected is justified on *any* economic basis is one which warrants consideration and up to now has defied our engineers. We have blindly presumed that what is required in the way of waste treatment by state or federal edict is required at any cost. This may be so, but the case may be strengthened by a proper economic evaluation. We should compute the total net benefits to the surrounding, immediate society of the specific waste-treatment alternative selected. Heretofore we have lamented that these benefits were intangible and hence immeasurable. (See Chapter 5.2 for more information about benefits.)

Benefits which are effected by water quality include (1) recreational use, (2) land use, (3) water withdrawal, (4) waste treatment, and (5) in-place water use.

16.1 Industrial Use of Municipal Sewage Plants

It is often possible and advisable for an industry to discharge its waste directly to a municipal treatment plant, where a certain portion of the pollution can be removed [6]. A municipal sewage-treatment plant, if designed and operated properly, can handle almost any type and quantity of industrial waste [5]. Hence, one possibility that should be seriously considered is the cooperation of industry and municipalities in the joint construction and operation of a municipal waste-water treatment plant. There are many advantages to be gained from such a joint venture:

1. Responsibility is placed with one owner, while at the same time the cooperative spirit between industry and municipality is increased, particularly if division of costs is mutually satisfactory.

2. Only one chief operator is required, whose sole obligation is the management of the treatment plant; i.e., he is not encumbered by the miscellaneous duties often given to the industrial employee in charge of waste disposal, and the chances of mismanagement and neglect, which may result if industrial production men operate waste-treatment plants, are eliminated.

3. Since the operator of such a large treatment plant usually receives higher pay than separate domestic plant operators, better-trained people are available.

4. Even if identical equipment is required, construction costs are less for a single plant than for two or more. Furthermore, municipalities can apply for state and/or federal aid for plant construction, which private industry is not eligible to receive.

5. The land required for plant construction and for disposal of waste products is obtained more easily by the municipality.

6. Operating costs are lower, since more waste is treated at a lower rate per unit of volume.

7. Possible cost advantages resulting from lower municipal financing costs, federal grants, and municipal operation can be passed on to the users and may permit higher degrees of treatment at a cost to each participant no greater than the cost for separate treatment at lower removal levels.

8. Some wastes may add valuable nutrients for biological activity to counteract other industrial wastes that are nutrient-deficient. Thus, bacteria in the sewage are added to organic industrial wastes as seeding material. These microorganisms are vital to biological treatment when the necessary BOD

reduction exceeds approximately 70 per cent. Similarly, acids from one industry may help neutralize alkaline wastes from another industry.

9. The treatment of all waste water generated in the community in a municipal plant or plants enables the municipality to assure a uniform level of treatment to all users of the river and even to increase the degree of treatment given to all waste water to the maximum level obtainable with technological advances.

10. Acceptance of the joint treatment project and relinquishment of individual allocations would give the municipality full control of the river's resources and permit it to use the capacity of the river to the best advantage for the public at large. The municipality has greater assurance of stream protection, since it has the opportunity for closer monitoring of effluent quality.

11. Public relations are good for municipality.

12. Land is generally more available.

13. No permit is needed.

Among the many problems arising from combined treatment, the most important is the character of the industrial waste water reaching the disposal plant. Equalization and regulation of discharge of industrial wastes are sometimes necessary to prevent rapid change in the environmental conditions of the bacteria and other organisms which act as purifying agents, to ensure ample chemical dosage in coagulating basins, and to ensure adequate chlorination to kill harmful bacteria before the effluent is discharged to a stream.

In recent years, two factors in particular have focused attention on the subject of combined treatment for sewage and industrial wastes: the increased interest in stream-pollution abatement and the phenomenal growth of industry in the postwar years, with the subsequent increase in demand for water.

Since most sewage plants use some form of biological treatment, it is essential for satisfactory operation that extremes in industrial waste characteristics be avoided and the waste mixture be: (1) as homogeneous in composition and uniform in flow rate as possible and free from sudden dumpings (shock loads) of the more deleterious industrial wastes; (2) not highly loaded with suspended matter; (3) free of

excessive acidity or alkalinity and not high in content of chemicals which precipitate on neutralization or oxidation; (4) practically free of antiseptic materials and toxic trace metals; (5) low in potential sources of high BOD, such as carbohydrates, sugar, starch, and cellulose; and (6) low in oil and grease content.

If the industrial-waste characteristics are such that the waste can be treated safely and effectively in the municipal sewage plant, there still remain two major considerations: a municipal ordinance which protects the treatment plant from any individual or industrial violation and sewer-rental charges which enable the municipality to defray the increased costs of construction and operation resulting from acceptance of the industrial wastes.

Combined municipal and industrial waste treatment is the most desirable arrangement and at the same time it is the most difficult to achieve. It is the author's contention that the difficulty is usually not a scientific one but rather one of human compatability and understanding. In numerous cases, the economics of the situation were overwhelmingly in favor of combined treatment and yet separate waste-treatment installations were finally used, primarily because of personality clashes and lack of sympathetic understanding.

What can municipalities do to assist industry in waste-treatment practices?

1. If the municipal treatment plant is new or being enlarged, it should be designed to serve the entire community, with *all* the existing and planned industry as members of a "corporation." Combined meetings, lectures, and actual visits to the treatment-plant site will enhance mutual understanding of the problems involved.

2. If the plant is already in operation, municipal officials should meet with industrial representatives and discuss the advantages and disadvantages of accepting the industrial wastes into the system. Adequate safeguards such as research, literature study, and pilot-plant experiments should precede administrative decisions, and any decision to accept the waste should be accompanied by a substantial and specifically detailed contract between the owner and user. Methods of sampling, analyses, charges, and waste characteristics should be clearly stated in this contract.

3. A municipality can purchase land, build a treat-

ment plant for its industry, float bonds, and receive rent from industry for use of the plant and amortization of the bonds. In this way the rental becomes an operating expense which industry can deduct from income *before* taxes rather than the long-term depreciation of capital assets involved in having its own plant. This is particularly attractive to industry.

4. Most important, perhaps, is the understanding a municipality must have for its industry. All the members of the city council should be in agreement that without industry municipal survival is doubtful and its growth potential is nil. An industry located within the city limits is contributing the maximum to a city in the form of taxes and intangible benefits. Although it is common for new industry to locate outside city limits, where it can purchase sufficient land at a reasonable cost for future expansion, these industries too are an indirect but valuable addition to the community.

5. A municipality can design its treatment plant so that it will handle an industry's waste without pretreatment by the industry. An ideal arrangement is to take the industry into the business of waste disposal as a "member of the municipal corporation." Industry should pay for this service but not at the same rate as an individual householder, since a large contributor deserves concessions solely on the basis of lower unit costs for larger volumes. In addition, industry's intangible benefits to the community should be assessed and its share of the capital cost reduced proportionately.

6. Municipal controls of the influent from the industrial plant are costly and difficult to establish. Instead, it is recommended that industry control its own effluent so that the "corporation" disposal system operates efficiently. When, and if, the system ceases to function as designed, a corporation meeting should be called to decide what measures should be taken to correct the situation. In this manner, the expenses of sampling and billing, as well as the ill feelings caused by policing, are eliminated.

7. A good corporation will continually try to improve its efficiency of operation by conducting research on new methods of treatment. Research has been proven to pay off in the long haul and the lack of it has often led to plant and process obsolescence. Industry, with its research experience, could well lead the way in this connection by supporting continued research in specific combined treatment processes.

8. The designing sanitary engineer should be selected by both the municipality and the industry for his competence and ability to work with equal ease with both groups. His fee should be on a lump-sum basis, approximating the sliding-scale percentage of estimated construction cost but not necessarily tied to this: rather than being penalized for reducing the capital costs (and hence his fees), he should be rewarded financially for economizing on construction and improving plant efficiency.

In summary, a municipality can assist its industry by encouraging mutual understanding, by embarking on a program of education, and by designing its plants to handle industrial waste. Other methods of assisting industry depend upon the formation of a "corporation treatment plant" with joint responsibility for efficient operation. Municipal ownership and reduction in charges based upon intangible as well as tangible industrial benefits should also be considered.

The Water Pollution Control Federation (4, 1976) published a rather detailed (34 pages) pamphlet on all the various aspects of joint treatment. The reader is directed to this pamphlet for the practical and economic aspects of this subject.

16.2 Municipal Ordinances

Although there are many types of municipal ordinances, all are designed to place an upper limit on the concentration of various constituents in waste. Sometimes this upper limit is zero, since any quantity of a certain pollutant would be detrimental to the plant or its component parts. In addition to their obligation to abide by municipal ordinances, many industries enter into separate contracts with the city. Generally, such contracts include: the obligation of the municipality to construct, operate, and maintain the treatment facilities and to finance the overall project by means of some type of bond; a declaration on the part of the industry as to the maximum quantity of flow, BOD, and solids; the percentage by volume of industrial waste as compared to municipal waste; the amount the industry will pay each year to cover operation and maintenance; provision for a penalty if stated limits are exceeded; and any other pertinent matters involving the joint usage of the treatment system.

The following are dangers of inadequate sewer-use control [4]: (1) explosion and fire hazards; (2) sewer clogging; (3) overloads of surface water (storm- and/or cooling-water pollution); (4) physical damage to sewers and structural damage to treatment plants; (5) interference with sewage treatment.

A comprehensive sewer ordinance [5] usually consists of the following principal parts: introduction; definition of terms; regulation requiring use of public sewers where available; regulations concerning private sewage and waste disposal where public sewers are not available; regulations and procedures regarding the construction of sewers and connections; regulations relating to quantities and character of waters and wastes admissible to public sewers; special regulations; provision for powers of inspectors; enforcement (penalty) clause; validity clause; and signatures and attest.

Since industrial wastes vary so greatly in character, only broad limits can be established in any model ordinance, and ordinances should always be based on recommendations of the consulting engineer. (See Table 16.1.) Most ordinances [5] provide for the control of waste substances other than sanitary sewage in the following ways:

1. They prohibit the discharge to the public sewers of flammable substances or materials that would obstruct the flow.

2. They state that industrial wastes will be admitted to the public sewers only by special permission of a stated municipal authority.

3. They ban all wastes that would damage or interfere with the operation of the sewage works, except when such wastes have been adequately pretreated, and even then their admission is to be at the discretion of a stated municipal authority.

4. They enumerate in detail, in a separate ordinance, the procedures outlined in (3).

5. They give detailed regulations to supplement the procedure in (3), stating specific limits for objectionable characteristics of industrial wastes.

A model ordinance [5] may spell out in detail the following regulations relating to quantities and character of water and wastes admissible to public

sewers:

Section 1. No storm water, roof runoff, cooling water, ground-water, etc., will be allowed in the sanitary sewer.

Section 2. Storm water or other uncontaminated drainage will be discharged to sewers that are designated *combined* or *storm sewers* only.

Section 3. No person shall discharge any of the following wastes to sanitary sewers except as hereinafter provided: (a) any liquid or vapor having a temperature higher than 150°F; (b) any waste containing more than 100 ppm by weight of grease; (c) any gasoline, etc., or other flammable or explosive liquid, solid, or gas; (d) any garbage that has not been properly ground; (e) any ashes, metals, cinders, rags, mud, straw, glass, feathers, tar, plastics, wood, chicken manure, or other interfering or obstructing solids; (f) any wastes having a pH less than 5.5 or higher than 9.0, or having other corrosive effects; (g) any toxic wastes that may be a hazard to sewage plant, persons, or receiving stream; (h) any suspended solids the treatment of which at the sewage plant may involve unusual expenditures; (i) any noxious gases.

Section 4. There shall be installations of interceptors for grease, oil, and sand, when necessary.

Section 5. These installations shall be maintained by owner.

Section 6. This section establishes the conditions pertaining to the admission of any wastes having (a) a 5-day BOD greater than 300 ppm, (b) more than 350 ppm suspended solids, (c) any of the quantitative characteristics described in Section 3, (d) an average daily flow greater than 2 per cent of the average daily flow of the city.

Section 7. Where preliminary treatment facilities are provided for any wastes, they shall be maintained by the owner at his own expense.

Section 8. When required, the owner of any property served by a sewer carrying industrial wastes shall install a suitable manhole for observation, sampling, and measuring.

Section 9. All measurements and analyses of the characteristics of waters and wastes referred to in Sections 3 or 6 shall be determined in accordance with standard methods [1].

Section 10. No statement contained in this article shall preclude any special agreement or arrangement between the city and any industry.

Table 16.1 Industrial contaminants and their general limiting values for discharge into municipal sewerage systems*.

Contaminant	Concentration generally limiting for municipal sewerage systems	Reason for limitation	If contaminant is excessive, the acceptable pretreatment generally required is
1. Flow	50% of municipal sewage flow	Causes sewage treatment system to react differently from its normal pattern as designed for municipal sewage. Unequalized or unproportioned industrial flow is especially troublesome	1. Equalization and proportioning 2. Recirculation and reuse within industry to reduce flow 3. Redesign sewage treatment plant to react more specifically to industrial waste
2. BOD 5, 220 °C	300 ppm	Exerts a disproportionately high percentage of oxygen-demanding organic matter to municipal wastewater	1. Change in industrial manufacturing process 2. Equalization 3. Biological pretreatment plant
3. Color	Visible in dilutions of 4 parts sewage to 1 part industrial waste	Color is normally not removed by domestic sewage treatment plants, will appear in the combined, treated effluent, will be readily detected and visually undesirable from an aesthetic standpoint.	1. Change in industrial manufacturing process 2. Chemical pretreatment to remove color 3. Equalization and/or proportioning
4. Suspended solids	350 ppm	Overloads disproportionately normal domestic sewage treatment plants	1. Change in industrial manufacturing process 2. Equalization 3. Sedimentation pretreatment plant
5. pH	5.5–9	Corrosion of sewers and treatment plant equipment causes a diminution or a malfunctioning of biological treatment units	1. Equalization 2. Neutralization 3. Change in industrial manufacturing process
6. Grease	100 ppm	Interferes with plant operating equipment—including aeration, primary sedimentation, etc. Overloads sludge-handling treatment units such as scum collection, digestion, and sludge-drying beds	1. Change in industrial manufacturing process 2. Install grease traps or remove pretreatment units
7. Heavy metals Cr, Sn, Pb, Zn Hg, Cu, Ni, etc.	1 ppm Cu, Cr, 5 ppm Zn, Ni	Inhibits biological action in municipal sewage units, such as activated sludge, trickling filters, and especially sludge digesters	1. Equalization 2. Chemical pretreatment and sedimentation pretreatment

Table 16.1 (*continued*)

Contaminant	Concentration generally limiting for municipal sewerage systems	Reason for limitation	If contaminant is excessive, the acceptable pretreatment generally required is
8. Nonorganics and other toxic chemicals	None so as to be toxic to bacteria serving the treatment plant or man or animals working in or near the sewage plant	Exhibit toxicity towards biological treatment units and cause health hazard to man and animals	1. Change industrial plant process 2. Use of advanced pretreatment wastewater techniques
9. Inflammable liquids, foaming agents, rags, solidifiable greases, ashes, metals, cinders, mud, straw, glass, feathers, tar, plastics, wood, chicken manure, etc.	None in such quantities that will cause either a hazard to the environment or a nuisance to the operation of the plant	Causes a nuisance and interferes with the normal operation of the domestic sewage-treatment plant	Removal by process change or physical means such as screening
10. Temperature	150°F	Hastens corrosion, drives out dissolved oxygen, volatilizes hazardous gases such as H_2S	1. Change in industrial process 2. Use of cooling water systems
11. Storm water	None resulting from direct connections or faulty sewer construction	Occupies valuable volume capacity of domestic and industrial sewers	1. Construct separate sewer 2. Use better construction procedures
12. Refractory organic matter	None	Contaminates the municipal sewage plant effluent for possible reuse downstream for water supplies	1. Change in industrial manufacturing process 2. Carbon adsorption pretreatment
13. Refractory mineral matter	Boron 0.7 ppm NaCl 1000 ppm	Contaminates the municipal sewage plant effluent for possible reuse for irrigation waters	1. Change in industrial manufacturing process 2. Pretreat industrial plant wastewater by membrane separation or distillation

*All industrial wastes can be treated in some manner at some cost with some relative effectiveness so as to render the contaminants suitably low in concentration or changed in character with the result that they may be discharged safely into the environment either directly or indirectly (through some municipal system). It is suggested that a literature survey be made of the effect of each contaminant at its expected level of concentration in the resulting domestic wastewater before acceptance or rejection of the industrial waste into any municipal system.

In cases where any doubt exists about the effect of the industrial waste based upon personal experience or literature survey, laboratory and/or field prototype studies should be made to ascertain the precise effect on municipal sewage-plant systems. Pretreated effluents from industry can be examined in a similar manner.

The exact values of each contaminant are based upon the best evidence available to the writer—exceptions can be found in each case. The user of this table must be prepared to render to it some flexibility in both allowable concentration and type of contaminants as more knowledge becomes available.

From Nemerow, N.L. Industrial Water Pollution Krieger Pub. Co. pg 198–199, 1987.

16.3 Sewer-Rental Charges

Sewer-rental charges are necessary to help meet the city's budget and to ensure that industry pays a fair share of the cost of disposing of its wastes. Several methods can be used to charge for sewer service: (1) an *ad valorem* tax on property, which is the traditional method in more than 80 per cent of U.S. communities and is successful in small towns and villages; (2) special assessments, with charges set according to front footage; (3) sewer-rental charges (approximately one-sixth of municipalities having treatment plants use this method); (4) special contracts negotiated with industry; (5) combination of two or more of the above methods. In many cases a municipality charges the industry or industries solely on the basis of water consumption. Although this may not always prove equitable, it has several advantages to the municipality and to the industries. First, the billing system is simplified, omitting the need for detailed and time-consuming cost procedures. Second, the system eliminates the need for measuring flows from the industries and their strength characteristics. Thus, the municipality treats its industries just as it does its householders, rather than as a "culprit."

In considering charges, fixed expenses such as operation, maintenance, and debt retirement should all be taken into account. A portion of each of these three costs can be charged to all the users of the sewer system, and the remaining portion to property owners having access to the system. This is done by itemizing the cost of each component unit of the sewer system and then allocating percentages of the annual cost of each unit to the users and the rest to the property owners. Total annual charges to users and property owners are determined from the summation of the unit costs. The total share allocated to property owners may now be prorated according to individual property valuations (or sometimes front footages). The user's share necessitates additional prorating based on the following waste factors: volume, suspended solids, BOD, and (sometimes) chlorine demand. This is carried out in the computations of user's share for each unit. If the unit is designed solely upon a volume basis, such as the main sewage pumps, the entire user's share is charged to volume contributors. On the other hand, for the sludge digester, 90 per cent of the cost may be charged to contributors of

suspended solids and 10 per cent to contributors of BOD. If the volume of sewage is based on water consumption and the supply is private (wells or river water), a meter is normally supplied by the industry for flow measurement.

When all users' charges attributed to volume, solids, and BOD are added, one obtains the total users' cost for each category. The total of the three in turn represents the users' share of the annual sewer costs, and the total of the users' and property owners' shares represents the complete annual sewer costs.

Schroepfer [8] uses the following example to illustrate a fair allocation of costs. The total annual cost of operating a sewage-disposal system in a certain town consists of:

(1) Fixed charges:
Intercepting sewers*	$ 35,000
Treatment plant*	75,000

(2) Operating and maintenance
costs	70,500
Total	$180,500

Table 16.2 shows the allocation of the fixed charges for the sewers and treatment plant, Table 16.3 the allocation of the operation and maintenance costs, and Table 16.4 the allocation of fixed and operational charges. Figures 16.1 and 16.2 illustrate the prorating of the fixed charges of the sewers and treatment plant.

Property owners' charges should be distributed according to assessed evaluation, which in the example under discussion is taken to be $20,000,000. Therefore, $2.88 per $1000 of property valuation will be charged to property owners. Users' charges depend on flow, solids, and BOD, as stated previously. Hence, the annual flow and the quantities of each type of these waste loads must either be determined after the first year's operation or estimated prior to establishing the users' charges for the year. The third column of Table 16.5 lists the unit rates obtained from the data given in columns 1 and 2.

---

*Capital investment: for intercepting sewers $700,000 and for treatment plant $1,500,000; debt retirement: 5 per cent per year (total interest and principal).

Table 16.2 Allocation of fixed charges (After Nemerow, N.L. Ind. Water Pollution Krieger Pub. Co. pg 201, 1987.)

Units	Total fixed charges, $	Chargeable to property owners		Chargeable to users, $	Users' share chargeable to					
					Volume		Suspended solids		BOD	
		%	$	$	%	$	%	$	%	$
Intercepting sewers	35,000	64	22,300	12,700	100	12,700				
Treatment plant										
Main pumping station										
Equipment	1,500	40.5	600	900	100	900				
Structures	1,250	64	800	450	100	450				
Screen and grit chambers	1,500	64	950	550	60	330	40	220		
Preliminary sedimentation tanks	4,500	40.5	1,800	2,700	85	2,300	15	400		
Trickling filters	30,000	25	7,500	22,500	10	2,250			90	20,250
Final sedimentation tanks	9,000	30	2,700	6,300	50	3,150			50	3,150
Receiving pumps	750	25	200	550					100	550
Chlorination tanks and equipment	2,000	35	700	1,300	40	520			60	780
Digester tanks and receiving filters	8,000	30	2,400	5,600			100	5,600		
Subtotal	58,500	30.3	17,650	40,850	24.2	9,900	15.2	6,220	60.6	24,730
Main control building	7,500	30.3	2,300	5,200	24.2	1,310	15.2	790	60.6	3,100
Plant water supply	2,500	30.3	800	1,700	24.2	410	15.2	260	60.6	1,030
Roads and grounds	2,500	30.3	800	1,700	24.2	410	15.2	260	60.6	1,030
Plumbing and heating	4,000	30.3	1,200	2,800	24.2	680	15.2	430	60.6	1,690
Total plant costs	75,000	30.3	22,750	52,250	24.2	12,710	15.2	7,960	60.6	31,580
Total fixed charges	110,000	41	45,050	64,950	39.2	25,410	12.2	7,960	48.6	31,580

Table 16.3 Allocation of operation and maintenance costs. (After Nemerow, N.L. "Ind. Water Pollution" Krieger Pub. Co. pg 201, 1987.)

Unit	Total operating and maintenance cost, $	Chargeable to property owners		Chargeable to users, $	Users' share chargeable to					
					Volume		Suspended solids		BOD	
		%	$	$	%	$	%	$	%	$
Intercepting sewers	2,200	60	1,300	900	60	500	40	400		
Main pumping station	9,200	17	1,600	7,600	100	7,600				
Preliminary treatment	6,700	50	3,400	3,300	50	1,700	50	1,600		
Secondary treatment	13,500	15	2,000	11,500	10	1,200			90	10,300
Effluent chlorination	5,200	15	800	4,400	10	400			90	4,000
Sludge disposal	17,500	5	900	16,600			100	16,600		
General	5,000	15	800	4,200	25	1,000	43	1,800	32	1,400
Supervisory	6,200	15	900	5,300	25	1,300	43	2,300	32	1,700
Collection and billing	5,000	15	800	4,200	25	1,000	43	1,800	32	1,400
Total	70,500	17.8	12,500	58,000	25.4	14,700	42.1	24,500	32.5	18,800

Table 16.4   Summary of the allocation of the fixed and operating charges. (After Schroepfer [8]).

Fixed charges	Chargeable to			
	Users		Property owners	
	%	$	%	$
Sewers	36	12,700	64.0	22,300
Treatment plant	69.7	52,250	30.3	22,750
Operation and maintenance costs	82.2	58,000	17.8	12,500
Totals		122,950		57,550
Averages	68.1		31.9	

(From Nemerow, N.L. "Ind. Water Pollution" Krieger Pub. Co. pg 202, 1987.)

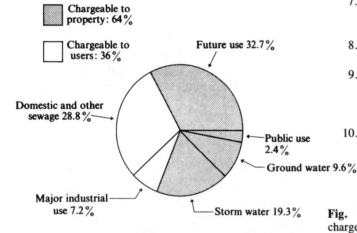

Chargeable to property: 64%

Chargeable to users: 36%

Future use 32.7%

Domestic and other sewage 28.8%

Public use 2.4%

Ground water 9.6%

Major industrial use 7.2%

Storm water 19.3%

Fig. 16.2 Allocation of fixed charges on the intercepting sewers. (After Schroepfer [8].)

References: Introduction

1. American Public Health Association, *Standard Methods for the Examination of Water, Sewage, and Industrial Wastes*, 10th ed., New York (1955).
2. California State Water Pollution Control Board, *A Survey of Direct Utilization of Waste Waters*, Publication no. 12, Sacramento, Cal. (1955).
3. Geyer, J. C., "The effect of industrial wastes on sewage plant operation," *Sewage Works J.* 9, 625 (1937).
4. Joint treatment of industrial and municipal wastewaters—A publication of the technical practice committee WPCF Washington, D.C. 1976.
5. *Municipal Sewer Ordinances*, Manual of Practise no. 3, Federation of Sewage and Industrial Wastes Association, Washington, D.C. (1957).
6. Nemerow, N.L. "Fiber Losses at Paper Mills: Effects on streams and sewage treatment plants" *Sewage and Ind. Wastes* 23, 880 (1951).
7. Nemerow, N. L., *Water Wastes of Industry*, Bulletin no. 5, Industrial Engineering Program, North Carolina State College, Raleigh (1956).
8. Schroepfer, G. M., "Sewer service charges," *Sewage Ind. Wastes* 23, 1493 (1951).
9. *The Treatment of Sewage Plant Effluent for Water Reuse in Process and Boiler Feed*, Technical Reprint T-129, Graver Water Conditioning Company, New York (1954).
10. Veatch, N. T., "Industrial uses of reclaimed sewage effluents," *Sewage Works J.* 20, 3 (1948).

Table 16.5   Calculation of users' charges based on three factors. (After Schroepfer [7].)

	Annual quantity	Total, $	Unit rate, $
Volume of flow	1,370 million gal	40,000	2.93/1000 gal
Suspended solids	3,647,000 pounds	32,460	0.89/100 lb
BOD	3,847,000 pounds	50,380	1.40/100 lb

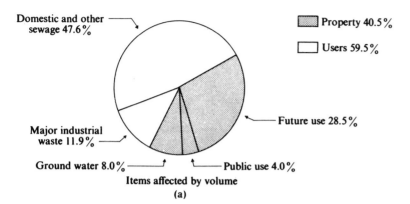

Domestic and other sewage 47.6%

Property 40.5%

Users 59.5%

Future use 28.5%

Major industrial waste 11.9%

Ground water 8.0%

Public use 4.0%

Items affected by volume

(a)

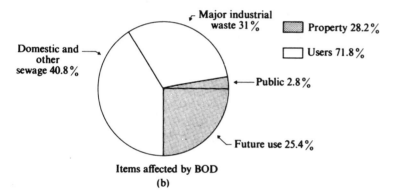

Major industrial waste 31%

Property 28.2%

Users 71.8%

Domestic and other sewage 40.8%

Public 2.8%

Future use 25.4%

Items affected by BOD

(b)

Fig. 16.3 Allocation of fixed charges on the treatment plant: (a) items affected by volume; (b) items affected by BOD; (c) items affected by suspended solids. (After Schroepfer [7].)

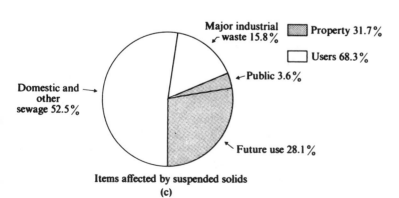

Major industrial waste 15.8%

Property 31.7%

Users 68.3%

Public 3.6%

Domestic and other sewage 52.5%

Future use 28.1%

Items affected by suspended solids

(c)

CASE HISTORY OF A PROJECT FOR JOINT DISPOSAL OF UNTREATED INDUSTRIAL WASTES AND DOMESTIC SEWAGE

For the purposes of our discussion, we shall consider the case of two relatively small municipalities containing 27 small industries (mostly tanneries) which require adequate and effective treatment of their wastes. The problem presents a challenge in engineering, economics, and administration.

16.4 Existing Situation

Cayadutta Creek rises in the central part of Fulton County in New York State, flows generally south for about 14 miles through the cities of Gloversville and Johnstown, and enters the Mohawk River at Fonda (Fig. 16.4). The total catchment area cover 62 square miles above Station 6. There are no official gauging stations on this stream but approximate flow data for a comparatively short time (1898–1900) are available. This creek has been characterized by an expert state hydrologist, a member of the U.S. Geological Survey, as similar to that of Kayaderosseras Creek, which is located in Saratoga County and drains into the Hudson River basin.

Ninety-one per cent of the population of the Cayadutta Creek catchment area is concentrated in the cities of Johnstown and Gloversville and the village of Fonda. In 1952, New York State cited this creek as "one of the most grossly polluted streams in the state." From within the city of Gloversville to the junction with the Mohawk River the stream is entirely unsuited for the support of fish life, whereas formerly it was trout water throughout its entire length. It has been stated (by M. Vrooman: *March 10, 1950, Report to City of Gloversville*) that the dry weather flow of Cayadutta Creek is higher than the average for streams in the state, owing to the nature of the watershed, the sandy soil, and the larger wooded area. He also stated that "the average daily flow of the Creek at the Gloversville sewage treatment plant is 17 million gallons and the low measured dry weather flow is 4.2 million gallons." These figures were evidently obtained from separate, independent, and unofficial flow measurements. The tanning industry

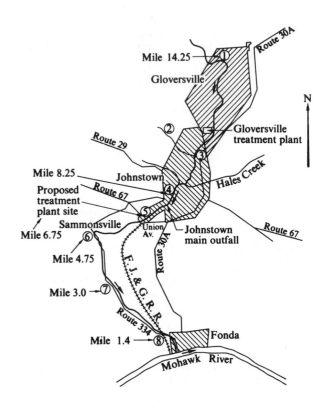

Fig. 16.4 Cayadutta Creek.

is an old one in American history and has a record of contributing to the damaging pollution of Cayadutta Creek. The National Tanners Association (private communication) estimates that the Fulton County area has been losing about one plant every four years. However, there has been more glove- and garment-leather demand as the population of the U.S.A. rises. They predict, therefore, that the overall demand for glove and garment leather (produced in

Fulton County) will continue slowly upward, but it will be met in fewer plants with increased production.

The sewage and wastes from the cities of Gloversville and Johnstown are discharged into Cayadutta Creek. In 1960, Gloversville had a population of 21,741 while that of Johnstown was 10,390 (Fig.16.5). Almost the entire urban population is served by public sewer systems but only half of the system is tributary to a sewage-treatment plant, which serves the people and industry of Gloversville. It consists of a bar screen, grit chamber, two antiquated Dortmund-type primary settling basins, a fixed-nozzle trickling filter, one final Dortmund-type settling basin, and some sludge-drying beds. The plant was built in the early 1900s and is incapable of handling more than half the waste water at the present flow rates.

The tanning industry retained a New York City consulting firm to represent their interests in this problem. The two cities retained a local consulting engineering firm, which in turn retained the present author to advise them on study procedures and solutions to the pollution problem in Cayadutta Creek.

16.5 Stream Survey

A stream survey is an essential part of any well-conceived waste-treatment study. Ideally, a survey designed to study the oxygen-sag curve should be carried out during extremely hot weather, extremely low stream flow, and typical high organic matter loading. It is seldom possible to conduct a stream survey under all of these "ideal" conditions. In this study we were particularly fortunate to collect stream samples during extremely low flow conditions—comparable to those which may be expected to occur for a seven-day period only once in ten years—while, at the same time, the municipal and industrial pollution loads were considered to be above average. Although stream temperatures were not high (11° to 14°C), these values are never very high, owing to the relatively cold mountain water diluting the wastes. For example, during the state survey of 1951, samples collected on August 22 and 23 at Stations 5 and 6 showed temperatures of only 15° to 19°C.

During October 1964, the creek was visited and examined at various locations and dissolved-oxygen values were determined in order to locate the sag

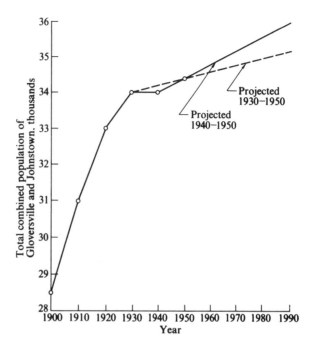

Figure 16.5

curve points. After an initial appraisal and a trial survey, the creek was sampled at the eight stations shown in Figure 16.3. After the first day samples were collected only from Stations 1, 5, and 6, on six days at

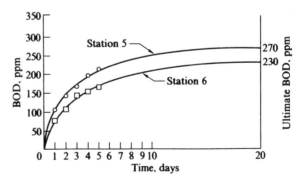

Fig. 16.6 Laboratory BOD's for Stations 5 and 6 at 20°C. Each point represents an average of seven samples collected from the creek on seven different days at different times of day, all during a drought flow period (October 8–18, 1964).

Table 16.6 Cayadutta Creek analyses during October 1964.

Station and milage	Date	Time	DO, ppm	Flow, cfs	BOD, ppm					Water temp., °C
					Day 1	Day 2	Day 3	Day 4	Day 5	
1. Bleeker St. Bridge, Gloversville (clean H_2O site) (14.25)	10/8		10.12						8.0	8
	10/12		9						8.5	13
	10/13		10.0						9.5	12
	10/14		10.5						10.0	10
	10/15		9						8.5	11
	10/17		9.5						9.3	12
	10/18		10.0						8.9	11
4. Main St., Johnstown (8.25)	10/8		1.0						310	12.5
	10/12	1:30	1.9						240	15
	10/13	10:15	0.5						290	13
	10/14	9:00	5.1						340	10
	10/15	4:00	2.7						280	12
	10/17	11:05	6.2						240	12
	10/18	1:30	6.0						30	12
5. Harding property below Johnstown (6.75)	10/8		2.75		140	230	260	290	340	11.5
	10/12	2:00	2.8		110	180	220	260	290	15
	10/13	10:25	5.0		100	130	130	140	150	14
	10/14	9:30	3.1		120	190	170	180	200	14
	10/15	4:30	3.0		100	140	160	210	220	13
	10/17	11:25	5.4		100	70	130	150	160	14
	10/18	2:05	5.5		70	70	100	140	140	15
6. Sammonsville Bridge (4.75)	10/8		2.2	59.4	130	200	270	300	330	13
	10/12	2:25	1.8	51	140	170	250	230	240	14
	10/13	10:45	1.1	44	40	70	60	80	140	12
	10/14	9:45	4.6	39	110	150	140	150	180	11
	10/15	4:50		39	40	110	150	160	160	12
	10/17	11:55	4.8	12.2	40	30	100	100	70	12
	10/18	2:40	4.8	12.2	40	30	50	50	40	13
	10/31			34						
	11/1			42						
7. Rt. 334, adjacent to Peresse Rd., Berryville Cross (3.0)	10/8		1.25						190	11
	10/12	2:45	0						200	14
	10/13	11:10	4.1						140	12
	10/14	10:05	5.0						170	11
	10/15	5:15	3.9						80	10
	10/17	12:15	1.8						70	11
	10/18	3:10	4.0						50	11
8. Rt. 334, 1 mile north of Fonda next to Cannarella house (1.4)	10/8		5.1						180	10.5
	10/12	3:05	3.5						90	13
	10/13	11:30	5.1						140	11
	10/14	10:20	5.1						160	11
	10/15	4:30	5.0						80	12
	10/17	12:45	3.6						100	11
	10/18	3:30	4.5						40	

Table 16.7 Time of flows from Station 5 downstream taken about one week before stream-sampling program in October 1964.

Site	Distance between points, mil	Time	Fall, feet
Start at Harding farm (Station 5)			
To old power dam	1	45 min	70
To bridge at Sammonsville (Station 6)	1	2 hr	20
To bridge at Fonda Ave.	1.75	2 hr	80
To railroad bridge	1.50	2 hr 40 min	55
Begin slack water of Mohawk River	1	2 hr 30 min	50
Total	6.25	9 hr 55 min	275

Table 16.8 Summary of 7-day sampling of Cayadutta Creek during dry period from 10/8/64 to 10/18/64. (From Nemerow, N. L. "Ind. Water Pollution" Krieger Pub. Co. pg 207, 1987.)

Station	Reading	Average of 7 samples	Range
1	DO, ppm	9.73	9–10.5
	Temp., °C	11	8–13
	5-day BOD, at 20°C, ppm	8.96	8–10
5	DO, ppm	3.93	2.75–5.5
	Temp., °C	13.8	11.5–15
	5-day BOD, at 20°C, ppm	214	140–340
	L	270	
		(projected factor 1.26)	
	1-day	106	70–140
	2-day	144	70–230
	3-day	167	100–260
	4-day	196	140–290
6	DO, ppm	3.17	1.1–4.8
	Temp., °C	12.4	11–14
	5-day BOD, at 20°C, ppm	166	40–330
	L	230	
		(projected factor 1.39)	
	1-day	77	40–140
	2-day	109	30–200
	3-day	146	50–270
	4-day	153	50–300
	*Flow, cfs†	36.7	12.2–59.4

*Time of travel between Stations 5 and 6 was 2 hr 45 min (0.115 days).
†Flow for 10/12/64 computed as arithmetic average of flow on 10/8/64. This flow represents approximate value of 7-day consecutive low flow likely to occur in Cayadutta Creek below Johnstown once in ten years: $\frac{62}{90} \times 20 = 13.8$ cfs for entire Cayadutta Creek (slightly less for below Johnstown). The 20 cgs in the equation is the minimum 7-day flow for Kayaderosseras Creek (*New York State Upper Hudson River Drainage Basin Survey Series Report no. 2*, p. 244).

different times during each day. Composite samples were analyzed for dissolved oxygen, BOD, and temperature and, in addition, the creek flow was measured on each sampling date at Station 6; these results are shown in Table 16.6. Flow times are shown in Table 16.7, a summary of the BOD and flow data in Table 16.8, the BOD curves for Stations 5 and 6 in Fig. 16.6, flow data from a similar, gauged creek in Table 16.8, the probability of the minimum flow data occurring in Table 16.10 and Fig. 16.7, and a multiple-regression technique analysis of the stream data in Tables 16.11 and 16.12.

Cayadutta Creek analysis. Using the multiple regression method described in Chapter 4, we take the following three equations. When solved simultaneously, these will yield the "best" equation, which relates the dissolved-oxygen sag to the BOD, flow, and temperature at the bottom of the sag (Station 6).

$$b_1 \Sigma X_1^2 + b_2 \Sigma X_1 X_2 + b_3 \Sigma X_1 X_3 = \Sigma X_1 Y, \qquad (1)$$

$$b_1 \Sigma X_1 X_2 + b_2 \Sigma X_2^2 + b_3 \Sigma X_2 X_3 = \Sigma X_2 Y, \qquad (2)$$

$$b_1 \Sigma X_1 X_3 + b_2 \Sigma X_2 X_3 + b_3 \Sigma X_3^2 = \Sigma X_3 Y. \qquad (3)$$

From Table 15.11 we can substitute numerical data in order to find b_1, b_2, and b_3. From Eq. (1) we obtain

$$364.78 b_1 + 4.58 b_2 + 125.62 b_3 = -43.78. \qquad (4)$$

Dividing Eq. (4) by 364.78 yields

$$b_1 + 0.01256 b_2 + 0.3435 b_3 = -0.1200. \qquad (5)$$

Multiplying Eq. (4) by 0.01255 gives

$$4.58 b_1 + 0.05748 b_2 + 1.5728 b_2 = -0.5494. \qquad (6)$$

We next apply Eq. (2) and obtain

$$4.58 b_1 + 10.68 b_2 + 0.14 b_3 = 3.02. \qquad (7)$$

Subtracting Eq. (6) from (7) yields

$$10.62252 b_2 - 1.4328 b_3 = 3.5694. \qquad (8)$$

Table 16.9 Minimum flow data of measured creek compared with that of Cayadutta Creek. (From Nemerow, N.L. "Ind. Water Pollution" Krieger Pub. Co. pg 208, 1987.)

Year	Minimum daily flow of Kayaderosseras Creek*, cfs	Equivalent minimum daily flow† of Cayadutta Creek (Station 6), cfs
1927	13	8.9
1928	18	12.4
1929	24	16.5
1930	19	13.1
1931	19	13.1
1932	20	13.8
1933	19	13.1
1934	20	13.8
1935	34	23.4
1936	20	13.8
1937	21	14.5
1938	20	13.8
1939	20	13.8
1940	23	15.9
1941	15	10.3
1942	23	15.9
1943	29	19.3
1944	21	14.5
1945	26	17.9
1946	20	13.8
1947	22	15.2
1948	18	12.4
1949	14	9.6
1950	23	15.9
1951	32	22
1952	31	21.4
1953	20	13.8
1954	23	15.9
1955	20	13.8
1956	32	22
1957	19	13.1
1958	20	13.8
1959	18	12.4
1960	25	17.2

*Hydrologically similar to Cayadutta Creek.

†Calculated by dividing the drainage area of Cayadutta Creek (62 mile²) by that of Kayaderosseras Creek (90 mile²) and multiplying by the known rate of flow for the latter, e.g.,

$$\frac{62 \times 13}{90} = 8.9.$$

Table 16.10 A normal probability distribution analysis of data (1927–1960) from Table 16.9.

Flow, cfs	Magnitude (M)	Plotting position*
8.9	1	0.0286
9.6	2	0.0572
10.3	3	0.0858
12.4	4	0.1143
12.4	5	0.1430
12.4	6	0.1715
13.1	7	0.2000
13.1	8	0.2290
13.1	9	0.2570
13.1	10	0.2860
13.8	11	0.3140
13.8	12	0.333
13.8	13	0.371
13.8	14	0.400
13.8	15	0.428
13.8	16	0.458
13.8	17	0.486
13.8	18	0.515
13.8	19	0.544
14.5	20	0.571
14.5	21	0.600
15.2	22	0.629
15.9	23	0.658
15.9	24	0.685
15.9	25	0.715
15.9	26	0.744
16.5	27	0.770
17.2	28	0.800
17.9	29	0.829
19.3	30	0.858
21.4	31	0.887
22.0	32	0.916
22.0	33	0.945
23.4	34	0.974

*Calculated from the formula $M/(N + 1)$, where M = magnitude in decreasing order of drought severity and N = number of values.

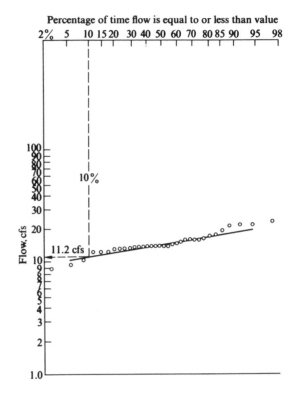

Fig. 16.7 Minimum flow of Cayadutta Creek and expected recurrence of this level.

Dividing Eq. (8) by 10.62252, we obtain

$$b_2 - 0.1349b_3 = 0.3360. \qquad (9)$$

Multiplying Eq. (4) by -0.3435, we get

$$-125.32b_1 - 1.573b_2 - 43.047b_2 = +15.038. \quad (10)$$

Next we multiply Eq. (8) by $+0.1443$:

$$1.533b_2 - 0.20675b_3 = 0.51506. \qquad (11)$$

Finally, we apply Eq. (3) and obtain

$$125.32b_1 + 0.04b_2 + 51.83b_3 = -20.55. \quad (12)$$

Adding Eqs. (10), (11), and (12), we have

$$8.5763b_3 = -4.99694, \qquad (13)$$

$$b_3 = -0.5826. \qquad (14)$$

From Eq. (9),

$$b_2 - (-0.07859) = 0.3360,$$
$$b_2 = +0.2574. \qquad (15)$$

From Eq. (5),

$$b_1 + (0.00323) + (-0.20012) = -(0.1200),$$
$$b_1 + 0.00323 - 0.20012 = -0.1200,$$
$$b_1 = 0.0769. \qquad (16)$$

Table 16.11   Summary of data required from Cayadutta Creek analyses October 1964 for Churchill method of analysis.

Date	DO, ppm		Drop in DO, ppm (Y)	BOD at sag, ppm	Temp. at sag, °C	Flow at sag, cfs
	Station 1	Station 6				
10/8	10.1	2.2	7.9	330	13	59.4
10/12	9.0	1.8	7.2	240	14	51
10/13	10.0	1.1	8.9	140	12	44
10/14	10.5	4.6	5.9	180	11	39
10/15	9.0	2.9	6.1	160	12	39
10/17	9.5	4.8	4.7	70	12	12.2
10/18	10.0	4.8	5.2	40	13	12.2

As a check, we substitute in Eq. (12):

$125.32 (0.0769) + 0.04 (0.2574) + 51.83 (-0.5826)$
$$= -20.55$$
$$9.6471 + 0.010296 - 30.1962 = -20.55$$
$$-20.5488 = -20.55$$
$$\bar{Y} = a + b_1\bar{X}_1 + b_2\bar{X}_2 + b_3\bar{X}_3$$

$a = \bar{Y} - (b_1\bar{X}_1 + b_2\bar{X}_2 + b_2\bar{X}_3)$
$a = 6.56 - [(0.0769 \times 9.35) + (0.2574 \times 12.4)$
$\qquad + (-0.5826 \times 3.92)]$
$a = 6.56 - [0.7190 + 3.1918 - 2.2838]$
$a = 6.56 - 1.6270$
$a = 4.9330.$

Table 16.12   Churchill analysis applied to Cayadutta Creek data.

Date	Dissolved oxygen		Drop in DO, ppm (Y)	BOD at sag, 1000/ppm (X_1)	Temp. at sag, °C (X_2)	Flow at sag, 100/cfs (X_3)	Y^2
	At Station 1, ppm	At Station 6, ppm					
10/8/64	10.1	2.2	7.9	3.03	13	1.68	62.41
10/12/64	9.0	1.8	7.2	4.17	14	1.96	51.84
10/13/64	10.0	1.1	8.9	7.14	12	2.27	79.21
10/14/64	10.5	4.6	5.9	5.56	11	2.56	34.81
10/15/64	9.0	2.9	6.1	6.25	12	2.56	37.21
10/17/64	9.5	4.8	4.7	14.29	12	8.20	22.09
10/18/64	10.0	4.8	5.2	25.00	13	8.20	27.04
Totals			45.9	65.44	87	27.43	314.61
Means			$\bar{Y}$	$\bar{X}_1$	$\bar{X}_2$	$\bar{X}_3$	
			6.56	9.35	12.4	3.92	$n\bar{Y}^2$
Corrected items*							301.21
Corrected totals							13.40

*n = number of samples (7).

Therefore, the Cayadutta Creek equation is

$$Y = a + b_1X_1 + b_2X_2 + b_3X_3$$

or

$$Y = 4.9330 + 0.0769X_1 + 0.2574X_2 - 0.5826X_3,$$

where X_1 = 5-day BOD (1000/ppm) 20° C, at Station 6
X_2 = temperature (°C)
X_3 = flow (100/cfs)
Y = DO drop from Station 1 to Station 6 (ppm).

To verify this stream equation, we can substitute our actual observed stream values for X_1, X_2, and X_3 and obtain a calculated Y value, which can then be compared with the observed value for accuracy:

Date (1964)	Observed Y (ppm)	Calculated Y (ppm)
10/8	7.9	7.53
10/12	7.2	7.72
10/13	8.9	7.25
10/14	5.9	6.70
10/15	6.1	7.01
10/17	4.7	3.37
10/18	5.2	5.42

During the October 1964 study the lowest flow was 12.2 cfs, the highest temperature 14°C, and the DO sag allowed was between 9.0 and 2.0 (i.e. 7.0) ppm. To find the BOD load at sag, we calculate as follows:

$$Y = 4.9330 + 0.0769X_1 + 0.2574X_2 - 0.5826X_3$$

$$7.0 = 4.9330 + 0.0769X_1 + 0.2574 (14) - 0.5836\left(\frac{100}{12.2}\right)$$

$$7.0 = 4.9330 + 0.0769X_1 + 3.6036 - 4.7773$$

$$3.2407 = 0.0769X_1$$

$$X_1 = 42.1417 = \frac{1000}{ppm}.$$

$$\text{ppm allowed} = \frac{1000}{42.1417} = 23.73 \text{ ppm}$$

The average BOD at sag was 166 ppm during the entire 7-day survey. Therefore, the BOD reduction required in the stream at Station 6 was

$$\frac{166 - 23.73}{166} \times 100 = 85.70\%.$$

Although the Streeter-Phelps method yielded

Table 16.12 *(continued)*

YX_1	YX_2	YX_3	X_1^2	X_1X_2	X_1X_3	X_2^2	X_2X_3	X_3^2
23.94	102.7	13.27	9.18	39.39	5.09	169	21.84	2.82
30.02	100.8	14.12	17.39	58.38	8.17	196	27.44	3.84
63.55	106.8	20.20	50.98	85.68	16.21	144	27.24	5.15
32.80	64.9	15.10	30.91	61.16	14.23	121	28.16	6.55
38.13	73.2	15.62	39.06	75.00	16.00	144	30.72	6.55
67.16	56.4	38.54	204.20	171.48	117.48	144	98.40	67.24
130.00	67.6	42.64	625.00	325.00	205.00	169	106.60	67.24
385.60	572.4	159.49	976.72	816.00	381.18	1087	340.40	159.39

$n\overline{YX}_1$	$n\overline{YX}_2$	$n\overline{YX}_3$	$n\overline{X}_1^2$	$n\overline{X}_1\overline{X}_2$	$n\overline{X}_1\overline{X}_3$	$n\overline{X}_2^2$	$n\overline{X}_2\overline{X}_3$	$n\overline{X}_3^2$
429.38	569.38	180.04	611.94	811.51	256.56	1076.32	340.26	107.56
−43.78	3.02	−20.55	364.78	4.58	125.62	10.68	0.14	51.83

values of k_1 and k_2 which gave a Fair's f of about 35,* the results are not dependable because of the variability of wastes from one moment to the next as well as the multiple entrances of wastes into the stream. The only reliable procedure for evaluating the oxygen-sag characteristics is to collect many stream samples under these critical conditions and statistically correlate the data in order to obtain a stream equation. This method is commonly referred to as the Churchill multiple-regression technique. The stream equation represents the line which best fits the data for the conditions under which the samples were collected. Projection of the line beyond this range of conditions is not recommended, but extrapolations to different conditions within the range of existing data can be made with a reasonable degree of certainty. The stream equation for Cayadutta Creek, developed by extensive analysis, can be used to compute the BOD reductions necessary to maintain a certain minimum dissolved-oxygen level at a given temperature. These calculations are given below and in Figs. 16.8 and 16.9.

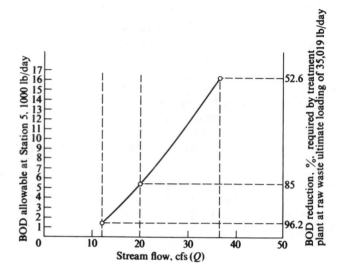

Fig. 16.9 Special design curve for computing treatment plant requirements at 12.4°C and 2 ppm DO at Station 5.

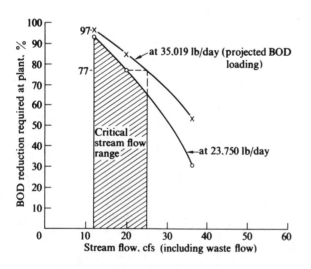

Fig. 16.8 BOD reduction required at Station 5, at 12.4°C and 2 ppm DO remaining and based on a waste discharge of 23,750 lb/day.

---

*$f = k_2/k_1$ = reaeration rate/deoxygenation rate.

To find the waste-treatment requirements, we use the stream equation developed during the low-flow* critical period for Cayadutta Creek during the October 1964 study. We obtain

$$Y = 4.9330 + 0.0769\ X_1 + 0.2574\ X_2 - 0.5826\ X_3,$$

when $X_1 = 1000/$ppm BOD

$\qquad X_2 = $ temperature $= 12.4$ (°C)

$\qquad X_3 = 100/$cfs $= \dfrac{100}{12}$ (see Table 15.10) $= 8.33$

$\qquad Y = $ DO sag from Station 1 to Station 6 $=$

$\qquad\quad = 9.73 - 2.00 = 7.73$

$7.73 = 4.9330 + 0.0769\ X_1 + 0.2574\ (12.4)$
$\qquad\quad - 0.5826\ (8.33)$

$7.73 + 4.85 - 4.9330 - 3.200 = \quad 0.0769\ X_1$

$12.58 - 8.133 = \dfrac{4.447}{0.0769} = X_1 = 57.75$

$$\text{ppm} = \frac{1000}{57.75} = 17.3 \text{ ppm at sag}$$

The BOD values for the seven days of the survey, at Station 5, were 340, 290, 150, 200, 220, 160, and 140 ppm, giving an average of 214 ppm. At Station 6, the values were 330, 240, 140, 180, 160, 70 and 40, with an average of 166 ppm. The percentage decrease in BOD between Stations 5 and 6 was

$$\frac{214 - 116}{214} = \frac{48}{214} = 22.5\%.$$

The BOD (in pounds) being discharged at Station 5 on the seven days of the survey was a total of 142,500 lb (see Table 16.14) and an average of 23, 750 lb/day.

Taking 23,750 × 77.5% = 18,400 lb/day as the BOD left at Station 6, with no treatment at 12 cfs, 12.4°C, and an allowable DO deficit at the sag point of 8.65 ppm (10.65–2.0), we obtain an allowable 17.3 ppm BOD at the sag or

$$\frac{17.3 \times 12}{1.54} \times 8.34 = 1122 \text{ lb/day.}$$

Since 23,750 − 0.225 = 18,400 lb/day, the BOD reduction required is

$$\frac{18,400 - 1122}{18,400} \times 100 = \frac{17,278}{18,400} = 93.8\%.$$

At 12.4°C and the same DO sag (7.73 ppm), but at the increased stream flow of 20 cfs, we obtain an allowable BOD at the sag of

$$Y = 4.9330 + 0.0769\, X_1 + 0.2574\, X_2 - 0.5826\, X_3$$

$$7.73 - 4.9330 - 3.2000 = 0.0769 X_1 - 0.5826 \left(\frac{100}{20}\right)$$

$$\frac{7.73 + 2.4130 - 8.1330}{0.0769} = X_1 = \frac{2.0100}{0.0769} = 26.05$$

$$\text{ppm} = \frac{1000}{26.05} = 38.4 \text{ ppm}$$

at 38.4 ppm $\times\ 8.34 \times \dfrac{20}{1.54} = 4155$ lb BOD at Station 6

$$\text{BOD reduction required} = \frac{18,400 - 4155}{18,400} \times 100$$
$$= \frac{14,245}{18,400} \times 100 = 77.5\%$$

At the same temperature (12.4°C) and the same DO sag (7.73 ppm), but at the average stream flow at the sag of 36.7 cfs (October 1964 survey), we obtain an allowable BOD at the sag of

$$Y = 4.9330 + 0.0769\, X_1 + 0.2574\, X_2 - 0.5826\, X_3$$
$$7.73 = 4.9330 \times 0.0769\, (X_1) + 0.2574\, (12.4°C)$$
$$- 0.5826 \left(\frac{100}{36.7}\right)$$
$$7.730 + 1.590 - 4.9330 - 3.2000 = 0.0769\, X_1$$
$$\frac{9.3200 - 8.1330}{0.0769} = \frac{1.1890}{0.0769} = 15.45 = X_1$$
$$\text{ppm} = \frac{1000}{15.45} = 64.8$$

at 64.8 ppm $\times\ 8.34 \times \dfrac{36.7}{1.54} = 12,850$ lb BOD at Station 6

$$\text{BOD reduction required} = \frac{18,400 - 12,850}{18,400} \times 100$$
$$= \frac{5,550}{18,400} = 30.2\%$$

Figures 16.8 and 16.9 indicate that to maintain 2 ppm of dissolved oxygen (a preselected safe value for this class of stream) at loadings of 23,750 pounds of BOD per day at the bottom of the sag at a temperature of 12.4°C, BOD reductions of 65 to 94 per cent at 23,750 pounds per day loading and 77 to 97 per cent at 35,019 pounds per day would be required for critical stream flows of 12 to 25 cfs.

16.6 Composite Waste Sampling

Waste samples were collected hourly for a 24-hour period from three sources, the Johnstown 30-inch sewer (main), the Johnstown 8-inch sewer (Tynville), and the Gloversville sewage treatment plant, on November 17 and December 3, 1964. Similar samples were collected for a 24-hour period on January 21, 1965, except that the 8-inch Johnstown sewer was eliminated as being relatively insignificant. Weirs were installed in the Johnstown lines to record the total flows from Johnstown as well as from Gloversville. Samples were collected and composited according to the rate of flow at the hour of sampling. A summary of the proportionate pollutional loads and volumes for these three days is shown in Table 16.13. Additional 24-hour composites from each line were collected according to flow and analyzed on February 18, March 30, April 21, and May 6, 1965 (Table 16.14).

Table 16.13   Summary of 24-hours sampling results.

Source	Date	BOD load				Flow		Suspended solids		
		lb/day	% in peak period*	% of daily total	mgd	Ratio of peak period to daily avg.	% of daily total	lb/day	% in peak period	% of daily total
Gloversville	11/17/64	13,350	45.2	67.5	3.80	1.38	66	27,550	80.8	87.5
	12/3/64	13,350	50.2	56.3	3.67	1.34	50	11,100	61.6	53
	1/21/65	23,400	41.1	63.1	3.96	1.35	60.5	21,600	25.8	63
Johnstown Main	11/17/64	6,400	44.8	32.2	1.92	1.55	33.4	4,100	57.7	12.5
	12/3/64	15,000	27.5	42	3.60	1.46	49	18,400	42.0	45
	12/1/65	13,600	43.2	36.9	2.58	1.33	39.5	12,800	37.0	37
Tynville	11/17/64	102.4	25.4	0.53	0.0449	1.24	0.6	32.9	56	
	12/3/64	223	25.2	1.4	0.0809	0.925	1	213	17.9	2.0
Totals	11/17/64	19,852			5.7649			31,683		
	12/3/64	28,573			7.3509			29,713		
	1/21/65	37,000			6.54			34,400		

*The peak period was taken to be 6.00 a.m. to 12 noon.

16.7 Composite Waste Analyses

The hourly data reveal several significant findings.

a) Slugs, which can be defined for this situation as instantaneous discharge of high volumes of waste, concentrated acid or alkali, or BOD, are apparently not a major problem. The flow increases by about 100 per cent of the daily average, for a period of about 12 hours during the daytime. The pH becomes alkaline (8 to 10) during the same period but returns to normal (7 to 8) during the twelve night hours. The BOD varies considerably during both day and night but is generally high and is confined to a range of 300 to 700 ppm from 6 a.m. until midnight. There is little pattern of discharge of BOD and no apparent practical gain as far as BOD is concerned from separate equalization basins. Since the flow and pH are largely higher during the entire daytime period, equalization to level out these factors would require very large basins. The cost of such units and the potential danger of septicity seem to the author to far outweigh the benefits derived from levelling the flow and pH. In this instance, it seems that the great numbers of varied tanneries themselves contribute to equalization of waste simply by their diversity.

b) The total BOD loads and flows given in Tables 16.13 and 16.14 can be examined more easily by referring to Table 16.15. The total flow measured averages 6.724 million gallons per day and contains an average of 23,442 pounds of 5-day, 20°C BOD and approximately 20,650 pounds of suspended solids. These values do not include any flows or loads not connected to the Gloversville sewage treatment plant or the 30-inch sewer outfall in Johnstown. The 8-inch sewer outfall in Johnstown, although measured and sampled at the beginning, contains less than 1 per cent of the total volume or BOD load and can therefore be considered insignificant in these surveys.

c) The maximum variations in flow and load from day to day were found to be 18 per cent from the average flow and 22 per cent from the average BOD. These variations are considered well within normal values and tend to substantiate the use of the average daily values given under the previous section in designing waste-treatment facilities.

Industrial production records during the sampling days (Tables 16.16 and 16.17) demonstrated that all major industries connected to the two major sewer systems were in operation at almost full capacity during these days. This provides some measure of

Table 16.14 Composite analyses (24-hr) of Gloversville and Johnstown waste water.

Characteristic	Feb. 18–19, 1965		March 30, 1965 32		April 21, 1965		May 6, 1965	
	Gloversville	Johnstown*	Gloversville	Johnstown	Gloversville	Johnstown	Gloversville	Johnstown
Flow, mgd								
24-hr average	3.41	3.05	3.05	4.88	3.63	2.84	4.3	2.07
6 a.m. to 12 noon average			4.73	6.33	4.29	3.00	5.9	2.92
6 a.m. to 2 p.m. average								
pH	9.4	8.3						
Total solids, ppm	3130	2450	2430	1970	2542	1840	3120	2962
Suspended solids, ppm	258	81	265	145	418	213	475	322
Volatile suspended solids, ppm			196	96	265	135	305	250
BOD (5-day, 20°C), ppm	405	385	300–435	285–330	371–435	371–386	485–520	540–585
BOD, lb/day	11,500	9800	(367)†	(307)	(403)	(378)	(502)	(562)
Settleable solids, ml/liter			20.0	5.5	13.0	4.5	14.0	16.0
On supernatant								
Suspended solids, ppm			97	70	128	86	172	135
Volatile solids, ppm			78	52	92	60	120	110
BOD, ppm			95–180 (285–330)	225–355	266–281	326–386	210–300	375–405
8-hr readings								
Suspended solids, ppm							690	423
Volatile solids, ppm							420	310
BOD, ppm							405–495 (450)	405–435 (420)
Total ash, ppm	2250	1570						
Total volatile, ppm	880	880						
Suspended ash, ppm	140	48						
Suspended volatile, ppm	118	33						
Suspended volatile, %	46	41						
Analysis of settled 2-hr sludge								
Total solids, %	1.54	0.88						
Total ash, %	34.4	37.5						
Total organics, %	63.6	62.5						

*30-inch sewer.
†Average values are given in parentheses.

Table 16.15   Summary of total loads for treatment.

Date	Total flow, mgd	Total BOD, lb/day	Total suspended solids, lb/day
11/17/64	5.765	19,852	31,683
12/3/64	7.351	28,573	29,713
1/21/65*	6.540	37,000	34,400
2/18/65	6.460	21,300	9,405
3/30/65	7.930	21,925	12,700
4/21/65	6.470	21,150	17,700
5/6/65	6.370	27,850	22,700
Average	6.724	23,442	20,650

*Since an unusually large percentage of deerskin was tanned, this day was not considered typical of even maximum normal operation and therefore it was excluded from the average.

assurance in using the average flow and BOD values obtained during this period. Table 16.18 shows industry's percentage of the total measured flow on these days. This reveals an industrial waste problem of about 50 per cent by volume when industry is operating near its rated capacity.

16.8 Laboratory Pilot-Plant Studies

To form more definite conclusions on the proper units to be included in the waste-treatment plant, certain small-scale laboratory studies were necessary, including sludge digestion and activated-sludge treatment.

Sludge digestion. A mixture of primary and secondary settled sludge was collected from the settling basins at the Gloversville treatment plant. A pilot digester, consisting of a glass container and a gas-collecting system maintained at 37°C, was set up in a private laboratory in Johnstown. The raw sludge sample selected was analyzed for organic matter at the start of the "batch" digestion period and again after 50 days of digestion, and gas volume measured almost daily (see Table 16.19).

Although this was a batch-type experiment, over the 50-day period 9.07 cubic feet of gas were produced per pound of volatile matter destroyed. Greater amounts of gas may be expected from a continuous digestion operation maintained at optimum environmental conditions. In this experiment, more gas would have evolved after an increased digestion period, but the rate of gas production did slow down

considerably after 50 days. Normal gas production for sewage sludge is about 15 cubic feet per pound of organic matter destroyed. Digestion experiments on a continuous basis and over a longer period would be needed to find whether an accumulated toxic effect exists. However, Vrooman and Ehle [16] reported successful digestion of this waste sludge.

Activated-sludge treatment. The apparatus used in the study consisted of an aeration tank with two mixers and three separate air-diffuser tubes fitted with porous stones. Air flow (cubic feet of air/hour) was measured by a previously calibrated rotameter. The tank was 23.5 in. long, 8.5 in. wide, and filled to a depth that gave an aeration volume of 6 gal.

Since settling was expected to be an integral part of a biological treatment plant of this type, various mixtures of settled tannery waste (1:1 mixture of beamhouse and tanyard wastes) and settled domestic sewage were added to the aeration tank in a semi-batch procedure to simulate continuous operation as closely as possible. The standard average aeration period of 6 hours was used and the waste mixture was added in three increments of 2 gallons each at 2-hour intervals. The tank contents (6 gallons of a mixture of tannery wastes and domestic sewage) were first settled for a 1-hour period; then 2 gallons of the supernatant were siphoned off and 2 gallons of the waste mixture added; the tank contents were aerated for 2 hours and settled for 15 minutes, 2 gallons of supernatant were withdrawn, and 2 more gallons of waste mixture added. This procedure resulted in the addition of 6 gallons of waste mixture in a period of 4 hours for a total aeration time of 8 hours and an average aeration period of 6 hours as shown below:

Time (hr)	Volume added (gal)	Aeration time (hr)
0	2	8
2	2	6
4	2	4
		Average: 6

The results obtained are summarized in Table 16.20 and shown graphically in Fig. 16.9. Each loading represents about one week's aeration data with samples being taken for analysis several times during this week of adaptation and acclimation.

Table 16.16 Industrial production during sampling days.

Company	November 17, 1964		December 3, 1964		January 21, 1965	
	Water used, gpd	% production*	Water used, gpd	% production	Water used, gpd	% production
Wood and Hyde Leather	200,000	100	225,000	100	225,000	100
Filmer Leather	41,310	50	41,310	50	41,300	50
Twin City Leather	120,000	50	120,000	100	120,000	100
Wilson Tanning	49,920	80	54,337	80	38,000	50
Leavitt-Berner Tanning	151,700	100	151,700	100	151,700	100
F. Rulison & Sons	76,300	100	76,300	100	76,300	100
Peerless Tanning	24,000	$33\frac{1}{3}$	48,000	$66\frac{2}{3}$	48,000	$66\frac{2}{3}$
Karg Bros.	266,000	68	254,000	65	386,000	90–95
Decca Records†	10,807	100	10,807	100	10,807	100
U.S. Rabbitt Tanning Co.†	10,000	100	7,000	75	7,000	75
Gloversville Continental Mill†	200,000	66	190,000	100	190,000	100
Independent Leather	134,000	60	107,000	40	161,000	60
Liberty Dressing	78,950	70–75			120,803	60
G. Levor		85		80		
Framglo Tanners (1)	500,000	100	500,000	100	500,000	100
Framglo Tanners (2)		80		80		80
Rebel Dye†	26,250	10	30,500	10	24,500	9
Lee Dyeing† (Johnstown)	175,000	40	1,000	0	1,000	0
Adirondack Finishing†	450,000	80	450,000	80	500,000	90
Crown Finishing (Maranco Leather)	42,352	100	42,000	100	42,352	100
Simco Leather	61,300	100	61,000	100	61,000	100
Johnstown Tanning	60,000	70	35,000	60	10,000	40
Napatan	21,072	100	21,000	80	21,000	100
Ellithorp Tanning	105,000	100	110,000	100	110,000	100
Johnstown Knitting†			100,000	45	100,000	75
Gloversville Leather						
Riss Tanning						
Industrial total, gpd	2,803,961		2,735,954		2,845,662	
Total flow, mgd	5.7649		7.3509		6.54	
Total BOD, lb/day	19,852		28,573		37,000	
Industrial portion of total water flow, %	48.7		37.2		43.4	

*Percentage of plant's total productive capacity.
†Figures are based upon yearly consumption (average figure).

Laboratory experiments verified that 65 to 75 per cent of this waste would degrade biologically even when loaded at the high rate of 95 to 115 pounds of BOD per 1000 cubic feet of aerator (see Fig. 15.9). Higher BOD reductions (75 to 85 per cent) were obtained with lower BOD loadings (60 to 82 pounds per 1000 ft³) and increased dilution of the tannery waste with domestic sewage.

These studies showed that the activated-sludge process or a modification of it could be utilized successfully in the overall treatment of the combined tannery and sewage wastes. Larger prototype field

Table 16.17   Water consumption related to production percentage.*

Industry	Water consumption, gal		Production percentage				
			Beamhouse		Tanning		
	From meter reading	From other sources	Type of skin	Rated potential	Compared 1/21/65	Operation potential	Type
Leather tanneries							
Wood and Hyde Leather	150,000	75,000	Burn – sheep	100	100	100	Combination
Filmer Leather	90,000	Pond (in future)	Horse, cow, jacks, deer	100	75	80	Combination
Twin City Leather	10,028		Sheep	100	100	100	Combination
Wilson Tanning	25,215		Sheep, goat, deer	0	0	33⅓	Combination
Leavitt-Berner Tanning (not contributing to sewage treatment plant)	140,000		Sheep, goat, deer	100	100	100	Combination
F. Rulison & Sons	60,150		Horse, cow	80	80	80	Combination
Peerless Tanning	21,766					75	Chrome
Karg Bros.	400,000		Pig, deer	100	100	100	Chrome
Independent Leather	230,000					100	Combination
Liberty Dressing	85,582		Goat, calf			45	Combination
G. Levor	Drinking water only	600,000	Pig, calf, goat	80		80	Combination
Framglo Tanners (1)	Drinking water only	400,000	Sheep, deer			100	
Framglo Tanners (2)	Drinking water only	300,000		80		None	
Crown Finishing (Maranco Leather)	26,465						
Simco Leather	45,030					100	Chrome, some comb. & veg.
Johnstown Tanning (1)	11,000					10	
(2)	0						
Ellithorp Tanning	248,000	From Levor				90	Chr. & comb.
Gloversville Leather	0	152,000				25	Chr. & comb.
Riss Tanning	27,970					33⅓	Chr. & comb.
Nonleather industries							
Rebel Dye	61,000 (8,145 ft³)		20				
Adirondack Finishing	474,310		80				
Lee Dyeing (Johnstown)	8,800 (1,175 ft³)		0				
Gloversville Continental Mill	208,000 gal		66⅔		Anticipate 330,000 gal due to dyeing technique but not necessarily increased production		
Johnstown Knitting	150,200 (20,178 ft³)		75				
Diane Knitting	10,590 ft³		75				
Decca Records	17,745 (2,366 ft³)		75				
Mohawk Cabinet			80				
U.S. Rabbitt Tanning Co.	1,000		0		Water running in tubs to keep them soaked. Should be disregarded		
Total industrial flow, gal	4,019, 262						
Total flow, gal	6,460,000						
Total BOD	21,300 lb/day						
Industrial flow (% of total)	62.5%						

*Data are for the 24-hr period from 6 a.m. on February 18 to 6 a.m. on February 19, 1965.

Table 16.18 Industrial waste flow.

Date	Total flow, mgd	Industrial flow (estimated by survey), mgd	Industrial portion of total flow, %
11/17/64	5.765	2.804	48.7
12/3/64	7.351	2.736	37.2
1/21/65	6.540	2.846	43.4
2/18/65	6.460	4.019	62.5
Average			ca.48

experiments would disclose whether these results can be projected directly to full-scale operation.

The present Gloversville treatment plant has experienced much difficulty due to its overloaded condition. However, from the best records available, when all the flow units of the plant were operating, about 58 to 60 per cent of the BOD was removed. This reduction in BOD also shows that biological degradation by trickling filtration is possible with this waste under full-scale field conditions. The exact degree of this oxidation could be determined more easily in a properly designed and operated field pilot plant.

16.9 Literature Survey

A study was also made of previous research work or reported practice dealing with combined treatment of domestic sewage and tannery waste by the activated-sludge process. The reports of Chase and Kahn [2], Braunschweig [1], Jansky [6], Thebaraj et al. [15], Snook [14], Hubbel [5], Pauschardt and Furkert [12], Kubelka [8, 9], Fales [3], Kalibina [7], Furkert [4], and Mausner [10] provide some evidence that the activated-sludge treatment process is feasible for tannery–sewage waste mixtures. This review of previous work tends to substantiate the biological pilot studies described above. Most of this reported work, however, has been on a research or pilot-plant basis or of a more sewage-diluted waste. There is a definite lack of reviews of full-scale biological treatment used on tannery and sewage waste mixed 50:50 (by volume).

Braunschweig [1] disproves the notion that chromium in the newer tannery processes interferes with

Table 16.19 Sludge digestion (laboratory study).*

Accumulated gas produced, cc	Days of digestion at 37° C
250	1
290	2
460	3
640	4
730	5
850	6
940	7
990	8
1040	9
1140	10
1140	11
1150	12
1190	13
1240	14
1290	15
1340	16
1380	17
1420	18
1480	19
1540	20
1600	21
1650	22
1700	23
1820	24
1900	25
1940	26
1980	27
1980	28
2010	29
2020	30
2030	33
2040	36
2050	40
2170	49
2200	50

*The results were analyzed as follows:

organic matter (raw sludge)	= 5.4533 gm
organic matter (after 50 days)	= 1.5542 gm
loss of organic matter	= 3.8991 gm (71.2%)
gas produced: total	= 2200 cc
per gram of organic matter destroyed	= 567 cc
per gram of volatile matter added	= 403 cc

To convert them from the metric system:

$2200 \times 0.000353 = 0.0777$ ft^3 of gas produced (total)

$\dfrac{3.8991}{454} = 0.00858$ lb of volatile (organic) matter destroyed

$\dfrac{0.00777}{0.00858} = 9.07$ ft^3 of gas per pound of organic matter destroyed

Table 16.20 Activated-sludge pilot laboratory studies.

Waste treated		BOD loading, lb/1000 ft³	Suspended solids under aeration, ppm	Air rate, ft³/lb BOD removed	BOD		
Origin	Quantity, gal				Influent, ppm	Effluent, ppm	Reduction, %
Sewage + Tannery waste mixture	5 1	60	2330	2450	239	44	81.6
Sewage + Tannery waste mixture	4 2	82.8	2221	1900	331	78	76.5
Sewage + Tannery waste mixture	3 3	114.7	2768	1735	459	165	64.1
Sewage wastes Gloversville Johnstown	5 1	70	3386	2000	280	39	86
Sewage wastes Gloversville Johnstown no. 1 Johnstown no. 2	4.015 2.112 0.044	73.8	2508	2116	295	68	76.9
Tannery waste mixture + Tap water	3 3	93.0	2646	1070	374	91	75.6

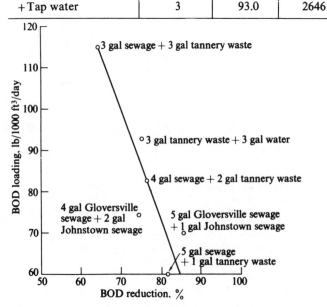

Fig. 16.10 Activated-sludge treatment: BOD reduction related to BOD loading. (From Nemerow, N.L. "Ind. Water Pollution" Krieger Pub. Co. pg 220, 1987.)

aerobic biological treatment. Jansky's work [6] is typical of that of the more recent supporters of activated sludge as a treatment method. Thebaraj *et al.* [15] found that all aerobic biological systems were effective and that the choice depended upon several economic and practical considerations. Fales [3] much earlier was of the same opinion as Thebaraj but pointed out the higher operating costs of activated sludge compared to trickling filtration. Furkert [4] also verifies our digestion studies and noted that except for the high H_2S content of the digester gas the composition of the gas is normal.

16.10 Conclusions from Study

The following specific conclusions and recommendations were made as a result of this study:

1. The stream survey was instrumental in providing evidence that secondary treatment of the combined industrial and sanitary wastes of the area is required and that 65 to 94 per cent BOD reduction will be need-

ed depending upon the dilution available in the stream. Use of the curves plotted in Figs. 16.6 and 16.7 would allow for a more precise selection of BOD reduction required for specific critical stream flows.

2. The existing dry-weather flow to be treated averages 6.724 million gallons per day with peaks of two to three times this rate; about 50 per cent of this flow originates from the industries of the area.

3. The combined area waste contains a daily average of 23,442 pounds of 5-day, 20°C BOD and 20,650 pounds of suspended solids. These loadings are affected considerably by the type of skin tanned, deerskin being an especially significant contributor of high BOD and solids loads.

4. Laboratory pilot studies demonstrated that the conventional activated-sludge treatment process is capable of reducing the BOD of the combined waste from 65 to 85 per cent (depending primarily upon the organic loading) at loadings ranging from 60 to 115 pounds of BOD per 1000 cubic feet of aerator capacity.

5. A digestion batch experiment yielded about 9 cubic feet of gas per pound of volatile matter destroyed and effected a 71 per cent reduction in organic matter.

6. A literature study confirmed the findings of the laboratory results—that the combined wastes of this type are amenable to biological oxidation.

7. Because of the unique nature of the volume and characteristics of the tannery–sewage waste mixture, as well as the size and cost of the project, field prototype studies should precede full-scale plant construction.

8. There should be additional laboratory research on development of improved methods of aerobic biological treatment of tannery wastes to allow for greater BOD reduction at higher BOD loadings.

The decision reached as a result of this study was that the cost was too high, the risk too great, and previous reported experience too slight for full-scale biological treatment to be recommended at the time. A prototype in the field—preferably at the site of the Gloversville treatment plant—was to be built and operated for about 6 months to obtain detailed data for the final design and to obtain greater certainty that the earlier findings were valid. This prototype should contain both trickling-filtration and activated-

sludge units (as well as provision for its modification). It should also allow for experimentation with series and parallel operation of the units and both diffused and mechanical aeration. Some sludge-digestion studies should be carried out over the entire period.

A schematic drawing of this field prototype is shown in Fig. 16.10. It consists of two sets of screens in series ($\frac{1}{2}$-inch openings followed by $\frac{1}{4}$-inch openings), pump, primary settling, trickling filter, aeration, and final settling. The plant began operation in early August 1965 and sampling was begun on August 16. Table 16.21 gives data for the first 7 weeks of operation. Table 16.22 shows how the prototype operating results influenced the final design parameters.

16.11 Overall Planning Study Conclusions

The conclusions reached through data collection, pilot and prototype plant studies, engineering evaluations, and reviews of design and operational experiences in major municipal sewage-treatment plants treating large amounts of tannery waste may be summarized as follows.

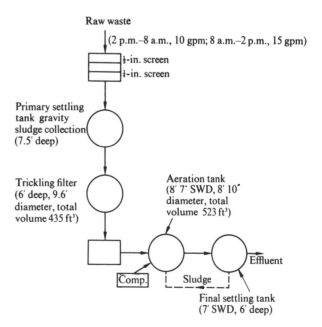

Fig. 16.11 Field prototype of the Gloversville-Johnstown joint treatment plant. (From Nemerow, N.L. "Ind. Water Pollution" Krieger Pub. Co. pg 221, 1987.)

Table 16.21 Prototype operating data.

Date	Raw waste			Primary effluent			Trickling-filter effluent			Final effluent			Under aeration	
	BOD, ppm	pH	SS,* ppm	BOD, ppm	pH	SS, ppm	BOD, ppm	pH	SS, ppm	BOD, ppm	pH	SS, ppm	SS, ppm	pH
8/16/65	655	7.6	603	448	7.3	248	353	7.3	251	93	7.2	139	1430	7.2
8/17/65	468	8.4	528	373	8.3	256	241	7.6	202	81	7.2	44	1603	7.3
8/18/65	468	9.1	578	380	9.0	170	230	8.5	262	73	7.4	75		
8/19/65	563	8.0	369	390	8.0	211	268	7.6	187	85	7.2	78	1850	7.3
8/20/65	493	8.5		408	8.7	281	256	8.3	166	73	7.4	84	2631	7.6
8/23/65	443	7.0	382	433	7.4	145	316	7.5	243	72	7.5	43	2313	7.4
8/24/65	555	8.1	393	425	8.1	184	330	7.8	250	90	7.4	89	2066	7.5
8/25/65	370	7.9	392	313	7.8	169	398	7.8	394	79	7.4	197	2487	7.5
8/26/65	408	8.3	454	360	7.8	243	231	7.5	168	102	7.3	127	2788	7.2
8/27/65	273	8.8	337	308	8.5	135	215	8.1	155	19	7.6	53		
8/30/65	298	7.9	347	210	8.1	111	124	7.5	116	25	7.4	47	1848	7.6
8/31/65	250	7.9	379	230	7.9	230	163	7.8	214	24	7.5	67	1934	7.4
9/1/65	423	7.8	460	260	7.9	292	256	7.8	260	54	7.5	160	854	7.6
9/2/65	483	8.6	550	408	8.4	350	435	8.4	360	96	7.5	166	2243	7.8
9/3/65	455	7.9	431	315	7.9	311	235	8.0	205	141	7.5	216	2775	7.6
9/9/65	563	8.1	486	418	8.2	82	275	7.9	87	76	7.4	264	2430	7.9
9/10/65	563	7.9	262	413	7.4	110	290	7.4	137	51	7.1	218	2500	7.4
9/13/65	563	8.4	329	453	8.1	154	290	8.2	142	112	7.4	180	2219	7.6
9/14/65	628	8.7	392	538	8.6	240	351	7.6	210	162	7.4	366	1899	7.4
9/15/65	658	7.6	356	568	7.6	257	349	7.5	178	69	7.0	22	2653	7.2
9/16/65	623	9.0	235	593	8.5	108	368	8.5	105	62	7.5	19	3025	7.6
9/17/65	635	7.6	284	460	7.6	60	348	7.5	96	75	7.1	44	3226	7.5
9/21/65	720	7.9	425	425	7.8	199	215	7.6	178	135	7.4	130	2300	7.5
9/22/65	530	7.2	435	450	7.0	167	204	7.0	166	55	7.0	338	2892	6.9
9/23/65	490	8.0	349	393	8.0	173	183	7.6	224	33	7.3	51	2790	7.5
9/24/65	730	8.7	288	523	8.5	192	170	8.3	115	111	7.5	140	2490	7.8
9/27/65	543	8.5	306	383	8.4	247	233	7.8	240	117	7.4	93	2326	7.6
9/28/65	480	8.1	403	420	8.0	197	289	7.8	193	127	7.6	124	2361	7.8
9/29/65	516		315	463		177	258		180	126		141	2664	
9/30/65	650	8.1	307	490	7.6	145	344	7.5	195	192	7.4	157	2096	7.4
10/1/65	435	7.8	312	363	7.6	162	280	7.6	229	158	7.2	241	1495	7.4

*SS = suspended solids.

1. *Degree of treatment*. Primary treatment by settling, followed by secondary treatment through biological processes, is required to meet New York State standards for plant effluent quality that may be accepted by the Cayadutta Creek under conditions of minimum dissolved oxygen content (at times of low flow and high temperature). The efficiency of treatment units and processes must be high, with an overall plant re-moval of approximately 85 per cent of the incoming BOD.

2. *Pretreatment at mills.*

a) Tanneries should remove fleshings, hair, hide pieces, and trimmings to make discharges transportable in gravity sewers. This can be accomplished by means found most efficient and economical,

Table 16.22   Prototype operating results and design parameters.

Unit and effect	Prototype results	Design parameters	Comment

Primary settling tank
  Suspended solids
    removal, %
  Surface settling
    rate
  5-day BOD removal
	49(75)*
	
	390(330)
	24(39)
		60
		
		800
		30
			Prototype tank construction and inherent limitations in small tanks resulted in lower settling efficiencies. Better results are expected in full scale tanks with scum- and sludge-removal facilities and improved hydraulic characteristics. Additional settling-tank efficiency could be obtained by using flocculating agents if needed.

Roughing filter, with loading of 150 lb of BOD$_5$/1000 ft^3	33% BOD$_5$ removal	30% BOD$_5$ removal	Performance of roughing filter established by test.

Aeration tank
  Process load-
  ing of 0.26 lb
  BOD/lb MLSS†
  (8-hr peak)
  Process loading
  of 0.4 lb BOD/lb
  MLSS (24-hr
  average)
	81% BOD$_5$ removal
	
	77% BOD$_5$ removal
		81% BOD$_5$ removal
		
		77% BOD$_5$ removal
			Performance of activated sludge system established by test.

Aeration tank (oxygen requirements, lb/day)	0.7 lb BOD$_{SR}$ + 0.02 lb MLSS	1 lb O$_2$/lb BOD$_5$ removed/day	Aerator capacity designed for mean peak 8-hr BOD loading with 25% present safety factor. Oxygen transfer ratio, $\propto$, to be determined in laboratory prior to final specifications on aerators.
Secondary settling tanks, surface settling rate (gal/ft²/day)	760 gal/ft²/day	600 gal/ft²/day (at peak 8-hr rate of 13.1 mgd)	Selection of design overflow rates not on basis of pilot plant results. Inclusion of skimming devices on secondary settling tanks due to experience with pilot plant.
Flotation (thickening of waste-activated sludge)	Waste sludge of less than 0.5%. Solids thickened to 5% or greater at loadings greater than 2 lb/ft²/hr	Design loading: 2 lb/ft²/hr	Review of prototype data indicates that the design loading is suitable and that this loading should be achieved without the use of chemical conditioning.

Sludge digestion tank	Digestion studies did not develop digestion rate curves.	Displacement time
		  Primary, 25 days
		  Secondary, 25 days
			Digestion studies were not conclusive.

*The data in parentheses give the results achieved by using polymer.   †MLSS = mixed liquor suspended solids.

including primary settling tanks and/or mechanical screening. Animal greases plus petroleum solvents should be removed at the tanneries.

  b) The glue factory should remove settleable solids to make discharges transportable in gravity sewers. This can be accomplished by means found most efficient and economical, including primary settling tanks and/or mechanical screening.

3. *Sewage-treatment plant: processes (general)*. The liquid wastes treatment will include (a) pretreatment by mechanical screening, grit removal, preaeration of grease for removal in primary treatment, and pretreatment for pH control and chemical coagulation (in future); (b) primary treatment by settling; and (c) secondary treatment by trickling filters (high-rate roughing filters), then through activated sludge, followed by settling, with provisions for chemical precipitation for more complete solids removal in the future, with discharge to the Cayadutta Creek where further treatment is by dilution and the oxygenation capacity of the stream.

Sludge treatment and disposal will include high-rate digestion with sludge-gas utilization followed by (a) dewatering by lagoons and disposal by approved landfill methods; (b) dewatering by vacuum filters and disposal by approved landfill methods; or (c) dewatering by vacuum filters, incineration (multiple hearth) and disposal of ash by landfill.

4. *Sewage treatment plant: processes (recommendations)*. The following treatment units and processes are specifically recommended and should be included in preliminary planning.

a) Three mechanically cleaned bar-rack screens to remove large debris from the flow

b) Two circular grit-removal units designed on surface overflow rate to remove grit and sand prior to primary settling (separation and washing of settled grit and organic matter by two hydrocyclone classifying devices). Prior to final design, consideration to be given to utilization of an aerated grit-removal unit

c) Disposal of screenings and grit in sanitary landfill

d) Grease removal by skimming in the primary settling tanks; grease flotation facilitated by aeration following or incorporated with grit-removal unit and immediately preceding the primary settling tanks

e) Possible future chemical application in the aeration structure for pH control and introduction of chemicals to aid precipitation of wastes, for short periods, at times of exceptionally low flows in Cayadutta Creek

f) Possible future addition of coagulating chemicals in the flow to the secondary clarifiers for "polishing" effluent and/or in the discharge from the secondary clarifiers for control of algal nutrients, if found necessary (structure provided for addition of coagulating chemicals in inflow)

g) Six rectangular primary settling tanks with mechanical sludge collectors and scum skimmers

h) Biological secondary treatment in two stages by two high-rate (roughing) filters, with stone or plastic media and rotating arm distributors, and activated-sludge treatment (in multiple units) in two sections, with mechanical aeration units directly powered by electric motors

i) Two circular secondary settling tanks with sludge- and scum-collection mechanisms, sludge collectors to be of the "vacuum cleaner" type

j) High-rate digestion provided through a primary digester followed by a secondary digester; floating covers on both digesters with gas-collection and holder facilities; gas utilization for heating of sludge and buildings; gas recirculation mixing provided in both digesters, with possible operation of either digester as primary

k) Dewatering of digested sludge and disposal of sludge cake by approved landfill methods

Final recommendations and determinations on sludge dewatering and disposal must consider the net annual costs, reflecting capital costs and operation, the physical problems involved in handling sludge as amounts increase over the years, and the future utilization of the land considered for landfill. Certain of these evaluations can be made from the engineering and cost points of view. However, the Cities of Gloversville and Johnstown and the Town of Johnstown, as well as the New York State Department of Health, must also consider future land use and future disposal of both sewage-treatment-plant wastes and municipal refuse. As a guide in the financial comparisons, the following is an estimate of the approximate net annual costs (capital plus operational) for the three possible methods.

A. Dewatering by lagoons and disposal by approved landfill methods $20,000
B. Dewatering by vacuum filters and disposal by approved landfill methods $50,000
C. Dewatering by vacuum filters, incineration (multiple-hearth), and disposal of ash by landfill $65,000

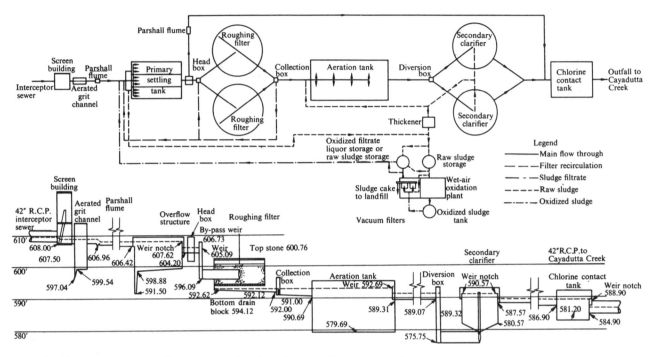

Fig. 16.12 Line diagram and hydraulic profile of the Gloversville-Johnstown joint waste-water treatment plant. (Courtesy Morrell Vrooman Engineers.)

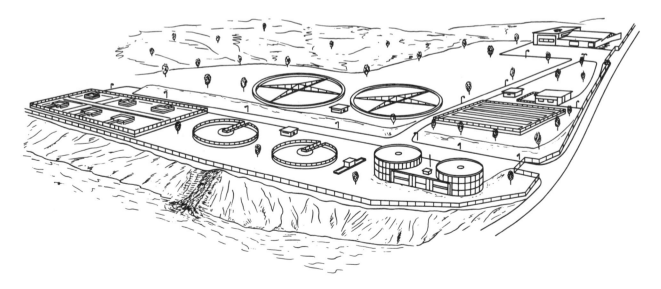

Fig. 16.13 General view of the Gloversville-Johnstown joint treatment plant. (Courtesy Morrell Vrooman Engineers.)

Dewatering at Fond du Lac, Wisconsin, was formerly achieved by vacuum filters. To reduce high dewatering costs, vacuum filtration was abandoned and replaced by evaporation and drainage in lagoons. Dewatered sludge removed from the lagoons is disposed of in landfill. This technique has been in operation since 1962. Sewage-treatment-plant loadings at Fond du Lac are only slightly less than those anticipated in our Gloversville–Johnstown design (6.4 million gallons per day including 3 million gallons per day of tannery wastes). Fond du Lac suspended solids are 173,000 pounds per week (compared to Gloversville–Johnstown 195,578 pounds per week).

Based on the above operational experience and consideration of the low annual costs compared to dewatering by vacuum filtration, dewatering by lagooning is recommended. In accordance with the requirements of the New York State Department of Health, the alternative to dewatering of the digested sludge by lagooning would be dewatering by vacuum filtration. For both methods, ultimate disposal would be by landfill. Both primary and secondary treatment units are shown schematically in Fig. 16.12 and in a general view in Fig. 16.13.

16.12 Solids Handling

The estimated quantity of raw solids to be handled at the plant is 170,000 pounds per week, of which approximately 70 per cent are of industrial origin, primarily tannery wastes.

A review of existing secondary waste-water treatment plants in the United States and Canada handling a large percentage of solids disclosed four plants that were treating tannery effluents in combination with municipal waste-waters. All four plants utilize digestion. Two dispose of the digested solids in liquid form on wastelands using tank trucks, one lagoons the digested solids, and one uses vacuum filtration, drying, and landfill. All of the plants are able to digest the solids effectively.

The problems with digestion of tannery–municipal solids have not been primarily chemical or biological but physical. Hair and scum have caused serious problems in the digesters themselves and in digestion-tank appurtenances. Extensive pretreatment, including fine screens, have been necessary at some plants to reduce such problems. The screens have in some cases introduced another problem, blinding.

The cities of Gloversville and Johnstown are surrounded by a rural area providing land for landfill of the final residue from the sludge-handling system. Both cities operate refuse landfill operations and own large areas of land designated for future landfill use. Therefore, landfill of the dewatered solids from the waste-water treatment plant could be accomplished.

A review of construction and operational costs indicated that digestion and dewatering in lagoons would provide the most economical solution to conditioning the solids. This solution was discussed extensively with the regulatory agencies and with the Gloversville–Johnstown Joint Sewer Board, who have responsibility for administering the project, but was eventually eliminated.

There was considerable interest in the wet-air oxidation system being used to condition and to destroy organic solids. This system was viewed as capable of treating industrial solids without possible upset by the changing chemistry of the leather industry. The Zimpro Division of the Sterling Drug Company had considerable experience in handling similar solids at South Milwaukee and in Kempen, Germany. The wet-air process offered the added advantage of producing a solid that was readily filterable. The filterability of either raw or digested solids has been a potentially troublesome and expensive feature of the operation.

The sludge-handling process finally selected was a low-pressure (300 psi), wet-air oxidation plant, vacuum filters for dewatering, and ultimate disposal of solids in landfill. The wet-air plant as designed will reduce the nonsoluble organic solids by 40 to 50 per cent. The high BOD filterate will be pumped to a holding tank and discharged to the head of the plant during low BOD load periods. This plant will have the capacity to handle the weekly solids loading in 5 days on a 16-hour schedule or in $3\frac{1}{2}$ days on a 24-hour schedule. It is basically one unit with several key items of equipment duplicated. The two vacuum filters will dewater cake from the oxidized-sludge holding tank. This tank will be equipped with overflow weirs and a sludge collector. The dense sludge pulled from the tank will be filtered during the 8-hour day shift, 5 days per week. This solution to the sludge-handling problem is not conventional but it fits the unique problem of these particular communities in their location and with their industries.

16.13 Final Design of the Gloversville—Johnstown Joint Sewage Treatment Plant

1. *Flow*

24-hr mean	9.50 mgd
8-hr mean peak	13.12 mgd
1-hr mean peak	16.22 mgd
1-hr maximum peak (to be used in hydraulic design of conduits)	19.3 mgd
Maximum hydraulic capacity through primary	30 mgd

2. *Process loading to sewage treatment plant*

BOD weekday 24-hr mean	35,019 lb/day
BOD weekday 8-hr mean	53,351 lb/day
Suspended-solids load	195,578 lb/week

3. *Mechanically cleaned bar screens*

Number of units	3
Width of screen channel	3 ft
Bar spacing	1-in. clear opening

   Velocity through screen at

5.0 mgd	1.2 ft/sec
9.5 mgd	1.6 ft/sec
30 mgd	2.7 ft/sec

4. *Grit-removal units*

Number of units	2
Type	Aerated
Dimensions	7 ft wide × 30 ft long × 7 ft deep
Particle size removed	100% of 0.2 mm at 20 mgd
Grit-cleaning devices (cyclone with screw classifier)	2 units
Grit disposal	Landfill

5. *Primary settling tanks*

Number of tanks	6

   Flows and surface settling rates

8-hr peak (basic design)	13.12 mgd; 800 gal/ft²/day
24-hr	9.50 mgd; 580 gal/ft²/day
Maximum peak hour	19.3 mgd; 1180 gal/ft²/day
Total surface area required	16,400 ft²
Surface area of each tank	2736 ft²
Tank dimensions	152 ft long × 18 ft wide × 8 ft deep

   Displacement time at

13.12 mgd	1.8 hr
9.50 mgd	2.5 hr

   Estimated removals

Suspended solids	60%
5-day BOD	30%

6. *High-rate (roughing) filters*

Number of units	2

   Process loading:

Peak 8-hr BOD	150 lb/1000 ft³
24-hr average BOD	98 lb/1000 ft³
Diameter of filters	165 ft
Depth of filters	6 ft

   Hydraulic loading

8-hr peak	315 gal/ft²/day
24-hr average	227 gal/ft²/day
Volume of filter media	250,000 ft³
Filter media	4-in. stone or plastic
Recirculation pumps	3–3500 gpm variable speed
Removal, 5-day BOD	30% at 150 lb/1000 ft³

7. *Aeration tank and equipment*

Number of tanks	1 with 2 compartments
Volume division of each tank	1/4 and 3/4
Tank dimensions	260 ft long × 130 ft wide × 13 ft deep
Total tank volume	439,000 ft³

   Process loading:

24-hr average (39 lb/1000 ft³)	0.25 lb BOD/lb MLSS at MLSS of 2500 ppm
8-hr peak (60 lb/1000 ft³)	0.39 lb BOD/lb MLSS at MLSS of 2500 ppm

Displacement time including
    33% recirculation
    24-hr flow, 9.50 mgd   6.3 hr
    8-hr flow, 13.12 mgd   4.5 hr
Aeration equipment
    Type   Mechanical
    Number of aerators   8
    Connected horsepower   800 hp
    Aerator oxygenation
      capacity (each)   310 lb/hr

8. *Secondary settling tanks*
    Number of tanks   2
    Diameter   120 ft
    Side water depth   10 ft
    Surface settling rate
      13.12 mgd flow   600 gal/day/ft²
      9.50 mgd flow   434 gal/day/ft²

9. *Chlorination equipment and contact tank*
    Number of tanks   1 with 2 compartments
    Dimensions   100 ft long × 60 ft wide (30 ft each) × 7 ft deep
    Detention time at 30 mgd   15 min
      9.5 mgd   47 min
      5.0 mgd   90 min
    Chlorinators
      Number   2
      Rating   4000 ppd
    Evaporators
      Number   2
      Rating   4000 ppd
    Residual analyzer
      Number   1

10. *Secondary return sludge pumps*
    Number of pumps   3
    Type   Variable speed
    Maximum capacity each   2500 gpm

11. *Waste sludge pumps*
    Number of pumps   3
    Type   Variable speed
    Maximum capacity each   200 gpm

12. *Sludge thickener*
    Number of units   2
    Type of thickener   Flotation
    Loading   2 lb/ft² /hr
    Operation   100 hr/week
    Total surface area required   264 ft²
    Total surface area provided   300 ft²

13. *Wet-air oxidation unit*
    Number of units   1
    Capacity   25 tons/day
    Operating pressure   300 psig
    Insoluble organic matter
      reduction   50%
    Operating volume and
      schedules
      170,000 lb/week   65% volatile content; 5% solids
      24 hr/day continuous
        operation $3\frac{1}{2}$ days/week
      16 hr/day operation   5 days/week
    Oxidized-sludge storage   120,000 gal for 1 day
    Duplicate items   Boiler High-pressure pump

14. *Raw-sludge holding tanks*
    Number of tanks   2
    Total holding capacity   7 days
    Tank proportions
      Side water depth   20 ft
      Diameter   42 ft

15. *Vacuum filtration*
    Number of units   2
    Filter area   400 ft² each
    Design filter rate
      Oxidized sludge   5 lb/ft²/hr
      Raw sludge   5 lb/ft²/hr
    Operating volume and
      schedules
      Oxidized sludge,
        115,000 lb/week   29 hr filter time/week
      Raw sludge,
        170,000 lb/week   43 hr filter time/week

16. *Ultimate disposal of oxidized filter cake*
Sludge to be landfilled in city refuse areas or other selected areas. Because of the character of oxidized-sludge cake, sludge need not be covered.

16.14 Estimated costs and financing

The estimated costs of the plant units described above are as follows:

	Amount
Site development	$300,000
Screen building and grit tanks	158,000
Primary settling tank	323,000
Roughing filters (two)	437,000
Aeration tank	645,000
Secondary settling tanks	310,000
Chlorine contact tank	67,000
Recirculation building	203,000
Overflow and Parshall flume structures	10,000
Sludge building, including thickeners, vacuum filters, and wet-air plant	1,056,000
Sludge and oxidized liquor holding tanks	77,000
Yard piping and conduits	279,000
Waterline to plant site and meter pit	32,000
Fencing	11,000
Administration building	228,000
Electrical contract	250,000
Subtotal	$4,386,000
Contingency (5%)	219,300
Total	$4,605,300

Table 16.23 Inconsistencies between theory of design and actual practice in design.

Situation	Theoretical solution	Actual practice	Reason for violating theoretical solution
Flow BOD	Measured as 5 to 7 mgd Measured as 20,000–25,000 lb/day	Designed for 9.5 mgd Designed for 35,019 lb/day	Addition of a large glue-manufacturing plant and another tannery; consideration of future loads
Sludge handling	Digestion plus lagooning was least costly and proved acceptable in the laboratory	Zimpro plus vacuum filtration plus landfill	Lack of State approval because of health hazards of lagooning digested sludge; lack of confidence in efficiency of sludge digestion
Charging for services	Incentive plan based upon unit costs for a pound of BOD and suspended solids and a gallon of waste	Percentage of water bill	Ease of charging; elimination of need to sample, police, and analyze industrial wastes; strong representation on joint Sewer Board of tannery industry
Equalization	Theory would indicate that it is needed because of great fluctuations in instantaneous flow and character	In practice, not required because the great number of tanneries and length of travel in sewers provide equalization	Great number of tanneries; increase in cost of construction and operation of equalization basin
Biological treatment	Theory and laboratory results indicate that activated sludge is an excellent method of reducing high BOD	Practice shows it more suitable to use a combination of roughing filters and activated-sludge treatment in series	Activated sludge computed to be about twice the cost*

*The costs are compared as follows:

	Roughing filter	Activated sludge
Fixed charge	$0.55/lb	$0.39/lb
Operating costs	$0.03/lb	$0.72/lb
Total costs	$0.58/lb	$1.11/lb

The project costs and the Federal and State grants are estimated as follows.

Interceptor sewers	$1,227,000
Waste-water treatment plant	4,600,000
Subtotal	5,827,000
All other costs and contingencies	1,212,000
Total	7,039,000
Less Federal and State grants	3,981,000
Net cost to community	3,058,000

It is estimated that the plant will have twenty full-time employees and that the annual operational and maintenance costs will be $262,000. The local communities will receive a reimbursement from the State of New York to the amount of one-third of this cost, leaving a local net cost of $175,000.

The estimated total annual costs for the project and their distribution between the cities are as follows:

Annual debt service (30 years at $4\frac{1}{2}\%$)	$220,000
Annual net operation and maintenance costs	175,000
Total	$395,000

Distribution

City of Gloversville (55%)	$217,250
City of Johnstown (45%)	177,750
Total	$395,000

The agreement between the cities of Gloversville and Johnstown calls for a 55/45 per cent split of capital costs and of operation and maintenance for three years. The division of operation and maintenance charges will be reviewed every three years, to reflect the results of samples collected and analyzed.

The final method of allocating costs to the users has not been fully established at this time, although certain principles have been established. The two cities would like to keep the rates in the cities the same, based on the volume of water used, probably with surcharges for industrial users. It is the aim of the Sewer Board and the cities to avoid a rate structure dependent upon repeated and critical sampling of industrial users. The average home-owner in the cities will pay for the service based upon his water usage and present estimates put the average annual cost at less than $20 per year per home.

15.15 Application of the Plan in Practice

There are many lessons which this case history serves to teach us. Several of the theoretical solutions to this problem had to be abandoned because of the situation existing in the social, economic, and governmental world of today. Table 16.23 describes five such inconsistencies along with the reasons for deviating from the theory.

It has been reported that the plant has been experiencing some grease problems due to ineffective pretreatment by some tanneries (1974–1975). However, over 90 per cent BOD reduction and about 85 per cent suspended solids reduction has been attained.

References: Case History

1. Braunschweig, T. D., "Studies on tannery sewage," *J. Am. Leather Chemists' Assoc.*, **60**, 125 (1965).
2. Chase, E. S., and P. Kahn, "Activated sludge filters for tannery waste treatment," *Wastes Eng.*, **26**, 167 (1955).
3. Fales, A. L., Discussion of paper by W. Howalt, "Studies of tannery waste disposal," *Trans. Am. Soc. Civil Eng.*, 1394 (1928); *Hide and Leather*, **75**, 48 (1928).
4. Furkert, H., "Mechanical clarification and biological purification of Elmshorn sewage, a major portion of which is tannery wastes," *Tech. Gemeindebl.* **39**, 285 (1936) and **40**, 11 (1937); *Chem. Abstr.*, **31**, 5076 (1937).
5. Hubbel, G. E., *Water Works Sewerage*, **82**, 331 (1935).
6. Jansky, K., "Tannery waste water disposal," *Kozarstvi*, **11**, 327 and 355 (1961); *J. Am. Leather Chemists' Assoc.*, **57**, 281 (1962).
7. Kalibina, M. M., "The application of the biological method of judging the efficiency of a purifying plant. Observations on the growth of organisms in an activated sludge tank . . .", *Chem. Abstr.*, **25**, 3422 (1931).
8. Kubelka, V. *Veda Vyzkum Prumyslu Kozedelnem*, **1**, 113 (1952).
9. Kubelka, V., "Principles of a final biological purification of tannery effluents," *Veda Vyzkum Prumyslu Kozedelnem*, **1**, 113 (1956).
10. Mausner, L., "Tannery waste water," *Gerber*, **41**, 1519 (1938).
11. Nemerow, N. L., and H. S. Sumitomo, *Pollution Index for Benefit Analysis*, to be published.
12. Pauschardt, H., and H. Furkert, *Stadtereinigung*, 411 and 427 (1936).

13. Research Report no. 10, Civil Engineering Department, Syracuse University, January 1969.
14. Snock, A., *Collegium*, **703**, 612 (1928).
15. Thebaraj, G. J., S. M. Bose, and Y. Nayudamma, "Comparative studies on the treatment of tannery effluents by trickling filter, activated sludge and oxidation pond systems," *Central Leather Research Institute, Madras, India*, **13**, 411 (1962).
16. Vrooman, M. and V. Ehle, *Sewage Ind. Wastes*, **22**, 94 (1950).

Questions

1. What specific alternatives are open to a municipal official concerning acceptance of industrial wastes? What are the alternatives open to an industrial plant manager when considering joint treatment?
2. List and discuss the 12 advantages of joint treatment.
3. Can all industrial wastes be treated in municipal sewage-treatment plants? What are the limiting concentrations of contaminants, the reasons for limitation, and the acceptable pretreatment required for excessive contaminants of each type?
4. What are some of the problems of combined treatment?
5. What can municipalities do to assist industries in waste-treatment practices?
6. What methods are used by municipalities for defraying costs of sewer services and sewage-treatment costs?
7. What method of charging an industry is advocated by both the American Society of Civil Engineers and the American Bar Association? What are its main advantages?
8. In the case history of joint treatment presented by the author, what are the pertinent background factors related to its solution?
9. Describe how the stream survey, composite waste sampling and analysis, and laboratory pilot studies were vital to the joint-treatment decision.
10. Why was a literature study necessary?
11. What were the overall planning conclusions?
12. What are some of the inconsistencies between theory and design used in this case study?
13. Does joint treatment of raw wastes represent a reasonable or optimum solution to the problem?

CHAPTER 17

JOINT TREATMENT OF PARTIALLY TREATED INDUSTRIAL WASTES AND DOMESTIC SEWAGE

Most industrial wastes contain only a few harmful constituents and the removal of these leaves the remaining wastes amenable to treatment along with domestic sewage. The well-trained waste engineer will recognize the situations which call for this approach, whereas the less-knowledgeable engineer will demand complete separation of all industrial wastes from city sewers and treatment plants. There is a fine line of distinction between a wholly untreatable waste and one containing only certain components which are untreatable when the waste is combined with domestic sewage. It is the objective of this chapter to define this difference more clearly. Again, it should be stressed that the concern of a municipality for its industries and, in return, that of industry for the effectiveness of the municipal treatment facilities are significant factors in the success of joint treatment.

AN EXAMPLE OF COMBINED TREATMENT FOLLOWING PRETREATMENT

A certain highly industrialized city had for 25 years been handling all the wastes from within its environs in its secondary-type treatment plant. This disposal plant was now overloaded and a decision had to be made concerning the treatment of industrial wastes. The author was employed by the municipality to advise its industries on treatment. Besides the overloaded conditions which reduced plant efficiency, other operating difficulties were being caused by grease which clogs the fixed nozzles or filters, feathers which clog the distributing devices and build up in the digester, toxic chemicals which inhibit bacterial action in the digester, build-up of grease and feathers which reduces or stops digestion, and excess floating solids which go over the effluent weir into the receiving stream. The plant produced inconsistent removal

efficiencies and was at least partially inoperative about 25 per cent of the time owing to the character and quantity of industrial wastes.

The existing plant contained a bar screen, grit chamber, circular rim-driven clarifier, fixed-nozzle trickling filter, secondary circular clarifier, separate sludge digester, and sand drying beds. The average daily flow was 3.88 million gallons per day, the suspended-solids load 9276 pounds per day, and the BOD load 11,533 pounds per day; 44.1 per cent of the average flow, 47.5 per cent of the average BOD, and 41.2 per cent of the suspended-solids load occurred during the nine daylight hours between 8 a.m. and 5 p.m.

The industries and their contributing pollution loads were as follows:

1. A dyeing and finishing mill for synthetic textiles contributed 33 per cent of the treatment-plant flow and 80 per cent of the total industrial flow, 25 per cent of the total plant BOD and 62 per cent of the total industrial BOD, but only 4 per cent of the total suspended solids.

2. A laundry contributed 6.2 per cent of the total plant BOD and suspended solids. This waste included rags and lint, which lead to stoppage of sewer lines, but it comprised less than 1 per cent of the plant flow.

3. A rendering plant contributed 4.5 per cent of the total plant BOD. There was a great deal of grease and odor in this waste, and 1.91 per cent of the flow and suspended solids came from it.

4. A poultry-plant waste consisted of blood, feathers, and paunch manure and contributed 1 per cent of the flow, 1.4 per cent of the BOD, and 1.7 per cent of the suspended solids of the total combined wastes.

5. A slaughterhouse and meat-packing plant con-

tributed 2.8 per cent of the BOD and 1.3 per cent of the suspended solids, but only 0.28 per cent of the flow.

6. An electrical-plating company contributed 2.2 per cent of the flow, 0.69 per cent of the BOD, and 0.87 per cent of the suspended solids. High chromium concentrations were found in several samples of these wastes.

7. The domestic customers contributed 84 per cent of the plant's suspended solids, 60.4 per cent of the flow, and 57 per cent of the BOD.

Overloaded plant capacity is a serious problem in many municipalities in this country. Precedent plays an important part in the solution to such a problem; people who have been disposing of their wastes in a certain way for a number of years will tend to resist any proposed change in overall methods. The initial method of plant financing will also tend to set a pattern for subsequent financing. Therefore, continued treatment in the same general system as in the past, but with improved efficiency, will be welcomed by all parties concerned. The task of designing waste-treatment facilities is difficult enough in itself, without attempting to convince the public that its approach has been, and continues to be, erroneous.

With these principles in mind, the author felt that the problem under discussion could best be solved by continued use of combined treatment, provided all other factors favored it. Since both the public and the private corporations concerned favored this approach and since it was more economical, the only major deterrent to its acceptance was the presence of toxic wastes and wastes which interfered with the proper operation of the plant, although only the chromium waste from the plating plant appeared to present any difficulty. Since this waste was a small percentage of the total flow, the author believed that a satisfactory arrangement could be made to eliminate it.

Three steps were now necessary to complete the technical solution to the problem: (1) to ascertain the capacity of the various existing treatment-plant units, (2) to reduce the incoming waste load to a minimum by proper pretreatment of industrial wastes at each individual factory, and (3) to reevaluate the present plant and suggest the additions required to handle the future waste load effectively.

17.1 Ascertaining Present Plant Capacity

Table 17.1 presents the load-ratings for each treatment unit in the present plant. A rated capacity of 100 per cent means that the unit is currently loaded to its maximum for optimum removal efficiency. A rating greater than 100 per cent means that the unit is overloaded by the percentage above 100 and a percentage less than 100 indicates that it is not being used to capacity. Since a minimum of 85 per cent efficiency was required to protect the receiving stream, it is clear that only the grit chambers (detritors) were adequate to handle even the present pollutional load.

Table 17.1   Capacity of present plant units.

Units	Rated capacity (average)*	Percentage of rated capacity being used
Detritors (two)	8.00 mgd	48.5
Primary settling basin	3.18 mgd	122
Secondary settling basin	2.00 mgd	194
Trickling filters (loading after primary settling, 9240 lb/day)	2210 lb/day (for 90% efficiency)	418†
	5600 lb/day (for 85% efficiency)	165†
	11,300 lb/day (for 80% efficiency)	82†
Digester	65,000 ft³	518
Sludge-drying beds	One loading per month of 9 in. of sludge (825 ft³ of sludge per day)	202
	2260 lb dry solids per day	246

* Based on normal accepted design loadings for the various units.
† Assuming 20 per cent reduction in primary settling basin.

17.2 Reducing the Incoming Load

Each industrial plant was visited and helped to assess the load it was discharging to the treatment plant. (At this point in any project, the engineer can suggest the design and construction of pretreatment methods discussed more fully in Chapters 6 through 10.) Methods of reducing or eliminating the undesirable wastes were proposed; all the industries were receptive to suggestions about changes in processes or disposal practices.

The textile industry, contributor of the largest volume, agreed to carry out investigations on the possibility of water reuse, but, after several months of laboratory study, no practical method of rendering the water reusable was found. Little change, therefore, could be expected in the load from the finishing mill. On the other hand, the hosiery-mill section of the textile operation made substitutions in soaps, detergents, and so forth, that considerably lowered the BOD issuing from this mill.

The rendering plant agreed to install new baffles in its existing grease tank trap and, in addition, to construct a second grease tank trap and new pipeline, which would eliminate the escape of grease into the sewer. This grease had been contributing significantly to the clogging of the fixed nozzles on the trickling filter at the treatment plant. The slaughterhouse and packing plant eliminated all blood from their discharge by separating the killing area completely from the rest of the plant. In this way, the blood wastes, which are used for fertilizers, glues, and animal feeds, were drained into large drums in the basement for separate disposal. All grease and floor drainings went to the grease tank before discharge to the city sewer, and orders were given for the grease tank to be opened and cleaned periodically. All paunch manure was "screw conveyed" out of the basement of the plant into tank trucks, which carried it to the country for sale as soil fertilizer. In this way, all solid manure was eliminated from the sewer.

The poultry plant presented another facet of the same problem. Feathers were clogging filter nozzles at the treatment plant and resisted settling in the sedimentation tank, because of their large surface area relative to density. Some feathers, however, after becoming thoroughly wetted, did settle and finally ended up in the digester, where they interfered with that process because they did not decompose. The poultry plant installed a series of three fine screens, which are cleaned daily, to remove feathers and other suspended matter which formerly escaped from the plant into the effluent sewer. Four cement gutters covered with perforated steel plates lead into the screens, with Wade-type drains (see any plumber's supply manual) placed at each end of the gutters as primary aids in keeping feather discharge to a minimum.

The laundry installed a metal frame with a heavy wire-screen cloth attached and a series of baffle plates in the catch basin located at the rear of the building. The cloth is changed as often as necessary and eliminates much of the suspended solids leaving the plant. Additional screens were installed within the plant to catch larger objects such as bits of cloth and towels.

The metal-plating plant installed automatic pH recorders on the plating-room effluent line. Ammonium hydroxide (NH_4OH) was added to stabilize the pH value at about 7.0, thus preventing excess acidity from entering the city sewer and ensuring elimination of slug acid discharges. In addition, a study was made of the plating-bath operations and precautions were taken against overflows of the metal baths (especially those containing expensive cyanides and chromium).

A small chemical company made minor discharge-line changes to eliminate storm water from the sewer. (The treatment of uncontaminated storm water in waste-treatment plants is usually an uneconomical procedure and can result in lowered plant efficiency.) They also agreed not to discharge any toxic chemical to the sewer unless adequately diluted. Two separators were used to facilitate the removal of these foreign constituents from the process waste waters. One was designed to remove all oily ester-type material originating from floor washings; the other is a two-stage system that is used for separation of all process water.

Frequent discussions with individual industries improved their attitudes toward pretreatment. Overall waste-treatment objectives were pointed out in these meetings and tours of the city waste-treatment plant greatly aided the industries to understand the problems involved.

17.3 Reevaluation of Present Plant and Suggestions for Additions

The third phase in the technical solution to this problem involved evaluating the final load reaching the city treatment plant after all industries had installed pretreatment devices. The wastes from each contributing industry were analyzed several times, over prolonged periods, to ascertain the pollution loads they contained. In addition, the total load reaching the city treatment plant was measured (Tables 17.2 and 17.3).

Table 17.2   Total loading of treatment plant.

Characteristic	Before pretreatment*	After pretreatment†
Flow, mgd	3.88	3.36
BOD, lb/day	11,533	10,130
Suspended solids, lb/day	9,276	8,129

* Average of 6 sampling days during June to September 1954.
† Average of 20 sampling days during June to September 1955.

The reader will notice, in Table 17.3, that certain plants' waste loads were increased after pretreatment. Although in theory this increase should not take place, it often occurs in practice, because of the variability of industrial operations. This set of figures, therefore, points out the need for numerous samplings over extended periods to ascertain precise loadings. The engineer should not be discouraged by a few results which may be somewhat misleading if interpreted directly. Moreover, a major objective in this particular case was the removal of nuisance wastes rather than reduction of loading. It may not necessarily follow that removal and reduction will occur simultaneously.

These new pollution-load measurements indicated a decrease of 1403 pounds of BOD per day since the installation of the pretreatment devices. At the average rate of $150 per pound of BOD for capital expenses of treatment, this represented a considerable saving. Also, this measured load included 322 pounds of BOD and 118 pounds of suspended solids from the packing plant, which would be removed when the packing plant's renovations were completed. The industries had eliminated many of the nuisances which existed before. The removal of feathers, blood, toxic chemicals, rags, and inflammable materials from the city waste waters resulted in increased efficiency of the existing treatment plant: even without further additions, the plant was able consistently to remove about 80 per cent of the BOD, 75 per cent of the suspended solids, and 66 per cent of the color of the total waste water, whereas previously it functioned irregularly and was only about 50 per cent efficient when it was operating. These results were obtained even though the plant was still obviously overloaded.

Additional plant capacity was advocated—in the form of increased clarifier volume, digester volume, filter-stone area, and final sludge-handling equipment—to achieve maximum removal of pollution and to protect the best usage of the receiving waters. These additions were designed and constructed; once they were in operation, efficiences of the operations rose to over 90 per cent. The new sections of the plant were built at a relatively low cost and operated at maximum efficiency, due in large part to the excellent cooperation of the industries and the city.

This example illustrates the importance of individual consultations with industry and the need for

Table 17.3   Results of pretreatment practiced by industries.

Plant	Flow, gpm		BOD, lb/day		Suspended solids, lb/day	
	Before*	After*	Before	After	Before	After
Synthetic dyeing and finishing mills	857	838	3048	2736	376	509
Laundry	50	100 x	719	222	576	433
Rendering plant	51	86 x	514	1995†	179	1323†
Poultry plant	28	80 x	162	187	159	122
Slaughterhouse and packing plant‡	6.95		322		118	
Plating plant	59	69 x	80	47	81	25

* Before and after pretreatment.
† Increase due to insufficient sampling before pretreatment rather than decreased efficiency.
‡ Improvements not completed at time of final survey.
x Increased flow was a result of insufficient data points "Before"

rendering small-factory wastes compatible with domestic sewage, as a prelude to combined treatment. Although alternatives such as separate treatment were not pursued, this solution was considered the best for all concerned because the precedent of joint treatment had been established in this community for a long time. However, the effort and time involved for the waste engineer who handles a problem this way should not be underestimated; unfortunately, because such an expenditure of time and energy is often not financially rewarding to the engineer, this thoroughness of approach is rarely attained in actual practice.

Questions

1. What industries require pretreatment before discharge into municipal sewers?
2. What is the objective of industrial-waste pretreatment?
3. What industries are involved in this case problem?
4. What steps are necessary to decide the optimum pretreatment required?
5. How is pretreatment actually achieved?
6. Was this solution to the disposal problem optimum or was some other alternative better?

CHAPTER 18

DISCHARGE OF COMPLETELY TREATED WASTES TO MUNICIPAL SEWER SYSTEMS

In many situations, an industry may find it advisable to discharge its wastes into the city sewer system even after complete treatment. It is, of course, an advantage to an industry to have someone else assume the responsibility for final handling and disposal of its residual liquid wastes, because there is always the possibility of a mishap at the plant which would make the wastes from the industry's treatment plant unacceptable for direct discharge to the receiving stream. The municipal treatment plant thus serves as an added protective device for the stream. It may also be more convenient for an industry to discharge by gravity into a nearby municipal sewer than to pump treated waste a long distance to a suitable location on a river. The attitude of the municipality and the cost of its sewer-service charges play an important part in an industry's decision to utilize the municipal sewer and treatment plant. Acceptance of industry as an integral part of a city's life and equitable charges for the use of city sewers encourage industry to make use of the city's sanitary facilities. Let us emphasize, once again, that the means of disposal of industrial wastes must be determined on an individual basis: each case is unique and often a single factor, such as location or precedent, can influence an industry's selection of the final stage of its waste treatment.

AN EXAMPLE OF THE DISCHARGE OF A COMPLETELY TREATED WASTE INTO A MUNICIPAL SEWER SYSTEM

For our discussion, we shall consider an industry which is erecting a new plant near its existing one, to manufacture electrical and mechanical business machines. The site for the new plant was selected for its convenience, setting and surroundings, and availability; but, as is often the case, at the time the site was selected little consideration was given to the waste-disposal problem. Though this oversight has been a common occurrence in the past, fortunately more and more companies are according waste treatment its proper importance when they select future plant locations, since they are beginning to realize that it can be a costly experience when an industry does not look into all aspects of waste disposal prior to selecting a site for a new plant.

The plant is located in a small town and is the only major industry. Production in the new plant will be essentially similar to that of the old, on an expanded basis. The existing plant manufactures about 1500 machines per day and discharges all its wastes untreated, through two holding lagoons, into a small creek. The new plant at the new location, however, will be served by the municipal sewer system, and the municipal primary-treatment plant is approximately 10 to 15 miles away from the old site. Effluent from this plant discharges by gravity to the only flowing stream in the area. The municipality has agreed in advance to construct the necessary additional sewer lines to the industrial site and the industry has agreed to pay for amortizing the bonds covering the sewer construction. The entire waste can flow by gravity from the industry to the municipal treatment plant. In fact, the topography of the area is such that all creeks, streams, and even underground waters drain toward the city. The city obtains its water supply from underground wells located near the city limits, on a direct line between the industrial-plant site and the municipal sewage-treatment plant. Should the industry decide not to use the municipal sewer facilities, it would have to pump its entire waste output about four miles, against a 200- to 300-foot elevation rise, into the receiving stream having adequate dilution.

The business-machine corporation needed to

determine the quantity and quality of the wastes being produced at the existing plant, so that it could estimate the wastes which would result from future production at the new plant and then determine the degree and type of treatment to be given those wastes before they entered the city waste-water treatment system.

18.1 The Sampling Program

The first step was to sample existing wastes. All the wastes from the existing plant were discharged into two pipelines. Both of these were weired and boxed for sampling and flow measurement; photographs of these weirs and boxes are shown in Figs. 18.1 and 18.2. The wastes from Pipeline 1 (Fig. 18.1) originated primarily in the plating room and those from Pipeline 2 (Fig. 18.2) came from the pickling room. Slug discharges and changes in production were included in the samples. Each line was sampled for a 24-hour period on each day of the week except Sunday. The days selected, however, were in different weeks over a two-month period in 1958, as follows: Monday, November 17; Tuesday, September 23; Wednesday, October 8; Thursday, November 6; Friday, October 24; Saturday, November 15. (See Table 18.4 for figures for the production of machines per plating line on these sampling days.)

Samples were collected from each weir box every half-hour during the 24-hour period, in volumes according to the rate of waste discharge at the time. Flows were recorded and samples composited each half-hour over the entire period of sampling, as illustrated graphically in Figs. 18.3 through 18.6. Batch dumpings of 55-gallon drums from the pickling and blacking rooms were also recorded during the sampling period (September 29, 1958, to November 18, 1958); the results are presented in Table 18.1.

18.2 Analyses of Wastes

The industry decided to construct its own laboratory and make its own analyses under qualified supervision; hence a chemical laboratory was designed and equipped by the author. All chemical analyses were carried out by qualified chemists employed by the industry. Metals were analyzed in accordance with the American Public Health Association's manual, *Standard Methods for the Examination of Water, Sewage, and Industrial Wastes*. The provisional method for heavy metals was used and a modified method for preliminary treatment of samples, to induce ionization of metals, was utilized.

The 24-hour composite samples from each production line were analyzed for pH, copper, chromium, zinc, cyanide, iron, and nickel. These analytica

Fig. 18.1   Pipeline 1 (plating-room wastes), showing weirs and boxes set up for sampling and measuring.

Fig. 18.2   Pipeline 2 (pickling-room wastes).

results are presented in Table 18.2. In addition, three 55-gallon barrels of waste from the pickling and blacking rooms were sampled and analyzed for cyanide only. These results are shown in Table 18.3.

18.3 Plant-Production Study

In order to ascertain the level of production during the sampling days, a detailed production study was made. With this information, one can establish the relation between production and waste discharge; it is then relatively easy to predict future waste volumes

and pollution loads resulting from increased plant production. Present normal production at the plant ranges from 1300 to 1700 machine units per day. The equivalent number of machines produced per plating line is presented in Table 18.4.

18.4 Suggested In-Plant Changes to Reduce Waste

An in-plant production survey was carried out to determine means of reducing metal impurities in waste effluents. This brought out the fact that the plant operates five plating lines (as shown in Table 18.4),

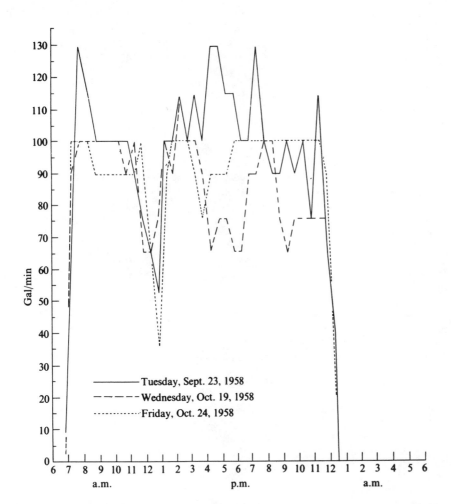

Fig. 18.3 Wastewater flow, Weir 1 (plating room).

each of which warrants some alteration from the standpoint of decreasing the amount of waste it contributes. Some of the more pertinent suggestions made by the author for changes are outlined in the following paragraphs.

1. Bright nickel cycle (see Fig. 18.7)

a) Add a high-pressure fog spray of water in the last 2-foot section of the nickel-plating line, to reduce the "drag-out" which results when a metal part is removed from a plating bath. Fog sprays utilize a fine spray of water under pressure to rinse the acids and metallic ions off the work (machine parts). A relatively small quantity of water is consumed by the spray, with the same net effect as a stream or bath of water.

b) Utilize a static rinse as make-up for nickel plate. Static (rather than continuous) rinses reduce the quantity of metal and acids lost in waste. As the static-rinse bath builds up in concentration of rinsed-off metals, it becomes increasingly less capable of rinsing the work. The rinse water can then be concentrated somewhat and returned to the plating bath as make-up water.

c) Use high-pressure jet sprays in the cold-water rinse and recirculate the cold-water rinse to a holding tank, from which it is bled as often as possible to the static-rinse tank as make-up water. There must be an overflow in the holding tank, leading to

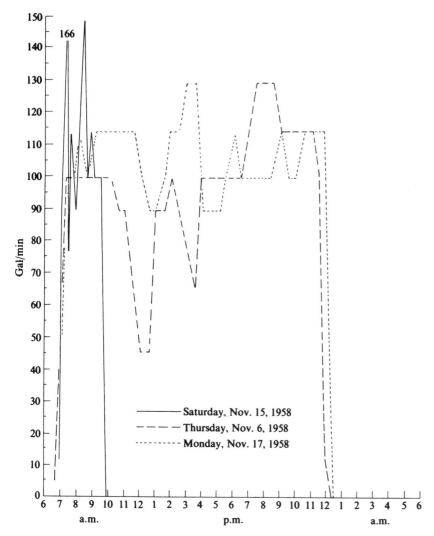

Fig. 18.4 Wastewater flow, Weir 1 (plating room).

the sewer. By this practice, wastewaters are used to the maximum before final loss to the sewer.

d) Replace the cyanide dip with an acid dip. Cyanides are not only detrimental to biological treatment of wastewater, but also hazardous and even lethal to humans, in the form of hydrogen cyanide gas. Even in small concentrations, cyanide compounds can be toxic to fish. One purpose of the cyanide dip is to brighten the work and to remove the last traces of contaminating matter: acid dips have proved to be adequate substitutes in many cases, as well as being more economical.

2. Zinc cycle (see Fig. 18.8)

a) Add a high-pressure fog spray of water in the last 2-foot section of the zinc-plating line, to reduce "drag-out."

b) Install a static rinse to replace the cold-water rinse and reuse static-rinse water as make-up for the zinc-plating bath.

c) Use small, high-pressure sprays in cold-water rinses and overflow each in the direction of the zinc-plating tank. (This utilizes the principle of counter-current flow.) If the flow of rinse water is small enough, it may serve eventually as make-up water

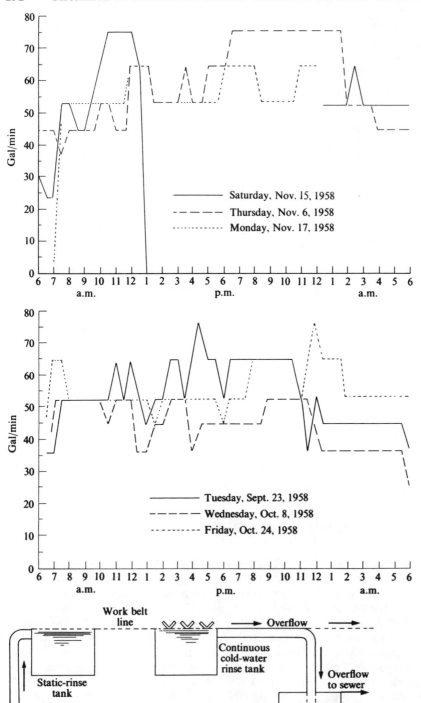

Fig. 18.5 Wastewater flow, Weir 2 (pickling room).

Fig. 18.6 Wastewater flow, Weir 2 (pickling room).

Fig. 18.7 Schematic diagram of nickel-cycle process in the business-machine plant (side view).

Table 18.1   Additional batch waste discharge.

Nickel strip from pickle room (Ni plus CN)		Cyanide from pickle room* (CN plus Fe)		Cyanide from blacking room (case hardening) (CN)	
N†	Date	N	Date	N	Date
1	9/29	1	10/13	6	10/7
1	9/30	1	10/18	6	10/21
2	10/3	2	10/24	7	11/13
1	10/6	1	10/31		
1	10/7	1	11/3		
1	10/8	1	11/7		
2	10/9	2	11/8		
2	10/13	1	11/13		
3	10/17	2	11/15		
1	10/20	1	11/15		
1	10/21	1	11/18		
2	10/24				
1	10/27				
2	10/29				
2	11/5				
1	11/6				
2	11/7				
1	11/15				

*Usually dumped Saturday.
†Number of 55-gallon barrels.

for the plating tank. An overflow pipe on the cold-water-rinse tank closest to the plating tank can serve as entrance to the sewer, if there is no room for an overflow in the zinc-plating tank.

3. Chrome cycle
a) Add a high-pressure fog spray of water just before racks move to the static rinse, to reduce "drag-out" from chrome-plating bath.
b) Return rinse water to the chromate plating tank, instead of discharging it to the sewer twice a week.

4. Combination copper and nickel cycle
a) Remove nickel part of cycle to new plant, where there will be a more modern automatic nickel cycle.
b) Design a new copper cycle so that no "drag-out" occurs across floors, where drippings are subsequently washed into the sewer. This is purely

a matter of clean housekeeping and is often over-looked by management.
c) Install a high-pressure fog spray of water near the end of the copper-cyanide plating tank to reduce "drag-out" prior to rinse.
d) Install a static rinse to follow the plating bath, instead of the present flowing rinse.
e) Use static rinse as make-up water for the plating bath.

5. Barrel nickel cycle
a) Install a static rinse following the acid nickel-plating tank to replace one of the two continuously flowing rinses in use at present.
b) Reuse static-rinse-tank contents as make-up for the acid nickel-plating tank.

6. Blacking room
Replace the molten-cyanide case-hardening tank with a carburonitriding electrical furnace. This will eliminate pollution due to cyanide from the case-hardening process.

7. Nickel stripping
Reduce nickel stripping to a minimum or eliminate it completely. This is a source of a great percentage of cyanide in the waste (see Tables 18.1 and 18.3).

18.5 City Wastewater Treatment Plant

The city possesses a primary-treatment plant with separate sludge-digestion facilities. The disposal plant contains the following units: 30-inch sewer-influent line; 1-inch bar screen (mechanically cleaned, 30-minute time switch); two auxiliary hand-cleaned bar screens (not used at present); screenings grinder (used three times a day, grindings returned to flow); two grit chambers (each closed off once a month for cleaning); Parshall flume for flow measurement; two horizontal settling basins (operated in parallel); chlorinator (750 lb/day capacity, prechlorination used in summer); two fixed-cover digesters (contents circulated 35 minutes per day, sludge withdrawn every two weeks); gas holder; gas boiler plus auxiliary oil boiler used only in winter (when, and if, digester gas fails); four vertical-type waste-water pumps (capacity: three 4-mgd, one 2-mgd at 100 per cent efficiency*); two

*Normal operating efficiency, 60 to 80 per cent.

Table 18.2   Analytical results of CN sampling* of two waste lines: metal content† and pH.

Date	Weir no. 1 (plating); average volume, 100 gal/min							Weir no. 2 (pickling); average volume, 50 gal/min						
	Cu	Cr	Zn	CN	Fe	Ni	pH	Cu	Cr	Zn	CN	Fe	Ni	pH
Mon., 11/17/58	3.2	12.4	15	17.9	1.1	5.5	7.6	0.8	0.28	<1	1.8	0.6	6.0	8.2
Tues., 9/23/58	2.8	32	7.5	11.4	2.5	6.3	9.4	1.6	0	<1	2.3	5.0	6.0	8.6
Wed., 10/8/58	3.8	22	7.0	20.8	7.6	5.0	9.8	0.8	0.2	<1	2.3	3.2	7.5	8.0
Thurs., 11/6/58	5.2	19.6	20	26.8	2.8	8.5	8.0	0.4	0.1	<1	3.1	2.6	13	8.6
Fri., 11/14/58	4.2	20	13.5	30.5	2.8	7.5	6.5	0.4	0	<1	1.3	4.8	11	7.8
Sat., 11/15/58	7.8	5.6	215	14.3	4.2	35	2.6	2.2	0.24	1.7	3.7	1.1	32	3.0
Average	4.5	18.6	46.3	20.3	3.5	11.3	7.3	1.0	0.12	<1.0	2.4	2.9	12.6	7.3
Range	2.8–7.8	5.6–32	7.0–215	11.4–30.5	1.1–7.6	5.0–35	2.6–9.8	0.4–2.2	0–0.28	1–1.7	1.3–3.7	0.6–5.0	6.0–32	3.0–8.6

*As 24-hour composites.
†In parts per million.

Table 18.3   Analyses of cyanide (in ppm) in pickling- and blacking-room barrels.

Sample number	Nickel strip	Blacking room	Pickling room
1	41,340	38,220	53,040
2	19,980	38,940	34,320
3		61,360	

sludge pumps (duplicate units pump every morning to digester). The effluent of the plant discharges to a Class B (best usage for bathing or recreation) stream.

Tables 18.5, 18.6, 18.7, and 18.8 present recent operating data for this plant. Careful consideration should be given to the sludge digesters, which appear to be amply designed for present loading conditions, providing a total volume of about 51,000 cubic feet. Active digestion should require about 17,500 cubic

feet, leaving 33,500 cubic feet for sludge storage—about 53 days of storage for the digested sludge. This may be of some concern to the operator during the long winter months at this plant location, when sludge cannot be pumped to open drying beds. The presence of toxic metals in the waste water would further aggravate this situation by retarding active digestion.

18.6 Toxic Limits for Metals

Literature on the subject reveals that even small concentrations of metals can deter sludge digestion. Table 18.9 shows that generally no more than 1 ppm of Cu, CN, or Cr and 2 to 5 ppm of Zn or Ni should be allowed in the sewage-plant influent. Dilution of plating wastes with municipal sewage often dictates the degree of pretreatment, if any, required to meet these limits.

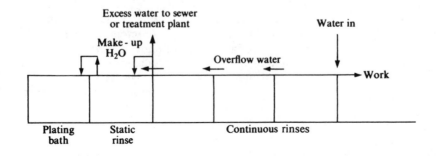

Fig. 18.8   Zinc-cycle process tanks in the business-machine plant (view from above).

18.7 Treatment for Industrial Wastes

Because of the difficulty in predicting the effects of these wastes on the municipal treatment plant, it was decided to recommend pretreatment by the industry in two stages. The first stage was to be installed immediately and the second one only if it is shown to be required.

Stage 1. Drag-out and excessive metal contamination in the wastes should be reduced as recommended in Section 18.4. All chromium, cyanide, and acid or alkaline industrial wastes should be segregated and piped separately from the plant to the industrial waste-treatment plant. A treatment unit of the flotation type should be constructed to remove oils and greases as well as solid matter. This tank would also serve as an equalizing device. The "float" from the flotator-equalizer can be burned rather than delivered to the

Table 18.4 Number of machine units per plating line on waste-sampling days.

Date	Nickel	Chrome	Zinc	Barrel nickel	Copper
Mon., 11/17/58	2400	940	1560	1650	740
Tues., 9/23/58	1990	1720	1080	1870	1340
Wed., 10/8/58	No production data available				
Thurs., 11/6/58	2290	1280	1670	1700	1500
Fri., 10/24/58	2330	1380	1640	1800	3840
Sat., 11/15/58	No production				

Table 18.5 Wastewater influent at city disposal plant.*

Month	1956		1957		1958		BOD, ppm	
	Average flow, mgd	Suspended solids, ppm	Average flow, mgd	Suspended solids, ppm	Average flow, mgd	Suspended solids, ppm	1957†	1958‡
January	5.25	138	5.52	66	5.20	95	24	25
February	5.19	135	5.39	78	5.06	87	29	108
March	8.47	70	6.06	82	7.31	45	30	77
April	8.67	53	6.36	72	8.30	75	32	328
May	6.39	81	5.55	95	7.12	124	39	132
June	5.33	80	4.98	89	6.47	59	44	88
July	0.71	94	4.69	112	5.23	98	42	80
August	4.07	129	4.26	99	4.57	99	64	142
September	3.86	128	3.94	174	4.44	90	44	131
October	4.17	139	3.59	135			71	
November	3.84	123	3.36	138			61	
December	4.83	92	4.58	150			74	
Average	5.40	105	4.86	108			46	

*Contains a considerable quantity of storm water.
†Samples collected more than once per month; values appear low owing to storm water diluting sewage.
‡Samples collected once per month; values appear low owing to storm water diluting sewage.

Table 18.6 Efficiency of sedimentation units at city disposal plant (1957).

Sanitary characteristic	Maximum	Minimum
Flow, mgd	6.78	3.28
Raw suspended solids, ppm	356	52
Final suspended solids, ppm	126	24
Removal of suspended solids, %	75	35
Raw BOD, ppm	73.5	24.5
Final BOD, ppm	58.5	14.7
Removal of BOD, % (average)	63	20

Table 18.7 Analysis of sludge solids at city disposal plant (1958).

Month	Raw sludge		Digested sludge	
	Total solids, %	Volatile matter, %	Total solids, %	Volatile matter, %
January	3.85	85	7.9	56
February	5.0	84	7.7	56
March	3.5	94	8.5	60
April	3.5	83	9.5	59
May	4.6	85	8.0	59
June	3.8	80	10.1	54.5
July	4.5	83	8.0	59
August	4.45	76	8.8	55
September	4.6	81	9.5	58

Table 18.8 Gas produced in digester at city disposal plant.

Year	Gas production total, ft³
1948	6,253,883
1949	6,475,236
1950	6,464,516
1951	6,350,273
1952	5,699,839
1953	6,220,937
1954	5,994,279
1955	5,964,300
1956	5,463,108
1957	5,562,321

concentration tank. Oils and greases can be either skimmed and burned or concentrated in other units.

Sludge from Stage 1 should be pumped to a concentration tank, from which it can be removed to dump areas by septic-tank cleaning trucks. The supernatant from the concentration tank should be returned to the beginning of the treatment plant's cycle.

Stage 2. If the effluent from the industry's treatment plant does not meet the standards set forth by the city or by the author in Section 18.6, the wastes should be further treated by the means outlined in the schematic drawing in Fig. 18.9. The detailed design of all treatment units should be prepared at the outset, and both stages submitted to the city and the state prior to construction of a new plant, with the agreement that the units in Stage 2 will be constructed only after the sampling and analysis program is completed and only if these analyses show Stage 2 treatment to be necessary. It should be stipulated that no wastes which exceed the concentrations listed in Table 18.9, when they reach the sewage treatment plant (Sampling Point 2) will be discharged into the city sewage system.

Sampling and analysis program. For a complete sampling and analysis program, the following samples should be collected:

Sampling Point 1. Daily composite of effluent from industrial plant, drawn from sampling manhole at industrial waste-treatment plant.

Sampling Point 2. Daily composite of sewage influent to city treatment plant to check degree of dilution available.

Sampling Point 3. Daily composite of raw sludge before it gets to digester at city treatment plant.

Sampling Point 4. Weekly composite of sludge in digester at city treatment plant to check the digester efficiency.

When these samples have been collected, analyses should be made for cyanide, zinc, chromium, and nickel. Volume of flow at Points 1, 2, and 3 should be recorded daily. A total organic-solids determination should be made daily on the composite raw sludge collected at Point 3.

Table 18.10, taken from Jones [4], indicates goals of a plating-waste treatment designed to protect fish life in a receiving stream.

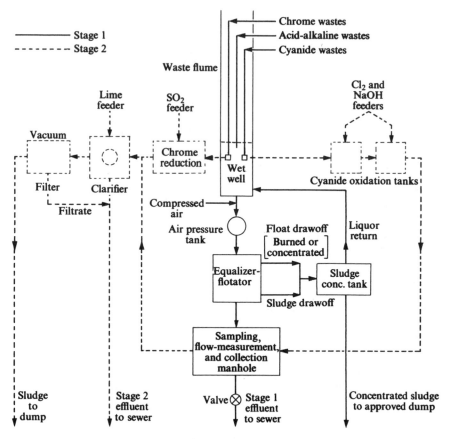

Fig. 18.9 Waste-treatment sequence, with Stage 1 units indicated by solid lines and Stage 2 units by dashed lines.

Table 18.9 Toxic limit for metals in raw sewage subject to sludge digestion.*

Metal	Reference								
	1	8	5†	3	10	2	7	6	9
Chromium	5.0	5.0	0.05			1.0		1.5	
Cyanide	2.0	1.0	0	0.1	1–1.6				
Copper	1.0	1.0	0.30	0.2		1.0	0.7		
Iron	5.0								
Zinc		5.0	0.3	0.3					> 5.0
Nickel			2.0						

*Concentrations given are in ppm. References are listed at end of this chapter.
† For streams and sewers.

Table 18.10   Lethal limits for metals as salts. (After Jones [10].)*

Salt	Fish tested	Lethal concentration, ppm		Exposure time, hr
Aluminum nitrate	Stickleback	0.1	Al	144
Aluminum potassium sulfate (alum)	Goldfish	100		12–96
Barium chloride	Goldfish	5,000		12–17
Barium chloride	Salmon	158		?
Barium nitrate	Stickleback	500	Ba	180
Beryllium sulfate	Fathead minnow	0.2	Be	96
Beryllium sulfate	Bluegill	1.3	Be	96
Cadmium chloride	Goldfish	0.017		9–18
Cadmium chloride	Fathead minnow	0.9		96
Cadmium (salt?)	Rainbow trout	3	Cd	.168
Cadmium nitrate	Stickleback	0.3	Cd	190
Calcium nitrate	Goldfish	6,061		43–48
Calcium nitrate	Stickleback	1,000	Ca	192
Cobalt chloride	Goldfish	10		168
Cobalt (salt?)	Rainbow trout	30	Co	168
Cobalt nitrate	Stickleback	15	Co	160
Copper nitrate	Salmon	0.18		?
Copper nitrate	Stickleback	0.02	Cu	192
Copper nitrate	Rainbow trout	0.08	Cu	20
Copper sulfate	Stickleback	0.03	Cu	160
Copper sulfate	Fathead minnow	0.05	Cu	96
Copper sulfate	Bluegill	0.2	Cu	96
Copper sulfate	Minnow	1.0	Cu	80
Copper sulfate	Brown trout	1.0	Cu	80
Cupric chloride	Goldfish	0.019		3–7
Lead chloride	Fathead minnow	2.4	Pb	96
Lead nitrate	Minnow	0.33	Pb	?
Lead nitrate	Stickleback	0.33	Pb	?
Lead nitrate	Brown trout	0.33	Pb	?
Lead nitrate	Stickleback	0.1	Pb	336
Lead nitrate	Goldfish	10		1–2
Lead nitrate	Rainbow trout	1	Pb	100
Magnesium nitrate	Stickleback	400	Mg	120
Manganese nitrate	Stickleback	50	Mn	160
Manganese (salt?)	Rainbow trout	75	Mn	168
Mercuric chloride	Rainbow trout	0.01	Hg	204
Mercuric chloride	Rainbow trout	0.15	Hg	168
Mercuric chloride	Rainbow trout	1.0	Hg	600
Nickel chloride	Goldfish	10		200
Nickel chloride	Fathead minnow	4	Ni	96
Nickel nitrate	Stickleback	1	Ni	156
Nickel (salt?)	Rainbow trout	30	Ni	168
Potassium chloride	Goldfish	74.6		5–15
Potassium chloride	Straw-colored minnow	373		12–29
Potassium nitrate	Stickleback	70	K	154

Table 18.10 (*continued*)

Salt	Fish tested	Lethal concentration, ppm		Exposure time, hr
Silver nitrate	Stickleback	0.004	Ag	180
Sodium chloride	Goldfish	10,000		240
Sodium chloride	Plains killifish	16,000		96
Sodium chloride	Green sunfish	10,713		96
Sodium chloride	Gambusia	10,670		96
Sodium chloride	Red shiner	9,513		96
Sodium chloride	Fathead minnow	8,718		96
Sodium chloride	Black bullhead	7,994		96
Sodium nitrate	Stickleback	600	Na	180
Sodium nitrate	Goldfish	1,282		14
Sodium sulfate	Goldfish	100		96
Strontium chloride	Goldfish	15,384		17–31
Strontium nitrate	Stickleback	1,500	Sr	164
Titanium sulfate	Fathead minnow	8.2	Ti	96
Uranyl sulfate	Fathead minnow	2.8	U	96
Vanadyl sulfate	Fathead minnow	4.8	V	96
Vanadyl sulfate	Bluegill	6		96
Zinc sulfate	Stickleback	0.3	Zn	204
Zinc sulfate	Goldfish	100		120
Zinc sulfate	Rainbow trout	0.5		64

*In this table the concentration values are all lowest at which definite toxic action is indicated by research. It must not be assumed that lower concentrations are harmless. Most of the data are for temperatures between 15° and 23° C. Exposure times have been approximated in some cases.

References

1. Coburn, S. E., "Limits for toxic wastes in sewage treatment," *Sewage Works J.* **21**, 522 (1949).
2. Connecticut State Water Commission, 8th Report (1938–1940), Hartford, Conn.
3. Dodge, B. F., and W. Zabban, "How a small electroplater can treat cyanide plating waste solutions with hypochlorite," *Plating* **42**, 71 (1955).
4. Jones, E., *Fish and River Pollution*, Butterworth, Washington, D.C. (1964), p. 74.
5. Kittrell, F. W., "Metal plating wastes in municipal sewerage systems," in Proceedings of 5th Southern Municipal and Industrial Waste Conference, 1956, at Chapel Hill, N.C., p. 216.
6. Pagano, J. F., R. Teweles, and A. M. Buswell, "The effect of chromium on the methane fermentation of acetic acid," *Sewage Ind. Wastes* **22**, 336 (1950).
7. Rudgal, H. T., "Bottle experiments as guide in operation of digesters receiving copper–sludge mixtures." *Sewage Works J.* **13**, 1248 (1941).
8. Rudolfs, W. (Chairman Research Committee), "Review of literature on toxic materials affecting sewage treatment processes, streams, and BOD determinations," *Sewage Ind. Wastes* **22**, 1157 (1950).
9. Sierp, F., and H. Ziegler, "Influence of zinc upon the purification of waste water," *Chem. Abstr.* **44**, 2152 (1950).
10. Whitlock, E. A., "The significance and treatment of cyanides in industrial waste waters," *Water Sanit. Engr.* **4**, 249 (1953).

Questions

1. Why is it advantageous to discharge completely treated industrial waste to the municipal sewage-treatment plant?

2. What play the most important role in industry's decision to use this alternative system?

3. What is the background of this case example?

4. What procedure was used in determining waste-treatment requirements at the new plant?

5. What ultimate treatment solution was selected?

CHAPTER 19

COMPLETE TREATMENT FOLLOWED BY DISCHARGE TO RECEIVING WATER COURSE—MOENCH TANNERY AND CLOSING OF GLUE PLANT

Industry has a moral, legal, and economic responsibility to consider waste treatment as an integral part of production costs. Therefore the existence or absence of a municipal plant, tax aid, or subsidy should not influence industry's decision to continue manufacturing. Waste treatment is a variable cost of doing business similar to the variable costs of labor, marketing, and raw materials.

Industry should seek to minimize production costs by selecting a proper plant site with lowest total cost of labor, taxes, raw material, water and other utilities, marketing, and waste treatment facilities.

Industry should attempt to treat waste by the lowest cost yielding a satisfactory effluent for the particular receiving stream. Considerable study, research, and pilot investigations should precede the selection of the optimum solution. Planning ahead will provide industry with time to make appropriate decisions. Conversely, lack of planning on minimizing waste-treatment costs may mean that industry could decide to cease production should it suddenly be confronted with the need for an immediate solution.

Closing the doors is one alternative always open to an industry facing higher production costs due to waste-treatment costs. More likely is a transfer of operations to another location, where (for one reason or another) total production costs are lower. Going completely out of business, in my opinion, is a result of a personal decision on the part of the owner—that is, the total cost of production exceeds revenue or is too high relative to revenue obtained from this same product.

We expect industrial production costs in the U.S. to rise due to a rather uniform acceptance of pollution-abatement costs as part of production costs. However, the advantages of a particular location and ingenuity in waste-treatment methods can have a considerable effect on the waste treatment portion of production costs. Industry is rapidly becoming aware of the importance of minimizing waste-treatment costs in production early enough so as not to be forced to select a "cease-production" alternative.

Here we present the actual case where one plant decided to close and another to completely treat its wastewater to remedy a polluted creek situation.

19.1 History of Problem

The Eastern Tanners Glue Division Plant of Peter Cooper Corporation, located on the bank of Cattaraugus Creek in the Village of Gowanda, New York, was approximately 18 miles from its discharge into Lake Erie. The Company produced 65,000 to 80,000 pounds of animal glue per day from varying proportions of tannery waste material such as limed hide fleshing, salted hide trimmings, and chrome-tanned leather splits and shavings. The availability of raw materials governs to a large extent the proportions of raw materials used in production from day to day.

The location of the Plant relative to Moench tannery and the village of Gowanda on Cattaraugus Creek is depicted in Fig. 19.1.

The manufacturing process entails the preliminary preparation of the raw materials (each in a somewhat different manner) before the actual glue extraction can be accomplished. This preliminary processing involves the washing out of lime, salt, and other impurities (large, unwanted proteins). After preliminary processing the stocks are sent to cooking kettles for glue extraction. (Flow diagrams of the three production schemes are shown in Fig. 24.20.)

The plant water supply may be separated into three classifications: (1) potable water, (2) condensing or

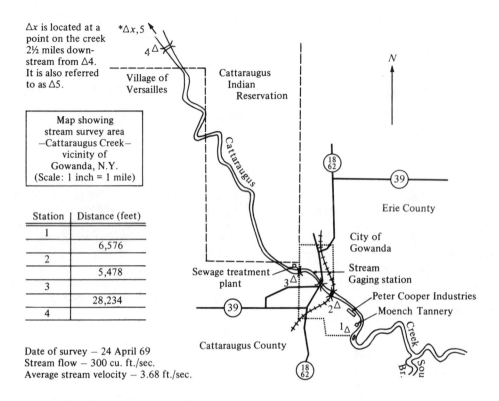

Δx is located at a point on the creek 2½ miles downstream from Δ4. It is also referred to as Δ5.

Village of Versailles

Cattaraugus Indian Reservation

*Δx,5

4 Δ

Cattaraugus

N

Map showing stream survey area —Cattaraugus Creek— vicinity of Gowanda, N.Y. (Scale: 1 inch = 1 mile)

Station	Distance (feet)
1	
	6,576
2	
	5,478
3	
	28,234
4	

Date of survey – 24 April 69
Stream flow – 300 cu. ft./sec.
Average stream velocity – 3.68 ft./sec.

Erie County

City of Gowanda

Sewage treatment plant

Stream Gaging station

Peter Cooper Industries

Moench Tannery

3 Δ

2 Δ

1 Δ

39

Cattaraugus County

South Br. Creek

Fig. 19.1 Cattaraugus Creek with relative locations of dilution and pollution.

cooling water, and (3) process water. The *potable water* is supplied from the Gowanda Village mains for offices, washrooms, and toilets. No cross connections exists between the potable water-supply system and the condensing or process systems. All domestic sewage from the plant is discharged in the Gowanda Municipal Sewerage System.

The *cooling water* is untreated creekwater used in the concentration of the light glue liquors by means of a triple-effect evaporator having a barometric condenser on the third-effect vapor body. This water is pumped directly from the Cattaraugus Creek through a screened intake located approximately 180 feet downstream from the process-water intake. About 2.4 mgd is required for this cooling. No carryover of solids from the third effect has ever been detected. The condenser water is discharged directly to the Cattaraugus Creek through an outfall sewer located approximately 130 feet downstream from the intake.

The *process water* intake exists at the southeastern upstream end of the Plant. The water is screened through a self-cleaning traveling water screen to remove trash, leaves, ice, and other large impurities before being pumped to the coagulation basin and clear well. About 4.5 to 6.5 mgd of this water are pumped, coagulated with alum, and used in the following processes: (1) boiler make up, (2) boiler water filter and softener, (3) noncontaminated cooling water, (4) cook-kettle make up, (5) flushing, floor wash, grease refining, and tankage, and (6) glue stock treatment. Only the last two processes (5 and 6) are discharged to waste treatment.

19.2 Waste Character and Treatment

The character of the waste is such that during the process-wash period from 7 a.m. to 5 p.m. the flow is at a maximum rate of about half of the 24-hour flow and at a high pH of 11–12. A large amount of the settleable solids consist of lime, dirt, hair, and grease particles, which are rapidly settleable or floatable. During the balance of the 24-hour period the wastewater is relatively low in both pH (2.5–6.0) and settleable solids. These acid wastes are principally dilute H_2SO_4 with dissolved chromium sulfate resulting from detanning of chrome splits, acid soluble proteins, and spent sulfurous acid. In 1965 the company put in certain preliminary and primary treatment to remove the high settleable-solids and scum content. The overall design by the company at that time was based on a maximum hourly flow of 5300 gpm and a 4.4 mgd maximum daily flow.

19.3 Preliminary and Primary Treatment

The process wastewater flows by gravity through two open channels to the manually cleaned bar screens to remove any relatively large foreign objects. The bar-screened effluent is distributed to four 7-ft diameter by 4 ft long Link Belt Company drum screens having a total effective screening area of approximately 269 ft². The screening media in all of the four units consists of 32–30 gauge, 316 stainless steel plates perforated with .077 in. diameter holes spaced on 7/64 in. staggered centers. The screenings are flushed from the drum screens into a Link Belt Grit Collector equipped with a surface skimmer and a dewatering screw conveyor. The dewatered screenings (or 3000 pounds per day) are returned to production.

The screened effluent from the four drum screens is distributed to two straight-line collector tanks (primary sedimentation) each 135 ft long, 20 ft wide, and 10 ft working depth (about 400,000 gallons total working capacity). Either of these basins can be operated independently in case repairs are required of the other.

Settled solids are collected in six hoppers and pumped to either one of two sludge dewatering and mixing basins having capacities of 525,000 gallons each. Floatables are periodically removed from the collector surface by means of manually-operated skimmers and pumped back to production or to one of the two mixing and dewatering basins.

Clarified effluent from the collectors is discharged over adjustable weir plates to collector troughs and piped to Cattaraugus Creek.

The sludge and floatables are further dewatered alternately in the two 525,000 gallon basins prior to vacuum filtration. A gantry crane installed in 1956 was used to produce a more uniform mixed sludge which could be vacuum filtered. From the basins the residue is pumped by a variable speed, positive displacement group having a capacity of 85–160 gpm to a filter conditioner, where a slurry of hydrated lime is added. The conditioned sludge is filtered over a Komline-Sanderson Coil Filter, 11 ft, 6 in. in diameter and 10 ft long, installed in 1962. The filter cake (35 per cent dry solids) is discharged to a pile on the plant property for removal to farmers as a fertilizer by company-operated dump trucks free of charge. The maximum capacity of the vacuum filter on this type sludge is said to be 8–10 tons per hour of filter cake.

The Link Belt and Komline Sanderson equipment were installed in October 1965 and have been operating satisfactorily since.

19.4 Existing Treatment Plant Efficiency

In an attempt to characterize the waste after primary treatment, a number of sampling and analysis programs were made. All of the samples taken were composited with flow, with the single exception of the diurnal sampling program, where grab samples were taken every half hour. Analyses were performed in accordance with standard methods.

The results of these analytical programs revealed that the composite waste, after existing primary treatment, was extremely variable in character. Five-day Biochemical Oxygen Demand values of the faw waste averaged 1500 mg/1, but ranged from 1000 to 3000. Chemical Oxygen Demand values varied from 2500 to 6000 mg/1 with an average of about 4000. Suspended-solid values averaged 1500 mg/1, with approximately 80 per cent being volatile suspended solids.

The diurnal variation of the waste was extreme, as illustrated in Figs. 19.2 and 19.3. As much as a three fold increase in BOD could be expected in only two hours and the pH of the waste was either above 11 or below 3 for most of the day. It became readily apparent that some form of equalization would be required for conventional treatment methods to be applicable.

As noted above, the major source of waterborne wastes is the milling process. The amount of material dissolved in milling (and hence discharged) is directly

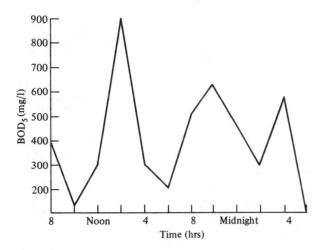

Fig. 19.2 Diurnal variation of biochemical oxygen demand.

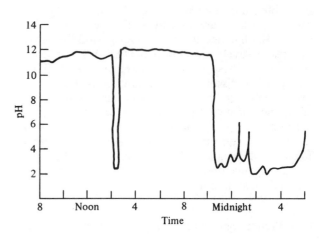

Fig. 19.3 Diurnal variation of pH.

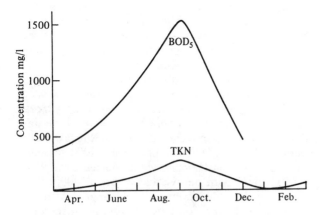

Fig. 19.4 Seasonal variation of biochemical oxygen demand.

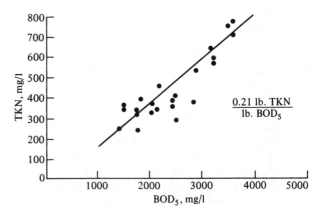

Fig. 19.5 Relationship between total Kjeldahl nitrogen and biochemical oxygen demand.

proportional to the water temperature and the conditioned of hides. This causes a marked variation in the daily average BOD discharged at different times of the year (Fig. 19.4). Figure 19.4 also illustrates the seasonal variation in total Kjeldahl nitrogen. In addition the relationship of total Kjeldahl nitrogen and five day BOD in the primary, wastewater effluent appears to be fixed as shown in Fig. 19.5.

19.5 Cattaraugus Creek Pollution Situation

From time to time Cattaraugus Creek showed visual evidence of contamination—that is, color, floating matter, turbidity, and odor. Of major significance were the periodic drops in oxygen levels in the stream below the Company's outfall. Despite several studies of the Cattaraugus Creek made by the author, which showed that an overall reduction of about 60 per cent BOD would maintain an oxygen concentration level of 4 P.M. at the critical sag point, New York State maintained in a letter (dated December 20, 1968) that "treatment on the order of 92 per cent removal of BOD would be satisfactory, although there was still some question of the allocation of the stream capacity among all users, Moench Tanning, Peter Cooper Corporation, and the Village of Gowanda. Nitrogen removal should also be considered in the Peter Cooper Corporation Studies." The stream's capacity and the allocation of treatment among the three major contributors of contamination were reevaluated by state officials. The following computations are from their letter of December 20, 1968, to the author.

Figure 19.4 shows the relationship between TKN and BOD_5 for the primary effluent. TKN values were approximately 21 per cent of BOD_5, a condition to be expected from a proteinaceous waste.

New stream parameters were calculated based on a critical stream temperature of 25°C and MA7CD flow of 60 cfs. The parameters are as follows:

$$K_2 = 2.58 \text{ day}^{-1} \text{ base } e$$
$$Kr = 0.88 \text{ day}^{-1} \text{ base } e$$

Considering the discharges from the three sources given above, it was found that the Class B section of Cattaraugus Creek will govern (minimum D.O. of 4.0 mg/l) the assimilative capacity. The assimilative capacity was calculated, with no capacity set aside for industrial growth, to be as follows:

Ultimate oxygen demand	Nitrogenous oxygen demand	NH_3 + Org. N	BOD_u
5760 lb./day	2970 lb./day	660 lb./day	2790 lb./day

Based on equal per cent treatment and allocations of the present discharge loadings, the allowable effluent loadings would be as follows:

Allowable effluent loadings				
	Peter Cooper	Moench Tanning	Gowanda STP	Total
BOD_u (#/day)	2500	255	35	2970
NH_3 + Org. N (#/day)	625	25	10	660
NOD #/day	2815	110	45	2970
UOD	5315	365	80	5760
Additional % treatment of BOD_u	93.9	93.9	93.9	93.9
Additional % treatment of NH_3 + Org. N	91.5	91.5	91.5	91.5

19.6 Corporate Relationship in the Community

Peter Cooper Corporation and Moench Tanning Company are the two major waste contributors in the Gowanda Village area. Both are old industries—the tannery is about 100 years old, and the glue plant is about 70 years old. The village owns and operates an old but quite efficient primary sewage-treatment plant. The Village (population about 3000 to 4000 persons) operates a rather strong mayoral-type government and a Village Board. The two industries employ about 500 people, which means that almost one person in each family is employed by one or the other of these industrial plants. In addition each family is indirectly effected by the activities of the industries.

However, to the Village-board members, the waste-treatment problems of the industries represent massive, incomprehensible, insurmountable areas of endeavor. It is the opinion of community that its industries can solve these problems and continue to operate and benefit the village community. In addition, there is apparently no interest on the part of the industries to cooperate with each other in waste-treatment processes. Peter Cooper corporation, the largest area liquid-waste contributor, has exhibited no inclination towards "joint" treatment of its wastes.

In a 1971 reevaluation of the capacity of the Cattaraugus Creek to assimilate wastes in the Gowanda area, the State made the following assessments:

| | Allowable loadings on the Creek in #/day | | | |
	Class D Class B –TOD–	NH₃ #/day	NOD #/day
Peter Cooper Corporation	6300 4525	463	2080

For a complete description of the characteristics and studies of Cattaraugus Creek, the reader is referred to another work of the author entitled Stream, Lake, Estuary, and Ocean Pollution, Chapter 10. Van Nostrand Reinhold Publishing Company, New York, 1991.

19.7 In-plant Studies

Since the glue manufacturing process is considered old and perhaps archaic, an in-depth assessment of production methods was deemed necessary. The primary objective of this phase of the waste-abatement program was to modernize plant production because it effects the volume and strength of wastewaters.

Samples of the Peter Cooper Corporation primary treated effluent were collected every 30 minutes and composited according to flow over a 24-hour period on 13 different days between July 30, 1968, and September 10, 1968. The untreated waste-flow averaged 3.331 mgd (see Table 19.1) and contained an average of about 1500 ppm BOD and 1400 ppm total suspended solids. Approximately 42–54 per cent BOD reduction is obtained by the existing treatment system as well as 65–80 per cent suspended-solids removal.

Normal production consists of 16 mills of fleshings, 16 mills of hide stock, and 24 mills of chrome stock per day. (See Table 19.1). The composite primary effluent averages a pH of 7–9 with settleable solids of from 5 to 63 ml per liter after 1 hour of settling.

Table 19.1 24-hour composite sampling of effluent (30 minute samplings)

Date (1968)	Total flow (Avg)		Production mills			Composite pH	Settleable solids 1 hour (ml/1)
	(GPM)	(MGD)	(Fleshings)	(Hide)	(Chrome)		
7/30–31	1880	2.710	(1)20	20	12	8.5	
7/31–1	1768	2.545	(1)20	20	12	9.0	12.0
8/5–6	2320	3.340	16	24	16	7.0	12.0
8/7–8	2650	3.815	12	24	20	8.0	8.0
8/12–13	2390	3.440	8	24	24	8.3	63.0
8/13–14	2410	3.475	8	24	24	8.5	38.0
8/14–15	2570	3.700	8	24	24	8.3	26.0
8/20–21	2400	3.580	8	24	18	8.5	27.0
8/19–20	2515	3.630	8	24	24	8.5	(2)124.0 35
8/29–30	1982	2.860	8	24	24	8.0	5.0
9/5–6	2325	3.350	8	24	24	8.5	2.3
9/9–10	2360	3.400	16	16	24	—	48.0
9/10–11	2400	3.460	16	16	24	—	22.0
Average		3.331					

(1) Plus 4 mills of sheep and goat trimmings.
(2) Before settling.

A. Findings

There are eight major waste-flows within the plant arising from the processing of the three raw materials of glue. These are shown in Table 19.2, which also shows the extreme differences in pH, color, BOD, COD, and suspended solids of each of the eight waste-flows. The samples reported in this table were collected according to flow and were composited over the entire waste-production period.

In Table 19.3 a comparison of the waste flows, BOD, COD, and suspended-solids loads is given for each of the three types of raw material.

Table 19.4 presents the waste-volume reductions, which were made possible by metered washing and controlled operation.

Table 19.2 Peter Cooper plant process wastes

BOD 5 day 20°C (ppm)	#/mill	Average Orig. total #/day	% of total BOD on a typical average day	COD (ppm)	COD BOD	COD #/mill	COD Avg. Orig. #/day	Suspended solids (ppm)	#/mill	Average #/day
7,300 (return 1200)	2310	36,960	55.0	14,046	1.93	4450	71,000	12,500	3,950	63
900	75	1,200	1.8	1,288	1.44	108	1,730	650	54	
2,220	408	6,528	9.7	2,977	2.45	1000	15,950	8,500	2,830	45
635	53	848	1.3	1,129	1.78	95	1,505	332	28	
1,195	30	720	1.1	5,658	4.73	142	3,400	2,500	63	1
(rerun 41 toxic) toxic Est.[1] 153	51	1,220	1.8	725	4.73	241	5,780	735	245	5
180	355	8,520	12.7	2,080	1.76	625	15,100	960	288	6
550	415	11,200	16.6	3,874	2,50	1035	27,900	100	27	
		67,196					142,365			120

*9 mills used 3 × per 24 hrs.
†est. 7000 gal./hr for 1½ hr.
1 est. based upon using a COD/BOD ratio of 4.73 as found in the No. 5, a similar, but stronger waste.

Table 19.3 Comparative volumes and loads

	Total flow Gallons/mill	(% of total)	Total BOD #BOD/mill	(% of total)	Total suspended solids #S.S./mill	(% of total)	Total COD #COD/mill	(% of total)
Fleshing	48,000	[1]23 [2]19	2385	[1]64.5 [2]61	4004	[1]33.5 [2]52	4996	[1]59.5
Hide stock (incl. salt mill)	82,000	[1]39 [2]33	876	[1]23.7 [2]22	2885	[1]40.0 [2]37	2130	[1]27.8
Chrome stock (incl. acid dump)	79,000	[1]38 [2]48	436	[1]11.8 [2]17	590	[1] 6.5 [2]11	1008	[1]12.7
Total of three processes Per Mill	[1]209,000 [2]248,500		3697		7485		7696	

(1) Based upon a ratio of 1:1:1 for mill production of fleshings, hide, and chrome stock.
(2) Based upon a ratio of 1:1:1.5 for normal mill production of fleshings, hide, and chrome stock.

Table 19.3   (*continued*)

Description of process	Flow of waste per mill average (orig. gal.)	Average number mills/day	Average original total flow per day	% of total flow on typical average day	pH	Color
Fleshing, wash out of lime	38,000	16	608,000	14.0	9.0	Grey
Fleshing, wash out of acid	†10,000	16	160,000	3.7	5.6	Light grey
Hide stock, wash out of lime	40,000	16	640,000	14.8	11.4	Light brown
Hide stock, wash out of acid	†10,000	16	160,000	˙3.7	5.8	Light grey
Chrome stock, direct drop of acid	3,000	24	72,000	1.7	0.6	Deep blue
Chrome stock, wash out of acid	40,000	24	960,000	22.2	1.3	Light blue
Chrome stock, wash out of lime	36,000	24	864,000	20.0	11.7	White
Hide stock, salt mill	32,000	*27	864,000	20.0	6.9	Dark grey
Total production	209,000		4,328,000	100.1		

Table 19.4   Reductions of flow by process change in washing out of lime and acid.

Type raw stock	Original or existing mill rate (gal./mill)	Anticipated or proposed new rate (gal./mill)	Reduction in volume			
			[1](%)	[2](%)	[3](%)	[4](%)
Fleshings						
Wash lime	38,000	25,000	34.3			
Wash acid	10,000	10,000	0			
Total.	48,000	37,000	23.0			
Hide stock						
Wash lime	40,000	20,400	49			
Wash acid	10,000	7,000	30			
Salt wash	32,000	32,000	0			
Total	82,000	59,000	28			
Chrome stock						
Wash lime	36,000	17,280	52			
Wash acid	40,000	28,800	28			
Direct acid drop	3,000	3,000	0			
Total	79,000	49,080	38.0			
Overall total	209,000	145,080	30.5	31.7	31.8	33.1

(1) Based upon 1:1:1 ratio of *fleshings* to *hide* to *chrome* stock in production.
(2) Based upon 1:1:1.5 ratio of fleshings to hide to chrome stock in normal existing operation.
(3) Based upon 25 mills of *hide* and 25 mills of *chrome* stock and none of fleshings to produce the same amount of glue as that produced by all 56 mills using 16 mills of fleshings.
(4) Based upon 20 mills of hide and 30 mills of chrome stock and none of fleshings to produce the same amount of glue as that produced by all 56 mills using 16 mills of fleshings.

Discussion of in-plant study for waste reduction:

1. The waste-flow at present from the Peter Cooper Corporation originates from chrome stock usage (48 per cent), from hide stock (33 per cent), and from fleshings (only 19 per cent).

2. As shown in Table 19.5, 79,000 gallons of waste per mill of chrome stock results in a production of 1550 pounds of glue or *51 gallons of waste per pound of glue*; 82,000 gallons of waste per mill of hide stock results in a production of 1500 pounds of glue or *54.7 gallons of waste per pound of glue*; 48,000 gallons of waste per mill of fleshings results in a production of 950 pounds of glue or *50.5 gallons of waste per pound of glue*. In other words, 50–55 gallons of waste are produced per pound of glue regardless of the type of raw stock (fleshings, hide, or chrome).

3. The BOD load at present from the Peter Cooper Corporation arises *61 per cent* from fleshings, *22 per cent* from hide, and only *17 per cent* from chrome stock.

4. 2385 pounds of BOD per mill of fleshings results in a production of 950 pounds of glue or *25 pounds of BOD per pound of glue*; 876 pounds of BOD per mill of hide stock results in a production of 1500 pounds of glue or only *0.28 pounds* of BOD per pounds of BOD per pound of glue. In other words about 0.3 to 0.6 pounds of BOD per pound of glue can be expected from chrome and hide stock and six times that amount for fleshings.

B. Directions for Waste-Volume and Strength Reductions

1. If Peter Cooper maintained its production constant in terms of pounds of glue produced but eliminated all production of glue using fleshings, little or no change in waste-flow would result; however, BOD, suspended solids and COD would decrease *80, 55,* and *71* per cent, respectively, when using hide stock in place of fleshings, and *89, 91,* and *88* per cent, respectively, when using chrome stock in place of fleshings.

2. In addition, extensive recent plant studies have shown that it is now feasible to reduce the flow of total waste from 30.5 to 33.1 per cent by more closely controlling the washing rates. This may or may not also result in a diminution of the pollutional loads, leaving more impurities in the product.

3. Preliminary laboratory results have shown that when the lime-wash wastes from hide stock and chrome stock are segregated, and held and treated with the spent, concentrated, acid-chrome waste, excellent precipitation, settling, and clarification of the waste results.

4. These studies show that if the use of fleshings were discontinued as a source of raw material and replaced with hide and/or chrome stocks, the improvement in the quality of the effluent over the present effluent would be the equivalent of secondary treatment of the present effluent. Still further improvement of the modified effluent (through elimination of fleshings) is believed possible through segregation of the spent acid-chrome-stock wastes and the controlled feed of such acid wastes to the alkaline lime-wash wastes to effect a co-precipitation and clarification of the mixed wastes, resulting in the removal of additional BOD with a resulting improvement in quality of final effluent.

19.8 Biodegradability of Glue-plant Wastes

A long-term biochemical oxygen demand was determined for the untreated glue-plant waste. The data

Table 19.5 Inplant studies, waste discharge as a function of raw material

Raw material	Flow		BOD$_5$		Suspended solids		COD	
	Gals Mill	Gals Lb glue	Lbs Mill	Lbs Lb glue	Lbs Mill	Lbs Lb glue	Lbs Mill	Lbs Lb glue
Chrome	79,000	51.0	436	0.28	596	0.40	1008	0.65
Hide	82,000	54.7	876	0.58	2885	1.92	2130	1.42
Fleshings	48,000	50.5	2385	25.00	4000	4.25	4558	4.80

obtained are plotted and shown in Fig. 19.6. The reader may note the similarity between the shape of this curve and that of normal domestic sewage. Glue-plant waste is certainly biodegradable with no apparent lag period and with an obvious nitrogenous stage of oxygen demand.

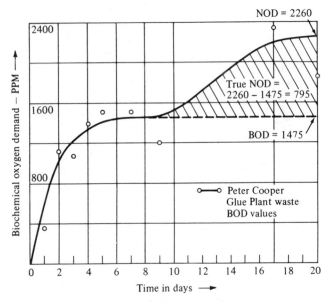

Fig. 19.6   BOD and NOD curves for Peter Cooper.

19.9 Pilot Plant Experimental Data

Three types of secondary treatment were studied in pilot plants to ascertain the potential degree of removal efficiencies: (A) Activated sludge, (B) Unox, and (C) Bio-Disc.

A. Activated Sludge Pilot Plant

The activated sludge pilot plant was operated as a single-stage unit with a detention time averaging 12 hours. The influent BOD was 1330 mg/l with an average effluent of 101 mg/l (92 per cent removal). TKN concentrations were reduced from 314 mg/l to 120 mg/l for an average removal of 62 per cent. The average F/M ratio during this period was 0.59 lbs BOD/lb of mixed-liquor suspended solids with values ranging from 0.42 to 0.86. A much lower F/M ratio was attempted by increasing the solids under aeration. However, continuing increases of the influent BOD

concentration resulted in a rather narrow band of F/M ratios. At the lowest F/M ratio of 0.42 with a low influent BOD of 735 mg/l, the effluent BOD was reduced to 38 mg/l, and the effluent TKN concentration was 26 mg/l. These were the best results recorded during the single-stage activated sludge pilot plant.

A two-stage activated sludge system was set up with detention times of 6 hours in each stage. The first stage was operated as a high rate system with F/M ratios between 0.5 and 1.5 lbs of BOD/lb of mixed-liquor suspended solids. The second stage was operated at a low loading rate to achieve maximum effectiveness in nitrification. The influent BOD during this period averaging 1550 mg/l was reduced to 207 ppm, (87 per cent removal). Influent TKN averages of 300 mg/l were decreased 62 per cent to an average of 114 mg/l. The F/M ratio over both stages was 0.65 lbs of BOD/lb of mixed-liquor suspended solids. The best results obtained were an effluent of 69 ppm TKN with a BOD concentration of 34 ppm.

This two-stage activated sludge system was modified to provide a detention time of 12 hours in the first stage, and thus a lower F/M loading. BOD concentrations decreased from 1266 mg/l to 138 mg/l for an average removal of 89 per cent. The average effluent ammonia concentration was 70 ppm and it is estimated that all of the TKN had been converted to ammonia. No TKN analyses are available for this run. On September 30, the effluent BOD was 111 ppm and the effluent ammonia concentration was 26 ppm. This was the last day of this case, and it is estimated that if this case had been pursued for an additional week to ten days to further acclimate the nitrifying bacteria, the results would have been acceptable. The average F/M ratio during this case was 0.32 and the individual values varied from 0.23 to 0.40.

In summary, a review of the three pilot plants using the air-fed activated sludge system shows that substantial nitrification can be accomplished by either the single-stage or the two-stage activated sludge system. However, in either system, the loading rates must be in the lower range so that net sludge production is minimal, thereby increasing the sludge age of the system. While ammonia concentrations were not lowered to the desired 18.5 ppm in the effluent during any of the days of operation, it is anticipated that effluent ammonia concentrations would be achieved by further operation at lower loading rates.

B. Unox Pilot Plant

A two-stage Unox pilot plant maintained high loadings for approximately one month in an effort to build up mixed-liquor suspended solids in both stages and to acclimate the nitrifying bacteria in the second stage. The system was operated most successfully with approximately 6 hours detention in each stage. The average influent BOD of 1873 mg/l was reduced 90 per cent to 42 mg/l. The average TKN of the influent was 343 ppm and the effluent TKN concentration averaged 92 ppm for an overall removal of 73 per cent. The average F/M ratios were 0.72 and 0.23 for the two stages. Since BOD values were taken on the effluent from stage one, we were able to calculate efficiencies of removal and loading factors on both stages independently. Mixed-liquor suspended-solids values in the oxygen system were maintained at much higher levels than in the air system mentioned earlier, while the detention times were approximately the same. Mixed-liquor suspended-solids values were 10,300 and 8,500 mg/l. Sludge age for stage one was 5.5 days and for stage two was 10 days. The lowest ammonia concentration was approximately 70 mg/l in the effluent from stage two. The average temperature during this time was approximately 25°C, which indicated that reduced temperature did not play an important part in reduced efficiency. Dissolved oxygen concentrations during this period averaged 15 mg/l and the pH of the mixed liquor averaged 6.0 and on the final day of operation dropped as low as 5.1. It was felt that the low pH in the mixed-liquor was partially responsible for the reduced efficiency of the ammonia removal.

A single-stage Unox pilot plant was again operated for approximately one month before collecting data. The initial data revealed substantial effluent nitrification occurring. The detention time in the system was 14.5 hours. The average BOD concentration in the influent was 1800 mg/l with an effluent of 13 mg/l for an average removal of 98 per cent. TKN concentrations were reduced 71 per cent from 333 mg/l to 98 mg/l with an average F/M ratio of 0.17. The last two days of operation under this case showed complete nitrification with effluent values of 20 mg/l and 10 mg/l respectively of TKN and ammonia. Effluent values of nitrate at this time were very high and revealed that approximately two-thirds of the TKN had been completely nitrified. As in the two-stage system,

dissolved oxygen concentrations were maintained very high, and averaged 15 mg/l. The pH of the mixed-liquor suspended solids in this single unit was approximately 6.0 and it dropped to only 5.9 on the final day of operation. It was felt that if the single-stage system had been operated for approximately one to two more weeks, complete nitrification would have occurred.

In reviewing the data from both the two-stage and the one-stage system, using oxygen, it became apparent that the single-stage process achieved much lower suspended solids concentrations in the effluent (15 mg/l compared with 50 mg/l for the two-stage system). Sludge volume indexes averaged approximately 35 for both stages of the two-stage system and 20 for the single unit. Due to the limited time of operation after complete acclimation of both systems no definite conclusions can be drawn as to which system would result in the highest efficiency of nitrification.

C. Bio-disc Pilot Plant

The feed rate of the Bio-disc pilot plant was initially set at 5 gph to accomplish a detention time of approximately 5 hours. The initial disc speed was 11 rpm. As the waste strength gradually increased, and dissolved oxygen concentrations in the unit dropped, a higher disc speed of 21 rpm was adopted. Neither case produced substantial degrees of nitrification. However, BOD removals averaged approximately 88 per cent.

The flow rate to the unit was then reduced to 2.5 gpm, thus providing a 10 hour detention. Influent BOD values during this period averaged 1475 mg/l with an effluent of 142 mg/l (average removal 90 per cent). TKN concentrations, however, were only reduced from 343 ppm to 232 ppm. Individual TKN removals ranged from a low of 9.3 per cent to a high of 56.4 per cent during this study. At this time, no conclusions have been reached regarding the reasons for this wide variation in results. In summary, the Bio-disc acted as a very efficient BOD removal process. Nitrification was difficult to obtain on this particular waste stream at the loadings employed and pilot plant configuration used.

19.10 Preliminary Design

While all the pilot plant nitrification data is not nearly as conclusive as desired, because of time limitation, it was necessary to proceed with preliminary design of treatment systems that could achieve the effluent concentrations required. Several alternative secondary treatment schemes were proposed. Each of the treatment schemes utilized the existing primary treatment plant and was arranged in such a way as to facilitate the use of several abandoned steel and concrete tanks on the plant site. Those treatment schemes seeming to have merit included:

1. Two-stage activated sludge, using air with detention times of 12 hours in each stage.
2. Two-stage activated sludge using oxygen with detention times of 6 hours in each stage.
3. Activated sludge using air with 36-hour detention (long-term aeration).
4. Single-stage activated sludge using oxygen with a detention time of 18 hours.
5. Bio-disc roughing with 6 hours detention, followed by activated sludge nitrification with a 12-hour detention.

The final choice of processes would seem to be one based mainly on economics including maximum reuse of certain existing tankage.

In reviewing the literature on nitrification it was found that temperature plays an important role in determining the rate of nitrification. Since the plant was located in a climate characterized by severe winter conditions, a considerable amount of thought was given to maintaining heat in the aeration tanks. It was discovered that a sizeable stream of hot clean water from a barometric condenser was being discharged continuously back into the receiving water at 112°F. Preliminary heat transfer calculations were made and it was determined that by routing this stream through a series of closed pipes in the aeration tanks, the temperature of the mixed-liquor could be maintained at 50°F or above at all times during winter operations. It is felt that such a system not only promotes nitrification, but also eliminates serious freezing conditions, which might affect the aerators.

19.11 Final Decision of Glue Plant

Cost analysis for secondary treatment revealed that all alternatives would result in a capital expenditure of between $1.8 and $4.1 million. The Bio-disc system may have resulted in the lowest annual total operating cost of about $67,000. The lowest capital cost was estimated to be $1.8 million for lagoons with mechanical aeration. After repeated board meetings of the Peter Cooper Corporation they arrived at their decision based largely upon the capital costs for treatment. The company decided that it could not afford to incur this additional environmental expenditure in order to continue to produce its current product, animal glue, at this location. The plant ceased production.

19.12 Characteristics of Tannery Waste

The Moench Tanning Company is located in the Village of Gowanda, New York, and immediately adjacent to and on the banks of the Cattaraugus Creek. It produces chrome-tanned upper shoe leather with saving of hair and flesh grease. The upstream creek water is relatively clean and flowing at a mean annual flow rate of 212 cfs as measured by a USGS a few miles downstream. The company is located about 1 mile upstream from the village center. The village possesses its own primary sewage treatment plant which operates efficiently but is running at about design capacity. The Peter Cooper Animal Glue Manufacturing Plant is located immediately downstream of the company.

The creek's best usage is that of fishing, and although repeated stream surveys have shown that about 50% removal of BOD would protect the stream for this purpose, New York State Department of Environmental Conservation has required, in general, a minimum of secondary treatment.

The wasteflow and strength characteristics of the tannery have been established by many measurements to average as follows:

Hides processed per day = 74,300 lb
Flow of wastewater = 253,000 gal/day
BOD (5 day, 20 C) = 1,323 mg/l
Total Kjeidahl Nitrogen = 180-445 mg/l
Suspended solids = 1,530 mg/l
Chromium = 0.1 mg/l Cr^{+++}
$\qquad\qquad$ < 0.07 mg/l Cr^{VI}
ph = 10.9
k_1 = 0.076

The typical flow of wastewater pattern during a given 24 hour period is shown in Figure 19.7. The tannery owns and operates a bar screen and two rectangular settling tanks operating in series. Each of these tanks are simply holding tanks and possess no provision for sludge removal. The effluent from the top of the second basin (as well as the sludge periodically) is discharged continuously into the creek.

19.13 Laboratory Prototype Results

A laboratory pilot plant was built and operated on a continuous basis during 1968. The results were published in 1969 [6]. The pilot plant consisted of a high solids biological aeration unit fed a continuously mixed, unsettled, composite of the Moench Tanning Company waste. In general, the findings were that the composite leather plant waste could be treated successfully *without primary settling* and *without segregation* of the beamhouse and tanhouse wastes. And furthermore, the findings showed that 92% BOD reduction could be obtained at elevated loadings over 200 lb of BOD per 1000 ft^3 per day. And, in addition, the suspended solids were separable effectively in the final settling tank. All of this was made possible by feeding a high pH waste and maintaining high suspended solids concentration (12-16,000 mg/l) in the biological aeration basin.

19.14 Field Pilot Plant Results

The separate system called for by the results of the laboratory prototype experiments was a relatively new one and untried in actual practice. Waste treatment at tanneries had not been practiced to any extensive degree before this time. And what systems had been constructed utilized separate tan and beamhouse treatment. It was decided, therefore, to construct and operate a field prototype at the site of the tannery. This plant would receive a small percentage of the tannery's actual wastewater (normally no less than 5% or 10 gpm, whichever was the smaller quantity) on a 24-hr daily basis, 7 days each week, and at least representative cold and warm weather seasons.

The field prototype treatment plant was operated over a six month period from the late fall of 1970 to early summer of 1971. The treatment system consisted of a waste collection and overflow chamber into which a centrifugal pump extracted a continuous flow of approximately 5% of the total plant wastewater flow. The pumped flow was directed to an equalizing basin the contents of which was kept in constant motion by a lighting-type mixer for a 24-

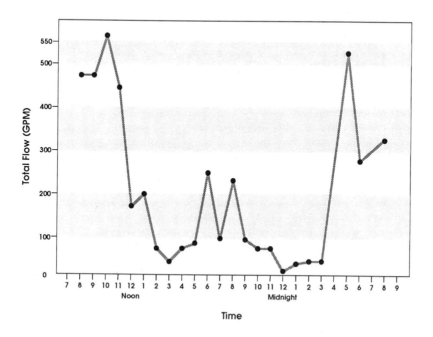

Fig. 19.7 Typical hourly total waste flow of Moench Tanning Company (January 16, 1968).

hour period. The equalized waste flowed continuously to a high-solids, biological aerator which was supplied air by means of a floating, mechanical aerator. Detention time in the biological unit was about 12 hr; MLSS 10–15,000 mg/l, and pH 8.3–9.0. The effluent was settled for two hours in a final sedimentation basin. The sludge was pumped to a compaction tank for final concentration prior to disposal by landfill.

Some difficulties were encountered during the operation of the prototype plant. For example, the floating mechanical aerator froze up due to the cold weather and was replaced with a compressed air system. Some solids clogging of air diffusers also occurred. These were problems which could be rectified in the design of the full-scale treatment facility.

These studies proved the laboratory results that an equalized chrome leather tannery waste could be treated effectively without prior sedimentation at a pH of greater than 10.5 by biological aeration. The BOD and suspended solids reductions, although slightly less than that obtained in the laboratory studies, averaged 80–90%.

It was agreed, then, to design a full-scale treatment plant for the entire plant wastewater.

19.15 Treatment Solution and Methods Selected

A plant serving only the tannery at the tannery plant site using combined treatment of equalized beamhouse and tanhouse wastes was selected as a solution to the Moench Tanning Company wastewater problem. This decision was based largely on three factors: (a) the waste could be effectively treated separately to remove excessive contaminants; (b) there existed no desire to cooperate on a combined plant on the part of the other two major contributors; and (c) a separate treatment system could be designed at the lowest capital cost to the tannery.

19.16 Design and Performance of Treatment Plant

The full-scale wastewater treatment facility was designed and constructed in 1971 and began operating in June of 1971. It was designed to contain equalization for one day's flow, followed by biological aeration with no pH adjustment of the pH 11 equalized influent for about 12 hr. The aeration basin effluent is settled and the sludge is withdrawn to a compaction tank from where it was periodically pumped out to a truck for conveyance to the company's sanitary landfill near the plant. Some sludge is recirculated from the compaction

tank to the biological aeration in order to keep a MLSS concentration of about 12,000 mg/l.

The results of the first 26 month average operation of the treatment plant is shown in Table 19.6. Some interesting loading parameters were also computed. They are presented in Table 19.7.

The MLSS averaged 11,825 mg/l of which 45% were

Table 19-6.   Treatment Plant Performance

	Plant Influent	Plant Effluent	Reduction (%)
pH	10.6	8.1	—
Temperature	68.2 F	—	—
Suspended solids, mg/l	2,763	314	88.7
BOD$_5$, mg/l	1,587	362	77.3

Table 19-7.   Loading Parameters

	Plant Influent	Plant Effluent
# BOD/1000 # HIDES	50.57	11.59
# BOD/1000 ft^3/day	139.2	—
F/M	0.200	—
# Suspended solids/1000 # HIDES	—	9.8

organic. The total Kjeldahl nitrogen in the effluent averaged about 150 mg/l.

The reduced efficiency in both BOD and suspended solids reductions, when compared to laboratory and field prototype experiments, was attributed to the following three problems:

1. A change in production processes by: (a) replacing dimethylamine sulfide with lime and sodium sulfide for unhairing; (b) introducing a grease plant effluent wastewater—resulting from processing fleshings—into the wastewater stream; (c) adding leather buffing waste to the effluent; and (d) an increase in hide processing from 70,000 to 100,000 lb/day.

2. A scale-up effect in the final sedimentation basin resulting in velocity currents and subsequent loss in solids over the weir and into the effluent.

3. The inability to pump sludge continuously from the bottom of the sedimentation basin and into the sludge concentration basin. Intermittent pumping at high rates caused disturbance with solids settling.

After all attempts had been made to rectify the effects which the changes in production made, the writer began to combat problems 2 and 3. Field studies made in August and September of 1974 established the relationship between the aeration basin pH and MLSS and the suspended solids in the plant effluent. These relationships are shown in Figure 19.8. Operational procedures were altered accordingly.

The final sedimentation basin was changed by adding a shallow baffle plate ahead of the effluent weir and the approach velocity of wastewater influent was reduced by better horizontal distribution.

A second sludge compaction basin was built so that sludge could be pumped on a continuous basis from the settling basin. One could be compacting sludge while the other received sludge.

Laboratory experiments were carried out to determine what coagulents could be used to enhance sedimentation of the MLSS entering the basin. It was found that 100 mg/l of ferric sulfate improved the removal of suspended solids to the point where consistent concentrations of less than 100 mg/l of suspended solids was present in the laboratory effluent. The coagulent had to be added to the aeration basin effluent and flocculated properly prior to final sedimentation. Two chemical feed tanks and a flocculating basin were constructed to serve this purpose.

All of these full-scale alterations in both the construction and operation of the treatment plants were completed by the fall of 1976. Since that time, the plant has given much better efficiency in BOD and suspended solids removal.

Creek Water Quality After Treatment

Cattaraugus Creek water quality has improved dramatically since the advent of the tannery waste treatment plant. A certain and significant part of this improve-

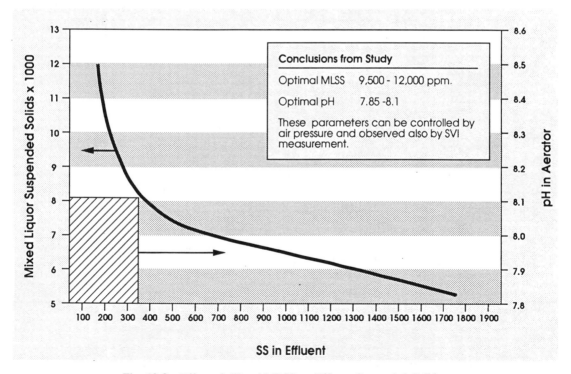

Fig. 19.8   Effect of pH and MLSS on Effluent Suspended Solids

ment is due to the closing of the major contributor of contaminants, the Peter Cooper Glue Manufacturing Plant.

As can be seen from the creek banks and bridges, the creek is now clear, oxygen abundant again, and supporting fish life to the extent never before realized. In fact, the tannery is now very careful and very conscious about its wastewater discharge since the recreational value of the creek has increased so greatly. The efficiency of the treatment plant is presented in Table 19.8.

Table 19-8. Treatment Plant Operational Efficiency Since 1976

	Plant Influent	Plant Effluent	Reduction (%)
pH	10.6	8.1	—
Temperature, °F	68.2	—	—
Suspended Solids, mg/l	2,763	131	95.3
BOD_5, mg/l	1,587	226	85.8

19.17 Significant Design and Operation Considerations

Diffused air system was selected finally because it preserved and conserved heat especially important during cold winter temperatures. Surface aerators were found to be not viable on activated tannery sludge in typical Western New York State climate.

Preliminary tannery waste treatment with Hydrasives was installed with screen sized in two units of 0.04 in. These reduced the BOD loading on the treatment plant by 8–12%. In addition, many clogging problems in treatment plant equipment were eliminated.

Positive displacement pumps were selected because they are more accurate and variable and were used as a check on the flow rate.

Several chemicals were tried to lower BOD, provide oxygen, and to enhance sludge dewatering but were not found acceptable; *i.e.*, (a) H_2O_2 could not be justified on a cost basis; (b) activated carbon was too costly and produced little filtration efficiency; and (c) several polymers were also found to be outperformed as sludge conditioners by both ferric chloride and ferrous sulfate.

Aluminum handrails and gratings are recommended to minimize maintenance.

Moench Tannery production operations were im-

proved by the optimization of performance of the waste treatment plant. This was possible as a result of the conservation practices which were forced on the company. Two examples of these are: (a) chromium recovery and reuse; and (b) sulfides recovery and reuse from the Beam House Liquors.

COST OF TREATMENT

Operational costs for 1977 were as follows:

Electricity	$24,000	*Does not include management nor administrative time; but includes all other labor costs for treatment plant including sludge dewatering and landfilling.
Direct Labor*	$30,000	
Chemicals	$65,000	
Maintenance	$25,000	
Total	$144,000	

*These costs do not include amortization of capital expenses. The total capital cost with improvements and modifications is approximately 1 million as of 1978. Production for 1977 ~ 15,000,000 ft. of hides (plus splits), thus waste treatment production costs were: 14,400,000¢/15,000,000 ft. of hide, or about 1¢/ft. of hide.

Evaluation of this Solution

We have seen and studied a very unusual waste treatment problem. It has been solved satisfactorily in what may be considered by many as an unconventional manner. The uniqueness of the problem and its solution can be summarized as follows:

- It was solved by separate treatment of the tannery plant waste rather than in combination with the other area-wide wastes which seemed on the surface to be more feasible.
- It was solved by combining the beamhouse and tanhouse wastes rather than treating them separately and differently.
- No primary sedimentation was used; instead, all wastes were equalized and kept in motion so as to prevent settling.
- No pH adjustment of the highly alkaline combined wastewater was utilized prior to biological treatment.
- A mixed liquor suspended solids concentration of from 9,500 to 12,000 mg/l was found to be optimum for maximum BOD reduction and suspended solids removal.
- Chemical coagulation following biological aeration was found to be effective and necessary to attain sufficient BOD and suspended solids reduction without resorting to additional tertiary treatment.

References

1. LaGrega, Michael D., "The animal glue industry: waste characterization and treatment," MS Thesis, Syracuse University, 1971.
2. Nemerow, Nelson, L., "Liquid wastes of industry: theories, practices and treatment," Reading, Mass.: Addision-Wesley, 1971.
3. "Standard methods for the examination of water and wastewater," American Public Health Association, 1971.
4. "Status of Unox nitrification," Internal Report, Union Carbide Co., May, 1971.
5. Wild, H. E., C. N. Sawyer, and T. C. McMahon, "Factors affecting nitrification kinetics," paper presented at the 43rd Annual Conference, Water Pollution Control Federation, Boston, Mass., 1970.
6. Emerson, D. B. and Nemerow, N. L. "High Solids, Biological Aeration of Unneutralized, Unsettled Tannery Wastes," *Proc. 24th Purdue Univ. Industrial Waste Conf.*, May 7, 1969.

Questions

A. For Glue Plant
1. Describe the background of this problem of waste treatment.
2. Is the waste treatable?
3. What appears to be the most difficult technical problem associated with treating this waste?
4. Summarize the steps which were taken to provide enough data to facilitate making a decision on waste treatment.
5. What were the real alternatives open to the company?
6. When does it become more prudent to cease all operation to avoid waste-treatment costs?
7. What decision did this industrial plant arrive at after all the facts were clarified?
8. Explain in your own words why you believe this plant chose the alternative it did.
9. Could the decision to close have been avoided as a reasonable course of action?

B. For Tannery
10. What are the significant pollutional characteristics of this tannery waste?
11. How was treatability ascertained?
12. Give at least six unique aspects used in the solution to this tannery waste?
13. What is the approximate cost of this tannery waste treatment (expressed as a percentage of production cost or sales cost of leather)?

C. For The Overall Problem
14. Does the Environmental Impact Statement analysis shown in Chapter 1–4 yield quantitative response comparable to that actually taken in this case?

CHAPTER 20

DISCHARGE OF PARTIALLY TREATED
INDUSTRIAL WASTES DIRECTLY TO STREAMS

Large industries located outside city limits often have water requirements so great that they must develop their own sources of water and likewise they must dispose of their own wastes. This is one price an industry must pay to obtain sufficient space and escape municipal taxes. There are scattered instances, of course, of cities extending their sewer lines to accept the wastes of a nearby industry; but usually a plant draws its process water from a nearby river or well and discharges its wastes to the same stream, after a careful analysis of the uses of the stream and its condition. Wastes from these large plants contain so much pollution that some treatment is required before discharge into the stream. Since treatment of large volumes of waste water is expensive, an industry should investigate many alternative methods of protecting the receiving waters. Although such studies require time-consuming survey, analysis, and evaluation, an industry cannot bypass these steps without spending an excessive amount of money for waste treatment. In short, the more the industry knows about its wastes, generally the lower the cost of the actual treatment.

AN EXAMPLE OF DISCHARGING WASTE TO A STREAM AFTER PARTIAL TREATMENT

A large textile mill is located on a river, in a small mill town with a small Imhoff tank providing sewage-treatment facilities. The stream receiving the mill waste has been classified C (having fish survival for its best usage). Two dams downstream from the mill cause impoundment of the stream and subsequent nuisance conditions. The mill takes its water supply from the stream just above the plant, where it maintains a reservoir containing about one billion gallons of water when full. The state has directed the industry to maintain the stream condition as classified, so it has little choice except to treat its wastes before discharging them directly to the stream. Since extensive treatment of such large quantities of waste as there are in this instance can be economic suicide in many cases, a great deal of planning, research, and analysis must be carried out in order to minimize costs and yet attain the best possible efficiency. The state has granted the mill a six-month period of grace, to set up a pilot plant to determine the feasibility of biological treatment and aeration as means of reducing the BOD of the mill waste.

With the above problems in mind, the textile mill asked the author to supervise the investigation and to analyze, interpret, and evaluate the results of the six-month pilot-plant study. In order to make positive recommendations to management, the author also made in-plant and river studies and held conferences with management to impress upon them the following facts.

Waste treatment must be considered an integral part of production. In manufacturing, the objective is to produce, with the least expense, an article which will satisfy the consumer. Similarly, in waste disposal, the objective is to reduce, with the least expense, the pollutional nature of the effluent, in order to maintain the standards of the receiving river. Waste treatment which costs more than it should or does more than is required is contrary to the laws of good business; similarly, inadequate waste treatment reflects poor business judgment, since public hostility breeds sales resistance and a bad corporate image. A plan for waste treatment in this instance must, therefore, include the minimum of aeration, neutralization, and retention which will still accomplish adequate reduction of the pollutional load on the river.

In addition to these three pollution-reducing

devices, certain microorganisms can be developed during aeration that will assist in decomposing the organic matter in the waste. But these bacteria require the proper environment, two important features of which are air and pH, and the maintenance of sufficient dissolved air and a near-neutral pH costs money. For instance, the amount of detention time required determines the size of the holding tanks.

20.1 Procedure

The pilot plant for the textile-mill waste consisted of a holding basin, an aeration tank, and a final settling basin (Figs. 20.1 to 20.4). It took six months of operation to decide whether these devices, could—at a reasonable cost—maintain the "fishing" classification of the river. The six-month study was divided into four separate investigations: (1) efficiency of the pilot plant; (2) characteristics of the receiving river; (3) substitution of soluble for insoluble sizing, to reduce the strength of the waste; (4) volumes and loads of individual waste lines. The pilot plant was operated continuously over the entire period. Rates of flow of air and waste were varied and BOD was used as the criterion of treatment efficiency. The results are presented in Table 20.1.

The river was sampled during a two-week period in mid-October and again for one day in November. The river flow was purposely maintained at about 15 mgd, the present minimum flow, by means of overflow control at the reservoir. The results are presented in Table 20.2. The oxygen-sag curve and BOD profile are plotted in Fig. 20.5. Monthly averages of the flow of the river for the last five years are given in Table 20.3.

Experiments in starch substitution were carried out by the author: carboxymethyl cellulose (CMC) and Penfer gum 300, two substitutes for starch, were compared with pearl starch, the sizing most commonly

Fig. 20.1   Waste-treatment engineer observing high pH of incoming waste.

Fig. 20.2 Side view of pilot plant built to test methods of treating textile-mill wastes.

used. The results are shown in Tables 20.4 and 20.5. Samples taken from the waste pipeline were collected and analyzed when the mill was using starch only and when it was using starch-substitute products. These results are presented in Table 20.6.

A 24-hour survey of all four waste lines and of the combined effluent was made on a normal operating day, August 22; the flows and BOD's of each are given in Table 20.7. There are four main wastes from this plant: dye, starch, kier (or bleach), and desize (Table 20.7). The desize waste, the main offender, contributes 60.8 per cent of the BOD load and 40.5 per cent of the flow. Since the other three wastes contribute smaller but significant flows and BOD loads, segregation does not appear feasible, especially because of the high BOD reduction required. The total flow measured by means of a weir box was 6.09 mgd on the test day. The total BOD load was between 32,000 and 35,550 pounds per day. The waste was highly colored, warm, and had a composite pH of about 11 or higher. It definitely needed to be treated for the stream standards for the river (Table 20.8) to be maintained. The quality standards for this class of river water are

described in Table 20.8. See Chapter 25.1 for characteristics of textile wastes.

20.2 River Studies

The oxygen curve of the river below the mill showed that the bottom of the sag occurred at the first dam (Station 3) (see Fig. 20.5). Below this point the river showed some recovery of oxygen. The plotted results illustrate that the critical stretch of the river was from the mill outfall to the first dam. This represents a distance of only about two miles and an average flow time of about 450 minutes ($7\frac{1}{2}$ hours). The BOD in this two-mile stretch diminished precipitately, indicating a rapid utilization of oxygen. The BOD of the river, determined from an 8-day average (Fig. 20.5), was 360 ppm just below the mill effluent and only 219 ppm at the dam. This means that a reduction of 141 ppm of BOD (61 per cent) was taking place in this critical stretch. If the *initial* river BOD could be reduced to 141 ppm and the organic matter reduced at the same rapid rate, theoretically there would be none left at the first dam. However, since organic matter

Fig. 20.3 Operator observing anemometer, which registers total air being fed to the wastes in aeration tank.

remaining after treatment probably will oxidize more slowly, some would remain at the first dam.

Deoxygenation rates of 0.3 to 0.6 per day can be attributed to the decomposition of the organic matter between the mill outfall and the first dam. Although reaeration rates cannot be accurately determined in this stretch because of the absence of any oxygen at Station 3, extremely high values are evident from calculations using certain assumptions. The use of these approximate K_1- and reasonable K_2-values resulted

in BOD reductions of 60 to 70 per cent, which were necessary to maintain the required 2 ppm of dissolved oxygen. This is expressed graphically in Fig. 20.5. A line parallel to the BOD curve was drawn so as to give zero BOD at a point just above the first dam. A BOD of about 120 ppm (rather than 360 ppm) is required at Station 2, which constitutes a BOD reduction of 67 per cent. Thus it can be shown, both numerically and graphically, that 60 to 70 per cent BOD reduction must be achieved to maintain 2 ppm of dissolved

Fig. 20.4 Aeration basin, viewed from above. Final settling basins are at the top of the picture.

oxygen at Station 3. The addition of clean, highly oxygenated dilution water during low flows can be construed as equivalent to a certain BOD reduction. For example, at normal deoxygenation rates, 1 pound of dissolved oxygen will be utilized during the first day by about 5 pounds of 5-day BOD. As stated before, the river studies were made at the present minimum flow of 15 mgd; any additional flow, carrying clean water, would naturally increase the total oxygen assets of the river. The following computations clarify, by the use of actual figures, the value of clean-water storage upstream and the resulting increased minimum flow available below the mill.

The effect of river dilution on BOD reduction is as follows. Keeping in mind the fact that there are 8.34 pounds of water in a gallon, with present river conditions (storage upstream, one billion gallons; summer minimum discharge, 15 mgd), we find

Table 20.1   Summary of six months of operation of pilot plant.

Treatment or characteristic of waste	Waste-flow rate, gpm						
	2	2	2	3	4	4	6
Holding-basin detention, hr	27	10	10	6	5.25	5.25	3
Aeration-tank detention, hr	12	12	12	8	6	6	4
Final settling basin detention, hr	8	8	8	5.5	4	4	2.75
Air rate, cfm	90	90	60	60	60	60	86–95
Air, ft³/lb BOD	5400	5400	3600	2650	2400	2065	2130
Number of 4-hr composites collected	68	16	18	19	22	16	30
BOD of raw waste at 20°C, ppm	834	840	798	842	750	872	822
BOD of primary settled waste							
ppm	374	710	711	701	700	755	730
% reduction	56.5	15.5	10.9	15.5	6.7	13.4	11.2
BOD of aerated waste							
ppm	210	196	232	333	346	569	419
% reduction	74.8	76.9	71.0	60.4	53.9	34.8	49.0
BOD of final settled waste							
ppm	198	193	194	330	350	583	402
% reduction	76.3	77.1	75.9	60.9	53.8	33.1	51.0
CO_2 used per day* prior to aeration, lb							
Minimum	0	0	3	12	27	0	43
Average	2	10	17	48	47	0	65
Maximum	10	62	52	121	76	0	94

*To reduce pH to about 9.0

15 mgd × 8.34 lb/gal × 8 ppm DO upstream
$$= \sim 1000 \text{ lb } O_2/\text{day}$$

available in the river. At the normal deoxygenation rate of 0.1 per day, this oxygen, exclusive of reaeration, will take care of a pollutional load containing about 3300 pounds of 5-day 20°C BOD (assuming only about 30 per cent of the 5-day BOD will be satisfied in the first day of critical flow time).

Changing the river conditions by increased dilution (storage upstream, 1.5 billion gallons; summer minimum discharge, 22.5 mgd), we have

22.5 mgd × 8.34 lb/gal × 8 ppm DO upstream
$$= \sim 1500 \text{ lb } O_2/\text{day}$$

available in the river. At normal deoxygenation rates of 0.1 per day, this oxygen, exclusive of reaeration, will take care of a pollution load containing about 5000 pounds of BOD. Therefore, instead of a 70 per cent reduction (2300/3300) being required to meet

stream standards, only 2300/5000 (about 45 per cent) will be necessary. In other words, this dilution results in approximately 24 per cent BOD reduction.

20.3 Pilot-Plant Results

The pilot plant, consisting of a combination of holding, aeration, and final-settling basins, showed BOD removals of 33 to 77 per cent. A total of 189 four-hour composite samples were collected and analyzed during the 6-month period. The percentage of BOD removal depended on the detention time of the mill waste in the holding basin, the quantity of aeration applied to the waste, and the degree of pH reduction (Table 20.1). Holding the waste for 10 hours and aerating it for 12 hours at 60 cfm resulted in an overall BOD reduction of 71 per cent. This would be an acceptable method of treatment without any other disposal devices. However, the cost of installing and maintaining equipment to produce such great amounts of air and CO_2 would

Table 20.2 Results of sampling of river receiving textile-mill wastes.

Sampling days*	Station number and location	Flow, mgd	Dissolved oxygen		Temper- ature, °C	BOD (20°C, 5-day), ppm	pH	Time of flow to next station, min	Accu- mulated time, min
			ppm	% saturation					
October 11–14†	1. Water plant	15.4	8.9	89	16	0.75	6.8	0	0
	2. Just below waste entry	15.4	3.7	42	22	400	10.0	75	75
	3. First dam	15.4	0	0	23	180	7.4	375	450
	4. Bridge	15.4	0.3	3.2	18.5	160	7.3	90	540
	5. Second dam	15.4	1.9	19	16	142	7.4	720	1260
	6. Bridge	15.4	0.3	3	15	122	7.4	75	1335
	7. Bridge	15.4	1.6	15.7	15	100	7.4	130	1465
October 18–21‡	1. Water plant	16.0	8.9	84	13	0.50	7.1	0	0
	2. Just below waste entry	16.0	3.8	41	20	320	10.3	75	75
	3. First dam	16.0	0.1§	0	19	258	7.9	375	450
	4. Bridge	16.0	0.0	0	17	202	7.6	90	540
	5. Second dam	16.0	2.4	23	13.5	171	7.4	720	1260
	6. Bridge	16.0	0.8	7.4	12	131	7.4	75	1335
	7. Bridge	16.0	3.6	33.3	12	145	7.5	130	1465
November 16	1. Just below waste entry	14	8.0	94	24	280	10.7	0	0
	2. Just below waste entry	14	4.6	52.3	22	240	10.4	30	30
	3. First dam	14	0.0	0	20.5	200	8.0	375	405
	4. Just below dam	14	0.0	0	18	190	7.2	60	465

*All figures are averages of four consecutive days of stream sampling.
†With chlorination of sewage from Imhoff tank below Station 1.
‡No chlorination of sewage from Imhoff tank below Station 1.
§One sample showed 0.4 ppm; three showed 0.0 ppm.

be prohibitive. The mill would do well to investigate the possibility of using boiler-flue gas as a source of CO_2 and giving less aeration time to the wastes. It was found that little or no CO_2 was needed prior to aeration when the waste was held for a 24-hour period. One experiment, using a 5-hour holding period and 6 hours of aeration without any neutralization with CO_2, gave about 35 per cent BOD reduction (Table 20.1). Several alternative plans for disposal of the wastes, which would achieve 60 to 70 per cent BOD reduction, were suggested, based on the pilot-plant results.

1) Lagoon 27 hr + substitution of soluble sizing on 40% of grey goods

BOD reduction: (56.5%) + (14.5%)* = 71%

2) Lagoon 27 hr + aeration 4 hr

BOD reduction: (56.5%) + (15%) = 71.5%

3) Holding basin 3 hr + aeration 4 hr + substitution of soluble sizing on 40% of grey goods

*Estimated from soluble-sizing reduction.

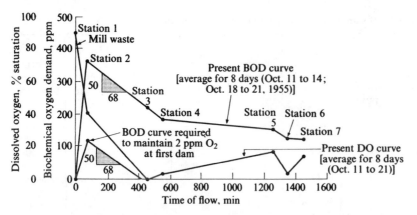

Fig. 20.5 Dissolved-oxygen and BOD profile of the river discussed in this chapter. The two downstream dams located at Sta.s 3 and 5 were originally built to retain water for small farm mills (later abandoned). Lawsuits ensued about the use and possible removal of Sta.3 dam.

BOD reduction: $(13.5\%) + (36.5\%) + (14.5\%)$
$$= 64.5\%$$

4) Holding basin 10 hr + aeration 12 hr

BOD reduction: $(10.9\%) + (60.1\%) = 71\%$

5) Holding basin 6 hr + aeration 8 hr + substitution of soluble sizing on 40% of grey goods

BOD reduction: $(15.5\%) + (44.9\%) + (14.5\%)$
$$= 74.9\%$$

6) Segregation* of kiering and desizing wastes 3 and 4 + holding lagoon 27 hr + aeration 4–12 hr

BOD reduction: $(56.5 \times 61\%) + (23.5 \times 61\%)$
$$= \sim 50\%$$

7) Increase upstream storage reservoir to 1.5 billion gallons, which means a minimum flow of 22.5 mgd during a 3-month summer drought period + build a rock-lined trench leading to biological-oxidation basins (27-hr detention)

BOD equivalent reduction: $(24\%) + (56.5\%)$
$$= 80.5\%$$

After a careful analysis of these methods, we see

---

*Segregation would reduce the volume of waste to be treated from about 6.2 mgd to 3.4 mgd, but aeration of the remaining wastes for up to 12 hours would result in an overall BOD reduction of only 0.80×61 per cent = about 50 per cent of total. Because of the necessity of obtaining 70 per cent reduction, segregation would therefore not be practical, unless some treatment were also given to both dye and starch wastes.

that there is great merit in using plan 7. Since there is an adequate amount of clean water available upstream (Table 20.3), it could be stored to supplement the river in the low-flow months, at a reasonable cost and with little or no future operating expense involved. Not only would this plan provide dilution of the river during the low-flow months, but detaining the waste for 27 hours in the biological-oxidation basins would result in a BOD reduction of 56.5 per cent. Adding this to the equivalent BOD reduction of 24 per cent provided by dilution, one obtains, for the total treatment, the equivalent of about 80 per cent reduction in BOD.

Table 20.3 Volume of flow of river receiving textile-mill wastes.

Month	Average daily flow, mgd				
	1954	1953	1952	1951	1950
January	167.2	73.4	54.7	44.5	73.8
February	55.0	98.5	70.0	52.9	62.5
March	71.9	80.0	151.1	76.4	70.6
April	77.1	48.4	93.8	62.7	68.6
May	56.0	69.6	56.0	37.3	49.3
June	37.7	67.5	38.8	48.2	68.0
July	26.0	30.9	30.5	25.2	63.1
August	18.6	26.5	40.8	28.3	35.9
September	10.6	23.8	27.7	34.2	48.4
October	8.3*	24.7	26.5	22.0	47.7
November	11.6*	27.7	32.0	36.8	38.1
December	23.3*	70.1	41.4	145.1	81.0
Daily average for year	46.9	53.4	55.3	51.1	58.9

*Estimated.

Table 20.4   Results of experiments in starch substitution.

| Sizing compound | Method of preparation | Laboratory BOD studies | | | |
| | | BOD (5-day, 20°C) | | BOD (10-day, 20°C) | |
		ppm	% red.	ppm	% red.
Pearl starch (100%)	Heated at 160°F for 4 hr with Rhozyme; final solution 0.1%	800		874	
Pearl starch (65%) and carboxymethyl cellulose (35%)	Same as above	336	58.3	525	40
Penfer gum 300 (100%)	Same as above	369	53.9	511	41.5
Pearl starch (65%) and carboxymethyl cellulose (35%)	Heated at 205°F for 20 min, no enzyme added; final solution 0.1%	283	64.6	265	69.7
Penfer gum 300 (100%)	Same as above	321	60.0	318	63.7

Storage of clean water was found to be economically feasible, as shown in Table 20.3, since storage for any 3-month period from January to July during the 5-year period would have yielded more water than required to maintain flows of greater than 22.5 mgd in the stream. During a 5-year period, no monthly average flow fell below 26 mgd between January and July.

20.4 Substitution of Soluble Sizing

A mixture of 65 per cent pearl starch and 35 per cent carboxymethyl cellulose (CMC) resulted in a 5-day 20°C BOD which was 64.6 per cent lower than that obtained by using pure pearl starch (Table 20.4). In addition, no enzymes were needed, since the sizing is easily washed off the cloth with warm water. Penfer gum of the 300 series showed a 60 per cent 5-day BOD reduction. Even after 10 days, the BOD values were 63.7 to 69.7 per cent lower than those obtained with pure starch. The approximate 5-, 10-, and 20-day 20°C BOD values were verified in additional experiments (Table 20.5). There did not appear to be any lag in the oxidation of the soluble sizes. Field studies (Table 20.6) substantiated the BOD results of the laboratory; however, these involved only one day of actual testing. There is no question in the mind of the author as to the advantages of using either a mixture of starch

Table 20.5   Additional results of starch-substitution experiments with 65% pearl starch and 35% carboxymethyl cellulose in a 0.1% solution.

BOD* at 20°C, ppm

5-day	10-day	15-day	20-day
386	483	600	450†

*Average of four bottles (2 dilutions, duplicate samples) each day.
†Insufficient samples within BOD range.

Table 20.6   Tests for BOD outlet of No. 4 waste line (desize).

| Date | Flow, mgd | BOD | | Sizing compound |
		ppm	lb/day	
Aug. 22–23	2.49	1040	21,600	Starch (100%)
Dec. 9	~2.49*	813	~16,900	Pure starch (60%); CMC, starch, and Penfer gum (40%)

*No exact flow was determined.

Table 20.7 Results of measurements of waste flow and BOD load for 24 hours (Aug. 22–23).

Waste line		Flow		BOD		
Number	Source	mgd	% of total	ppm	% of total	lb/day
1	Dye	1.735	28.2	320	13.02	4,635
2	Starch	0.950	15.45	440	9.81	3,485
3	Kier (bleach)	0.971	15.80	720	16.40	5,830
4	Desize	2.490	40.50	1040	60.80	21,600
Total (sum of 4 lines) as measured individually		6.146				35,550
Totals of effluent as measured		6.09		630		32,000

and CMC or Penfer gum of the 300 series. A definite BOD reduction is to be obtained by using them. However, the higher cost of the soluble sizes and an increased percentage of rejects have acted as deterrents to the changeover. If these objections could be overcome, less pollutional waste would be discharged.

The following is a list of recommendations.

1. Provide storage for 1.5 billion gallons of water upstream. This should provide a 3-month low-flow minimum of 22.5 mgd during summer operating periods.
2. Transport the entire waste of the mill through a shallow, rock-lined trench to a biological-oxidation basin, for a minimum of 24 hours of storage. The basin should be no deeper than 4 feet and constructed so as to minimize short-circuiting. The plant complied with both recommendations. Figure 20.6 shows the shallow rock-lined trench carrying the waste to the biological-oxidation basin and Fig. 20.7 shows the approach to the first oxidation basin, before it was filled.
3. Assign a full-time qualified engineer (preferably with water- and waste-treatment experience) to the position of waste-treatment supervisor. This man should keep *daily* records of the pollutional characteristics of both the raw waste and the oxidation-basin effluent, as well as the river conditions. These data should be kept on file at the mill and copies forwarded to the water-pollution control board each month.

Table 20.8 Standards of quality for fishing (C) classification.*

Items	Specifications
Floating solids, oils, settleable solids, sludge deposits	None [permitted] which are readily visible and attributable to sewage, industrial ... or other wastes, and which measurably increase the amounts of these constituents in receiving waters, after opportunity for reasonable dilution and mixture [of the waters] with the wastes discharged thereto
pH	May range between 6.0 and 8.5, except that swamp waters may range between 5.0 and 8.5
Dissolved oxygen	Not less than 2 ppm
Toxic wastes, deleterious substances, colored or other wastes, heated liquids.	[Not permitted if,] alone or in combination with other substances or wastes, [they are] in sufficient amounts or at such temperatures as to be injurious to fish survival or impair the waters for any other best usage, as determined by the Water Pollution Control Authority for the specific waters which are assigned to this class

*In accord with state regulations, these waters are suitable for fish survival, industrial and agricultural uses, and other uses requiring water of lower quality.

Fig. 20.6 Textile-mill waste traveling in a shallow, rock-lined trench to the biological-oxidation basin, where it will be held for a minimum of 24 hours.

4. Continue to investigate means of overcoming the drawbacks to the use of soluble sizes. The extra BOD reduction obtained by the use of these sizes may be needed in the near future, for additional protection to the stream.

5. Continue to operate the pilot plant, as time and personnel permit, to obtain additional information on minimum air and CO_2 requirements for secondary treatment.

6. Attempt to maintain continuous flow through the first dam at all times.

The first three recommendations should be carried out immediately.

Without the use of carbon dioxide to acidulate the wastes, BOD reductions were less. When the raw, un-neutralized finishing-mill waste was held for only 5 hours and aerated for only 6 hours, an average of 28.5 per cent BOD reduction occurred.* That almost one-third of the BOD was oxidized at a pH of between 11 and 12 was quite a revelation. Some questions existed about the nature of the oxidation: mainly, whether it was chemical or biological. When the flow through the pilot plant was reduced (Table 20.9) to a holding time of 10 hours and an aeration period of 12 hours, a BOD reduction of 38.8 per cent occurred, despite the fact that the initial pH averaged 11.3. In earlier experiments at the same rate of flow (Table 20.1), 17 pounds of CO_2 per day were required to give a BOD reduction of 71 per cent. To install equipment to supply CO_2 in adequate amounts for this mill would cost approximately $150,000. In addition, it would cost about $275 per day for power and fuel to generate the CO_2. This is a considerable expense, even for a textile mill as large as this one. However, in many cases a mill can eliminate the high neutralization cost by providing sufficient detention time and sacrificing some efficiency; also, as mentioned previously, the use of flue gas for neutralization is often an inexpensive way to achieve extra BOD reduction.

---

* Average of Tables 20.1 and 20.9.

Fig. 20.7 Biological-oxidation basin built to hold textile wastes for 24 hours. Note shallowness: it is not over four feet in depth. Smokestacks of textile mill can be seen in background.

Table 20.9 Summary of additional results of pilot-plant study.

Treatment	Waste-flow rate	
	4 gpm	2 gpm
Holding-basin detention, hr	5.25	10
Aeration-basin detention, hr	6	12
Air rate		
cfm	60	60
ft³/lb BOD	2370	4650
Number of 4-hr composites	34	48
BOD of raw waste, ppm	760	765
BOD reduction (5-day, 20°C)		
Primary settling tank, %	6.7	8.3
Aeration tank, %	22.1	38.8
Final settling tank, %	19.1	37.5
CO_2 used, lb/day	0	0

Table 20.10 Environmental conditions during aeration.*

Number of 4-hr composites averaged	48
Temperature, °F	
Maximum	70
Minimum	55
Mean	64
pH level	
Raw	11.3
Primary settled	11.1
Aerated	10.0
Final settled	9.8
Plate counts in aeration basin, total bacteria/ml	
Maximum	14,000,000
Minimum	10,800
Mean	5,351,853

*Waste flow, 2 gpm; no pH adjustment.

The pH dropped only slightly, from 11.3 to 11.1, in the holding basin, but dropped more rapidly to 10.0 in the aerator (Table 20.10). The reaction of CO_2 in the air with caustic alkalinity in the waste, producing carbonates, and the CO_2 given off by bacterial action on organic matter, could have been responsible for the lowered pH. Total bacterial plate counts (similar to those made in water analyses) showed that an average of somewhat over 5 million microorganisms per milliliter (as reported by the mill chemist) were living in the aeration basin. The numbers varied from as high as 14 million to as low as 10,000. This concentration of bacteria and a BOD reduction of about 40 per cent signify that bacteria can survive, and apparently metabolize organic matter, at a pH as high as 11.3. The possibility of utilizing bacteria for oxidizing dissolved organic matter at elevated pH values should not be overlooked.

On the basis of these findings, a full-scale treatment plant was constructed, complying with the major recommendations of the author.

Questions

1. When can you discharge a partially treated industrial waste directly into the stream?
2. What does this case study entail?
3. What steps were taken to meet stream standards in a manner economical to the mill?
4. How was the stream-water-quality problem solved?
5. Describe the special combination of factors which, in this case, led to the alternative finally selected.

CHAPTER 21

DISCHARGE OF COMPLETELY TREATED WASTES
TO STREAMS OR LAND

Complete treatment of wastes prior to direct discharge to a receiving stream is gradually receiving more and more consideration. The amount of dilution water in streams is not increasing and, on the other hand, pollution loads unfortunately *are* increasing. With the population explosion and industrial expansion, we can expect more extensive waste-treatment requirements. At present, complete treatment is required only in special instances and in the case of the large, wet industries—for example, textiles, pulp and paper, steel, and chemicals.

There is some doubt as to what is meant by the expression "complete treatment." It is generally conceded that complete treatment refers to secondary treatment; that is, the removal of about 85 to 90 per cent of the BOD by a combination of physical, biological, and/or chemical means. According to this definition, one is removing only two polluting constituents: suspended solids and dissolved organic matter (including colloidal solids). Does this definition, then, imply that the removal of *any* two forms of pollution—such as color and suspended matter, oils and alkalinity (high pH), or acids and organic matter—also constitutes complete treatment? The author doubts that this is the original meaning of the term; and in these days when "complete treatment" is insufficient and certainly not complete in some cases, a reevaluation of our terminology is in order. For example, an industry may have little or no dissolved organic matter in its waste and yet be required to remove two or more other forms of pollution. In the author's mind, this also constitutes complete treatment, as the term is currently defined. The expression "complete treatment" will hardly be satisfactory, with its present definition, when the public begins to accept and include "tertiary treatment" in its thoughts on the subject. Tertiary treatment presently provides for the removal of three or more forms of contamination: suspended solids, dissolved organic solids, and dissolved inorganic solids. True complete treatment would remove refractory solids as well.

An industry requiring complete treatment for its waste usually discharges a large volume of waste and is located outside, and some distance from, a municipality, on a stream requiring the maintenance of high standards of water quality. This author prefers to consider "complete treatment" as that which renders waste waters reusable for industrial and (in some cases) municipal water supplies. This normally will mean a fairly complete removal of all suspended, dissolved, and colloidal solids, including both inorganic and organic fractions. Since this is, at present, rarely practiced, we are forced to accept as a definition of "complete treatment" the removal merely of a major portion of the suspended solids and dissolved organic matter.

AN EXAMPLE OF COMPLETE WASTE
TREATMENT BY A FIRM PRIOR TO DIRECT
DISCHARGE INTO THE RECEIVING STREAM

21.1 The Problem

Townsends, Inc., an integrated poultry operation, consists of a hatchery, feedmill, soybean mill, and poultry-processing plant located about two miles east of Millsboro, Delaware. It is owned privately by the Townsend family, and the raising and processing of chickens is their main business. The waste problem is at the poultry-dressing plant. This plant, built in 1957, is located about 50 yards from Swan Creek, a tributary of the tidal Indian River. The relative locations of the plant, town, and receiving waters are shown in Fig. 21.1. Of special significance is the location of the Millsboro extended-aeration sewage-treatment plant which discharges into the Indian River about 3 miles above the confluence with Swan

Creek. The poultry-plant waste from Townsends is discharged after screening and ineffective flotation treatment into Swan Creek about 1 mile upstream of the confluence with the Indian River. The proximity of these waste discharges to the shellfish area only $2\frac{1}{2}$ miles below Swan Creek is a major concern to the regulatory authorities. Although the main portion of the town of Millsboro (Fig. 21.1) is served by the extended-aeration sewage-treatment plant followed by chlorination, many of the homes along Route 24 are individually served with septic tanks and well-water supplies located in relatively sandy soils. The underground disposal in sandy areas of sewage or wastes may represent some danger to these water supplies. During each of the last three years many areas of the Indian River Bay have had to be closed periodically during the summer for cleaning because of bacterial contamination. Coliform standards have been set at 70/100 ml for shellfish and at 1000/100 ml for swimming.

In 1956, the Delaware Water Pollution Commission concluded in their *Indian River Drainage Basin Survey* that:

1. A portion of the fresh-water flow within the Indian River watershed originates from swampy and marshlike areas which have a decided effect upon the chemical and physical composition of the runoff waters. These waters are generally high in iron and color, low in turbidity, suspended solids, and dissolved oxygen, and acid in pH.

2. Average dry-weather flow in this basin area is approximately 0.25 cfs per square mile.

3. Small tributaries predominate in the drainage basin. The only surface supply location with sufficient volume for either domestic or industrial use is at Millsboro dam.

4. Studies made by the Delaware State Water Pollution Commission when the former owner of the plant was in operation clearly indicated that dry cleaning of manure solids, coagulated blood, and feather removal followed by satisfactory removal of settleable solids with heavy disinfection will effectively and satisfactorily protect state waters downstream from this plant.

Millsboro (the closest and most significant municipal co-polluter in this case) is located on the Indian River, 13 miles from the ocean and is one of the prin-

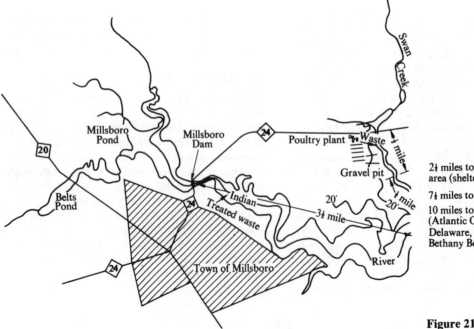

2¼ miles to shell-fishing area (sheltered coves)

7¼ miles to Indian River Bay

10 miles to Indian River Inlet (Atlantic Ocean) with Lewes, Delaware, on the north and Bethany Beach on the south

Figure 21.1

cipal towns of the Indian River County of Delaware. It is a distributing point for carloads of poultry feed and coal for the broiler chicken industry. The Commission's historical survey revealed that as late as 1956 the pool at the base of Millsboro Dam was still noted for its herring run in April and May; in good seasons as many as a million have been taken in a few weeks. At times crabbers brought thousands of soft-shell crabs to Millsboro for shipment alive in boxes filled with wet grass. The alternate opening and closing of Indian River Inlet prior to 1938 nearly ruined the industry, though a few soft-shell crabs were still shipped from there during the periods of transformation. The new inlet revived the market for crabs, fish, oysters, and clams taken in Indian River. The main body of the Indian River from Millsboro to the ocean is tidal, with an elevation of less than 10 feet at Millsboro. This, the flattest stretch of the area, yields a slope of only 0.7 foot per mile.

In 1956 about one-third (18,600 acres) of the Rehoboth and Indian River Bays, which receive the poultry waste, was utilized for oyster cultivation at an annual "take" of $800,000. At the same time an additional $237,500 revenue resulted from the growing and harvesting of clams in these bays. Each year as much as $250,000 is spent in the shore-line areas between Lewes and Bethany Beach for fishing tackle, bait, and other small items associated with the sport. Boat rentals have been estimated to bring $96,000 per year. The Indian River is a vital link to the tremendous menhaden fishing industry in the mid-Atlantic states. In 1953 the U.S. Fisheries Statistics Report stated that Lewes, Delaware, was the nation's leading fishing port poundwise with landings of about 363 million pounds, consisting almost entirely of menhaden. This catch had a reported value of $4,117,000. Duck hunting is also estimated to contribute about $25,000 per season and muskrat trapping about $15,000. Despite the value of the shell-fishing, fishing and hunting industries, the 1956 Delaware Report stated that "there is little doubt that bathing and swimming is a primary interest in this drainage basin area." They were referring to the areas of Rehoboth Beach, Lewes, Rehoboth Bay, Dewey Beach, Indian River Bay, and Bethany Beach.

The foregoing information led the Delaware Water Pollution Commission in 1956 to conclude that "this entire basin must, of necessity, be classified as an unusually clean water area which has as its major interests bathing, swimming, boating, sports fishing, commercial fishing, shellfish, wildlife, recreation, and seasonal real estate."

21.2 Stream Studies

The Delaware State Water Pollution Commission conducted many studies of the Indian River and Inlet Bay areas during 1952–55. Figure 21.2 shows the drainage basin and the location of the sampling points (described in Table 21.1).

The State of Delaware Water Pollution Commission investigated the quality of the Indian River (Fig. 21.1), which extends from the Millsboro Dam to the vicinity of Oak Orchard, on July 1 and 28, 1953. The results are shown in Tables 21.2 to 21.3. Fresh-water flow in the Indian River Basin was determined during two periods, May 1, 4, and 5, 1953, and May 5, 1955. These results are shown in Table 21.4. One may note that in Tables 21.2 and 21.3 samplings were taken as near to high tide as possible. Thus the increased volume of dilution water from the bay might tend to minimize the effects of pollution. It may also be noted that the water temperatures were high—a fact which is not considered abnormal since the Indian River Bay is broad and quite shallow. The high dissolved oxygen (although the upper reaches near the Millsboro Dam are relatively low) may indicate that little or no pollution is present.

The reader should recognize the scarcity of meaningful analytical data on the sanitary characteristics of the receiving stream. The evidence for pollution comes from instances of fish deaths rather than direct stream analytical measurements. However, coliform bacteria counts have been run on many samples of the Indian River at the sampling points shown in Fig. 21.3. Typical data on coliforms at some of these points collected as late as 1964 are shown in Table 21.5. These data indicate that considerable attention should be given to the bacteriological quality of the receiving water, since these are primarily recreational and fishing waters.

21.3 State Decision

The stream data illustrate a lack of positive evidence on the effect of organic loading, especially from the poultry plant. They do show bacterial contamination

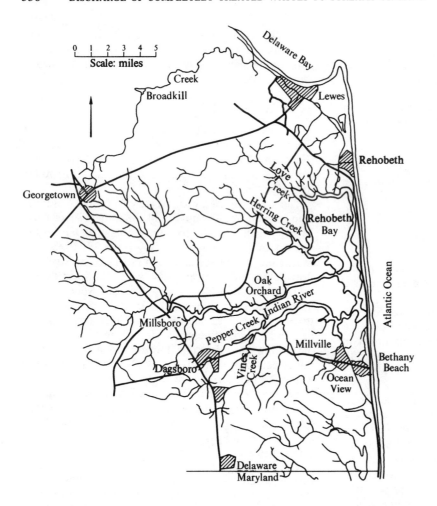

Fig. 21.2 Delaware Water Pollution Commission Survey of the Indian River drainage basin.

in recreational, fishing, and shellfish-producing waters. The author had to decide whether to recommend a complete stream survey to determine the exact degree of treatment required for the poultry wastes. He decided against this survey for the following important reasons: (1) it would be costly and time-consuming with no apparent financial support available from the poultry processer or the state; (2) the state commission had already decided that a high degree of treatment was required (and only this would be approved) in order to protect the valuable resources of the receiving waters downstream.

21.4 Poultry-Waste Characteristics

During the normal 8- to 11-hour working day at the poultry plant 9000 to 10,000 chickens weighing $3\frac{3}{4}$ pounds each are processed every hour. The processes and their associated wastes are summarized in Fig. 21.4. A separate septic-tank sewage-disposal system serves the 225 plant employees. The chickens are not force-fed (a procedure of fattening before killing to produce more weight) at the plant and dry removal is practiced. Although the killing room is separated from the rest of the processing operation and blood

Table 21.1 Indian River basin sampling stations.

Description	Miles from Indian River inlet	Description	Miles from Indian River inlet
Assawoman Canal, Ocean View	5.50	Indian River	13.31
White Creek tributary, Ocean View	6.20	Pepper Creek	13.45
White Creek, Millville	6.34	Pepper Creek	13.50
Indian River	6.59	Iron Branch at railroad	13.52
Indian River	7.16	Vines Creek, Frankford	13.70
Lewes–Rehoboth Canal jetty	7.60	Indian River at Millsboro	14.10
Indian River	7.73	Iron Branch near Millsboro	14.40
Indian River, mouth of Island Creek	8.30	Vines Creek	14.65
Blackwater Creek	8.46	Vines Creek	14.70
Indian River	9.01	South tributary of Iron Branch	14.80
Stokely Cannery, Rehoboth Beach	9.40	South tributary of Iron Branch	14.82
Vines Creek	9.43	Lewes–Rehoboth Canal Bridge	14.86
Lewes–Rehoboth Bridge, Route 41	9.45	Vines Creek near Frankford	14.90
Love Creek	10.00	Vines Creek	14.95
Unity Branch in Fairmont	10.16	Shoals Branch at Betts Pond	15.20
Indian River	10.20	Famys Branch near Millsboro	15.50
Chapel Branch in Angola	10.24	Betts Pond at Route 113	15.70
Indian River	11.00	Vines Creek near Millsboro	15.90
Vines Creek	11.70	Roosevelt Inlet	16.59
Indian River	11.74	Cow Bridge Branch	17.62
Indian River	11.80	Stockley Branch	17.90
Pepper Creek	11.82	Wood Branch near Morris Millpond	18.75
Love Creek	11.95	Deep Branch near Morris Millpond	18.85
Swan Creek near Millsboro	12.10	Cow Bridge near Morris Millpond	19.60
Indian River	12.57	Wood Branch near Georgetown	20.07
Iron Branch near Millsboro	12.58	Wood Branch near Georgetown	21.40
Vines Creek, Frankford	13.30	Wood Branch, Georgetown	22.20

Table 21.2 Indian River water quality, Millsboro to Oak Orchard *

Sampling station, miles from inlet	Time	Temperature, °C	D.O., ppm	D.O. saturation, %	Salinity as NaCl, ppm	Depth, ft
Millsboro Dam (14.10)	1:00 p.m.	27.8	4.65	60.7	3,500	
13.31		27.5	5.5	71.0	3,200	
12.57		30.5	7.8	105.7	2,400	
11.74		29.5	7.3	97.3	2,500	
11.00	1:30 p.m.	30.5	7.35	101.8	4,500	5
10.20		30.0	7.5	105.5	6,800	
9.01		30.0	7.4	105.5	8,050	
8.30	2:00 p.m.	30.0	8.3	131.0	16,300	
8.55		31.0	8.6	137.0	16,450	2
11.80		31.5	8.65	139.5	16,450	1.5
7.73		30.0	8.7	136.6	16,700	

* High tide (from U.S. Geological Survey Table) 2 p.m. at Indian River inlet. Water throughout stretch being studied, quiescent before 1:30 p.m., choppy after.

Table 21.3 Quality of water in the Indian River from Millsboro to Oak Orchard on July 28, 1953 (high tide).

Sampling station, miles from inlet	Time (p.m.)	Temperature,°C	Air temperature, °C	D.O., ppm	D.O. saturation, %	Salinity as NaCl, ppm
Millsboro Dam (14.10)	2:00	17.5	25.0	3.5	17.1	7,100
13.31	2:05	28.0		5.0	68.6	8,200
12.57	2:10	30.0		11.8	170.0	9,000
11.74	2:20	30.0		8.3	123.0	11,800
11.00	2:28	30.0		7.9	118.0	12,200
10.20	2:35	30.0		7.7	118.0	14,300
9.01	2:45	29.5		7.7	120.0	18,200
8.30	2:55	30.0		7.6	124.0	20,000
7.73	3:04	28.5		8.3	134.0	21,400
7.16	3:12	28.0	26.0	8.5	138.0	23,000
6.59	3:22	27.5		8.5	139.0	28,000

is scooped out of the killing-floor area and disposed of with the screenings, the film which collects on the walls is washed into the sewer at the end of each working day. The feathers, which constitute about 14 per cent of the raw chicken weight, are sold for rendering for about $16 per ton; the offal, making up 16 per cent of the weight of the chicken, is sold for about $21 per ton. The processing waste-water is screened through

Table 21.4 Fresh-water volume within Indian River basin.

Station (miles from inlet)	Flow, mgd	
	May 1, 4, and 5, 1953	May 5, 1955
11.95	10.25	4.04
10.24	7.2	2.80
10.16	13.0	4.03
12.10	6.20	3.58
14.10	111.1	39.40
14.40	3.54	1.55
14.80	3.02	1.60
11.82	1.59	1.63
13.45*		3.60
11.70	3.68	
13.70	0.36	
Total	159.37	60.60

*Upstream from 11.82 and not added into total.

four Sweco vibrating screens which are cleaned daily with alkali (1 pound/day) to keep them clean of feathers.

The Delaware Water Pollution Commission carried out composite analysis of the poultry-plant effluent on July 21 through 24, 1964. The results are shown in Table 21.6. The BOD averaged about 630 ppm with total nitrogen (mostly organic) of about 60 ppm and a slightly alkaline pH (7–8). Suspended solids were about 200 to 600 ppm, mostly organic.

Although quite accurate water-flow records are kept at the poultry plant and indicate a consumption of about 800 gallons per minute (500,000 gallons per day or an average of 5 gallons per bird), for additional information the plant effluent was weired and measured every half-hour during a typical operating day on January 27, 1965 (Table 21.7). Some reduction in process wash water may be achieved by closer control and using higher pressure nozzles. The flow rate, however, must be approved by the Department of Agriculture, which supervises cleanliness within poultry-processing plants.

21.5 The Solution

Since the Indian River and its receiving bays were already contaminated in the 1950s before the poultry plant began operations in 1957 and since the poultry-plant waste was also found to be highly pollutional

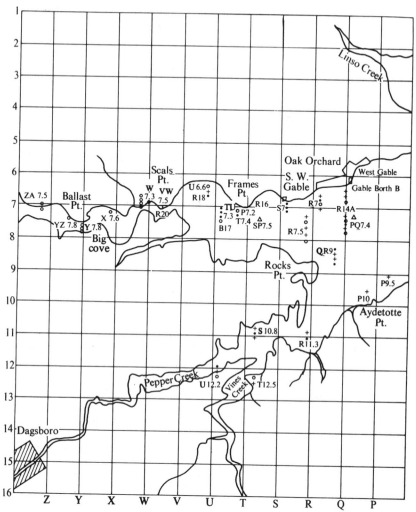

Figure 21.3

(both from analysis and stream observations), adequate treatment of the waste was necessary. From Tables 21.6 and 21.7 the total BOD load was computed to be 2550 pounds—equivalent to a population of 12,750 persons. The major question was what type of treatment should be used to protect the best uses of the stream. Obviously, the major concern is bacterial contamination, so that chlorination of the poultry-plant waste would be a minimum requirement. Chlorination in the presence of 2550 pounds of BOD and the other suspended and floating matter normally found in poultry-plant effluents would be difficult and costly. Organic matter reacts rapidly with chlorine and the chlorine necessary would be expected to cost well over $250 per day. Therefore, more economical means for removing a major portion of the organic matter prior to chlorination were demanded.

A two-stage, oxidation-pond treatment system was chosen to perform the task because of the low construction and operation costs compared with other biological treatment systems. The first stage consists of a baffled, high-rate, deep pond to allow sedimentation of heavy solids, flotation of grease or feathers which escape preliminary treatment by the screens, and bacterial degradation of the organic matter. This pond is 595 feet long, 109 feet wide, and 8 feet deep;

Table 21.5 Selected data on coliforms at stations in Indian River.

Date	Sampling point*	Coliform count, MPN/100 ml	Salinity, ppm
8/14/61	U6.6	790	14,500
7/25/62	QR9	430	23,600
7/29/63	VW7.3	430	
7/25/62	W7.3	4,600	17,700
8/8/62	W7.3	2,400	
8/21/62	W7.3	11,000	18,600
8/8/62	W7.3	2,400	
7/15/64	T7.2	430	
10/13/64	ST7.5	430	
	ST7.5		
7/24/62	R7	430	21,600
6/21/61	S7	1,600	
7/24/62	R7	930	22,500
7/15/64	R7.5	430	

*See Fig. 21.3.

30 over-and-under baffles on 15-foot centers cover the middle 435 feet. The second stage is a shallow photosynthetic pond designed to remove more organic matter and convert inorganic phosphates and ammonia nitrogen to an algal mass. It is 635 feet long, of nonuniform width, and about 2 feet deep, covering an area of about 212,000 square feet. The effluent from this two-stage treatment is chlorinated before discharge into Swan Creek. Detailed drawings of this plant are shown in Fig. 21.5. The area chosen for this two-stage treatment plant was predominantly sandy.

Four tests in the area confirmed that the soil was about 99.5 per cent inorganic matter (stable at 900°C) and only about 5 to 6 per cent moisture.

A rough cost estimate of $83,738 was given to Townsends by the author on May 25, 1965 (Table 21.8) and some minor revisions were made in the original plans on July 22, 1965. On August 25, 1965, George and Lynch Construction Company signed a contract with Townsends, Inc., for construction of the treatment plant at a cost of $90,000 and a construction period of about 45 days. Some photographs of the treatment plant during construction are shown in Fig. 21.6. The plant was officially inaugurated on March 14, 1966, although operation actually began about January 1, 1966.

21.6 Results

Some details of the design and operation of the facility were presented in *Poultry Meat* (August 1966). In a letter dated January 30, 1967, Mr. Donald J. Snyder, manager of the Dressed Poultry Division of Townsends, stated, "The system has been working very well and has never given us any trouble... and we are happy to have anyone inspect the system if they should care to."

Samples were collected and analyzed by the Delaware Water Pollution Commission on April 7, June 29, and December 6, 1966, and March 23, 1967. These results are shown in Table 21.9.

From these four samples it was apparent that the treatment facilities were operating satisfactorily at a BOD loading of about 1390 lb/day. The loading on

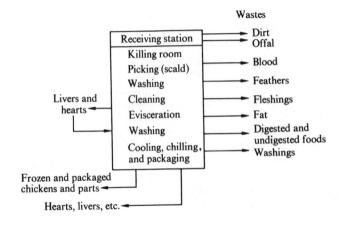

Fig. 21.4 Flow sheet of poultry-processing plant.

Table 21.6 Sanitary characteristics of the poultry-plant effluent as reported by the Delaware State Water Pollution Commission.

Characteristic*	Date (July 1964)					
	21	21†	22	22	23	24
Sample no.	638	640	642	672	674	676
5-day BOD††	425	1200	395	800	500	457
Chloride, ppm	64	74	37	55	45	24
COD††	1710	3690	3250	1590	2700	2780
Total N	62.4	91.8	57.4	60.2	57.7	59.3
Organic N	56.3	80.2	54.3	54.3	53.2	55.7
NH_3N	6.1	11.6	3.1	5.9	4.5	3.6
NO_2-N	0.076	0.018	0.09			
Acidity				28	5	20
Total alkaline (as $CACO_3$)				48	73	39
pH			7.6	6.9	8.1	7.4
Total suspended solids			360	606	254	204
Suspended volatile solids			360	584	244	180
Suspended ash			0	22	10	24
Total solids			801			
Total volatile solids			482			
Total ash			319			

*All results are given in milligrams/liter unless otherwise indicated.
†Plant washdown during sample collection on this day led to unusual results.
††An additional plant effluent sample was composited and analyzed on 3/17/65 and found to contain 418 ppm of BOD and 880 ppm of COD.

the first basin is

$$\frac{1390}{(595 \times 109)/43,560} = 1390/1.49 = 935 \text{ lb BOD/acre.}$$

At a daily waste-flow rate of about 530,000 gallons the detention time in this first basin is 7.35 days:

$$\frac{595 \text{ ft} \times 109 \text{ ft} \times 8 \text{ ft} \times 7.5}{11 \text{ hrs/day} \times 800 \text{ gal/min} \times 60 \text{ min/hr}}$$

This unusually high loading resulted in a BOD reduction of about

$$\frac{313 - 87}{313} \times 100 = 72.5 \text{ per cent.}$$

The second basin handled a BOD loading of 385 lb/day or

$$\frac{385}{212,000/43,560} = 385/4.87 = 79 \text{ lb/acre}$$

and effected an addditional BOD reduction of

$$\frac{87 - 79}{87} \times 100 = 9.2 \text{ per cent}$$

when the algae are not removed from the effluent and

$$\frac{87 - 26}{87} \times 100 = 70 \text{ per cent}$$

when the algae are filtered out of the final effluent. Detention time in the second basin averages about 6 days. No attempt is made to remove algae from the final effluent but the effluent is withdrawn slightly below the surface.

The overall BOD reduction obtained during the first year of operation (based upon only four samples) was about

$$\frac{313 - 26}{313} \times 100 = 92 \text{ per cent}$$

Table 21.7 Poultry-plant effluent flow on January 27, 1965.*

Time	Flow, gpm	Time	Flow, gpm
6:00 a.m.	645	12:30 p.m.	645
6:30	1190	1:00	645
7:00	800	1:30	645
7:30	1020	2:00	645
8:00	800	2:30	645
8:30	645	3:00	525
9:00	1190	3:30	380
9:30	800	4:00	352
10:00	1020	4:30	380
10:30	1020	5:00	408
11:00	645	5:30	380
12 noon	645	6:00	408
		6:30	380

*Average rate was 674 gpm or 40,400 gal and chickens were processed on this day at the rate of 8570 per hour, so that $\frac{40,440}{8,570} = 4.7$ gal/chicken were used.

when the algae were filtered from the final effluent and

$$\frac{313 - 79}{313} \times 100 = 75 \text{ per cent}$$

when the algae cells were left in the final effluent. Although scum removal in the first basin was frequently required in 1966, no operating difficulties or nuisance resulted from an overall plant BOD loading of

$$1390/(1.49 + 4.87) = 1390/6.36$$
$$= 219 \text{ lb BOD/acre/ day.}$$

The preliminary results point out some other interesting phenomena, for example, that the expected rise in pH in the second pond was coupled with a corresponding reduction in phosphates and coliform bacteria. Total coliform counts in the chlorinated effluent approximate 10/100 ml and apparently meet current shellfish standards. Although it is too early

Table 21.8 Rough cost estimate of waste-treatment system for poultry-processing plant.

Item*	Cost per item, $
Pumps (2)	3,708
Installation and delivery of pipeline	500
Pipeline at $7.50/ft for 955.5 ft	7,180
Chlorinator (duplicate of existing one)	1,500
Excavation for two basins at $0.50/yd³	30,000
Cement of soil cement at $4.50/barrel	15,800
Wood at $200/mbf	10,000
Poured concrete at $25/yd³	2,500
Steel at $0.20/lb	10,000
Concrete block at $0.50/unit	1,100
C.I. pipe and fittings	750
Chlorination shack	200
Flagstone	500
Total	83,738

*These figures do not include some additional items which should be considered by the company such as

Fencing, especially for the no. 1 basin;
Ditching around basins to prevent groundwater intrusion;
Seeding of the birms to prevent erosion;
Landscaping to improve the aesthetic appearance of the system.

to formulate any firm and final conclusions, one can observe that elevated BOD loadings were handled in a properly designed two-stage, oxidation-pond treatment plant system and produced satisfactory operating results.

Continued sampling of the treatment plant facilities on March 29, June 14, and July 26, 1967, yielded the results shown in Table 21.10. Excellent BOD reduction continues, in the range of 85 to 90 per cent. In addition, coliform bacteria counts are less than 10/100 ml, which is acceptable for discharge into water used primarily for shellfish cultivation.

This example shows how a large poultry plant discharging about half a million gallons of waste per

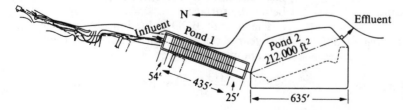

Fig. 21.5 Diagram of the two-stage, oxidation-pond treatment system designed for the poultry-processing plant.

Table 21.9

Characteristic	Influent pond no. 1			Effluent pond no. 1			Effluent pond no. 2			
	4/7/66	6/29/66	3/23/67	4/7/66	6/29/66	3/23/67	4/7/66	8/29/66	12/6/66	3/23/67
Physical										
Color, units			380			175				220
Turbidity, units			110			60				55
Dissolved oxygen, mg/liter		3.84			0			5.18	2.5	
Temperature, °C		21.5			24			25	7.0	
Minerals										
pH	6.4	6.4	6.9	6.8	6.7	6.8	7.4	8.7	7.0	7.1
Acidity (CaCO$_3$), ppm	31	26	42	41	55	50	27	0	52	36
Alkalinity (CaCO$_3$), ppm	38	32	16.4	135	143	155	138	8	144	153
Hardness (CaCO$_3$), ppm		80	65		69	69		73		68
Chloride (Cl), ppm			82			84				83
Nitrogen balance (mg/liter as N)										
Total Kjeldahl N	98.3	54.3	42.5	15	31.7	33.3	17	22.9	30	28.0
Organic N	89.9	45.4	28	4.2	8.7	5.3	9.0	17.4	4.5	8.0
NH$_3$-N	8.4	8.9	14.5	10.8	23	28.0	8.0	5.5	25.5	20.0
NO$_2$-N	0.47	2.44	0.013	<0.02	0	0	<0.02	0	0	0
NO$_3$-N	4.3	2.98	0.4	<2	0.1	0.24	0.38	0.4	0.5	0
Waste analyses										
BOD, mg/liter	300	380	260	70	86	105	97(u)* 24(f)	83(u) 27(f)	65	70
COD, mg/liter	600	560	370	185	150	150	196	270(u) 190(f)	130	120
Total PO$_4$, mg/liter	8.6	0.74	10	11.4	2.8	9.9	7.4	1.6	9.7	9.2
Ortho PO$_4$, mg/liter										
Methylene blue alkyl benzene sulfonate		16			1.9			1.5		
Solids balance										
Settleable solids, mg/liter	1.2	2.0	0.6	<0.1	<0.1	<0.1	0.2			<0.1
Total suspended solids, mg/liter		338	148		70	50		218	52	52
Volatile suspended solids, mg/liter		326	140		64	50		208	52	52
Total solids, mg/liter		592	470		318	332		56.9	29.7	323
Total volatile solids, mg/liter		362	238		125	118		33.5	207	110
Bacteriological analysis										
Total coliform/100 ml	2.5×10^6	6×10^6		0.2×10^6	9×10^4		5.4×10^4	3×10^3		
Fecal coliform/100 ml	6×10^4	1.2×10^6		5×10^4	1×10^4		6×10^3	1×10^3		
Fecal streptococci/100 ml	1.3×10^6	2.0×10^6		3.2×10^5	4.5×10^4		6.3×10^4	$<1 \times 10^3$		

*Unfiltered = u; filtered = f.

a, b

Fig. 21.6 Two views of oxidation basin no. 1: (a) down length, showing scum collection area in foreground and baffles in background; (b) on diagonal across basin in baffled area.

Table 21.10   Continued analyses of treatment plant.

Characteristic	3/29/67			6/14/67		10/31/67		7/26/67	
	Influent	Effluent basin no. 1	Effluent basin no. 2	Influent	Final chlorinated effluent	Influent	Final chlorinated effluent	Effluent	Final chlorinated effluent
pH	6.4	6.7	7.1	6.4	8.3	6.0	6.9	7.2	7.1
Acidity (CaCO₃), ppm	29	66	35	30	5	36	34	32	35
Alkalinity (CaCO₃), ppm	35	152	134	40	121	29	110	135	135
Hardness (CaCO₃), ppm	49	72	73	130	180	93	86		
Chloride (Cl), mg/liter	98	84	84	149	124	160	135		
Total Kjeldahl N	49.8	32	30.5			87.2	30.8	30.5	
Organic nitrogen	49.8	5.0	10			74.0	7.3	106	
NH_3–N	0	27	20.5	23.6	21.2	13.2	23.5	19.9	
NO_2–N	0.042	0	0	1.1	0.41	0.39	0.10	0.120	
NO_3–N	6.8	0.43	0.14	50.0	6.40	4.3	1.3	0.05	
BOD, ppm	340	100	55	365	39	470		38	30
COD, ppm	420	125	90	280	160	560	110		
Total PO_4	10	17	14	13.5	12.2	25	13	1.3	<0.1
Settleable solids, ml/liter				2.5		1.0	<0.1	0.3	
Total suspended solids, ppm	292	68	72	220	110	274	80	54	68
Total solids, ppm	727	443	350	776	479	820	459	523	514
Color	395	180	195	115	115	86	79	40	95
Turbidity	162	62	54	21	36	115	40	16	40
Temperature, °C								27	27
Coliform bacteria, 100 ml							<10		<10

day solved its pollution problem in a satisfactory manner. It was forced by circumstances to provide the equivalent of secondary treatment but did so at a cost of less than $100,000 capital expense. Adequate screening, followed by two-stage oxidation utilizing over-and-under contact baffles, and final chlorination gave 85 to 95 per cent BOD reduction. The cooperative spirit exhibited by both the plant and the regulatory authority, combined with some engineering innovations in design, resulted in success.

21.7 An Example

The Growers and Packers Cooperative (Gro-Pac), formed in 1936, was originally located in North Collins, New York. Its wastes were discharged to the North Collins Municipal Sewage Treatment Plant. In February 1965 the municipal treatment plant experienced operating difficulties due to an overload from the plant of high flows and odors from the cannery. At that time, the cannery's waste was classified as a readily oxidizable organic waste which could be biologically treated. Because of waste-related problems with the municipal treatment plant, Gro-Pac decided in 1966 to construct a new plant in Eden, New York, a few miles north of North Collins. The new facility was completed in 1967 and its owners contracted with a local engineer to design an industrial waste treatment facility at their new location.

As a rule, wastes were treated in the plant's three-pond lagoon system. The treated wastes were then discharged into a drainage ditch, which finally discharged into Rythus Creek. During the 1967 packing season (a 10-week period), the following problems occurred:

1. There were offensive odors coming from the ponds.
2. There was the possibility that Rythus Creek's was being polluted.
3. There was over-production and excessive hydraulic loading.
4. Discharge into the drainage ditch was visible.
5. Maintenance of treatment facility was inadequate.

Because of these problems, the state health department issued a letter to Gro-Pac requesting that the problems be corrected before the 1968 packing

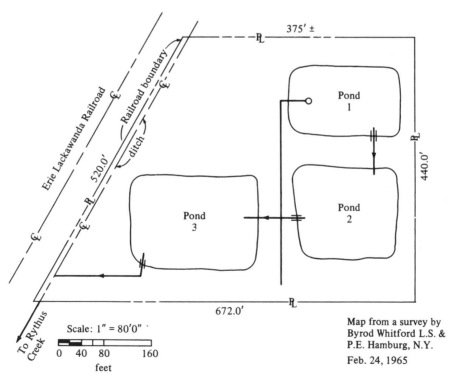

Map from a survey by
Byrod Whitford L.S. &
P.E. Hamburg, N.Y.
Feb. 24, 1965

Fig. 21.7

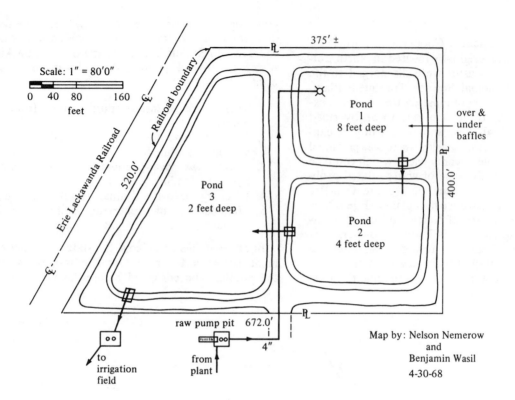

Fig. 21.8

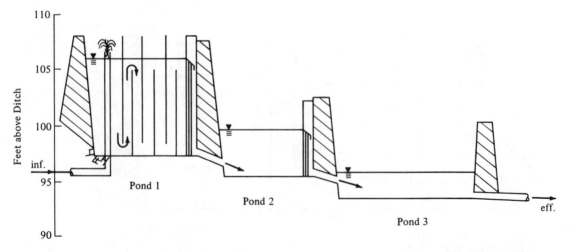

Scale:
Vertical: 1″ = 5′0″
Horizontal: 1″ = 80′0″

By Nelson L. Nemerow
and
Benjamin Wasil
4-30-68

Fig. 21.9

season began or an operating permit would not be issued.

The composite wastes from this plant contained BOD in the range of 200–1000 ppm, COD in the range of 500–2200 ppm, suspended solids in the range of 48–1200 ppm, and pH in the range of 4.2 and 6.6.

Basis of Design

The decision for the design of the new treatment facility was based on the following constraints:

1. The waste produced was a readily-oxidizable organic waste.
2. Adequate (but not abundant) land was available.
3. No sewers or municipal treatment facilities were located in the immediate area.
4. Rythus Creek, the receiving stream, was used mainly as irrigation water.
5. The plant operated each year during the packing season (10 weeks) only.

It was decided after reviewing three previous studies by Nemerow [1, 2, 3,] to use a three-pond system of lagoon treatment, with over-and-under baffles in the deepest of the three ponds. After lagoon treatment this waste was sprayed to irrigate a rye-grass field. Under ideal conditions the sprayed effluent reached the creek only after percolating through the soil substrata. In case of surface runoff only the three-pond-lagooned effluent would reach the creek.

The three studies [1, 2, 3] had shown that over-and-under baffles increased the BOD removal efficiencies of normally operated lagoons. This waste-treatment facility as designed and built is shown in Figs. 21.7, 21.8, and 21.9.

Essentially these three ponds acted as high-rate oxidation basins. The ponds were designed to operate at 8, 4, and 2 foot depths, respectively. Pond 1 was expected to operate as anaerobic in its lower levels and aerobic near the surface. The products of anaerobic decomposition from the settled solids were expected to diffuse from the bottom to the higher levels, where more oxygen would be available. The over-and-under baffle system (33,000 feet^2 of surface area) was used to enhance bacterial, protozoal, and slime growth. The baffles served both to effect complete mixing and to increase the contact between bacterial growth and

organic matter. The detention time was short enough (8 days) to prevent complete anaerobiosis.

Pond 2 was shallower and since its effluent would be of higher quality and clarity more oxygen could be supplied to the oxygen deficient waste by algal growth and surface reaeration. However, the detention time in this pond was also short (4 days) to ensure an adequate oxygen balance. Some residual solids were expected to settle out in this pond also.

In Pond 3 no settled solids were expected. The major emphasis was placed upon algal production in order to supply sufficient oxygen for final oxidation of the remaining decomposible products. The detention time in this pond was designed to provide 6 days under which aerobic conditions should have been sufficient for rather complete (80–90 per cent) BOD reduction. This highly treated waste would then be sprayed on a rye-grass field and tile-underdrain system, which would finally reach Rythus Creek. This additional phase of treatment (spray irrigation) would produce a highly polished and clarified effluent which would minimize its impact on Rythus Creek.

A total detention time in the system of 18 days was estimated if the bottoms and sides of the ponds were sealed with bitumastic or clay and if the ponds were constructed and operated as designed.

The first packing season for the new treatment facility was in 1968 lasting for a period of 10.5 weeks. Prior to this, the following modifications were made:

1. An over-and-under baffle system covering 33,000 ft^2 of surface area was used.
2. Pond 3 was enlarged to double its original size.
3. The spray-irrigation area was regarded and replanted with rye grass.
4. Several banks of distribution pipes were added to the spray-irrigation system.
5. A collection weir and effluent sump were constructed for Pond 3.

The total cost of the treatment facility including the above modifications was $55,000.

Sampling

Samples of the raw waste and three-pond effluents were collected as soon as they became available, as were upstream and downstream samples of Rythus Creek—a potential receiving stream for drainage

Table 21.11   1968 packing-season average analytical results of waste-treatment system

Sample location	BOD (ppm)	BOD (per cent reduction)	COD (ppm)	COD (per cent reduction)	Normal pH range	Suspended solids (ppm)	Suspended solids (per cent reduction)
Raw waste	429	0	1004	0	4.2–6.6	304	0
Pond 1 effluent	276	35.7	191	81	4.5–6.3	48	84.3
Pond 2 effluent	201	53.2	343	65.8	5.2–6.9	47	85
Pond 3 effluent	135	68.7	127	87.2	5.4–7.0	69	77.3
Rythus Creek: When flowing upstream	3.4	—	7.6	—	5.6–7.9	7	—
When flowing downstream	3.8	—	14.1	—	5.8–8.4	13	—

from the spray-irrigation land. Samples were analyzed for BOD, COD, pH, and suspended solids.

Analytical Results

Although the results varied during each packing season as well as from year to year, the overall average results are shown in Tables 21.11–21.15.

Operational Problems and Corrective Measures

During the first year of operation (1968 packing season) a few of the over-and-under baffles in Pond 1 did not remain in position; they floated to the surface and tilted on their sides, which resulted in an inefficient flow distribution throughout Pond 1. In addition, the stilling basin approach to the effluent weir of Pond 1 leaked on the sides all the way to the bottom, allowing bottom anaerobic sludge to escape through the weir stops into the remaining two ponds. In addition, apparently the last baffle was placed too close to the effluent weir and, under existing circumstances, did not provide sufficient settling prior to weir discharge.

Since some of the oxygen-deficient matter was scoured from the bottom of Pond 1 at times, faint odors of hydrogen sulfide existed near the pond's banks. The following steps were immediately taken to correct the situation:

1. The effluent weir stilling basin was filled and sealed with clay so that all the effluent from Pond 1 was discharged only over the surface of the weir.

2. An attempt was made to "right" those baffles which had erupted from their soil anchor.

While the first remedial measure was successful, the floating-baffle problem was not corrected until the 1969 packing season. Although some baffles were placed back in their original vertical position and weighted with concrete filled cans, this task was not completed until the ponds could be emptied and the baffles placed in concrete supports during the summer of 1969.

Because of an unanticipated scouring of the bottom of Pond 1, other remedial measures were taken to prevent odors from reaching the objectionable or nuisance level. These measures included the following:

1. Feeding compressed air into the screened influent pumped to Pond 1.
2. Adding $NaNO_3$ (100 #/day) to the raw screened waste prior to pumping it to Pond 1.
3. Recirculating (by means of a small pump) some of the contents of Pond 3 back into Pond 1.

Even though the ponds were all black and extremely low in dissolved oxygen, odors and hence nuisance conditions were averted during the first packing season (1968) by applying the above-mentioned measures.

As corrections were made in Pond 1, the compressed air being fed to the raw-waste-screen chamber and the recirculation of Pond 3 contents to Pond 1 were discontinued in 1969 with no apparent deleterious effects on the efficiency of the treatment system. However the feeding of sodium nitrate was continued

Table 21.12   1969 packing-season average analytical results of waste-treatment system

Sample location	BOD (ppm)	BOD (per cent reduction)	Normal pH range	SS (ppm)	SS (per cent reduction)
Raw waste	312	0	4.4–5.6	310	0
Pond 1 effluent	197	37.5	6.1–6.3	362	—
Pond 2 effluent	130	63.5	6.2–6.5	200	35.5
Pond 3 effluent	84	74	6.5–6.6	120	61
Rythus Creek: When flowing upstream	—	—	—	—	—
When flowing downstream	—	—	—	—	—

Table 21.13   1970 packing-season average analytical results of waste-treatment system

Sample location	BOD (ppm)	BOD (per cent reduction)	Normal pH range
Raw waste	470	0	4.8–7.2
Pond 1 effluent	176	63	6.5–7.2
Pond 2 effluent	121	75	6.9–7.3
Pond 3 effluent	70	85	6.9–7.9
Rythus Creek: When flowing upstream	—	—	—
When flowing downstream	—	—	—

since analyses indicated a possible deficiency of nitrogen in the raw waste.

Residents in the area were canvassed periodically during the canning season to discover whether they had encountered objectionable odors. None were noted in 1969; however, no data on the odors were available in 1970–71. Slight odors could be detected periodically if one stood on the dike between Pond 1 and Pond 2. These odors were later found to be caused by sulfur in the water supply and the sulfur bacteria converting the sulfur to hydrogen sulfide. Those odors were present only under ideal wind and waste-loading conditions and almost vanished as one moved away from the treatment area. There were no odors present

Table 21.14   1971 packing-season average analytical results of waste-treatment system

Sample location	BOD (ppm)	BOD (per cent reduction)	SS (ppm)	SS (per cent reduction)	Normal pH range
Raw waste	615	0	454	0	5.1–6.8
Pond 1 effluent	274	555	126	50	6.0–7.2
Pond 2 effluent	209	66	66	86	6.8–7.4
Pond 3 effluent	160	75	70	85	6.8–7.5
Rythus Creek: When flowing upstream	—	—	—	—	—
When flowing downstream	—	—	—	—	—

Table 21.15   BOD and efficiency over the period 1968–71

	Packing season							
	1968		1969		1970		1971	
	BOD		BOD		BOD		BOD	
Sample location	(ppm)	(per cent reduction)	(ppm)	(per cent reduction)	(ppm)	(per cent reduction)	(ppm)	(per cent reduction)
Raw waste	429	0	312	0	470	0	615	0
Pond 1 effluent	276	35.7	197	37.5	176	63	274	55.5
Pond 2 effluent	201	53.2	130	63.5	121	75	209	66
Pond 3 effluent	135	68.7	84	74	70	85	160	75

at any time in the spray irrigation field area. Some odors, however, did exist near a truck used to load and haul away the leaves and bean snips screened out of the waste. After bringing this to the attention of the cannery owners, the situation was promptly corrected by emptying the truck more often and washing it thoroughly between loadings.

Gro-Pac personnel were extremely cooperative during the entire period of operation. They assisted in sample collection, made the necessary structural corrections to Pond 1, and cooperated fully in operating the ponds at their proper levels, and in changing waste-treatment operating procedures immediately as instructed.

Results of Treatment

The waste-treatment system, a three-stage lagoon system of three varying depth basins, effected removals of 68–85 per cent BOD and 60–85 per cent suspended solids over a four-year operational period (1968–1971). This reduction was ample to (1) prevent nuisances and (2) to allow for final disposal of the effluent by spray irrigation with no soil clogging, ponding, odors, or contamination of receiving waters. The first pond accomplished the most of the reduction of contaminants as expected. The last pond contained dissolved oxygen, an elevated pH, some algal growth, and low suspended solids. Raw-waste BOD, COD, and suspended solids were about

450, 1000, and 300 ppm, respectively, and were reduced by the series pond treatment to values of approximately 110, 127, and 70 ppm, respectively.

Flow Measurement

Because of the erratic pattern of discharge of raw waste from the cannery and because of the resulting discontinuous discharges from each pond, accurate flow measurements during the average 10-week packing season were not available.

However, the raw-waste pump rated at 400 gpm was operated approximately 50 per cent of the time for 18 hours of the day. This amounted to a daily hydraulic loading of 216,000 gallons per day.

At four times during the 1968 packing season, the flow leaving Pond 1 was measured during the 18 hours of operation; this measurement averaged about 175 gpm. This was in good agreement with the estimated raw-waste flow if one assumes a 12 per cent loss in Pond 1 due to evaporation and exfiltration.

The total waste load was therefore computed on this flow basis (0.216 mgd). This load would be 450 ppm BOD $\times$ 8.34 $\times$ 0.216 mgd or 810 pounds of BOD per day on the three-pond system. Since the three-pond area contains 3.1 acres, the unit loading on this entire lagoon system is 260 pounds of BOD per acre per day.

In comparison with loadings applied to domestic-sewage-lagoon systems of 20–50 pounds of BOD per acre per day, these were roughly 10 times as high. Three reasons for the relatively acceptable efficiency at the elevated loadings are proposed: (1) a readily oxidizable wastewater; (2) a properly designed three-stage-lagoon system and (3) the use of baffles to obtain additional biological growth surface area.

Conclusions

The following conclusions were reached from this case treatment:

1. A three-stage biological-oxidation-pond system was used to successfully treat bean-cannery waste during short packing seasons.
2. No permanent nuisances were created in the area surrounding the plant. Temporary odorous conditions were corrected by $NaNO_3$ feeding and aeration of the lagoon influent in the raw-pump wet-well chamber.
3. Efficient operation of this plant was greatly hampered in the earlier stages by construction and structural failures.
4. After all the structural problems were corrected, the plant operated at a considerably higher efficiency—an increase from 68 per cent to 85 per cent BOD reduction.
5. The three-pond system with over-and-under baffles in the first pond effected seasonal average BOD, COD, and suspended solids removals of 78, 87.3, and 77 per cent, respectively. The estimated BOD loading was 260 pounds per acre per day.
6. The pond-system-treated effluent was successfully spray irrigated on a rye-grass field without nuisances such as odors, ponding, or contamination of the potential receiving streams. The irrigation-field loadings were 72,000 gallons per acre per day over the 3 acre field and a BOD load of 67 pounds per acre per day was applied. These values are well within acceptable hydraulic loads and BOD loading did not seem to be the critical design factor.

References

1. Nemerow, N. L., "Accelerated Waste Water Pond Pilot Plant Studies," Advances in Biological Waste Treatment, Proceedings of 3rd Conference Biological Waste Treatment, Manhattan College, April 1960, Pergamon Press Ltd., London (1963).
2. Nemerow, N. L., "Poultry processing waste treatment at Millsboro, Delaware", Proceedings 22nd Industrial Waste Conference, Purdue University, May 1967.
3. Nemerow, N. L., "Baffled biological basins for the treatment of poultry plant wastes," *WPCFJ.* **41** (9), 1602 (1969).

Questions for Problem 1

1. When is it necessary from a technical standpoint to completely treat an industrial waste prior to discharge into a stream?
2. What does this particular problem in waste treatment entail?
3. Why was this wastewater so difficult to treat?
4. Why was it important to remove such a high percentage of the organic matter?
5. Why wasn't a river study carried out?
6. Describe the unique treatment system used to obtain our objectives in this case.
7. What was the major limiting constraint placed by the industry on waste treatment in this case?
8. What combination of factors allowed the author to comply with this constraint and still obtain the treatment required?

Questions for Problem 2

1. What is the major difference in background factors in this problem from the first one in this chapter?
2. What is the major difference in technical character of this waste from the first waste?
3. What is the purpose of the baffling in the oxidation ponds in both problems?
4. What combination of conditions led to the solution selected by the author to this problem?
5. What are the purposes of the relatively deep first basin and shallow third basin?
6. What is necessary for final disposal of treated industrial wastewater onto the land?

CHAPTER 22

COMPLETE TREATMENT OF INDUSTRIAL WASTE PLUS EFFLUENT REUSE

22.1 Introduction to Mathura Refinery Problem

The Government of India decided to set up a large oil refinery in the Northwest region to meet the growing petroleum products demand of the region. The techno-economic studies conducted established that the Mathura area was the most suitable location for the refinery, the principal considerations being central location within the demand area, proximity to both Broad Gauge and Metre Gauge Railway Lines as well as the National Highway, and availability of land.

The Mathura Refinery has a nominal capacity of processing 6 million tons of crude per annum, out of which 3 million tons are imported crude and 3 million tons are Bombay High Crude. The processing is in cycles. The imported crude has a sulphur content of less than 2%. The sulphur content of Bombay High Crude is less than 0.2%. It went into operation in 1982.

22.2 Manufacturing Processes

The Mathura Refinery has been designed for maximization of middle distillates and consists of the following processing units:

Unit	Capacity in '000 tons per annum
(1) Atmospheric Distillation Unit with Desalter	6,000/7,000
(2) Vacuum Distillation Unit	2,300
(3) Fluid Catalytic Cracking Unit	1,000
(4) Vis Breaker	1,000
(5) Bitumen	500
(6) Merox Units for:	
SR LPG	70
Cracked LPG	145
Visbreaker Naptha	70
ATF/KERO (Aviation Fuel/	1,500
Kerosene)	
FCC/LSR Gasoline	300
(Fluidized Catalyic Cracker)	
(7) Sulphur Recovery Unit	10

In addition to these process units, the refinery will have the following facilities:

1. Crude and Product Storage Tanks
2. Product Despatch facilities by Pipeline, Rail and Road
3. LPG Bottling facilities.
4. Bitumen Drum filling facilities.
5. TEL Blending facilities.
6. Thermal Power Station. 37.5 MW Capacity
7. Effluent Collection and Treatment Facilities.
8. Water Treatment Facilities
9. Other auxiliary facilities such as:
   -Pump Stations
   -Laboratory
   -Warehouses
   -Workshop, etc.

22.3 Effluents and Characteristics

The effluents emanating from the Process Units and other facilities can be divided into two categories: those to be discharged to the surface and those to be discharged to atmosphere. Only the former will be considered in this book. The basic effluents to be discharged to the surface are the liquid effluents in the Refinery. The waste water streams emanating from various places are shown in Tables 22.1 and 22.2.

The refinery effluent leaves the boundary limits of the plant and joins with the treated effluent from the Refinery Township (consisting of 800 private homes) at a point downstream of Brahmandghat, ensuring adequate mixing in the Yamuna River. There is no habitation from the discharge point for about 15 to 20 kms downstream. This discharge point is about 40 kms upstream

Table 22.1 Characteristics of Effluent Streams

Description of Waste	Flow (M³/hr)	pH	Temp. (°C)	BOD (ppm)	COD (ppm)	Hydro-carbon (ppm.)	Phenol (ppm.)	Sus-pended solids (ppm)	Total Salt con-tent (ppm)	Ammoni-cal Ni-trogen (ppm)	NaSH, NO₂S (ppm)	H₂S (ppm)
1	2	3	4	5	6	7	8	9	10	11	12	13
1. INDUSTRIAL SEWAGE (a) Main Process Units	108	7–8	Ambient	400–700	700–1100	6000	4–10	400–800	1400–3000	50–60	—	—
(b) FCC Unit	10		-do-	50		1000						
(c) Offsite areas	30		-do-	50		1000						
(d) Storm Water from Process Units	1000 Max Inter-mittent	7–8	-do-	50		1000	2	200				
(e) Storm Water from FCC/Merox and Offsites.	1050 Max Inter-mittent		-do-	50		500		200				
2. Storm Water Drain from Tank Farm	2190 Max	7–8	-do-	50		500		200				
3. Crude Oil Tank Drain	100 M³/Day					20,000		800	4500–5500			
4. Electric Desalter Drain	60	7.5–8.5	40°C	250–350	500–700	150	2–4	400–1200	3500–4500	20–30	200	4
5. Domestic Sewage	40	6.5–7.5	Ambient	200								

Table 22.2

Description	Quantity/ Discharge	Frequency	pH	BOD (ppm)	COD (ppm)	Hydrocarbon (ppm)	Phenol (ppm)	Sulphides as H₂S (ppm)	Total Sulphur (ppm)	Total Alkalinity as NaOH	RHS as 'S' (ppm)	H₂S as 'S' (ppm)
1	2	3	4	5	6	7	8	9	10	11	12	13
1. Spent Caustic from Naphtha Caustic Wash	30M³	Once or twice in a week	14	75,000	85,000	3000	5000—10,000	25,000—50,000	35000	5%		
2. KERO/ATF Merox Coalescer Water wash	0.55M³/hr	Continuous	12–13					Traces			(Contains Naphenic Acids)	
Caustic prewash	2.75M³/hr 76M³/hr	-do- Once in 9 days	12–14	(50% available NaOH spent as Sodium Napthanates)				Traces		0.2–0.4%		
Fixed Bed Reactor Caustic	69M³	Once in 5 days	14	(Total N - 8 Wt ppm)				250	200	6.9%	1.0	1.0
4. SR LPG Merox Caustic Prewash	2.16M³	Once in a month		(80% of available NaOH spent as Na₂S of NaSH)								
Visbreaker Naphtha Merox Caustic Prewash	1.73M³	Once in a month		(80% of available NaOH spent as Na₂S & NaSH)								
Fixed Bed Reactor	2.05M³	Once in 90 days	14	(Total N - 8 Wt ppm)				250	200	6.9%	1.0	1.0
5. Cracked LPG Merox Caustic Prewash	4.23M³	Once in a month		(80% of available NaOH spent as Na₂S & NaSH)								
Extractor	0.41M³	Daily						Traces	(Traces of thiosulphate, Disulphide, Mercaptans)			
6. FCC/LSR Gasoline Caustic Prewash	1.82M³	Once in a month		(80% of available NaOH spent as Na₂S & NaSH)								
Caustic Settler	13.64M³	Once in 3 months		(Acid Oil Content (Phenols & Cresols) 7% Vol.)					2000	1%		
Sour Water	15.5M³/hr	20 days in a year		(Ammonia 2400 ppm,		80	2400—4000 Cyanides-	3200—4000 Traces				6000

of the city of Agra water supply works, which is the nearest municipal consumer downstream. Agra is a tourist city in which the Taj Mahal resides.

The treatment facilities that are being provided are:

1. Effluent Treatment Plant (ETP)
2. Neutralization facility for DM Plant Effluent
3. Sour water stripper

A schematic diagram showing the sources, collection, and treatment of waste water is shown in Figure 22.1.

22.4 Effluent Treatment Facilities

The wastewater streams entering the effluent treatment plant are collected in four sumps depending on their characteristics as shown in the flow diagram of ETP given in Figure 22.2.

The incoming streams are collected in different sumps as shown below:

1. Alkaline Waste — Two concrete, rectangular tanks of 1000 M^3 each.

2. Industrial Sewage/ Storm Water (Process Units) Salty Effluent, Electric Desalter Drain. — Two sumps of 220 M^3 and 160 M^3 capacity.

3. Storm Water Drain — One sump of 425 M^3 capacity.

4. Domestic Sewage — One sump of 40 M^3 capacity.

22.5A Process description: Alkaline waste

The Alkaline Waste Stream which flows intermittently from the units is stored in two tanks. These tanks are provided with oil skimming capabilities to remove any free floating oil. The alkaline waste from these tanks is fed at a constant rate to the mixing sump where it is diluted by process unit cooling tower blowdown. The alkaline stream flow is regulated between 8–18 M^3/hr and the cooling tower blow down varies between 136–156 M^3/hr. There are provisions to divert water from a guard basin to a mixing sump in lieu of a cooling tower blow down. The average flow of the mixed stream is taken as 150 M^3/hr when the sour water stripper is shut down (for 20 days in a year). During this period, sour water will be directly routed to the mixing sump. The effluent from the mixing sump is pumped to the API oil separator. This API separator has two channels out of which one is used at a time, the other being a standby. The oil skimmed from the API separator and alkaline waste storage tank flows to slop oil sump. The sludge from API Separator goes to the oily sludge sump.

The effluent from the API separator goes to the pH adjustment chamber where a 10% concentration of sulphuric acid is added to bring down the pH value of the stream to 9.5. At this pH all the sulphides exist in the form of NaHS or NA_2S and not as hydrogen sulphide and as such there is no possibility of liberation of H_2S into the atmosphere. In the reaction chamber, chlorine is dosed to react with sulphides and precipitate sulphur. Ferrous sulphate is used as flocculant and is added in the flocculator. It then goes to the clarifier cum thickener for clarification as well as thickening of flocs formed in the flocculator. The sludge collected here goes to the sludge conditioning sump. The overflow goes to the equalization pond where it mixes with process and other wastes.

22.5B Oily Water

The industrial waste sump consists of two parts. From one part the water is pumped to a splitter box and from the other it is pumped to the guard basin. From the splitter box, all the water goes to the API oil separator. During normal operation and during rains when the flow is more, the excess quantity is diverted to a guard basin. The storm water which is collected in another sump is pumped to the guard basin. The water collected in the guard basin is pumped back to the API oil separator for treatment during lean periods (when there are no rains and the flow is low) at the rate of 150 M^3/hr.

The API Oil Separator consists of four channels, out of which three are used at a time, the fourth being a standby. The design load for each channel is 3700 M^3/day (based on dry weather flow—7500 M^3/day and 3600 M^3/day of storm water). The skimmed oil from the API separators goes to the slop oil sump and sludge to oily sludge sump. The effluent water from here goes to an equalization pond.

The overflow from equalization pond flows by gravity to high rate trickling filter.

Domestic Sewage

The Domestic Sewage is collected in a separate sump.

After treatment in a screen and grit chamber, it is routed to the trickling filter. There are two units of screen and grit chamber; one is used while the other serves as a standby.

The recirculation stream from the effluent of the final clairifer also joins the inlet to the trickling filter. Full provision for recirculation up to 100 percent is made, including that for storm water flow. There also is a provision to add urea and phosphoric acid as nutrients at the inlet of trickling filter.

Activated Sludge Process

The effluent from the trickling filter is transferred to an aeration tank by pump. The sludge recirculation stream is also pumped to the inlet of the aeration tank separately. The aeration tank consists of three channels with three aerators per channel. The removal efficiency of the aeration tank for BOD, Phenols and sulphides is 90 percent.

Both the trickling filter and aeration basins have by-pass arrangements to be used in case of breakdown/ maintenance. The overflow from the aeration tank goes to a final clarifier from where the bio-sludge is removed to the bio-sludge sump. The clarified effluent then flows to a final guard pond. Part of the clarified effluent is pumped to the trickling filter as recirculation. In the event that recycling is required, there is a provision to send water from the guard pond to the oily process wastewater sump for retreatment. The effluent from the guard pond is routed to a holding pond (having two days holding capacity) for further oxidation. From here the effluent joins the stormwater channel.

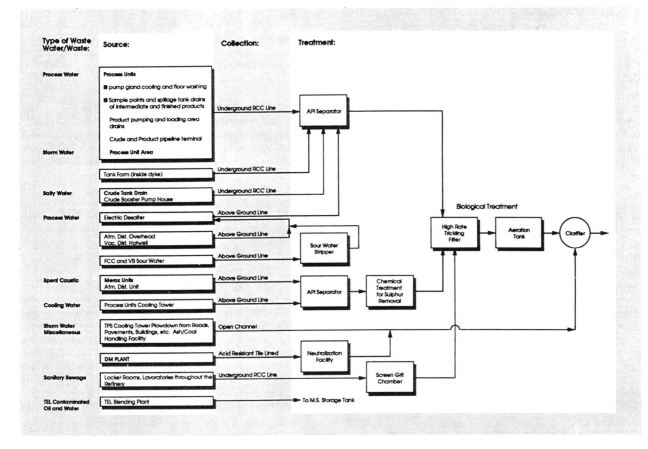

Figure 22.1   Schematic diagram showing source collection and treatment of liquid effluents flow diagram.

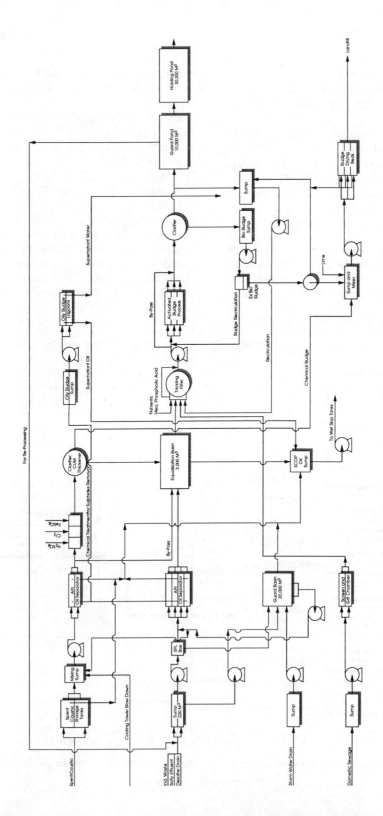

Figure 22.2 Mathura Refinery, schematic flow diagram of wastewater treatment plant.

Slop Transfer

The skimmed oil collected in the slop oil sump is pumped out to two wet slop tanks (each with a 500 kl capacity). The slop oil, after initial preparation in wet slop tanks, is pumped to the main slop tanks. After further preparation, it is reprocessed in an visbreaker unit. (There is a provision to reprocess it in an atmospheric distillation unit also).

Sludge Disposal

The oily sludge is pumped out to oily sludge lagoons consisting of two channels with a capacity of 6,000 M^3 each. There is provision to send the oil that may get separated in the oily sludge lagoon to a slop oil sump.

The excess biological sludge is thickened in a sludge thickener. The thickened sludge goes to a sludge conditioning sump where it mixes with the chemical sludge. Here lime is added for better sludge settling. The conditioned sludge is pumped out to sludge drying beds. The overflow from the bio-sludge thickener and underdrain from the sludge drying beds are collected in a sump and pumped to the trickling filter recirculation sump. The overall treatment scheme for both the domestic and industrial waste is shown in Figure 22.3.

Design Considerations for ETP

Design Flows	M^3/hr	M^3/day
Alkaline Waste after dilution with Cooling Tower blowdown	150	3600
Dry Weather flow of industrial waste water	315	7500
Storm Water	150	3600
Domestic Sewage	40	1000
Total:	655	15,700

The Waste Caustic Treatment facilities are designed for 150 M^3/hr flow.

The API Separators for Industrial Waste Water is designed for 465 M^3/hr.

The screen & grit chambers are designed for a maximum flow of 80 M^3/hr.

The trickling filter and aeration tank are designed for a flow of 655 M^3/hr.

The design load for chemical section is :

BOD	2500 kg/day	700 mg/lit.
Hydrocarbon	100 kg/day	27 "
Alkalinity as NaOH	6600 kg/day	1831 "
Sulphides as S	1200 kg/day	333 "
Phenols	350 kg/day	97 "

The effluent from chemical treatment section shall confirm to the following :

pH	7.0–8.5 Max.
Sulphides	40 mg/lit. Max.
Suspended Solids	30 " "

22.6 River Standards

The present river, groundwater, and coastal water quality standards are given in Tables 22.3 & 22.4.

22.7 Effluent Disposal

The treated effluent from the holding pond joins the storm water channel and the combined streams form the effluent outfall channel. Before this channel leaves the refinery boundary limits, an oil trap has been provided. The treated effluents from the refinery township also join this channel. The estimated quantity of treated waste water that is discharged into Yamuna River via this brick lined open effluent outfall channel is of the order of 3 MGD. An inspection road is provided along the channel. The point of discharge of treated waste water into the river water has been so selected as to ensure proper mixing. It has been ascertained that the extent of dilution of the waste water, even when there is minimum flow in the river, will be of the order of 1:10. This discharge point is over 40 kms upstream of the Agra Water Supply Works, which is the nearest Municipal Consumer Down Stream.

The treated effluent from the treatment plant in the refinery, as well as the combined effluent (including treated effluent from the township), conforms to the limits specified by the Indian Standard Specifications for Industrial Effluents to be discharged into inland surface waters (IS-2490).

There is a proposal to use the treated effluent for irrigation purposes. The National Environmental Engineering Research Institute has been entrusted with studies for the utilisation of treated effluents for irrigation.

In this case there would be no impact upon the river. The main criteria for the effluent use as irrigation water will be the tolerance of the agricultural products for the treated wastewaters.

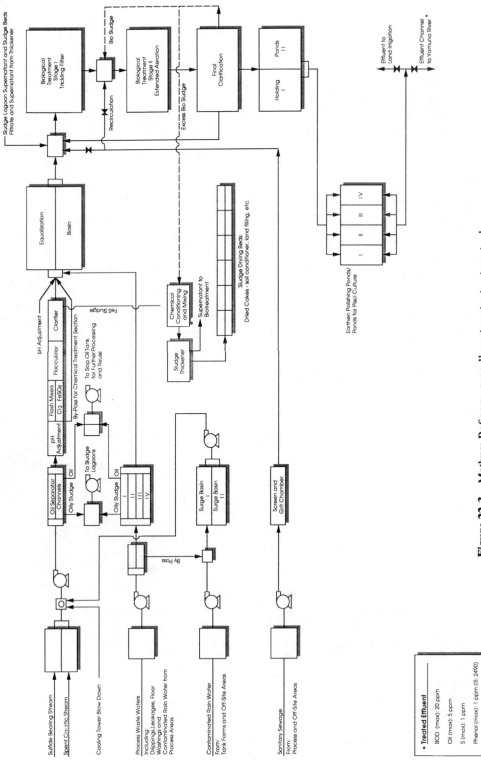

Figure 22.3 Mathura Refinery, overall wastewater treatment scheme.

Biological Treatment Stage I Trickling Filter

Bio Sludge

Biological Treatment Stage II Extended Aeration

Final Clarification

Holding I   Ponds II

Recirculation

Excess Bio Sludge

Sludge Lagoon Supernatant and Sludge Beds Filtrate and Supernatant from Thickener

Equalization Basin

pH Adjustment

Clarifier

Flocculator

Flash Mixers Cl₂ FeSO₄

pH Adjustment

Oil Separator Channels

Fe S Sludge

To Slop Oil Tank for Further Processing and Reuse

By-Pass for Chemical Treatment Section

To Sludge Lagoons

Oil

Oily Sludge

Oily Sludge

Oil

I II III IV

Chemical Conditioning and Mixing

Supernatant to Biotreatment

Sludge Thickener

Sludge Drying Beds:
Dried Cakes - soil conditioner, land filling, etc.

I II III IV

Earthen Polishing Ponds/
Ponds for Pisci Culture

Effluent to
Land Irrigation

Effluent Channel
to Yamuna River *

Sulfide Bearing Stream

Spent Caustic Stream

Cooling Tower Blow Down

Process Waste Waters
Including:
Drippings,Leakages, Floor
Washings and
Contaminated Rain Water from
Process Areas

By Pass

Surge Basin I   Surge Basin II

Contaminated Rain Water
From:
Tank Farms and Off-Site Areas

Screen and
Grit Chamber

Sanitary Sewage
From:
Process and Off-Site Areas

* **Treated Effluent**

BOD (max): 20 ppm

Oil (max): 5 ppm

S (max): 1 ppm

Phenol (max): 1 ppm (IS: 2490)

Table 22.3 I.S.I. Standards for Discharge of Industrial Effluents (All values except pH, temperature and radioactive materials are in mg/l)

Characteristics	Tolerance limits for industrial effluents discharged		
	Into Inland surface waters (IS:2490–1974)	Into Public sewers (IS:3306–1974)	On Land for irrigation (IS:3307–1965)
BOD, 5 days 20°C	30	500[a]	500
COD	250	—	—
Suspended solids	100	600[b]	—
Total dissolved solids (inorganic)	—	2100*	2100
pH	5.5–9.0	5.5–9.0	5.5–9.0
Temperature °C	40	45	—
Oil and grease	10	100	30
Phenolic compounds	1.0	5.0	—
Cyanides	0.2	2.0	—
Sulphides	2.0	—	—
Fluorides	2.0	—	—
Total residual chlorine	1.0	—	—
Insecticides	Absent	—	—
Arsenic	0.2	—	—
Cadmium	2.0	—	—
Chromium (Hexavalent)	0.1	2.0	—
Copper	3.0	3.0	—
Lead	0.1	1.0	—
Mercury	0.01	—	—
Nickel	3.0	2.0	—
Selenium	0.05	—	—
Zinc	5.0	15	—
Chlorides	—	600*	600
Boron	—	2*	2
Sulphates	—	1000*	1000
Per cent sodium	—	60	—
Ammoniacal nitrogen (N)	50	50	—
Radioactive materials Alpha emitters, μc	10^{-7}	10^{-7}	10^{-9}
Beta emitters, μc	10^{-6}	10^{-6}	10^{-8}

Table 22.4 IS: 1968–1976 Tolerance Limits for Industrial Effluents Discharged into Marine Coastal Areas

Characteristic	Tolerance Limit
Copper (As Cu), mg/l, Max	3.0
Lead (as Pb), mg/l, Max	1.0
Chromium, mg/l, Max	1.0
Cadmium (as Cd), mg/l, Max	2.0
Mercury (as Hg) mg/l, Max	0.01
Nickel (as Ni), mg/l, Max	5.0
Zinc (as Zn), mg/l, Max	5.0
Total suspended solids, mg/1, Max	
a) For process wastewaters	100
b) For cooling water effluent	Total suspended matter content of influent cooling water plus 10%
Particle size of :	
a) Floatable solids, Max	3 mm
b) Settleable solids, Max	850 microns
pH value	5.5 to 9.0
Temperature, Max	45°C at the point of discharge
Biochemical Oxygen Demand (5 days, at 20°C), mg/l, Max	100
Oils and grease, mg/l, Max	20
Phenolic compounds (as C_6H_5OH) mg/l, Max	5.0
Cyanides (as CN), mg/l, Max	0.2
Sulphides (as 5), mg/l, Max	5
Alpha emitters, μc/ml, Max	10^{-8}
Beta emitters, μc/ml, Max	10^{-7}
Residual chlorine, mg/l, Max	1
Arsenic (as As). mg/l, Max	0.2
Selenium (as Se), mg/l, Max	0.05
Ammoniacal Nitrogen (as N), mg/l, Max	50
Chemical Oxygen Demand, mg/l, Max	250
Pesticides :	
a) Organo-phosphorus compounds (as P) mg/l, Max	1
b) Chlorinated hydrocarbons (as Cl) mg/l, Max	0.02
Fluorides (as F), mg/l, Max	15

[a]Subject to relaxation or tightening by local authority.
[b]Relaxable to 750 by the local authority.
*These requirements shall apply only when after treatment the sewage is disposed of on land for irrigation.

CHAPTER 23

COMBINED INDUSTRIAL AND MUNICIPAL WASTES FOR TREATMENT AND EFFLUENT RECHARGE INTO GROUNDWATER FOR LAND PRESERVATION

23.1 Problem: The Rivers Jala and Spreca

Three major pollutors contribute to the excess contamination of the Jala and Spreca Rivers in Central Yugoslavia: (1) the municipality of Tuzla, Yugoslavia, (2) the industries of the town of Tuzla, and (3) the HAK Chemical Industries I and II. Their respective locations are shown in Figures 23.1 and 23.2.

All three major wastewaters are presently discharged directly and untreated into the Jala river, which flows into the Spreca river at Lukavac at the outlet of the Modrac Accumulation Lake created by the dam 17 years ago.

The Spreca River flows for 60 kms to the confluence of the Bosna River. The Jala and Spreca rivers are classified as Class III; however, neither river currently meets these qualities.

The major problems facing Tuzla and its associated industrial plants are:

1) should separate or combined treatment of the three major wastes be utilized?

2) can stream standards be maintained by the treatment selected?

3) what treatment(s) are required?

4) how shall the sludge(s) be disposed of?

Tuzla means "salt" in the Yugoslavian language for the major industry upon which the town was founded and based on several hundred years ago. Salt is still being produced in Tuzla, and the waste mineral salts are still being discharged and untreated into the two rivers. Liquid salt is also still being pumped out of the ground for manufacturing salt product, resulting in gradual subsidence of the ground in the surrounding area. This has resulted in cracking of walls in buildings. Many houses

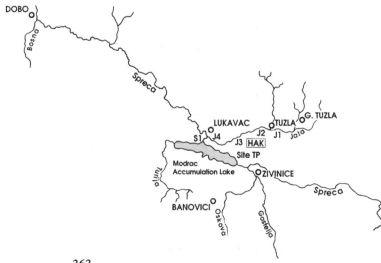

Fig. 23.1 Catchment area of the River Spreca, showing location of river monitoring points.

and buildings have already been demolished for safety reasons and people and facilities relocated at great expense and emotional stress to other areas around Tuzla.

23.2 Data Available

The following characteristics have been made available by the Sodaso Research Institute in Tuzla from their analyses of the three major sources of wastes:

(1) Tuzla waste is typically domestic.
(2) The town industries are mainly typical small organic food manufacturers such as dairy, meat, yeast, beer, laundry, garages, hospitals, etc.
(3) The HAK I chemicals of propylene oxide, polyols, detergents (hard) and considerable

amounts of dissolved solids. HAK II chemicals are largely toluidine, nitrophenol, and ODCB. The design period is for the year 2000.

1-Tuzla has a 1980 population of 90,000, of which 60,000 are connected to the sewerage system. In the year 2000, a total of 120,000 P.E. will be connected— 100,000 to a combined system and 20,000 to a separate system. The characteristics of Tuzla town (1) and industries (2) of the town wastes are given in Table 23.1. The HAK I chemical industries characteristics are shown in Table 23.2. The HAK II chemical industrial component characteristics are shown in Table 23.3. In Table 23.4 the HAK II total effluent characteristics after neutralization are given.

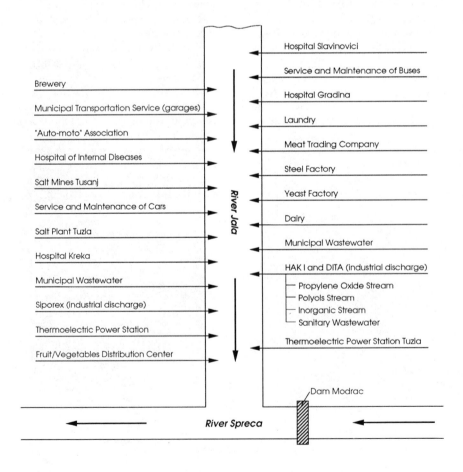

Fig. 23.2   Schematic layout of the main polluters in the River Jala Basin

Table 23.1   Town of Tuzla and Town Industries Characteristics

Type of Waste	Volume (M³/d)	BOD (Kg/d)	COD (Kg/d)	SO4 (Mg/l)	CL (Mg/l)	P-P04 (Mg/l)	N-NH3 (Mg/l)	NO3-N (Mg/l)	Oils & Fats (Kg/d)
Domestic	30,000	7,200	18,000						
Beer Mfg	470	873	1,253						
Yeast Mfg	411	3,058	3,900	900	1170	95	800	400	
Milk Mfg	400	360	650						
Meat Mfg	161	193	386						64
Auto Services	2,880								
Service And	30	1.5	5.0						
Laundry	50	64	103						
Hospitals	800								
Fruit & Veg.	320	480	1,440						
Total Tuzla	35,521	12,421.5	26,217						

Table 23.2   HAK I chemical industries characteristics

Characteristic			
Flow	(M³/day)	4,349	
	(M³/hr avg)	243	
pH	(min)	7.6	
	(min avg)	9.5–10.7	
	(min avg)	10.4–12.1	
Temp (°C)	(max	47–65	
	(min)	35–58	
COD (ppm)	(avg daily)	2,375	
COD (Kg/day)	(avg daily)	10,531	
BOD (ppm)	(avg daily)	718	
BOD (Kg/day)	(avg daily)	3,188	
ECH (ppm)		5–30 (epichlorohydrine)	
PCH (ppm)		<4–43.8 (propylene chlorohydrine)	
PDH (ppm)		25.5–230 (propylene dichlorohydrine)	
PO (ppm)		22.5–121 (propylene oxide)	
PCH2 (ppm)		<4–15 (propylene chlorodihydrine)	
DCDIPE (PPM)		30–69 (dichloridepropiline)	
Propyleneglycol (ppm)		1570 (biodegradable)	
Ca + + (ppm)	(average)	10,904	
Ca + + (Kg/day)	(average)	47,421	
Hydroxide (ppm)		953 (to neutralize to ph 8.3 following prelim. neutralization by industry)	
Cl-(ppm)	(average)	21,500	
Cl-(Kg/day)	(average)	93,500	
Total Suspended (Mg/l) Solids		1,449 (following preliminary sedimentation by Industry)	

Table 23.3   HAK II industrial component characteristics

Type of Waste	Volume (m³/d)	Toluidine (kg/d)	pH	Toluene (kg/d)	DNT (kg/d)	NaCl (kg/d)	Na₂CO₃ (kg/d)	(NH₄)₂SO₄ (kg/d)	NH₄NO₃ (kg/d)	NaNO₃ (kg/d)	NaNO₂ (kg/d)
Streams from Incineration	55.9	28.9	~7	No salts; may contain excessive NO₃ after stack washing.							
Effluent from Toluene Extraction	134.2	0		134.4	19.2	21.6	26.4	55.2	732	88.3	1452
Effluent from Neutralization (m³/d)	(3A) 29.3	(3B) 8.0		(2) 450		(7) 120					
ODCB (ppm)	200	—		10		—					
NaCl (ppm)	80,000	20–24,000		600		900					
Na₂CO₃ (ppm)	50,000	160–240,000		—		—					
NaOH (ppm)	150,000	60–85,000		—		—					
BOD (kg/d)	0	0		68		—					
COD (kg/d)	0	0		90		20 (ppm)					
Susp. Solids (ppm)	0	0		184		110					
pH	All effluents between 12.5–13.0										

Table 23.4   HAK II chemical industries final effluent following neutralization

Flow (m³/d)	-	798.9
COD (kg/d)	-	681
BOD (kg/d)	-	227
Susp. Solids (kg/d)	-	97
pH	-	7–9
Toluidine (kg/d)	-	28.9
DNT (kg/d)	-	19.2
ODCB (kg/d)	-	11.9
Toluene (kg/d)	-	134.4
Total Organics (kg/d)	-	194.4
NaCl (kg/d)	-	10,249.8
Na₂CO₃ (kg/d)	-	3,411.4
NaNO₃ (kg/d)	-	1,452.0
NaNO₂ (kg/d)	-	88.8
NH₄NO₃ (kg/d)	-	732.0
(NH₄)₂SO₄ (kg/d)	-	55.2
Total Inorganics (kg/d)	-	15,989.2

Location of Proposed Town of Tuzla Treatment Plant

Six hectares of land have been reserved next to the power plant for this facility (Fig. 23.1). More area is available for extension. A limit on maximum building height of 30 meters has been set, but a minimum distance of 5 meters beneath the cables must be observed and no inflammable materials may be used. The Jala river overflows its banks approximately once every 100 years.

Discharge Requirements for Tuzla Sewage Treatment Plant

According to the local water authority responsible for the water quality of the Tuzla region, the effluent of the combined sewage treatment plant has to meet the following requirements:

BOD— 20 mg/1 (average)
SS— 30 mg/1 (average)

In addition, legally prescribed maximum allowable concentrations are given for two poisonous components are given:

epichlorohydrine   0.01 ppm
propandiol   10 ppm

The Institute for Chemical Engineering in Tuzla has

Table 23.5 Recommended MDK Values of Organic Impurities from HAK-I Wastewaters (Propilen-Oxide) and Calculated Concentrations for Common Treatment and the Rivers Jala and Spreča Waterstreams.

Component	Legally Prescribed MDK III Class (mg/l)	Recommended MDK for III Class (mg/l)	Recommended MDK for Activated Sludge (mg/l)	Estimated Average Conctn. in Wastewater HAK-I (mg/l)	Calculated Conctn. in the Jala Watercourse (89 1/s) (mg/l)	Calculated Conctn. if HAK-I is Discharged into the Spreča (mg/l)	Conctn. in Common Treatment 24,000 m³/d (mg/l)	Conctn. in Common Treatment 40,000 m³/d (mg/l)
Epichlorhidrine (ECH) Propilenhlorhidrine (PCH)	0.01	0.02	13	11.4(5–20)	5.2	0.09	1.84	1.10
Propilenhlorhidrine (PCH)	—	0.004	30	25.2(0)	12.8	0.2	4.00	2.45
Propilendichloride (PDC)	—	0.09	35	16.2(5–10)	8.2	0.13	2.61	1.57
Sekundarnibutilhloride (SBH)	—	1.5	90	-.(2–5)	2.5	0.04	0.8	0.48
Propilenoksid (PO)	—	1.6	no norm.	93(88)	47.2	0.73	15	9.0
Propionaldehyde (PA)	—	1.0	"	103(103)	55.3	0.86	17.?	10.6
Propandiol (PG)	10	5.5	"	312(380)	158	2.5	50	30.3
Dichloridedopropiletel (DCDIPE)	—	—	—	(30–67)(9)	15–34	0.12–0.27	4.8–10.8	2.9–5.6

*The recommended MDK is given on basis of "Study of toxicity," ITEN Sarajevo, 1980.

formulated recommendations for the maximum concentrations allowable in class III surface water for the most important components present in the waste water HAK I and HAK II, on a basis of their own experiments and literature. These figures are given in tables 23.5 and 23.6.

Both the Jala and the Spreca rivers have to meet the standards of class III (Table 23.1). Further, a waste water amount of four times the dry weather flow has to be treated in a primary sedimentation tank and an amount of 1.5 times the dry weather flow has to be treated biologically. The term, dry weather flow, expresses the average daily flow divided by 24 hours. The waste water from HAK I and HAK II always has to be treated completely biologically.

23.3 A Review of Legal Regulations and the Categorization of the Jala Watercourse

According to the legal regulations, the decree on the classification of waters and the decree on the categorization of water-courses, respectively, surface, ground and lake waters are divided into four classes (1-IV) with limit-values of quality indicators as set out in Table 23.7.

The decree on maximum allowable concentrations (MDK) or radionucleides and other hazardous substances in inter-republic watercourses, international waters and coastal sea waters, establishes MDK values for individual substances for Class I, II, & IV waters, respectively. Where several toxic and hazardous substances occur simultaneously in the waters, the MDK values must be reduced proportionately to the number and types of substance present. However, for the major proportion of contaminants in the effluents from chemical industries based on chlorine (and including the manufacture of propylene oxide and polyols), MDK values have not been established and thus need to be determined by means of toxicity tests.

The decree specifies that the indicators are applied to surface watercourses at their lower volume equivalent to 95% guaranteed flow and also to the guaranteed minimum volume of artificial storage reservoirs.

In accordance with the decree on categorization of watercourses, it was specified that the waters within the catchment areas of the rivers Spreca and Bosna should correspond to the following classes.

Table 23.6 MDK Values for Watercourse and Biology of Impurity Components from HAK-II (TDI) Wastewaters and Calculated Concentrations for Common Treatment.

Component	Recommended MDK for III Class Watercourse (mg/l)	Recommended MDK for Biological Treatment (mg/l)	Estimated Conctn. in HAK-II Wastewaters (Original) (mg/l)	Estimated Conctn. in HAK-II Wastewaters after Pre-treatment. (mg/l)	Calculated Conctn, in the Jala (mg/l)	Calculated Conctn. if it is Discharged into the Spreča (mg/l)	Conctn. in Common Treatment 24,000 m^3/d (mg/l)	Conctn. in Common Treatment 40,000 m^3/d (mg/l)
Dinitrotoluen DNT	0.5	0.5	339	18.7	2.5	0.04	0.79	0.48
Mononitrotoluen MNT	0.01	0.01	26.3					
Orthotoluendiamin OTD	0.001	<1	94					
m-toluendiamin MTD	0.001	<1	62					
Toluidini	<1		26.8	28.2	3.16	0.06	1.2	0.72
0-nitrofenol	0.06	0.6						
M-nitrofenol	0.06	3.0	12.3	<1	0.13	0.002	0.04	0.02
P-nitrofenol	0.025	0.4						
O-dichlorobenzeneODCB	0.02	0.2	9.6	7.2	0.96	0.015	0.3	
Nitrati (KaO N)	15	—	3190					
Nitriti (KaO N)	0.5	—	330					
Total diluted materials	1500	—	25,000					
Suspended materials	80		—					
pH	6–9	6.5–8.5	10–11	6.5–8.5				
Toluene	25	—	0	131.2	17.5	0.27	5.6	3.3

*MDK data is given on basis of "Study of possibilities for purification of HAK-II wastewaters." IHI, 1980.

Table 23.7   Categorization of Surface, Ground and Lake Waters

	Class I	Class II	Class III	Class IV
Suspended solids, max.	10 mg/l	30 mg/l	80 mg/l	—
Total dissolved Solids, max.	350 mg/l	1000 mg/l	1500 mg/l	—
Dissolved oxygen, min.	8 mg/l	6 mg/l	4 mg/l	0.5 mg/l
5 - day Biochemical oxygen demand, max.	2 mg/l	4 mg/l	7 mg/l	—
pH value	6.8–8.5	6.8–8.5	6–9	—
Saprobity degree (Liebmann)	Oligosaprobic	Beta-mesaprobic	Alpha mesaprobic	—
Productivity degree	Oligotrophic	Eutrophic	—	—
Most probable number Coliform bacteria per 100 ml, max	200	6000	—	—
other characteristics:	without visible waste material, color or noticeable odor	without visible waste material color or odor	without visible waste material	without visible waste material

River Spreca:
—from source to the Modrac reservoir   Class II
—the Modrac reservoir   Class II
—from Modrac reservoir to the
confluence with river Bosna   Class III

River Jala:
—from source to the new sewage conveyor
of Tuzla   Class II
—from new sewage converyor of Tuzla
to confluence with river Spreca   Class III

River Bosna
—from confluence of river Miljacka
to confluence with river Sava   Class III

This means that, in accordance with the above decrees, the river Jala should correspond to Class II upstream of monitoring point J2 and to Class III from J2 to J4, while the Spreca should correspond to Class III at point S1.

23.4 Alternative Solutions

The following three alternative solutions for abating the pollution of the Jala and Spreca rivers are realistic considerations:

1. Separate treatment of all the three major wastewaters: (1) Tuzla town sewage with its associated small industries, (2) HAK I chemical and (3) HAK II chemical with three separate discharges into the receiving water.
2. Combined treatment of HAK I and HAK II chemical wastes and separate treatment of Tuzla town wastes with two separate discharges into the receiving water.
3. Combined preliminary and primary treatment of HAK I and HAK II chemical wastes with subsequent combination with Tuzla town wastes for secondary treatment and discharge of one effluent into the receiving water.

23.5 Laboratory Studies

Laboratory prototype studies were carried out on (a) Tuzla wastes, (b) mixed with Tuzla effluent in 20–50 and 40–60 percent volume ratios. HAK I and II wastes and a mixture of 14% Hak I and II wastes with 86% Tuzla wastes by volume. Biological aeration was the secondary treatment used since it is more amenable to small-scale study in the laboratory. After considerable difficulty in deriving acclimated sludge seed–effective biological treatment was obtained after primary sedimentation and neutralization of HAK wastes in (b) and

(c) systems (See Figures 23.4 and 23.5). Sludges formed were removed from the treatment systems prior to biological aeration. Sludge loadings of 0.4 kgs COD per kg. MLSS (Mixed Liquor Suspended Solids) were maintained in all systems. The major difficulty encountered during biological treatment was an elevated and increasing sludge volume index—an indication of bulking sludge. The sludge volume indices were from 100 to 300 in the Tuzla waste treatment system and over 400 in the combined system(c) (See Figure 23.5).

23.6 Mixing of Waste Waters

Beside organic components, the wastewater from HAK I and HAK II contains large amounts of dissolved matter. In the waste water from HAK I, Cl^- and Ca^{++} are the most important ions; in the waste water from HAK II, these are $SO_4^=$ and $CO_3^=$. If the waste waters from HAK I and HAK II are mixed, formation of $CaCO_3$ and $CaSO_4$ will take place. Even if the waste waters from HAK I and HAK II are neutralized before mixing, the solubility products of $CaCO_3$ and $CaSO_4$ will still be exceeded, due to the high Ca^{++} concentration.

In the mixture of domestic waste water, HAK I and HAK II waste waters, the formation of $CaCO_3$ will also occur at pH > 6. It is important to note that it has been assumed that the domestic sewage contains 300 ppm of $CO_3^=$.

The following discussion is to acquaint the reader with some of the problems of pre-treatment in both the case of separate treatment of industrial waste water and that of mixed treatment of domestic sewage and industrial waste water.

In this connection the formation of $CaCO_3$ is especially significant. If the domestic wastewater and the industrial waste water are treated separately, the formation of $CaCO_3$ and possibly of $Ca SO_4$ will only take place in the industrial waste water. In this case there are two possibilities:

a.) The waste waters from HAK I and HAK II are pretreated and neutralized separately. Then both waste waters are mixed, either before transportation to the waste water treatment plant, or at the site of the treatment plant (Figure 23.3). During the mixing $CaSO_4$ will form as follows:

$$Ca^{++} + SO_4^{--} \rightarrow CaSO_4 \downarrow$$

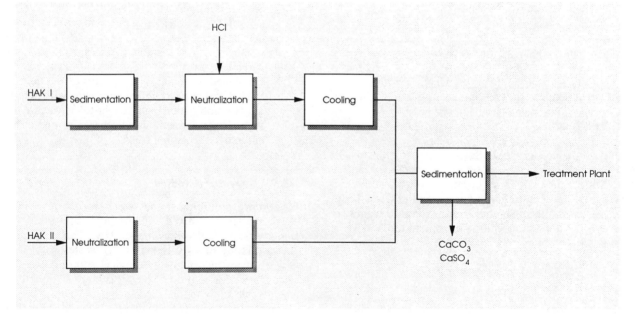

Fig. 23.3 Separate netralization before mixing. Separate treatment of domestic and industrial waste water.

The maximum solubility of SO_4^{2-} at 7800 ppm Ca^{2+} is approximately (Fig. 23.3) 500 ppm.

About 3000 kg Ca SO_4/day will be removed. Beside that, Ca CO_3 will be formed, as follows:

$$Ca^{2+} + 2\,HCO_3 \rightarrow \underline{Ca\,CO_3^{\downarrow}} + CO_2 + H_2O$$

The maximum Ca CO_3 production in this case is 1800 kg/d. After mixing, (Fig. 23.4) reaction and sedimentation, the waste water has a considerable CO_2 concentration.

During aeration in the waste water treatment plant this CO_2 will partly be stripped and reaction 2 will move further to the right, resulting in the occurrence of more $CaCO_3$ in the aeration tank.

b.) If the waste waters from HAK I and HAK II are mixed before neutralization (Figure 23.4), besides the formation of $CaSO_4$ mentioned under a), the following reaction takes place:

$$Ca^{2+} + CO_3^{--} \rightarrow Ca\,CO_3 \downarrow$$

The maximum Ca CO_3 production will now be 3600 kg/d. After neutralization and cooling, the waste water is transported to the waste water treatment plant. Al-

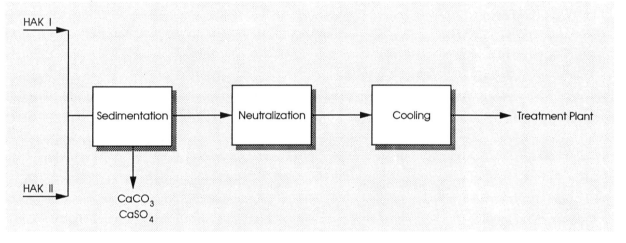

Fig. 23.4   Mixing HAK I and HAK II before neutralization.

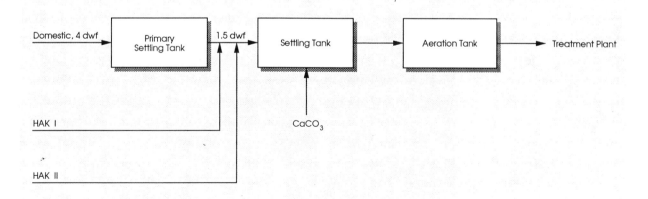

Fig. 23.5   Mixed treatment of industrial and domestic sewage.

though the waste water contains almost no $CO_3^=$, the Ca^{2+} concentration is still very high, so that one has to plan on the formation of some $CaCO_3$. The amount will be less than in the possibility mentioned under a.

The formation of some $Ca\,CO_3$ in the aeration tank is unavoidable in view of the presence of some thousands ppm's Ca^{2+}. This method results in less $CaCO_3$ formed in the aeration tank.

If the pre-treated and neutralized waste waters from HAK I and HAK II are mixed with domestic waste water, according to Figure 23.5, account must be taken that $CaCO_3$ will form as follows:

$$Ca^{2+} + 2HCO_3^- \rightarrow CaCO_3^\downarrow + CO_2 + H_2O$$

Mixing HAK I and HAK II waste water first and thus removing some $CaCo_3$ will not change this situation because the Ca^{2+} concentration will decrease slightly compared to its total value, and domestic sewage itself contains enough carbonate to continue the reaction.

The consequence of this mixing is that the $CaCO_3$ formed must be removed.

During aeration CO_2 will be partly stripped and some $Ca\,CO_3$ will be formed in the aeration tank. Also, in this case the formation of $Ca\,CO_3$ in the aeration tank will be unavoidable due to the high Ca^{2+} concentration.

The preceding discussion is depicted in Figures 23.3 and 23.4.

23.7 Special Significant Recommendations of Consultant

1. Utilize high solids, biological aeration of combined Tuzla municipal and industrial wastes with no prior neutralization.
2. Burn, recover, and reuse the solid waste of HAK 1, and HAK 2, for fresh calcium oxide.
3. Reuse final effluent from combined wastewater treatment plant for groundwater recharge to assist in abating land subsidence still occurring in Tuzla as a result of salt pumping from the ground.
4. Reuse combined treatment plant waste sludge for argicultural purposes after adequate lab and field testing and public education.
5. Conduct laboratory anaerobic digestion studies of combined plant sludge to determine potential inhibition or effectiveness of the process.

In this case we take advantage of the high quantity of solids which will arise from treating the industrial wastes. In addition we propose to reuse both the sludges resulting from treatment and the effluent following complete treatment: the sludge reuse for agricultural purposes and the effluent recharge into the groundwater to prevent land subsidence. The solution, then, is a unique combination of innovative treatment and optimum reuses of both sludge and effluent.

CHAPTER 24

LACK OF COMPLETE COORDINATION BETWEEN INDUSTRIAL PRODUCT MANUFACTURING AND WASTE TREATMENT

24.1 Introduction

Unfortunately, there are many cases in which the industry makes the proper decision concerning waste treatment and the consultant provides a reasonably adequately designed waste treatment plant, and yet the treatment plant does not provide the proper environmental protection for the receiving water and surrounding environment.

There are many reasons for this situation. Most are primarily involved with the lack of coordination between production by industry and operation of its waste treatment facilities. The difficulty is due to improper interpretation of production quantity and the type by consulting engineers, and poor or no concern by management for *operation* of the waste treatment units.

This case history is presented here in order to assist engineers, consultants and plant production managers, in order to avoid problems after presumably making a potentially feasible design decision and monetary expenditure.

Hopefully, future waste treatment plants will be operated to perform more effectively as designed.

A Montreal, Canada, consulting engineer was retained by the prime production plant engineer to provide a preliminary study for evaluating waste treatment and, once accepted, to provide plans and specifications for treatment of all wastes from a whey processing plant.

The Vermont Whey Plant Incorporated, a whey processing plant, is located on Bovat Road in the town of Georgia, Vermont, about 3/4 miles east of the junction of State Route 7 and Interstate 89, and on Town Road No. 31.

24.2 Processes

The plant was conceived and designed originally in 1975–1976 to produce sweet whey (from cheddar cheese) for about 12 hours a day and acid whey (from cottage cheese) for about 8 to 10 hours, with the intervening 2 to 4 hours between each of the productions were to be devoted to cleanup. In 1978, after construction (1977), it was reported that the plant was in more or less continuous operation for the entire year and in processes 6 or 7 days per week.

The plant consists of standard whey processing facilities, although with special modifications due to the use of a "hot" whey process. This requires that raw whey be concentrated within 12 hours of the time of generation at the cheese processing plant. This severely limits the amount of the holding time that is available and requires careful scheduling and virtually continuous operation in order to insure that the majority of whey can be processed into an edible grade whey. The fresh whey is received in tank trucks and stored in silos. It is then condensed from about 6.5 percent to approximately 42 percent solids in the first stage of a two stage evaporation process. The first stage consists of 5 effects and 11 passes. Whey is then "finished" in the second stage consisting of 3 effects and 4 passes. In this stage, whey would normally be condensed from 42 percent to approximately 50 percent solids. The whey concentrate is then held in crystalizers to crystalize the lactose and then dried in a cyclone dryer. The dried whey is bagged in 50 pound packages and shipped.

The plant is provided with typical supporting facilities such as boilers, air compressors, cooling towers, as well as a cleaning system using both acid and caustic.

24.3 The Problem

Since the time when the plant was completed in 1977 and operation attempted in June of that year, it has experienced numerous difficulties with the processing equipment and cleaning systems. It did not begin actual operation until May 1978, and was closed because of these difficulties and unsuitable market conditions, in September of 1979. These have caused major losses of whey, both processed and unprocessed, along with cleaning chemicals, to the treatment plant, and have also been responsible for a use of water considerably in excess of design expectations.

However, there is also a possibility that the 1975 design loading expectations were lower than they should have been, and the original performance expectations were higher than what they should have been. The treatment plant was designed by the original engineering firm on the basis of data available at the time. See data and computations given in Table 24.1.

Parameters in Table 24.1 vary with loading, biological solids variation and seasonal air temperature. The expected BOD_5 concentration in effluent from the stabilization basin in normal operation is 25–30 mg/l, or less. Based on results in Canada (Winnipeg area), Minnesota, Michigan and Montana plants (case studies), similar results have been obtained with similar BOD_5 loadings and treatment operation.

Data on rapid sand and/or mixed media (sand and anthracite) filtration of treated wastewater effluent and chlorination shows BOD_5 reductions of 50% in applied values of 20–40 mg/l BOD. In our case at Georgia, the strength of the treated effluent from Pond No. 3, the stabilization basin, should be 20 mg/l or less for each of the BOD and suspended solids values, before the filtration process. If desirable and/or necessary, alum coagulation in the stabilization basin can be readily carried out by the application of the alum solution from a rowboat. Such treatment has been successfully done in stabilization ponds.

The aeration equipment proposed is the Polcon Helixor, a submerged static unit of polyethylene fabrication. A diagram of oxygen transfer values made by Dr. H. Paulson of Iowa State University is attached (Figure 24.1). From the curves there has been added the oxygen transfer line for the proposed basin depth in the Georgia plant (13′6″), the value of 2.25 lbs. of oxygen per hour at an applied air rate of 26 cubic feet per min-

ute to each aerator. This value is increased to 2.5 lbs. oxygen per hour due to the result of further transfer curves made from data where the aerators are spaced at 10-foot centres or less.

Thus, the oxygen transfer rate is, in terms related to power:

1.9 lbs per horsepower-hour or
2.55 lbs oxygen per kilowatt-hour.

24.4 The Treatment Plant

The general flow pattern of the treatment plant consists of raw sewage pumps, equalization, two stage aerated lagoons, a waste stabilization pond, sand filtration and chlorination with disposal of the treated liquid by percolation in the surrounding soil. A nutrient feed system is available.

The General Design as Found by the Reviewing Engineer After Plant Construction in Early 1979.

The raw waste lift station wet well (about 6,000 gallons) contains 2 (325 gpm) pumps. When both pumps are in operation at the same time, the total output is limited to 425 gpm.

The equalization basin is 35,000 gallons to provide some dampening of the variation in pH due to acid and caustic wash cycles, as well as to extract some of the heat from the raw waste by means of a heat exchange coil system. It was also initially intended to provide some separation of large suspended solids, and was equipped with 2 air diffusers to be used only to prevent septic conditions during times of zero or no wastewater flow. A V-notch weir with a flow recorder and proportionate waste sampler has been added on the discharge end of the equalization basin. A bypass allows for direct flow to either aeration basin.

Aeration basin 1 is earthen, sealed with bentonite clay on all inside walls and bottom, with approximate surface dimensions of 110′ × 150′ and bottom of 40′ × 80′. The normal water level is 13′ 6″ with 2′ freeboard and the side slope of 2.5:1. This basin is equipped with 45 stationary diffusers (Polcon Corp) each 5′ high. Aeration basin 2 is essentially the same with only 32 Polcon diffusers.

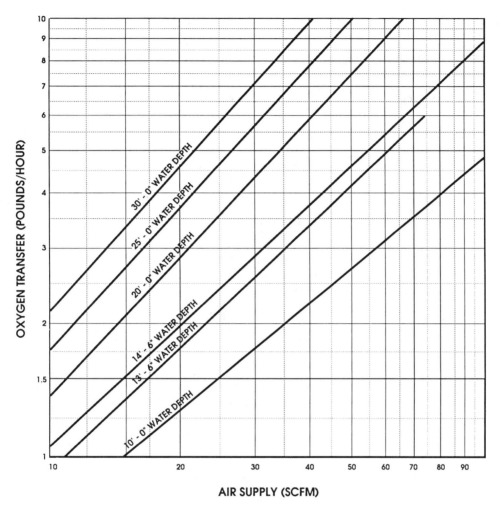

AIR SUPPLY (SCFM)

Tested at: 10' - 0" and 14' - 6" SWD
Extrapolated to: 20' - 0", 25' - 0" and 30' - 0" SWD

D.O. Level	0 ppm
Temperature	20 c
Helixor Diameter	12 inches
Helixer Length	5 feet
CoION	0.5 mg/l

Source: University of Iowa, by Dr. W. Paulson

Figure 24.1

Table 24.1 Vermont Design Factors *Whey Drying Plant-Wasted Water

Effluent Sources:

1. Truck Washing (32 trucks)	a) First hot water bursts (15 secs)-	To Process Stored	
	b) Alkaline wash cycle (C.I.P.)		
	c) 2 burst rinses after alkali wash 2 × 15 secs at 150 GPM × 32	2600 Gals	
	d) Iodine rinse bursts (15 Secs) 1 × 15 secs at 150 GPM × 32	1300 "	
	e) Final cold burst rinse (15 secs) 1 × 15 secs at 150 GPM × 32	1300 "	
		5200 Gals.	
2. Storage silos (3 silos/day)	a) Pre-rinse hot (20 secs)	To Process 250 Gals	
	b) 2 Hot rinses (20 secs at 120 GPM 2 × 120 × .33 × 3		
	c) 2 Hot rinses (30 secs)at 120 GPM 2 × 120 × .5 × 3	360 "	
	d) Iodine rinse burst (30 secs) 1 × 120 × .5 × 3	180 "	
		790 Gals.	

3. 40% Whey Tanks
& Crystallizers
(10 tanks)

Same Process as # 2

but Volume is $\dfrac{10}{3} \times 800$

	2500 "	
	3290 "	

4. Low Con Evap.	a) 10 minute pre rinse at 100 GPM	To Process 500 Gals	
	b) Caustic Rinse C.I.P.		
	c) Hot rinse 15 minutes at 100 GPM	1500 "	
		2000 "	
5. Hi Con Evap.	Assume same as Low Con	2000 "	
6. Dryer Wash (1 per week)	a) First rinse to whey dry process		
	b) Ave. Anhydro, Niro, other data	3000 "	
7. Line Cleaning	Assume same as 40% tanks & Crysta.	3300 "	
8. Miscellaneous	Hose Stations, Floor washings Est.	10,000 "	

9. Boiler Make-up Water as follows:

a) Low con evap.	10,000 lbs/hr	
b) Hi " "	1,500 "	
c) Deaerator Steam	4,000 "	
d) F.O. Heating	300 lbs/hr	
e) C.I.P. "	2,000 "	
	17,800 "	for 20 Hrs.
356,000/8.33 = 43,000 Gals =		48,000 Gals

10. Sanitary Wastewater - Personal use

30 people × 40 gals = 1200 gals say	1,500 "	
11. Pump Seal Water 15 gal/min × 1440 min.	22,000 "	
12. Contingency - Extra water	25,000 "	
Total	125,000 "	

13. Evaporation of Whey - Condensate Water

a) From steam effects
$$70,000 \times \tfrac{1}{4} = 17,600 \text{ lbs/hr for } 20$$
$$= 352,000 \text{ lbs} \times \frac{1}{8.33} = 42,000 \quad ''$$

b) From other effects
$$70,000 \times \tfrac{3}{4} = 52,500 \text{ lbs/hr for } 20$$
$$= 1,050,000 \text{ lbs} \times \frac{1}{8.33} = 126,800 \text{ Gals}$$

Here cooling tower pick-up is
$$\frac{12 \times 10^6}{1 \times 10^3 \times 8.33} \times 20 \text{ hours} = 29,000 \quad ''$$
$$\text{net} \quad 98,000 \quad ''$$

Therefore boiler make-up - condensate
$$= 48,000 - 42,000 = 6,000 \text{ say} \quad 10,000 \quad ''$$

J. Miller expects to sewer 65,000 gals from condensate

Total of	98,000	''
Expected Sewered Water Items 1–12	125,000	''
Item 13 Boiler net make-up	10,000	''
Condensate Volume	65,000	''
	200,000	''

SAY 210,000 Gals = 150 USGPM

WASTE LOADING RATE

ITEM	BOD (lbs)	WATER (gals)
1. Truck washings (St. Albans)	93	5,200
2. Storage Silos (var. sources)	20	1,000
3. 40% tanks ('' '')	66	3,300
4. Low Con Evapor ('' '')	80	2,000
5. Hi '' '' ('' '')	80	2,000
6. Dryer 1 time/week	250	3,000
7. Line Cleaning	75	3,300
8. Miscellaneous hose Stations, floor washings	10	10,000
9. Boiler Make-up	0	48,000
10. Sanitary Wastewater	6	1,500
11. Pump Seal Water (Cond)	6	22,000
12. Contingency Water	0	25,000
13. Condensate Water (Reused)	200	65,000
14. Occasional C.I.P. Wash Liquid	50	2,000
	936	193,300

Expected BOD $= \dfrac{936 = \text{mg/l} \times 8.3 \times 193,300}{1 \times 10^6}$

$\text{mg/l} = \dfrac{936 \times 1 \times 10^6}{8.3 \times 193,300} = 580 \text{ mg/l}$

Assume spill or waste whey

Therefore 40% increase $= 580 \times 1.4 = 810 \text{ mg/l BOD}_5$

Therefore Design for 810 mg/l BOD_5 at 210,000 gals/day, with the process facility considered adequate to efficiently reduce a BOD_5 loading of 40–50% in excess of 810 mg/l.

EFFLUENT TREATMENT DESIGN
VERMONT WHEY DRYING PLANT

Anticipated Waste Water Flow   210,000 U.S.G./Day
      "   "   "   BOD$_5$ Load   810 mg/l

TREATMENT PROCESSES:

1. Sump in main building basement with access for solids removal
2. Effluent pump to discharge line,
   6" diameter stainless wastewater line - 700'
3. Reinforced conrete Equalization Basin, 4 hour average flow retention with heat pick and heat dispersion system (coils)
4. Aeration Pond (1 of 2) of 4–5 days average flow retention (8 days winter volume)
5. Second Aeration Pond same size as item #4
6. Stabilization Basin of 10 days average flow retention (18 days in winter)
7. Rapid mixed media filtration
8. Flow volume measurement
9. Chlorination
10. Ground recharge by Basin percolation

Winter operations are 55% maximum of summer volumes. Therefore all basin capacities have 1.83 or greater retention periods.

Biological Treatment system is extended aeration, a modification of the activated sludge process.

Anticipated pH value average   6.5
      "   Temperature   120°F (49°C)
      "   possible susp. solids 1000–1200 mg/l

Anticipated BOD reduction in 1st Basin =   85%
   15% of 810 mg/l loading   = 121 mg/l
Expected reduction in 2nd Aeration Basin = 50%
   50% of 121 mg/l applied   = 60 mg/l
Anticipated Polishing or Nitrification - Stabilization 50%
   50% of 60 mg/l applied = 30 mg/l BOD (mainly soluble)

The waste stabilization pond is an L-shaped, unaerated basin with the short leg having a bottom width of 50' and a top width of 80' and the same dimensions of the long leg 75' and 105'. The liquid depth is 6 feet.

All waste from the stabilization pond flows by gravity into a small sump in the blower and filter building designated as pump house No. 2 and is pumped through 2 pumps and a sand filter, chlorinated, and discharged to one of 4 percolation ponds. Each pump is rated at 150 gpm.

A 72 ft.^{2} sand filter of the dual media type—containing 9" of anthracite above about 20" of sand (0.8 mm diam and U.C. of 1.0)—receives the stabilization pond effluent. The filter is backwashed by a high-capacity pump (rated at 750 gpm).

A separate chlorination room is provided to chlorinate the treated waste with a Fisher and Porter unit at a rated capacity of 1 pound per hour.

Four percolation ponds are provided with surface dimensions of about 85' × 290' and bottom dimensions of about 20' × 225' with 13 feet of SWD and 2 feet of freeboard. The basins are earthlined (with *no* bentonite added for sealing) to permit percolation.

A nutrient storage tank (7,000 gallons) is provided for anhydrous ammonia.

Three Godrey type air compressors are provided with a nominal rated capacity of 1120 CFM at STP. They are designed to operate at 1765 rpm and can theoretically produce up to 10 psig pressure. Each blower is fitted with a 50 hp. motor drawing 62 amps under 3 phase, 460 volt service at 40°C.

Waste Treatment Plant Layout

The planned waste treatment plant layout is shown in Figure 24.2. The plant was built exactly in accordance with the working drawings provided by the consulting engineer, with one error of a 6″ Dia. pipe rather than an 8″ dia. pipe, leading to the pump sump from the stabilization bacin.

The $ 64 Questions

Does the design once accepted by the State of Vermont Water Resources and Conservation Agencies "clear" the consulting engineer from any claim of negligence? If so, why did the waste treatment plant malfunction? And who is to blame? And, finally, most important to the preservation of the environment, how was the problem solved?

Verification of Acceptable Design

The consulting engineer designed this treatment plant originally (contract letter of December 7, 1974), to achieve an effluent quality of 10 mg. per liter BOD_5, 10 mg. per liter suspended solids, and 2 mg. per liter of phosphorous. The plant was approved by the Vermont Agency of Environmental Conservation and the Vermont Whey Pollution Abatement Authority, in a certificate of Compliance on Dec. 27, 1978, and again on March 9, 1979.

It was designed for an average BOD_5 of 810 mg. per liter and 210,000 of wastewater per day as found in the previous section. The entire plant was designed on the basis of the evaporator capacity - a nominal output load of *100,000 pounds of whey powder* per day or an in put load of raw liquid (6.5 % solids) of about 1,538,000 pounds of liquid whey.

Reasons for Malfunctioning of Waste Treatment Plant

After validating the correctness of the consultant's design of the waste treatment plant, based upon his knowledge of production plans, we must seek reasons for the ineffectiveness of the treatment units. The coordination of waste treatment design and plant efficiency depends upon the smooth working relationship between industrial plant production and waste treatment plant operation. Any disturbance in this delicate relationship will produce malfunctioning of pollution control units.

In this case, let's examine the whey production as one of the two vital components of the relationship. In 1974, when this waste treatment plant was designed, the production plant was not built. Nor were any typical plants of this exact type in operation anywhere in the world. Therefore, this plant had to be designed on the best production data (and hence waste load and character) available to the consulting engineer.

Originally, the production plant was conceived and designed (according to the consulting engineer) for about 12 hours of dried sweet whey from cheddar cheese as the predominant product, and 8 to 10 hours of dried acid whey. The intervening 2 to 4 hours were devoted to clean up. Switching from acid whey to sweet whey requires a careful cleanup of the equipment so as not to interfere with the start of the sweet whey cycle. Therefore, the effluent treatment process was designed to handle *regular* washings of equipment and also an additional cleaning each day after the acid whey dry processing shift. As shown in Figure 24.2, the concentrated acid and caustic streams in the wash cycles were to be returned to their original holding tanks, not only to protect the treatment plant, but also to conserve chemicals for reuse several times. All first rinse waters of equipment in the first phase of clean–up, or clean–in–place, were to be sent to holding tanks (Figure 24.2) built for that purpose. This was planned to avoid treating raw whey, concentrated whey, partially crystallized whey, and product whey from the drier. The contents of these holding tanks were to be subsequently fed to the product process stream for reuse. Provision was also built into the system to dispose of rejected whey in various stages of processing to be trucked off the site and not sent to the treatment system. Presumably, when production is proceeding smoothly and as planned these "built-in" safeguards whould be in operation. However, many plant production problems developed mainly due to the use of the "hot whey" drying process to save heat energy in drying. This process demands rapid utilization of the hot whey as it is received from the cheese plants, as well as continuous and uninterrupted cheese plant manufacturing. Since this was not always possible, bacterial contamination during storage in the silos resulted in spoiled raw material. In addition, improperly operating clean-in-place systems, feed pumps, blocked-up crystallizers, and faulty dryer operation all contributed to production interruptions. The net result of these production problems was mal-

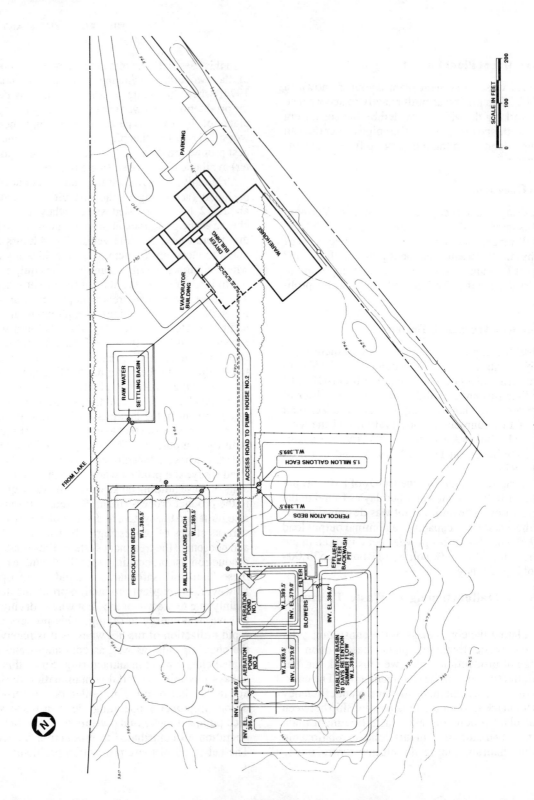

Figure 24.2 Vermont Whey Pollution Abatement Authority, edible whey plant.

functioning of the treatment plant because:

1. Dumping of tanker trucks of raw, spoiled or unsuitable whey to the sewer,
2. Dumping of spoiled crystallizer contents (concentrated, partly-evaporated whey),
3. Frequent cleaning of evaporator system with all rinses going to sewer due to faulty evaporator operation,
4. Cleaning of dryer equipment and hosing to sewer, more frequently than anticipated,
5. Dumping to sewer of concentrated acid and caustic due to improper clean-in-place acid and caustic feed control and recirculation,
6. Regular dumping of hydraulic fluid from main pump in basement of the main building to the main building to the effluent sewer, such liquid (allegedly) being very toxic to wastewater organisms.

24.5 Design Limitations and Reasons for Malfunctioning

Operating difficulties within the plant started almost immediately with large discharge of solids pumped to the treatment facility. With frequent cleaning and washing, the wastewater volumes were peaking out and with the increased pumping capacity in the plant, and the unwarranted aerators in operation in the equalization basin, very little settling of solids took place in the basin. Solids were then carried directly into the aeration basins. The combinations of both porduction problems and ensuing operating difficulties are accentuated by:

1. Several sources of wastewater and solids being sewered due to inadequate technology by operating personnel, unfamiliarity with equipment, some faulty operation, etc.
2. No wastewater treatment plant operator
3. No laboratory facility for wastewater analysis.

The crux of the operating difficulties has been the operation of the equalization basin which was intended also as a settling basin for solids.

In addition to production units and waste–treatment operation problems, the winter whey drying production is actually about twice that predicted and designed for by the consulting engineer. This occurred during the season when biological activity is usually reduced because of less than optimum temperatures.

Therefore, misses in predicting production quantity and performance of systems utilized, as well as absence of adequate waste treatment operation, resulted in inefficient treatment and adverse environmental impact.

24.6 Resolution of the Problem

One must investigate the following information and procedures to determine factual data upon which to base reasonable conclusions.

1. Was the waste treatment plant designed and constructed in a professional engineering manner?
2. With proper treatment plant operation, would the treatment plant operate satisfactorily?
3. Were additional steps taken by the designing engineer to avoid adverse environmental impacts?
4. Was the whey production plant designed and operated as planned by both management and the waste treatment plant consulting engineer?
5. Was the owner (management) cooperative to the point of assisting the designing engineer in avoiding adverse environmental impacts?

Obviously, the answer to one or more of the above questions must have been "no," since the waste treatment plant was forced to handle unacceptable problems.

However, the findings when one examines each of the questions will reveal the source(s) of the problem(s). One further question, then, is also pertinent:

6. What would have been done—and at what cost—to counteract the environmental problems?

1. There are several criteria which can be used to ascertain whether the treatment plant was designed in a professional manner.

A. *Overall system* selected—of equalization, aerated lagoons in two stages, waste stabilization pond, sand filtration, and chlorination followed by soil percolation of the effluent—is certainly a suitable one for this type of wastewater and with that amount of land avaiable and for a discharge which may eventually reach the near by Lake Arrowhead Mountain.

B. *Detention time* in the biological units—*10 days* for aerated lagoons and *10 days* more for the stabilization basin. These are based upon maximum flows expected during summer months. They would appear normal when compared to times used for other dairy or sewage-type wastewaters.

C. Could the whey waste be expected to *biodegrade at a rate* at least as fast as normal domestic wastewater? And did the engineer take into account in his design the biodegradability? The answer to both questions is apparently, yes. Belschner used a k_1 value of 0.375 which he obtained by substitution in the following equation by aiming at an

$$t = \frac{E}{2.3 \, k_1 \times (100 - E)}$$

efficiency of BOD reduction, E, in the first aerated lagoon of 85 percent and a detention time, t, of 4 days. A Minnesota Report verified the use of k_1 of 0.37 for summer conditions for similar treatment of sewage with some dairy wastes present.

He also used W.C. Boyle's "Kent Cheese Study" of the Chicago area as a guide, and presumed that all cheese plant wastes are the same or similar and should respond to aerated lagoon treatment in like manner. It would have been better if Belschner could have obtained an effluent sample from a similar plant in order to measure its biodegradability. However, no other plant in operation in North America produces dried whey product from alternate use of both sweet and acid whey. The engineer did visit one in Solaipa, France, to observe its operation. However, that plant did not have a waste treatment plant. It would have been desirable if the engineer obtained effluent samples from this plant and determined k_1 values by laboratory analysis. Apparently, he was satisfied that the waste would respond as predicted. It is not always easy to collect samples and run the longterm BOD analyses necessary for determining k_1 when in a foreign country, a long way from your home base.

D. Was the lagoon *BOD loading* reasonable?

Loading = 810   mg/1 − 8.34 × .21   MGD = 1418 #BOD/Day

AREA = $\dfrac{2 \times 110' \times 150'}{43,560 \text{ ft.}^2/\text{acre}}$ = ~0.76 acres

$\therefore$

$\dfrac{1418}{0.76}$ = ~1866# BOD / day / acre

depth = 13.5 feet or 222,750, ft^3 or 222.75 1,000 ft^3

Loading = 8.38 # BOD/ 1,000 ft^3 a low and acceptable organic loading on aerated lagoons. However, the above computations were based upon the "expected?" BOD concentration of 810 mg/l and a summer flow of 0.210 mdg. Obviously, if this BOD value was exceeded, the aerated lagoons might be "strained" to remove the orgaic matter. Some data supplied to this reviewer of August 1978 analyses of equalization basin contents showed BOD values of 1250 to 4300 ppm, although the flow was quite consistent with the 0.210 mdg planned and designed. The mean BOD during this period was about twice that designed for and, therefore, the loadings on the lagoons would have been doubled or about 17 pounds/1,000 ft^3—still not an excessive value—to be handled without unusual difficulty if not for the presence of sludge solids in addition.

24.7 Sludge Production and Removal

The engineer provided no *automatic* and continual system for removal of sludge. Two sources of sludge will occur as a result of treatment:

1. direct contribution from production wastes both normally expected and abnormal or unusual spills.
2. indirect growth of sludge as a result of biological activity in the aerated lagoons.

Dr. Belschner reasoned that positive means of sludge removal would not be necessary on a regular basis from the second (2) source because the loading was so low that all the growth (as a result of biodegradation) would be stabilized and ashed during the long retention period. He also provided for manual removal from the "equalization" basin during spills. This, to our thinking, is not feasible, since it (1) depends too much on operator initiative, (2) defeats the purpose of equalization, and (3) does not really account for accidental or purposeful production spills which occur so rapidly as to defy detection until the sludge has passed over into the aerated lagoons. He also provided a sump collection basin within the production plant for the purpose of retaining solids from "spills" before they were pumped to the equalization basin of the treatment plant. Once again, Dr. Belschner relied on production plant personnel to enhance his treatment plant's performance. Apparently, repeated and numerous spills of both dissolved organics and sludge solids with little or no concern by production personnel and absence of any treatment plant operating personnel proved too much for the treatment plant to handle.

This phase of Dr. Belschner's design, in our opinion,

was the weakest area of the treatment plant. Sludge from both sources named above inundated and fouled up the plant operation.

24.8 Sufficient Aeration Capacity

1,418 # BOD/day should require aeration capacity of

$$1418 \times \frac{1,000 \text{ ft}^3 \text{ air}}{\# \text{ BOD} / \text{day}} = 1,418,000 \text{ ft}^3 \text{ air/day}$$

or $\dfrac{1,418,000}{1440 \text{ min/day}} = \underline{\sim 1,000 \text{ cfm}}$

Three air compressors are provided by the consulting engineer capable of providing 1,120 cfm. Therefore, ample air has been provided for the load expected— 1,418 # BOD/day. The latter was computed by the engineer, allowing for a generous 40% increase in BOD due to "spills or waste whey."

There is some confusion as to whether each aeration basin is provided with this aeration capacity (giving a total air capacity of 2240 cfm) or whether these 3 compressors supply all the air for the plant (giving 1120 cfm).

2. Even with the satisfactory waste treatment plant operation as it was designed, I believe that solids buildup would have eventually clogged the sand filters resulting in surface overflows.

A spray irrigation system for the effluent from the stabilization basin would have been more effective, since the ashed solids would have been returned to the land and carried into the soil with no adverse environmental impact. It also would not have been possible— even with good treatment plant operation—to completely eliminate all "spill" sludge solids from the production plant reaching the heart of the waste treatment plant. Thus, a further deterioration in the effectiveness of the total plant system as designed could be expected.

3. Some additional steps apparently were taken by the consulting engineer to avoid or eliminate troubles with the waste treatment plant.

A. He attempted to advise the production plant operating personnel in preventing spills and purposeful discharges from reaching the waste treatment plant. These discharges were to be collected separately and removed by empty whey tank trucks to external land disposal areas.

B. He objected to construction of the sump pit collection basin in the whey plant reduced to half its capacity. The shorter detention period would tend to diminish opportunities to capture spill solids before settling and being automatically pumped to the waste treatment plant.

C. He pleaded for an additional sum of money presumably based upon the premise that the construction contract price was about $100,000 less than that amount originally allocated. It is presumed that this money would be used to ameliorate the solids disposal problem.

D. He requested, apparently, a qualified waste treatment plant operator and laboratory facilities at the plant for analyses and supervision of the waste treatment plant operation.

4. The Whey production plant apparently was not designed and operated as initially planned, and as represented to the consultant by management.

In the first place, only sweet whey was processed, and acid whey was not received or processed. The "hot whey" process did not operate as predicted and considerable waste batches occurred. The retention in the collection sump within the plant was reduced to half when built, providing for less opportunity for "spill solids" separation. All these production deviations from planned operations led to *overloading of the waste treatment plant*.

5. Evidently, for reasons not completely known, the management was not cooperative with the designing engineer in (1) either providing sufficient additional money for correcting problems which developed, or for (2) altering production procedures to eliminate or counteract spills to the waste treatment plant. Most treatment facilities require some alteration after construction to correct oversights or to compensate for changes in production. Without these changes, the plant could not be expected to operate properly.

6. The *sand filtration unit* should be replaced by a spray irrigation system. *Primary sludge* clarification unit should preceed the aerated lagoons to remove "excess" solids reaching the plant from production practices if these spills cannot be avoided. The capital cost of thse two additions would be about $100,000, but more precise costs for these modifications and additions should be obtained in the local area and at the time when contracts are to be let.

Part 4 | MAJOR INDUSTRIAL WASTES

INTRODUCTION

The purpose of Part 4 is to provide the reader with a fairly complete list of references to the majority of publications concerning industrial wastes. The origin, character, and methods of treatment of the major types of industrial waste are described. It is not the purpose of this section, however, to provide the reader with details of each and every waste or method of treating it, but rather to give a condensation of existing information and guide to the literature. The author believes that solving waste-treatment problems should demand not an ability to memorize details but rather the application of scientific principles and judgment to the solution of practical problems. Although the reader may have to search the literature for the detailed information he needs, he should be capable of applying this information once he has absorbed the contents of this text.

The following table provides the reader with a brief summary of the major liquid wastes, their origin, characteristics, and current methods of treatment, which are described in the last six chapters of this text. It should be useful as a quick reference, but in no way can it be considered complete for any specific industry.

The authors have divided industrial wastes roughly into six major classifications—apparel, food and drugs, materials, chemicals, energy and non-point sources—and has devoted a chapter to each category, with the exception of the energy industry, which merits two chapters because the wastes from energy industries such as steam power plants and coal processing are one thing and those from atomic-energy plants are quite another. Nuclear wastes present such unique problems that the author felt they merited a chapter to themselves.

Summary of Industrial Waste: Its Origin, Character, and Treatment

Industries producing wastes	Origin of major wastes	Major characteristics	Major treatment and disposal methods
Apparel [Chapter 25]			
Textiles	Cooking of fibers; desiring of fabric	Highly alkaline, colored, high BOD and temperature, high suspended solids	Neutralization, chemical precipitation, biological treatment, aeration and/or trickling filtration
Leather goods	Unhairing, soaking, deliming, and bating of hides	High total solids, hardness, salt, sulfides, chromium, pH, precipitated lime, and BOD	Equalization, sedimentation, and biological treatment
Laundry trades	Washing of fabrics	High turbidity, alkalinity, and organic solids	Screening, chemical precipitation, flotation, and adsorption
Dry cleaning	Solvent cleaning of clothes	Condensed, toxic, organic vapors	Recovery and reuse
Food and Drugs [Chapter 26]			
Canned goods	Trimming, culling, juicing, and blanching of fruits and vegetables	High in suspended solids, colloidal and dissolved organic matter	Screening, lagooning, soil absorption or spray irrigation
Dairy products	Dilutions of whole milk, separated milk, buttermilk, and whey	High in dissolved organic matter, mainly protein, fat, and lactose	Biological treatment, aeration, trickling filtration, activated sludge
Brewed and distilled beverages	Steeping and pressing of grain; residue from distillation of alcohol; condensate from stillage evaporation	High in dissolved organic solids, containing nitrogen and fermented starches or their products	Recovery, concentration by centrifugation and evaporation, trickling filtration; use in feeds; digestion of slops

Summary of Industrial Waste: Its Origin, Character, and Treatment (*Continued*)

Industries producing wastes	Origin of major wastes	Major characteristics	Major treatment and disposal methods
Meat and poultry products	Stockyards; slaughtering of animals; rendering of bones and fats; residues in condensates; grease and wash water; picking of chickens	High in dissolved and suspended organic matter, blood, other proteins, and fats	Screening, setting and/or flotation, trickling filtration
Animal feedlots	Excreta from animals	High in organic suspended solids and BOD	Land disposal and anaerobic lagoons
Beet sugar	Transfer, screening, and juicing waters; drainings from lime sludge; condensates after evaporator; juice and extracted sugar	High in dissolved and suspended organic matter, containing sugar and protein	Reuse of wastes, coagulation, and lagooning
Pharmaceutical products	Mycelium, spent filtrate, and wash waters	High in suspended and dissolved organic matter, including vitamins	Evaporation and drying; feeds
Yeast	Residue from yeast filtration	High in solids (mainly organic) and BOD	Anaerobic digestion, trickling filtration
Pickles	Lime water; brine, alum and turmeric, syrup, seeds and pieces of cucumber	Variable pH, high suspended solids, color, and organic matter	Good housekeeping, screening, equalization
Coffee	Pulping and fermenting of coffee bean	High BOD and suspended solids	Screening, settling, and trickling filtration
Fish	Rejects from centrifuge; pressed fish; evaporator and other wash water wastes	Very high BOD, total organic solids, and odor	Evaporation of total waste; barge remainder to sea
Rice	Soaking, cooking, and washing of rice	High BOD, total and suspended solids (mainly starch)	Lime coagulation, digestion
Soft drinks	Bottle washing; floor and equipment cleaning; syrup-storage-tank drains	High pH, suspended solids, and BOD	Screening, plus discharge to municipal sewer
Bakeries	Washing and greasing of pans; floor washings	High BOD, grease, floor washings, sugars, flour, detergents	Amenable to biological oxidation
Water production	Filter backwash; lime-soda sludge; brine; alum sludge	Minerals and suspended solids	Direct discharge to streams or indirectly through holding lagoons
Cane sugar	Spillage from extraction, clarification, etc. Evaporator entrainment in cooling and condenser waters	Variable pH, soluble organic matter with relatively high BOD of carbonaceous nature.	Neutralization, recirculation, chemical treatment, some selected aerobic oxidation
Agriculture	Variable origin depending upon exact source; agriculture chemicals, irrigation return flows, crop residues, and liquid and solid animal wastes	Highly organic and BOD detergent cleaning solutions	Biological oxidation basins; some composting and anaerobic digestion; land application
Palm oil	Mechanical extraction of crude oil, refining of crude palm oil to edible oil	High BOD, COD, solids and total fats and low pH	Neutralization, coagulation, floatation, filtration, and biological conversion

Industries producing wastes	Origin of major wastes	Major characteristics	Major treatment and disposal methods
Materials [Chapter 27]			
Pulp and paper	Cooking, refining, washing of fibers, screening of paper pulp	High or low pH, color, high suspended, colloidal, and dissolved solids, inorganic fillers	Settling, lagooning, biological treatment, aeration, recovery of by-products
Photographic products	Spent solutions of developer and fixer	Alkaline, containing various organic and inorganic reducing agents	Recovery of silver; discharge of wastes into municipal sewer
Steel	Coking of coal, washing of blast-furnace flue gases, and pickling of steel	Low pH, acids, cyanogen, phenol, ore, coke, limestone, alkali, oils, mill scale, and fine suspended solids	Neutralization, recovery and reuse, chemical coagulation
Metal-plated products	Stripping of oxides, cleaning and plating of metals	Acid, metals, toxic, low volume, mainly mineral matter	Alkaline chlorination of cyanide; reduction and precipitation of chromium, lime precipitation of other metals
Iron-foundry products	Wasting of used sand by hydraulic discharge	High suspended solids, mainly sand; some clay and coal	Selective screening, drying of reclaimed sand
Oil fields and refineries	Drilling muds, salt, oil, and some natural gas; acid sludges and miscellaneous oils from refining	High dissolved salts from field; high BOD, odor, phenol, and sulfur compounds from refinery	Diversion, recovery, injection of salts; acidification and burning of alkaline sludges
Fuel oil use	Spills from fuel-tank filling waste; auto crankcase oils	High in emulsified and dissolved oils	Leak and spill prevention, flotation
Rubber	Washing of latex, coagulated rubber, exuded impurities from crude rubber	High BOD and odor, high suspended solids, variable pH, high chlorides	Aeration, chlorination, sulfonation, biological treatment
Glass	Polishing and cleaning of glass	Red color, alkaline nonsettleable suspended solids	Calcium-chloride precipitation
Naval stores	Washing of stumps, drop solution, solvent recovery, and oil-recovery water	Acid, high BOD	By-product recovery, equalization, recirculation and reuse, trickling filtration
Glue manufacturing	Lime wash, acid washes, extraction of nonspecific proteins	High COD, BOD, pH, chromium, periodic strong mineral acids	Amenable to aerobic biological treatment, flotation, chemical precipitation
Wood preserving	Steam condensates	High in COD, BOD, solids, phenols	Chemical coagulation; oxidation pond and other aerobic biological treatment
Candle manufacturing	Wax spills, stearic acid condensates	Organic (fatty) acids	Anaerobic digestion
Plywood manufacturing	Glue washings	High BOD, pH, phenols, potential toxicity	Settling ponds, incineration
Metal container	Cutting and lubricating metals, cleaning can surface	Metal fines, lub, oils, variable pH, surfactants, dissolved metals	Oil separation, chemical precipitation, collection and reuse, lagoon storage. Final carbon absorption

Summary of Industrial Waste: Its Origin, Character, and Treatment (*Continued*)

Industries producing wastes	Origin of major wastes	Major characteristics	Major treatment and disposal methods
Petrochemicals	Contaminated water from chemical production and transportation of second generation oil compounds	High COD, T.D.S., metals, COD/BOD ratio, and cpds. inhibitory to biol. action	Recovery and reuse, equalization and neutralization, chemical coagulation, settling or flotation, biological oxidation
Cement	Fine and finish grinding of cement, dust leaching collection, dust control	Heated cooling water, suspended solids, some inorganic salts	Segregation of dust-contact streams, neutralization and sedimentation
Wood furniture	Wet spray booths and laundries	Organics from staining and sealing wood products	Evaporation or burning
Asbestos	Cleaning and crushing ore	Suspended asbestos and mineral solids	Detention in ponds, neutralization and land filling
Paint and inks	Solvent-based rejected materials scrubbers for paint vapors; refining and/or removing inks	Contain organic solids from dyes, resins, oils, solvents, etc.	Settling ponds for detention of paints, lime coagulation of printing inks
Chemicals [Chapter 28]			
Acids	Dilute wash waters; many varied dilute acids	Low pH, low organic content	Upflow or straight neutralization, burning when some organic matter is present
Detergents	Washing and purifying soaps and detergents	High in BOD and saponified soaps	Flotation and skimming, precipitation with $CaCl_2$
Cornstarch	Evaporator condensate or bottoms when not reused or recovered, syrup from final washes, wastes from "bottling-up" process	High BOD and dissolved organic matter; mainly starch and related material	Equalization, biological filtration, anaerobic digestion
Explosives	Washing TNT and guncotton for purification, washing and pickling of cartridges	TNT, colored, acid, odorous, and contains organic acids and alcohol from powder and cotton, metals, acid, oils, and soaps	Flotation, chemical precipitation, biological treatment, aeration, chlorination of TNT, neutralization, adsorption
Pesticides	Washing and purification products such as 2,4-D and DDT	High organic matter, benzenering structure, toxic to bacteria and fish, acid	Dilution, storage, activated-carbon adsorption, alkaline chlorination
Phosphate and phosphorus	Washing, screening, floating rock, condenser bleedoff from phosphate reduction plant	Clays, slimes and tall oils, low pH, high suspended solids, phosphorus, silica and fluoride	Lagooning, mechanical clarification, coagulation and settling of refined waste
Formaldehyde	Residues from manufacturing synthetic resins and from dyeing synthetic fibers	Normally high BOD and HCHO, toxic to bacteria in high concentrations	Trickling filtration, adsorption on activated charcoal
Plastic and resins	Unit operations from polymer preparation and use; spills and equipment washdowns	Acids, caustic, dissolved organic matter such as phenols, formaldehyde, etc.	Discharge to municipal sewer, reuse, controlled-discharge
Fertilizer	Chemical reactions of basic elements. Spills, cooling waters, washing of products, boiler blowdowns	Sulfuric, phosphorous, and nitric acids; mineral elements, P, S, N, K, Al, NH_3, NO_3, etc. Fl, some susp. solids	Neutralization, detain for reuse, sedimentation, air stripping of NH_3, lime precipitation

Summary of Industrial Waste: Its Origin, Character, and Treatment (*Continued*)

Industries producing wastes	Origin of major wastes	Major characteristics	Major treatment and disposal methods
Toxic chemicals	Leaks, accidental spills, and refining of chemicals	Various toxic dissolved elements and compounds such as Hg and PCBs	Retention and reuse, change in production
Mortuary	Body fluids, washwaters, spills	Blood salt, formaldehydes, high BOD, infectious diseases	Disch. to municipal sewer holding and chlorination
Hospital-Res. Labs.	Washing, sterilizing of facilities, used solutions, spills	Bacteria, various chemicals radioactive materials	Disch. to mun. sewers; holding and biol. aeration in large
Chloralkali wastes	Electrolytic cells, making chlorine and caustic soda	Mercury and dissolved metals	In-plant control, aeration, and adsorption
Organic chemicals	Various chemical productive processes	Varied types of organic chemicals	Biological degradation, in-plant control, process modification
Energy [Chapter 29]			
Steam power	Cooling water, boiler blow-down, coal drainage	Hot, high volume, high inorganic and dissolved solids	Cooling by aeration, storage of ashes, neutralization of excess acid wastes
Scrubber power plant wastes	Scrubbing of gaseous combustion products by liquid water	Particulates, SO_2, impure absorbents or NH_3, $NaOH$, etc.	Solids removal usually by settling, pH adjustment and reuse
Coal processing	Cleaning and classification of coal, leaching of sulfur strata with water	High suspended solids, mainly coal; low pH, high H_2SO, and $FeSO_4$	Settling, froth flotation, drainage control, and sealing of mines
Nuclear power and radioactive materials [Chapter 30]			
	Processing ores; laundering of contaminated clothes; research-lab wastes; processing of fuel; power-plant cooling waters	Radioactive elements, can be very acid and "hot"	Concentration and containing, or dilution and dispersion
[Chapter 31]			
Non-point sources	Dirt, dust, combustion prod. runoff, salt runoff, organic matter runoff	Various but largely mineral and organic matter	Sealing sources, holding and treating by various means

CHAPTER 25

THE APPAREL INDUSTRIES

The apparel industry may be subdivided into three classifications: textiles, leather goods, and laundry trades. Each of these is concerned with wearing apparel—shirts, suits, shoes, work clothes, and so forth.

TEXTILE WASTES

Textile mill operations consist of weaving, dyeing, printing, and finishing. Many processes involve several steps, each contributing a particular type of waste, e.g. sizing of the fibers, kiering (alkaline cooking at elevated temperature), desizing the woven cloth, bleaching, mercerizing, dyeing, and printing. Textile wastes are generally colored, highly alkaline, high in BOD and suspended solids, and high in temperature. Wastes from synthetic-fiber manufacture resemble chemical-manufacturing wastes and their treatment depends on the chemical process employed in the fiber manufacture. Equalization and holding are generally preliminary steps to the treatment of those wastes, because of their variable composition. Additional methods are chemical precipitation, trickling filtration, and, more recently, biological treatment and aeration. The textile industry has long been one of the largest of water users and polluters and there has been little success in developing low-cost treatment methods, which the industry urgently needs to lessen the pollution loads it discharges to streams.

25.1 Origin and Characteristics of Textile Wastes

The sources of polluting compounds are the natural impurities extracted from the fiber and the processing chemicals which are removed from the cloth and discharged as waste. It is necessary for the industrial waste engineer to have a working knowledge of the various processes which produce the wastes and which vary with the particular material. The materials can be subdivided into three groups: cotton, wool, and synthetic fibers.

Cotton. Raw cotton is carded, spun, spooled and warped, slashed (filled with starch), drawn, and woven or knitted into cloth before being sent to the finishing mill. No water-borne pollution originates in this sequence of operations, since they are all mechanical processes, except slashing. In slashing, the warp thread is sized with starch to give it the tensile strength and smoothness necessary for subsequent weaving. The starches used for sizing are cellulose derivations. The sized cloth, referred to as "grey goods," contains 8 to 15 per cent slashing compound, which must be removed in the finishing operation. The grey goods are desized to allow further wet processing, kiered to remove natural impurities, bleached to render them white, mercerized to give the fabric luster, strength, and dye affinity, printed or dyed, and finally filled or sized again, to make them more resistant to wear and smoother to the touch. In addition, some goods are waterproofed, with aluminum acetate or formate mixed with gelatin and a dispersed wax. Each of these processes may involve many steps and may be carried out simultaneously with different machines in different parts of the mill. Figure 25.1 based on a 1967 survey [19] summarizes the operations involved in cotton textile finishing.

Masselli and Burford [68] found that the major wastes and their respective BOD loads resulting from cotton finishing are as shown in Tables 25.1 and 25.2. Brown's findings [3] indicate that starch waste constitutes about 16 per cent of the total volume of the waste produced, 53 per cent of the BOD, 36 per cent of the total solids, and 6 per cent of the alkalinity. Caustic waste constitutes about 19 per cent of the total volume, 37 per cent of the BOD, 43 per cent of the total solids, and 60 per cent of the total alkalinity. General waste is com-

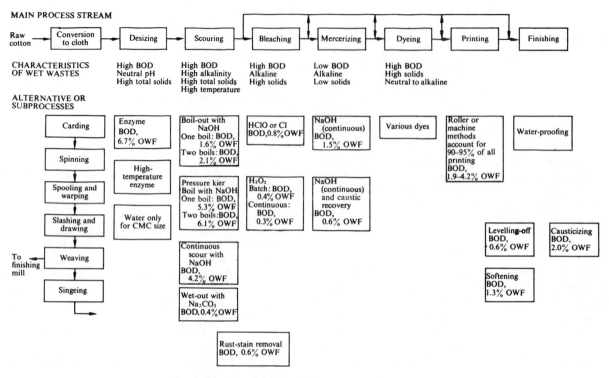

Fig. 25.1 Cotton-textile finishing process flow chart.

Table 25.1 Pollutional loads contributed by various textile processes.
(After Masselli and Burford [68].)

Department	Process	lb BOD/1000 lb cloth*	% of total
Desizing		53	35
Scouring	Either {pressure kier, first scour	53	16
	{pressure kier, second scour	8	1
	Or continuous scour	42	15
	Average	47	
	Subtotal (scouring)		32
Dyeing		0.5–32	15–30
Printing	Color-shop wastes	12	7
	Wash after printing, with soap	17–30	17–30
	Wash after printing, with detergent	7	7
	Subtotal (printing)		15–35
Bleaching	Hypochlorite bleach	8	3
	Peroxide bleach	3	1
Mercerizing		6	1
Total		125–250	

*Approximately 800 to 1000 lb of impurities are discharged in the waste per 1000 lb of cotton processed.

Table 25.2 BOD contributed in the dye process. (After Masselli and Burford [68].)

Process	lb BOD/1000 lb cloth
Vat dye, continuous	18
Vat dye, jig	32
Naphthol, jig	14
Direct, jig	0.5
Sulfur, jig	31

posed of wastes from all other processes (washing, bleaching, dyeing, and finishing) and constitutes 65 per cent of the total volume, 10 per cent of the BOD, 21 per cent of the total solids, and 34 per cent of the total alkalinity.

Wool. Wool wastes originate from scouring, dyeing, oiling, fulling, carbonizing, and washing processes. Practically all the natural and acquired impurities in wool are removed by scouring it in hot detergent–alkali solutions. Because of the high pollutional content of these scouring wastes, some wool is scoured with organic solvents; the grease-laden solvent is then recovered by distillation, leaving behind a recoverable wool grease. During the dyeing process, the hot dye solution is generally circulated by pumps through the wool, which is packed in a removable metal basket suspended in a kettle. In oiling, the carding oil is usually mixed with water and sprayed on the wool. This oil, usually olive oil or a lard–mineral oil mixture, varies from 1 to 11 per cent of the wool by weight. Oiling increases the cohesion of the fibers and aids in the spinning, but all of the oil has to be washed out of the cloth later in the finishing process.

Fulling is the process whereby the loosely woven wool from the loom is shrunk into a tight, closely woven cloth. In most plants, soap mixed with soda ash and a sequestering agent is used in the fulling process; excess fulling solution must be squeezed and/or washed out of the fabric. Carbonizing is a process utilizing hot concentrated acids to convert vegetable matter in the wool into loose, charred particles, which are mechanically crushed and then shaken out of the cloth in a machine called a "duster." In addition, piece dyeing, bleaching, and return fulling may take place, but these operations usually involve a small percentage of the total cloth processed.

A summary of alternative processes for finishing wool is given in Fig. 25.2, based on a recent government survey [1967, 90].

The actual wool-fiber content in "grease wool," as taken from the sheep's back, averages only 40 per cent, the remaining 60 per cent being composed of natural impurities such as sand, grease, suint (dried sheep perspiration), and burrs. Consequently, when $2\frac{1}{2}$ pounds of grease wool are scoured, only 1 pound of scoured wool is obtained; in other words, for every 1000 pounds of wool scoured and produced, 1500 pounds of impurities are discharged to waste. In addition, 300 to 600 pounds of process chemicals are discharged. In terms of BOD, 200 to 250 pounds of BOD are discharged per 1000 pounds of scoured wool produced.

Wool scouring and finishing mills produce a composite effluent having a pH of 9 to 10.5 and containing approximately 900 ppm BOD, 3000 ppm total solids, 600 ppm total alkalinity, 4 ppm total chromium, and 100 ppm suspended solids. The waste is brown in color and mainly colloidal in nature. The major BOD source is the wool grease and suint removed in the scouring and the soap used in fulling and washing. Approximately 70,000 gallons of water are used to process each 1000 pounds of wool, but this waste is only slightly affected by plain sedimentation. Most woolen mills in the U.S. are dyeing and finishing mills, which purchase scoured wool. In such mills the contribution of wool waste to the BOD in the plant waste is negligible. About 24 per cent of the BOD in a woolen mill's waste originates from the dye process, 75 per cent from the wash after fulling, and only 1 per cent from neutralizing after carbonizing. Tables 25.3, 25.4, and 25.5 detail the waste characteristics.

Synthetic fibers. Synthetic fibers are essentially composed of pure chemical compounds and have no natural impurities. Because of this, only light scouring and bleaching are necessary to prepare the cloth for dyeing. Processing of the fibers and cloth is readily done on the conventional machinery used for cotton and wool. At present, the major synthetic fibers are rayon, acetate, nylon, Orlon, and Dacron. Rayon is chiefly composed of regenerated cellulose; acetate is a cellulose–acetate fiber; "nylon" is the generic term for any long-chain synthetic polymeric amide; Orlon is a trade name for synthetic, orientable fibers from

MAIN PROCESS STREAM

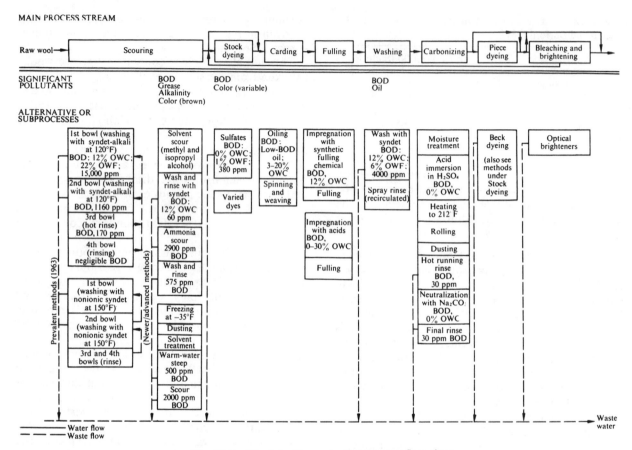

Fig. 25.2   Wool-textile production process flow chart.

polymers containing a preponderance of acrylic units, the newest of which are the acrylonitriles and ethyl or methyl acrylate; Dacron is a polyester fiber manufactured from ethylene glycol and terephthalic acid. All the pollution from treatment of these fibers originates in the various scouring and dyeing chemicals used to process them. (See Figs. 25.3 and 25.4 based on a U.S. Department of the Interior survey [1967, 19].) The volumes and BOD's of such compounds are presented in Tables 25.6, 25.7, and 25.8.

Flax (like cotton) is grown, harvested, deseeded, bundled, loaded on racks, and retted either in closed 'tunnels' or in open basins. The 'closed tunnel' retting system utilizes an aerobic fermentation process while the 'open basin' depends upon aerobic microbiological activity and, in fact, may actually use compressed air to promote faster retting. After retting for one to eight

days (depending on temperature and aerobiosis used), the flax bundles are removed and dried, usually in tee-pee shapes in open fields for five days. These swollen flax tubes can easily be separated from the bast fibres inside. The fibres can be bundled and sent to the spinning mill for subsequent treatment, if necessary, before weaving sililar to that of cotton. Meanwhile, the flax retting wastewater is highly polluting and possesses high oxygen consumption, suspended solids, and brown color, and low pH (4 to 6). Flax retting actually dissolves 25% of the raw fibre weight. Raw flax itself undergoes about 16 steps in preparing it for spinning. Only 11% of the raw flax actually becomes linen and 14% becomes tow (scrap or small fiber flax).

25.2 Treatment of Textile Wastes

Masselli *et al.* [12] emphasize certain preliminary

Table 25.3 Inventory of process chemicals and BOD in woolen-mill wastes. (After Masselli *et al.* [12]).

Process chemical	Chemical composition	Use	% OWF used*			Concentration in effluent, ppm	BOD*	
			Scouring and carding	Finishing	Total		% OWC	% OWF
Soap	Fatty-acid soap	Scouring, fulling	2.1	5.5	7.6	152	155	11.7
Soda ash	Na₂CO₃	Scouring, fulling	14.2	2.8	17.0	340	0	0
Quadrafos	Na₆P₄O₁₃	Washing		0	0.5	10	0	0
Pine oil	Pine oil	Washing	0.5	0	0.5	10	108	0.5
Paragon 500	?		0.5	0	0.5	10		
Proxol T	Mineral oil, plus nonionic emulsifier	Carding	0.5	0	0.5	10	20	0.1
Acetic acid, 84%	CH₃COOH	Dyeing		1.2	1.2	24	62	0.7
Olive sub C3	Oil	Spinning	0.4	0	0.4	8		
Sulfuric acid	H₂SO₄	Carbonizing, dyeing	0	0.2	0.2	4	0	0
Chrome mordant	Na₂Cr₂O₇ + (NH₄)₂SO₄	Dyeing	0	0.4	0.4	8	0	0
Chrome	Na₂Cr₃O₇	Dyeing	0	0.6	0.6	12	0	0
Glauber salt	Na₂SO₄	Dyeing	0	0.4	0.4	8	0	0
Monochlorobenzene	C₆H₅Cl	Dyeing	0	0.2	0.2	4	3	0
Nopco 1656	Soluble fatty ester	Spinning	0	0.2	0.2	4	12	0
Iversol	Blend of soaps, solvents, and detergents	Washing	0	1.6	1.6	32	60	1.0
Rinsol	Detergent	Washing	0	2.9	2.9	58	72	2.1
Supertex E	Fatty-acid soaps, solvent cresylic acid	Washing	0	0.2	0.2	4	25	0.1
Wool finish B	High carbohydrates and enzymes	Finish	0	2.3	2.3	46	57	1.3
	Added impurities (subtotal)		18.7	18.5	37.2	748		17.5
	Natural impurities (grease, suint, dirt)		150.0	0	150.0	3000	16.7	25.0
	Grand total		168.7	18.5	187.2	3748		42.5

*Per cent OWC is BOD inherent in chemical, based on its weight; per cent OWF is BOD due to the chemical, based on weight of wool.

practices to reduce the quantity and strength of textile wastes: good housekeeping, closer process control, process-chemical substitution, and recovery. They admit that even the best housekeeping practices will reduce the BOD load by only 5 to 10 per cent; however, closer control of cotton kiering, sizing, and the amount of chemicals used in the various other processes may reduce pollution loads up to a maximum of 30 per cent. The author also agrees with the

statement of Masselli and Burford that no treatment plant should be planned until serious consideration has been given to pollution reduction through chemical substitution. These authors list cotton- and woolen-mill processes where substitution would be effective.

Cotton mills. (1) Substitution of low-BOD synthetic detergents (syndets) (0 to 20 per cent BOD*) for soap (140 to 155 per cent BOD). The maximum reduction is approximately 35 per cent, in plants where considerable soaping is done. (However, a disadvantage

*Pounds of BOD per pound of chemical, that is, 0 to 0.02 lb BOD/lb detergent.

Table 25.4   Comparison of pollution of woolen processes. (After Masselli *et al.* [12].)

Process	Scour and finish mill, BOD			Finish mill, BOD		
	% OWF*	% of total†	% reduction‡	% OWF*	% of total†	% reduction§
Method I						
Scour with soap	25.0	55.4				
Stock dye with acetic acid	4.9	10.9		4.9	24.4	
Card with 100% BOD oil						
Full with soap	15.0	33.3		15.0	74.6	
Wash with soap						
Neutralize after carbonizing	0.2	0.4		0.2	1.0	
Total	45.1			20.1		
Method II						
Scour with 12% BOD detergent	22.1	74.6				
Stock dye with ammonium sulfate	0.9	3.0		0.9	12.0	
Card with 20% BOD oil						
Full with 12% BOD detergent	6.4	21.6		6.4	85.3	
Wash with 12% BOD detergent						
Neutralize after carbonizing	0.2	0.7		0.2	2.7	
Total	29.6		34	7.5		63
Method III						
Solvent scour, recover grease only	0	0				
Wash out suint salts and dirt with 12% BOD detergent	10.0	58.9				
Stock dye with ammonium sulfate	0.9	5.3		0.9	12.9	
Card with 3% BOD oil						
Full with 12% BOD detergent	5.9	34.7		5.9	84.3	
Wash with 12% BOD detergent						
Neutralize after carbonizing	0.2	1.2		0.2	2.9	
Total	17.0		62	7.0		65
Method IV						
Scour with methyl and isopropyl alcohols; recover grease and suint	0	0				
Wash out dirt with detergent	1.0	19.6				
Stock dye with ammonium sulfate	0.9	17.6		0.9	22.0	
Card with 3% BOD oil						
Full with 12% BOD detergent	3.0	58.8		3.0	73.2	
Wash with 12% BOD detergent						
Neutralize after carbonizing	0.2	3.9		0.2	4.9	
Total	5.1		89	4.1		80

*Based on oven-dry wool.
†Based on the total of that particular method.
‡Based on the total of Method I (45.1% OWF).
§Based on the total of Method I (20.1% OWF).

Table 25.5   Analysis of waste from woolen mill. (After Masselli *et al.* (12).)

Method	Alkalinity*			Solids			
	pH	$CO_3^=$, ppm	HCO_3^-, ppm	Total, ppm	Fixed, ppm	Volatile, ppm	BOD, ppm
Grease scour, 1st bowl, soap-alkali	9.7	4870	7340	64,448	19,133	45,315	21,300
Grease scour, 1st bowl, detergent–Na₂SO₄	8.0	0	6442	60,593	19,889	40,012	15,400
Grease scour, 2nd bowl, soap–alkali	10.4	9153	2214	25,624	15,131	10,493	4,780
Grease scour, 2nd bowl, detergent–Na₂SO₄	8.3	16	463	6,368	2,086	4,478	1,160
Grease scour, 3rd bowl, soap–alkali	9.7	355	154	1,129	555	574	255
Grease scour, 3rd bowl, detergent–Na₂SO₄	7.3	0	75	1,609	525	1,083	170
Stock dyeing, acetic acid	7.3	18	803	3,855	2,248	1,266	2,182
Stock dyeing, ammonium sulfate	6.7	0	194	8,315	3,782	4,533	379
Wash after fulling, 1st soap, soap used for fulling	10.0	2117	584	19,267	4,771	14,489	11,455
Wash after fulling, 1st soap, detergent used for fulling	9.7	380	60	4,830	977	3,853	4,000
Neutralization following carbonizing, 1st running rinse	2.2	0	0	2,241	193	1,048	28
Neutralization following carbonizing, 1st soda–ash bath	8.5	517	2788	9,781	9,559	222	28
Optical wool bleaching in dye kettles	6.0	0	281	908	376	532	390

*No free hydroxide present in any waste samples.

to substituting syndets for soap is their persistence in streams and underground water supplies.) (2) Substitution of steam ranges for oxidation of dyes, in place of dichromate–acetic acid baths (5 to 15 per cent reduction). (3) Use of less caustic in kiering (10 to 20 per cent BOD reduction, 10 to 30 per cent caustic reduction). (4) Use of low-BOD dispersing, emulsifying, leveling, etc., agents in place of high-BOD agents (5 to 15 per cent reduction). (5) Substitution of low-BOD sizes (carboxymethyl celluloses, 3 per cent; hydroxymethyl cellulose, 3 per cent; polyacrylic acid, 1 per cent; polyvinyl alcohol, 1 per cent) for the high-BOD starch (50 to 70 per cent) now widely used. This could theoretically reduce the total BOD from a cotton mill by 40 to 90 per cent. (6) Replacement of acetic acid in dyeing with an inorganic salt, such as ammonium sulfate or chloride (0 per cent BOD).

Woolen mills. (1) Replacement of soap used in scouring by low-BOD detergents (maximum BOD reduction possible is 5 per cent). (2) Replacement of Na₂CO₃ with detergent–Na₂SO₄ mixture, to reduce high alkalinity of waste. (3) Replacement of carding oils (100 per cent BOD) with mineral oils with nonionic emulsifiers (20 per cent BOD); BOD reductions of approximately 10 per cent will be effected in scouring-and-finishing mills and 25 per cent in finishing mills. (4) Replacement of the soap used for fulling and wash after fulling with low-BOD detergents; 15 to 30 per cent BOD reductions may be obtained; H₂SO₄ may also be substituted for soap for fulling. (5) Use of ammonium sulfate instead of acetic acid will reduce the total plant BOD 5 to 10 per cent in a scouring-and-finishing mill.

It should be noted that, if a so-called "no-BOD" compound is substituted for a "high-BOD" compound, other difficulties may arise. For example, when soluble sizes are used to replace starch, the sizes may persist for many miles in a river. In due time they may find their way into water supplies, interfere with treatment, and because of their resistance to treatment may be found in a domestic water supply. This should not cause plants to eliminate substitution of these sizes as a preliminary step in waste treatment, but rather cause them to examine water uses and treatment methods carefully. In some cases, where effluent water does not find its way into the domestic water supply or where adequate water-treatment methods are used, substitution of soluble sizes in the textile industry may prove invaluable. Recovery of

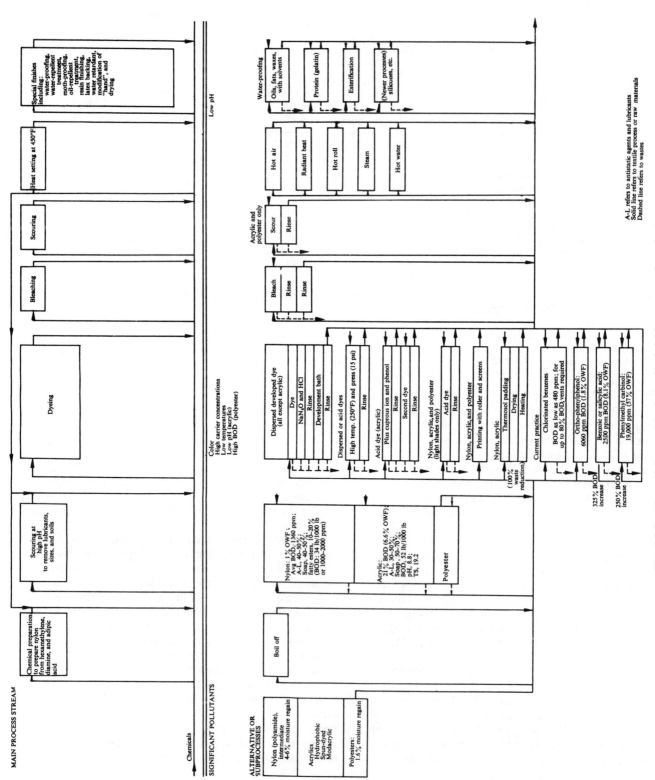

Fig. 25.3 Noncellulose synthetic-textile finishing process flow chart.

MAIN PROCESS STREAM

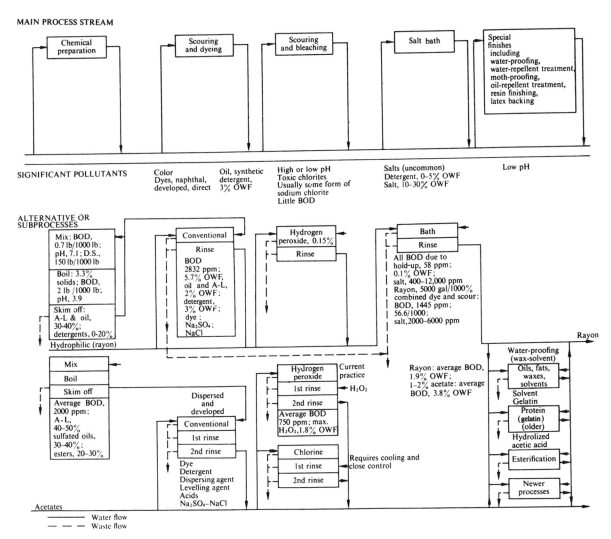

Fig. 25.4 Cellulose synthetic-textile finishing process flow chart.

certain materials should be considered by all mills, since some 200,000 tons of cotton impurities—such as waxes, pectins, and alcohols—are now being dumped into sewers each year [12]. Caustic soda and slashing starch are examples of recoverable chemicals. Many mills are already recovering caustic soda, but mainly so that the remaining waste can be treated more effectively by biological means. About half the caustic soda is found in the contaminated kiering liquors, and the other half comes from mercerizing. Dialysis [17] and evaporation have been used to

purify and recover this caustic soda. Since the kier liquors are relatively dilute solutions of caustic soda, which also contain many other colloidal and suspended impurities such as gums, pectins, and hemicelluloses, it has not been found practical to dialyze or evaporate the waste in order to recover caustic soda.

Masselli [11] calculates that 400,000 tons of glucose might be recovered annually, solely from the starch–desize wastes from our textile mills. This figure was based on an average addition of 10 to 15 per cent

Table 25.6 BOD loads and concentration from processing of various fibers. (After Masselli *et al.* [12].)*

Fiber	gal/1000 lb	BOD, % OWF	Average BOD, ppm
Rayon	5,000	5.0–8.0	1200–1800
Acetate	9,000	4.0–6.0	500–800
Nylon	15,000	3.5–5.5	300–500
Orlon	25,000	10.0–15.0	500–700
Dacron			
o-phenylphenol	12,000	15.0–25.0	1500–2500
Monochloro-benzene	12,000	3.0–5.0	300–500
Benzoic acid	12,000	60.0–80.0	6000–8000
Salicylic acid	12,000	50.0–70.0	5000–7000
Phenylmethyl carbinol	12,000	40.0–60.0	4000–6000
Cotton	70,000	12.5–25.0	220–600
Wool	70,000	40.0–60.0	700–1200

*Based on actual plant surveys, with the exception of data for Dacron.

starch during the sizing process. Recovery of glucose would not only be of economic benefit to the industry but would also reduce the BOD load to be treated by 45 to 94 per cent. Steam produced from the evaporation of desize wastes could be used in the mill.

Lanolin in wool grease from woolen-mill wastes has often been recovered through solvent extraction. A cleaning-solution solvent, such as carbon tetrachloride or benzene, is generally used. A potential supply of 50,000 to 100,000 tons of wool grease exists in our woolen mills, and BOD reductions of 20 to 30 per cent are effected by grease recovery.

Suint can also be recovered by a new alcohol-extraction process and sold for detergent manufacture or potassium salts. Recovery of suint results in an additional 20 to 30 per cent BOD reduction and 20,000 to 40,000 tons of suint can be produced in this manner in the U.S. per year.

Soap is also a valuable product recoverable from woolen-mill wastes. Although little soap recovery is practiced in this country, it could result in 30 to 70 per cent reductions in the BOD of wool-scouring and finishing-mill wastes and the recovered fat might be rendered or used as a fuel source.

It therefore appears that recovery, with its inherent large BOD reductions, is an important step in any waste-treatment plan. However, if recovery, chemical substitution, process control, and good housekeeping practices are not sufficient to eliminate pollution, additional waste-treatment methods must be utilized.

Table 25.7 BOD contribution of process chemicals used in finishing of synthetic fibers. (After Masselli *et al.* [12].)

Fiber	Process chemical*		BOD†		
			% OWC	% OWF	% total
Acetate	Scour and dye				
	Antistat-lubricant			1.5	44
	2% sulfonated oil		52	1.0	31
	1% syndet‡		5	0.1	2
	2% aliphatic ester		41	0.8	24
	2% softener		0	0	0
		Total		3.4	
Nylon	Scour				
	Antistat-lubricant			1.5	29
	1% soaps		150	1.5	29
	1% fatty esters		55	0.6	11
	Dye				
	2–4% sulfonated oils		56	1.7	32
		Total		5.2	

Table 25.7   (*continued*)

Fiber	Process chemical*	BOD†		
		% OWF	% OWF	% total
Dacron	Scour			
	Antistat-lubricant		1.5	9
	1% nonionic syndet	5	0.1	0
	Dye			
	4% acetic acid (84%)	58	2.3	13
	10% o-phenylphenol	138	13.8	78
	Other possible carriers			
	6% p-phenylphenol			
	40% benzoic acid	165	(66.0)	(94)
	40% salicylic acid	141	(56.4)	(94)
	30% phenylmethyl carbinol	150	(45.0)	(92)
	Total		17.7	
Rayon	Scour and dye			
	Antistat-lubricant		1.5	50
	3% syndet	14	0.4	14
	2% soluble oil	53	1.1	36
	10–30% common salt	0	0	0
	Total		3.0	
Orlon	Scour			
	Antistat-lubricant		1.5	12
	2% soaps	150	3.0	24
	0.5% syndet	0	0	0
	3.0% formic acid	20	0.6	5
	Second dye			
	1% wetting agent	14	0.1	1
	3% phenolic compounds	200	6.0	48
	3% copper sulfate			
	2% hydroxy ammonium sulfate	4	0.1	1
	Scour			
	2% syndet	0	0	0
	1% pine oil	108	1.1	8
	Total		12.4	

*Per cent figure before process chemical indicates amount used (OWF).
†Per cent OWC is BOD inherent in chemical, based on its own weight; per cent OWF is the BOD due to the chemical based on weight of the wool; per cent of total is the chemical's contribution to the total BOD load.
‡Syndet indicates synthetic detergent.

25.3 Final Waste Treatment

Generally, chemical coagulation and biological treatment are the chief methods used for final removal of excess BOD, though both processes have their limitations, as pointed out in Chapters 11 and 13. Alum, ferrous sulfate, ferric sulfate, or ferric chloride are used as coagulants, in conjunction with lime or sulfuric acid for pH control. Calcium chloride has also been found effective in such procedures as coagulating

Table 25.8   Estimated pollution load from the processing of various fibers. (After Masselli et al. [12].)*

Fiber	Natural impurities	Sizes, oils, antistats	Scouring	Dyes, emulsifiers, carriers, etc.	Special finishes, waterproof, etc.	Total
Cotton	3–5	0.5–10.0	0.5–6.0	0.2–8.0	0.2–8.0	4.4–37.0
Greasy wool	20.0†–30.0	0.2–9.0	1.5–15.0‡	0.5–10.0	0.2–8.0	21.9†–72.0
Scoured wool	1.0–2.0	0.2–9.0	1.0–15.0‡	0.5–10.0	0.2–8.0	2.9–44.0
Rayon	0	0.5–6.0	0.5–5.0	0.2–5.0	0.2–8.0	1.4–24.0
Acetate	0	0.5–6.0	0.5–5.0	0.2–5.0	0.2–8.0	1.4–24.0
Orlon	0	0.5–6.0	0.5–5.0	0.5–10.0	0.2–8.0	1.7–29.0
Nylon	0	0.5–6.0	0.5–5.0	0.2–5.0	0.2–8.0	1.4–24.0
Dacron	0	0.5–6.0	0.5–5.0	3.0–60.0	0.2–8.0	4.2–78.0

*These estimates are based on surveys and processing methods described in the literature. All results give the per cent OWF.
†If grease and suint are removed by solvent extraction, this load may be reduced to approximately 2 per cent.
‡High values include soap used for fulling also.

wool-scouring wastes. Each coagulant, when used with a specific waste, has its own optimum isoelectric point (pH for maximum coagulation) which must be determined experimentally. Some wastes are more easily coagulated with one chemical than another; others cannot be coagulated with any known economical coagulant. Masselli [12] presents data (Table 25.9) to assist the waste engineer in estimating the BOD reduction possible by chemical coagulation of a particular waste. Figure 25.5 shows schematically the major treatments used today for cotton-finishing wastes.

The author [14–16] gives five ways that textile-dye wastes must be treated before discharge to a stream: (1) equalization, (2) neutralization, (3) proportioning, (4) color removal, and (5) reduction of organic oxygen-demanding matter. It was found that $Al_2(SO_4)_3 \cdot 18H_2O$ (alum) completely removed the apparent color from a sewage–dye-waste mixture and also reduced the BOD by 63 per cent. The dosage of alum required was 200 ppm at the existing pH of 8.3 and 140 ppm at a pH of 7.0.

Chamberlain [5] reported that, instead of coagulating dye wastes chemically, he used chlorine, in the form of chlorinated copperas, to oxidize or bleach many dyes and to remove BOD from sulfur dyes. Chlorine can also be added at the same time

Table 25.9   BOD reductions through chemical coagulation of certain wastes. (After Masselli et al. [12].)

Chemical waste	BOD reduction, %
Soap*	90
Phenol	0
Glucose	0
Starch	57
Gelatin*	65
Glue*	33
Emulsified mineral oil	80
Sulfonated castor oil	82
Sulfonated vegetable oil	44
Coconut oil*	92
o-phenylphenol	0
Salicylic acid	17
Benzoic acid	8
Acetic acid	8
Oxalic acid	86
Sodium acetate	0
Alum–wax emulsion	85

*The following wastes required more than 3 lb of alum as coagulant per 1000 gal: soap, 5 lb; gelatin, 18 lb; glue, 5 lb; coconut oil, 10 lb.

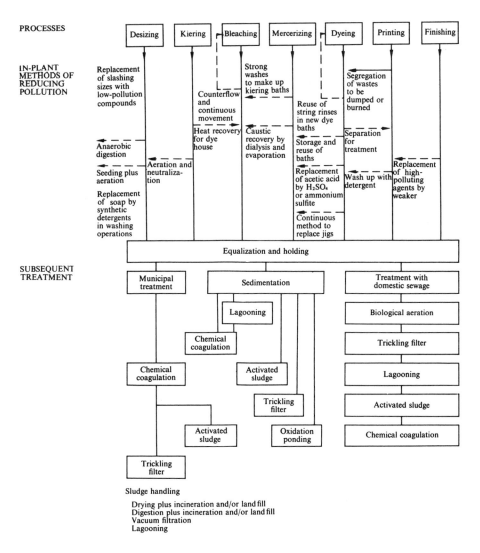

PROCESSES

| Desizing | Kiering | Bleaching | Mercerizing | Dyeing | Printing | Finishing |

IN-PLANT
METHODS OF
REDUCING
POLLUTION

Replacement of slashing sizes with low-pollution compounds

Anaerobic digestion

Seeding plus aeration

Replacement of soap by synthetic detergents in washing operations

Aeration and neutralization

Counterflow and continuous movement

Heat recovery for dye house

Strong washes to make up kiering baths

Caustic recovery by dialysis and evaporation

Reuse of string rinses in new dye baths

Storage and reuse of baths

Replacement of acetic acid by H_2SO_4 or ammonium sulfite

Continuous method to replace jigs

Segregation of wastes to be dumped or burned

Separation for treatment

Wash up with detergent

Replacement of high-polluting agents by weaker

Equalization and holding

SUBSEQUENT
TREATMENT

Municipal treatment

Sedimentation

Treatment with domestic sewage

Lagooning

Biological aeration

Chemical coagulation

Trickling filter

Chemical coagulation

Activated sludge

Lagooning

Activated sludge

Trickling filter

Activated sludge

Oxidation ponding

Chemical coagulation

Trickling filter

Sludge handling

Drying plus incineration and/or land fill
Digestion plus incineration and/or land fill
Vacuum filtration
Lagooning

Fig. 25.5 Cotton-textile finishing waste-treatment flow chart.

as the chemicals, to aid in coagulation and color removal, as well as in the final stage of waste processing. The chlorine requirements are normally 100 to 250 ppm. The action of chlorine is primarily one of oxidizing organic dyes to colorless end-products.

Of the biological means of treating textile wastes, trickling filtration, activated sludge, and dispersed-growth aeration have been most successful. Trickling filtration is generally desirable from the standpoint of flexibility, lower operating costs, and capability of handling shock loads of wastes. Activated-sludge

treatment gives greater BOD reduction, but entails large units to provide the long detention periods (12 to 48 hours) usually needed and also requires highly qualified supervision. Dispersed-growth aeration generally gives somewhat lower BOD reduction than activated-sludge treatment, but does away with the sludge problem; it also takes a minimum of operation and maintenance.

The initial pH of the waste is a controlling factor in the efficiency of any biological treatment. Optimum BOD reductions are obtained when the pH is between

7 and 9, but some BOD reduction takes place when the pH is between 9 and 11, the extent in this range depending on the character of the waste and the equalization it receives prior to aeration. Little or no BOD reduction occurs when the pH exceeds 11.5. Since methods of controlling the pH can be expensive, the cost factor may limit the acceptability of biological treatment. The pH is normally lowered by the addition of acid (H_2SO_4), compressed gas (CO_2), or flue gas. The first two methods are quite effective but relatively expensive. However, flue gases, which usually contain 12 to 14 per cent carbon dioxide, can be used to lower the pH of caustic solutions and, once the capital expense of a pipeline, scrubber, and blower has been overcome, the operating costs are low and the operation is practical. (The principles of neutralization with flue gas were discussed in Chapter 8.) Neutralization is not only feasible, by means of flue gas, but necessary in the biological treatment of alkaline textile wastes. The Beaches [1]

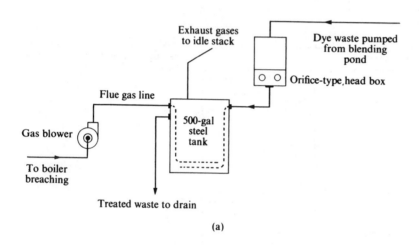

(a)

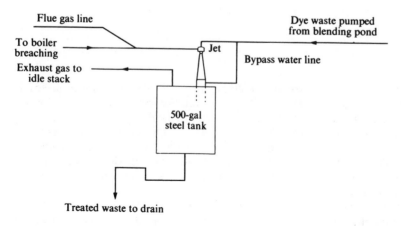

(b)

Fig. 25.6 Gas pilot plants: (a) blower flue; (b) jet flue. (After Beach and Beach [1].)

showed that certain dye wastes had a pH ranging from 2 to 11 and contained up to 30 ppm sulfides. Various methods of passing flue gas into the wastes were studied and they found that, by using a commercial fume scrubber operating on the aspirator principle, the pH could be reduced from 9.0 to 6.1 and 98 per cent of the H_2S eliminated. Their pilot plants are diagrammed in Figs. 25.6 and 25.7. Other workers have also found flue-gas treatment feasible [4, 6, 7, 8, 9, 13, 18].

Dispersed-growth aeration is gaining acceptance as a method of treating highly dissolved organic textile wastes. As pointed out in Chapter 13, laboratory results for dispersed-growth aeration of cotton-kiering liquors, enzyme-desizing wastes, and starch-rinse wastes have indicated satisfactory treatment. Pilot-plant results of this treatment on finishing-mill

wastes were so good [48] that a full-scale treatment plant, based on these results, was designed.

A combination of trickling filters and activated sludge was used successfully in treating a 40 to 60 per cent mixture of textile-finishing-mill waste and domestic sewage [20]. Another laboratory research project [2] illustrated that, at a high loading of 2.73 pounds of BOD per cubic yard of stone and at a relatively high pH of 10.5, a 58 per cent reduction was obtained by trickling filtration. These results indicate that it may be feasible to treat biologically a highly alkaline sewage–waste mixture without prior neutralization. Such a procedure might mean lower efficiency of treatment, but this disadvantage would be offset by considerable savings in chemical costs to the municipalities and industries involved. Also, the reduction in pH of the filter effluent (from 10.5 to 9.1)

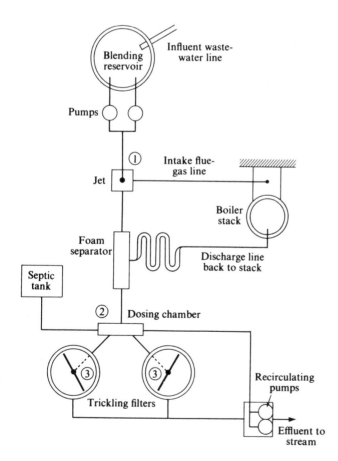

Fig. 25.7 Diagrammatic sketch of a waste-water treatment plant. At (1), the pH is adjusted and the sulfides are removed; at (2), dye waste is blended with settled sewage; and at (3), the combined waste is treated on trickling filters to lower the BOD. (After Beach and Beach [1].)

will produce a higher degree of efficiency in any subsequent biological treatment unit. Therefore, the trickling filter can be used to advantage as a "roughing" or preliminary biological treatment. The 42.5 per cent color removal obtained by this process is a desirable secondary result. A complete activated-sludge system [63] was used in one case to treat the waste from a small bleachery, which had an influent pH of 11.4 and a BOD of 950 ppm.

These examples of how wastes from textile mills can be successfully handled should do much to encourage engineers and textile-plant managers to keep trying to improve their methods of waste disposal.

References

1. Beach, C.J. and M.G. Beach, "Treatment of alkaline dye wastes with flue gas" Proceedings of 5th Southern Municipal and Industrial Waste Conference, April 1956, page 162.

2. Sanitary Engineering Division, Research Report No. 1, "Biological treatment of highly alkaline textile mill waste-sewage mixture," *Proc. Am. Soc. Engrs.* **81**, Paperno., 750(1955).

3. Brown, J.L., Jr., "Combined treatment, textile waste and domestic sewage," Proceedings of 6th Southern Municipal and Industrial Waste Conference, April 1957, page 179.

4. Brown, K.M., and S. Kurtis, "Solved difficult waste disposal problem," *Petrol. Refiner* **29**. 111(1950).

5. Chamberlain, N.S., "Application of chlorine and treatment of textile wastes," Proceedings of 3rd Southern Municipal and Industrial Waste Conference, March 1954, page 176.

6. Curtis, H.A., and R.L. Copson, "Treating alkaline factory waste liquors, such as kier liquor from treating cotton with caustic soda," U.S. Patent No. 1802806, April 1931.

7. Jung, H., "Purifying water," French Patent No. 767586, 20 July 1934.

8. Jung, H. "Chemical treatment of sewage," *Gesundheit Ing.* **69**, 305, (1948).

9. King, J.C., "A solution to highly alkaline textile dye wastes—flue gas treatment," Proceedings of 4th Southern Municipal and Industrial Waste Conference, April 1955.

10. McKinney, R.E., J.M., Symonds, W.G. Shifrin and M.

11. Vezina, "Design and operation of a complete mixing activated sludge system," *Sewage Industrial wastes* **30**, 287, (1958).

11. Masselli, J.W. and M.G. Burford, "Pollution reduction program for the textile Industry," *Sewage & Industrial Wastes* **28**, 1273 (1956).

12. Masselli, J.W., N.W. Masselli, and M.G. Burford, "A Simplification of Textile Waste Survey and Treatment," New England Interstate Water Pollution Control Commission (1959).

13. Murdock, H.R., "Stream pollution alleviated-processing sulfur dye wastes," *Ind. Eng. Chem.* **43**, 77A (1951).

14. Nemerow, N.L., "Textile dye wastes," Proceedings, 1st Southern Municipal and Industrial Waste Conference, March 1952, page 165.

15. Nemerow, N.L., "Textile dye wastes," Proceedings of 7th Industrial Waste Conference, Purdue University, May 1952, page 282.

16. Nemerow, N.L., "Textile dye wastes," *Chemical Age* **66**, 887 (1952).

17. Nemerow, N.L. and W.R. Steele, "Dialysis of caustic textile wastes," Proceedings of 10th Industrial Waste Conference, Purdue University, May 1955, page 74.

19. U.S. Department of the Interior, "The Cost of Clean Water," Industrial Waste Profile No. 4, Textile Mill Products, Washington, D.C. (1967).

18. Steele, W.R., "Application of flue gas to the disposal of caustic textile wastes," Proceedings of 3rd Southern Municipal and Industrial Waste Conference, March 1954, page 190.

20. Hazen, R., "Pilot plant studies on treatment of textile wastes and municipal sewage," Proceedings of 6th Southern Municipal and Industrial Waste Conference, April 1957, Page 161.

21. Krofta, M., et al., "Development of Low-Cost Flotation Technology and Systems for Wastewater Treatment," Proceedings of 42 Purdue Industrial Waste Conference, 1987. Page 185.

22. Shaul, G.M. et al., "Fate of Azo Dyes in the Activated Sludge Process," Proceedings of 41 Purdue Industrial Waste Conference, 1986. Page 603.

23. Brower, G. R. and G. D. Reed, "Economical Pretreatment for Color Removal from Textile Dye Wastes," Proceedings of 41 Purdue Industrial Waste Conference, 1986, Page 612.

24. Bergenthal, J. F. et al., "Full-scale demonstration of textile dye wastewater reuse," Proceedings of 40 Purdue Industrial Waste Conference, 1985, Page 165.

25. Anderson, G. K. et al., "Ozonolysis of Wastecoaters Using a Spinning Disc Reactor," Proc. 39, Purdue Ind. Waste Conf., 1984, Page 163.

26. Wilson, F. and P. H. King, "Treatment of Wood Scour Wastewater: Recent Trends in New Zealand," Proc. 38 Purdue Ind. Waste Conf., 1984, Page 193.

27. Anderson, G. A. et al., "Biological Treatability Study of Wastewater from a Nylon Fibers and Plastics Facility," Proc. 38 Purdue Ind. Waste Conf., 1983, Page 201.

28. Jiajun, H. et al., "A Study on Reuse of Water in A Woolen Mill," Proc. 38 Purdue Ind. Waste Conf., 1983, Page 211.

29. Junkins, R., "Case History: Pretreatment of Textile Wastewater," Proc. 37 Purdue Ind. Waste Conf., 1982, Page 139.

30. Kertell C. R. and G. F. Hill, "Textile Dychase Wastewater Treatment: A Case History," Proc. 37 Purdue Ind. Waste Conf., 1982, Page 147.

31. Kremer, F. et al., "Energy and Materials Recovery Options for the Textile Industry," Proc. 37 Purdue Ind. Waste Conf., 1982, Page 157.

TANNERY WASTES

Tannery wastes originate from the beamhouse and the tanyard. In the beamhouse, curing, fleshing, washing, soaking, dehairing, lime splitting, bating, pickling, and degreasing operations are carried out. In the tanyard, the final leather is prepared by several processes. These include vegetable or chrome tanning, shaving, and finishing. The finishing operation includes bleaching, stuffing and fat-liquoring, and coloring. The discharge from a tannery averages 8000 to 12,000 gallons of waste per 1000 pounds of wet, salted hide processed. The waste averages 8000 ppm total solids, 1500 ppm volatile (organic) solids, 1000 ppm protein, 300 ppm NaCl, 1600 ppm total hardness, 1000 ppm sulfide, 40 ppm chromium, 60 ppm ammonium nitrogen, and 1000 ppm BOD. It has a pH of between 11 and 12 and normally produces a 5 to 10 per cent sludge concentration because of the lime and sodium sulfide contents. The generally accepted procedure for waste treatment is equalization, sedimentation, trickling filtration, or activated-sludge treatment. The latter two biological treatments generally reduce the BOD by 85 to 95 per cent and the sulfide by 100 per cent.

25.4 Origin and Characteristics of Tannery Wastes

Tanning is the act of converting animal skins into leather. For a detailed discussion of the composition of animal skin, the reader is referred to the work of Masselli et al. [1]. The dry matter of the skin is almost entirely protein, of which 85 per cent is collagen. The skin also contains minor amounts of lipids, albumins, globulin, and carbohydrates. The preliminary processes prepare the hide protein (mainly collagen) so that all undesirable impurities are removed, leaving the collagen in a receptive condition to absorb the tannin or chromium used in tanning.

Curing involves dehydration of the hide by drying it with salt or air in order to stop proteolytic enzyme degradation. Fleshing removes the areolar (fatty) tissues from the skin by mechanical means. Washing and soaking remove the dirt, salts, blood, manure, and nonfibrous proteins and restore the moisture lost during preservation and storage. Unhairing is accomplished by the use of lime, with or without sodium sulfide; this makes the skins more attractive and more amenable to the removal of trace protein impurities. Lime splitting separates the skin into two layers: one is the more valuable grain layer; the other, the lower or flesh side, is called the "split." Bating prepares the hide for tanning by reducing the pH, reducing the swelling, peptizing the fibers, and removing the protein-degradation products. Bating is generally accomplished with ammonium salts and a mixture of commercially prepared enzymes (predominantly trypsin and chymotrypsin). The bating bath renders the grain silky, slippery, smoother, and more porous, increases its width, and diminishes its wrinkles. Pickling generally precedes chrome tanning and involves treatment of the skin with salt and acid to prevent precipitation of the chromium salts on the skin-fibers. Degreasing removes natural grease, thus preventing formation of metallic soaps and allowing the skin to be more evenly penetrated by tanning liquors.

Chrome tanning is used primarily for light leather, while vegetable tanning is still preferred for most heavy-leather products. The process of chrome tanning is of shorter duration and produces a more resistant leather. Vegetable tanning produces leathers which are fuller, plumper, more easily tooled and embossed, and less affected by body perspiration or changes in humidity. In the United States mainly

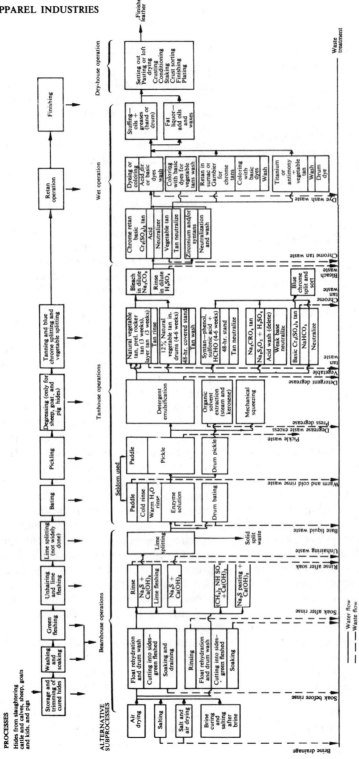

Fig. 25.8 Leather tanning and finishing process flow chart.

quebracho and an extract of chestnut wood are used. Bleaching with dilute Na_2CO_3, followed by H_2SO_4, gives the leather a lighter and more uniform color before dyeing. The process of incorporating oils and greases into the tanned skins is called stuffing and fat-liquoring, and makes the hides soft, pliable, and resistant to tearing. Dyeing to produce the final colored leather product is usually done with basic dyestuffs.

A recent study by the Department of the Interior [1967, 2] revealed changes in the processing of leather, which are shown in a detailed flow chart (Fig. 25.8).

An equalized tannery waste, including rinses, is high in total solids (6000 to 8000 ppm) of which about half (3000 ppm) is NaCl. It contains about 900 ppm BOD, 1600 ppm total hardness, 120 ppm sulfide, 1000 ppm protein, and 30 to 70 ppm chromium. Of importance to the industrial-waste engineer are the high BOD, hardness, sulfide, chromium, and sludge content. About one gallon of this waste is produced for each pound of hide received by the tannery. Masselli et al. [1] also state that for every 1000 pounds of wet, salted hide, there is a BOD load of 76 pounds, 52 per cent of which is discharged in the unhair waste, 20 per cent in the soaks, and 13 per cent in the delime and bate wastes.

Wide fluctuations in the nature of tannery wastes, due to intermittent dump discharges, make these wastes difficult to treat, especially in combination with municipal sewage. Protein and other material extracted from the hides are estimated to produce 50 to 70 per cent of the BOD load and process chemicals are estimated to produce 30 to 50 per cent.

The average composition and waste contribution of major tannery wastes are presented in Tables 25.10 and 25.11, after the findings of Masselli et al. [1]. In Table 25.12 the annual consumption of process chemicals for a cattle-skin tannery is presented, showing, among other things, that the total of process chemicals (including curing salt) used per year is 14,080,000 pounds—approximately 61 per cent of the weight of hides. About 71 per cent of this total, or about 440 pounds of chemicals for each 1000 pounds of hide received by the tannery, is discharged to waste. Salt constitutes 57 per cent of the chemicals in this waste.

25.5 Treatment of Tannery Wastes

In 1952 there were 443 tanneries in the United States, approximately 60 per cent of them located in the North-east. (The number had decreased to about 250 by 1967.) Treatment of tannery wastes is limited in most cases to equalization and sedimentation, although some chemical coagulation and sludge digestion are also practiced. Equalization is necessary to minimize the wide fluctuations in the composition of the waste caused by intermittent dump discharges of strong liquors, and sedimentation is necessary

Table 25.10 Cattlehide tannery survey, average composition and contribution of strong liquors. (After Masselli et al. [1].)

Process	Volume		BOD				Sodium chloride, ppm	Total hardness, ppm	Protein, ppm	Total solids, ppm	Volatile solids, ppm
	gal/day	% of total	ppm	lb/day	% of total	lb/1000 lb hide					
Soaks	73,100	42	2,200	1310	20	15*	20,000	670	1,900	30,000	3,600
Unhair	27,200	16	15,500	3510	52	40*	18,000	25,000	22,700	78,000	18,000
Relime	27,200	16	650	147	2	2*	3,500	25,000		20,300	2,500
Delime and bate	17,600	10	6,000	880	13	10†	10	4,100	4,300	15,000	8,800
Pickle	9,800	6	2,900	237	4	3†	47,000	2,400		79,000	7,200
Chrome tan	8,500	5	6,500‡	425	6	8†	26,000	1,800		93,000	13,000
Color and fat-liquor											
First dump	5,100	3	2,000	85	1	3§				16,000	8,000
Second dump	5,100	3	2,200	93	1	3§	250	2,600		9,500	4,900
Total	173,600			6687							

*Based on wet, salted hide.
†Based on fleshed, split hide, after relime.
‡Estimated at 50 per cent concentration of volatile solids.
§Based on chrome-tanned leather.

Table 25.11   Pigskin tannery survey; average composition and contribution of strong liquors. (After Masselli *et al.* [1].)

Process	Volume gal/day	Volume % of total	BOD ppm	BOD lb/day	BOD % of total	BOD lb/1000 lb hide	Sodium chloride, ppm	Total hardness, ppm	Protein, ppm	Total solids, ppm	Volatile solids, ppm
Soaks	3,000	19	2,400	60	8	17*	35,000			28,000	2,300
Unhair	4,000	26	14,000	467	61	70†	5,700	38,000	18,400	55,000	12,900
Delime and bate	4,000	26	4,400	147	19	23†	640	4,200	1,600	14,000	7,400
Pickle	700	5	4,200	25	3	9†	80,000			98,000	12,000
Degrease											
Kerosene layer	340	2		(1210)‡		435§					
Brine layer	800	5	2,600	17	2	7	100,000			110,000	2,300
Vegetable tan	30		24,000	6	2§	1				93,000	25,000
Chrome tan	600	4	2,300††	12	2	5§	51,000			80,000	4,600
Color and											
fat-liquor											
First dump	1,000	6	490	4	1	1**	410			3,950	890
Second dump	1,000	6	3,950	33	4	8**	135			3,980	3,030
Total	15,470			771							

*Based on wet, salted hide.
†Based on fleshed hide after soaking (30 per cent of flesh removed).
‡Calculated from kerosene (53 per cent BOD) only; not included in total.
§Based on pickled hide.
**Based on leather.
††Estimated at 50 per cent concentration of volatile solids.

because of the large volumes of sludge (5 to 10 per cent) present in the waste. If secondary treatment is necessary, trickling filters and activated-sludge systems can be used (when the pH has been reduced to 9.0 and the hardness to 200 ppm) to produce an 85 to 95 per cent BOD reduction. The evolution of H_2S from the waste, as a result of contact with another acid waste, should be prevented. When discharging tannery waste to a municipal treatment plant, provisions must be made to remove hair and fleshings (usually by screening) and to avoid deposition of scale in the sewer line. A schematic description of tannery waste treatment is given in Fig. 25.9).

There have been three comprehensive studies of the treatment of tannery wastes. The first was made by the Leather Chemists' Association on Stream Pollution [3]. It found that all 32 plants reporting used primary sedimentation, but only five employed secondary treatment. Similarly, spent tans were lagooned and released at periods of high stream flow by only five of the plants. Although the U.S. Public Health Service [4] recommends chemical precipitation as one of the treatments to be used on combined tannery wastes, the survey [5] showed that only one plant used chemical precipitation or applied its effluent to a filter bed. The Tannery Waste Disposal Committee of Pennsylvania [6] found that mixing of wastes and sedimentation removed 85 per cent of the

suspended solids and 40 per cent of the BOD. This committee recommended equalization and settling for at least eight hours, then trickling filtration and sedimentation. As an intermediate step it recommended the addition of coagulants to the filter effluent. In a more recent study, Haseltine [7] found combining tannery wastes with domestic sewage feasible. The tannery waste is equalized separately for three to four days; it is then mixed with twice its volume of sewage, aerated, and settled. The sludge is concentrated, dewatered on vacuum filters, and incinerated.

Masselli *et al.* [1] noted that treatment of tannery wastes is limited in most cases to equalization and sedimentation. Many tanneries discharge without treatment directly into coastal waters or to a municipal sewage-treatment plant. They recommend that, when treatment is required, the strong liquors should be segregated and the dilute rinses discharged directly without treatment. The wastes should be screened to prevent damage to, or clogging of, pumps and pipelines. Holding basins or storage tanks should be used to provide equalization for at least one day's flow of waste. When sedimentation is called for, it is necessary to provide mechanical sludge removal followed by centrifugation or vacuum filtration. The dewatered sludge can be finally disposed of by sand or cinder-bed filtration or by lagooning.

Table 25.12 Annual consumption of process chemicals in cattlehide tannery. (After Masselli *et al.* [1]).

Process chemical	lb used	% BOD	lb BOD	ppm in waste
Sodium chloride	1,368,000	0	0	684
Sodium chloride (used in curing 19% OWH)	4,408,000	0	0	2200
Lime	2,470,000	0	0	1235
Sodium sulfide (62% Na$_2$S, 25% S$^-$)	981,000	40	392,000	490
Sulfuric acid	350,000	0	0	175
Soda ash	161,000	0	0	80
Oropon [95% (NH$_4$)$_2$SO$_4$]	144,000	5	7,200	72
Calcium formate	88,000	12	11,000	44
Lactic acid (30%)	77,000	32	25,000	38
Sodium formate	56,000	2	1,100	28
Sterizol	42,000			20
Ammonium chloride	20,000	0	0	10
Chemicals absorbed by hide*				
Tanolin R (16% chromium)	1,670,000	0	0	626†
Tamol L	729,000	0	0	36
D-1 oil	37,000	83	3,100	2
Other oils (total of 12)	650,000	80	52,000	33
Quebracho	146,000	5	700	7
Soyarich flour	100,000			5
Tanbark H	88,000	11	1,000	4
Titanium dioxide	88,000	0	0	4
Ade 11 tan	38,000			2
Gambade	156,000	4	600	8
Maratan B	136,000			7
Methocel	20,000	6	120	1
Orotan TV	30,000	5	150	2
Semisol glue	37,000			2
Upper tan	28,000			1
Totals	14,080,000		494,300	5343

*Absorption estimated at 90%, only 10% discharged to waste. Pounds BOD and ppm in waste are based on 10% of the pounds used.
†Based on 75% discharged to waste.

Some important information pertaining to the results of primary treatment of tannery wastes, including settling, lagooning, chemical precipitation, and sludge digestion, is provided in references 8, 9, 10, 11, 12, 13, 14, 15, 16, 17, 18, 19, 20, 21, 22, 23, and 24. Results of secondary treatment of tannery wastes are given in references 25, 26, 27, 28, 29, 30, 31, and 32.

Under proper conditions, the activated-sludge process developed on a pilot-plant scale by Chase and Kahn [1955, 33] was able to render tannery wastes entirely suitable for discharge into streams. An effluent was produced which was practically clear, free from objectionable odor, and perfectly stable. During a period of over four months, between March and July 1916, the following reductions were obtained: albuminoid ammonia, dissolved, 80 per cent, suspended, 69 per cent; free ammonia nitrogen, 65 per cent;

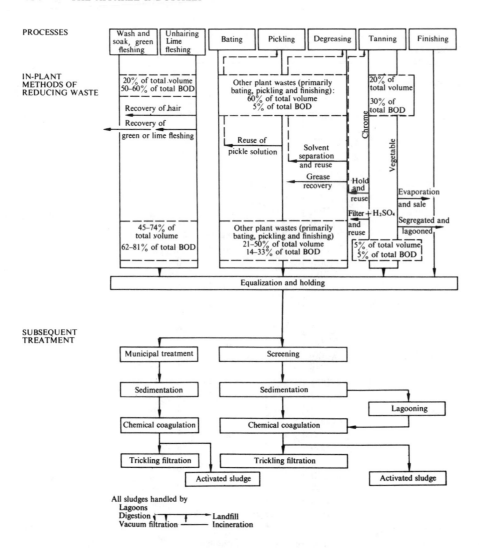

Fig. 25.9 Leather tanning and finishing waste-treatment flow chart.

total suspended solids, 78 per cent; fats (by ether extraction), 79 per cent.

Owing to the variability of the wastes it was found necessary to equalize their flow and quality, in order to preserve the bacterial activity of the sludge and to make the wastes more amenable to treatment. Milk of lime was added periodically to avoid an acid effluent from the equalizing tank and to maintain the alkalinity level of the wastes above the minimum required for good clarification.

Return activated sludge (sludge which is returned to the aeration tank from the final settling tank after activated-sludge treatment), as high as 50 per cent of the influent, was found to be quite economical when compared with the relative quantities of air required per gallon of wastes treated by aeration of activated sludge. Within certain limits it was found that the rate of air supply and the period of aeration could be considered inversely proportional, with economic advantages favoring longer periods of aeration.

The pilot-plant operation indicated that, with the most economical use of air, the volume of sludge to be disposed of would probably not exceed 10,000 gallons per million gallons of wastes treated. It could be dried on sludge beds, if applied in shallow doses not over 10 inches deep, because the activated-sludge process produced no objectionable odors and had no fly problems (such as occurred with trickling filters); furthermore, aeration renders the sludge more stable and less likely to create malodors when drying.

Gates and Lin [1966, 34] studied in two pilot plants the possibility of using an anaerobic lagoon with an oxygen cover for treating tannery wastes. They obtained 88.5 per cent reduction in BOD at organic loadings of about 1000 pounds of BOD acre/day. The final effluent, although colored, was clear and odorless. Most of the reduction occurred in the anaerobic section—which included sludge digestion—the aerobic lagoon being utilized primarily for odor control.

Emerson and Nemerow [25] (1969) recently reported and published results of laboratory and pilot-plant experiments on biological treatment of equalized combined beamhouse and tanhouse wastes. The treatment without neutralization by high-solids (15,000 ppm suspended solids) aeration, at loadings of 140 pounds of BOD per 1000 ft³ with a detention time of 12 hours, yielded 90 per cent BOD reduction.

References

1. Masselli, J. W., N. W. Masselli, and M. G. Burford, *Tannery Wastes*, New England Interstate Water Pollution Control Commission (1958).

2. U.S. Department of the Interior, *The Leather Industry— The Cost of Water*, Profile Report no. 6 (1967).

3. "A condensed report by the Tannery Waste Disposal Committee of Pennsylvania to the Sanitary Water Board on the treatment of tannery wastes," *J. Am. Leather Chemists' Assoc.* 26, 70 (1931).

4. U.S. Public Health Service, *Tannery Wastes*, Ohio River Survey, Supplement D (1943), p. 1218.

5. "Tannery Waste Disposal Committee of Pennsylvania, a report," *J. Am. Leather Chemists' Assoc.* 26, 70 (1931).

6. Reuning, H. T., "Report of stream pollution committee," *J. Am. Leather Chemists' Assoc.* 38, 292 (1943) and 39, 378 (1944).

7. Haseltine, T. R., "Combined treatment, tannery wastes and sewage, Williamsport, Pa.," *Sewage Ind. Wastes* 30, 65 (1958).

8. Eldridge, E. F., "Report on sanitary engineering projects," *Mich. State Univ. Eng. Exp. Sta. Bull.* 67, 32 (1936), 83, 3 (1939), and 87 (1939).

9. Fales, A. L., "Treatment of industrial wastes from paper mills and tannery on Neponset river," *Ind. Eng. Chem.* 21, 216 (1929).

10. Foster, W., "Chrome tannery treatment plant, description," *Sewage Ind. Wastes*, 24, 927 (1952).

11. Harnley, J. W., R. F. Wagner, and H. G. Swope, "Treatment at Griess-Pfleger Tannery, Waukegan, Ill.," *Sewage Works J.* 12, 771 (1940).

12. Howalt, W., and E. S. Cavett, "Studies on tannery waste disposal," *Proc. Am. Soc. Civil Engrs.* 53, 1675 (1927); *Trans. Am. Soc. Civil Engrs.* 92, 1351 (1928).

13. Hubbell, G. E., "Tannery waste disposal at Rockford, Mich.," *Water Works Sewage* 82, 331 (1935).

14. U.S. Public Health Service, *Industrial Waste Guide*, Ohio River Pollution Survey, Supplement D (1943).

15. Künzel-Mehner, A., "Treatment with ferric chloride," *Sewage Works J.* 17, 412 (1945).

16. Maskey, D. F., "Study of tannery waste disposal," *J. Am. Leather Chemists' Assoc.* 36, 121 (1941).

17. Porter, W., "Operating problems from tannery wastes, Ballston Spa, N.Y.," *Sewage Works J.* 21, 738 (1949).

18. Reuning, H. T., "Tanning wastes," *J. Am. Leather Chemists' Assoc.* 42, 573 (1947).

19. Reuning, H. T., and R. F. Coltart, "An effective tannery waste treatment plant," *Public Works Mag.* 78, 21 (1947).

20. Riffenburg, H. B., and W. W. Allison, "Treatment of tannery wastes with flue gas and lime," *Ind. Eng. Chem.* 33, 801 (1941).

21. Rosenthal, B. L., "Treatment of tannery waste and sewage mixture on trickling filters," *Sanitalk* 5, 21 (1957).

22. Vrooman, M., and V. Ehle, "Digestion of combined tannery and sewage sludge," *Sewage Works J.* 22, 94 (1950).

23. Warrick, L. F., and E. J. Beatty, "Treatment with domestic sewage," *Sewage Works J.* 8, 122 (1936).

24. Wimmer, A., "The sewage from the tannery city Backnang," *Sewage Works J.* 9, 529 (1937).

25. Alsop, E. C., "Purification of liquid tannery waste by forced oxidation," *J. Am. Leather Chemists' Assoc.* 7, 72 (1912).

26. Eddy, H. P., and A. L. Fales, "The activated-sludge process in treatment of tannery wastes," *Ind. Eng. Chem.* 8, 548 (1916).

27. Loveland, F. A., *J. Am. Leather Chemists' Assoc.* 7, 474 (1912).

28. Mohlman, F. W., "Treatment of packing-house, tannery, and corn-products wastes," *Ind. Eng. Chem.* 18, 1076 (1926).

29. Power, R. M., *Sanitalk* 5, 19 (1957).

30. Rosenthal, B. L., "Treatment of tannery wastes by activated sludge," *Sanitalk*, 6, 7 (1957).

31. Sarber, R. W., "Tannery waste disposal," *J. Am. Leather Chemists' Assoc.* 36, 463 (1941).

32. Sutherland, R., "Tanning industry," *Ind. Eng. Chem.* 39, 628 (1947).

33. Chase, E. S., and P. Kahn, "Activated sludge and filters for tannery waste treatment," *Wastes Eng.* 26, 167 (1955); *J. Am. Leather Chemists' Assoc.* 50, 366 (1955).

34. Gates, W. E. and S. Lin, "Pilot plant studies on the anaerobic treatment of tannery effluents," *J. Am. Leather Chemists' Assoc.* 61, 10 (1966).

35. Emerson, Dwight B. and N. L. Nemerow, "High solids biological aeration of unneutralized unsettled tannery wastes," *Purdue Industrial Waste Conference Proceedings* (1969).

36. Rao, T. D. and T. Viraroghavan, "Treatment of Tannery Effluent," Proc. 41 Purdue Ind. Waste Conf., 1986, Page 204.

37. Cheda, P. V. et al., "Joint Wastewater Management for a Cluster of Tanneries in Kanpur," Proc. 39 Purdue Ind. Waste Conf., 1984, Page 151.

38. Bailey, D. G. et al., "Anaerobic Treatment of Tannery Waste," Proc. 38 Purdue Ind. Waste Conf., 1983, Page 673.

39. Porter, J. R. and T. A. Doane, "Odor Control at a Western Michigan Tannery and Wastewater Treatment Plant," Proc. 37 Purdue Ind. Waste Conf., 1982, Page 163.

40. Suddath J. L. and E. L. Thackston, "Improved Color Removal in Spent Vegetable Tanning Liquors," Proc. 36 Purdue Ind. Waste Conf., 1981, Page 801.

41. Chang I. L. and N. S. Zaleiko, "UV-Oxidative Degradation of Protein-Containing Wastewaters-Tannery Unhairing Waste," Proc. 36 Purdue Ind. Waste Conf., 1981, Page 814.

42. Huber, C. V. and T. A. Doane, "A Case Study-Tannery Meets EPA Pretreatment Standards," Proc. 35 Purdue Ind. Waste Conf., 1980, Page 95.

LAUNDRY WASTES

The laundry industry is a service—not a manufacturing—industry and therefore no specific subdivisions exist. According to a statement issued by the American Laundry Institute in January 1961, professional laundries have become the nation's largest personal-service industry, with an annual sales volume of more than $1,600,000,000. The industry processes more than 5 billion pounds of laundry per week, including more than 50 million men's shirts. Laundry wastes originate from the use of soap, soda, and detergents in removing grease, dirt, and starch from soiled clothing. The waste has a high turbidity and alkalinity and a readily putrescible organic content with a BOD of 400 to 1000 ppm. The usual method of treatment is chemical precipitation, after adjustment of pH by dilution or chemical addition. If secondary treatment is required, laundry wastes may be oxidized readily on trickling filters. The activated-sludge process is sometimes used, but it is not as satisfactory as the trickling filter. The spread of laundromats in non-sewered areas is creating especially critical problems.

25.6 Origin and Characteristics of Laundry Wastes

Wastes originate from the washing of clothes, which are usually placed in a double cylinder with water, soap, and other washing agents. Rotation of the inner perforated cylinder (the outer cylinder is stationary) produces the agitation necessary to free or dissolve the impurities (dirt) from the fabrics. A detailed discussion of commercial laundry methods is given by Smith [1]. The amount of alkali used (and hence the washing formulas) varies with the type and amount of soil content; the present tendency is to use alkalis, such as sesqui- or orthosilicate, which have low buffering values.

The U.S. Public Health Service [2] estimates laundry water consumption, and hence waste production, as four gallons per pound of clothes. Boyer [3] describes the highly putrescible character of laundry wastes as strongly alkaline, exceedingly turbid, highly colored, and containing large quantities of soap, soda ash, grease, dirt, dyes, and scourings from cloth; its BOD will be on average twice that of domestic sewage and at times it will be five times as great. Rudolfs [4] presents analyses (Table 25.13)

Table 25.13 Comparison of the sanitary characteristics of commercial and domestic laundry wastes.* (From Rudolfs [4].)

Analysis	Commercial	Domestic
pH	10.3	8.1
Total alkalinity, ppm	511	678
Total solids, ppm	2114	3314
Volatile solids, ppm	1538	2515
BOD, 5-day, ppm	1860	3813
Oxygen consumed, ppm	868	1045
Grease, ppm	554	1406

*"Commercial" presumably refers to large-scale operations such as linen service for hotels and restaurants, while "domestic" refers to laundries processing home apparel [author's note].

of both commercial and domestic laundry wastes. The U.S. Public Health Service in its survey [2] indicated that most laundry waste waters have compositions within the limits shown in Table 25.14. (The reader will note some discrepancies between the pollution loads given in Tables 25.13 and 25.14.)

Table 25.14 Typical laundry waste water composition [2].

Analysis	Ranges of values
pH	9.0–9.3
Alkalinity above pH 7.0, as Na_2CO_3, ppm	60–250
Total solids, ppm	800–1200
BOD, 5-day, ppm	400–450

Eckenfelder and Barnhart [5] studied the treatment of wastes from laundromats and small laundry operations. Most installations contain between 25 and 35 machines, each using 25 to 30 gallons of water per washing cycle. Of this water, 22 gallons were hot (140°F) and 8 gallons cold, so that the average temperature of the waste discharged was 100°F. An average waste-water volume of 50,000 gallons per week per installation can be expected. Approximately 100 pounds of commercial detergent are used per week. The characteristics of the composite waste are summarized in Table 25.15.

Table 25.15 Analysis of typical laundromat effluent (24-hr composites). (After Eckenfelder and Barnhart [5].)

Characteristic	Range of values
Turbidity*	208–300
COD, ppm	344–445
Detergent, as ABS†, ppm	50–90
pH	7.0–8.1
Suspended solids, ppm	140–163

*Based on an arbitrary scale setting, with pure water equal to zero.
†Alkyl benzene sulfonate.

25.7 Treatment of Laundry Wastes

In 1944 Gehm [6] came to the following conclusions concerning treatment of laundry wastes.

1. To remove about 75 per cent of oxygen-consuming solids and grease, laundry wastes can be treated most economically by acidification with H_2SO_4, CO_2, or SO_2, followed by coagulation with alum or ferric sulfate. Coagulation by other salts and lime may be effective in some cases, but is usually too costly.

2. Laundry waste can be effectively treated by trickling filtration or by the activated-sludge process, with long aeration periods.

3. Sludge obtained can be dried directly on sand beds, digested anaerobically, or filter-pressed. As a final recovery, soap or the dried sludge can be reclaimed.

4. After chemical coagulation, laundry wastes can be further purified by biological filtration or activated-sludge treatment.

5. Domestic sewage can contain laundry waste up to 20 per cent of its volume when being treated by the activated-sludge process and any amount of laundry waste when being treated by biological filtration.

Both Eliassen and Schulhoff [7] and the Florida State Board of Health [8] demonstrated that flotation produced better results than sedimentation. This is undoubtedly due to the relatively large percentage of emulsified grease in these wastes.

Eckenfelder and Barnhart [5] concluded that through a combination of physical adsorption, with seven parts of carbon to one part of detergent, and chemical coagulation, with 100 grains per gallon of alum, it is possible to remove almost all of the anionic synthetic detergent in wastes from laundromat operation. They also found that settling for a period of four hours will result in a sludge containing one to two per cent solids.

Andres *et al.* [9] present three methods of treatment for laundromat wastes prior to their discharge into ground water: 1) Separmatic treatment, with pressure diatomaceous-earth filters; 2) Lansing treatment—a flotation process, using pH reduction, followed by air flocculation and floating of coagulated sludge; 3) activated carbon, alum, and soda ash coagulation as proposed above [5]. Although the three systems, used either in combination or separately, may remove 85 to 95 per cent of suspended solids, BOD, and synthetic detergents, none appears to effect a significant reduction in dissolved solids.

The U.S. Public Health Service [2] summarized the most recent information available in 1956, stating that appreciable waste reduction could be accomplished by avoiding overuse of washing agents and by controlling washer loads. Most commercial laundry wastes are discharged directly to municipal sewage systems and treated together with domestic sewage. Considerable experimental work on the separate treatment of laundry wastes with trickling filters has been reported in the literature. In practice, some laundry wastes are being treated by chemical flocculation and sedimentation, with further purification obtained by lagooning and sand filtration.

Sigworth [10] concludes that the aesthetically objectionable foam and flavor from syndets (synthetic detergents) in potable water supplies can be controlled at the treatment plant by carbon dosages in the range of one to two parts (per million) of carbon for each part of syndet formulation. He states that "of all presently known water-purification processes, activated carbon is the only tool which will assure complete success." Table 25.16 gives the dosages of activated carbon required to remove the taste and foam characteristics of five typical detergents.

Current treatment methods include the following systems: (1) natural methods of disposal—into deep wells, fissures in cavernous limestone strata, and artificial lagoons—although there are serious objections to each of these methods; (2) trickling filtration; (3) the activated-sludge process; (4) chemical precipitation; (5) sand filter and high-rate trickling filter; (6) acidification with H_2SO_4, CO_2, or SO_2, followed by coagulation with alum and ferric sulfate; (7) the activated-carbon method and diatomaceous-earth filter. Spade [1962, 11] discusses the basic features of this last method and the properties and characteristics of diatomaceous earth. Since the treatment does not use coagulation and settling basins, it makes possible a great saving in space, and also includes completely automatic filtration. How-

Table 25.16   Foam and flavor studies on synthetic detergents. (After Sigworth [10].)

Synthetic detergent	Amount producing characteristic		Carbon dosages necessary to treat 25 ppm concentration of detergent	
	Noticeable foam, ppm	Unpalatable water, ppm	To correct foam, ppm	To correct flavor, ppm
A	10.0	17.5	34	43
B	2.0	10.0	44	20
C	500.0	20.0	0	50
D	1.0	8.0	25	50
E	5.0	15.0	30	40

ever, its main disadvantage is the inability to handle large flows; the maximum economical range at present seems to be 40,000 gal/day. If the ability to handle greater flows can be achieved, consideration may be given to developing diatomaceous filters as polish units for the final effluent of sewage-treatment plants. Typical analyses for the Maric diatomaceous-earth filter and the Bruner clear stabilizing-filter systems are given in Table 25.17.

Table 25.17 Comparison of the Bruner and Maric filter systems (After Spade [1962, 11].)

Filter system	Waste characteristic*				
	BOD	D.O.	Detergents, as ABS	Total solids	Suspended solids
Bruner					
Raw	162	1.6	38	563	100
Final	28	5.2	1.1	1020	2.0
Maric					
Raw	132	4.7	35	340	220
Final	12	3.8	3.0	16	12

*All results are given as parts per million.

25.8 Dry Cleaning Wastes

In 1986 dry cleaning sales were about $3.5 billion at 32,000 retail and 225 industrial clothing cleaning plants (Seitz 12, 1987). Although cleaning tonnage has increased, it is claimed that solvent tonnage has decreased: Perchloroethylene, known as perc, is used in 75–80 percent of the dry cleaning plants, Valclene (a fluorocarbon solvent) is used in only about 2 percent of them, while the use of kerosene type, Stoddard solvent, is used by the remaining 18 percent. Recycling of perc and Stoddard solvent are commonly reused by the industry. The petroleum-based solvents are also sold back to the automotive parts-cleaning services. Most dry cleaning establishments recycle the solvent by filtering it through either a bank of disposable cartridge filters (using various media including activated carbon) or by distillation. In 1987 perc cost about $3.30 per gallon when purchased in 52 gallon drums, while Valclene cost $16 per gallon. However, because of the nature of Valclene which effects the ozone layer, its use will diminish in the future. Labor and energy still represent about 99 percent of the cleaning costs of a suit with the solvent cost only 1 percent. Recycling, then, is important not for the process economics, but primarily for environmental protection.

References

1. Smith, R. B., *Washroom Methods and Practice in the Power Room Laundries*, Moore-Robbins Publishing Co., New York (1948).
2. "Industrial waste guide to the commercial laundering industry," *U.S. Public Health Serv. Bull.* 509 (1956).
3. Boyer, J.A., "The treatment of laundry wastes," Texas Agr. Exp. St. Bull. 42, 1 October 1933.
4. Rudolfs, W., *Industrial Wastes*, Reinhold Publishing Corp., New York (1953), p. 471.
5. Eckenfelder, W.W., and E. Barnhart, "Removal of Synthetic Detergents from Laundry and Laundromat Wastes," N.Y. State Water Pollution Control Board, Research Report No. 5, March 1960.
6. Gehm, H. W., "Volume, characteristics, and disposal of laundry wastes," *Sewage Works J.* 16, 571 (1944).
7. Eliassen, R., and B. Schulhoff, "Laundry waste treatment by flotation," *Water Works Sewerage* 90, 418 (1943).
8. Florida State Board of Health, "Experimental pilot plant studies treatment wastes," *Wastes Eng.* 24, 512 (1953).
9. New York State Water Pollution Control Board, "Effect of Synthetic Detergents on the Ground Waters of Long Island," Research Report No. 6. by C.W. Lauman, Inc. and Suffolk County Health Dept., June 1960.
10. Sigworth, E. A., "Synthetic detergents and their correction with activated carbon," *J. North Carolina Section Am. Water Works Assoc.*; in Proceedings of 40th Meeting of Water Pollution Control Association, 1960, p. 45.
11. Spade, J. F., "Treatment methods for laundry wastes," *Water Sewage Works* 109, 110 (1962).
12. Seitz, W. "Dry Cleaners' Prosperity Bypasses Perc," Chemical Business, Jan. 1987, page 7.
13. Van Gilo, G. J. et al., "Treatment of Emulsified and Colloidal Industrial Wastewater Using a Combined Ultrafiltration," Proc. 39 Purdue Ind. Waste Conf., 1984, p. 269.
14. Jeng, F. T. and C. J. Shih, "Treatment of Laundry Wastewater from a Nuclear Power Plant by Reverse Osmosis," Proc. 39 Purdue Ind. Waste Conf., 1984, p. 281.

CHAPTER 26

FOOD INDUSTRIES

26.1 Introduction

Food-processing industries are those whose main concern is the production of edible goods for human or animal consumption. Processing plants included in this group are: (1) canneries, (2) dairies, (3) breweries and distilleries, (4) meat-packing and -rendering plants (including poultry plants and animal feedlots), (5) beet-sugar refineries, (6) pharmaceutical plants, (7) yeast plants, and (8) miscellaneous plants, producing such foods as pickles, coffee, fish, rice, soft drinks, bakeries, and water production. The production processes usually consist of the following steps: cleaning the raw material, removal of inedible portions, preparation of the foodstuff, and packaging. The wastes to be considered are: spoiled raw material or spoiled manufactured products; rinsing or washing waters; condensing or cooling waters; transporting waters; process waters; floor- and equipment-cleaning liquids; product drainage; overflow from tanks or vats; and unusable portions of the product.

The characteristics of food-processing wastes exhibit extreme variation. The BOD may be as low as 100 ppm or as high as 100,000 ppm. Suspended solids, almost completely absent from some wastes, are found in others in concentrations as high as 120,000 ppm. The waste may be highly alkaline (pH 11.0) or highly acidic (pH 3.5). Mineral nutrients (nitrogen and phosphorus) may be absent or may be present in excess of the (BOD/N) or (BOD/P) ratio necessary to promote good environmental conditions for biological treatment. Similarly, the volume of wastes may be almost negligible in some industries, but reach one or more million gallons per day in others.

Food-processing wastes usually contain organic matter (in the dissolved or colloidal state) in varying degrees of concentration, so that biological forms of waste treatment are indicated. Since these wastes differ from domestic sewage in general characteristics and in particular by their higher concentrations of organic matter, pretreatment is required to produce an equivalent effluent. In addition, one or more of the following adjustments are frequently necessary to provide the proper environmental conditions for the microorganisms upon which biological treatment depends: continuous feeding, temperature control, pH adjustment, mixing, supplementary nutrients, and microorganism population adaptation.

Among the aerobic or anaerobic biological treatments available, the major and more effective methods make use of activated sludge, biological filtration, anaerobic digestion, oxidation ponds, lagoons, and spray irrigation. The loadings of the biological units must be carried out with care, since many of the wastes contain high concentrations of organic matter. Quite frequently long periods of aeration or high-rate two-stage biofiltration is required to produce an acceptable effluent. The type of treatment selected will depend on the following aspects: degree of treatment required, nature of the organic waste, concentration of organic matter, variation in waste flow, volume of waste, and capital and operating costs.

CANNERY WASTES

The canning industry is one of the most important to the people of the United States because (1) its total annual retail value today (1967) exceeds \$5 billion for canned food and \$3 billion for frozen food and (2) it utilizes great quantities of water—in 1964 the "pack" of 944 million equivalent cases of canned and frozen fruits and vegetables required 76 billion gallons of water. The industry is extremely diversified; about 200 plants in the United States can or freeze literally dozens of different raw products. Cannery wastes are classified according to the product being processed, its season of growth, and its geographic location. Since the harvesting and processing periods of the three main groups of products—vegetables, fruits, and citrus fruits—are short, many canneries are designed to process more than one product. Wastes from these plants are primarily organic and result from trimming, juicing, blanching, and pasteurizing of raw materials, the cleaning of processing equipment, and the cooling of the finished product. The four most common and effective methods of treatment are: discharge to municipal treatment plant, lagooning with the addition of chemical stabilizers, soil absorption or spray irrigation, and anaerobic digestion.

26.2 Origin of Cannery Wastes

Figure 26.1 gives a fairly detailed schematic illustration of the canning and freezing processes for both fruits and vegetables. Peas, beets, carrots, corn, squash, pumpkins, and beans are among the vegetables which produce strong wastes when processed for canning. Since the preparation for processing differs with each vegetable, the methods used should be studied individually, but there is little other difference in cannery procedures, and hence the origin of all vegetable wastes is similar. The process waste usually consists of: wash water; solids from sorting, peeling, and coring operations; spillage from filling and sealing machines; and wash water from cleaning floors, tables, walls, belts, and so forth.

Among fruits, the processing of peaches, tomatoes, cherries, apples, pears, and grapes presents the most common problems of waste discharges. Waste flows may originate from lye peeling, spray washing, sorting, grading, slicing and canning, exhausting of condensate, cooling of cans, and plant cleanup. Other wastes originate from specific operations not necessarily common to all fruit processing.

The three main citrus fruits—oranges, lemons, and grapefruit—are usually processed in one plant, which produces canned citrus juices, juice concentrates, citrus oils, dried meal, molasses, and other by-products. Liquid wastes from citrus-fruit processing include cooling waters, pectin wastes, pulp-press liquors, processing-plant wastes, and floor washings. Citrus-cannery waste is a mixture of the peel, rag, and seed of the fruit, surplus juice from the washing operations, and blemished fruits.

Table 26.1 Volume and characteristics of cannery wastes. (From Sanborn [1].)

Product	Volume per case, gal	5-day BOD, ppm	Suspended solids, ppm
Asparagus	70	100	30
Beans, green or wax	26–44	160–600	60–85
Beans, lima	50–257	189–450	422
Beans, baked	35	925–1440	225
Beets	27–65	1580–5480	720–2188
Carrots	23	520–3030	1830
Corn, cream style	24	623	302
Corn, whole-kernel	25–70	1123–6025	300–4000
Peas	14–56	380–4700	272–400
Mushrooms	6600*	76–390	50–242
Potatoes, sweet	3500*	295	610
Potatoes, white	†	200–2900	990–1180
Pumpkin	20–42	2850–6875	785–3500
Sauerkraut	3	6300	630
Spinach	160	280–730	90–580
Apples, sauce	†	1685–3453	
Apricots	57–80	200–1020	260
Tomatoes, whole	3–15	570–4000	190–2000
Tomatoes, juice	38–100	178–3880	170–1168

*Per ton.
†Not given.

26.3 Characteristics of Cannery Wastes

The volume and characteristics of waste waters vary considerably from one plant to the next, and within the same plant from day to day. Data presented by Sanborn [1] illustrate the variability of the wastes after screening (Table 26.1). Eckenfelder [2] gives additional data on the characteristics of cannery wastes (Table 26.2).

Citrus-cannery waste is a slick, slimy, nonuniform mass, with a moisture content of about 83 per cent. A complete breakdown of the wastes of a cannery processing 700 tons per day of oranges, lemons, and grapefruit and producing 0.7 mgd of waste containing 6 tons of BOD is given by Ludwig et al. [3] (Table 26.3).

Wakefield [4] presents data on Florida citrus-plant wastes, which we reproduce in Table 26.4. Sanborn [1] also shows variation in the BOD of cannery wastes within the same plant canning different fruits and vegetables (Table 26.5).

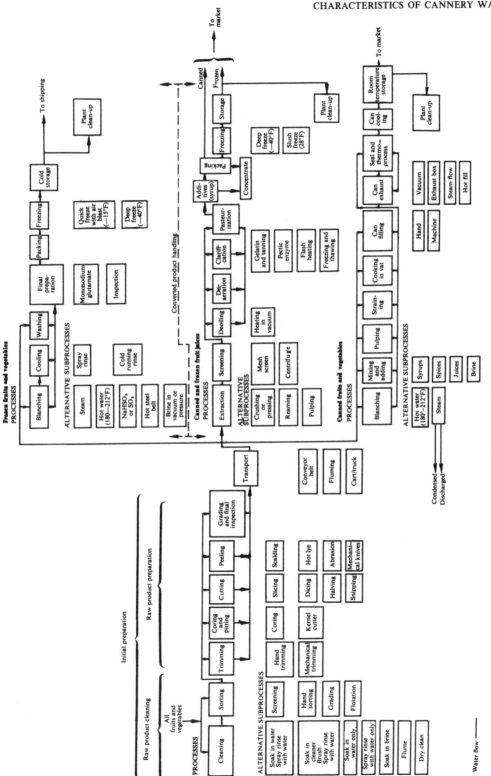

Fig. 26.1   Process flow chart for preparation of canned and frozen fruits and vegetables.

Table 26.2 Cannery-waste characteristics. (From Eckenfelder [2].)

Product	Flow, gal/case	BOD, ppm	Suspended solids, ppm
Tomatoes	4.5–78.0	616–1870	550–925
Corn	30–116	885–2936	530–2325
Green beans	104.5	93	291
Green beans and corn	99.5	270	264
Mixed vegetables	12.2	750	593
Pears	32.4–42.5	238–468	340–637
Peaches	37.5	1070	250
Apples	26.8	1600	300
Cherries	16.0	800	185

Table 26.3 Composition of citrus wastes. (From Ludwig et al. [3].)

Type of waste	Waste flow, gpd	BOD, ppm	Suspended solids, ppm
Cooling water	285,000	100	765
Pectin wastes	225,000	2720	1790
Pulp-press liquor	120,000	9850	780
Processing-plant waste	40,000	3230	3400
Floor washings	30,000	970	685
Composite wastes		2100	7200

26.4 Treatment of Cannery Wastes

Canning is a highly seasonal business, and hence the treatment of cannery wastes presents unique problems. Orlob et al. [1965, 5] present a detailed description of their six categories of cannery-waste treatment: (1) in-plant modifications, (2) preliminary treatment, (3) chemical treatment, (4) biological treatment, (5) land disposal, and (6) municipal treatment. The selection of the type of treatment most suitable for a particular plant must be guided by two sets of considerations: one comprising the standard aspects of volume and character of wastes and treatment required, the other taking into account the unique conditions of number and duration of packing periods. Cannery wastes are most efficiently treated by screening, chemical precipitation, lagooning, and spray irrigation. Digestion and biological filtration are also used, but to a lesser extent.

Screening is a preliminary step, designed to remove large solids prior to the final treatment or discharge of the waste to a receiving stream or municipal waste-water system. Only slight reductions in BOD are realized by screening. Mechanically operated screens (mesh size varying from 12 to 30) of either the rotating or vibrating type are used. Typical screening loads are about 40 to 50 pounds per 1000 gallons of waste water. Vibrating screens will produce solids having a moisture content between 70 and 95 per cent, depending on the product. The waste retained on the screens is disposed of in various ways: it can be spread on the ground, used for sanitary fill, dried and burned, or used to supplement animal feed.

Chemical precipitation, in conjunction with other treatment methods, is used to adjust the pH and to reduce the concentration of solids in the wastes. It has been quite effective for treating apple, tomato, and cherry wastes. Ferric salts or aluminate and lime have produced 40 to 50 per cent BOD reductions. Dosage rates amount to about 5 to 10 pounds of lime, plus 1 to 8 pounds of ferrous sulfate or alum, for each 1000 gallons of screened waste. Chemical precipitation produces about 10 to 15 per cent by volume of sludge, which will normally dry on sand beds in about a week without producing an odor.

Treatment in lagoons involves biological action (both aerobic and anaerobic), sedimentation, soil absorption, evaporation, and dilution. Some engineers advocate lagooning as the only practical and economical treatment of cannery wastes when adequate land is available. Lagoons in which aerobic conditions are not maintained give off unpleasant odors and provide a breeding ground for mosquitoes and other insects. To eliminate odors, $NaNO_3$ is used, at a dosage equal to 20 per cent of the applied oxygen demand (normally 20 to 200 pounds per 1000 no. 2 cases). However, using $NaNO_3$-treated lagoons for complete treatment may be impractical from the standpoint of cost because of the large volumes of wastes involved. Also, with strong wastes such as pea wastes, odors may still persist even with the $NaNO_3$ treatment. Surface sprays have been used to reduce the fly and other insect nuisance and in some cases to combat the odors arising from these lagoons. Seepage

Table 26.4 Citrus wastes. (From Wakefield [4].)

Plant or process	Flow, gpd	BOD, ppm	lb BOD/ 1000 cases	Suspended solids, ppm
Juice	158,610–813,200	182	12.7–43.1	25–85
Sectioning of grapefruit	211,700–420,260	873–945	384–887	124–140
Concentration of orange juice (average of four plants)	2,396,500	82	57.1*	27

*Per 1000 gal of concentrate.

Table 26.5 Cannery waste variation. (After Rudolfs [1].)

Product	Origin of waste	BOD, ppm
Peas	Pea washer	3,700
	Blancher overflow	13,815
	Blancher dump	34,490
	Ensilage-stack liquor	35,000–78,000
Corn	Corn washer	2,800
	Whole-kernel washer	7,000
	Ensilage-stack liquor	22,000–33,000
Kidney beans	Soak water	10,500
	Blancher	3,600
Cherries, sour	Pitter drippage	38,000–55,000
	Pit flume water	950–3,330
Grapefruit sections	Fruit-wash water	20–110
	Peeling table	38,080
	Sectioning tables	2,480
	Exhaust-box overflow	1,000
	Floor-wash water	4,000
	Peel-bin drippage	50,000

must also be considered, especially when lagoons are located near underground sources of water.

Spray irrigation is another economical and unobjectionable method that can be used whenever the cannery waste is nonpathogenic and nontoxic to plants. Its use is primarily limited by the capacity of the spray field to absorb the waste water. High BOD reductions may be expected as the waste per-colates through the vegetation and soil. Some spray-irrigation performances are given in Table 26.6. Use of ridge-and-furrow irrigation or absorption beds is limited to soils of relatively high water-absorbing capacity. Permanent pasture grasses are apparently able to handle a heavier organic load that alfalfa can. Wastes should be screened before spraying, although comminution alone has been used successfully in conjunction with spray irrigation.

Oxygen-demanding materials in cannery wastes can be removed by biological oxidation. When the operation is limited by seasonal conditions, it is difficult to justify capital investment for bio-oxidation facilities. However, in many instances cannery wastes can be combined with domestic sewage, and then bio-oxidation processes provide a practical and economic solution. High-rate trickling filters have reduced the BOD of pea, green-bean, and tomato wastes by as much as 97 per cent, and cider, apple, cherry, tomato, and citrus wastes have also been successfully treated. BOD loadings to filters having a removal rate of 90 per cent range from 0.5 to 2.0 pounds per cubic yard per day. Activated-sludge treatment has also been used to produce a clear, odorless citrus-waste effluent with at least 90 per cent BOD reduction. Mixed cannery wastes and pea and carrot wastes have been handled by conventional activated-sludge plants effecting 91 to 95 per cent BOD reduction on wastes whose raw BOD ranged from 1350 to 1500 ppm. At a BOD loading of 1.7 to 2.5 pounds of BOD per day per pound of sludge, detention times vary from three to five hours.

In a study of canning wastes in which the author was involved [1965, 5], Orlob sought to provide

Table 26.6   Spray irrigation performance. (After Eckenfelder [2].)

Product	Pump rate, gpm	Total area sprayed, acres	Rate of application, gpm/acre	Average application, in./day	Average loading	
					lb BOD/ acre/day	lb suspended solids/ acre/day
Tomatoes	{1000 500	5.63 6.4	178 86	2.96 0.70	413 155	364 139
Corn	350	2.28	153.5	3.35	864	500
Asparagus and beans	253	0.9	282	3.5	22.5	356
Tomatoes, corn, and lima beans	430	9.18	43.8	0.375	40.5	14.7
Lima beans	430	6.65	65	0.375	65	46
Cherries	216	2.24	96.5	3.61	807	654

an insight (not heretofore evident in the literature) into the interrelationship of the many technical and economic considerations which govern decisions between alternative methods of coping with the cannery-waste treatment problem. This revealing study (1) characterized the physical, chemical, and biochemical in-plant waste streams and composite flows resulting from processing of peaches and tomatoes; (2) evaluated the technical and economic feasibility of in-plant separation and/or treatment of cannery-waste flows; and (3) developed engineering-economic systems for cannery-waste treatment and/or disposal.

Water used for rapid cooling of cans after sterilization by heat in the canning industry is sometimes recirculated with intermittent cooling and chemical treatment. The treatment given must take into account the fact that some tins may burst or develop leaks during heating or cooling processes. Such leaks can contaminate the water with fruit or vegetable products. Also leaking cans may possibly draw water in during cooling operations. Therefore, it is also necessary that the recirculated water is sterilized in order to avoid subsequent contamination of their contents.

References: Cannery Wastes

1. Rudolfs, W., *Industrial Waste Treatment*, Reinhold Publishing Corp., New York (1953).

2. Eckenfelder, W. W., *et al.*, "Study of fruit and vegetable processing waste disposal methods in the eastern region," Final Report, New York State Canners' Association, September 1958.

3. Ludwig, H. F., G. W. Ludwig, and J. A. Finley, "Citrus by-product waste at Ontario, Calif.," *Sewage Ind. Wastes* 23, 1254 (1951).

4. Wakefield, J. W., "Semitropical industrial waste problems," in Proceedings of 7th Industrial Waste Conference, Purdue University, May 1952, p. 495.

5. Orlob, G. T., F. G. Agardy, R. C. Cooper, N. L. Nemerow, and R. G. Spicker, *Cannery Waste Treatment Utilization and Disposal—A Literature Review*, California State Water Quality Control Board and Water Resources Engineers, Inc. (1965).

6. White, T. E. et al., "Anaerobic Treatment of Apple Romance and Wastewater," Proc. 43 Purdue Ind. Waste Conf., 1988, P. 551.

7. Blanc, F. C. and O'Shaughnessy, J. C., "Static Pile Composting of Cranberry Receiving Wastes and Processing Residues, Proc. 43 Purdue Ind. Waste Conf. 1988, P. 569.

8. Blanc, F. C. et al., "Design and Operation of a Cranberry Wastewater Treatment System," Proc. 41 Purdue Ind. Waste Conf., 1986, P. 497.

DAIRY WASTES

In 1963, 16 million cows in the United States produced approximately 127 billion pounds of milk. This was distributed among five related industries as follows: 52 billion pounds of fluid milk, including 1.5 billion pounds of cottage cheese; 34 billion pounds of butter; 14.4 billion pounds of cheese; 11.9 billion pounds of ice-cream and frozen desserts; and 10.8 billion pounds of condensed and powdered milk. The plants handling or processing milk and milk products may be roughly classified as follows: receiving stations, bottling plants, cheese factories, creameries, condenseries, and dry-milk and ice-cream plants.

The receiving station serves as a collection point for raw milk from the farmers. Here milk-cans are emptied into a weighing vat, the milk is sampled, and it is then loaded into tank cars or trucks for shipment to processing plants. Waste-producing operations are washing and sterilizing of cans, vats, tanks, cooling equipment, and floors. At the bottling plant, the raw milk is dumped, sampled, weighed, clarified, filtered, preheated, pasteurized, cooled, and poured into glass or paper containers. Waste-producing operations here include washing of bottles, cases, cans, processing equipment, and floors. Figure 26.2 illustrates this process.

A typical milk-receiving station* now receives about 25 per cent of its milk in conventional 10-gallon milk-cans and the other 75 per cent in the new 2500-gallon pickup tank cars. About $2\frac{1}{2}$ gallons of rinse water are required per 10-gallon can in the total cleaning process. When cans are processed, the rinse water leaves the station in a steady discharge. For cleaning the pickup tanks, an automatic prerinse of about 200 gallons per truck is followed by a rinse of about 500 gallons, which is reused for 5 to 10 trucks before being discharged. An additional 500 gallons of cold-water rinse per truck concludes the tank cleaning. Thus, a similar volume of rinse water per gallon of milk received is used for the old-type cans and the new tank cars.

A cheese factory receives whole milk, cream, or separated milk, which is weighed, preheated, filtered, pasteurized, and cooled. A 1 to 3 per cent lactic-acid bacterial culture is sometimes used to attain the proper pH. It then is placed in cheese vats to which rennet, acid, or other souring agents are added. This causes a separation of casein in the form of a curd. The whey is then withdrawn, and the cheese washed with water. Other ingredients, such as cream, may be added (depending on the final product desired) and the cheese is shaped and packaged for sale (see Fig. 26.3).

A creamery processes whole milk, sour cream, and/or sweet cream into butter and other products (Fig. 26.4). When whole milk is received, it is centrifuged to separate the cream from the milk; the cream is churned into butter, while the separated milk may be processed into other dairy products, either for human consumption or for animal feed.

In the condensery, whole milk or other dairy products are evaporated to obtain a concentrated product. Unsweetened milk is the most important product and is produced by preheating whole milk and then evaporating and homogenizing it. Cans of evaporated milk are filled and sealed by machine, after which they are sterilized. Sweetened condensed milk is made in much the same manner, except that sugar is added. Other products include condensed nonfat milk, whey, and buttermilk.

A dry-milk plant uses atmospheric drying, vacuum drying, or spray drying to produce powdered whole or nonfat skimmed milk. The ice-cream plant, employing various formulas, combines milk and milk products with cream in certain proportions with flavorings, sugar (or other sweetening agent), and a stabilizer. The resulting mixture is homogenized, pasteurized, and cooled, after which fruits, nuts, or candies may be added, and the entire mixture frozen.

26.5 Origin and Characteristics of Diary Wastes

Dairy wastes are made up, for the most part, of: various dilutions of whole milk, separated milk, buttermilk, and whey from accidental or intentional spills; drippings allowed to escape into the waste through inefficient design and operation of process equipment; washes containing alkaline or other chemicals used to remove milk and milk products, as well as partially caramelized materials, from cans, bottles, tanks, vats, utensils, pipes, pumps, hot wells, evaporating coils, churns, and floors; and process washes of butter, cheese, casein, and other products.

Although dairy plants are found in most communities, there is considerable variation in the size of plants and in the type of product they manufacture. Table 26.7 shows the average composition of milk constituents, as reported by three different investigators. Dairy wastes are largely neutral or slightly

---

*Private communication to author.

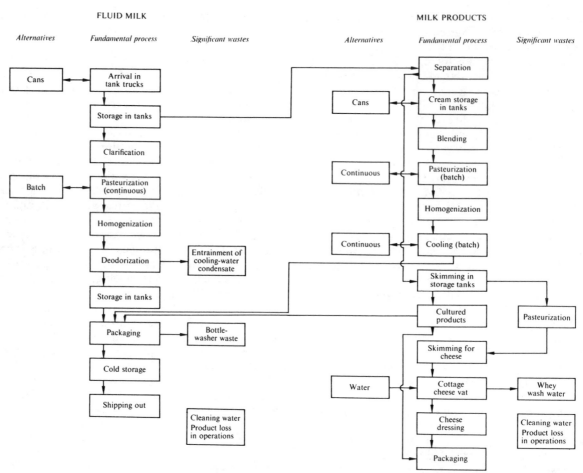

Fig. 26.2   Process flow chart for fluid-milk preparation.

alkaline, but have a tendency to become acid quite rapidly, because of the fermentation of milk sugar to lactic acid. Lactose in milk wastes may be converted to lactic acid when streams become devoid of oxygen, and the resulting lowered pH may cause precipitation of casein. Cheese-plant waste is decidedly acid, because of the presence of whey. Milk wastes contain very little suspended material (except the fine curd found in cheese waste) and their pollution effects are almost entirely due to the oxygen demand which they impose on the receiving stream. Heavy

black sludge and strong butyric-acid odors, caused by decomposing casein, characterize milk-waste pollution. Roughly, 100 pounds of whole milk will result in about 10 pounds of BOD. The average composition of milk, milk by-product, and cheese waste is given in Table 26.8.

26.6 Treatment of Dairy Wastes

Milk-plant wastes are generally high in dissolved organic matter, contain about 1000 ppm BOD, and

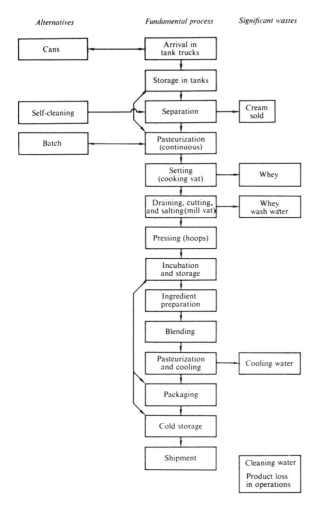

Fig. 26.3 Process flow chart for the preparation of cheese (natural and processed).

are nearly neutral in pH. Since these wastes are mainly composed of soluble organic materials, they tend, if stored, to ferment and become anaerobic and odorous. Therefore they respond ideally to treatment by biological methods. Aerobic processes are most suitable, but the final selection of a treatment method hinges on the location and size of the plant. The six conventional methods generally used which are most effective are: (1) aeration, (2) trickling filtration, (3) activated sludge, (4) irrigation, (5) lagooning, and (6) anaerobic digestion.

Because there is a wide variation in the flow rate and strength of milk wastes, holding and equalization are desirable to provide a uniform waste for treatment. Aeration is desirable, either as a means of treatment in itself or as pretreatment before biological processes. Aeration for one day often results in 50 per cent BOD reduction and eliminates odors during conversion of the lactose to lactic acid. Higher BOD reductions (50 to 80 per cent) have been obtained by aerating milk waste in the presence of some seed material, and following this with a period of settling. High-rate recirculating filters are quite commonly used for treating milk wastes. Some two-stage filters

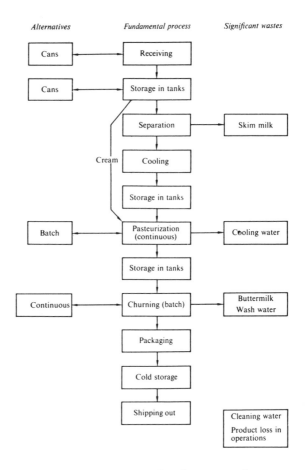

Fig. 26.4 Process flow chart for creamery butter.

Table 26.7   Composition of whole milk.*

Investigator	Water	Fat	Protein			Lactose	Ash
			Casein	Albumin	Total		
Roberts [101]	87.25	3.80			3.50	4.80	0.65
Eldrige [20]	87.30	3.60			3.80	4.50	0.80
Van Slyke	87.10	3.90	2.5	0.7	3.2	5.10	0.70

*Results are given as percentages.

Table 26.8   Average composition of milk, milk by-products, and cheese wastes.

Characteristics	Whole milk, ppm	Skim milk, ppm	Butter-milk, ppm	Whey, ppm	Process wastes, ppm	Sep-arated whey, ppm
Total solids	125,000	82,300	77,500	72,000	4516	54,772
Organic solids	117,000	74,500	68,800	64,000	2698	49,612
Ash solids	8,000	7,800	8,700	8,000	1818	5,160
Fat	36,000	1,000	5,000	4,000		
Soluble solids					3956	54,656
Suspended solids					560	116
Milk sugar	45,000	46,000	43,000	44,000		
Protein (casein)	38,000	39,000	36,000	8,000		
Total organic nitrogen					73.2	1,300
Free ammonia					6.0	31
Na					807	648
Ca					112.5	350
Hg					25	78
K					116	1,000
P					59	450
BOD, 5-day	102,500	73,000	64,000	32,000	1890	30,100
Oxygen consumed	36,750	32,200	28,600	25,900		

yield greater than 90 per cent BOD reduction, while single-stage filters loaded at the rate of about one pound of BOD per cubic yard (1610 pounds of BOD per acre · foot) yield about 75 to 80 per cent BOD reduction.

The activated-sludge process has proved a successful method for the complete treatment of milk wastes. This system employs aeration to cause the accumulation of an adapted sludge. The flora and fauna in the active sludge, when supplied with sufficient air, oxidize the dissolved organic solids in the waste. Excess sludge is settled out and subsequently returned to the aeration units. There is some indication that the treatment can be carried out without wasting any sludge, although this requires an aeration period sufficient to "burn up" most of the excess sludge. Properly designed plants which provide ample air for handling the raw waste plus returned sludge are not easily upset, nor is the control procedure difficult. Operation costs, however, may be higher than those involved in trickling filtration. Both continuous-flow and batch-operated activated-sludge treatment plants have removed 90 to 97 per cent of the BOD. Generally, a tank volume of about 80 to 100 gallons is needed for each pound of BOD in the waste under treatment. Since milk wastes average about 1000 ppm BOD, the volume of returned sludge is usually 6 to 7 times the raw-milk flow.

References

1. Swampalli, R. Y. and Bauman, E. R., "RBC Kinetics in Treating Domestic and Industrial Diary Wastewater Under Low and High Organic Loading Conditions," Proc. 41 Purdue Ind. Waste Conf., 1986, P. 445.

2. Sabkowicz, A.M., "Celrobic Anaerobic Treatment of Dairy Processing Wastewater," Proc. 41 Purdue Ind. Waste Conf., 1986, P. 459.

3. Backman, R. C. et al., "The Treatment of Dairy Wastewater by the Anaerobic Up-flow Packed Bed Reactor," Proc. 40 Purdue Ind. Waste Conf., 1985, P. 361.

4. Doran, F. J. et al., "Nitrogen Transformations and Reduction by Ridge and Furrow Land Treatment of Dairy Waste." Proc. 40 Purdue Ind. Waste Conf., 1985, P. 405.

5. Martin, J. H. and R. R. Zall, "Dairy Processing Wastewater Bioaugmentation—An Evaluation of Effectiveness," Proc. 40 Purdue Ind. Waste Conf., 1985, P. 351.

BREWERY, DISTILLERY, WINERY, AND PHARMACEUTICAL WASTES

The fermentation industries include breweries and distilleries, manufacturers of alcohol and certain organic chemicals, and some parts of the pharmaceutical industry, such as producers of antibiotics. Fermentation has been defined as the decomposition of complex organic substances into material of simpler composition, under the influence of nitrogenous organic substances called ferments. The transformation of grape juice into wine, the manufacture of alcohol from molasses, and the use of yeast in dough to make bread are familiar examples of fermentation.

Two main types of raw material are used for producing alcohol or alcoholic products: starchy materials, such as barley, oats, rye, wheat, corn, rice, and potatoes, and materials containing sugars, such as blackstrap and high-sugar molasses, fruits, and sugar beet. The process of converting these raw materials to alcohol varies somewhat, depending on the particular raw material and the desired alcoholic product. For instance, flavor is of prime importance in the manufacture of beer, and this concern accordingly influences the process used. Manufacturers of distilled products, on the other hand, are more concerned with alcohol yield. The process of brewing and distilling consists of: (1) conversion of malt to a finely divided state in a malt mill; (2) preparation of the mash by mixing malt with hot water and, in some cases, with raw grain; (3) transformation of starches to sugar by the action of malt; (4) draining and washing the "sweet" water from the mash to fermentation tanks; (5) fermentation of sugars to alcohol by yeasts; (6) cooling, skimming, and clarification of the fermented liquor; and (7) locking in casks (if used for beer); storing in vats (if used for alcohol).

Two types of yeast are used for fermentation: "bottom yeast," used in beer manufacture, and "top yeast," used in top-fermented beers and whiskey mash. The latter type is also used almost exclusively in the manufacture of commercial compressed (bakers') yeast. The production of yeast, whether it be intended for fermentation purposes or for direct sale for baking purposes, is accompanied by the formation of by-product wastes of a highly pollutional character.

The growth of biological and pharmaceutical plants was greatly accelerated during and after World War II. Manufacture of new products, particularly antibiotics, has greatly increased the waste-disposal problem of this industry.

26.7 Origin of Brewery, Distillery, Winery, and Pharmaceutical Wastes

The brewing of beer has two stages, malting of barley and brewing the beer from this malt. Both these operations are carried on in the same plant. The malting process consists of the following steps: (1) grain is removed from storage and screened; (2) screened grain is placed in a tank and steeped with water to bleach out the color; (3) grain is then allowed to germinate, while air and water are introduced to stimulate growth of enzymes to be used for inoculum; (4) the grain malt is removed after five to eight days of aeration to the dryer, where it is dried for about four days, to a predetermined moisture content; (5) the finished malt, after the sprouts have been screened out, is stored and aged in large elevators.

The malting process produces two major wastes: those arising from the steep tank after grain has been removed, and those remaining in the germinating drum after the green malt has been removed. In the actual brewing process, considerable water is required, most of which is used for cooling purposes. Brewery wastes are composed mainly of liquor pressed from the wet grain, liquor from yeast recovery, and wash water from the various departments. After the distillation of the alcohol process, a residue remains which is referred to as "distillery slops," "beer slops," or "still bottoms."

During production of wines, considerable quantities of both solid and liquid wastewaters accumulate. Liquid wastes originate mainly from three sources— spillage, cleanup water, and cooling water. The primary wine residues are grape stems from stemming and crushing, pomace (pressing wastes), and crude lees (fermentation residues); wine lees and spillage from storage cellars; stillage or distilling residues from brandy stills; heads and fusel oil fractions separated from fermentation columns; and liquid wastes from the cellar washing operations.

There are several sources of wastes in a distillery. Of major concern are the "dealcoholized" still residue and evaporator condensate, when the stillage is evaporated. Minor wastes include redistillation residue and equipment washes. In the manufacture of compressed yeast seed, yeast is planted in a nutrient solution and allowed to grow under aerobic conditions until maximum cell multiplication is attained. The yeast is then separated from the spent nutrient solution, compressed, and finally packaged. The yeast-plant effluent consists of: (1) filter residues resulting from the preparation of the nutrient solutions, (2) spent nutrients, (3) wash waters, (4) filter-press effluents, and (5) cooling and condenser waters.

Pharmaceutical wastes originate principally from the spent liquors from the fermentation processes, with the addition of the floor washings and laboratory wastes. Liquid wastes from pharmaceutical plants producing antibiotics and biologicals can be classed as: (1) strong fermentation beers, (2) inorganic solids, such as diatomaceous earth, which are utilized as a precoat or an aid to the filtration process, (3) washing of floors and equipment, (4) chemical waste, and (5) barometric condenser water from evaporation.

26.8 Characteristics of Brewery, Distillery, Winery, and Pharmaceutical Wastes

Malt wastes were shown [1] to contain an average of 72 ppm of suspended solids and 390 ppm of BOD, when 6996 bushels of barley were processed per day. The volume of wastes averaged 75 gallons per bushel, or 524,700 gallons per day. The solids were mainly organic and high in nitrogen, indicating considerable protein material. A major portion of the solids were in solution, as indicated by the low suspended-solids content.

A summary of individual wastes from a typical bourbon distillery is given in Table 26.9. Table 26.10 shows the composition of some distillery "slops" [2]; all the slops have an acid reaction and contain from 2 to 7 per cent total solids. Beer slops from the fermentation of rye contained 3.3 per cent total solids, of which 91 per cent was volatile. Other data on the solids concentration and BOD content of fermentation wastes are given in Table 26.11.

Yeast wastes consist primarily of the spent nutrient (although only 20 per cent of the wastes by volume, they account for 75 to 80 per cent of the total BOD). They are brown, have the typical odor of yeast, and are highly hygroscopic. The solids are almost entirely dissolved and colloidal; the suspended-solids content is seldom above 200 ppm. The composition of this waste, according to Trubnick and Rudolfs [13], is given in Table 26.12. The average characteristics of a fermentation process-plant waste are shown in Table 26.13.

The solid waste from pressing (pomace) consists of organic materials such as stems, skins, seeds, and tissues of grapes. Solid waste from filtering operations is predominantly inorganic, diatomaceous earth. Both the soluble and insoluble wastes occurring in the wastewaters are primarily organic, consisting of components of wine and grapes such as organic acids, sugars, alcohols, pectins, proteins, etc. An unpublished EPA report gives standard effluent levels for wine and brandy wastes. (See Table 26.14.)

Wastes from pharmaceutical plants producing penicillin and similar antibiotics are strong (high BOD) and generally should not be treated with

Table 26.9   Summary of distillery waste. (From Ruf et al. [1].)

| Source of waste | Volume, gal | Suspended solids | | pH | BOD, ppm | Population equivalent |
		Total, ppm	Volatile, ppm			
Cooker condensate	54,000	7	6	7.2	5.4	14
Cooker wash water	2,750				1,370	185
Redistillation residues	1,000			7.7	1,700	84
Floor and equipment wash						227
Evaporator condensate	29,490	35	30	4.5	375	540
Thick slop to feed house	35,235	50,000	48,000	4.3	20,000	35,300

Table 26.10   Composition of distillery slops. (From Buswell [2].)

Item	Spirit type	Bourbon type	Molasses	Apple brandy
pH	4.1	4.2	4.5	3.8
Total solids, ppm	47,345	37,388	71,053	18,866
Suspended solids, ppm	24,800	17,900	40	50
BOD, ppm	34,100	26,000	28,700	21,000
Total volatile solids, ppm	43,300	34,226	55,608	16,948

Table 26.11   Solids and BOD of fermentation wastes. (From Boruff [3].)

Fermentation waste	Solids, %	BOD, ppm
Brewery-press liquor	3	10–25,000
Yeast plant	1–3	7–14,000
Industrial alcohol	5	22,000
Distillery slops	4.5–6	15–20,000

Table 26.12   Composition of spent nutrient of yeast plant. (From Trubnick and Rudolfs [13].)

Characteristic	Concentration or value
Total solids, ppm	10,000–20,000
Suspended solids, ppm	50–200
Volatile solids, ppm	7,000–15,000
Total nitrogen, ppm	800–900
Organic nitrogen, ppm	500–700
Total carbon, ppm	3,800–5,500
Organic carbon, ppm	3,700–5,500
BOD, ppm	2,000–15,000
Sulfate, as ppm SO_4	2,000–2,500
Phosphate, as ppm P_2O_5	20–140
pH	4.5–6.5

domestic sewage, unless the extra load is considered in the design and operation of the treatment plant. Some wash waters range as high as 14,000 ppm BOD, and average values of combined wastes are 2500 to 5000 ppm. Brown [5] reports five main pharmaceutical wastes and their characteristics as follows: (1) strong fermentation beers: small in volume but having 4000 to 8000 ppm BOD; (2) inorganic solids: waste slurry with little BOD; (3) washings of floor and equipment: large percentage of total volume and BOD from 600 to 2500 ppm; (4) chemical waste: solution of solvents which exert a substantial BOD when

Table 26.13

Characteristic	Concentration or value
BOD, ppm	4,500
pH	6–7
Total solids, ppm	10,000
Settleable suspended solids, ml/liter	25

diluted with other wastes; (5) barometric-condenser water: resulting from solids and volatile gases being mixed with condenser water, causing 60 to 120 ppm BOD.

Wastes from the production of fine chemicals and antibiotics, including vitamins B_1, B_2, and B_{12}, streptomycin, lysine, sulfaquinazoline, nicarbazin, and glycamide, possess the following pollutional characteristics [1962, 18].

Characteristic	General antibiotic wastes	Specific antibiotic wastes	
		Terramycin	Penicillin
BOD, ppm	1500–1900	20,000	8,000–13,000
Suspended solids, ppm	500–1000	10	
pH	1–11	9.3	2–4

26.9 Treatment of Brewery, Distillery, Winery, and Pharmaceutical Wastes

The principal pollutional load from a distillery is stillage, the residual grain mash from distillation columns. As much of this as possible is recovered by the industry as a by-product for manufacturing animal feed or for conversion to chemical products. Without such recovery, the population equivalent of distillery wastes, based on BOD, would be about 50,000 for each 1000 bushels of grain mash. Screening the dried grains reduces the population equivalent to 30,000. Complete stillage recovery makes possible a population equivalent of only about 2500 and a large-volume, but weak, waste. Dried yeast and dried spent grains are two valuable by-products recovered from these fermentation-waste slops, and much research is being devoted to recovering additional valuable materials from them. If additional BOD

Table 26.14   Standard effluent levels for the wine and brandy industry

Type plant	BOD	Suspended solids	COD	pH
Winery without stills (#/ton of grapes crushed)	0.31	0.12	0.47	6–9
Winery with stills (#/ton grapes crushed)	1.20	NA	1.80	6–9

removal is required after recovery processes have been carried out, the residual waste can be treated effectively by trickling filtration or anaerobic digestion. Centrifuging has also been used to concentrate distillery slops. Trickling filtration has effected BOD removals of 60 to 98 per cent [4, 6, 7, 8], while anaerobic digestion removes 60 to 90 per cent of the BOD [9, 10, 5, 11, 12]. Trubnick and Rudolfs [13] investigated the treatment of compressed-yeast wastes, with the results shown in Table 26.15.

Although by-product recovery of winery residues is possible, the short (2–3 month) operating season acts against this solution. Land disposal has been objectionable because of odors and plugging of the soil. However, very shallow ponds (less than 6 inches in depth) or spray irrigation systems have been used successfully. Aerobic lagoons were selected in Hammondsport, New York at the Taylor Wine Company (Tofflemire, 1970).

Anaerobic digestion and controlled aeration have both been used to reduce the BOD of pharmaceutical wastes by approximately 80 per cent. Effluents from such treatments can be further processed in sand

Table 26.15   Treatment efficiences for fermentation wastes. (From Trubnick and Rudolfs [13].)

Treatment process	Average BOD reduction, %
Electrodialysis	28
Chemical treatment	10
Anaerobic digestion	83
Activated sludge	30
Trickling filter	72

filters, to produce an effluent with about 35 ppm of BOD. An interesting biofiltration plant [6], consisting of two aerators, two clarifiers, and two high-rate filters, has given greater than 90 per cent BOD reduction and 65 per cent suspended-solids removal. The initial waste contains 5700 ppm BOD in a flow of 90,000 gallons per day. Milcher [1962, 14] and Winar [1967, 15] recommended deep well injection while McKinney [1962, 16], Dazai [1966, 17], and Howe [1962, 18] report success with activated-sludge treatment of antibiotic wastes. At times the only possible treatment of antibiotic wastes is by evaporation and incineration. Because bulk rubbish is usually handled in addition to process wastes at a pharmaceutical plant, incineration is readily carried out. Residues from penicillin and other antibiotics can be dried and used in stock food. A vacuum-dried mycelium from the manufacture of penicillin can be digested to produce methane; at the same time this reduces the organic matter by about 55 per cent. The traces of antibiotics and mystery factors in spent mycelium permit sale of these waste as animal-growth-promoting agents.

References: Brewery, Distillery, Winery, and Pharmaceutical Wastes

1. Ruf, H. W., L. F. Warrick, and M. S. Nichols, "Malt house waste treatment studies in Wisconsin," *Sewage Ind. Wastes* 17, 564 (1935).

2. Buswell, A. M., "Treatment of 'beer slop' and similar wastes," *Water Works Sewerage* 82, 135 (1935).

3. Bonacci, L. N., and W. Rudolfs, "Electrodialysis treatment," *Sewage Ind. Wastes* 14, 1281 (1942).

4. Davidson, A. B., "Designing a distillery waste disposal plant," in Proceedings of 5th Industrial Waste Conference, Purdue University, November 1949, p. 159.

5. Brown, J. M., "Treatment of pharmaceutical wastes," *Sewage Ind. Wastes* 23, 1017 (1951).

6. Howe, R. H. L., and S. M. Paradiso, "Miracle drug waste and plain sewage treatment by modified activated sludge and biofiltration units," *Wastes Eng.* 27, 210 (1956).

7. Pitts, H. W., "Treatment problems," *Sewage Ind. Wastes* 27, 970 (1955).

8. Roberts, N., and J. B. Hardwick, "Pilot-plant studies on distillery waste," in Proceedings of 6th Industrial Waste Conference, Purdue University, February 1951, p. 80.

9. Buswell, A. M., "Anaerobic fermentation plants," in Proceedings of 5th Industrial Waste Conference, Purdue University, November 1949, p. 168.

10. Davidson, A. B., and H. B. Brown, "Rapid anaerobic digestion studies," in Proceedings of 7th Industrial Waste Conference, Purdue University, May 1952, p. 142.

11. Pearson, E. A., D. F. Feuerstein, and B. Onodera, "Treatment and utilization of winery wastes," in Proceedings of 10th Industrial Waste Conference, Purdue University, May 1955, p. 34.

12. Tatlock, M. W., "Treatment of yeast products wastes," in Proceedings of 3rd Industrial Waste Conference, Purdue University, May 1947, p. 111.

13. Trubnick, E. H., and W. Rudolfs, "Treatment of compressed yeast effluents," in Proceedings of 4th Industrial Waste Conference, Purdue University, September 1948, p. 109.

14. Melcher, R. R., "Pharmaceutical waste disposal by soil injection," *Biotechnol. Bioeng.* 4, 147 (1962).

15. Winar, R. M., "The disposal of waste water underground," *Ind. Water Eng.* 4, 21 (1967).

16. McKinney, R. E., "Complete mixing activated sludge treatment of antibiotic wastes," *Biotechnol. Bioeng.* 4, 181 (1962).

17. Dazai, M., T., Higashikara, M. Ogawa, and H. Ono, "Treatment of industrial wastes by activated sludge. VIII: Treatment of antibiotic wastes, report 2," *Kogyo Gijutsuin Hakko Kenkyusko Kenkyu Hokoko* 27, 77 (1965); *Chem. Abstr.* 64, 17243 (1966).

18. Howe, R. H. L., "Complete biological treatment of antibiotic production wastes," *Biotechnol. Bioeng.* 4, 161 (1962).

19. Schlott, D. A. et al., "Design, Construction and Start-up of an Anaerobic Treatment System for Pharmaceutical Wastewater." Proc. 43 Purdue Ind. Waste Conf., 1988, P. 651.

20. Wolf, D. J. and D. K. Emerson, "Effects of Reactor Configuration on Operation of a Pharmaceutical Waste Treatment System." Proc. 43 Purdue Ind. Waste Conf., 1988, P. 661.

MEAT PACKING, RENDERING, AND POULTRY-PLANT WASTES

The meat industry has three main sources of waste: stockyards, slaughterhouses, and packinghouses. Animals are kept in the *stockyards* until they are killed. The killing, dressing, and some by-product processing are carried out in the *slaughterhouse*, or abattoir. To obtain the finished product—namely, the fresh carcass, plus a few fresh meat by-products such as hearts, livers, and tongues—the following operations are performed in the slaughterhouse. The animals are stuck and bled on the killing floor (cattle being stunned prior to sticking). Carcasses are trimmed, washed, and hung in cooling rooms. Livers, hearts, kidneys, tongues, brains, etc., are sent to the cooling rooms to be chilled before being marketed. Hides, skins, and pelts are removed from the cattle, calves, sheep, and pigs, and are salted and placed in piles until they are shipped to tanners or wool-processing plants. Viscera are removed and, together with head and feet bones, are sent to the rendering plant; other bones are shipped to glue factories. Many slaughterhouses are equipped to render their own inedible offal into tallow, grease, and tankage; other independent rendering plants convert inedible poultry, fish, and animal offal and waste products into animal feed and grease. This is accomplished by cooking the inedibles at a high temperature for several hours. The cooked material is pressed to remove the grease and the pressings are ground for feed. Thus, slaughterhouse wastes are produced on the killing floor, in the process of dressing the carcass, in rendering operations, in the hide cellar, and in the cooling room.

Although *packinghouses* also perform some of the operations carried out in abattoirs, they are primarily concerned with the production of salable products. Carcasses are trimmed, cleaned, and cooled as in the slaughterhouse, but the packinghouse processes some of the meat further by cooking, curing, smoking, and pickling. Packinghouse operations also include the manufacture of sausages, canning of meat, rendering of edible fats into lard and edible tallow, cleaning of casings, drying of hog's hair, and some rendering of inedible fats into grease and inedible tallow. In addition, the packinghouse is equipped to process, to varying degrees, the by-products of the slaughterhouses. Blood is usually collected, coagulated, dried, and finally formed into edible and inedible products. Tanning, wool palling, and the manufacture of glues, soaps, and fertilizers are usually carried out in separate plants. Packinghouse wastes, therefore, issue from various operations on the killing floor, during carcass dressing, rendering, bag-hair removal and

processing, casing, and cleaning, in the making of tripe, during the manufacture of by-products such as glue, soap, and fertilizer, and from laundering of linens and uniforms of the packinghouse workers.

Figure 26.5 is a schematic illustration of the operation of meat-processing plants.

Poultry processing differs from the processing of other meats enough to warrant a separate discussion. The operations of the poultry industry consist, in general, of the following steps: (1) the processor furnishes baby chicks and feed to the grower; (2) the grower, after about six weeks, sends the broilers to the processor; and (3) the processor prepares and markets the broilers. Figure 26.6 depicts the steps involved in processing poultry.

Broilers are delivered to the processing plants and are hung live by their feet on a moving chain which delivers them to the killing table, where their throats are slit. The blood usually flows into a trough and thence to drums for storage. (In some plants, the blood is sprayed with a coagulant and the residue shoveled into drums, while the supernatant goes to the sewers.) As the endless chain continues through the plant, the birds are mechanically plucked, washed, cleaned, washed again, and finally removed from the chain. At certain locations along the chain, drums are placed for storing feathers, heads, feet, offal, and the scrap-waste products, since it is important not to allow any of this waste to be flushed into the sewer. The dressed bird is then cut, frozen, or refrigerated, depending on the operation of the day. Rendering plants in the area offer a ready market for offal, feet, heads, scraps, and even blood, which they convert into tankage for feed additives or process as pet food or fertilizer. In spite of all the apparently profitable methods for disposal of these poultry wastes, a large percentage unfortunately ends up in the sewers.

26.10 Origin and Characteristics of Meat-Packing Wastes

Stockyard wastes contain excreta, both liquid and solid. The amount and strength of the wastes vary widely, depending on the presence or absence of cattle beams (horns), how thoroughly or how often manure is removed, frequency of washing, and so

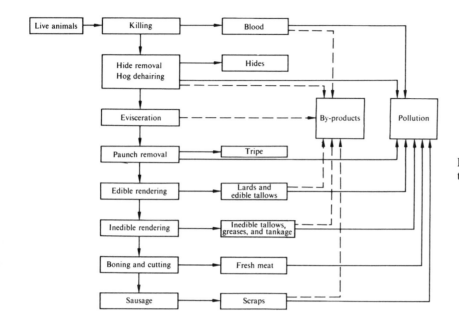

Fig. 26.5 Fundamental processes in the meat-packing industry.

forth. An analysis of wastes from one Chicago stockyard is given in Table 26.16. Another study of these same wastes showed a volume of 623,000 gallons per day for a 27-acre section of the yard and an average BOD of 100 ppm (population equivalent of 3100) [1].

Table 26.16 Stockyard waste [1].

Characteristics	Concentration, ppm
Total suspended solids	173
Volatile suspended solids	132
Organic nitrogen	11
Ammonia nitrogen	8
BOD	64

In a recent [1967, 2] survey, five processes were singled out for analysis, because of their potential impact on the waste load of the meat-packing industry. Some are actually by-product processes, but in an industry well known for "utilizing all parts but the 'squeal,'" by-product recovery has essentially become part of the process.

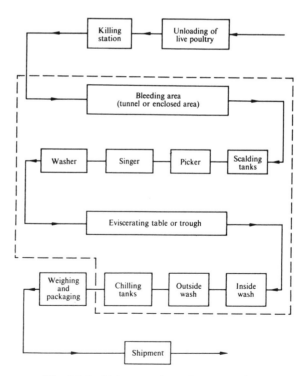

Fig. 26.6 Major steps in poultry processing.

1. *Blood recovery* is an "all or nothing" situation: either it is recovered or it escapes to the sewer. Recovery means a 42 per cent reduction in the gross waste load of a meat-packing plant. In 1966 over 95 per cent of the industry, on a live-weight basis, was recovering blood. Blood is a rich source of protein and, hence, for all but the very small plant it is economically rational to recover it. The very small plant, which does not render (inedible products), which does not produce tankage, and which is not located in an area where it can sell raw blood to others, will probably dump blood in the sewer.

2. *Paunch material* is a source of pollution problems if it is dumped into the sewer, as the total-solids concentration becomes so large that it interferes with the efficient workings of traditional waste-treatment methods.

3. *Edible rendering* can be highly pollutional, depending upon the method. The worst offender is wet rendering without evaporating tank water. This process is the oldest type and is not being adopted by new plants. If tank water is evaporated waste loads are cut in half. Newer methods of rendering, including dry rendering and low-temperature rendering, cut waste loads by 60 per cent.

4. *Inedible rendering* similarly is either a dry or a wet process. Wet rendering must be followed by evaporation of tank water in order to cut waste loads in half. Both forms of dry rendering, batch and continuous, will produce waste loads 60 per cent lower than those from a wet rendering system without evaporation of tank water. Continuous dry rendering is the latest method in terms of technology.

5. *Cleanup* by water from high-pressure hoses has been and continues to be the general practice in the meat-packing industry. Pollution loads could be substantially reduced by the use of dry cleanup prior to the wet cleanup. This could also mean greater recovery of scraps for inedible rendering, instead of hosing them into the sewer. Another effect of dry cleanup is the reduction of waste-water volume. Data have indicated a high, direct correlation between water use per 1000 pounds (live weight) killed and waste load. Decrease in waste-water volume seems to be accompanied by utilization of dry cleanup and, hence, lower waste load per unit of product as well as lower waste-water volumes per unit of product.

Slaughterhouse processes are centered about the killing floor. The wastes produced here have a deep reddish-brown color, high BOD, and contain a considerable amount of suspended material. Blood, being highly nitrogenous, decomposes readily. In addition, the wastes contain varying amounts of manure, hair, and dirt. Analysis of several samples of killing-floor waste taken from an average slaughterhouse showed an average BOD of 2000 ppm and a total nitrogen content of about 500 ppm at a flow of 5000 gallons per day.

The content of individual process wastes from a packinghouse is shown in Table 26.17. Table 26.18 presents the characteristics for combined slaughterhouse and packinghouse wastes.

Meat-plant wastes are similar to domestic sewage in regard to their composition and effects on receiving bodies of water. However, the total organic contents of these wastes are considerably higher than those of domestic sewages. On the other hand, the danger from pathogenic organisms in packing- and slaughterhouse wastes is slight when compared

Table 26.17 Packinghouse wastes.

Source	Suspended solids, ppm	Organic nitrogen, ppm	BOD, ppm	pH
Killing floor	220	134	825	6.6
Blood and tank water	3,690	5400	32,000	9.0
Scalding tub	8,360	1290	4,600	9.0
Meat cutting	610	33	520	7.4
Gut washer	15,120	643	13,200	6.0
Sausage department	560	136	800	7.3
Lard department	180	84	180	7.3
By-products	1,380	186	2,200	6.7

Table 26.18 Slaughterhouse and packinghouse wastes.

Type of kill	Volume per animal, gal	Suspended solids, ppm	Organic nitrogen, ppm	BOD, ppm	Population equivalent per animal
Mixed	359	929	324	2240	40.2
Cattle	395	820	154	996	19.6
Hogs	143	717	122	1045	7.5
Mixed	996	457	113	635	30.7
Cattle	2189	467		448	49.2
Hogs	552	633		1030	28.6

with domestic sewage. In the absence of adequate dilution, the principal deleterious effects of meat-plant wastes are oxygen depletion, sludge deposits, discoloration, and general nuisance conditions.

Rendering-plant wastes depend on the process used. If the "wet" process is used, a liquor waste containing high concentrations of organic matter, especially nitrogenous compounds, may result. Some plants evaporate the liquor and mix the residue with the product of the rendering process. Although "dry" rendering does not result in a liquor, there is a small amount of drainage and press liquor, most of which is returned to the rendering vat. Wash waters from the rendering plant may add considerable pollution material to the waste.

The total liquid waste from the poultry-dressing process contains varying amounts of blood, feathers, fleshings, fats, washings from evisceration, digested and undigested foods, manure, and dirt. The manure from receiving and feeding stations and blood from the killing and sticking operations contribute the largest amount of pollution from the process. The composition of poultry-plant wastes is given in Table 26.19. Some recent data on poultry-plant characteristics before and after various treatments are given by Vasuki and Sabis [1967, 3] (Table 26.20).

26.11 Treatment of Meat-Packing Wastes

In-plant recovery practices, screening, flotation, and biological treatment are the major methods used to treat meat-plant wastes. Some practices aimed at reducing the quantity or strength of wastes from meat plants are profitable, at least in larger plants.

Table 26.19 Composition of poultry-plant wastes.

Characteristic	Concentration
Volume	3.26 gal/bird
Total solids	26.6 lb/1000 birds
Suspended solids	15.3 lb/1000 birds
Settleable solids	9.4 lb/1000 birds
Grease	1.3 lb/1000 birds
BOD, 5-day	30.0 lb/1000 birds

Among them are recovery of blood and grease, and utilization of tank waters and tankage-press liquors. Grease recovery is usually accomplished by means of baffled basins or traps on waste lines. Practically all plants recover the major portion of the blood from the killing operation. While separate disposal of paunch manure is seldom profitable, it is nonetheless desirable; even in plants where this is done, a certain amount of the liquid waste reaches the sewers. The extent to which waste loads from packinghouses are reduced by plant practices depends to some extent on whether or not such methods cost less than handling the waste load in a municipal disposal system. The latter cost, in turn, usually depends on the degree of difficulty the municipality encounters in removing the pollution at its treatment plant.

The most common methods used for treatment of meat-plant wastes are fine screening, sedimentation, chemical precipitation, trickling filters, and activated sludge. Screening by rotary wire-mesh screen removes coarse materials such as hair, flesh, paunch manure, and floating solids. Removals of 9 per cent of

Table 26.20   Characteristics of poultry-plant waste before and after treatment.

Characteristic	Plant 1		Plant 2		Plant 3		Plant 4		Plant 5		Plant 6	
	Influent	Effluent	Influent*	Effluent	Influent	Effluent	Influent	Effluent	Influent†	Effluent	Influent	Effluent‡
BOD	30.6§	28.8	37.0	11.6	38.5	Treated in municipal plant	17.7	5.8	28.9	Secondary treatment recently started	25.5	5.0
Organic nitrogen		2.6	6.1	2.3			1.8	1.3	1.9		2.7	0.8
Ammonia nitrogen		0.3	0.7	2.2				0.3	0.2		1.2	1.3
Nitrite nitrogen		0.003	0.010	0.001			0.006	0.004	0.006		0.06	0.006
Nitrate nitrogen	0.4	0.7	0.4	0.1					0.2		0.9	0.1
Suspended solids	23.1	22.0	29.9	17.2	24.4		15.1	17.2	12.6		15.2	7.3
Volatile suspended solids	21.7	20.4	29.3	15.9	23.4		14.5	15.1	11.9		14.6	7.1
Total solids	41.5	41.6	72.4	48.4			36.0	34.3	29.8		42.7	33.1
Volatile total solids	33.0	28.4	52.9	29.3			20.6	18.6	22.0		21.6	13.8
pH	6.4	6.4	7.3	7.1	7.3		7.5	7.5	6.4		6.5	7.5
Kill rate (birds/day)	52,000		75,000		80,000		43,500		76,800		80,000	

*After settling.   ‡After flotation.
†After screening.   §Results are given in pounds per 1000 birds, except for the pH.

the suspended solids on a 20-mesh screen and 19 per cent on a 30-mesh screen have been reported. Sedimentation in Imhoff tanks is also satisfactory, being capable of removing 63 per cent of the suspended solids and 35 per cent of the BOD, with a 1- to 3-hour detention period. Trickling filters, at rates varying from 0.6 to 1.0 mgad, can give 81 to 90 per cent removal, with no accompanying nuisances. Activated-sludge treatment produces a satisfactory effluent after about 9 hours of aeration, at a rate of 3.5 cubic feet of air per gallon of waste. Experiences with double filtration in Mason City, Iowa, and Wells Forge, North Dakota, showed overall BOD reductions in excess of 95 per cent.

Few meat-plant wastes are chemically coagulated, because of the high costs. However, one plant uses $FeCl_3$ to reduce the BOD from 1448 to 188 ppm and the suspended solids from 2975 to 167 ppm. Operating costs were at one time reported to be $68 per million gallons. However, the effluent was sold for irrigation, grease was recovered, and sludge was utilized, thus reducing the net cost to about $25 per million gallons.

Chlorine and alum, if used in sufficient quantities, appreciably reduce the BOD and color of rendering-plant wastes and make possible improved clarification. Once again the chemical cost is high, but the BOD of raw wastes ranging from 1500 to 3800 ppm can be reduced to 400 to 600 ppm.

A trickling filter operating in conjunction with an air-flotation unit produced satisfactory results (61 per cent BOD reduction) at a loading of 2.6 pounds of BOD per cubic yard.

Poultry-plant wastes should and do respond readily to biological treatment; if troublesome materials such as feathers, feet, heads, and so forth, are removed beforehand, satisfactory biological treatment is attainable. Treatment facilities for one poultry-dressing plant's wastes include stationary screens in pits, septic tanks, and lagoons; an overall removal of 93 percent of the BOD was reported to result from the use of these measures. The author has designed a two-stage oxidation-basin treatment system which provides 85 to 90 per cent BOD reduction.

References: Meat-Packing, Rendering, and Poultry-Plant Wastes

1. U.S. Public Health Service, Industrial Waste Guide to the Meat Industry (1954).
2. Miller, P. E., "Poultry wastes," in Proceedings of 6th Industrial Waste Conference, Purdue University, February 1951, p. 176.
3. Vasuki, N. C. and W. R. Sabis, "In Plant control of poultry waste discharges," 1st Mid-Atlantic Industrial Waste Conference Univ. of Delaware Nov. 1967 (1963)
4. Norcross, K. L. et al., "Start-up and Operation Results from SBR Treatment of a Meat Processing Wastewater," Proc. 42 Purdue Ind. Waste Conf., 1987, P. 475.
5. Young, K. S., "Performance of World's Largest Cych'cal Activated Sludge Process Treating Combined Municipal/Packing House Wastewater," Proc. 42 Purdue Ind. Waste Conf., 1987, P. 483.
6. Cooper, R. N., "Irrigation of Pasture with Meat-Processing Plant Effluent," Proc. 42 Purdue Ind. Waste Conf., 1987, P. 491.

26.12 Feedlot Wastes

Large-scale livestock operations have removed animals from pasturage and now handle large numbers in small confinement areas (feedlots), where feed and water are brought to the livestock. Poultry, cattle, and swine are the major animals involved. Loehr [5] gives the average characteristics of these wastes as obtained from Hart and Turner [1] (Table 26.21). It is reported that, on a BOD basis, one dairy cow is the equivalent of 20 to 25 persons, one beef animal, 18 to 20 persons, one hog, 2 or 3 persons, and that 10 to 15 chickens are the equivalent of one person. Cassell [2] reports that in 1963 the American poultry industry produced approximately 63 billion eggs for consumption, from a hen population of about 365 million birds. It is also estimated that about 5 million tons of organic material was produced by the laying-hen industry. The farmer has traditionally disposed of chicken manure by spreading on land. This is no longer desirable because the attendant odors and flies offend nearby residents and vacationers. Chicken manures vary from 10 to 15 per cent total solids, of

Table 26.21   Average characteristics of the surface liquid-manure lagoons. (From Loehr [6].)

Characteristic	Poultry manure					Dairy manure			Swine manure		
Loading, lb volatile solids/ft³	0.001	0.004	0.006	0.010	0.016	0.004	0.008	0.01	0.005	0.01	0.014
pH	7.6	7.0	7.3	7.1	7.3	7.4	7.0	7.0	6.9	6.7	6.3
Total solids, mg/liter	600	1600	5500	4000	8000	2200	7000	7000	1500	2200	3000
Volatile solids, % of TS	55	45	45	50	60	70	75	75	60	60	65
BOD_5, mg/liter	40	100	300	300	1500	500	500	1200	500	500	1000
Alkalinity, mg/liter	400	1300	5500	4800	7500	1300	2500	3500	700	2000	3000

which about 75 per cent is organic matter with a BOD of about 30,000 ppm. Cassell [2] reports a wet weight of 0.1 to 0.4 pound per hen, a wet volume of 6 to 8 cubic inches per hen, and about 0.01 to 0.015 pounds of BOD per hen.

Cassell found chicken manure conditioned with anionic-cationic polyelectrolyte mixtures can be effectively dewatered and the volume reduced by as much as 65 per cent. In laboratory experiments, he found that the following conditions were necessary to digest chicken manure anaerobically:

pH	7.5
Volatile acids	1500 mg/liter
Alkalinity	10–12,000 mg/liter
NH_3–N	1500 mg/liter
Detention time	20 days
Loading	0.088 lb volatile solids/ft³/day
Temperature	35°C
Na^+	0.018 molar

Loehr found that treatment and disposal of animal wastes are complicated by the nature of the wastes, the volume to be handled, the lack of interest of the livestock producer in waste treatment, and the proximity of the suburban population. In the past, most of the wastes have been recycled through the soil environment with a minor release to the nearby waters. He found that anaerobic treatment in lagoons offers a possible approach for handling the tremendous quantities of manure that originate from confinement feedlot operations, but admits that this is not the complete answer to the problem. In combination with units to treat the effluent from the lagoons, anaerobic lagoons may be a useful *part* of the process for treating livestock and feedlot wastes that have a high solids content.

Clark [4] and Dornbush and Anderson [5] give data on the character of effluent from two-cell anaerobic lagoons treating hog and poultry feedlot wastes (Table 26.22). In a recent publication, Wadleigh [1968,

Table 26.22   Effluent quality of field livestock waste lagoons.

Parameter	Type of livestock		
	Swine	Swine	Poultry
Loading rate,* lb volatile solids/day/1000 ft³	0.36–3.9	10–15	4–11
pH	6.7–8.0		6.8–7.9
Total solids, mg/liter		1000–1500	
Volatile solids, mg/liter	850–2330	400–800	
Volatile acids, mg/liter	72–528		
Alkalinity, mg/liter	1120–2220		
BOD_5, mg/liter			320–1350
COD, mg/liter	940–3850	500–1300	590–2550
Total nitrogen, mg/liter		150–350	113–290

*Note: lb/1000 ft³ × 16 = g/m³.

6] provides information on the amount of solid and liquid wastes produced by livestock in the United States (Table 26.23 and 26.24). Liquid wastes are reported to amount to over 400 million tons per year.

The New York State Health Department has reported on the treatment of duck-farm wastes from Suffolk County, representing 6,250,000 ducks (60 to 70 per cent of the nation's total). The waste waters from these farms contain two major objectionable impurities, manure and waste grain. Each of the 21 farms with treatment facilities has an aerated lagoon with 5-day detention, two or three sedimentation tanks in parallel with 12-hour detention (only one is used at a time while the others dewater and dry), and a chlorine contact chamber.

References: Animal Feedlot Wastes

1. Hart, S.A. and M.E. Turner, "Lagoons for livestock manure," *J. Water Pollution Control Federation* 37, 1578 (1965).
2. Cassell, E.A., "Studies on chicken manure disposal I," New York State Health Dept., Laboratory Studies Research Report No.12, 1966.

Table 26.23 Production of wastes by livestock in the United States. (From Wadleigh [1968, 9].)

Livestock	U.S. animal population (1965), millions	Solid wastes, gal per capita/day	Total production of solid waste, million tons/yr	Liquid wastes, gal per capita/day	Total production of liquid wastes, million tons/yr
Cattle	107	23,600	1004.0	9000	390.0
Horses	3	16,100	17.5	3600	4.4
Hogs	53	2,700	57.3	1600	33.9
Sheep	26	1,130	11.8	680	7.1
Chickens	375	182	27.4		
Turkeys	104	448	19.0		
Ducks	11	336	1.6		
Total			1138.6		435.4

Table 26.24 Population equivalent of the fecal production by animals, in terms of biochemical oxygen demand (BOD). (From Wadleigh [1968, 6].

Biotype	Fecal, gal per capita	Relative BOD per unit of waste	Population equivalent
Man	150	1.0	1.0
Horse	16,100	0.105	11.3
Cow	23,600	0.105	16.4
Sheep	1,130	0.325	2.45
Hog	2,700	0.105	1.90
Hen	182	0.115	0.14

3. Loehr, R.C., "Effluent quality from anaerobic lagoons treating feedlot wastes," *J. Water Poll. Control Fed.* 39, 384, (1967).

4. Clark, C.E. "Hog waste by lagooning," *J. Sanit. Eng. Div. Am, Soc. Civil Engrs.* 91, 27, (1965).

5. Dornbush, J.N., and J.R. Anderson, Lagooning of live-stock wastes," Proc. 19th Ind. Waste Conf., Purdue Univ. Eng. Ext. Series, Bull. No. 117, 1964, page 317.

6. Wadleigh, C.H., "Wastes in Relation to Agriculture and Forestry," U.S. Dept. Agr., Misc. Pub. No. 1065, March 1968.

BEET-SUGAR WASTES

The process of extracting sugar from beets is essentially the same in all factories in the United States. The sugar "campaign" usually starts in the fall and lasts from 60 to 100 days, operating 24 hours a day during this "on" season. The majority of factories operate only what is known as the "straight house," in which sugar is extracted to the point at which a heavy molasses is obtained. Some factories also operate Steffen houses, in which powdered lime is added to the beet molasses to precipitate the sugar as calcium sucrate [1]. A typical beet-sugar plant conducts the following operations [2]: (1) beets are weighed, unloaded, screened, and washed; (2) beets go to slicers and diffusers, and the resulting pulp is sent to pulp presses and then to storage, from which raw juice is withdrawn; (3) product (raw juice) is put through first and second carbonation tanks, where lime and CO_2 are added; (4) in the sulfidizer, SO_2 is added; (5) product is filtered and the cake is removed; (6) residue is evaporated; (7) the evaporated liquor is centrifuged to draw off sugar; (8) sugar is dried and stored. Southgate [3] presents a typical flow sheet of these processes (see Fig. 26.7).

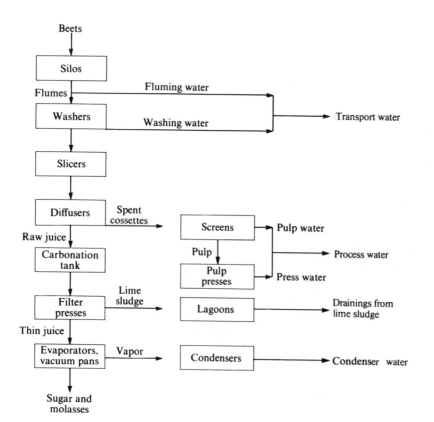

Fig. 26.7 Flow sheet showing discharges of waste waters from a beet-sugar factory. (After Southgate [3].)

26.13 Origin and Characteristics of Beet-sugar Wastes

There are five sources of waste water at a beet-sugar plant which employs the Steffen process: (1) the flume (transport) water, which is used to wash the beets and to transport them from stockpiles in the factory to the site where they are to be processed; (2) the process waste water, consisting of (a) the battery-wash water, from the operation of flushing the exhausted (de-sugared) cossettes (sliced beets) from the diffusion battery cells, and (b) the pulp-press water, from the partial dewatering of the exhausted pulp; (3) the lime-cake or lime-slurry residue from the carbonation process; (4) the condensate from the multiple-effect evaporators and vacuum pans used to concentrate the sugar solution; (5) the Steffen waste, resulting from the extraction of sugar from the straight-house molasses by the Steffen process.

Flume wash waters vary considerably in content of soil, stones, beet leaves, roots, and dissolved solids. Pulp-press water is high in organic material and suspended solids, as is the lime-cake slurry. Condenser water may contain organic matter entrained with the vapors from the last effect. The flume water represents about 72 per cent of the total beet-sugar waste volume; the BOD of this water, however, is comparatively low, about on a level with domestic sewage. Data collected by Eldridge [18] give the composition of beet-sugar factory wastes (Table 26.25); similar data by Pearson and Sawyer [4] are presented in Table 26.26. Rogers and Smith [5] found the general analysis to be as presented in Table 26.27.

The Steffen process consists of diluting the molasses to a specific concentration and treating it with enough powdered lime to precipitate calcium sucrate,

Table 26.26 Beet-sugar wastes. (After Pearson and Sawyer [4].)

Characteristic	Flume water	Process waste water
Suspended solids, ppm	400	1300
Volatile solids, %	35	75
Total solids, ppm		3800
BOD, ppm	200	1600
COD, ppm	175	1500
Protein-N, ppm	10	65
NH_3-N, ppm	3	15
Sucrose, ppm	100	1500
Volume, gal/ton beets	2000–3000	325

which can be removed from the liquor by filtration, and the sugar recovered by treating with carbon dioxide.

26.14 Treatment of Beet-sugar Wastes

Elimination of all unnecessary water usage and reuse of waste waters should precede any actual treatment in waste-treatment plant is uneconomical. Some ferable, because large volumes are discharged during a relatively short period of the year, so that investment in waste-treatment plant is uneconomical. Some factories have had satisfactory experiences with the reusing of flume water. Battery wash water and pulp-press water may be successfully eliminated by the use of somewhat more costly and less efficient diffusers, known as rack continuous diffusers. The lime-cake drainage may be eliminated by "dry" transportation from the pressure filters.

Any actual treatment of beet-sugar wastes is accomplished mainly by the use of lagoons. With

Table 26.25 Beet-sugar factory wastes. (After Eldridge [1].)

Characteristic	Flume water	Process water	Lime-cake drainage	Steffen waste
Volume, gal/ton beets	2200	660	75	120
BOD, ppm	200	1230	1420	10,000
Total solids, ppm	1580	2220	3310	43,600
Suspended solids, ppm	800	1100	450	700
Dissolved solids, ppm	780	1120	2850	42,900

Table 26.27 Beet-sugar wastes. (After Rogers and Smith [5].)

Characteristic	Amount or value
BOD, ppm	445
Total solids, ppm	6470
Suspended solids, ppm	4920
pH	7.9
Alkalinity, ppm	250

lagooning and land spraying, BOD reductions of 50 to 67 per cent have been obtained [6, 7, 5]. Coagulation has also been used [1, 4] and may be followed by sedimentation and/or biological filtration, if conditions require a higher degree of purification.

Hopkins *et al.* [6] found that, if total beet-sugar wastes were discharged uniformly across the upper end of 5-acre shallow lagoons, with a detention time of about one day, virtually all suspended solids, 55 per cent of the concentration of BOD, and 63 per cent of the pounds of BOD were removed. This procedure also reduced the alkalinity by 69 per cent, completely eliminated nitrate nitrogen, and reduced ammonia nitrogen by 94.3 per cent. Coliform-type bacteria increased, but phosphates were unchanged. Water loss was 3.27 acre-feet per day, of which 0.18 acre-foot was due to evaporation and 3.09 acre feet due to infiltration.

Five methods have been receiving attention recently:

1. Absorption by sawdust, calcined coke, coke, and slag is effective in removing organics and sugar found in sugar wastes; for example, with sawdust, 96 per cent of organics and 50 per cent of sugar can be removed [1966, 9].

2. Total reuse of waste water is being included in the design of new plants [1964, 9; 1966, 10].

3. Aerobic oxidation ponds may give up to 90 per cent overall BOD reduction when they follow anaerobic digestion (of cane-sugar wastes) [1966, 11].

4. Activated-sludge treatment has given 95 per cent BOD removal in laboratory scale units [1963, 12], and 70 per cent in pilot-plant operations when mixed with sewage [1964, 13].

5. The following patented process [1965, 14] represents a novel approach. Highly alkaline wastes, separated from the solids, are sparged with CO_2, the precipitated $CaCO_3$ is separated, and the pH is further reduced with mineral acid. The BOD of the liquor after clarification is reduced by fermentation with yeast, the CO_2 evolved is used for $CaCO_3$ precipitation and excess yeast produced is used for forage.

26.15 Cane Sugar Wastes

Origin. Sugar cane is shipped to the factory and passed through a series of cutters that prepare the cane for grinding. Juice is extracted from the sugar cane, leaving a fibrous residue called bagasse, which can be used as fuel for the boilers or can be disposed of as solid waste. Lime is then added to the extracted juice to raise its pH and to prevent the inversion of the sucrose molecule to glucose and fructose. Lime also helps to clarify the juice along with the addition of bentonite or some other coagulant aid. The juice is then pumped through the heaters and the clarifiers to the multiple-effect evaporators. The sludge from the clarification process which can either be used as a fertilizer or disposed of, goes to the vacuum filters. The juice in the evaporators is boiled under vacuum and a concentration of 30 to 40 per cent water is reached. This syrup is then passed through the vacuum pans where the sugar is crystallized. Centrifuges are used to separate the crystallized sugar from the residual components of the syrup and molasses. This molasses can be reused for further crystallization of sugar. The exhausted molasses can be used for hog feeding, alcoholic fermentation, or other purposes [15].

Characteristics. The cane sugar wastes may vary according to the plant location, water availability and other factors. Guzman [16] (1962) has divided these wastes into three categories in (1) cooling and condenser wastes, (2) the solid waste from the filter cake, and (3) concentrated wastes from spillage, scum leaks, washings, cleanings, and wastes containing lubricants and oil from machinery.

The wastewater from cooling and the condenser is generally low in BOD content and high in volume, whereas the concentrated wastes are generally low in volume and high in organic matter content. Parashar [17] (1969) has divided the sugar factory effluents into two categories, namely, high-pollution load and non-pollution load. High-pollution load effluents contain a BOD of 2000 to 3000 ppm. During the cleaning periods, the wastes discharged are much more concentrated than those discharged during normal plant operation. The volume of cooling and condenser waters varies considerably with the availability of water near the factory site. Water is recirculated at plants where it is not readily available. In one Puerto Rican factory, the volume of these waters was about 10 mgd and contained 14 pounds of BOD for every ton of cane ground. The major source of organic pollutants in these wastes are the small particles of sugar that gain access to the cooling and condenser waters by entrainment in the evaporators and vacuum pans. The filter-cake waste is the result of the clarigication of juice; it is generally used as a fertilizer. There is again much variation in the concentrated wastes. These wastes may be acidic or alkaline and at times the BOD may exceed 10,000 mg/l.

Treatment. The high concentration of the carbohydrates in the cane-sugar wastes makes them difficult to treat. High oxygen demand is exerted by sugar and its by-products; molasses has the highest BOD of 900,000 mg/l. The conventional treatment methods become difficult as volatile organic acids, which inhibit the activity of the microorganisms during biological oxidation, are formed. These acids can be neutralized by lime or other relatively inexpensive alkalis. However, the consumption of neutralizing agents is relatively high. The BOD of the cooling and recirculating waters can be reduced to low concentrations by some chemical methods, such as flocculation and oxidation; but they may not be economically feasible. Accelerated anaerobic digestion of the filter cake has resulted in average BOD reductions of 78 per cent. Oxidation ponds can also be used for treating such wastes. In one of the sugar factories in India, high-pollution load effluent of BOD content of 2000 to 3000 ppm is first subjected to aerobic oxidation and then to the action of micro-flora developed in rapid trickling biofilters (Parashar, 1969). The treated effluent is diluted with 10 times its volume of non-polluted water before discharge.

References: Beet- and Cane-Sugar Wastes

1. Eldridge, E. F., *Industrial Waste Treatment Practice*, McGraw-Hill Book Co., New York (1942), p. 84.

2. Rudolfs, W., *Industrial Wastes*, Reinhold Pub. Co., New York (1953), page 473.

3. Southgate, B.S., *Treatment and Disposal of Industrial Wastes*, H.M.S.O. London (1948) p. 317

4. Wintzell, T., and T. Lauritzson, "Return of beet sugar factory waste water," *Sugar* 39, 26 (1944) and 40, 28 (1945).

5. Rogers, H. G. and L. Smith, "Beet sugar waste lagooning," Proc. 8th Ind. Waste Conf., Purdue Univ., May 1953, p. 136.

6. Hopkins, G., "Evaluation of broad field disposal of sugar beet wastes, *Sewage Ind. Wastes* 28, 1466 (1956).

7. Porges, R., and G. Hopkins, "Broad field disposal of beet sugar waste," *Sewage Ind. Wastes* 27, 1160 (1955).

8. Brandon, T. W., "Reuse of process water," *Sewage Ind. Wastes* 20, 360 (1948).

9. Eldridge, E. F., and F. R. Theroux, "Chemical treatment," *Sewage Ind. Wastes* 7, 769 (1935).

10. Eldridge, E. F., and F. R. Theroux, "Steffens waste, spray drying studies," *Sewage Ind. Wastes* 9, 533 (1937).

11. Allen, L. A., A. H. Cooper, and M. C. Maxwell, "Microbiological problems in manufacture of beet sugar wastes," *Sewage Ind. Wastes* 21, 942 (1949).

12. Black, H. H., and G. N. McDermott, "Industrial waste guide," *Sewage Ind. Wastes* 24, 181 (1952).

13. Eldridge, E. F., " Full-scale experimental plant results, Michigan," *Sewage Ind. Wastes* 9, 531 (1937) and 10, 913 (1938).

14. Brandon, T. W., "Waste waters from beet sugar factories," *Intern. Sugar J.* 49, 98 (1947).

15. Haupt, H., "By-product recovery," *Sewage Ind. Wastes* 8, 350 (1936).

16. Guzman, R. M., "Control of cane sugar wastes in Puerto Rico," *J. Water Pollution Control Federation* 34, 1213 (1962).

17. Parashar, D. R., "Treatment of sugar factory effluents of high pollution load-self oxidation and use of water hycinth," *Proc. Annu. Conv. Sugar Tech. Ass. India* 37, 381–391 (1970); *Indian Sugar* 18, 879 (1969).

18. Kenda, W. et al. "A Systems Approach to Effluent Abatement by Hawau's Sugar Cane Industry." Proc. Cornell Argicultural Waste Management Conference, 1973. P. 161.

19. Lee, C. Y. et al. "Waste Reduction in Table Beet Processing." Proc. Cornell Waste Mgmt. Conf., 1973. P. 48.

20. Dasgupta, A. "Anaerobic Digestion of Lignocellulogic Residues from Sugar Cane Processing," Ph.D. dissertation, 1983. University of Miami, Florida.

MISCELLANEOUS FOOD-PROCESSING WASTES

Wastes from the manufacturing of such foods as coffee, rice, fish, pickles, beverages, baked goods, candy, and drinking water, are somewhat less prevalent than other food-processing wastes described in this chapter, but are nonetheless important and some study of them is advisable.

26.16 Coffee Wastes

The major part of our coffee comes from South America and almost all of it used to be processed in the USA, but an increasing amount of raw coffee is now being ground in the producing country for processing. Since washed coffee of good quality receives the highest price in the United States, some of the coffee beans are washed before shipment; that is, the ripe berry is picked and milled in a process which requires the use of water. This process is differentiated from that of the "dry" coffee, in which the berry is picked from the tree and the hull removed by dry milling. The water requirement for washing is about 260 gallons per 100 pounds of finished coffee, so, a significant pollution problem may exist from the washing.

The coffee cherry (the de-hulled berry) is dumped into a receiving vat from which it is conveyed to the pulpers by water. During this conveyance, stones and other debris are separated by means of traps and floaters; unsound cherries are diverted to separate pulpers. The pulper removes the skin and a large proportion of the flesh from the coffee bean. The hulled bean is then transported by water to a fermentation vat, where it is allowed to remain and ferment in a moist state, the excess water being drained off and reused if necessary. The period of fermentation may be as short as twelve hours or as long as two days. Fermentation is necessary before all the flesh of the cherry can be removed from the parchment which immediately surrounds the silver skin of the coffee bean. The protopectin in the flesh is insoluble and cannot be washed or scrubbed off; no machine has been developed that can satisfactorily remove the unfermented flesh and undried parchment from the bean. Because the protopectin adheres to the tough parchment, the bean cannot be satisfactorily used unless fermentation of the bean itself occurs. The fermentation process makes it possible to dry the bean in a clean parchment which guarantees its purity.

The theory of fermentation in coffee processing is as follows:

1) Protopectin $\xrightarrow{\text{Protopectinase}}$ Pectin,

2) Pectin $\begin{cases} \xrightarrow[\text{Pectinase}]{\text{Pectase}} \end{cases}$ $\begin{cases} \text{Pectic acid} + CH_3OH \\ \text{Galacturonic acid} \\ \quad + CH_3OH, \end{cases}$

3) Pectic acid + Calcium $\longrightarrow$ Calcium pectate (a soluble gel).

These products, some of which are soluble, are readily removed from the parchment of the coffee bean by washing. After fermentation, the beans are washed and conveyed to the drying patios, again by water. They are screened from the water and spread out to dry in the sun for several days; some plants use mechanical driers in addition to sun-drying. When the beans have dried sufficiently to ensure color and flavor, they are milled to remove the parchment, then graded, sacked, and shipped to the markets.

The main uses of water (and the origin of wastes) in coffee mills are (1) to transport beans to pulpers, (2) to transport pulp to a hopper or pile, (3) to transport beans to fermentation vats, (4) to wash fermented beans, (5) to transport fermented beans to drying patios, (6) miscellaneous uses, such as acting as a trap for stones and as a method of separating "floaters," for hydraulic classification of beans, and as boiler water.

The elements which make up the coffee cherry or bean are shown in Table 26.28. The four major wastes from processing the coffee bean are pulp, pulping waste, fermentation wash water (tank water), and parchment.

The pulp is the waste that is potentially most troublesome, but it is generally recovered and used for fuel or fertilizer. When the fresh pulp is stored in open piles its sugar attracts flies, and when it begins to

Table 26.28   Composition of coffee bean.

Component	Average %
Water	9–12
Ash	4
Nitrogen	12
Cellulose	24
Sugars	9
Dextrin	1–15
Fat	12
Caffetannic acid	8–9
Caffeine	0.7–1.3
Nitrogen-free extract	18
Essential oil	0.7
Water-soluble material	25.3

ferment a foul, repulsive odor emanates from it. The pulping water contains a relatively high amount of settleable solids and, since it contains sugar and other soluble materials, is highly pollutional. The fermentation (tank) wastes contain a great many colloidal geis of pectin and other products. This is a relatively weak waste, compared with the pulping waste water, and is relatively stable and inoffensive. The parchment

from the dry milling of the dried bean has no significance as a waste product, since it is nearly pure cellulose and is generally utilized as a fuel for the steam boilers which provide power for the mills. Table 26.29 presents the average sanitary characteristics of wastes from three coffee plants using different amounts of water. In Table 26.30, the volumes of water and the BOD of coffee wastes is given [1].

Table 26.29   Coffee wastes.

Source	BOD, ppm	Settleable solids, ppm	Total solids, ppm	Suspended solids, ppm
Pulp	47,000			
Fermentation wastes	1,250–2,220	660–700	4260	2060
Pulping waste	1,800–2,920	60–127	4960	848
Combined pulp and fermentation wastes	6,150–134,000	160	3220	

Table 26.30   Average results of examination of waste waters from processing of coffee (1946). (After Brandon [1].)

Waste water	Volume, gal/ton clean coffee	Proportion of total volume, %	BOD (3-day at 26.7°C), ppm	Proportion of total BOD, %
Pulping wastes				
Pulp water	4,490	34⎫	2400	45⎫
Main-tank effluent	2,220	17⎬57	3900	35⎬85
Repasser-tank effluent	840	6⎭	1450	5⎭
Tank-washing wastes				
First tank	280	2⎫	2800	4⎫
Second tank	270	2⎬5	1300	1⎬6
Repasser tank	165	1⎭	1900	1⎭
Channel washing				
Main	4,700	35⎫38	40	8⎫9
Repasser	440	3⎭		1⎭
Total	13,445*	100		100

*Usually 20,000 gal/ton of coffee.

Horton [2] reports the characteristics of fermentation wash-water wastes (Table 26.31) and depulping wastes (Table 26.32).

Horton [2] proved that biological filters with high rates of application and recirculation provided the most effective method of treating coffee wastes (Table 26.33) and proposed the effluent be used for irrigation. He also concluded that one hour of sedimentation reduced the BOD by only 16 to 29 per cent, but the BOD of mixed coffee wastes can be reduced a maximum of 50 per cent by chemical coagulation. Brandon [1] presents results and costs for treatment of coffee wastes by various methods (Table 26.34) and states that in most coffee mills screening and primary sedimentation of the wastes is justified, since the dried pulp recovered is valuable as both a fuel and a fertilizer and has the following composition: protein, 1.3 per cent, fiber 19.7 per cent, nitrogen-free extract 50.1 per cent, ash 9.0 per cent.

Table 26.31   Coffee fermentation wastes.* (After Horton et al. [3].)

Characteristic	Minimum, ppm	Maximum, ppm	Mean, ppm
BOD	295	3600	1700
pH	4.1	5.5	4.5
Turbidity	250	4000	1750
Suspended solids	235	2385	900
Total solids	885	3140	2100

*Based on 30 different samples of effluents.

Table 26.32   Coffee depulping wastes.* (After Horton et al. [3].)

Characteristic	Minimum, ppm	Maximum, ppm	Mean, ppm
BOD	3,280	15,000	9,400
pH	4.1	4.7	4.4
Turbidity	1,500	4,000	2,900
Suspended solids	625	1,055	790
Total solids	10,090	12,340	11,300

*Based on 12 samples on different days.

Table 26.33   Treatment of combined (fermentation and depulping) coffee wastes by recirculation through a biological filter.* (After Horton et al. [3].)

Rate of recirculation, mgd	Recirculations per hour	Settled waste, ppm	5-day BOD					
			After settling treatment for			Reduction after treatment for		
			2 hr, ppm	4 hr, ppm	6 hr, ppm	2 hr, %	4 hr, %	6 hr, %
20	5	2200	600	550	250	72.7	75.0	88.6
40	5	2450	920	650	420	62.4	73.5	82.9
40	8	2800	950	700	400	66.1	75.0	85.7
40	10	2951	900	700	450	69.5	76.3	84.7
60	5	2850	960	690	380	66.4	75.8	86.7

*Based on results of 5 tests.

Table 26.34 Comparative capital costs of alternative schemes for treatment of waste waters from processing of coffee. (After Brandon [1].)

Waste waters treated	Method of treatment	Reduction in polluting character, %	Reduction in total pollution, %	Capital cost for factory per ton of clean coffee per day, pounds sterling
Tank effluents and washings	Seepage pits		46	45
Pulp water and channel washings	Biological filtration	94	51	1000
Total			97	1045
Mixed wastes	Fermentation 12 days	85	85	580
	12 days plus biological filtration	94	14	480
Total			99	1060
Mixed wastes	Biological filtration to produce effluent with BOD of about 100 ppm	94	94	1850
	Biological filtration to produce effluent with BOD of about 40 ppm	98	98	2500

References: Coffee Wastes

1. Brandon, T. W., "Treatment and disposal of waste waters from processing of coffee," *East African Agr. J.* 14, 179 (1949).
   Brandon T. W., "Coffee waste waters, treatment and disposal," *Sewage Ind. Wastes* 22, 142 (1950).
2. Horton, R. K., M. Pachelo, and M. F. Santana, "Study of the treatment of the wastes from the preparation of coffee," paper presented at Inter-American Regional Conference on Sanitary Engineering, Caracas, Venezuela, Sept. 26–Oct. 2, 1946.

26.17 Rice Wastes

In the preparation of edible rice, large volumes of waste are produced in the soaking, cooking, and washing processes. The volumes of waste produced average approximately 60,000 gallons per ton of raw rice handled. About 12 to 14 per cent of this volume comes from the soaking and an equal amount from the cooking process. The remaining 75 per cent results from washing and draining of the rice.

Heukelekian [1] gives an analysis based on the average of a number of composite samples (Table 23.35). Since most of the BOD is in the form of

Table 26.35 Characteristics of composite rice waste. (After Heukelekian [1].)

Characteristic	Amount or value
pH	4.2–7.0
Total solids, ppm	1460
Ash solids, %	20.5
Suspended solids, ppm	610
Ash in suspended solids, %	10.8
Total nitrogen, ppm	30
Phosphates, ppm	30
BOD, ppm	1065
Starch, ppm	1200
Reducing sugars, ppm	70

colloidal and soluble materials, settling effects only a 29 per cent reduction and is not recommended because of the thin and variable volumes of sludge. Sixty per cent BOD reduction has been obtained by using 2000 ppm of lime as a coagulant; digestion can yield a BOD reduction of over 90 per cent, at a loading of 0.02 to 0.10 pounds of BOD per cubic foot per day; and dispersed-growth aeration, with a waste-to-seed ratio of 14.5:1 and adequate nitrogen addition, has been found to produce more than 90 per cent BOD reduction.

Reference: Rice Wastes

1. Heukelekian, H., "Treatment of rice water," *Ind. Eng. Chem.* **42**, 647 (1950).
2. Naito, S., "Water treatment in the sake industry," *Chem. Abs.* **76**, (103521u) (1972).
3. Noshiro, K. "Water pollution in the sake industry," *Chem. Abs.* **76**, (103518y) (1972).

26.18 Fish Wastes

While fishing, which is an important industry in the United States, employs 130,000 or more commercial fisherman and includes a total estimated catch of over 900 million, it is a far more important industry in Japan, Peru, Soviet Union, China, and Norway.

The production of oil, meal, fish solubles, and other materials from fish constitutes a sizable industry in the United States. On the East and Gulf Coasts, these materials are primarily produced from menhaden, a nonedible fish, and on both the East and the West Coast they are by-products of the sardine-canning industry.

A typical flow sheet of a fish-processing plant is given in Fig. 26.8. (The Iw numbers refer to sources of waste waters.) (Iw1) Boat storage-compartment wastes occur when fish are pumped out of the fish-storage compartments of the boats. (Iw2) Spent fish-transfer water is kept in tanks of about 10,000-gallon capacity, which are emptied daily. Impurities in this waste water consist largely of blood, particulate, dissolved fish solids, fish scales, and oily scum.

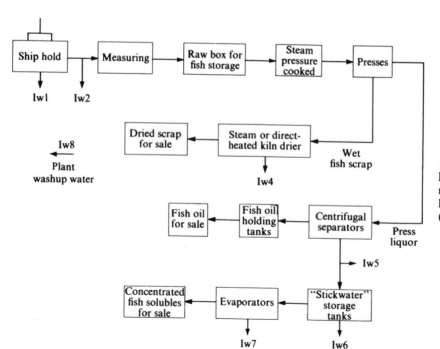

Fig. 26.8 Simplified flow sheet of menhaden fish-processing plant. The letters Iw refer to industrial waste. (After Paessler [2].)

(Iw3) Raw-box-leakage waste occurs when liquids escape through cracks in the raw-fish storage box, because of the weight of the fish, and also when liquids escape at the point where the conveyors take out the fish. This waste has a high concentration of blood and dissolved solids. (Iw4) Dryer-kiln deodorizer spray water occurs when water is sprayed continuously during processing, to absorb fumes from the drier kilns. This spraying is a common practice in all plants and may be of importance in air-pollution control. (Iw5) "Stickwater" wastes have a BOD which varies from 33,800 to 112,000 and represents more than 90 per cent of the total BOD of the waste from a menhaden plant. Its quantity, however, is limited, and it is discharged early in the season when water temperatures are low. Essentially it is composed of the water-base reject liquid from the centrifuges. (Iw6) "Stickwater" storage-tank wastes consist of spillages due to overflows when filling the tanks and "bottoms" that must be disposed of when the tanks are cleaned. (Iw7) Evaporator wastes occur when the evaporators are boiled out with water, rinsed, boiled again with caustic soda, and rinsed again, normally after each run of fish. Cooling water and some condensate are also continuously discharged from the evaporators. (Iw8) Washing wastes occur from cleaning of presses, floors, tanks, centrifuges, and other equipment. Table 26.36 contains the average composition of these wastes [1].

Chemical coagulation, because of low BOD removals and excessive sludge formation, is not recommended for this waste. A satisfactory method of treatment has been to combine all individual wastes with the stickwater waste (Iw6) and evaporate them. Barging the wastes to sea has been suggested as an alternative to evaporation.

Table 26.36   Composition of fish wastes. (After Paessler [2].)

Waste	BOD, ppm	Total solids, ppm	Total volatile solids, ppm	Grease, ppm
Iw1	42–265	15,576–20,606	2,489–3,394	
Iw2	3,050–67,205	18,421–64,857	5,912–46,907	1,314–17,234
Iw3	30,500–32,500	46,741–61,760	29,533–46,247	10,655
Iw4	120–300	14,171–18,949	1,906–7,957	45
Iw5	56,333–112,500	33,597–79,200	12,609–66,406	4,226–24,387
Iw6	47,063	52,998	45,483	18,157
Iw7	200–8,043	13,756–16,260	1,695–12,389	16–329

References: Fish Wastes

1. Matusky, R.E., J.P. Lawler, and T.P. Quirk, "Preliminary process design and treatability studies of fish processing wastes," Proc. 20th Industrial Waste Conf., May 1965.
2. Paessler, A.H., "Waste waters from menhaden fish oil and meal processing plants," Proc. 11th Ind. Waste Conf., Purdue Univ., May 1956, page 371.
3. Lanting, J. et al., "Thermophilic Anaerobic Design of Coffee Wastewater," Proc. 43 Purdue Ind. Waste Conf., 1988, P. 513.
4. Ripley, L. E. et al., "Bench-scale Evaluation of Anaerobic Contact Process for Treating Ice Cream Novelty Wastewater," Proc. 43 Purdue Ind. Waste Conf., 1988, P. 561.
5. Krofta, M. et al., "Treatment of Seafood Processing Wastewater by Dissolved Air flotation, Carbon Adsorption, and Free Chlorination," Proc. 43 Purdue Ind. Waste Conf., 1988, P. 535.
6. Allen, J. D. et al., "Full-Scale Operation of an Activated sludge Plant Treating Corn Wet Milling Wastewater," Proc. 41 Purdue Ind. Waste Conf., 1986, P. 505.

26.19 Pickle and Olive Wastes

These wastes arise when cucumbers or other vegetables are converted—by a combination of aging, chemical, and seasonings—to pickles. Since pickle factories produce either sweet or sour pickles, the wastes will therefore differ primarily in the degree of concentration of sugar. Four major wastes may be expected from a plant producing both types of pickle: (1)

lime waste, from "sweetening" the wooden vats during the winter months; (2) brine waste, from fermenting the pickles for two to three months in a 10 to 20 per cent brine solution; (3) alum and turmeric waste, from the solution which causes swelling and coloring of the pickles; (4) syrup wastes, from the solution of vinegar and sugar used for seasoning and packing the pickles. Each of these contributes pollution of a different type. Combinations of two or more make the stream-pollution problem difficult to remedy.

The brine and the alum and turmeric wastes are normally discharged almost continuously in large plants. The intermittent discharge of the lime waste or the acid syrup waste determines the pH and organic-matter content of the final wastes. Screening, neutralization, and equalization are minimum treatments for pickle wastes. Good housekeeping practices are a necessity in the prevention of excessive pollution. Sudden shock loads on treatment plants or receiving streams should be avoided at all costs. The reader is urged to consult Chapter 18 for details of the characteristics and treatment of pickle wastes. In the olive-curing process, lye and heat are used for the alkaline hydrolysis (saponification) of oils which are contained in the olives. This process produces glycerol and the metal salts of organic acids. These acid salts are long-chain carbohydrates with one or two points of unsaturation. Enzymatic oxidation of these compounds is slow for the aerobic microorganisms and therefore olive canning wastewater is not particularly amenable to biological treatment.

References: Pickle and Olive Wastes

1. Barnes, G. E., and L. W. Weinberger, "Internal housekeeping cuts waste treatment at pickle packing plants," *Wastes Eng.* **29**, 18 (1958).
2. Haseltine, T. R., "Biological filtration of kraut and pickle wastes," *Water Sewage Works*, **99**, 161 (1952).
3. Nemerow, N. L., *Pickle Waste Disposal*, private report to pickle company, 1952 (from files of the author).
4. Rice, J. K., "Water management reduces wastes and recovers water in plant effluents," *Chem. Eng.* **74**. no. 20 125 (1966).
5. Cortinovis, D., Water and Wastes Engineering **12**, 23, June (1975).

The following references have been published since 1970.

1. Cranfield, D., "Cucumber brining and salt recovery," *Rpt for Whitefield Pickle Co.* Montgomery, AL (1972).
2. Etchells, J. L., "Collected works of the U.S. food fermentation laboratory," *Dept. of Food Sci., NC State Univ.*
3. Henne, R. E. and J. R. Geisman, "Recycling spent cucumber pickling brines," *Proc. 1973 Cornell Agr. Waste Mgt. Conf.* (1973).
4. Lowe, E. and E. L. Dirkee, "Reconditions brine to cut pollution," *Food Engineering* **43**, 50–51 (1971).
5. Mercer, W. A., J. Naagdenberg, and J. W. Ralls, Reconditioning and reuse of live processing brines," *Proc. First Nat'l Symp. Food Processing Wastes* (1970).
6. Mercer, W. A. and J. W. Ralls, "Reduction of salt content of food processing liquid waste effluent," *EPA Water Pollut. Contr. Research Series* (12060 DXL) (1971).
7. Popper, K., "Possible uses of the mini-flow filter," *Proc. First Nat'l Symp. Food Processing Wastes* (1970).

26.20 Soft-Drink Bottling Wastes

Soft-drink bottling wastes result from the production of nonalcoholic beverages, both carbonated and non-carbonated. Wastes are produced from washing of bottles, production of syrup, treatment of water, and washing of floors. The wastes are usually highly alkaline, have a slightly higher BOD and suspended-solids content than domestic sewage, and are discharged to the sewer with or without screening.

The wastes from the bottle-washer are highly alkaline, since the washes consists of a series of alkaline detergent baths. Although, for reasons of economics as well as waste reduction, labels are used less than they were in the past, there are still large amounts of suspended solids resulting from straws, cigarette butts, paper, and other refuse left in the bottles. This foreign matter, plus leftover drinks in dirty bottles, is the major cause of the high BOD concentration. Wastes from the cleaning of floors, syrup-mixing and storage tanks, syrup filters, spillage, and so forth, are intermittent, and are not considered major sources of BOD and suspended solids. Wastes from water treatment will differ widely according to the quality required and the quality of the incoming water.

The characteristics of carbonated-beverage wastes, taken from Porges and Struzeski [6] and Besselievre [2] are presented in Table 26.37. Porges and Struzeski also observed [6] that, in 1954, 4643 bottling plants in the United States produced over one billion cases of soft drinks, valued at well over one billion dollars [3]. During that year, the per capita consumption in this country was more than 155 bottles. Analyses of several typical wastes from this giant industry are presented in Table 26.38.

Most soft-drink bottling plants are located near centers of population, so that discharge to the municipal sewer system appears to be the best means of waste disposal. Screening of wastes from the bottle-washer, to remove foreign matter left in bottles and labels if used, is sometimes practiced as a means of solids removal. To reduce the volume of waste, some plants reuse final rinse water from the bottle-washer for prerinsing the dirty bottles, or for other uses. Removal of waste drink and debris from the bottles and removal of labels before washing yields a pronounced reduction in BOD and suspended solids in

Table 26.37   Carbonated-beverage wastes.

Characteristic	Reference	
	6	2
pH	10.8	
Phenolphthalein alkalinity, mg/liter	150	
Total alkalinity, mg/liter	290	
5-day BOD, mg/liter	430	
Suspended solids, mg/liter	220	
Waste volume, gal/1000 cases	10,600	15,000
5-day BOD, lb/1000 cases		1,500
Suspended solids, lb/1000 cases		200

the waste water. The small amounts of waste resulting from this operation can be disposed of in various other ways than letting them escape to the sewers. The remaining wastes, although they have a high pH and alkalinity, have little or no undesirable effect on most municipal sewage-treatment processes.

Table 26.38   Five-day BOD, total solids, acidity, and pH of carbonated beverages. (After Porges and Struzeski [6].)

Beverage	5-day BOD, ppm	Total solids, ppm	Acidity		pH
			Mineral, ppm	Total, ppm	
Coca-Cola	67,400	114,900	244	1526	2.4
Pepsi-Cola	79,500	122,000	248	1466	2.5
Mission Orange	84,300	141,300	570	1579	3.0
Wagner Lift	64,600	110,800	316	2253	3.4
Tom Collins, Jr.	66,600	106,900	353	1246	3.2
Canada Dry quinine water	64,500	101,300	1181	3150	2.4
Average	71,200	116,200	490	1870	

References: Soft-Drink Bottling Wastes

1. Besselievre, E. B., "Industrial wastes, a community problem; the effects of certain types of wastes on city utilities," *Public Works Mag.* **83**, 74 (1952).
2. Besselievre, E. B., *Industrial Waste Treatment*, McGraw-Hill Book Co., New York, (1952), p. 107.
3. "Industrial statistics," in *1954 Census of Manufacturers*, Vol. 2, Bureau of Census, U.S. Dept. of Commerce, Washington, D.C.
4. Medbury, H. (ed.), *The Manufacture of Bottled Carbonated Beverages*, American Bottlers of Carbonated Beverages, Washington, D.C. (1945).

5. Morgan, R., *Beverage Manufacture*, Atwood and Co., London (1938).
6. Porges, R., and E. J. Struzeski, "Wastes from the soft drink bottling industry," *J. Water Pollution Control Federation*, **33**, 167 (1961); Proceedings of 15th Industrial Waste Conference, Purdue University, May 1960, p. 331.

26.21 Bakery Wastes

There are two types of bakery processes. The first is a dry-baking operation, such as bread baking, where the only waste waters are the floor washings and some specialized machinery wastes. The mixing vats and baking pans and sheets are all cleaned dry, floors are swept, and breadcrumbs are recovered. The waste water is low in BOD and suspended solids, with the main contaminants being flour and some grease.

The second type of baking operation—the production of cakes, pies, doughnuts, cookies, etc.—is distinctly different in both operation and waste characteristics. Pans and trays have to be washed and greased after each baking, which results in a very strong waste, with BOD values from 3000 to 5000 ppm and suspended-solids content from 2000 to 3000 ppm. The major contaminants are grease, sugar, flour, fruit washings, and detergents.

Strong bakery wastes are biologically treatable, with activated sludge giving good results. Because of the high carbon character of the waste, bulking or a light filamentous sludge is often produced. Grease traps are often inefficient, because detergents emulsify the fats, and flotation is a recommended practice to remove suspended solids. Acid treatment of small-quantity, high-strength wash waters can reduce BOD values by as much as 80 per cent. To date, little has been done in the way of research on bakery wastes, but flotation and centrifugation have been used with apparently good results.

References: Bakery Wastes

1. Conner, W. R., and M. J. Perry, "Treats liquid wastes more efficiently," *Food Eng.* **40**, 92 (1968).
2. Mulligan, T., "Bakery sewage disposal," *Bakers Dig.* **41**, 81 (1967).
3. Grove, C. S., and D. Emerson, "Laboratory pilot-plant studies for treatment of bakery wastes," Report (part I) to the Ebinger Baking Company, Brooklyn, N.Y., Oct. 27, 1968.

26.22 Waster-Treatment-Plant Wastes

In the past, the waste from water-treatment plants has not been considered an industrial waste. This may have had some basis in regard to municipal facilities, but certainly wastes from industry-owned water-treatment plants are industrial wastes. In addition, the water-treatment field is so large today that it should be considered an industry within municipal government.

Recent statements by both state and federal officials have indicated that even municipal plants will be required to treat their waste water, because federal and state legislation does not distinguish between sources of pollution. Cost of handling and treating wastes from water-treatment plants should be considered as a fundamental part of water-treatment costs. The sources of wastes in treatment plants are: (1) filter backwash water, (2) lime and lime soda sludge, (3) brine from cation exchange and sodium zeolite softeners, and (4) alum sludge.

Common methods of treatment being used are:

1. Direct discharge to
   a) sanitary sewers (controlled (high flow), uncontrolled, or sludge discharge)
   b) deep well disposal
   c) lagoons

2. Reclamation and reuse
   a) recovery of alum by sulfuric acid
   b) recalcination of lime
   c) recycle filter backwash after sedimentation to head of plant

3. Physical treatment methods
   a) centrifugation
   b) vacuum filtration
   c) sand drying beds
   d) flash drying
   e) thickening and sedimentation
   f) freezing

At present the most popular method is direct discharge to streams, with lagooning the most commonly used "treatment method."

26.23 Agricultural Wastes

Introduction. Agricultural waste results from the production of food and fiber and can have a detrimental effect on the environment when not properly managed. This includes liquid and solid animal wastes, crop residues, loss of agricultural chemicals, and irrigation return flows. All of these wastes, except for irrigation return flows, which are a problem mainly in the arid Southwest, exhibit variability in their characteristics due to modifications of agricultural techniques. These modifications are in response to changes in climate, terrain, soil characteristics, and socio-economic conditions between farming regions.

Origin. The characteristics of livestock wastes are a function of the animal, the type, digestibility, and composition of its feed ration, and the type of feeding operation. Livestock feces consists chiefly of undigested cellulose fibers. It also contains protein, potassium, calcium, magnesium, phosphorus, residue from digestive fluid, mucus, bacteria, and any foreign material, such as dirt, that was ingested while the animal was fed. Feces may also contain inorganic feed additives given to increase weight. Copper salts from commercial pig feed supplements are good examples of this.

Livestock wastes, or manure, encompass, (a) fresh excrement including both liquid and solid portions, (b) total excrement with bedding materials, (c) solid material after the liquid has been drained or evaporated, or (d) only the liquid drainage. Each of these has different properties, and definition of the waste by type and a detailed analysis of its composition are prerequisites to the design of any management or treatment system.

Characteristics. Poultry wastes contain 75–80 per cent moisture, 15–18 per cent volatile solids, and 5–7 per cent ash. Manure excreted daily represents about 5 per cent of the body weight of the bird. BOD has been shown to range from 8500–40,000 mg/l.

Daily swine waste production is a function of the type and size of animal, feed ration, temperature and humidity within their housing, and water volume used to wash floors. Quantities of feces and urine excreted increase with the weight and food uptake of the animal. Swine and poultry wastes are higher in nitrogen and phosphorus than dairy or beef cattle manures.

Dairy cattle production produces two wastes, manure from the animals and liquids from the cleaning of milking parlor equipment. Manure production averages about 86 lbs/animal/day, representing 7–8 per cent of body weight. Urine makes up 30 per cent of this on a weight basis.

Milk sanitation regulations require cleaning of the milking parlor floors and all equipment after each milking session. A chloride cleaning solution is used for milking equipment and fresh water is used to flush floors. Total volumes produced are a function of the size of the milking parlor and the total length of transfer lines, rather than the number of animals milked. Overall characteristics of the waste requiring treatment depends on the farm management–i.e., whether the two wastes are treated separately or combined.

Treatment. Treatment of livestock wastes is by biological means, normally employing lagoons, ponds, and oxidation ditches. These methods have been shown to be most efficient at least costs. Anaerobic lagoons are the most commonly employed treatment method.

Alternative measures of waste handling include composting, reprocessing wastes to feed supplements, anaerobic digestion for methane production and, the most common alternative method, land disposal for crop fertilization.

Land application rates for manure must be consistent with the ability of the land to assimilate the waste—i.e., both the hydraulic and organic load. Immediately following application incorporation of the wastes into the soil should be practiced to minimize the chance for loss via surface runoff. Care should also be taken to prevent the accumulation of nitrates, and other inorganic salts in the soil that could leach to ground or surface waters.

Agricultural chemical loss results from the transport, by surface runoff, of fertilizers, pesticides, and herbicides from fields to surface waters. Prevention of fertilizer loss can be achieved by applying only needed amounts at the optimum time for use by the crop. Methods such as side dressing of fertilizers meet these requirements.

Reduction in pesticide and herbicide losses can be achieved by proper timing of the application, using only the minimum required amounts that give effective control of the pest or weed. Soil conservation measures will also reduce losses as most of these compounds are lost while being absorbed into soil particles.

Crop residues present a problem only when their buildup results in organic material or nutrients being leached from them by surface runoff. Rapid incorporation into the soil after harvest will prevent this.

Irrigated agriculture is the largest user of our water resource, especially in the arid Southwest. There, irrigation water percolates through the soil, dissolves out weathered minerals, and returns to the river with a higher than original salt concentration. This impairs the beneficial use of the water by others. It then becomes a non-point pollutant and is considered under Section 31.2.

The problem is compounded further by evapotranspiration concentration of salts in the root zone, which necessitates increasing water volume that must be applied to maintain salt balance in the soil. Hence, increasing applied volumes increases leaching and concentrates more salts in the return flow.

Major sources of return flows that cause salinity problems are canal seepage and deep percolation. Average seasonal canal losses can vary from 10–50 per cent of the total diversion. When soils along the canals are high in residual salts, the salt pick up by canal seepage can easily exceed that leached from irrigated lands. Canal lining is normally used to prevent this when the costs can be justified.

Reduction of the salt buildup in deep percolating waters can be achieved by efficient use of applied irrigation water. Methods for increasing efficient use are those those which control the timing and volume of applications to maximize plant use of water and minimize evaporation losses. Automatic control structures for furrow irrigation systems and traveling sprinkling systems, coupled with night or late afternoon applications, will accomplish these objectives.

The main problem with return flows, however, is that return-flow quality is an economic externality. The harmful practices of one irrigator become a cost to the next down-stream user, while improving return-flow quality has no direct benefit to the individual irrigator and can have direct costs. No incentive exists for individual action, making this a problem which must be approached on a regional scale.

26.24 Palm Oil Wastes

Crude palm oil production involves only mechanical extraction of the fresh fruit bunches, as shown in Figure 26.9. The palm fruit is sterilized, digested and the resulting extraction of oil is clarified. (Maheswaran, 1, 1985).

About 5 cubic meters of palm oil effluent are released for every ton of palm oil produced. A typical analysis of crude palm oil wastewater is shown in Table 26.38.

Treatment:

Treatment of the crude palm oil wastes is anaerobic digestion (with or without recirculation) and then followed by aerobic treatment either by oxidation ditch, activated sludge or aerated lagoon. The effluent CAN be clarified with the sludge dried on a sand bed.

Palm Oil Refining

Palm oil refineries are also major souces of pollution. Refining is primarily a process of fractionation for separating the liquid (olein) and solid (stearin) of the palm oil. Fractionation is done dry (by chilling), with detergent, or solvent and detergent. The wastewater characteristics vary widely for the three systems. Dry refining is least polluting. The other two systems yield effluents with high BOD and solids as well as fats. The treatment facilities employed by refineries range from simple neutralization to complete coagulation, dissolved air flotation, filtration and biological conversion.

Table 26.39   Typical Analysis of Palm Oil Mill Wastewater (After Maheswaran 1985)

	mean (p.p.m.)
B.O.D. (3 day, 30°C.)	25,000
C.O.D.	53,630
Total Solids	43,365
Suspended Solids	19,020
Oil and Grease	8,370
Ammoniacal-nitrogen	35
Total Nitrogen	770
pH	3.8–4.5

References: Palm Oil

1. Maheswaran, A., "The food processing industry and the environment with emphasis on palm oil production" UNEP Industry and Environment, Oct/Nov/Dec. 1985, page 2.

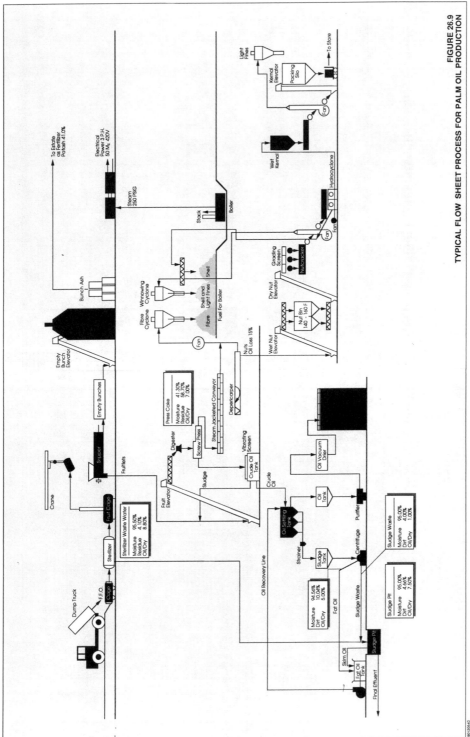

FIGURE 26.9

TYPICAL FLOW SHEET PROCESS FOR PALM OIL PRODUCTION

Figure 26.9   Schematic flow diagram of palm oil mill (After Maheswaren 1985)

CHAPTER 27

THE MATERIALS INDUSTRIES

In this text, industries producing materials are differentiated from food-processing industries in that the products are nonedible; from the apparel industries in that the products are not articles of clothing; and from the chemical industries in that the products are not specific chemicals or associated products. These industries fall into four groups: (1) wood fiber (including pulp and paper wastes and photographic wastes); (2) metal (steel and other metal wastes, metal-plating and iron-foundry wastes); (3) liquid-processing (oil-refinery, rubber, glass-manufacturing, and naval-stores wastes); and (4) special materials (glue, wood preserving, candles, and plywood).

WOOD FIBER INDUSTRIES

The pulp- and paper-making industry is the fifth largest in our economy and ranks third in its rate of expansion [1]. The per capita demand for its products is nearly 400 pounds. More than 1000 mills in 40 states produce a wide variety of paper and paperboard products, which total more than 30 million tons annually. Water is used by this industry at a rate of 4000 mgd, a usage equivalent to that of a population of about 40 million. (These figures are as of 1956.)

The manufacture of paper, like the manufacture of textiles, can be divided into two phases: pulping the wood and making the final paper product. The raw materials generally used in the pulping phase are wood, cotton or linen rags, straw, hemp, esparto, flax, and jute or waste paper. These materials are reduced to fibers which are subsequently refined, sometimes bleached, and dried. At the paper mill, which is often integrated in the same plant with the pulping process, the pulps are combined and loaded with fillers; finishes are added and the products transformed into sheets. The fillers commonly used are clay, talc, and gypsum. The four main types of pulp used are groundwood, soda, kraft (sulfate), and sulfite.

Fiber industries, therefore, produce two main wastes, namely pulp-mill and paper-mill wastes. Pulp-mill wastes come from grinding, digester cooking, washing, bleaching, thickening, deinking, and defibering. These wastes contain sulfite liquor, fine pulp, bleaching chemicals, mercaptans, sodium sulfides, carbonates and hydroxides, sizing, casein, clay, ink, dyes, waxes, grease, oils, and fibers. Treatment of these wastes is primarily by recovery of chemicals, equalizing, sedimentation, controlled dilution, coagulation, lagooning, biological oxidation, evaporation, spray drying, burning, and some fermentation. Paper-mill wastes originate in water which passes through the screen wires, showers, and felts of the paper machines, beaters, regulating and mixing tanks, and screens. The paper-machine wastes (white waters) contain fine fibers, sizing, dye, and other loading material. Wastes are treated by recovery of the white water, settling and vacuum filtration, flotation, and chlorination. Figure 27.1 is a schematic flow design for pulp- and paper-mill processes.

27.1 Pulp- and Paper-Mill Wastes

Origin of pulp- and paper-mill wastes. The major portion of the pollution from papermaking originates in the pulping processes. Raw materials are reduced to a fibrous pulp by either mechanical or chemical means. The bark is mechanically or hydraulically removed from wood before it is reduced to chips for cooking. Mechanically prepared (groundwood) pulp is made by grinding the wood on large emery or sandstone wheels and then carrying it by water through screens. This type of pulp is low-grade, usually highly colored, and contains relatively short fibers; it is mainly used to manufacture nondurable paper products such as newspaper. The screened bark effluent contains fine particles of bark and wood and some dissolved solids. Additional sources of waste from

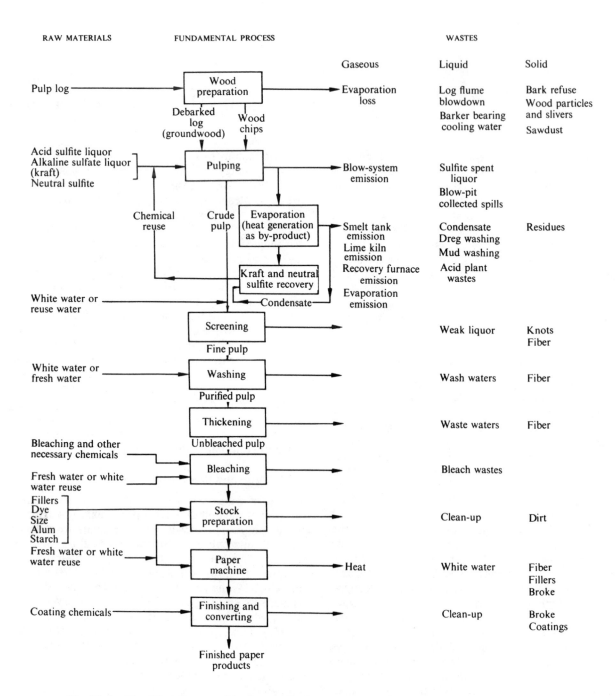

Fig. 27.1   Simplified diagram of fundamental pulp and paper processes. (Prepared for the F.W.P.C.A.)

wood preparation are the pressing of rejects prior to burning and floor drainings.

Chemically prepared pulps, as compared with mechanically prepared ones, are made by the soda, sulfate (kraft), or sulfite process; the semichemical process, which we will discuss later, is also used today. (Sulfate pulps, incidentally, do not bleach readily and therefore are generally used for brown, or other-colored, wrapping paper.) In all these methods the wood is prepared, as in the making of groundwood, by reduction to chips and screening to remove dust. The chemical processes differ from one another only in the chemical used to digest the chips.

Softwoods, such as poplar, are usually treated by the soda process. To a digester holding about four cords of chips, a mixture of soda ash (Na_2CO_3) and lime [$Ca(OH)_2$] is added and the total contents are boiled under steam pressure for about ten hours. This digestion decomposes or separates the binding, noncellulosic materials, such as lignins and resins, from the fiber, but it is a rather harsh treatment for the fibers and consequently weakens them. Coniferous woods are used in the preparation of both kraft and sulfite pulps. The sulfate process calls for a shorter digestion period of about five or six hours, with a mixture of sodium sulfide, hydroxide, sulfate, and carbonate. The lignin and noncellulosic materials are dissolved, leaving a stronger fiber for paper formation. Sulfite pulp is made by cooking with calcium bisulfite at over 300°F and 70 pounds of steam pressure.

After digestion, chemically prepared wood pulps are blown into a closed blow pit, where the black liquor is allowed to drain to the sewer or to recovery processes. The drained pulp is then washed. These wash waters may then be wasted, reused, or sent through recovery operations, while the washed pulp is passed through some type of refining machine to remove knots and other nondisintegrated matter. A cylindrical screen, called a decker, revolving across the path of the pulp partially dewaters it, after which it is passed to bleach tanks, where it is mixed in a warm, dilute solution of calcium hypochlorite or hydrogen peroxide. The dried, bleached pulp is then ready for sale or delivery to the paper mill.

The papermaking process involves first a selection of the appropriate mixture of pulps (wood, rag, flax, jute, straw, old newspaper, and so forth). The pulp mixture is disintegrated and mixed in a beater, to which are added various fillers and dyes, to improve the quality of the final paper product, and sizing, to fill the pores of the paper. The beater is essentially an oblong tank equipped with a rotating cylinder, to which are attached dull knives to break up the knotted or bunched fibers and cause a thorough mixing of the entire contents of the tank. Sometimes the pulps are washed in the "breaker beater" prior to the addition of the chemicals. The washing initially produces a rather strong waste, which is progressively diluted as the washing proceeds. After beating, the pulp is usually refined in a jordan, a machine that consists of a stationary hollow cone with projecting knives on its interior surface, fitted over a rapidly rotating adjustable cone having similar knives on its outside surface. This machine cuts the fibers to the final size desired. The pulp then passes to stuffing boxes, where it is stored, mixed, and adjusted to the proper uniform consistency for papermaking. Finally, the pulp is screened to remove lumps or slime spots, which would lower the quality of the final paper.

Next, the pulp is evenly distributed from a headbox over a traveling belt of fine wire screening, known as a fourdrinier wire, and carried to rolls. A small portion of the water contained in the pulp passes through the screen, while the longer fibers are laid down as a mat on the wire. A considerable portion of the fine fibers and some fillers also pass through the screen wire with the water. Because of its color, this waste water is called "white water." The paper mat passes through a series of rolls as follows: a screen roll to eliminate inequalities at the end of the wire, a suction roll to draw out more water, press and drying rolls to rid the paper of most of the remaining water, and finally finishing rolls (calenders), which produce the final shape of the paper. The final products are used for many purposes—printing paper, newspaper, wrapping paper, parchment paper, writing paper, blotting paper, tissue paper, and impervious food-wrapping paper.

The chief sources of waste at the pulp mills are the digester liquors and the chief sources at the paper mills are the beaters and paper machines. Fiber losses generally average 3 per cent or less. In so-called "closed systems," where white water is recirculated and reused, it is possible to reduce fiber losses to 0.1 per cent. However, even this low level of loss is significant, because of the large quantities of fibers processed per day. Table 27.1 gives some idea of the

Table 27.1   Average waste discharge per ton of paper product.*

Product	Waste, gal
Pulp mills	
Groundwood	5,000
Soda	85,000
Sulfate (kraft)	64,000
Sulfite	60,000
Miscellaneous paper	
No bleaching	39,000
With bleaching	47,000
Paperboard	14,000
Strawboard	26,000
Deinking used paper	83,000

*From *Industrial Waste Guide, Ohio River Pollution Control Survey*, Supplement D, U.S. Public Health Service (1943).

volumes of wastes discharged per ton of product.

With softwood timber steadily becoming scarcer, the use of hardwoods for pulp has increased during recent years. Another pulping method, known as semichemical pulping, has been developed chiefly for hardwoods. It is carried out by digestion under relatively mild chemical conditions, which softens but does not fully pulp the wood, and subsequent actual reduction to pulp by mechanical means. The product is used mainly for container board and coarse wrapping paper. The name "semichemical" has been applied mainly to cooking methods employing neutral sodium sulfite. However, other cooks resulting in slightly acid or basic pH values and utilizing semi-sulfate and semisoda, are also sometimes referred to as semichemical pulping.

Characteristics of pulp- and paper-mill wastes. Since the four types of pulping produce somewhat different wastes, each should be considered separately. Ghem [2] presents the general characteristics of groundwood wastes (Table 27.2). The Ohio River study [3] shows typical analyses of all types of pulp and paper wastes (Table 27.3). Gehm [2] also gives

Table 27.2   Typical analysis of wood preparation wastes.

Characteristic	ppm
Total solids	1160
Suspended solids	600
Ash (suspended solids)	60
Dissolved solids	560
Ash (dissolved solids)	240
BOD, 5-day	250

Table 27.3   Typical analytical results for pulp- and paper-mill wastes.

Product	BOD, ppm	Suspended solids, ppm
Pulp		
Groundwood	645	
Soda	110	1720
Sulfate (kraft)	123	
Sulfite	443	
Miscellaneous paper		
No bleach	19	452
With bleach	24	156
Paperboard	121	660
Strawboard	965	1790
Deinking used paper	300	

Table 27.4   Characteristics of kraft-mill wastes.

Characteristic	Maximum	Minimum	Average
pH	9.5	7.6	8.2
Total alkalinity, ppm	300	100	175
Phenolphthalein alkalinity, ppm	50	0	0
Total solids, ppm	2000	800	1200
Volatile solids, %	75	60	65
Total suspended solids, ppm	300	75	150
Volatile solids, %	90	80	85
BOD, 5-day, ppm	350	100	175
Color, ppm	500	100	250

the analyses of 24-hour composite samples of the combined effluent from modern unbleached kraft (sulfate) mills (Table 27.4).

Moggio [4] states that present-day kraft mills, efficiently operated, discharge an effluent containing no more than 100 pounds of the Na_2SO_4 equivalent of the cooking liquor per ton of pulp. He also claims

water-use figures of 20,000 to 30,000 gallons for unbleached kraft and 40,000 to 60,000 gallons for bleached kraft (per ton of pulp). The effluent character varies somewhat, depending on bleaching practices: suspended solids range from 20 to 60 ppm and are primarily fiber (about 0.5 per cent of the total product); dissolved-solids concentrations range from

Water sources

Water outlets

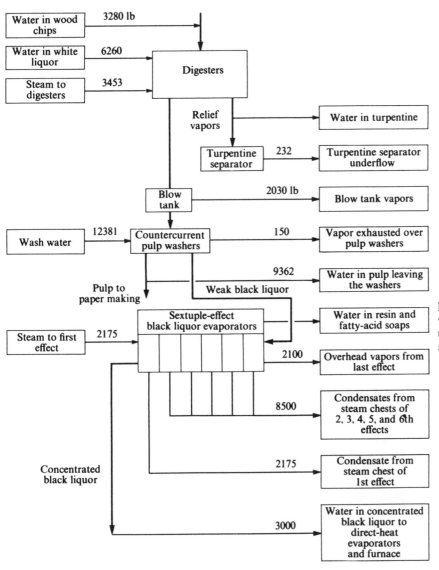

Fig. 27.2 Process flow diagram of water balance of kraft pulp manufacturing process (quantities in lb/ton of air-dry pulp). (After McDermott [5].)

1000 to 1500 ppm, of which about 60 per cent is ash; BOD values range from 100 to 200 ppm, or 20 to 40 pounds per ton of product. The effluent is coffee-colored and has a color value of about 500. Typical flow diagrams of kraft pulp-manufacturing processes and water balances are given in Figs. 27.2 and 27.3.

Gehm [2] states that, since the soda pulping process is very similar to the kraft one, the effluents are similar. The only essential difference is that either NaOH alone, or a lower concentration of Na₂S, is used in the soda process for cooking the wood. Soda cooks, then, will have a lower sulfur concentration.

Spent sulfite liquors average approximately 300 gallons per ton of pulp. Spent-sulfite-liquor waste has been described as a highly corrosive, dilute solution, containing about one-half of the solids in the pulp wood [6]. These solids may contain as much as 65 per cent lignosulfonic acid, 20 per cent reducing sugars, 8.4 per cent sugar-sulfur-dioxide derivatives, and 6.7 per cent calcium. The solids concentration of spent sulfite liquor drawn from the digesters may vary from 6 to 16 per cent and it may contain from 400 to 600 (or more) pounds of BOD per ton of pulp. The sulfur compounds possess an "immediate oxygen demand" which accounts for about 11 per cent of the 5-day BOD. The sugars (hexoses and pentoses) repre-

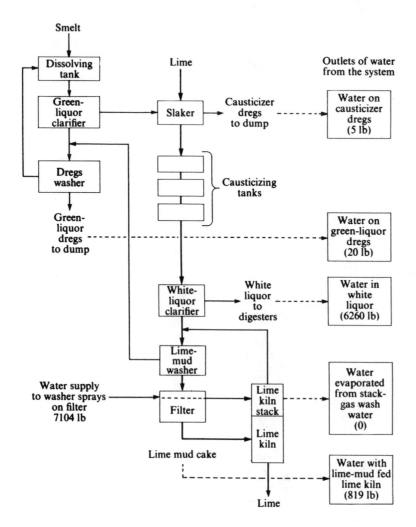

Fig. 27.3   Water balance in chemical reconstituting system (quantities in lb/ton of air-dry pulp). (After McDermott [5].)

Table 27.5 Relationship of major components of sulfite waste liquor [6].

Component	% Total solids, bone-dry basis
Lignin	51.6
Sugars	16.9
Sulfur	9.15
Calcium	4.5

sent about 65 per cent of the BOD. Although lignins account for over half the solids in this waste, they contribute little to the BOD. Haskins [cited in reference 6] describes sulfite waste liquor as a highly complex liquid, containing 10 to 12 per cent solids. The major components of the liquor are free and combined SO_2, volatile acids, alcohols, acetone, furfural, sugars, and lignin. A more detailed analysis of these wastes is presented in Tables 27.5 and 27.6.

An analysis of a typical semichemical-pulping mill

Table 27.6 Composition of a typical sulfite pulp-plant waste.

Component	Digester liquor, ppm	Blow-pit liquor, ppm
Total solids	111,100	38,700
Volatile solids	101,000	34,000
Ash	10,100	4,700
Calcium	3,990	1,550
Total sulfate	31,200	8,620
BOD, 20-day	42,900	

waste was given by Crawford in 1947 [7]. This mill used the semisoda cook on pine and mixed hardwoods and the neutral cook on tannin-extracted chestnut chips (Table 27.7). Vilbrandt [8] reports a breakdown in the analysis of solids and BOD components of semichemical soda-pulping waste (Table 27.8). Rudolfs and Nemerow [9] present an analysis of typical white-water waste from chipboard mills (Table 27.9). The pulping of raw materials other than wood results in somewhat different waste analyses. Bloodgood [10] reports the analyses of strawboardmill waste based on an eight-month daily test survey (Table 27.10).

Rudolfs and Nemerow [11] present the important sanitary characteristics of rag-, rope-, and jute-cooking liquors and of beater wash-water wastes (Table 2.11), and a further chemical analysis of the cooking liquors (Table 27.12). A Wisconsin survey of pulp and paper mills in 1946 [12] showed the wide variation in characteristics of all types of pulping and paper-machine wastes (Table 27.13). A recent survey [1968, 13] gives the solids, BOD, and pH waste loads and waste-water flows in various types of mill (Table 27.14).

Treatment of pulp- and paper-mill wastes. Pulp- and paper-mill wastes are treated in the following manner: (1) recovery, (2) sedimentation and flotation to remove suspended matter, (3) chemical precipitation to remove color, (4) activated sludge to remove oxygen-demanding matter, (5) lagooning, for purposes of storage, settling, equalization, and sometimes for biological degradation of organic matter.

Paper-mill waste treatment is one of the most developed and investigated systems. A great deal of research and many pilot-plant investigations are being currently

Table 27.7 Major wastes in semichemical board production [7].

Characteristic	Globe-digester blow	Washer effluent	Machine effluent
Volume, gal/day	24,000	2,000,000	864,000
Total solids, ppm	102,300	4,593	658
Fixed solids, ppm	35,000	1,547	166
Volatile solids, ppm	67,300	3,046	492
Total solids, ton/day	10.8	38.3	2.4
Color, ppm	165,000	12,000	500
Per cent of total	13.8	84.7	1.5
BOD, 5-day, lb/day	1,940	16,440	230

Table 27.8 Solids and BOD analysis of semichemical soda-pulping waste [8].

Component	Total solids, %	BOD, %
Pentosans	5.7	18.9
NaHCO₃	11.5	0.0
NaC₂H₃O₂	26.1	56.7
Sodium formate	6.8	4.2
Lignin	14.9	—*
Hydrolyzable fraction of suspended solids other than pentosans	5.4	—*
Total	70.4	79.2

*Doubtful 5-day BOD value.

Table 27.9 Analysis of white-water waste from chipboard mill [9].

Characteristic	Value or concentration
pH	7.0
Alkalinity, ppm CaCO₃	118
Suspended solids, ppm	840
Fixed solids, %	11.9
Volatile solids, %	88.1
Total solids, ppm	2180
Ash, %	19.3
Volatile total solids, %	80.7
Dissolved solids, % of total	61.5
BOD, ppm*	100–400

*From Rudolfs and Axe [277].

Table 27.10 Analysis of total strawboard waste [10].

Characteristic	Maximum daily	Minimum daily	8-month average
Total solids, ppm	4900	2600	3691
Total volatile solids			
ppm	4100	1100	2502
%			67.7
Total suspended solids, ppm	2190	484	1369
Suspended volatile solids			
ppm	1590	320	909
%			66.5
BOD, ppm	1372	409	955

Table 27.11 Rag, rope, and jute-mill wastes [11].

Characteristic	Jute		Rope		Rag	
	Cook	Wash	Cook	Wash	Cook	Wash
pH	12	11.2	10.8	8.7	12	8.1
Alkalinity, ppm CaCO₃	2850	574	18,000	350	31,655	264
BOD, ppm	3381	385	12,125	1250	29,225	526
Total solids, ppm	7200	1300	45,000	1100	96,000	2200
Total volatile solids, %	56	56	56	64	64	54
Dissolved solids, %	82	65	95	59	93	73
Total nitrogen, ppm	126	6	157	21	1,270	44
Ratio of BOD to total nitrogen	2.7:1	65:1	88:1	60:1	23:1	12:1

Table 27.12   Breakdown of jute, rope, and rag cooking-liquor solids soluble in cold water [11].

Type of cook waste	Total solids in cold water, ppm	Volatile matter, %	Components of volatile matter in cold water, % of volatile matter					
			CO_2	Grease	Lignin	Volatile acids, as HAC	Complex polysac-charides	Protein, total N × 6.25
Jute	5,540	68.8	30.0	0	21.8	40.9	10.9	4.5
Rope	37,000	54.3	25.4	0	39.4	14.8	10.6	6.4
Rag	85,800	58.4	24.6	7.6	24.3	10.4	3.6	3.8

carried out and reported on. Actual treatment equipment is installed only after exhaustive study of all other possibilities, since the cost of treatment is considered high in relation to the cost of the product produced. Thus, economic limitations have forced the industry to place emphasis on recovery rather than treatment.

Recovery processes in the paper mill involve the use of "save-alls," in either closed or partly closed systems. These save-alls are installed not only as a waste-treatment measure, but also as a conservation measure to recover fibers and fillers. The main types are based on filtration, sedimentation, or flotation processes. Filtration devices are usually some variation of a revolving, cylindrical, perforated screen or filter that removes the suspended solids in the form of a mat, which is subsequently scraped off the drum and returned to the paper-making stock system. Conical or other sedimentation tanks are also often used to separate the suspended matter by difference in specific gravity. All the principles of sedimentation discussed in Chapter 10 apply to these treatment units.

Table 27.13   Results of 1946 survey of wastes at Wisconsin pulp and paper mills [12].

Type of mill	Waste, gal/ton of product	Solids, lb/ton of product			Fiber loss, % of production	5-day BOD*
		Fixed suspended	Volatile suspended	Total soluble		
Paper						
Book	13,071	28.8	22.5	44.6	0.9	40
Tissue	23,048	5.7	30.4	113.4	1.5	47
Wrapping	28,432	17.9	31.2	146.8	1.6	56
Bond	20,461	18.7	57.2	108.6	2.8	116
Glassine	48,727	5.3	32.2	111.4	1.6	59
Paperboard	5,600	0.9	10.9	26.3	0.5	11
Black wadding	57,400	13.2	86.3	232.9	4.3	158
Pulp						
Rag and deinked	49,796	157.6	188.2	911.6	6.1	403
Kraft	64,838	33.6	61.2	329.8	3.1	451
Sulfite	50,470	3.7	37.6	2413.8	2.0	2857
Groundwood	2,302	0.15	11.2	7.0	0.6	20

*Population equivalent, persons/ton of product.

Table 27.14 Waste loads and waste-water quantities in typical pulp and paper mills.

Process	Waste load, in lb/ton of product										Waste-water quantities, gal/ton	
	Suspended solids		Dissolved solids		Total solids		BOD		pH			
	Range*	Mean	Range	Mean	Range	Mean	Range	Mean	Range	Mean	Range	Mean
Wood preparation	1.9–40	9		4	4–50	13	2–10	3	6.5–8.0	7.0	1,000–10,000	3,400
Pulping												
Groundwood												
Sulfate (kraft)												
Blow tower	(3.7)	4		17	(21.0)	21	(1.3)	1	(12)	12.0	(1,000)	1,000
Dirty condensate	0–0.5	0.1		4	6–11	7	6.5–9.0	8	9.5–10	10.0	950–1,900	1,200
Evaporator ejector	0.06–0.2	0.1		2	1–3	2	1.6–4.5	3	9–10	9.5	290–640	300
Causticizing waste	2.2–5.7	5		96	46–240	101	8.0–10.5	9	9–11.0	10.0	600–9,600	2,500
Green dreg	(1.0)	1		21	(22)	22	(1.0)	1	(12)	12.0	(200)	200
Floor drain	0.5–10	6		1	11.0–11.5	11	0.3–1.7	1	11.6–12	12.0	340–580	400
Subtotal		17		141		164		23	9.5–12.0			5,600
Sulfite												
Blow tower	0.42–1.9	1		246	36–348	247	29–194	116	2.2–2.9	2.7	1,840–1,950	1,900
Condensate	0.05–0.2	0.1		47	18–87	47	48–71	66	2.3–3.1	2.6	750–1,700	1,100
Uncollected liquor	0.3–43	21		84	50–515	105	50–61	53	2.2–2.6	2.4	2,000–10,000	7,500
Acid plant wastes	(5)	5	(5)		(10)	10			(1.2)	1.2	(300)	300
Boiler blowdown	(2)	2			(22)	22	(0.05)	0.05			(100)	100
Subtotal		29		382		411		235	1.2–2.9	11.0		10,900
Semichemical												
Blow tower	(2)	2	(6)	6	(8)	8	(1)	1		4.0	(1,000)	1,000
Condensate	(0.1)	0.1	(2)	2	(2)	2	(3)	3		3.5	(2,000)	2,000
Recovery system	(9)	9	(111)	111	(150)	150	(8)	8			(2,000)	2,000
Uncollected liquor	(11)	11	(29)	29	(40)	40	(18)	18	2.5–4.0	2.5	(2,000)	2,000
Subtotal	(22)	22		148		200		30				7,000
Deinking (all sources)†							11–25				9,700–36,000	
Pulp screening												
Groundwood												
Sulfate (kraft)	5–8	4		58	60–63	62	10–18	14	9–10	10.0	900–9,600	3,600
Sulfite	1.7–14	8		19		27	22–10.7	8	5.4–5.7	5.6	1,700–14,300	6,000
Semichemicals												
Deinking												

Table 27.14 (continued)

Process	Range	Mean	Range	Mean	Range	Mean	Range	Mean	Range	Mean	Range	Mean
Pulp washing and thickening												
Groundwood (no washing)	9–14	11		44	51–107	75	22–46	33	5.0–6.25	6.0	4,800–10,000	7,500
Sulfate (kraft)	10–30	15		127	94–180	142	10–35	25	8.9–9.4	9.0	3,000–11,000	7,000
Sulfite	6.5–9.0	8		123	68–1037	131	7.4–34.0	18	2.4–3.9	2.9	1,800–15,000	7,500
Semichemical	0.9–6.0	3		90	42–141	93	10–42	24	7.0–7.9	7.4	2,400–7,800	5,400
Deinking												
Bleaching												
Groundwood	14–124	60	92–280	180	216–294	240	8–88	30	2.9–6.8	2.9	12,000–32,000	4,000
Sulfate (kraft)	4–44	15	126–409	205	131–415	220	17–44	25		3.8	9,000–30,000	19,000
Sulfite												15,000
Semichemical												
Deinking		6		119		125		12		2.2		5,500
Paper-making‡												
General	10–166	46	21–425	73	31–591	119	3–80	16	4.3–6.9	5	5,700–40,000	13,000
Related products												
Newsprint	20–60	40					10–12	15			37,000	
Uncoated groundwood												
Coated printing paper												
Uncoated book paper	47–100	30		66		116	15–40	16			8,000–28,000	14,000
Fine paper		.73		80		153					9,000–40,000	18,000
Coarse paper	10–30	20					10–25	15			2,000–29,000	10,000
Special industrial paper	200–400	300					140–170	155			20,000–100,000	
Sanitary and tissue paper	50–100	50		150		200	15–30	22			8,000–37,000	14,000
Total mill effluent (integrated pulp and paper mills)												
Bleached sulfate and paper	50–200	170	150–1130	640	200–1300	810	30–220	120			39,000–54,000	45,000
Unbleached sulfate and paper§		50		460		510						27,000
Bleached sulfite and paper	40–100	100	560–1600	1040	600–1700	1140	235–430	330			40,000–70,000	55,000

*Single pieces of data are entered under the "Range" column in parentheses. The mean values shown are not truly statistical averages; they are considered to be probable average values based on the available data.

†The deinking process includes pulping, screening, washing, and thickening.

‡The waste waters from paper-making include those from stock preparation, paper-machining, and finishing and converting operations.

§Data for integrated unbleached sulfate pulp and paper mills are generated by subtracting the data for bleaching from those for the integrated bleached sulfate pulp and paper mill.

In flotation recovery units, also discussed in Chapter 10, the suspended fibers and other solids are removed in the form of a mat floating on the surface of the tank. This is a very efficient method for certain fibers which have a natural tendency to float in a suspension, being buoyed up by minute bubbles of air dissolved in the fibrous waste. The air is usually forced into the waste water under a pressure of about 45 pounds per square inch and released in an open flotator tank under atmospheric pressure, or under a slight vacuum. Recovery efficiencies are often better than 95 per cent suspended solids recovered. Recovery of the clarifier white water is achieved by recirculating this water into the beaters, head boxes, and showers. One difficulty encountered with recirculation of clarified white water has been slime growths, both in the mixture and on equipment. This greatly reduces the paper machine rates and lowers the value of the paper produced. Chlorination, organic mercurials, and environmental controls (pH and temperature) are used to control these growths [9].

Sulfite pulp mills use various methods of recovery. Equipment has been developed in which the sulfite waste liquor can be burned to produce enough steam to run the evaporator. (This process does not produce a salable by-product, but merely eliminates the waste problem.) Complete evaporation of the sulfite waste liquor produces both a fuel which can be burned without an additional outside fuel supply and a salable by-product—used in making core binder, insecticides and fungicides, linoleum cement, road binder, road-bank stabilizer, ceramic hardener, boiler compounds, synthetic vanillin, and other useful by-products. However, the main problem associated with these by-products has been the fact that the market in this country cannot absorb more than 5 to 10 per cent of them. Since evaporation is costly, owing to the low initial concentration of solids in the sulfite waste liquor and the boiler-scaling difficulties encountered, operation of such a process is limited to mills close to users of waste liquors.

In addition to those by-products obtained by evaporating sulfite waste liquor, other valuable by-products are obtainable by other processes. The liquor may be fermented to produce ethyl alcohol; about 40 liters of alcohol can be produced per ton of dry solids. This process reduces the BOD of the liquor by utilizing the simple sugars alone. Acetone and butyl alcohol can also be produced from the waste, with an overall BOD reduction of about 82 per cent. However, in 1950 only one mill in the United States manufactured ethyl alcohol from sulfite waste liquor, and the situation remains similar in 1969. The drawback of this process is that it costs more to manufacture alcohol from the sulfite-liquor waste than from blackstrap molasses or ethylene.

Another product of fermenting the liquor is yeast for cattle feed. In 1948 a plant was built in Wisconsin to produce yeast fodder by this method and it achieved a 60 to 70 per cent BOD reduction. Unfortunately, the market was found to be very limited, because of competition with brewers' yeast. Laboratory experiments have been used to produce torula yeast for stock feed, with a resulting reduction of 40 per cent in BOD; 350 pounds of yeast are obtained from a ton of waste solids.

Recovery is also practiced in kraft mills. The black liquor (spent cooking liquor) is processed by evaporation and incineration, in order to recover chemicals and to utilize the heating value of the dissolved wood substances. During the recovery process, Na_2SO_4, with or without added sulfur, is added to replace the relatively small proportion of chemicals lost in the various steps of the process. Following these additions and the incineration, the smelt is dissolved in water to form "green liquor." The chemical compounds in the green liquor are converted to the desired cooking chemicals by the addition of lime, so as to form of "white liquor" and a lime mud consisting chiefly of $CaCO_3$. The white liquor is returned to the pulping operation as the cooking liquor and the lime mud is calcined to form calcium oxide, which is reused in converting other green liquor to white liquor. By-product recovery of turpentine, resin, and fatty acids also aids in reducing the strength of kraft waste-water effluents. Maximum recovery of these by-products may result in effluents in which chemical toxicants are no longer a significant factor, as far as stream pollution is concerned. The turpentine is recovered from the digester relief gases, which also contain small quantities of $(CH_4)_2S$, dimethyl sulfide, methyl mercaptan, and ketones. The black liquor also contains recoverable quantities of sodium salts, rosin, and fatty acids, which separate during the concentration and cooking of the black liquor. This material is called "crude sulfate soap"; after it has been skimmed from the black liquor, it is treated with acid to form tall oil. The resin and fatty acids are further refined and have a variety of applications in industry.

Sedimentation and flotation. These treatments are achieved through the use of save-alls, which have been described earlier. Although save-alls are used on paper-machine white waters primarily for purposes of clarification (for white-water reuse) and fiber recovery, they may still be considered as part of process equipment. Removal of the fiber naturally results in decreased loss of solids to the sewer, and therefore effluents of lower pollutional strength. Another type of equipment which has resulted in lower sewer losses of black liquor is the foam trap or foam breaker, which prevents foam from spilling over and carrying the entrained liquor into the sewer.

Sedimentation is the usual method of total and final treatment of paper-mill effluents, the save-alls being restricted to usage within the mill. In spite of the use of in-plant save-alls, there are losses to the drains which cannot be prevented. Modern concrete or steel sedimentation tanks, usually circular, have been installed at many paper mills, where they give good service. Diameters vary from about 20 to 120 feet; solids discharged have a concentration of 4 to 15 per cent, with an average of 6 to 8 per cent. The fillers present in the effluent make final clarification difficult; effluents usually contain at least 30 ppm suspended solids, unless upflow-type clarifiers are used.

Chemical precipitation to remove colloids and color. The use of chemicals to treat paper-mill waste has generally been avoided, since it increases the quantity of sludge which must be disposed of. However, some mills have used this method. An Indiana mill, using sulfite pulp, treats all wastes with alum. The sludge is dried on beds, and the effluent is recirculated to the process. A Michigan mill uses chemical precipitation and obtains a BOD reduction of 64 per cent—somewhat higher than usual.

The Howard process is a precipitation, with lime as a coagulant, in three stages to a final pH of 11. In the first stage, calcium sulfide settles out and is returned as a slurry to the cooking-liquor make-up. In the second stage, lignin is precipitated and converted to a cake on a rotary filter. In the third stage, settling removes any colloidal material that is left when the pH is raised to 11. About 40 per cent BOD removal is

obtained, although claims for somewhat higher efficiencies have been made. The process is one of recovery and not actual treatment, even though chemical precipitation is used. The lignin is used both as a fuel and in the manufacture of plastics, the production of tannins, as an antiscale or antifoam agent in boilers, and in manufacturing synthetic vanilla. The Strellenert process is similar to the Howard process, except that: the precipitant is gypsum, and the liquors are heated to 160°C in pressure vessels; sulfur dioxide is evolved and reused in cooking liquors; lignin is precipitated and used for fuel, mainly in countries outside the United States.

The problem of color removal from kraft-mill effluents has become increasingly pressing. It is possible to remove color by treating the effluents with high dosages of hydrated lime, but this results in large volumes of hydrous sludges, which are extremely difficult to dewater. This has been a major hindrance to solving the color-removal problem. Because of the high lime dosages required, economic considerations have dictated the recovery of calcium from the sludges for reuse. A method of sludge treatment involving carbonation and heat is being investigated for calcium recovery.

A study of various methods of chemical treatment of rope-mill wastes [11] showed that, in general, most of the coagulants causing good clarification also produced excessive quantities of sludge and effected relatively little BOD reduction. A three-stage chemical treatment using $FeCl_3$, H_2SO_4, and alum in successive stages resulted in a total BOD reduction of 50 per cent, and an accumulated sludge volume of 27 per cent, with a turbidity reduction greater than 82 per cent, and a practically colorless effluent. All other single coagulants or combinations of chemical coagulants produced lower BOD reductions.

Activated-sludge treatment. Aerobic biological processes have been most successful on kraft-mill wastes. Whether this is due to the characteristics of this particular waste, or to the fact that the pollution problems of kraft mills are more pressing than those of other paper or pulp mills, or to more progressive viewpoints on the part of kraft-mill administrators, is still questionable; one modified version, dispersed-growth aeration, has been successful on a laboratory scale in treatment of specialty-paper-mill wastes. In recent years a promising accelerated treatment for kraft wastes has been developed, embodying the principle of activated sludge as used for domestic-sewage treatment, which requires the addition of nitrogen salt. Its greatest shortcomings are, however, the capital and operating costs involved when the process is applied continuously.

A Virginia kraft mill utilizes this process on its normal 16-mgd flow, containing an average BOD of 140 ppm. Aeration for three hours is provided, with a 25 per cent return of sludge. Based on average concentration, the BOD loading of the aerator is 56 pounds per 1000 cubic feet, and mixed-liquor concentrations vary between 0.2 and 0.3 per cent. Air requirements approximate one cubic foot per gallon of waste. Nitrogen and phosphorus nutrients are added as needed, to maintain one pound of available nitrogen and phosphorus for each 20 and 75 pounds of BOD respectively. During its initial operation, in June 1955, this plant accomplished an 85 per cent BOD reduction, while treating 60 per cent of the wastes of the kraft mill.

Experiments with aeration, in rag, rope, and jute mills, involved treating their cooking liquors and wash-water wastes with sludge and dispersed-growth seeds under optimum conditions, and showed these methods of treatment to be effective [11]. Aeration in the presence of sludge floc of mixtures of cooking liquors and the wash water from the first hour of washing appeared to be slightly more effective (but less dependable) than aeration in the presence of dispersed growth, as far as reduction of BOD was concerned. Important factors affecting the rate of oxidation of the wastes are pH, nutrients (N and P), seed adaptation, air supply, and temperature. At optimum conditions, BOD reduction of mixed wastes with a raw BOD concentration of about 2000 ppm, in the presence of dispersed-growth seed, was 78 to 96 per cent; with sludge floc, 90 to 98 per cent after 24 hours of aeration, depending primarily on speed and efficiency of adaptation of seed. Foaming of rag-mill wastes produced an operating difficulty during the aeration process.

Lagooning. The major method of pulp- and paper-mill final waste treatment is lagooning. Ponds, the most widely used form of lagoons, can be used for storage or as effluent-stabilization devices, with storage ranging from 10 days to 10 months. When ponds are used for stabilization purposes, retention periods of 10 to 30 days are common. Because of the tremendous size of ponds, their use depends on availability of land. Also, utilization of these ponds is usually seasonal and depends, as well, on the rate of flow of the receiving stream, with retention of wastes during low water flow and discharge during high flow.

A new kraft mill in Georgia handles its wastes in alternating settling basins, followed by two oxidation lagoons in series. The settling basins receive fiber-bearing waters after passage through flotation save-alls. Strong, nonfiber-bearing wastes are collected in an equalizing basin, from which they are pumped at controlled rates and blended with the effluent from the settling basins. The main object of this type of treatment is to remove the last traces of suspended matter and proportion the equalized waste to the receiving stream according to the flow. However, only limited BOD removals are obtained with lagoons, and odors and contamination of well waters must be taken into consideration.

High-pressure oxidation. The Zimmerman process (see Chapter 14) has been used successfully for treating sulfite-mill wastes in Norway [14], but economic factors must be studied carefully before such a high investment in capital equipment can be made in this country. The atomized suspension technique, utilizing a similar principle, has shown promise for treatment of concentrated, organic-waste liquors [15, 16]. Swedish mills appear to be more advanced in their use of this method of treatment than United States mills.

The current trends in waste treatment in the pulp and paper industry may be summarized in ten areas as follows:

1. In-plant changes have been successful in reducing both strength and quantity of waste. Extensive research is being undertaken on re-use and by-product recovery.

2. Disposal of spent liquor by deep well disposal or by conversion to salable products has been reported.

3. Sedimentation, flotation, and thickening: there has been considerable use of moving screens. In all of these processes, more and more attention is being paid to fiber recovery.

4. Chemical coagulation: contact flocculation, alum or ferrous salts, and activated silica have proved effective.

5. Solids handling: vacuum filtration, centrifugation, thickening, straining, pressing, incineration, wet-air oxidation, and landfill have been used.

6. Treatment of receiving waters: the paper industry, perhaps more than any other, has been emphasizing consideration of such techniques as: impoundment, intermittent storage at low flow, diffusion of effluent, mechanical aerators within the stream, etc.

7. Biological treatment: activated sludge (and all modifications), trickling filtration, aerated lagoons, and anaerobic treatment are all being utilized, with the emphasis on activated sludge and aerated lagoons.

8. Irrigation disposal: excellent results have been reported on a number of crops.

9. Color removal: activated carbon adsorption.

10. Foam separation.

References: Pulp- and Paper-Mill Wastes

1. "Industrial wastes forum," *Sewage Ind. Wastes* **28**, 654 (1956).

2. Rudolfs, W., *Industrial Waste Treatment*, Monograph Series no. 118, Reinhold Publishing Corp., New York (1953), p. 195.

3. U.S. Public Health Service, *Industrial Waste Guide*, Ohio River Pollution Control Survey, Supplement D, Appendix X (1943) p. 1193.

4. Moggio, W. A., "Control and disposal of kraft mill effluents," *Proc. Am. Soc. Civil Engrs.* **80**, separate no. 420 (1954).

5. McDermott, G. N., "Sources of wastes from kraft pulping and theoretical possibilities of reuse of condensates," in Proceedings of 3rd Southern Municipal and Industrial Waste Conference, March 1954, p. 105.

6. "Rayonier's new sodium base process," *Paper Mill News* **82**, 12 (1959).

7. Crawford, S. C., "Lagoon treatment of kraft-mill wastes," *Sewage Ind. Wastes* **19**, 621 (1947).

8. National Council for Stream Improvement, *Report on Semichemical Wastes*, Report no. 24, New York (1949).

9. Rudolfs, W., and N. L. Nemerow, "Some factors affecting slime formation and freeness in board-mill stock." *Tech. Assoc. Pulp Paper Ind.* **33**, 7 (1950).

10. National Council for Stream Improvement, *Treatment of Strawboard Wastes*, Report no. 15, New York (1947).

11. Rudolfs, W., and N. L. Nemerow, "Rag, rope, and jute wastes from specialty paper mills," *Sewage Ind. Wastes* **24**, 661, 765, 882, and 1005 (1952).

12. Warrick, L. F., "Pulp and paper industry wastes," *Ind. Eng. Chem.* **39**, 670 (1947).

13. Palladino, A. J., "Deinking waste treatment plant design," *Sewage Ind. Wastes* **23**, 1419 (1951). evaporation," Canadian Patent no. 562250, 26 August 1958.

14. Zimmerman, F. J., "New waste disposal process," *Chem. Eng.* **65**, 117 (1958).

15. Gauvin, W. H., "Application of the atomized suspension technique," *Tech. Assoc. Pulp Paper Ind.* **40**, 866 (1957).

16. Pinder, K. L., and W. H. Gauvin, "Atomized suspension techniques," *Ind. Wastes* **4**, 26 (1959). (1952).

17. Bialkowsky, H. W., and P. S. Billington, "Pilot studies, effect of waste discharge, prediction," *Sewage Ind. Wastes* **29**, 551 (1957).

18. Billings, R. M., "Stream improvement through spray disposal of sulphite liquor at the Kimberly-Clark Corp., Niagara, Wis.," Purdue University Engineering Extension Series, Bulletin no. 96, 1958, p. 71.

19. Bishop, F. W., and J. W. Wilson, "Integrated mill waste treatment and disposal, description," *Sewage Ind. Wastes* **26**, 1485 (1954).

20. Bjorkman, A., "Recovery in the cellulose industry," *Svensk Papperstid.* **61**, 760 (1958); *Inst. Paper Chem. Bull.* **29**, 877 (1959).

21. Black, H. H., "Spent sulfite liquor developments," *Ind. Eng. Chem.* **50**, 95A (1958).

22. Black, H. H., and V. A. Minch, "Industrial waste guide, wood naval stores," *Sewage Ind. Wastes* **25**, 462 (1953).

23. Blandin, H. M., *et al.*, "Forum discussion, paper

mill wastes," *Sewage Ind. Wastes* **19**, 1108 (1947).

24. Blandin, H. M., K. H. Holm, and B. F. Stahl, "Workshop on paper manufacturing wastes," in Proceedings of 2nd Industrial Waste Conference, Purdue University, January 1946, p. 155.

25. Bloodgood, D. E., "Development of a method for treating strawboard wastes," *Tech. Assoc. Pulp Paper Ind.* **33**, 317 (1950).

26. Bloodgood, D. E., "Disposal methods of strawboard mill wastes," *Sewage Ind. Wastes* **19**, 607 (1947).

27. Bloodgood, D. E., "Tenth Purdue Conference highlights industrial waste problems," *Ind. Wastes* **1**, 33 (1955).

28. Bloodgood, D. E., and G. Erganian, "Characteristics of strawboard mill wastes," *Sewage Ind. Wastes* **19**, 1021 (1947).

29. Bloodgood, D. E., *et al.*, "Strawboard wastes, laboratory studies," *Sewage Ind. Wastes* **23**, 120 (1951).

30. Blosser, R. O., "BOD removal from deinking wastes," Purdue University Engineering Extension Series, Bulletin no. 96, 1958, p. 630.

31. Blosser, R. O., "Solids removal practices in the paper industry," in Proceedings of 7th Ontario Industrial Waste Conference, June 1960, p. 121.

32. "BOD reduction by heat hydrolysis," Sulfite Waste Research Report, Technical Bulletin no. 29, National Council for Stream Improvement, New York (1949).

33. Boyer, R. A., "Sodium-base pulping and recovery," *Tech. Assoc. Pulp Paper Ind.* **42**, 356 (1959).

34. Boyer, R. A., and S. R. Parsons, "Operation of the 1957 experimental sodium-base recovery plant at Consolidated Waste Power and Paper Company," *Tech. Assoc. Pulp Paper Ind.* **42**, 565 (1959).

35. Brookover, T. E., "A paper board mill's attack on stream pollution," *Tech. Assoc. Pulp Paper Ind.* **28**, 74–79 (1945).

36. Brookover, T. E., "The paper industry and stream pollution," *Paper Trade J.* **131**, 26 (1950).

37. Brown, H. B., "Water conservation," *Sewage Ind. Wastes* **29**, 1409 (1957).

38. Brown, H. B., "Effluent disposal problems and pollution abatement measures of the southern pulp and paper industry," *Southern Pulp Paper J.* **53**, 1–3, October 1955.

39. Brown, W. G., "Market potential for protein concentrate produced from fermentation of spent sulfite liquor," Technical Bulletin no. 110, National Council for Stream Improvement, New York (1958).

40. Brown, R. W., D. T. Jackson, and J. C. Tongren, "Semichemical recovery processes and pollution abatement," *Paper Trade J.* **143**, 28 (1959).

41. Buehler, H., Jr., "Waste treatment in a paper mill," in Proceedings of 12th Industrial Waste Conference, Purdue University, May 1957.

42. Burbank, N. C., and C. D. Eaton, "Pulp and paper mill waste treatment," *Tech. Assoc. Pulp Paper Ind.* **41** (Supplement 195*A*) 6 (1958).

43. Buswell, A. M., and F. W. Sollo, "Methane fermentation of a fiber board waste," *Sewage Works J.* **20**, 687 (1948).

44. Calise, V. J., and R. J. Keating, "Some economic aspects of white water treatment in pulp and paper mills," in Proceedings of 9th Industrial Waste Conference, Purdue University, May 1954.

45. Callaham, J. R., "Alcohol recovery from sulfite waste liquor," *Sewage Ind. Wastes* **16**, 388 (1944).

46. Carpenter, C., and C. C. Porter, "Waste water utilization by clarification," *Tech. Assoc. Pulp Paper Ind.*, **28**, 147–151 (1945).

47. Cawley, W. A., "Sphaerotilus in streams," *Sewage Ind. Wastes* **30**, 1174 (1958).

48. Cawley, W. A., and C. C. Wells, "Lagoon system for chemical cellulose wastes," *Ind. Wastes* **4**, 37 (1959).

49. Cederquist, K. N., "Some remarks on wet combustion of cellulose waste liquor," *Svensk Papperstid.* **61**, 114 (1958).

50. Krofta, M. and L. K. Wang, "Total Closing of Paper Mills with Reclamation and Drinking Installations," Proc. **43** Purdue Ind. Waste Conf., 673, (1988).

51. Ferguson, A. M. D. et al., "On-site Anaerobic Treatment Proves Higher Design Loading Rate at Lake Utopia Paper Limited," Proc. **43** Prudue Ind. Waste Conf., 689, (1988).

52. Bagchi, A., "Improving Stability of a Paper Mill Sludge," Proc. **42** Purdue Ind. Waste Conf., 137, (1987).

53. Knocke, W. R. et al., "Treatment of Pulp & Paper Mill Wastewaters for Potential Water Reuse," Proc. **41** Purdue Ind. Waste Conf., 421, (1986).

54. Khan, M. Z. A., "Conducting of Pulp and paper Sludge Using Direct Slurry Freezing," Proc. **41** Purdue Ind. Waste Conf., 429, (1986).

55. Naylor, L. M. and Schmidt, E. J., "Effects of Papermill Wood Ash on Chemical Properties of Soil," Proc. **41** Purdue Ind. Waste Conf. 437, (1986).

56. Velasco, A. A. et al., "Full Scale Anaerobic-Aerobic Bi-

ological Treatment of a Semichemical Pulping Wastewater." Proc. **40** Purdue Ind. Waste Conf., 297, (1985).

57. Roberts, K. L. and J. Schoolfield, "Removal of Color from Kraft Pulp Mill Effluent," Proc. **40** Purdue Ind. Waste Conf. 305, (1985).

58. Cocci, A. A. et al., "Pilot-Scale Anaerobic Treatment of Peroxide Bleachery Waste, *Paper Machine Effluent and Waste Activated Sludge," Proc. **40** Purdue Ind. Waste Conf., 335, (1985).

59. Jackson, M. L., "Evaluation of Wastewater Treatment Process by a New Technique," Proc. **40** Purdue Ind. Waste Conf., 343, (1985).

27.2 Builders Paper and Roofing Felt Manufacturing Waste

Because felt manufacturing is a relatively small industry, not much attention has been given to its specific wastes. The waste generated from builders paper and roofing felt manufacturing plants consist of oil, grease, high pH, and large quantities of suspended solids. These pollutants are developed during the saturating and the deadening of the paper.

Treatment. Possible waste disposal solutions are (a) set settleable solids limitations; (b) restrict "bleed-off" wastewater containing a heavy load of settleable solids without some form of primary treatment; (c) increasing recycling of wastewater; and (d) send wastes to municipal treatment plants.

References

1. Anonymous, "Paving and roofing materials point source category," Federal Register **40**, 31190, September 1975.

2. Morris, D. C., "Effects of wastewater recycle in a paperboard mill," *J. Water Poll. Contr. Fed.* **45**, 9 (1973).

3. Quarles, J. "Builders paper and roofing felt segment on the builders paper and board mills," *Federal Register* **39**(No. 91), Part III, May 1974.

27.3 Photographic Wastes

Waste water from a large-scale film-developing and printing operation consists of spent solutions of developer and fixer, containing thiosulfates and compounds of silver. The solutions are usually alkaline and contain various organic reducing agents. Treatment usually consists of silver recovery carried out by the industry and subsequent treating of developer waste in combination with domestic sewage. Studies of two such plants* showed that the effect of developer waste on the treatment of sewage is insignificant if the ratio of this waste to sewage is comparatively low.

The Eastman Kodak Corporation† estimates the value of silver bullion recovered from photographic-processing waste at about $1.29 per troy ounce. It informs its processing plants that $800 worth of silver can be obtained by processing of the silver wastes from 100,000 rolls of black and white film and $2400 from a similar quantity of Kodacolor-X film. These monetary recoveries are based upon 100 per cent recovery efficiency and do not include any of the normal costs of waste treatment.

There are three methods generally available for silver recovery: metallic replacement, electrolysis, and chemical precipitation. Metallic replacement involves bringing spent hypo solutions into contact with a metal surface such as steel stampings, zinc, steel wool, and/or copper. After 66 to 99 per cent recovery of silver is completed, the metal is removed from the tanks, dried, and sold to the refiner. The electrolytic method consists of placing a cathode and an anode in the silver-bearing hypo solution. When an electric current is passed between these electrodes, silver plates out on the cathode. The permissible current density depends on whether sodium or ammonium hypo is used in the bath and on the acidity, silver, and sulfite concentration in the solution. Turbulence in the solution furnishes a continuous supply of silver-laden fixer solution to the cathode. To handle 50 gallons of solution containing 19 ounces of silver, 32 square feet of cathode area are required, along with compressed air agitation. Twenty hours of such operation will remove 98 per cent of the silver and can recover about 600 troy ounces of silver before the cathode has to be desilvered or replaced. Several compounds can be used in chemical precipitation. However, one common method utilizes a combination of sodium hydroxide and sodium sulfide to achieve

---

*"Treatment Data for Photographic Wastes," *Public Works* **85**, 104 (1954).

†*Recovering Silver from Fixing Baths*, Pamphlet No. J-10 (3-66 minor revision), Eastman Kodak, Rochester, N.Y.

the precipitation of silver sulfide. The sludge is dried and sold to the most convenient refiner. Typical quantities of precipitating chemical required are 1 ounce of sodium hydroxide (2 lb of NaOH per gallon) and 1 ounce of sodium sulfide (2 lb of Na_2S per gallon) for each gallon of waste fixer.

Additional references (Since 1962)

1. Anonymous, "Photo processors meet pollution control requirements," *Graphic Science* 12, 15–17 (1970).
2. Dannenberg, R. O. and G. M. Potter, "Silver recovery from coated photographic solutions by metallic displacement," *U.S. Bur. Mines, Rep. Invent.* 7117, 22 (1968).
3. Fritz, H., and F. Meier, "Treatment of wastewater from fiber developing plants," *Chem. Zetig.* 95, 467–471 (1971).
4. Geisler, F., "Continuous recovery of silver halides and silver," *Germany* 71, 442 (1970).
5. Hennessy, P. V., D. G. Rosenberg, and R. G. Zehnpfenning, "Photographic laboratory waste treatment," *Water and Sewage Works* 115, 131–135 (1968).
6. Hillis, M. R., "Electrolytic treatment of effluents," *Effluent Water Treatment J.* 9, 647–654 (1970).
7. Kay, M., F. J. Quinn, N. W. Marshal, and H. Meikle, "Recovery and pollution control by R. O.," *J. Soc. Motion Picture and Television Eng.* 81, 461–464 (1972).
8. Mohanrao, G. J., K. R. Krishnamurthi, and W. M. Deshpande, "Photofilm industry wastes; pollution effects and abatement," *Proc. 3rd Int. Conf. Water Pollution Resources, Munich* 1, 18–205 (1966).
9. Mahoney, J. G., M. E. Rowley, and L. E. West, "Industrial waste treatment opportunities for reverse osmosis," *Membrane Sci. and Tech.—Industrial Biological and Waste Treatment*, Flinn, J. E. (ed.), New York: Plenum Press (1970).
10. Robles, E. G. Jr., "Sulfite and thiosulfate in water or photographic waste waters," *U.S. Nat'l Tech. Info. Service*, Reprint No. 752528, 8 (1973).
11. Ullmicher, W., "Reclaiming silver halides from photographic waste water," Ger. Pat. 1258733 (cl. Go3c), Jan. 11, 1968, Appl. July 21, 1962.
12. Wing, B. A. and W. M. Steinfeldt, "A comparison of stonepacked and plastic packed trickling filters," *J. Water Poll. Contr. Fed.* 42, 255–264 (1970).
13. Zehnpfenning, R. G., "Possible toxic effects of photographic laboratory wastes discharged to surface water," *Water and Sewage Works* 115, 136–138 (1968).

METAL INDUSTRIES

Metal wastes include wastes from refining mills, plating mills, and parts washing and encompass a wide range of materials. For example, there are wastes not only from the manufacturing of steel, but from many other metals (copper and aluminum, to name two); wastes are produced from renewing surfaces on used metallic parts, such as airplane engines prior to their return to service; the coating of one metal with another, for protective purposes, for example, the plating of silverware or business machines, should be included as an intermediate process. The wastes from all three sources are similar, in that they possess various concentrations of metallic substances, acids, alkalis, and grease. They are characterized by their toxicity, relatively low organic matter, and greases.

27.4 Steel-mill Wastes

In modern steel production, there are three types of steel manufacturers: integrated producers, minimills, and specialty-steel mills. The first start with iron ore and coal and end up with many shapes of steels and amount to about 70% of the U.S. market. The minimills reprocess scrap steel, usually into some low-quality products, make up the remainder. The specialty steel mills, although similar to minimills, are smaller and manufacture much more expensive items. Both the minimills and specialty-steel mills have been able to take advantage of new technology for refining and casting. Some of the new technologies being used or considered include the following.

Direct ironmaking is a one-step production of molten iron from iron ore and coal. It promises major reductions in capital and operating costs over those in integrated mills. Other innovations center on transforming molten steel in one step to finished product, thus freeing capital, energy and labor costs associated with sequential operations.

Continuous casting can be used to produce steel sheets a few millimeters thick rather than rolling them laboriously from ingots of slabs. Direct ironmaking consists of a vessel in which iron ore and coal are charged along with an injection of oxygen. The molten iron produced is similar to that made in a blast furnace of the integrated mill. Coal serves the dual purpose of a reducing agent and fuel to melt the iron. Szekely (8, 1987) gives a schematic comparison of the three types of steel production as shown in Figure 27.8.

Origin of steel-mill wastes. Steel-mill wastes come

mainly from the by-product coke, blast-furnace, rolling-mill, and pickling departments. Wastes contain cyanogen compounds, phenols, ore, coke, limestone, acids, alkalis, soluble and insoluble oils, and mill scale. Wastes are treated by recirculation, evaporation, benzol extraction, distillation, sedimentation, neutralization, skimming, flotation, and aeration.

The "by-product" coke process. Coal is heated in the absence of air to produce coke and other products. This may be done in an integrated steel mill or in a separate plant proximate to, but not at the same site as, the steel mill. The cooking process evolves a gas, the further processing of which leads to the major wastes from this process; tar and ammonia are its main constituents. The following products are obtained from the burning in retorts of a ton of coal: coke, 1300 to 1525 pounds; $(NH_4)_2SO_4$, 17 to 26 pounds; tar, 5 to 12 gallons; gas, 10,800 to 11,300 cubic feet; phenol, 0.1 to 2.0 pounds; light oil, 2 to 3 gallons; naphthalene, 0.5 to 1.2 pounds.

The major wastes from preparation of the coke product itself come from the quench tower, where the hot coke is deluged with water. The coke dust present in this quenching water is called "breeze" and is commonly recovered from the water. A schematic drawing of the coking process is presented in Fig. 27.4.

The blast furnace. The wet scrubbing of blast-furnace gas evolves waters laden with flue dust. The wet scrubbers are downflow water sprays which clean the dust from the upflowing gases, an operation which is usually an intermediate stage between dry (or cyclone) dust separation and final electrostatic precipitation of the remaining fine particles. Secondary gas washers or precipitators are periodically cleaned by flushing with water, thus adding to the flow of discolored water.

The pickling process. Before applying the final finish to steel products, the manufacturer must remove dirt, grease, and especially the iron-oxide scale which accumulates on the metal during fabrication. Normally this is carried out by immersing the steel in dilute sulfuric acid (15 to 25 per cent by weight). This process, known as "pickling," produces a waste called "pickling liquor," composed mainly of unused acid and the iron salts of the acid (Fe^{+++} and Fe^{++}). The acid reacts with the iron salts, forming $FeSO_4$. As the acid is used, it becomes weaker and must be renewed. However, at a certain point, the concentration of $FeSO_4$ increases to such a degree that it inhibits the action of even a high concentration of sulfuric acid. At this point, the pickling liquor must be discharged and replaced by a fresh batch of sulfuric acid. It is this pickling-liquor waste which has received such wide publicity from an industrial-waste standpoint.

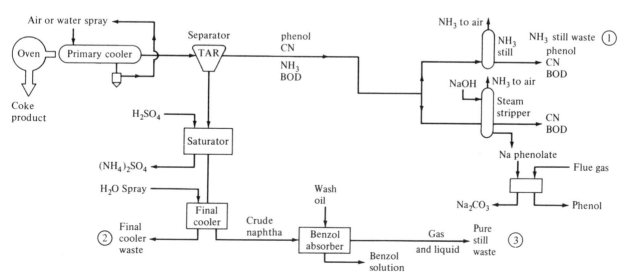

Fig. 27.4 Three major coke plant wastes.

A recent report by the U.S. Department of the Interior [1967, 4] gives the following pertinent data on the steel industry. Gross water intake in 1964 was 3815 billion gallons; gross water use, including recirculation and reuse, was 5510 billion gallons. Overall water reuse was thus 42.1 per cent. Cooling water reuse is estimated to have been 52.2 per cent and process water reuse 18.1 per cent. Water uses by the industry are summarized in the following table, for the year 1964, in billions of gallons.

The general trend in the steel industry is toward the use of subprocesses which will produce products of lighter unit weight at increasingly high speeds with minimum manual operation. Production units tend to become larger in order to realize economies of scale.

Characteristics of steel-mill wastes. Important wastes from the by-product coke phase of steel-mill operation come from the ammonia still, the final cooler, and from the pure still, where products such as benzene, toluene, and xylene are made from the crude naphthalene. Phenol and oxygen-demanding matter are the primary contaminants. A summary of the major constituents of these coke-plant wastes is given in Table 27.15

The blast furnace wet-scrubber effluent contains flue-dust solids, from washing the gas, composed of iron oxide, alumina, silica, carbon, lime, and magnesia. The amount of each constituent, in comparison with the total quantity of dust, varies with the type of ore used in the furnace, conditions of the furnace lining, the quality of coke used, the number of furnaces in blast, the amount of air being blown, and the regularity and thoroughness of dumping and flushing of dry dust catchers [2]. Fe_2O_3 comprises about 70 per cent, and silica about 12 per cent, of the flue-dust content. Some pertinent analyses of the physical characteristics of flue-dust wastes are given in Table 27.16 [2].

The amount of waste pickling liquor per ton of steel product depends on the size and type of plant. Steel production in the United States in 1948 was more than 11 million tons, with an estimated 600 million gallons of pickle-liquor waste [3], or about 55 gallons per ton of steel; whereas in Germany the figure ranges from 25 to 200 gallons per ton. One factor that increases the volume is that, since the steel products must be rinsed in water after they leave the pickling tank to remove all trace of acid, the rinse or wash water eventually becomes quite acidic and must also be discarded. The volume of rinse water is 4 to 20 times that of the actual pickling liquor, although naturally it is far more dilute. Wash waters contain from 0.02 to 0.5 per cent H_2SO_4 and 0.03 to 0.45 per cent $FeSO_4$, as compared with 0.5 to 2.0 per cent H_2SO_4 and 15 to 22 per cent $FeSO_4$ in the pickling liquors. Thus, H_2SO_4 and $FeSO_4$ in these ranges of concentration are the major contaminants in the wastes from pickling and washing of steel.

Industry water uses in 1964, billions of gallons [4].

Manufacturing process	Process water	Cooling water
Blast furnaces	276	586
Open-hearth furnaces	1	491
Basic oxygen furnaces	8	37
Electric furnaces	1	63
Hot-rolling mills and related	468	468
Cold mills and related	264	
Coke plants	6	632
Sanitary uses, boilers, etc.		254
Blowers, condensers, etc.		1955
	1024	4486

Treatment of steel-mill wastes. The primary method of treatment of by-product coke-plant wastes is to use recovery and removal units with high efficiencies, phenol being the main contaminant recovered. The BOD can be reduced by about one-third by the practice of recirculation and reuse of contaminated waters, and by-product recovery may be undertaken for profit in the case of such materials as ammonium sulfate, crude tar, naphthalene, coke dust, coal gas, benzene, toluene, and xylene. Quench water is usually settled to remove coke dust, and the supernatant liquor from the settling tanks is reused for quenching. Gravity separators are used to remove free oil from the wastes

Table 27.15 Analyses of by-product coke-plant wastes [4].

Characteristic	Source of wastes			
	Ammonia still	Final cooler*	Pure still	Combined
BOD, 5-day, 20°C	3974	218	647	53–125†
Total suspended solids, ppm	356		125	89‡
Volatile suspended solids, ppm	153		97	
Organic and NH₃-N, ppm	281	14	20	
NH₃-N, ppm	187		10	
Phenol, ppm	2057	105	72	6.4§
Cyanide, ppm	110			
pH	8.9		6.6	

*No recirculation.
†Depending on compositing technique.
‡Average of 11 daily 24-hour composites, including coke breeze.
§Single-catch sample.

Table 27.16 Flue-dust content of wet-washer effluents [2]

Characteristic	Value or concentration
Suspended solids content	
Range, ppm*	500–4500
Per cent by weight passing 100-mesh sieve	86–99
Per cent by weight passing 200-mesh sieve	74–97
Temperature, °F	100–120
pH	6–8
Specific gravity	3–3.8

*1200 ppm is average at Fairless Steel plant.

from benzol stills, since the emulsified oils are generally not treated and without separation the free portion of the oil would thus reach the sewers. Final cooler water is also recirculated, to reduce the amount of phenol being discharged to waste. Phenol is recovered primarily to prevent pollution of streams and to avoid the nuisance of taste in water supplies. Phenols may be removed by either conversion into nonodorous compounds or recovery as crude phenol or sodium phenolate, which have some commercial value. The conversion may be either biological (activated sludge or trickling filtration) or physical (ammonia-still wastes used to quench incandescent coke, a process which evaporates the NH₃). Although certain concentrations of phenol (0 to 25 ppm) may be handled by biological units, dilution with municipal sewage is a good idea, since this provides a buffering and diluting medium. The Koppers dephenolization process [1] lowers the phenol content by 80 to 90 per cent in ammonia-still wastes. The process, as shown in Fig. 27.5, is essentially a steam-stripping operation, followed by mixing in a solution of caustic soda and renewing pure phenol with flue gas.

In treating flue dust, sedimentation, followed by

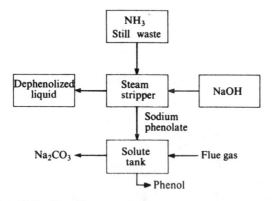

Fig. 27.5 The Koppers dephenolization process, using steam, caustic soda, and flue gas.

thickening the clarifier overflow with lime to encourage flocculation, has been found most effective for removing iron oxide and silica. Ninety to 95 per cent of the suspended matter settles readily and does so within a one-hour period, the resulting effluent having less than 50 ppm suspended solids. Primary and secondary (lime-coagulated) thickened sludges are also obtained, which can then be lagooned without creating nuisances. Henderson and Baffa [33] give details of a typical blast-furnace waste-treatment process (Fig. 27.6).

The treatment of pickling liquor is a problem of considerable magnitude. For most *small* steel plants, the recovery of by-products from waste pickling liquor is not economically feasible and they neutralize the liquor with lime. However, some companies do obtain by-products from this waste, namely: (1) copperas and $FeSO_4 \cdot H_2O$; (2) copperas and H_2SO_4; (3) $FeSO_4 \cdot H_2O$ and H_2SO_4; (4) $Fe_2(SO_4)_3$ and H_2SO_4; (5) Fe^{+++} and H_2SO_4; (6) iron powder; (7) Fe_3O_4 for polishing or pigments; (8) Fe_3O_4 and $Al_2(SO_4)_3$. These are described in more detail by Hoak [5].

The recently developed Blaw-Knox-Ruthner process for the recovery of sulfuric acid involves the

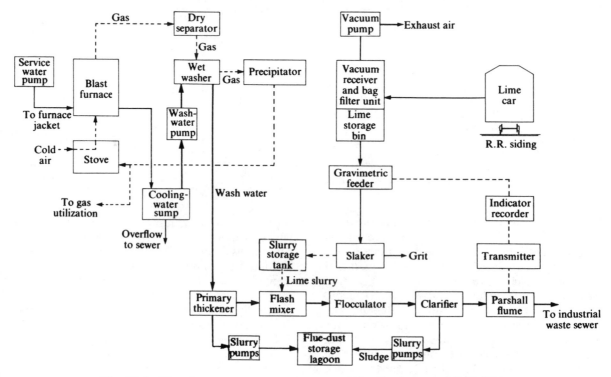

Fig. 27.6   Blast-furnace waste treatment process. (After Henderson and Baffa [2].)

concentration, by evaporation, of waste pickling liquor before it is discharged to a reactor, where anhydrous hydrogen chloride gas is bubbled through it, reacting with the ferrous sulfate to produce H_2SO_4 and $FeCl_2$. The ferrous chloride is separated from the sulfuric acid (which is returned to the pickling line) and is converted to iron oxide in a direct-fired roaster. This liberates HCl, which is recovered by scrubbing and stripping and is then recycled to the reactors. This process, shown in Fig. 27.7, has been successfully demonstrated in a pilot plant cooperatively run by many of the large steel producers [3].

Neutralization of pickle-liquor waste with lime is costly, because there is no saleable end-product and there is a voluminous, slow-settling sludge which is difficult to dispose of. Neutralization takes place in four stages: (1) formation of ferric hydrate with a pH below 4, (2) formation of acid sulfate, (3) formation of the ferrous hydrate with a pH between 6 and 8, and (4) formation of the normal sulfate. Calcium and dolo-

mitic lime are the least expensive neutralizing agents, caustic soda and soda ash being too expensive for such a purpose. Even with the cheaper chemicals, Hoak [5] concludes that the overall cost of neutralization ranges from \$5 to \$10 per 1000 pounds of acid from the pickling operation. Much research has been done on reducing the cost of neutralization by increasing the basicity of the neutralizing chemicals and by various methods of decreasing sludge volumes. However, research should be directed toward finding new methods of treatment, rather than relying on neutralization only.

Three areas of change in the pickling-waste problem are: (1) improvements in the treatment of waste from pickling with H_2SO_4; (2) a new HCl pickling operation; and (3) a new dry descaling operation.

New treatment methods for H_2SO_4 pickling include *deep well disposal*, which costs \$500,000 (on average) per installation and \$1/1000 gallons to operate, with a removal efficiency of 85 per cent (based on the fact

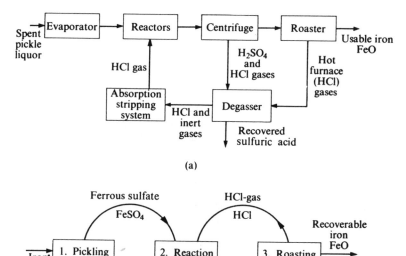

(a)

(b)

1. Pickling: $FeO + H_2SO_4 \rightarrow FeSO_4 + H_2O$
2. Reaction: $FeSO_4 + 2HCl \rightarrow FeCl_2 + H_2SO_4$
3. Roasting: $FeCl_2 + H_2O \rightarrow 2HCl + FeO$

Fig. 27.7 Blaw-Knox-Ruthner process for recovery of acid from spent pickle liquor: (a) process flow diagram; (b) chemistry of process. (Courtesy Blaw-Knox Co.)

that the rinse water is not treated and a small percentage of the pickle liquor in the well may escape as a pollutant), and *ion exchange*, which has a removal efficiency of 80 per cent [1965, 6] and costs as yet undetermined.

Hydrochloric acid pickling differs from H_2SO_4 pickling in the basic chemistry of the pickling action. Hydrochloric acid readily dissolves all the various oxides of iron in the scale, yet reacts relatively slowly with the base metal. The dissolved solids in the HCl pickle liquor are far below saturation concentration and the steel is left clean and free of crystals or insoluble slime. Sulfuric acid on the other hand acts at a high reaction rate with the parent metal and "blows" off oxides on the strip. Because of this, more scale-breaking is required before pickling. The benefits of HCl pickling are: easier regeneration of acid; no over-pickling and more flexibility on the line; elimination of the secondary scale breaker; higher pickling speeds; and a 20 per cent reduction in waste-water volume. The one disadvantage is the increased cost of HCl over H_2SO_4; however, on the whole it is definitely more desirable. Hydrochloric acid wastes are treated by deep well disposal and by neutralization. The capital costs for neutralization are $1 million for a plant with a capacity of 100,000 gallons per day and the operating cost are $20 per 1000 gallons of waste neutralized, with a removal efficiency of 80 per cent (considering the calcium salt residual and the fact that rinse waters are not neutralized). A third method of treatment is regeneration, which consists of the following processes: the pickle liquor is pumped to a spray roaster where water and free HCl in the pickle liquor are quickly driven off; the crystal descends inside the roaster while an increased temperature gradient roasts it, producing iron oxide and hydrogen

Treatment of three major wastes

NH₃ still (1)		Final cooler (2)	Pure still (3)
BOD	3974	218	647
suspended solids			125
CN	110		
NH₃	187		
ORGANIC	281	14	
+			
NH₄N	2057	105	72
PHENOL pH	8.9		
1. Recover NA phenolate with NAOH during steam distillation. Phenol is subsequently obtained by passing flue gas through. 2. Recirculate and reuse the contaminated NH₃ still wastewater for; (a) quenching (b) water sprays for primary and final coolers 3. Recover $(NH_2)SO_4$ by making acid with H_2SO_4 4. Biological treatment can be used to remove BOD and possibly to oxidize CN to $CO_2 + NO_2$.		1. Recirculate final cooler water. 2. Combine the blowdown from this cooler with residual NH₃ still waste and subject both to biological treatment.	1. Settle to remove suspended solids and free oil. 2. Send supernatant (which contains some emulsified oils and BOD) to biological treatment if is a very serious consideration in ultinate disposal.

chloride; the iron oxide is collected from the bottom of the roaster; some iron oxide is discharged in the gas and a cyclone is used to collect it; finally, the dry hydrogen chloride is recovered as HCl. For this method, the capital costs are $4 million for a plant with a capacity of 100,000 gallons per day; operation and maintenance costs are $8.80 per 1000 gallons of waste treated.

Abrasive descaling used on cold-rolled strip is done on two machines. The first uses steel spheres about 0.01 inch in diameter; the second uses even smaller angular grit. Both abrasives are cleaned continuously and recirculated [1966, 7].

Since coke plant wastes are so potentially polluting the characteristics of the three major wastes of Fig. 27.4 are given on p. 552 with suggested treatments for each.

References: Steel-Mill Wastes

1. American Iron and Steel Institute, *Annual Statistical Report* (1949).
2. Henderson, A. D., and J. J. Baffa, "Waste disposal at a steel plant: treatment of flue dust waste," *Proc. Am. Soc. Civil Engrs.* 80, separate no. 494, (1954).
3. Rathmell, R. K., "Pickle liquor treatment," *Finishing Abstr.* 8, 289 (1966).
4. U.S. Department of the Interior, *Blast Furnaces and Steel Mills*, Industrial Waste Profile, Vol. III, no. 1 Washington, D.C. (1967).
5. Hoak, R. D., "New developments in the disposal and utilization of waste pickling liquors," in Proceedings of 2nd Industrial Waste Conference, Purdue University, January 1946.
6. Poole, D. E., "Republic's continuous reclamation of HC1 pickling at Gadsden, Alabama," *Iron Steel Eng.* 42, 4160 (1965).
7. "Trends in steel pickling and waste acid treatment," *33 Mag. Metal Prod.* 4, 65 (1966).
8. Szekely, J. "Can Advanced Technology Save the U.S. Steel Industry?" *Scientific American.* 257, 34, (1, July 1987).
9. Stolzenburg, T. R. et al., "Analysis & Treatment of Reactive Waste: A Case Study in the Ductile Iron foundry Industry," Proc. **40** Purdue Ind. Waste Conf., 133, (1985).
10. Greenfield. J.H. and R. D. Neufeld, "Quantification of the Influence of Steel Industry Trace Organic Substances on Biological Nitrification," Proc. **36** Purdue Ind. Waste Conf., (1981).
11. Boyle, W.C. et al., "Leachate Tests on Selected Foundry Cupola Dusts and Sludges," Proc. **36** Purdue Ind. Waste Conf., 784, (1981).
12. Osantowski, R. and R. Hendriks, "Treatment of Steel Plant Blast Furnace Effluent Using Physical/Chemical Techniques," Proc. **35** Purdue Ind. Waste Conf., 725, (1980).

27.5 Other Metal-Plant Wastes

Processors of several other metals besides steel are significant waste contributors. Among these are brass, copper, gold, and aluminum plants, which are similar to steel mills in that impure metal is purified, worked, and fabricated into final usable products.

Brass and copper. The brass and copper industry produces plate, sheets, and strips by rolling operations, rods and wire by extrusion and drawing operations, and tubes by piercing or extrusion and drawing. Among the principal alloys used in these processes is what is called normal brass ($\frac{2}{3}$ copper and $\frac{1}{3}$ zinc, with small amounts of tin and lead). The molten metal comes from an electric furnace and is poured into molds of various sizes and shapes, to produce billets or bar castings for further operations. The bar castings are rolled into plates, sheets, and strips. After a certain amount of rolling, the metal becomes "hard" and must be annealed before further rolling, after which pickling is required to remove the resulting oxide scale or stain. Similar procedures must be carried out with rods, in order to form wire and tubes from billets. All processes require annealing followed by pickling.

Annealing is done in oil-fired furnaces; the alternate heating and cooling causes a rather heavy oxide scale to form on the surface of the metal. This scale must be removed before the rods, wire, and tubes are drawn, to prevent damage to the dies and sheets and so that the scale is not embedded in the final product. This is done by pickling in a bath of 5 to 10 per cent H_2SO_4 by volume. Stains, particularly on the finished product, are removed in a "bright dip" solution of 5 to 10 per cent H_2SO_4 and up to 0.5 pound per gallon of sodium dichromate. The metal leaving the pickling bath and bright-dip tank is washed with fresh water, which eventually overflows to waste. When the concentration of dissolved metal is too great for economical operation, the bath is dumped as waste. The frequency depends on the metal composition, length of pickling time, and amount of metal pickled; bright dips may be dumped daily, while pickle baths are usually dumped monthly. These two wastes constitute the main waste problems of this industry. Wise *et al.* [16] give the composition of samples collected from eight pickle tubs (Table 27.17) and bright-dip tanks (Table 27.18) during a brass-plant survey.

The most important methods of treatment appear to be precipitation of the metals as hydroxides in alkaline solutions or utilizing ion exchangers to recover valuable metals. Electrolysis is also sometimes used to recover or regenerate pure metals. The Bureau of Mines reports that high-grade alumina can be recovered economically from mineral-waste solutions such as copper-mine waste water. This method eliminates the cost of mining, crushing, and leaching the rock matter; thus, by-product recovery not only helps to solve a waste problem but also makes the new product more economical.

Copper mining and concentrating—typical of that done at the Cyprus Mining Company—produces three significant wastes as shown in Fig. 27.9.

Gold. Figure 27.10 illustrates the processing of gold bullion from gold-rich ores and the associated wastes.

Gold is found either in Placer form-free gold in veins or streams derived from the original ores after millions of years of erosion and degradation-or as an ore (alode) deposit. The latter form requires a great deal of capital (about $15,000 (1978)/ton of ore) and operating costs for refining while the former only requires stream separation of heavy, pure gold from other lighter gravels.

Aluminum. The production of various aluminum products by the Alcoa Company* illustrates typical processes of the aluminum industry and their associated wastes.

At the mines, power shovels scoop bauxite (the raw material of aluminum) out of the ground and dump it

---

*The author wishes to thank the Alcoa Company for the information in this section.

into railroad cars or trucks that haul it to a shipping point or directly to an alumina plant. Some bauxites are crushed, washed to remove some of the clay and sand waste, and dried in rotary kilns prior to shipment. Other bauxites are crushed and dried only, while bauxite from the Caribbean Islands is usually only dried. Bauxite delivered directly to an alumina plant is generally transported in its raw state.

In one manufacturing method, the Alcoa-Bayer process, the finely ground bauxite is fed into a steam-heated unit called a digester and a caustic solution made from soda ash and ground lime is added. This mixture is heated under pressure, whereupon the alumina dissolves, but the impurities do not. The mixture then flows through pressure-reducing tanks into a filter press, where cloth filters hold back the solids but allow the alumina-containing solution to pass through. The remaining solids, known as "red mud," are discarded. The liquid solution next passes through a cooling tower and into tall, silo-like tanks, called precipitators. Small amounts of alumina in crystal form are stirred into the liquid to serve as "seed particles" that stimulate the precipitation of solid alumina as the solution cools. This is aluminum oxide, chemically combined with water, and in this form is known as hydrated alumina. The hydrated alumina crystals that have formed in the solution are settled out and removed and the weakened caustic soda solution is returned to the digesters, where it is strengthened, mixed with new bauxite, and used over again.

The ore used in this process should be of high quality, which means that it should contain a low percentage of silica. Silica is a problem in bauxite refining because it combines with alumina and soda to form an insoluble compound that is filtered out with the red mud; thus valuable quantities of alumina and soda are lost. To recover this unintentionally discarded alumina and soda, and to make possible the use of ores containing substantial percentages of silica, the Alcoa Combination process was invented. In this process, the filtered-out red mud is treated to recover the alumina and soda that have combined with the silica. Limestone and soda ash are added and the mixture is heated to form a clinker-like "sinter." This sinter is treated with water which dissolves out the alumina–soda compound produced in the sintering step. The solution of alumina and soda in water is then returned to the digesters at the start of the Alcoa-

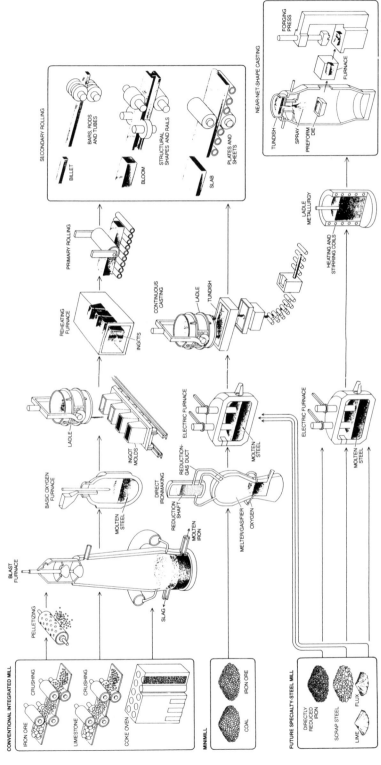

MANUFACTURE OF STEEL PRODUCTS varies markedly according to plant type: integrated mill, minimill or specialty-steel mill. Operations at integrated mills encompass both "front end" and finishing processes. The front end includes the preparation of coke, iron ore and limestone for charging a blast furnace, where the ore is reduced (stripped of oxygen) and heated to yield liquid iron and slag (a waste product), as well as the refining operations that convert the iron into steel. Finishing refers to the casting, reheating and rolling operations that transform the molten steel into shaped products such as rods, tubes, structural beams, plate and sheet. Minimills generally dispense with most of the front end of steelmaking and begin with steel scrap, but they could soon supplement the scrap with molten iron from direct-ironmaking plants. Direct ironmaking represents a substantially smaller investment than a blast furnace does and relies on iron ore and coal (rather than coke) as raw materials. The alternative is to buy directly reduced iron: iron from modern plants that reduce iron ore without melting it. Minimills were among the first steel mills to exploit continuous casting: the casting of billets, blooms or slabs directly from the molten steel. The process bypasses the casting of ingots and many of the reheating and rolling steps still carried out at some integrated mills. Specialty-steel mills are also primarily scrap-based, but unlike minimills, which tend to produce low-grade steels for a local market, they produce high-quality alloyed steels to customers' specifications. A future specialty-steel mill might impart still higher quality to its products by further refining and purifying the steel-alloy melt in special ladle-metallurgy stations and then casting the product in virtually its final shape.

Figure 27.8 Three Prevalent Steel Productions (after Szekely, 1987).

Source: After Sz. Ekely, August 8, 1987.

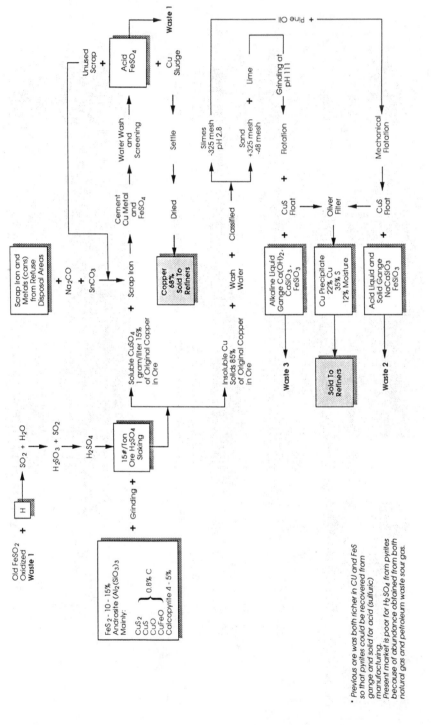

Figure 27.9   New Cyprus Mining Company copper mining and concentrated wastes flow diagram.

* Previous ore was both richer in CU and FeS
so that pyrites could be recovered from
gange and solid for acid (sulfuric)
manufacturing.
Present market is poor for H₂SO₄ from pyrites
because of abundance obtained from both
natural gas and petroleum waste sour gas.

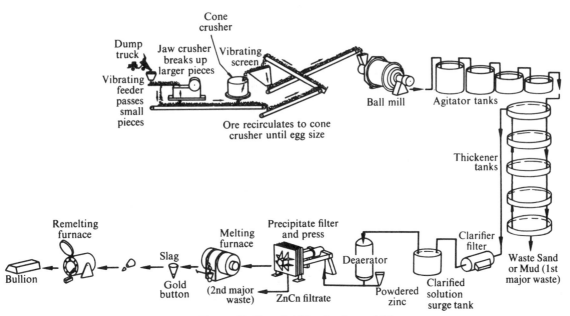

Figure 27.10 Gold Production and Wastes

Table 27.17 Pickle-bath wastes [16].

Acid and metal	gm/liter
H$_2$SO$_4$	59.7–163.5
Cu	4.0–22.6
Zn	4.3–41.4
Cr	0–0.56
Fe	0.1–0.21

Table 27.18 Bright-dip wastes [16].

Acid and metal	gm/liter
H$_2$SO$_4$	5.6–85.8
Cu	6.9–44.0
Zn	0.2–37.0
Cr^{+6}	4.3–19.1
Cr (total)	13.5–47.7
Fe	0.03–0.36

Bayer process.

In the final stages of alumina production, the hydrated alumina from the precipitators is filtered, washed, and dried in a long, slowly-revolving kiln at 1800°F to remove both the moisture associated with the hydrated alumina and the chemically combined water that is part of its crystalline structure. Hydrated alumina goes into the kiln with about 35 per cent combined water content, and comes out as a dry, white powder—commercially pure alumina.

It is generally conceded that there is about a 50%

conversion efficiency from bauxite to aluminina. The alumina can then be used also to produce electricity at a conversion rate of 4 KHW for each pound of alumina. This is shown in Figure 27.11.

Ingot, and scrap aluminum are converted into mill products—sheet, plate, bar, rod, wire, tube, extruded, and rolled shapes—that become the raw material for the finished goods manufactured by thousands of other companies.

Annodizing of aluminum produces a metal-acid and alkali wastes as shown in Fig. 27.12.

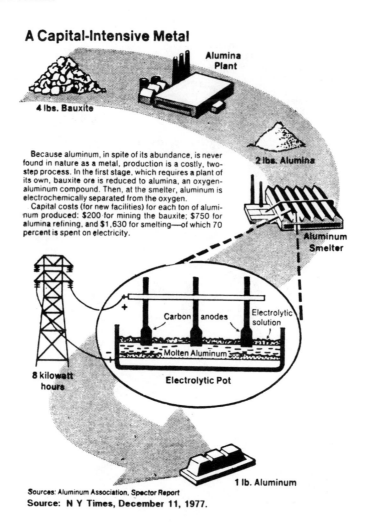

A Capital-Intensive Metal

4 lbs. Bauxite

Alumina Plant

2 lbs. Alumina

Aluminum Smelter

Because aluminum, in spite of its abundance, is never found in nature as a metal, production is a costly, two-step process. In the first stage, which requires a plant of its own, bauxite ore is reduced to alumina, an oxygen-aluminum compound. Then, at the smelter, aluminum is electrochemically separated from the oxygen.

Capital costs (for new facilities) for each ton of aluminum produced: $200 for mining the bauxite; $750 for alumina refining, and $1,630 for smelting—of which 70 percent is spent on electricity.

8 kilowatt hours

Carbon anodes   Electrolytic solution

Molten Aluminum

Electrolytic Pot

1 lb. Aluminum

Sources: Aluminum Association, Spector Report
Source: N Y Times, December 11, 1977.

Figure 27.11

References: Other Metal-Plant Wastes

1. Czensny, R., "Copper, toxicity to fish," *Sewage Ind. Wastes* 7, 760 (1935).
2. Griffith, C. R., "Lagoons for treating metalworking wastes," *Sewage Ind. Wastes* 27, 180 (1955).
3. Hill, H., "Effect on activated sludge process, nickel," *Sewage Ind. Wastes* 22, 272 (1950).
4. Hupfer, M. E., "Brass mill waste treatment," *Sewage Ind. Wastes* 29, 45 (1957).
5. Mitchell, R. D., *et al.*, "Brass and copper wastes, effect on sewage plant design, Waterbury, Conn.," *Sewage Ind. Wastes* 23, 1001 (1951).
6. McDermott, G. N., A. W. Moore, M. A. Post, and M. B. Ettinger, "Effects of copper on aerobic biological sewage treatment," *J. Water Pollution Control Federation* 35, 227 (1963).
7. McGarvey, F. X., R. E. Tenhoor, and R. P. Nevers, "Cation exchangers for metals concentration from

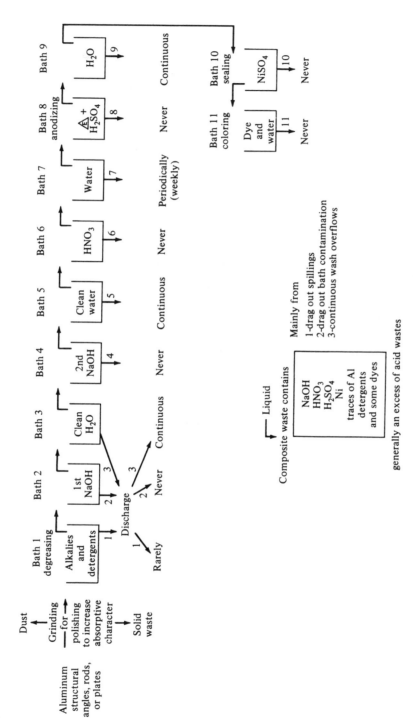

Fig. 27.12 Aluminum annodizing wastes.

pickle rinse waters," *Ind. Eng. Chem.* **44**, 534 (1952).

8. Pomelee, C. S., "Beryllium production wastes, toxicity studies," *Sewage Ind. Wastes* **25**, 1424 (1953).

9. Pirk, G. W., "Copper reclamation and water conservation," *Sewage Ind. Wastes* **29**, 805 (1957).

10. Rudgal, H. T., "Copper, effects on sludge digestion, Kenosha, Wis.," *Sewage Ind. Wastes* **18**, 1130 (1946).

11. Sanderson, W. W., and A. M. Hanson, "Colorimetric determination," *Sewage Ind. Wastes* **29**, 422 (1957).

12. Ullrich, H., "Treatment of waste-waters from aluminum surface treatment plants," *Chem. Abstr.* **58**, 7713 (1963).

13. Wise, W., "Treatment of metal processing wastes," *Sewage Ind. Wastes* **18**, 761 (1946).

14. Wise, W., "Character, treatment and disposal of wastes," *Sewage Ind. Wastes* **20**, 96 (1948).

15. Wise, W., *et al.*, "Composition of wastes, treatment," *Sewage Ind. Wastes* **20**, 772 (1948).

16. Wise, W., B. F. Dodge, and H. Bliss, "Brass and copper industry," *Ind. Eng. Chem.* **39**, 632 (1947).

The following references have been published since 1970.

1. Anonymous, "Brass mill wastes treatment," *Water and Wastes Eng.* **7**, b12 (1970).

2. Anonymous, "Metals affect enzymes in fish," *Sci. Newsletter* **97**, 459 (1970).

3. Anonymous, "Spectrum," *Environment Mag.* **13**, 23 (1971).

4. Anonymous, "Steel from aluminum wastes?" *Chem. Eng.*, 88, September 1971.

5. Anonymous, "Report on the Environment—1972," New York: *McGraw-Hill* (1972).

6. Bender, M. E., W. R. Matson, and R. A. Jordon, "On the significance of metal complexing agents in secondary sewage effluents," *Environmental Sci. and Tech.* **4**, 520–521 (1970).

7. Brant, M. V., D. C. Bone, and E. F. Emley, "Fumeless in-line degassing and cleaning of liquid aluminum," *J. of Metals* **23**, 48–53 (1971).

8. Ford, J. H. and F. E. Rizzo, "Liquid phase cementation of copper," *J. of Metals* **23**, 41–44 (1971).

9. Gurnham and Assoc. Inc., Chicago, "Industrial waste study; the basic nonferrous metals industry," Contract No. 68-01-0019, *Interior Rpt.*, 1–14, April 1971.

10. Haver, F. P. and M. M. Wong, "Recovery of copper, iron and sulfur from chalcopyrite concentrate using a ferric chloride leach," *J. of Metals* **23**, 25–29 (1971).

11. McCaull, J., "Building a shorter life," *Environment Mag.* **13**, 2–15 (1971).

12. Murs, G., "Waste water handling, aluminum industry," Personal communication with Director of Laboratories, *Alcan Ind.*, Oswego.

13. Scheiner, B. J., R. E. Lindstrom, and T. A. Henrie, "Processing refractory carbonaceous ores for gold recovery," *J. of Metals* **23**, 37–40 (1971).

14. Schmidt, F. A., R. M. Bergman, O. N. Carlson, and H. A. Walbalm, "Molybdenum metal by the bomb reduction of MoO_3," *J. of Metals* **23**, 38–44 (1971).

27.6 Metal-plating Wastes

Origin of metal-plating wastes. After metals have been fabricated into the appropriate sizes and shapes to meet customers' specifications, they are finished to final product requirements. Finishing usually involves stripping, removal of undesirable oxides, cleaning, and plating. In plating, the metal to be plated acts as the cathode while the plating metal in solution serves as the anode. The total liquid wastes are not voluminous, but are extremely dangerous because of their toxic content. The most important toxic contaminants are acids and metals, such as chromium, zinc, copper, nickel, tin, and cyanides. Alkaline cleaners, grease, and oil are also found in the wastes.

There are two main sources of waste from plating operations, each one distinctive in its volume and chemical nature: (1) batch solutions; and (2) rinse waters, including both nonoverflowing reclaimable rinses and continuous overflow rinses. Various stripping and cleaning operations may precede the plating processes. The Ohio River Valley Water Sanitation Commission (ORSANCO) [1] lists the major types of wastes originating from the stripping, cleaning, and plating of metal parts as follows:

1. *Proprietary solutions.* Most metal-finishing plants use solutions prepared according to manufacturers' formulas. These are mainly cleaners or plating-process accelerators of various types. The exact chemical composition of each solution should be obtained from the manufacturer.

2. *Cyanide concentrates.* This includes cyanide plating solutions and cyanide dips with relatively

high concentrations of cyanide. Since this chemical is one of the most toxic to both fish and other aquatic life, as well as man, even low concentrations in wastes are extremely dangerous and to be avoided at all costs.

3. *Cyanide rinse water* originates from the rinsing of cyanide-plated or -dipped metal parts.

4. *Concentrated acid and pickling wastes* originate primarily from stripping and cleaning of metal.

5. *Strong acid rinse waters* arise from rinsing after acid dips, pickling solutions, and strong acid process solutions.

6. *Chromates* originate from both plating and rinsing of metals that have been treated with chromate solutions to give them a durable protective finish. Since chromium, like cyanide, is toxic even in very low concentrations, chromium wastes are segregated and treated to remove all the chromium.

7. *Concentrated alkalis* are found in spent alkaline cleaning solutions—usually containing soaps, oils, and suspended solids—which are dumped periodically.

8. *Other wastes requiring treatment.* In most metal-finishing plants there are wastes which contain metal compounds, oils, soaps, and suspended solids, which can be treated by chemical precipitation and pH adjustment.

9. *Waste waters not requiring treatment.* These include cooling water and other waters unchanged in quality, which may be discharged without treatment.

Characteristics of metal-plating wastes. Most stripping baths are acidic in nature and consist of solutions of sulfuric, nitric, and hydrochloric acid, but alkaline baths containing sodium sulfide, cyanide, and hydroxide may also be used. Usually the chemicals in the stripping solution are present in concentrations of less than 10 per cent. Cleaning is carried out by organic solvents, pickling, or alkaline cleaning compounds. The organic-emulsion cleaners are petroleum or coal-tar solvents coupled with an emulsifier. Alkaline cleaners consist of sodium hydroxide, orthophosphate, complex phosphates, silicates, carbonates, some organic emulsifiers, and synthetic wetting agents.

Burford and Masselli [2] present a table showing the concentration of chemicals used in common plating baths (Table 27.19) and a flow chart for the most common plating baths (Table 27.20).

Cyanide salts are desirable, since they are good oxide solvents and in zinc plating they yield a brighter, less porous, galvanized plate. However, acid zinc sulfate is also being used in plating baths because it is said to conduct the current with less resistance than zinc cyanide.

The character and strength of plating wastes vary considerably, depending on plating requirements and type of rinsing used. The total plant waste may be either acidic or alkaline, depending on the type and quantity of baths used. A preponderance of cyanide or alkaline cleaning baths is likely to result in a highly alkaline pH, while the opposite may be true for chromate baths. The amount of stripping done is also important, since this operation contributes a highly acid waste to the plant mixture. Table 27.21 presents typical plating-waste concentrations obtained from seven different plants. The total volume of wastes from metal-plating plants, usually expressed as gallons per finished number of metallic units, varies even more than the characteristics. In most metal-finishing plants, the volume of waste is less than 0.5 mgd. Since most plants use excessive chromates for plating, the concentration of chromium in chromium-plating bath waste will usually be several times the concentration of other metals in other baths (Table 27.19).

Treatment of metal-plating wastes. The methods used for disposal of waste from plating operations can be divided into two classes: (1) modifications in design and/or operation within the manufacturing process to minimize or eliminate the waste problem; (2) installation of a chemical (sometimes physical) treatment plant to destroy or remove toxic and objectionable materials in plating-room effluents.

Many recommendations for modifications in design and operation to reduce wastes have been suggested. The Ohio River Valley Water Sanitation Commission [3] has published a guide for these practices. Additional modifications include: (1) installing a gravity-fed, nonoverflowing emergency holding tank for toxic metals and their salts; (2) eliminating breakable containers for concentrated material; (3) designing special drip pans, spray rinses, and shaking mechanisms; (4) reducing spillage, drag-out leak to the floor, or other losses, by curbing the area and discharging these losses to a holding tank; (5) using high-pressure fog rinses

Table 27.19   Common plating baths. (After Burford and Masselli [2].)

Bath formulas	Metallic + cyanide concentrations, ppm	Rinse concentration, ppm	
		0.5 gph drag-out*	2.5 gph drag-out*
Nickel			
40 oz/gal nickel sulfate	82,000 Ni	171 Ni	855 Ni
8 oz/gal nickel chloride			
6 oz/gal boric acid			
Chromium			
53 oz/gal chromic acid	207,000 Cr	431 Cr	2155 Cr
0.53 oz/gal sulfuric acid			
Copper (acid)			
27 oz/gal copper sulfate	51,500 Cu	107 Cu	535 Cu
6.5 oz/gal sulfuric acid			
Copper (cyanide)			
3.0 oz/gal copper cyanide	12,400 Cu	2.8 Cu	14 Cu
4.5 oz/gal sodium cyanide	28,000 CN	58 CN	290 CN
2.0 oz/gal sodium carbonate			
Copper (pyrophosphate)			
4 oz/gal copper (as proprietary mix)	30,000 Cu	62 Cu	310 Cu
29 oz/gal sodium pyrophosphate			
0.4% ammonia (by volume)			
Cadmium			
3.5 oz/gal cadmium oxide	23,000 Cd	48 Cd	240 Cd
14.5 oz/gal sodium cyanide	57,700 CN	120 CN	600 CN
Zinc			
8 oz/gal zinc cyanide	33,800 Zn	70 Zn	350 Zn
5.6 oz/gal sodium cyanide	48,900 CN	102 CN	510 CN
10 oz/gal sodium hydroxide			
Brass			
4 oz/gal copper cyanide	21,000 Cu	44 Cu	220 Cu
1.25 oz/gal zinc cyanide	5,250 Zn	11 Zn	55 Zn
7.5 oz/gal sodium cyanide	47,500 CN	99 CN	495 CN
4 oz/gal sodium carbonate			
Tin (alkaline)			
16 oz/gal sodium stannate	53,000 Sn	110 Sn	550 Sn
1 oz/gal sodium hydroxide			
2 oz/gal sodium acetate			
Silver (cyanide)			
4 oz/gal silver cyanide	24,600 Ag	51 Ag	255 Ag
4 oz/gal sodium cyanide	21,800 CN	45 CN	225 CN
6 oz/gal sodium carbonate			

*Drag-out is the amount of solution carried out of the bath by the material being plated and the racks holding the material. Rinse rate is assumed to be 4 gpm.

Table 27.20 Flow chart for some common plating baths. (After Burford and Masselli [2].)

Copper plating	Nickel plating	Chrome plating	Zinc plating
Electrocleaner (cathodic)	Electrocleaner (cathodic)	Electrocleaner (cathodic)	Electrocleaner (cathodic)
Running rinse →	Electrocleaner (anodic)	Running rinse →	Running rinse →
Hydrochloric acid dip (5%)	Running rinse →	Sulfuric acid dip	5% sulfuric acid dip
Running rinse →	5% sulfuric acid dip	Running rinse + spray →	Running rinse →
Copper cyanide "strike"	Running rinse →	Chrome solution	Zinc cyanide solution
Running rinse →	Bright nickel solution	Recovery rinse	Running rinse →
Running rinse →	Running rinse →	Mist spray rinse	Spray rinse →
Copper pyrophosphate solution	Soap dip	Running rinse →	Brightener still dip (HNO₂)
Running rinse →	Hot running rinse →	Hot still dip	Running rinse →
Hot rinse (slow overflow) →	Drying oven	Running rinse →	Running rinse →
Drying oven		Hot rinse (slow overflow) →	Hot water dip (slow overflow) →
		Drying oven	Drying oven

*Flow sheets for common types of conveyorized electroplating. (Wastes overflowing to final effluent are indicated by an arrow.)

Table 27.21 Plating-waste concentrations. (From Burford and Masselli [2].)

Plant	pH	Cu, ppm	Fe, ppm	Ni, ppm	Zn, ppm	Chromium, ppm +6	Chromium, ppm Total	Cn, ppm
A	3.2	16	11	0	0	0	1.0	6
A	10.4	19	3	0	0	0	0.5	14
B	4.1	58	1.2	0	0	204	246	0.2
C	2.8	11		0.2		3	7	1.2
D	2.0	300	10	0	82	0	0	0.7
E	2.4	35	8			555	612	1.2
E	10.7	14	4	19		32	39	2.0
F	10.5	6	2	25	39			10
G	11.3	18	18	26		36		15
G	11.9	23	21	32		95		13

rather than high-volume water washes; (6) reclaiming valuable metals from concentrated plating-bath wastes; (7) evaporating reclaimed wastes to desired volume and returning to plating bath at rate equal to loss from bath; and (8) recirculating wet-washer wastes from fume scrubbers.

Treatments of plating wastes by chemical and physical means are designed primarily to accomplish three objectives: removal of cyanides, removal of chromium, and removal of all other metals, oil, and greases.

The treatment of cyanides, although mostly accomplished by alkaline chlorination, is being carried out by no less than ten methods [156]: (1) chlorination (gas), (2) hypochlorites, (3) ClO_2, (4) O_3(ozonation), (5) conversion to less toxic cyanide complexes, (6) electrolytic oxidation, (7) acidification, (8) lime-sulfur method, (9) ion exchange, and (10) heating to dryness.

Chromium-bearing plating wastes are normally segregated from cyanide wastes, since they must be reduced and acidified (to convert the hexavalent chromium to the trivalent stage) before precipitation can occur. Although it is possible to precipitate the chromium directly in the hexavalent form with barium chloride, this method is not widely used. The removal of other metals such as Cu, Zn, Ni, Fe, and greases is usually accomplished by neutralization followed by chemical precipitation. To give the reader a better working knowledge of the processes, the three most

widely used treatment methods are discussed in the following paragraphs.

Alkaline chlorination. The treatment of cyanide-bearing wastes by alkaline chlorination involves the addition of a chlorine gas to a metal-plating waste of high pH. Sufficient alkalinity, usually $Ca(OH)_2$ or NaOH, is added prior to chlorination to bring the waste to a pH of about 11, thus ensuring the complete oxidation of the cyanide. Violent agitation must accompany the chlorination, to prevent the cyanide salt of sodium or calcium from precipitating out prior to oxidation. The presence of other metals may also interfere with cyanide oxidation, because of the formation of metal cyanide complexes. Extended chlorination may be necessary under these conditions. The probable reaction with excess chlorine in the presence of caustic soda has been expressed as

$$2NaCN + 5Cl_2 + 12NaOH$$
$$\rightarrow N_2 + 2Na_2CO_3 + 10NaCl + 6H_2O.$$

About 6 pounds each of caustic soda and chlorine are normally required to oxidize one pound of CN to N_2. Sometimes a full 24-hour chlorination period may be required to effect complete oxidation. Schematic drawings of both continuous and batch treatment of cyanide wastes are presented in Figs. 27.13 and 27.14.

Reduction and precipitation. Chromium-plating-waste treatment by reduction and precipitation involves reducing the hexavalent chromium (Cr^{+6} as chromic

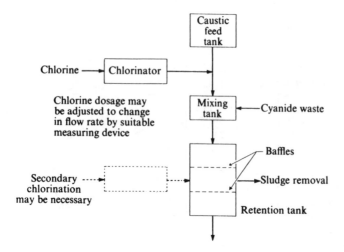

Fig. 27.13 Continuous chlorination of cyanide wastes. (After ORSANCO [1].)

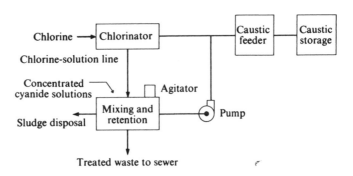

27.14 Batch chlorination of cyanide wastes. (After ORSANCO [1].)

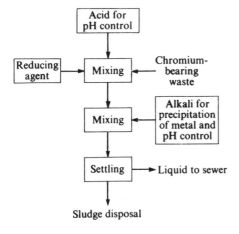

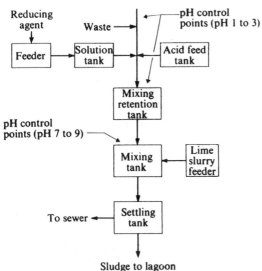

Fig. 27.15 Reduction and precipitation of chromium. (After ORSANCO [1].)

acid or chromates) in the waste to the trivalent stage (Cr^{+++}) with reducing agents such as $FeSO_4$, SO_2, or $NaHSO_3$. Sufficient free mineral acid should also be present to combine with the reduced chromium and to maintain a residual pH of 3.0 or lower, which will ensure complete reaction. When the reduction is complete, an alkali (usually lime slurry) is added, to neutralize the acid and precipitate the trivalent chromium. The following chemical reactions, using ferrous sulfate, illustrate this method of treatment:

$$H_2Cr_2O_7 + 6FeSO_4 + 6H_2SO_4$$
$$\rightarrow Cr_2(SO_4)_3 + 3Fe_2(SO_4)_3 + 7H_2O,$$
$$Cr_2(SO_4)_3 + 3Ca(OH)_2 \rightarrow 2Cr(OH)_3 + 3CaSO_4,$$
$$Fe_2(SO_4)_3 + 3Ca(OH)_2 \rightarrow 2Fe(OH)_3 + 3CaSO_4.$$

Figure 27.15 diagrams these reactions. One part per million of chromium usually requires about 16 ppm of copperas, 6 ppm of sulfuric acid, and 9.5 ppm of lime and produces about 2 ppm of chromic hydroxide and 0.4 ppm ferric hydroxide sludges, as well as almost 2 ppm of calcium sufate (some of which is also precipitated).

Neutralization. Treatment of other metal, oil, and grease-bearing wastes by neutralization and precipitation usually involves recombining the wastes with previously oxidized cyanide and reduced chromium wastes for subsequent and final treatment. If the combined waste is acid, an alkali (usually 5 to 10 per cent lime slurry) is added to neutralize and precipitate the metals. The floc produced is large and quite heavy, and hence the velocity of flow is decreased after adequate flocculation has occurred. The waste is then allowed to settle. Sludge is removed and usually

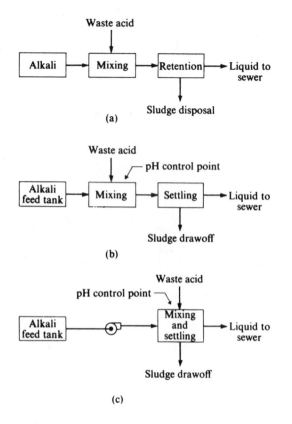

Fig. 27.16 (a) Acid neutralization; (b) continuous acid neutralization; (c) batch acid neutralization. (After ORSANCO [1].)

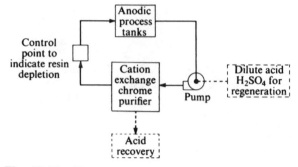

Fig. 27.17 Chrome purifier and recovery system. (After ORSANCO [1].)

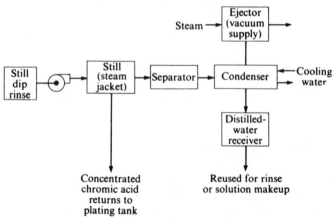

Fig. 27.18 Chrome acid recovery by vacuum evaporation. (After ORSANCO [1].)

lagooned, since this is the most economical treatment for the slow-drying, relatively innocuous, metal sludges. The processes involved in treating the final wastes are shown in Fig. 27.16.

Sulfide precipitation with Na_2S or $NaHS$ is an effective alternative to lime precipitation treatment to remove heavy metals. It is possible to attain effective precipitation of certain metals, such as As, Cu, Cd, and Hg, even at very low pH, low detention time in the reaction tank because of the high reactivity of sulfides and the feasibility of selective metal recovery.

Recovery practices are mainly those involving ion exchange and evaporation. The use of ion exchangers

is only an application of water-softening methods, and its best application is in the treatment of rinse water following plating operations, so that little or no foreign contamination other than the recoverable metal is present. Rinse water is passed through beds of cationic and anionic resins selected for the particular application, and the deionized water is recycled through the rinse tank. The ion beds must be regenerated periodically. The regenerating solution containing the concentrated metal salts may require further treatment prior to its reuse in plating operations. Figure 27.17 illustrates the use of an ion exchanger for chromium reovery.

Chrome, nickel, and copper acid-type plating solutions may be reclaimed from the rinse tank by evaporation in glass-lined equipment; the concentrated solution is then returned to the plating systems. The water condensed from the steam is reused in the rinse tank following the plating tank, to eliminate build-up of natural water salts. This process has proved effective for recovering valuable metal salts. The high initial cost for equipment is more than recovered, not only in the value of waste treatment, but also by recovery of metals, especially when volumes of metallic wastes are large. ORSANCO [156] shows a schematic arrangement for evaporating chromium wastes (Fig. 27.18).

References: Metal-Plating Wastes

1. Metal-Finishing Industry Action Committee, *Methods for Treating Metal Finishing Wastes*, Ohio River Valley Water Sanitation Commission, January 1953.
2. Rudolfs, W., *Industrial Waste Treatment*, Reinhold Publishing Corp., New York (1953), p. 289.
3. Ohio River Valley Water Sanitation Commission, *Plating Room Controls for Pollution Abatement*, (1951).
4. Ying, W. C. et al., "Removal of Hypophosphite and Phosphite from Electroless Nickel Plating Baths", Proc. **43** Purdue Ind. Waste Conf., 699, (1988).
5. Kane, J. E., "The use of Innovative Treatment Technologies to Upgrade Performance of an Existing Pretreatment System to meet New Discharge Standards," Proc. **43** Purdue Ind. Waste Conf., 707, (1988).
6. Semmens, M. J. and Y. Y. Chang, "Selective Cyanide Recovery from Wastewater Containing Metal Cyanide Complexes," Proc. **43** Purdue Ind. Waste Conf., 711, (1988).
7. Johannes, R. D. et al., "Electroplating/Metal Finishing Wastewater Treatment: Practical Design Guidelines," Proc. **43** Purdue Ind. Waste Conf., 727, (1988).
8. Brooks, C. S., "Nickel Metal Recovery from Metal Finishing Industry Wastes." Proc. **43** Purdue Ind. Waste Conf., 847, (1988).
9. Bricka, R. M. and M. J. Culliname, Jr., "Comparative Evaluation of Heavy Metal Immobilization Using Hydroxide and Xanthate Precipitation." Proc. **42** Purdue Ind. Waste Conf., 809, (1987).
10. Kavanaugh, D. J. and A. R. Boyce," A Continuous Closed Loop Regeneration System For a Chromic/Sulfuric Acid Etchant Bath," Proc. **42** Purdue Ind. Waste

Conf., 873, (1987).
11. Bowers, A. R. et al., "Iron process for Treatment of Cr (VI) Wastewaters." Proc. **41** Purdue Ind. Waste Conf., 465, (1986).
12. Edwards, J. D. and J. W. Cammarn, "A Case History of Removing Cadmium from Electroplating Wastewater Using Calcium Sulfide." Proc. **41** Purdue Ind. Waste Conf., 474, (1986).
13. McFadden, F. et al., "Nickel Removal from Nickel Plating Wastewater Using Iron, Carbonate, and Polymers for Precipitation and Coprecipitation," Proc. **40** Purdue Ind. Waste Conf., 417, (1985).
14. Low, W. L. and A. F. Gaudy, Jr., "Removal of Cadmium at High and Low Dosages," Proc. **40** Purdue Ind. Waste Conf., 431, (1985).
15. Lindsay, M. J. et al., "Sodium Borohydride Reduces Hazardous Waste," Proc. **40** Purdue Ind. Waste Conf., 477, (1985).
16. Ramirez, E. R. and J. P. VanDillen, "Engineering and Design of a Wastewater Treatment System for a Vibrobot Plating Operation," Proc. **40** Purdue Ind. Waste Conf., 483, (1985).
17. Hunter, J. S., "Performance of Buoyant media Filter in Metal Plating Shop Wastewater Treatment," Proc. **40** Purdue Ind. Waste Conf., 459, (1985).
18. Sloan, F. J. et al., "Removal of Metal Ions from Wastewater by Algave," Proc. **39** Purdue Ind. Waste Conf., 423, (1984).
19. Mayenkar, K. V. et al., "Removal of Chelated Nickel from Wastewaters," Proc. **39** Purdue Ind. Waste Conf., 457, (1983).

27.7 Motor Industry Wastes

Origins of motor industry wastes. Stamping plants, body, and final assembly operations contribute 23 per cent of the water intake and use 70 per cent of the manpower of the motor vehicle equipment industry [1967, 1]. In the stamping operation, which produces major body parts, the metal (normally strip or sheet steel) is cut to size and then stamped into the desired shape by large hydraulic presses. Portions of the stamped parts are normally welded together in the stamping operation. From here the parts are sent to the body manufacturing facility. In conventional industry terminology "body" refers to the passenger enclosure from the fire wall back and does not include the front-end parts such as the front fenders and hood.

In the body-assembly plant, the body is first constructed from stamped metal parts in the body shop and is then treated and painted in the paint shop. The exterior and interior trim, produced in parts plants, are added in the trim shop.

From the body-assembly operations, the completed body goes to the assembly plant. First, the chassis, wheels, and power train (engine, transmission, etc.) are assembled from parts produced elsewhere; this assembled chassis is joined to the already assembled body; finally, the front-end parts (fenders, hood, etc.) are added. These last items are stamped in a conventional stamping plant and are normally painted in the final assembly plant. The stamping plants are separate from the body and final assembly operations.

Characteristics of motor industry wastes

Stamping plants. These operations produce no significant liquid processing wastes, since only small amounts of water are used directly in processing. However, large amounts of oils (both lubricating and hydraulic) are used, and in many cases some of these find their way into the sewer system. Because these oils originate from a variety of points in the plant and because their introduction into the sewer system is of a miscellaneous nature, the concentration in the plant effluent can vary widely. Amounts of extractable material from 50 mg/liter to several thousand mg/liter have been encountered. The flow of contaminated process water will be, as noted above, quite small, varying from about 2000 to 10,000 gallons per day.

Large amounts of cooling water are used in welding systems. Recirculation is widely practiced, so in most cases the discharge will be restricted to cooling system blowdown. This can, however represent the major part of the plant water discharge. For typical stamping plants the cooling-system blowdown can vary from 25 to 150 gallons per minute. Powerhouse water (boiler blowdown, boiler-water pretreatment system blowdown, etc.) also represents a source of contaminants. Processes and significant wastes from body stamping and assembly are shown in Fig. 27.19.

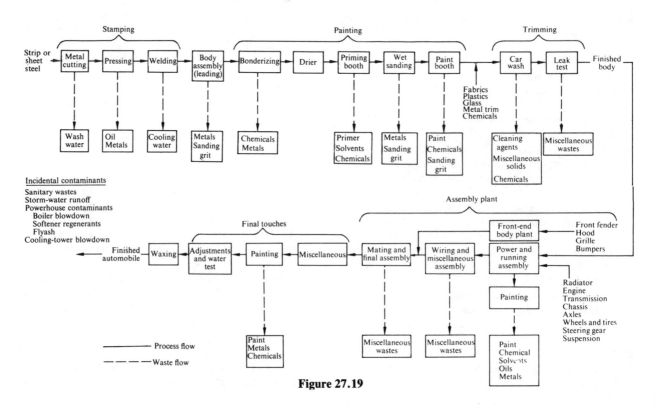

Figure 27.19

Assembly plants. The waste waters discharged from final assembly plants, body-assembly plants, or combined operations are of the same general type, that is, organic waste waters containing suspended solids. These solids originate primarily from the painting and paint-sanding operations. In addition, heavy metals such as zinc and chromium that originate in metal-treating (bonderizing) operations may be present, as well as powerhouse and cooling water, although the amount of this latter waste relative to the overall plant discharge will be considerably less than with stamping plants.

Because the waste waters are primarily organic in nature and contain suspended solids they are similar to sanitary waste water, but the organic content (as measured by BOD and COD tests) and suspended solids will normally be higher than in typical sanitary wastes.

The processes which produce liquid wastes are essentially uniform throughout the industry and do not vary between small and large plants. There are no new processes anticipated that will materially add to the pollution load, but one new process, electrostatic painting, may in the future actually effect a reduction in pollution loading. In a typical painting operation, a water curtain is used in the paint booth to entrap overspray, which otherwise could present an environmental pollution problem. This water is discharged (after a significant amount of reuse) and is a major source of contaminants, primarily organic materials and suspended solids. With electrostatic painting there is less overspray, in fact the water curtain can sometimes be eliminated entirely. It appears unlikely that this technique will be widely used in the foreseeable future to replace existing equipment, but it will probably be tried where new facilities are constructed. Thus, it will probably not make a significant contribution within the next decade, but may become increasingly important thereafter.

The average amount of waste produced per 100 cars for body assembly and final assembly is summarized as follows:

	lb/100 cars
Chemical oxygen demand (COD)	1007.77
Biochemical oxygen demand (BOD)	322.33
Hexavalent chromium (CrO$_4$)	4.50

Trivalent chromium (CrO$_4$)	2.08
Zinc (Zn)	1.12
Suspended solids	360.30
	gal/100 cars
Flow	201,958

Treatment of motor industry wastes

Stamping plants. Stamping-plant treatment systems are of several types. If the cooling water and powerhouse water do not require treatment and if the concentrated waste can be collected separately, the treatment system will usually consist of a batch system for removal of oil and suspended solids. Alternatively, incineration has been used for the concentrated, oil-containing waste water.

If, on the other hand, the cooling water and powerhouse water require treatment or if the concentrated waste water cannot be separated from the general plant collection system, an end-of-line facility for removal of suspended solids and oil is dictated. The overall efficiency for the removal of suspended solids and oil is in the range of 85 to 95 per cent.

Assembly plants. Typical waste-treatment facilities for assembly plants incorporate chemical clarification followed by conventional biological treatment, such as the activated-sludge process. Provision for reducing hexavalent chromium is incorporated in the chemical treatment, and trivalent chromium and other heavy metals are removed in the clarification step. The removal of heavy metals including chromium is essentially complete. Normal efficiencies for removal of organic material (as measured by BOD and COD tests) are in the range of 80 to 95 per cent and for suspended solids in the range of 85 to 95 per cent. Several modifications of this general approach can be used and these are detailed in a report.*

Since the waste waters from assembly operations are basically organic in nature, they can be sent to municipal sewage-treatment facilities. However, it is usually necessary first to adjust the pH and remove the heavy metals, including chromium. Also, since the concentrations of COD, BOD, and suspended solids

---

*The Cost of Clean Water, Motor Vehicles and Parts, Industrial Waste Profile no. 2, Vol. III, U.S. Department of the Interior, Washington, D.C. (1967).

are usually higher than in conventional sanitary waste, pretreatment to remove excess suspended solids (including some organic material) is normally provided. After such pretreatment, it is the general practice of the industry to discharge to municipal systems where possible rather than to provide secondary biological treatment on the plant site.

References

1. Anderson, E., "Small pumps play a big role in auto plant," *Water and Wastes Eng.*, E-4, September 1970.
2. Anonymous, "18 billion pound problem confronts auto makers annually," *Refuse Removal J.* 12, 18 (1969).
3. Anonymous, "G. M. stages pollution show and tell show," *Iron Age* 205, 23, April 1970.
4. Anonymous, "How C & NW fights water pollution," *Railway Age* 166, 26–27 (1969).
5. Balden, A. R., "Industrial water management at Chrysler Corp.—1969," *J. WPCF* 41, 1912–1922 (1969).
6. Besselievve, P. E., *The Treatment of Industrial Wastes*, New York: McGraw-Hill 247–263 (1969).
7. Delow, J. S., "Industrial requirements for effective pollution control—automobile industry," *Electrochem. Soc.*, Princeton, NJ, 29–39 (1972).
8. EPA, "Interim guidelines for the automobile and aircraft industries" (1975).
9. EPA, "Reference guide for wastewaters from the automobile industry," *EPA Contract No. 68-01-0028* (1975).
10. Fisco, R., "Plating and industrial waste treatment at the Fisher body plant," *Proc. 25th Ind. Waste Conf.*, *Purdue Univ.* (1970).
11. General Motors Corp., "Recovering phosphoric acid from plastic conditioning solution," *U.S. Pat. 3,506,397*; *Metal Finishing Abs.* 12, 218 (1970).
12. Greene, J. R., "Guide lamp combines treatment of water and wastes," *Water and Wastes Eng.* 6, E22–23, September 1969.
13. Hill, G. B., "Complete removal of chromic acid waste with aid of instrumentation," *Plating* 56, 172–176 (1969).
14. Hubbell, J. W., "New wastewater treatment plant reuses its oil sludge," *Civ. Eng.* 40, 81, (1970).
15. Nemerow, Nelson L., "Motor industry wastes," *Liquid Waste of Industry*, 431–433, Reading, Mass.: Addison-Wesley (1971).
16. Poole, D. A., "Design of waste treatment facilities for auto industry," *Proc. 14th Ont. Ind. Waste Conf.*, Niagara Falls (1967).
17. Schott, C. C., "Junks—melt'em down," *Steel* 163, 489 (1968).
18. Weidemann, C. R., "Control considerations in washing, painting, and soluble oil removal," *Metal Progr.* 98, 66–67 (1970).

27.8 Iron-Foundry Wastes

The waste from most small gray-iron foundries is dry, being composed of solids from molding and core sands and fly ash. Foundries produce castings from molten metal, which are then machined to final specifications. The major waste resulting from the formation of rough castings from molten metal is used sand. Often the sand is conveyed to a disposal site by flushing with water, in a procedure similar to that used in conveying flue dust from steel mills and coal dust from coal-mining operations. Disposing of the used sand is a difficult problem, since it requires considerable land area and working time. Foundry sand, normally new Ottawa sand, possesses specific sieve analyses and is quite expensive, so its recovery ought to be considered. The waste material from used molds consists of about 85 to 90 per cent sand, the remainder being clay, sea coal, and so forth. The suspended solids concentration varies between 2500 and 5000 ppm.

Most waste-treatment measures include some method of sand reclamation. One processor [5] recommends a filtering unit to remove excess water from the sand. The sand then enters a drier, which contains a firing unit at the end where the sand enters. The processor states that the performance of the reclamation unit has been most satisfactory, not only from an operating standpoint, but also as an investment, showing immediate returns. Typical screenings of the sand as purchased and the sieve analyses of the sand after reclamation are given in Table 27.22. The analyses of the reclaimed sand after screening show that it is comparable to the new sand.

Another waste-treatment system, set up according to the flow diagram in Fig. 27.20, consists of primary sedimentation prior to reuse of the water. The excess over that required for reuse is further clarified by chemical treatment and sedimentation. This effluent is subjected to a flotation treatment before final dis-

Table 27.22 Comparative sieve analyses of new and re-claimed sand [5].

Sieve mesh	Ottawa sand, new	Reclaimed sand
20		0.08
30	0.04	0.30
40	2.04	5.84
50	32.30	41.08
70	44.90	36.44
100	16.80	13.26
140	2.84	1.68
200	0.20	0.68
270	0.30	0.02
Pan	0.04	0.02

posal, and the sludge is dewatered by filtration. With this method, it was found that, although the sand waste contained 3760 ppm suspended solids, this figure was reduced to 363 ppm after primary settling, 30 ppm after the clarifier, and 10 ppm after flotation. The 13 per cent solids sludge, filtered at a rate of 13 pounds per square foot per hour, produced a cake of ¼-inch thickness and a filtrate with 900 ppm suspended solids.

Bader [1] describes a new foundry with a modern industrial-waste treatment system utilizing gravity separation of oils, a holding basin, air flotation, and sludge disposal (Fig. 27.21). The air-flotation equipment provides the secondary treatment necessary

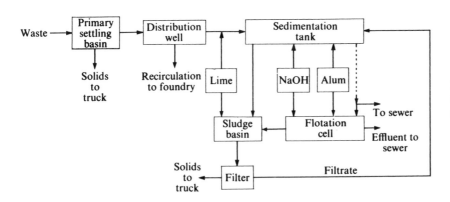

Fig. 27.20 Flow diagram of recovery of water and treatment of residual gray-iron foundry waste.

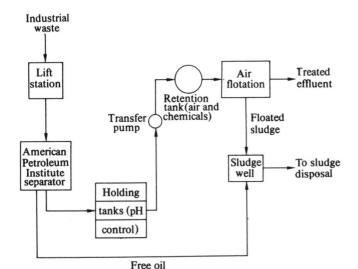

Fig. 27.21 Schematic plan of industrial waste system and description of various stages of waste treatment. This system handles all the liquid wastes except santiry, dust collection, and cooling.

because the separator (designed by the American Petroleum Institute) does not completely remove emulsified oil, small particles of free oil, and suspended oil. The system pressurizes the entire waste flow and consists of a transfer pump, retention tank, flotation unit, and related air-supply and chemical-feed accessories. Compressed air, alum, and a coagulant are added to the retention tank to improve removal efficiencies. After treatment, floated sludge is discharged to a sludge well and treated liquid is drawn from the bottom of the flotation compartment and either discharged to the spent-cooling-water sewer or returned to the holding tank for additional treatment.

References: Iron-Foundry Wastes

1. Bader, A. J., "Complete waste treatment system designed for new foundry," *Plant Eng.* 118 (1968).
2. Chapple, H., "Wet foundry sand reclamation," *Ind. Wastes* 1, 121 (1911).
3. Hartman, C. D., "Waste water control at a midwest finishing mill near Lake Michigan," in Proceedings of the 11th Ontario Industrial Waste Conference, June 1964, p. 85.
4. Hartman, C. D., *et al.*, "Waste water control at Midwest Steel's new finishing mill," *Iron Steel Eng.* 40, 182 (1963).
5. "Industrial waste treatment in foundry industry," *Mod. Castings* 44, 471 (1963).
6. Hathaway, C. W., *et al.*, "Treatment of machine shop and foundry wastes," *Sewage Ind. Wastes* 26, 1363 (1954).
7. Hathaway, C. W., *et al.*, "Treatment of gray iron foundry waste water," *Ind. Wastes* 1, 166 (1956).

The following references have been published since 1972.
1. Handwerk, R. J., "Recycling water effectively," *Foundry* 100, 40 (1972).
2. Krueger, G. N., "Planning your caster: It's water and pollution control facilities," *Iron and Steel Eng.* 49, 73 (1972).
3. Nobse, W. and D. Wystinch, "Practical and economic comparison of the most common metal finishing waste treatment systems," *Plating* 59, 126 (1972).

LIQUID-MATERIALS INDUSTRIES

The major liquid-materials industries are here categorized as oil, rubber, and glass manufacturing. We shall discuss the waste problems of each one separately in the pages which follow.

27.9 Oil-Field and Refinery Wastes

Origin of oil-field and refinery wastes. Oil wastes can be classified into those originating from (1) oil production and (2) oil refining. Wastes result from pumping, desalting, distilling, fractionation, alkylation, and polymerization processes; they are of large volume and contain suspended and dissolved solids, oil, wax, sulfides, chlorides, mercaptans, phenolic compounds, cresylates, and sometimes large amounts of dissolved iron. Treatment is by scrubbing with flue gas, evaporation, flotation, mixing, aeration, biological oxidation, coagulation, centrifugation, and incineration.

Generally, petroleum contains about 85 per cent carbon and 12 per cent hydrogen. The remaining 3 per cent is composed of small amounts of oxygen, nitrogen, and sulfur. Some products and by-products of oil refining are gasoline, kerosene, lubricants, gas oil and fuel oil, wax, asphalt, petroleum coke, and miscellaneous materials such as petrolatum and insecticides.

Crude oil is refined by fractional distillation to separate the various hydrocarbons, by application of heat and pressure (with or without catalysts) to alter the molecular structure of some of the distillation products, and by chemical and mechanical treatment of various fractions or products to remove impurities. The crude oil is first passed through a pipe still into a fractionating tower where the lighter products—gasoline, kerosene, and gas oil—are taken off and condensed. The gasoline and kerosene then pass through tanks in which sulfuric acid, caustic soda, plumbite, and water washes are applied to remove impurities. The gas oil, after certain proprietary treatment, is stored for sale as light fuel oil. Distillation of the remainder is continued in the fractionating tower, to take off lubricating or wax distillates. After all products have been taken off as distillates, the residue in the vacuum chamber is used for the manufacture of asphalt.

Wastes from oil fields are drilling muds, salt water, free and emulsified oil, tank-bottom sludge, and natural gas. Of these, salt water (brine) presents the most difficulty. Many oil-bearing strata have brine-bearing formations directly over or under them. Pumping rates are controlled and some of the areas of the well are sealed off to prevent these briny waters from seeping into the oil. However, this can never be completely accomplished, and brine is often pumped out of the well with the oil. The brine and oil must then be separated by gravity and the brine disposed of.

Wastes from oil refineries include free and emulsified oil from leaks, spills, tank draw-off, and other sources; waste caustic, caustic sludges, and alkaline waters; acid sludges and acid waters; emulsions incident to chemical treatment; condensate waters from distillate separators and tank draw-off; tank-bottom sludges; coke from equipment tubes, towers, and other locations; acid gases; waste catalyst and filtering clays; special chemicals from by-product chemical manufacture; and cooling waters. Oils from leaks and spills can amount to as much as 3 per cent of the total crude oil treated. The treatment of oils with alkaline reagents to remove acidic components and the sweetening processing of oils to convert or remove mercaptans produce a series of alkaline wastes which give off obnoxious odors and present difficult, and costly, waste-disposal problems; as do the acid sludges resulting from sulfuric-acid treatment of oils. A typical flow sheet of the petroleum-refining process, showing the above wastes, is given in Fig. 27.22.

Characteristics of oil-refinery wastes. Like most industries, oil refineries use enormous quantities of water. Virtually every refinery operation, from primary distillation through final treatment, requires large volumes of process and cooling waters. The demand is estimated by the National Association of Manufacturers to be 770 gallons per barrel of crude oil. In 1952 Giles [1] estimated that the oil refineries used 5400 mgd of water (based on a national crude-oil figure of 7 million barrels per day). This volume of water used was second only to that of the steel industry and represented about 20 per cent of the total indus-

trial consumption in the United States and slightly less than 50 per cent of municipal needs.

Giles also gave data on waste characteristics of a typical oil refinery (Table 27.23). Analyses of oil-field brine wastes in four major oil-producing states are given in Table 2.24. The Ohio River Survey [2] gives a more complete analysis of the mineral content of five Kansas oil-field brines (Table 27.25).

A study by the U.S. Department of the Interior [1967, 3] describes the petroleum refinery "as a complex combination of interdependent processes and operations, many of which are complex in themselves." The reader is referred to Fig. 27.22 for a flow diagram of a typical refinery and to this report for complete and detailed descriptions.

Although the sheer volume of wastes from oil refineries looms as so large a problem, the American Petroleum Institute reports that 80 to 90 per cent of the total water used by the average refinery is for cooling purposes only and is not contaminated except by leaks in the lines. The combined refinery wastes, however, may contain crude oil, and various fractions thereof, and dissolved or suspended mineral and organic compounds discharged in liquors and sludges from the various stages of processing. The oil may appear in waste waters as free oil, emulsified oil, and as a coating or suspended matter, though ordinarily not in proportions greater than 100 ppm. However, floating oil is visible even when present in very small concentrations, because of its ability to spread in very thin (0.000003 inch), nondestructible layers. The American Petroleum Institute has classified waste constituents according to the refinery units from which they are released (Table 27.26). Weston [4] gives the sanitary characteristics of 20 typical refinery wastes (Table 27.27).

Treatment of oil-field and refinery wastes. Several methods of disposing of brine wastes have been used, including solar evaporation of impounded brine, controlled diversion of brine into surface waters, recovery of mineral salts, and injection of the brine into subterranean formations; in spite of all these, there is sometimes no alternative to the payment of damages to downstream riparian owners through claim-adjustment associations. The only procedure finding wide use and acceptance for other than

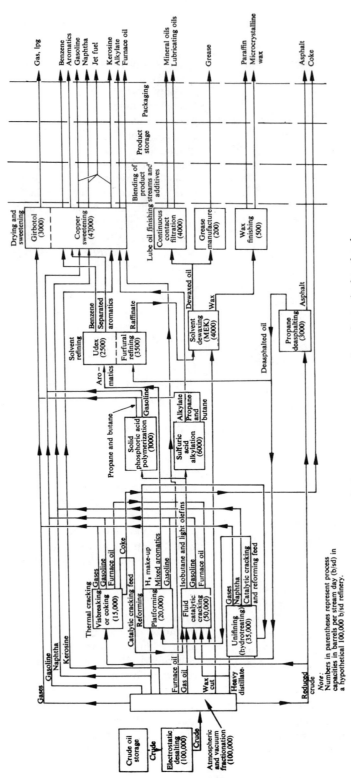

Fig. 27.22 Subprocess series representative of a typical technology.

Table 27.23 Data for typical oil refinery. (After Giles [1].)

Production, water, and waste character	Value
Crude run, bbl/day	20,500
Water usage, mgd	16
Effluent water quality	
Oil content, bbl/day	16
Phenolics, lb/day	380
BOD, population equivalent	31,500

Table 27.24 Oil-field brines.

State	Total solids, ppm	Ca and Mg, ppm	Na and K, ppm	Cl, ppm
Illinois	113,000	50,000	38,000	69,000
Kentucky	251,000	220,000	57,000	137,000
Oklahoma	236,000	130,000	77,000	145,000
Texas	69,000	1,800	24,000	40,000

Table 27.25 Mineral analyses of oil-field brine samples [2].

Radical	Concentration range, ppm
Calcium	1507–12,888
Magnesium	346–4290
Sodium	8260–63,275
Bromide	32–633
Carbonate	0
Bicarbonate	43–644
Sulfate	0–1578
Chloride	12,750–127,220
Total solids	25,210–248,600

coastal fields is disposal by injection. A drawback is that brines must be pretreated prior to subterranean injection to prevent equipment from being corroded and to prevent plugging of the sand; it often becomes apparent that the pretreatment required may be so extensive and costly that it is better to discharge the brines elsewhere after some other pretreatment. Chemical coagulation of brine wastes with alum is also

practical prior to discharging them into surface waters.

The main remedial measures for reducing refinery wastes are: (1) reduction of oil leakage by preventive maintenance of pipelines and equipment; (2) preventing formation of oil emulsions or, where these exist, isolation and separate treatment; (3) removal of floating oil in separators located as close to the original source of waste as possible; (4) isolation and separate treatment of objectionable wastes, e.g. with American Petroleum Institute (API) separators, which reduce the BOD to 5 to 10 ppm, provided emulsions are not present.

Acidification of caustic wastes (with H_2SO_4) removes some of the objectionable compounds. Acid sludges may be used as a source of fuel or to produce by-products such as oils, tars, asphalts, resins, fatty acids, and chemicals. Some refineries recover sulfuric acid from the acid sludges for their own use. Giles [48] gives the reduction of sanitary constituents of refinery wastes after chemical and biological treatment (Table 27.28).

Weston [5] points out that the method of handling petroleum-industry product losses and by-products is an economic problem in which, in some cases, the choice will lie between utilization of the wastes for by-products and proper industrial-waste treatment prior to disposal.

Beychok in an excellent new text [6] on refinery wastes presents a considerable amount of basic theories and practical formulations, tables, and graphs, as well as calculations to enable the engineer to predict composition and quantities of contaminants in various wastes. He discusses in detail three major methods of treatment: (1) in-plant pretreatment, (2) API and similar separators, and (3) secondary treatment following the use of separators. He also includes a much-needed chapter on cost data. The reader is urged to refer to this text for a comprehensive treatment of petrochemical processing and wastes.

The U.S. Department of the Interior [1967, 3] reports that the waste-treatment methods applicable to petroleum refineries can be divided into five types: physical, chemical, biological, tertiary, and special in-plant methods. (1) Physical methods include gravity separators, air flotation (without chemicals), and evaporation. Gravity separators (API and earthen

Table 27.26 Oil refinery waste: classification of substances found in wastes from the oil refinery industry as prepared by the American Petroleum Institute.*

Refinery unit	Native solutes and those present in end-products	Solutes resulting from chemical reactions	Naturally occurring emulsoids and suspensoids, and those persisting in end-products	Suspensoids and emulsoids resulting from chemical and physical actions
1. Oil storage	Organic sulfur compounds† Acids: H_2S, CO_2, organic acids Inorganic salts: NaCl, $MgCl_2$, Fe and Al compounds, $CaCl_2$ $(NH_4)_2S$, etc.		Suspended matter in tank bottoms Insoluble salts, SiO_2, $Al_2(SiO_3)_3$, S, finely divided substances Asphaltic compounds (in some cases)	
2. Distillation A. Straight distillation	Organic nitrogen compounds‡	Inorganic salts, sulfites, acid sulfites, Na_2CO_2, $(NH_4)_2S$, Na_2S, sulfates, acid sulfates	Insoluble organic and inorganic salts, S compounds, sulfonic and naphthenic acids, and insoluble mercaptides	Oil-water emulsions from steam in towers, etc.
	Organic sulfur compounds†	Acids and alkalis: H_2S, NaOH, NH_4OH, $Ca(OH)_2$		Soaps
	Phenol and like compounds	$(NH_4)_2SO_4$, $(NH_4)_2S$, NH_4Cl		Waxy emulsions
	Naphthenic acids			Oxides of metals
B. Cracking and distillation	Same as 2.A	Same as 2.A with the addition of phenols and phenolic compounds	Suspended coke	Same as 2.A
			Insoluble salts, FeS, and SiO_2	
3. Treating A. Sweetening, sulfuric acid, neutralization	Organic sulfur compounds† Organic nitrogen compounds‡ Naphthenic acids	Organic sulfur compounds Phenolic and sulfonate compounds Weak H_2SO_4 and other acid solutions	Suspended matter: Pbs, S, S compounds Acid and alkaline sludges Polymers and resins	Waxy emulsions Oil-water emulsions Inorganic salts: PbS, $CaSO_4$, $CaHPO_4$

Table 27.26 (*continued*)

Refinery unit	Native solutes and those present in end-products	Solutes resulting from chemical reactions	Naturally occurring emulsoids and suspensoids, and those persisting in end-products	Suspensoids and emulsoids resulting from chemical and physical actions
	Phenylates	Weak alkaline solutions Soaps Inorganic salts: $CaCl_2$, Na_2CO_3, Na_2SO_4, NaCl Oxides dissolved in alkaline solutions as PbO, CuO, etc.		Soaps Oxides: PhO, Fe_2O_2
B. Clay	Same as 3.A	Organic sulfur compounds Phenolic and sulfonate compounds Weak H_2SO_4 and other acid solutions Inorganic salts: $CaCl_2$, Na_2CO_3, Na_2SO_4, NaCl	Suspended clay, earth Polymers and resins	Suspended clay, earth SiO_2, H_2SiO_3, $Al(OH)_3$
4. Recovery A. Gas purification and recovery	Organic sulfur compounds† Organic nitrogen compounds‡	Inorganic salts: sulfates, acid sulfates, sulfites, acid sulfites, FeS, $(NH_4)_2S$, Na_2CO_3, Na_2S Mercaptides	Insoluble S compounds and mercaptides	Suspended Fe and S compounds

Table 27.26 (continued)

B. Acid recovery	Sulfonates Mineral acids Organic nitrogen compounds‡	Inorganic salts, and H³SO₄, SO₂, SO₃ Organic esters	Organic suspensoids: tars Some sulfur compounds§ Some nitrogen compounds**	Acid sludges
5. Miscellaneous A. Cooling- and boiler-water treating		Inorganic salts: BaCl₂, NaCl, NaHCO₃, Na₂SiO₃, CaCl₂, MgCl₃, Na₂CO₃, Na₂SO₄, Na₂HPO₄, CaHPO₄, etc.		Insoluble and colloidal compounds: CaCO₃, BaCO₃, Ca(OH)₂, Mg(OH)₂, Ba(OH)₂, Ca₃(PO₄)₂
B. Fire protection		Inorganic salts: NaHCO₃, Na₂SO₄, Al₂(SO₄)₃ Organic compounds		

*Asphalt and lubricating-oil units have not been included in the separate headings. Straight-run distillation involves the heavier oils, also sulfuric-acid treating embraces that of lubricating oils as well as light oils.

†Under this caption are included mercaptans, dialkyl sulfides, sulfonic acids, some alkyl and aryl sulfides, etc. Only a few of these compounds will be found in any one type of oil.

‡Under this caption are included amines, some amides, quinolines, and pyridines. Only a few of these compounds will be found in any one type of oil.

§Some alkyl sulfides, thiophenes, etc. Not all of these compounds are found in any one type of oil.

**Quinolines, some amides, pyridines, and some aryl amines. Not all of these compounds are found in any one type of oil.

Table 27.27   Characteristics of typical refinery wastes. (After Westom [112].)

Characteristic	Type of waste					
	Water layer	Water layer	Milk-water emulsion	Emulsion	Emulsion	Emulsion
Source of waste	Slop oil treatment	Slop oil treatment	Water wash	Bar condenser	Jet vacuum pump	Desalting
Type of unit	Spent caustic and heat, plant scale	Vacuum pre-coat filter, pilot scale	Treating	Combination unit	Lube oil vacuum still	Desalting unit
Quantity of waste			1 bbl/bbl product			
Acidity, ppm			15,313	59.5	520	739
Alkalinity, ppm				4.1	225	2.0
Ammonia, ppm						
BOD, ppm	5660–14,440	77–153	7,900	72.3	425	404
COD, ppm	22,000–56,000	500–1360	86,775			865–3031
Odor threshold				1.0		1.8
Oil, ppm	4900–10,300	37–130	31,600	236	94.3	32–713
pH	10–10.2	6.9–7.7	9.79	7.22	7.03	9.26
Phenol, ppm				2.3		4.1
Sulfide, ppm				1.3		
Suspended solids, ppm	60–940	30–139				
TLM, 24-hr*						

Table 27.27 (continued)

Characteristic	Type of waste					
	Condensate	Condensate	Condensate	Acid	Acid	Spent caustic
Source of waste	Low sulfur gas separator	Viscous breaker gas separator		Unit sewer	Unit sewer	Alkylate wash
Type of unit	Combination unit	Combination unit	Light oil recovery	Alkylation unit	H₂SO₄ sludge conversion unit	Chemical manufacturing
Quantity of waste						0.21 lb NaOH used per lb product
Acidity, ppm	1518	2963	69–13,175	1,105–12,325	1140–10,050	
Alkalinity, ppm	500	130–999	3–8350			46,250
Ammonia, ppm	408	3040	55–9500	1.2	2–13	256
BOD, ppm	1204	7239	214–16,255	31	10–272	3,230
COD, ppm	2.5	4.5	3.5	1251	910	
Odor threshold	3	2.5	6–230	3.7	3–5	1.5
Oil, ppm	8.5	7.85	5.0–9.2	131.5	124	10
pH	0.06	156	0–213	0.6–1.9	1.71	12.8
Phenol, ppm	600	1500	T–5000			50
Sulfide, ppm						2
Suspended solids, ppm						253
TLM, 24-hr*			14	0.4	3	

Table 27.27 (*continued*)

Characteristic	Spent caustic	Spent caustic	Carbolate	Alkaline	Special chemicals	Special chemicals	Refinery separator inlet
							Type of waste
Source of waste	Catalytic polymerization	Naphtha wash	Naphtha wash	Shut-down	Unit sewer	Unit sewer	
Type of unit		Treating	Fluid catalytic cracking unit	Caustic methanol sweetening	Alkylation sulfonation	Detergent drying	
Quantity of waste	0.8 lb NaOH used per bbl product						9.6 mgd
Acidity, ppm					18–245	150	0–188
Alkalinity, ppm	209,330	80,020	247,900		1.8	0.2–12.6	0–92
Ammonia, ppm					28–151	8–1,180	96–501
BOD, ppm	8,440	51,154	363,600		259–5382	2,585–51,350	
COD, ppm	50,350	144,120	901,200	371–299,000	1.2		3900–6500
Odor threshold	T–12				13–1310		
Oil, ppm			13 +	8–58	4.6	40–7,750	3.25–5.32
pH	12.9	13.4		9.5–12.5	0–7.4	7.37–9.31	
Phenol, ppm	22.2		309,300				2.4–6.2
Sulfide, ppm	3,060		0–3,380				0.5–7.9
Suspended solids, ppm	54–279	23,312	678				131–678
TLM, 24-hr	0.04						33

* Approximate 24-hour median tolerance limit for bluegill sunfish, expressed as percentage concentration of waste.
† Waste represents combined discharges from a complete refinery, with the exception of naphtha-treating wastes.

Table 27.28 Reduction in refinery effluent contaminants by secondary treatment [1].

Characteristic	Reduction, %	
	Chemical flocculation	Biological treatment
Odor	95	90
Turbidity	93	85
Oil	90	75
Suspended solids	65	45
BOD	50	85
Phenolics	0	90

basins), which are used in practically all refineries, are designed primarily for removal of floatable oil and settleable solids. They remove 50 to 99 per cent of the separable oil and 10 to 85 per cent of the suspended solids, as well as BOD, COD, and phenol, at times to a substantial degree depending on the influent wastewater characteristics. Air flotation without chemical addition obtains comparable results. Pollutant removals by evaporation ponds are very high, but application of this method is severely limited by location, climate, and land-availability considerations. (2) Chemical methods (coagulation-sedimentation and chemically assisted air flotation) are more effective in oil and solids removal, particularly in respect to emulsified oil. (3) Biological methods include activated sludge, trickling filters, aerated lagoons, and oxidation ponds. In general these treatment processes require waste-water pretreatment to remove oil and remove or control other conditions (such as pH and toxic substances). The activated-sludge process is the most effective for removal of organic materials (the main purpose of biological treatment); expected removal efficiencies are 70 to 95 per cent for BOD, 30 to 70 per cent for COD, and 65 to 99 per cent for phenols and cyanides. (4) Tertiary treatment to date has been limited to activated carbon and ozonation, which are effective in removing taste and odor and refractory organic substances from biologically treated waste waters. (5) The most important in-plant treatment methods are sour water stripping, neutralization and oxidation of spent caustics, ballast-water treatment, slop-oil recovery, and temperature control. These measures substantially reduce the waste loadings in the influent to general refinery treatment facilities and are necessary to ensure reasonable performance of these facilities.

References: Oil-Refinery Wastes

1. Giles, R. N., "A rational approach to industrial waste disposal problems," *Sewage Ind. Wastes* 24, 1495 (1952).
2. 78th U.S. Congress, First Session, House Document 266, "Industrial waste guide to the oil industry," *Ohio River Pollution Control Survey*, Supplement D, Appendix IX p. 1175, (1943).
3. "Oil pollution of water supplies," *J. Am. Water Works Assoc.* 58, 3 (1966).
4. Rudolfs, W., *Industrial Waste Treatment*, Reinhold Publishing Corp., New York p. 427, (1953).
5. Weston, R. F., "Waste disposal and utilization problems of the petroleum industry," in Proceedings of 1st Industrial Waste Conference, Purdue University, p. 98, November 1944.
6. Beychok, M. R., *Aqueous Wastes from Petroleum and Petrochemical Plants*, London: John Wiley.
7. Rhee, C. H. et al., "Removal of Oil and Grease in the Hydrocarbon Processing Industry," Proc. **43** Purdue Ind. Waste Conf., 143, (1988).
8. Rebhun, M. and N. Galil, "Biotreatment Inhibition by Hazardous Compounds in an Integrated Oil Refinery," Proc. **43** Purdue Ind. Waste Conf., 163, (1988).
9. Krafta, M. et al., "Development of Low-Cost Flotation Technology and Systems for Wastewater Treatment," Proc. **42** Purdue Ind. Waste Conf., 185, (1987).
10. Mueller, J. A. et al., "Nitrification in Refinery Wastewater Treatment," Proc. **40** Purdue Ind. Waste Conf., 507, (1985).
11. Sprehe, T. G. et al., "Process Considerations in Land Treatment of Refinery Sludges," Proc. **40** Purdue Ind. Waste Conf., 529, (1985).
12. Diesterweg, G. et al., "Tower-Biology and its Application for the Nitrification/Denitrification of Ammonia Rich Wastewater," Proc. **40** Purdue Ind. Waste Conf., 535, (1985).
13. Chou, C. C. and R. A. Swatloski, "Biodegradation of Sulfolane in Refinery Wastewater," Proc. **37** Purdue Ind. Waste Conf., 559, (1982).
14. Christiansen, J. A. and P. W. Spraker, "Improving Effluent Quality of Petrochemical Wastewaters with Mutant Bacterial Cultures," Proc. **37** Purdue Ind. Waste Conf., 567, (1982).

15. Kramer, G. R. et al., "Electrolytic Treatment of Oil Wastewater," Proc. **34**, Purdue Ind. Waste Conf., 673, (1979).

16. Meiners, H. and G. Mazewski, "Design Start-Up and Operation of a Refinery Treatment System," Proc. **34** Purdue Ind. Waste Conf., 710, (1979).

27.10 Petrochemicals

The petrochemical industry has been defined by the EPA for effluent guideline purposes as "the production of second generation petrochemicals (i.e., alcohols, ketones, cumene, styrene, etc.) or first generation petrochemicals and isomerization products (i.e., BTX, olefins, cyclohexane, etc.) when 15 per cent or more of refinery production is as first generation petrochemicals and isomerization products."

The petrochemical industry is very diverse and complex. The number of companies involved in the manufacture of chemicals which start with raw materials such as petroleum, coal, natural gas, and miscellaneous sources of carbon, has risen in the last 60 years.

From the basic raw materials a variety of petrochemicals are produced.

Raw materials
Petroleum
Coal
Natural Gas
Miscellaneous sources of carbon

First generation of petrochemicals

Alkenes	Ethylene
Olefins	Propylene
Paraffins	Butylene
Aromatics	Higher olefins

Basic intermediates
Ethylene Oxide
Dichloroethane
Ethyl Benzene
Acetaldehyde
Ethyl Chloride

Products

Synthetic detergent bases	*Solvents*	*Fuel additives*
Non-ionic detergents	Trichloroethylene	Tetraethyl lead
Ethanolamines	Perchloroethylene	Ethyl bromide
	Methylchloroform	

Plastics & resins	*Synthetic fiber bases*
Polyethylene glycol	Acetic anhydride
Polyvinyl chloride	Ethylene glycol
Polyethylene	Ethanolamines
Polystyrene	Acrylonitrile
Polyester	Acetic acid
Synthetics	

In the production of petrochemicals, a chemical reaction is usually encountered, where water is produced, becoming one of the waste products. Water is also used as a medium of transport for the chemical reaction. In both cases the water is heavily contaminated and it is an added waste stream not common in other industries.

Purification of organic compounds is achieved using several distillation columns. The bottom stream from the column represents the concentrated impurities that have been separated from the purified product and poses another problem. This stream is commonly incinerated for disposal.

Petrochemical products have an extraordinarily degrading effect on the environment.

Characteristics. As can be expected from such a wide range of products, the wastewater has varying characteristics.

1. High concentration of COD
2. High total dissolved solids
3. High COD/BOD ratio
4. Compounds inhibitory to biological treatment
5. High frequency of spillage
6. Heavy metal contamination due to catalyst used
7. Product revision causing variations in waste characteristic

Treatment. Chemical recovery, water reuse, and recycle of wastewater for cooling towers is common practice prior to treatment. This not only reduces the volume for treatment but it is an added benefit due to the high cost of chemicals. One practice unique to the petrochemical industry which decreases waste treatment is the generation of the influent of one process from the waste stream of a second process. In some cases the waste has to be transported overland for several miles, but even with the added cost of transportation this process still reduces waste treatment costs for the industries involved.

Once all of the above techniques for waste reduction have been exhausted, the remaining waste stream must be treated. Pretreatment includes equalization, pH adjustment, flocculation/coagulation, primary settling, floation, and oil separation.

Biological treatment uses trickling filters, activated sludge, aerated lagoons, oxidation ponds, and anaerobic digestion. Physical-chemical treatment uses activated carbon.

New processes of petrochemical waste treatment include the following.

The hydrocyclone process involves using a solid-liquid separation device similar to that used in the more widely known gas-solid cyclone. Other hydrocyclone methods use a filtration screening mechanism which catches particles, imbeds them, and holds them with a porous media that permits fluid to pass through. Hydrocyclone processes use the force of gravity and actually "part" the solids from the fluids.

Another process includes adsorption with molecular sieves, which are suspended in a slurry that comes in contact with the waste stream. One advantage to this process is its high gas-phase efficiences at high loadings. Langmuir adsoprtion isotherms versus linear adsorption isotherms result in increased efficiencies.

Photo-oxidation engineers have always been interested in the ultraviolet photolysis; but, because it involves great expense, it has never been applied. Dye-sensitized aerobic photo-oxidation, which is energized by visible light, is inexpensive.

Reverse osmosis-investigation is delving into the types of material used for membranes and the application of pressurized systems.

Ultrafiltration is a process where feed is pumped through a membrane system; materials pass through the membrane under applied hydrostatic pressure. The pore structure of the membrane acts as a filter, passing small solutes and retaining larger suspended material.

During the pervaporation process the contaminant is concentrated by passing the solute through a membrane. Molecules must diffuse through the liquid film. Once diffused these particles must be adsorbed and removed from the downstream surface of the film and introduced into the gas stream above the membrane.

1. Anonymous, "Hard to treat waste problem is solved," *Water and Wastes Eng.* **10**, D19, January 1973.
2. Anonymous, "Petrochemicals get roasted," *Chem. Eng.* **8**, 32, February 1974.
3. Baker, C. D. *et al.*, "Recovering para-cresol from process effluent," *Chem. Eng. Proc.* **69**, 77, August 1973.

4. Brumotts, V. A., *et al.*, "Granular carbon handles concentrated wastes," *Chem. Eng. Proc.* **69**, 81 August 1974.

5. Day, R. W., "Hydrocyclone in process and pollution control," *Chem. Eng. Proc.* **69**, 67 September 1973.

6. Desai, S. V., *et al.*, "Pervaporation for removal of dilute contaminants from petrochemical effluent streams," *AICHE Symp. Ser.* **135**, 169 (1973).

7. Eden, H. P., "Wastewater treatment and recovery through reverse osmosis," *AICHE Symp. Ser.* **135**, 167 (1973).

8. Edwards, V. H., *et al.*, "Removal of 2,4 D and other persistent organic molecules from water supplies by reverse osmosis," *Am. Water Works Assn.* **66**, 610 (1974).

9. Fichel, R. G. "Continuous adsorption, a chemical engineering tool," *Am. Inst. of Chem. Eng. Symposium Ser.* **135**, 64 (1973).

10. Ford, D. C., *et al.*, "Pollution control in new petrochemical complex," *Environ. Sci. and Tech.* **7**, 906 (1973).

11. Goldsmith, R. L., *et al.*, "Ultrafiltration of soluble oil waste," *J. Water Poll. Contr. Fed.* **46**, 2183 (1974).

12. Guisti, D. M., *et al.*, "Activated carbon adsorption of petrochemicals" *J. Water Poll. Cont. Fed.* **46**, 947 (1974).

13. Hamada, M. F., *et al.*, "Organics removed by low pressure R.O.," *J. Water Poll. Contr. Fed.* **45**, 2146 (1973).

14. Hoffman, T., *et al.*, "Stimulation of a petroleum refinery waste treatment process," *J. Water Poll. Contr. Fed.* **45**, 2321 (1973).

15. Keilman, D. J., *et al.*, "Biological treatment of wastewater from a polyolefin plant," *AICHE Symp. Ser.* **135**, 172 (1973).

16. Lawson, C. T., *et al.*, "Limitations of activated carbon adsorption for upgrading petrochemical effluents," *AICHE Symp. Ser.* **136**, 577 (1974).

17. Maijaji, Y., *et al.*, "Biological treatment of industrial wastewater by using nitrate as an oxygen source," *Water Resources* **9**, 95 (1975).

18. Mills, K. J., "Today's toxic hazards require advanced measurement techniques," *Proc. Eng.* **86** (1975).

19. Minor, P. S., "Organic chemical industry's wastewater," *Environ. Sci. and Tech.* **8**, 621 (1974).

20. Mohler, G. F., *et al.*, "Bio-oxidation process saves water," *Hydrocarbon Proc.* **52**, 84 (1973).

21. Prengle, H. W., Jr., *et al.*, "Recycle wastewater by ion exchange," *Hydrocarbon Proc.* **54**, 173 (1975).

22. Price, K. S., *et al.*, "Brine shrimp bioassay and seawater BOD of petrochemicals," *J. Water Poll. Cont. Fed.* **46**, 63 (1974).

23. Reiner, R. E., *et al.*, "Pure O_2 activated sludge treatment of a petrochemical waste," *27th Purdue Ind. Waste Conf.*, Purdue University, 840 (1972).

24. Sargent, J. W., *et al.*, "Light energized oxidation of organic wastes," *J. Water Poll. Cont. Fed.* **46**, 2547 (1974).

25. Selmar, H., "Substitution of toxic materials," *Plating* **61**, 332 (1974).

26. Simons, W. T., "Slurry bed adsorption with molecular sieves," *Chem. Eng. Proc.* **70**, 70 (1974).

27. Stevens, B. W., *et al.*, "Recovering organic material from wastewater," Chem. Eng. **82**, 84 (1975).

28. Waggy, G. T., *et al.*, "Identification and control of petrochemical pollutants inhibitory to anerobic process," *USEPA Office of Monitoring and Research*, EPA-R2-73-194 (1973).

29. Wallenbrunk, R., "Wastewater reuse and inplant treatment," *AICHE Symp. Ser.* **135**, 153 (1973).

27.11 Fuel and Lub-Oil Wastes

Origin of wastes. Fuel-oil wastes are also a problem at the distribution stage. Two major types normally occur in metropolitan areas.

The first fuel-oil waste is found at refueling stations for diesel trains. A typical diesel engine may contain several fuel tanks on each side. These tanks are filled by 3-inch hose-pipes under pressure. Although most hose-pipes are fitted with automatic shutoff valves that close when each fuel tank is full, this procedure is usually too slow and inefficient for the operators. In some cases the hoses are not removed from the tanks until the fuel oil has reached the very top and overflows. The oil spills onto the surrounding ground and over a period of time saturates the soil. Subsequent rains may carry this oil into a nearby receiving stream.

The second oil waste originates from automobile and truck crankcase oil. The dirty crankcase oil drained out at garages or terminals was formerly collected and resold to refineries for reprocessing. Since the Federal Government recently eliminated its financial support to the refineries for fuel-oil reprocessing, most refineries are not too enthusiastic about collecting, transporting, and reprocessing this oil. In fact, garages, instead of receiving some payment for the old oil, are now often forced to pay the refineries to take the oil away. In some instances this cost may force collectors of crankcase oil to let these wastes find their way to drainage ditches and creeks.

Waste characteristics. Although the wastes are obviously various dilutions of oil in water, no specific analyses have been officially published.

Waste treatment. The only effective treatment is retention of the oily wastes for shipment to refineries. When the oils have been contaminated with water, skimming often precedes collection for transportation away from the site. Lub oils can be heated, coagulant added, and centrifuged prior to reuse.

27.12 Rubber Wastes

Origin of rubber wastes. Rubber is not one substance, but many. The principal types of rubber are: (1) natural rubber—all rubberlike materials produced by coagulation of the rubber-plant sap (latex); (2) synthetic rubber, made by copolymerization of butadiene and styrene (GR-S) or isoprene and butadiene with small amounts of isobutylene (GR-1) for non-oil-resisting rubbers and neoprene-type oil-resisting rubbers (polymers of chloroprene); (3) scrap rubber, a mixture of discarded rubber items and residues from manufacturing processes; (4) rubberlike plastics, including a group of nonrigid ones, both thermoplastic and thermosetting. Rubber latex is usually distributed to manufacturers of finished rubber products as a milky colloidal emulsion of sugars, resins, and protein constituents. The composition of rubber varies widely with the source, but the useful constituent is rubber hydrocarbon, which makes up 90 per cent of plantation rubber. It is now universally accepted that the unit structural group in the rubber molecule is $(C_5H_8)_n$,

$$
\begin{array}{ccc}
H & & H \quad H \\
C = C & - & C = C, \\
H \quad | & & H \\
\quad HCH & & \\
\quad H & &
\end{array}
$$

with the configuration which is 2-methyl-butadiene-1, 3, commonly known as isoprene. Naturally, the most important characteristic of rubber is its high modulus of elasticity, the combination of strength, flexibility, and resilience.

Wastes from the production of rubber have a high BOD, taste, and odor; the problems they present vary considerably, depending on the plant site, the raw material used, and the number of intermediary products. Rubber-manufacturing wastes may be divided into four general classes: (1) steel-products wastes, (2) rubber-commodities wastes, (3) reclaimed-rubber wastes, and (4) synthetic-rubber wastes.

To explain these classifications: the same company that produces the rubber product may also manufacture steel rims for wheels, metal parts for mechanical rubber goods, stainless-steel-lined beverage containers, and so forth. As a result, these rubber wastes would include zinc- and brass-plating wastes, and other metal-manufacturing wastes.

The making of rubber commodities involves washing, compounding, calendering, and curing processes, followed by the actual manufacturing of all kinds of rubber products. The wastes include a large volume of washing waters with the impurities exuded by the crude rubber.

Reclamation of used rubber is accomplished by shredding the old rubber and discharging it on a travelling belt, where it passes under a magnet which removes any bits of metal. The ground rubber and fabric, freed of metal, are subjected to a caustic treatment under high temperatures for several hours, which destroys the fabric and frees the rubber. The recovered rubber is then washed, dried, milled, strained, and refined for reuse.

Synthetic rubber evolves from a process whereby butadiene is mixed with some other monomer such as styrene or acrylonitrile plus a catalyst, in a soap solution, to produce synthetic latex. Coagulation of the latex either in an acid-brine solution or with alum follows, after which the latex is washed, dried, and baled. Waste from the synthetic-rubber plant consists of whatever coagulated rubber escapes, plus the acid and saline liquid, and occasional batches of materials that will not polymerize properly.

Characteristics of rubber wastes. Schatze [1] presents some of the sanitary characteristics of reclaimed and synthetic rubber wastes (Table 27.29). Black [2] and Rostenbach [3] give other analyses of typical rubber wastes (Tables 27.30 to 27.32).

Table 27.29 Characteristics of reclaimed and synthetic rubber wastes. (After Sechrist and Chamberlin [4].)

Characteristic	Reclaimed rubber	Synthetic rubber
Total solids, ppm	16,800–63,400	1900–9600
Suspended solids, ppm	1,000–24,000	60–2700
Oxygen consumed, ppm	3,600–13,900	75–4500
BOD, ppm	3,500–12,500	25–1600
Chlorides, ppm	130–2,000	90–3300
Hydroxide alkalinity, ppm	0–2,700	
pH	10.9–12.2	3.2–7.9

Table 27.30 Analysis of typical wastes from the synthetic-rubber industry. (After Black [2].)

Plant area	Approximate daily discharge, mgd	pH	Total solids, ppm	Suspended solids, ppm	5-day BOD, ppm	Odor concentration
1. Butadiene waste	1.90	2.8	300	27.6	2550	16,100
Styrene waste	4.62	6.2	150	4.5	180	690
Copolymer plant						
3. Process waste	2.34	4.3	5580	12.3	69	62
4. Recovery and reactor	0.39	8.0	570	23.6	492	8,760
5. Main sewer*	2.63	7.0	6530	46.0	168	930
6. Main outfall, institute†	119.40	5.4	270	15.5	81	460

*Represents the mixture of 3 and 4.
†Represents the mixture of 1, 2, and 5, plus large quantities of condenser water.

Table 27.31 Wastes from combined operations of three plants: butadiene, GR-i, and RD-S [3].

Characteristic	Value or concentration
pH	7.6
Oil, ppm	9
Suspended solids, ppm	79
BOD, ppm	78
Dissolved oxygen, ppm	2.0
Flow, gpm	2000

Table 27.32 Wastes from combined operations of three plants: butadiene, styrene, and GR-S [3].

Characteristic	Value or concentration
pH	8.6
Oil, ppm	18
Settleable solids, ml/liter	1.4
Odor threshold	16
Flow, gpm	
Styrene plant	190
GR-S plant	735
Butadiene plant	1120
Total	2045

Study of these tables reveals that rubber wastes are highly objectionable because of their high BOD and odiferous nature. The odors are detectable even at extremely low concentrations and make water unpalatable for several hundred miles downstream from a rubber plant.

Treatment of rubber wastes. The most common means of treatment practiced today are aeration, chlorination, sulfonation, and biological methods; coagulation, ozonation, and treatment with activated carbon are also used.

Black [2] presents the results of aeration achieved by spraying rubber wastes at a pressure of 10 psi (Table 27.33). Increase in aeration pressure above 10 psi improved the efficiency. Spray aeration was less effective than diffused-air aeration, and both methods were less effective with styrene wastes than with butadiene wastes. Chlorination has been used to reduce the phenolic constituents of rubber wastes [2, 4], and sulfonation of styrene waste can yield an almost odorless waste [2] if sufficient time is given for the

reaction (Table 27.34). In comparison with aeration, chlorination, and sulfonation, which are valuable mostly for their capability as to odor removal, biological treatment of rubber wastes is a method which affords the greatest reduction of BOD (Tables 27.35 and 27.36). However, addition of nitrogen and phosphorus, or mixture with domestic sewage in a ratio of one to three, is required for efficient biological treatment of rubber wastes.

In certain cases rubber wastes are treated by trickling filtration. Culver [1963, 5] stated that combined treatment of rubber wastes with municipal sewage could be achieved by the activated-sludge process and a BOD removal of 85 per cent was obtained. Morzycki [1966, 6] reported that sewage containing synthetic latexes was purified by coagulation with CaC_2, $MgSO_4$, $Al_2(SO_4)$, $FeCl_3$, and $Fe_2(SO_4)_3$; optimum results were obtained in experiments involving $Al_2(SO_4)_3$, and iron salts and BOD was reduced by 84 to 87 per cent. Ruchhoft [7] mentioned that the coagulation method is of little value because less removal of taste and odor is

Table 27.33   Effect of spray aeration on butadiene and styrene wastes and their dilutions [2].

Material	Concentration, %	Temperature of treatment, °C	Odor concentration		
			Initial	After bubble aeration	After spray aeration
Butadiene	100	24	1200	100	250
	4*	25.5	64		16
	4*	50	64		16
	4	50	16		8
	4	50	8	1	8
	2	24	32		8
	1	50	2	None	2
	1	24	4		4
	1	50	8		4
	1*	50	16		8
	1*	24	16		8
	0.5	24	8		4
Styrene	100	24	600	500	600
	4	24	64	4	16
	1	24	16		4
	0.5	50	8		4

*Duplicates except for temperature of treatment.

Table 27.34 Odor concentrations obtained after treating butadiene and styrene wastes with Na_2SO_3 and Na_2S [2].

Waste and original odor level	Days treated	Treated with Na_2SO_3			Treated with Na_2S		
		100 ppm	250 ppm	500 ppm	100 ppm	250 ppm	500 ppm
Butadiene (4100)	1	4100	4100	4100	4100	4100	4100
	3	4100	4100	4100	8200	8200	8200
	4	4100	4100	4100	4100	8200	8200
	5	4100	4100	4100	4100	4100	4100
	6	4100	4100	4100	4100	4100	4100
	7	4100	4100	4100	4100	4100	4100
	10	1000	2000	2000	4100	4100	4100
	13	500	1000	1000	2000	2000	2000
	17	250	250	250	500	500	500
Styrene (1000)	1	128	64	64	65	6	128
	2	64	32	32	32	32	64
	3	32	32	32	32	32	32
	5	16	16	16	8	16	32
	6	16	16	16	8	8	16
	7	8	16	16	8	8	8
	8	4	16	8	4	4	8
	9	4	8	8	4	4	4

Table 27.35 Treatment of composite rubber waste by the use of an experimental trickling filter. (After Black [2].)

Feed applied to filter	Total time, hr	Total amount applied		pH		5-day BOD, ppm		BOD removed, %
		liters	mgd	Influent	Effluent	Influent	Effluent	
50% composite rubber waste and 50% sewage	0	0	0	6.3		340		
	16.5	14.0	1.49	6.9	7.8	340.5	25.3	92.6
	20.0	16.0	1.01	6.9	7.7	404	18.2	95.5
	23.0	18.0	1.16	6.9	7.8	398	39.6	90.1
	39.5	42.0	2.56	6.7	7.9	359	35.2	90.2
	42.5	43.0	0.59	6.9	7.8	252	28.8	88.6
	45.5	44.0	0.59	6.9	7.7	436	46.2	89.4
	68.0	58.0	1.10	6.9	7.7	195	32.5	83.3
	72.0	60.0	0.88		7.7	230	19.3	91.6
Fresh sewage only	24		1.0	7.4	7.5	292	205.5	29.6
	48		1.0	7.7	7.5	189	26.9	85.8
	72		1.0	7.7	7.6	263*	34.5	86.9
100% neutralization of composite rubber waste	2			7.3	7.9	328	72	78.0
	5			7.2	7.9	392	35	91.1
	7	4.0	1.01	7.3	8.4	385	71.2	78.9
	24	15.0	0.81	7.4	8.3	445	35	92.1
	31	18.0	0.99	7.4	8.4	484	43.8	91.0

*4-day BOD

Table 27.36   Effects of adding butadiene and styrene wastes to activated sludge. (After Black [2].)

Type of waste	Concen-tration of waste, %	Feed*	pH	Initial		4-hr effluent		24-hr effluent	
				Suspended solids, ppm	BOD, ppm	BOD, ppm	Reduction, %	BOD, ppm	Reduction, %
Neutralized	3.0	C	7.7	2364	320.0	27.2	92	10.7	96
		B	7.7	2196	371.1	40.3	91	13.6	96
		S	7.7	2244	310.0	47.4	85	12.2	96
Neutralized	6.0	C	7.5	2176	167.2	50.8	70	12.9	92
		B	7.7	2128	405.3	53.1	81	12.4	95
		S	7.4	2176	161.3	69.1	57	9.5	94
Neutralized	12.0	C	7.1	1828	185.3	59.2	68	13.1	93
		B	7.7	2128	405.3	131.4	67	20.0	95
		S	7.1	1940	168.4	95.6	43	8.8	95
Neutralized	24.0	C	7.1	1712	167.4	53.5	68	17.8	89
		B	7.7	2080	674.7	412.5	39	53.8	92
		S	7.3	1668	187.1	146.8	22	15.6	92
Sample and fresh-filtered sewages	0.0	C	7.7	1720	301.7	108.1	64	6.7	98
		B	8.5	2384	307.6	72.8	76	9.5	97
		S	7.7	1672	301.2	91.9	70	63	98
Sample and fresh-filtered sewages	0.0	C	7.5	1696	151.7	69.5	54	11.8	92
		B	8.4	2416	152.6	78.4	49	6.7	96
		S	7.5	1748	151.5	93.8	38	7.5	95
Unneutralized	3.0	C	7.3	1644	189.3	45.0	76	15.5	92
		B	6.6	2320	259.2	26.7	90	5.7	98
		S	7.1	1716	195.1	44.6	77	11.2	94
Unneutralized	6.0	C	6.9	1492	71.2	27.9	61	25.3	65
		B	4.7	2244	220.6	29.7	86	11.2	95
		S	6.7	1576	93.8	35.0	37	16.2	83

*C, control; B, butadiene; S, styrene.

obtained; ozone treatment is also not valuable, but activated carbon completely removes taste and odor, provided 100 ppm of carbon is applied.

References: Rubber Wastes

1. Schatze, T. C., "Effect of rubber wastes on sewage treatment process," *Sewage Works J.* 17, 497 (1945).

2. Black, O. R., "Study of wastes from rubber industry" *Sewage Works Journal* 18, 1169, (1946).

3. Rostenbach, R. E., "Status report on synthetic rubber wastes," *Sewage Ind. Wastes* 24, 1138 (1952).

4. Sechrist, W. D., and N. S. Chamberlain, "Chlorination of phenol-bearing rubber wastes," in Proceedings of 6th Industrial Waste Conference, Purdue University, November 1951, p. 396.

5. Mills, R. E., "Progress report on the bio-oxidation of phenolic and 2,4-D waste waters," in Proceedings of 4th Ontario Industrial Waste Conference, June 1957, p. 30.

6. Morzycki, J. et al. "Effluents from sewage contaminated with latex" *Chem. Abstracts* 64, 3192 (1966).

7. Ruchhoft, C. C., *et al.*, "Synthetic rubber waste disposal." *Sewage Ind. Wastes* 20, 180 (1948).

27.13 Glass-Industry Wastes

In the manufacture of optical and other specialty glass, polishing produces a waste containing detergents, finely divided iron from the polishing process, and a considerable quantity of glass particles, many of which are microscopic in size. These materials form a quasi-emulsion with the following characteristics: (1) brick-red to scarlet color, (2) low BOD, (3) alkaline reaction, (4) nonsettleable solids, (5) resistance to acid cracking, and (6) resistance to alum coagulation.

McCarthy [2] showed that coagulation with calcium chloride produced a clear supernatant, when 250 ppm were used. This method reduced the total solids from 1080 to 3 ppm, the BOD from 40 to 28 ppm, and the color from 900 to 35 ppm. No other coagulant produced comparable results.

References: Glass-Industry Wastes

1. Durham, R. W., "Disposal of fission products in glass," Paper no. 57-*NESC*-54, read at 2nd Nuclear Engineering and Science Conference, March 1957, at Philadelphia, American Society of Mechanical Engineers, New York.
2. McCarthy, J . A., "Coagulating rouge with calcium chloride," *Public Works Mag.* **85**, 170 (1954).

The following references have been published since 1969

1. Anonymous, "Industrial waste study of glass and concrete industry," *Commerce Bus. Daily* **16**, February 24, 1971.
2. Helbing, C. H., *et al.*, "Recycle and reuse of plant effluent at the PPG Industries works No. 50, Shelbyville, Indiana," *Proc. 26th Ind. Wastes Conf.*, Purdue University, 358 (1971).
3. Kohler, R., "Reiningung von quarz- und glasschleifereiabwaessern mittels aktivierter kieselsaeure als fickungshilfsmittel," *Wasser, Luft une Betrieb* **14**, 446–8 (1970).
4. Telle, A. J., "Control of emissions from glass manufacture," *Amer. Ceramic Soc. Bulletin* **51**, 637 (1972).
5. Venis, E., "Glass-making plant," *Chartered Mech. Eng.* **15**, 102 (1969).

27.14 Metal-Container Wastes

Very little has been written about the treatment of container manufacturing wastes. Some information does exist, however, on the treatment of wastes originating from the integral parts of the manufacturing process, namely, metal finishing, paint manufacturing, and container coating. The characteristics of wastes and waste treatment vary greatly with each step. Significant reductions can be accomplished "in-plant" long before end of pipe treatment must be considered.

Origin.   Container fabrication wastes contain metal fines, which can be settled out, and lubricating and shop oils, which respond to biological treatment and can therefore be discharged into municipal sewers.

The cleaning and surface pretreatment of the container is usually accomplished in baths and often is accompanied by rinses, as in the metal finishing industry.

Characteristics.   The characteristics of these baths are generally noted as extremes of pH, dissolved heavy metals and metal salts, phosphates and chromates, and limited BOD exerted by surfactants solvents and oils. New materials provide alternatives to the treatment of these wastes, having minor and major effects on the process. Substitution of nonphosphate cleansers to reduce part of the phosphate loading and nonchromium conversion coatings to eliminate chromates offer some improvements. The replacement of sheet metal coil stock with precoated stock produces the following changes in the manufacturing procedures (and sources of wastes): No lubricating oils are required; a water spray rinse replaces the cleaning solution bath; and, conversion coatings are eliminated.

Less drastic alterations on the process can help to minimize bath wastes. These alterations include methods in practice for a considerable time (e.g., spill prevention, good housekeeping, sealing floor drains, spray and fog rinsing, etc.). Only the recent refinements deserve mention here: (1) extend bath life with automatic replenishment of spent chemicals; (2) use countercurrent rinsing to force contaminant concentrations to first stage; (3) avoid dragout (that is, avoid loss of bath solutions during mechanized removal); improve rinsing with agitation, aeration, or sonic vibration; and (4) include floors and foundations in leak surveys to protect groundwater supplies.

Treatment. Eventually treatment of the spent bath wastes must be considered. Either an API separator or centrifuge can be employed for removal of oils with the latter offering greater efficiencies per unit area and flexibility for varied flow conditions. Oil emulsions have been successfully treated from various tanks at a metal finishing plant with a "portable" platform that combines heating, chemical addition, and reaction tank prior to centrifuging.

Highly acidic cleaning solutions can be mixed with chromium bath waters to prepare the latter for reduction and precipitation. Alkaline solutions require neutralization in acid mix tanks and, depending on phosphate and trace metal content, can either be settled and sewered or combined with phosphating solutions for further treatment.

The combination of precipitation and lime is commonly used for removal of phosphates and reduced chrome from solutions. In a comparison of total system costs, alum and ferric chloride were found to be less expensive than lime for both capital and operating costs categories. Metal finishing operations in proximity to a steel foundry would do best to consider combining bath wastes with steel pickle liquor wastes which contain ample quantities of the iron salts ($FeCl_3$). Another waste material (solid) which would reduce the chemical costs of precipitation is fly ash, as found in boiler flue gases.

Activated carbon adsorption has been used for the removal of both phosphates and chromium, especially when quantities of wastes are low (high concentrations). The adsorbed material may be removed for recovery or disposal depending on the chemical used for regeneration. Ion exchange can handle bath wastes on a much larger scale and can recover waste products in an equally pure state. A large amount of ion to be exchanged or high frequency of regeneration indicates the need for a moving-bed ion exchanger that can provide continuous regeneration.

A unique process of solvent extraction claims to make phosphate bath treatment profitable. Phosphate sludges are dissolved in acid and subjected to "selective extraction of metals with a specific acid solvent." Both the solvent and the metals are recovered in separate form with sulphuric acid, leaving a pure phosphate solution from which TSP can be crystalized. The cost of the major reagents is approximately $100/ton, whereas the market value for the by-products is $250/ton.

The method used for deriving pure water from salt water has been suggested as a method for chromium recovery. The bath mixture is subjected to freezing temperatures resulting in an ice and chromium solution slurry. Counter-washing in a vertical closed cylinder forces an ice plug up the column and the chromium solution downward for separate recovery.

The treatment of bath wastes can be avoided by concentration and containment of the bath wastes in either metal drums or lagoons. Sensing devices for pH and conductivity could be placed in the lagoon effluent channels to warn of any toxic escape into the environment or municipal sewer systems.

The container decorating and coating processes generate wastes from the cleaning of equipment and the scrubbing of stack gases from baking ovens. These wastes are composed of the raw materials used in printing inks (paints) and coatings—i.e., inorganic or organic pigments, oil or resin binders, heavy metals and metal salts, solvents, latex, driers, and brocides. The Federal Government EPA guidelines stated recently that the paint and ink formulating industry shall permit *no* discharge to navigable waters, a testament to the strength and toxicity of these wastes.

The trend in the decorating industry is toward using water reducible paints and coatings. This substitution will reduce the BOD that the petroleum solvents exert on receiving waters, both from paint formulations and equipment cleaning solutions. The numerous acrylics, epoxies, and polymers on the market allow any petroleum solvent paint to be duplicated in appearance and application with current machinery. The elimination of solvents entirely from paint formulations is available in 100 per cent solids thermosetting polymers and powder coatings.

Precoated stock that eliminates the need for base coat application and hi-gloss base coats that make overprint varnish unnecessary reduce pollution by eliminating the steps that contribute to it. Spray-on interior coatings produce far less waste from spillage and clean-up than their roll-on predecessors. Finally, scrubbing wastes can be avoided through substitution of an ultraviolet-cured finish rather than an oven-baked finish.

Many in-plant measures can be instigated similar

to metal finishing methods, but here the emphasis is placed on collection and reuse of cleanup water. Sumps or holding tanks in equipment areas would contain clean-up wastes for sedimentation prior to recycling. Volume reduction is achieved through high pressure cleaning. Concentrated spent solutions could then be drummed, barged, or injected into deep wells. In operations where solvent contamination must be confronted, a recovery system employing a vertical thin film evaporator can be used to separate coating or ink solids from their solvents for total reclaim.

Application of conventional methods of suspended-solids removal has met with only limited success. Finally divided pigment, resin, and metallic particles are highly resistant to sedimentation. Clarification should only be attempted in conjunction with acid treatment and/or chemical flocculation, pressure aeration, frothing, and skimming. Both supernatant and sludge can be further treated by aeration and in drying lagoons correspondingly. Rather than attempt sedimentation, some industries have applied filtration to remove suspended-solids wastes.

Removal of inorganic dissolved solids must be accomplished by some sort of physical or chemical treatment—generally by evaporation. Carbon adsorption is employed in a few operations, but is far more suitable for a low-volume operation. When heavy metal removal is the objective, chemical treatment or filtration should be considered.

Collection and containment is an attractive alternative when quantities are limited. Drums of spent materials could be returned to ink formulating industries where recovery or treatment operations occur on a much larger scale.

References

1. Anonymous, "Better lineup for can coatings," *Chemical Wk.*
2. Anonymous, "Lone Star adopts ultra lightweight seamless steel can," *Brewers Digest.*
3. Anonymous, "Solvent extract recovers phosphate and metals from sludge," *USBM Report RI*, 7662, USDI (1972).
4. Anonymous, "Largest treatment plant for metal finishing," *Metal Finishing* **49**, 36, September 1973.
5. Anonymous, "Water reducible coatings solves problems," *Ind. Finishing* **49**, August 1973.
6. Anonymous, "Treating spray painting wastewater," *Plant Eng.* **48**, June 1972.
7. Anonymous, "Sorption wins phosphoric acid from finish wastes," *Chem. Eng.* **79**, June 1972.
8. Anonymous, "Solid state logic: Electronic ally for beer drinkers, cardlor," *Plant Eng.* **27**, August 1973.
9. Barrett, W. J. and G. Mornair, "Waterborne wastes of paint and inorganic pigment industries," *USEPA*, 670/2–74–030 March 1974.
10. Carrol, J. "How to minimize pollution from metal prefinishing lines," *Ind. Water Eng.* **11**, June 1974.
11. Chalmers, R. K., "Treatment of wastes from metal finishing," applications of new concepts of physical chemical wastewater treatment, September 1972.
12. Cheremisinoff, N. and Y. H. Habib, "Removal technology for Cadmium, Chromium," *Water and Sewage Works* **119**, 46 (1972).
13. Church, F. L., "Coors goes all aluminium," *Modern Metals* **27**, January 1972.
14. Church, F. L., "Can makers cleaning up energy hogging image," *Modern Metals* **30**, 45 (1974).
15. Emmerling, J., "Economic recovery of waste solvent provided by system permits reuse of bottons, avoids pollution," *Chem. Proc.*, Chicago, Ill. **38**, April 1975.
16. EPA, "Upgrading metal finishing facilities to reduce pollution," EP 710 M 56/No. 1/974, August 1974.
17. Gadomski, R. R. and A. V. Gimbrone, "An evaluation of emission and control technologies for the metal decorating processes," *Air Pollution Control Assoc.*
18. Jarmuth, R. A., "Regeneration of chromated aluminum deoxidizers," *USEPA*, ORD, EPA 660/2–73–023, December 1973.
19. Kuman, I. J. and N. L. Clesceri, "Phosphorus removal in wastewater, a cost analysis," *Water and Sewage Works* **120**, March 1973.
20. Pipes, W. O. "Aluminium hydroxide effects on wastewater treatment process," *Water Poll Contr. Fed. J.* **45**, April (1973.
21. Rakowski, L. R., "Schlitz blitz," *Modern Metals* **29**, April 1973.
22. Schrantz, J., "Pollution compliance with water reducible coatings," *Ind. Finishing* **49**, 10 (1973).
23. Schneider, H. J. and R. L. Price, "Pollution: Cope with it or avoid it," *Ind. Finishing* **48**, 110 (1972).
24. Schrantz, J. "Rock Island's arsenal: Zero discharge," *Ind. Finishing* **49**, June 1973.
25. "Water reuse in industry; Metal finishing" *Mech. Eng.* **95**, 29 (1973).
26. Zievers, J. F. and C. J. Novotny, "Curtailing pollution from metal finishing," *Environ. Sci. and Tech.* **7**, 209 (1973).

27.15 Porcelain Enameling Industry

This industry applies a glass-like coating to a base metal such as aluminum, cast iron or steel. It will improve chemical and abrasive resistance of the coated product, increase its thermal and electrical resistance, and usually enhance its appearance.

The base metal is coated either by spraying or dipping into a "slip." A slip typically contains color oxides, chemical additives, water, and a type of "frits." A frit is an alkali-silicate, glassy material supplemented by the addition of metals, usually cadmium, barium, and lead. Following the coating the porcelain is baked at about 1000°F. to perfect the bond. A typical slip formula is given in Table 27.37.

Table 27.37 Formulation of a Commonly-Used Porcelain Enameling Slip (After Donahue, Assoc. 1981)

	Amt. in Slip (%) by Weight	Metal Content (% by Wt.)
Frit	59.1	12.5%Pb; 5%Ba
Color oxide 1	0.7	53%Pb; 25%Sb
Color oxide 2	0.2	40%Cr$_2$O$_3$
Boric acid	1.8	35% CR$_2$O$_3$
KOH	1.5	
Na$_2$SiO$_3$	1.2	
K$_2$SiO$_3$	2.4	
Si	9.5	
H$_2$O	23.58	

Spraying the slip on the enamel product is the more common method used. About one gallon is required to apply one 2 mil coat to 200 ft^2 of aluminum surface. About half (50%) of the slip is lost on overspray. This spray, when trapped in a background curtain or water spray, constitutes the largest and most significant quantity of metal wastes in the aluminum enameling industry. About 11% of the slip by weight is comprised of metallic elements of which lead (Pb) is the major contaminant in overspray waste.

A mass water balance of the porcelain enameling process on aluminum is shown in Fig. 27.18. A typical waste sludge analysis (D in Fig. 27.23) is given in Table 27.38. A typical analysis of non-porcelain enameling metal preparation process waste stream (F in Fig. 27.23) is presented in Table 27.39. All these data was made available by *Donohue Associates in 1981.

There are two problems: successful recycle or disposal of the sludge waste, D, and acceptable discharge of the metal waste stream, F, to a municipal wastewater treatment plant. Donohue Associates suggest the following possible solutions to these problems:

1. Use of alternate raw materials
2. Recycling of the sludges
3. Removal and disposal of objectionable wastes
4. Encapsulation and fixation of these metals
5. Conditioning and placing the sludges in on-site segregated landfills.

Donohue Associates concluded that two processes were technically feasible and economically justifiable for proposed disposal of the sludges, D.

1. Treatment with polyelectrolytes and discharge into on-site or municipal landfills
2. Construction of segregated landfills on-site to receive the untreated sludges.

*Small Scale Reclamation/Disposal of Toxic Porcelain Enameling Waste Sludges NSF Project No. ISP 8009922 Donohue Assoc. Feb. 27, 1981

27.16 Naval-Stores Wastes

Naval stores are manufactured by refining oleoresinous materials from pine wood. Among the products are wood rosin, wood turpentine, pine oil, dipentene, and other monocyclic hydrocarbons. Gum naval stores are obtained by simple steam distillation of resinous material taken from living pine trees. Wood naval stores, obtained from cut pine wood and stumps, can be subdivided into products obtained by (1) destructive distillation, (2) sulfate distillation or (3) steam distillation. Because of the nature of the raw material, gum refining is carried out on a seasonal basis. Wood (southern pine), on the other hand, is refined all year round.

Stumps that have been rotting on the ground for many years are conveyed to a washer where dirt, sand, and loose or rotted wood are removed. The clean stumps are fed directly to grinding equipment, which delivers wood chips about one inch long and $\frac{3}{16}$ inch thick for extraction. The extraction process is carried out in large pressure vessels at about 280°F and at a pressure of 80 psi. A solvent, usually a narrow-range petroleum fraction (200 to 400°F naphtha), is pumped

Table 27.38   Characteristics of waste sludges (mg/l)

	TSS	pH	Pb	Cr	Cd	Se	Ba	Zn
Untreated Mill Room Sludge I (grab sample)	2,920	10.26	97.5	1.06	4.0	0.1	53	
Untreated Mill Room Sludge II (1 week composite)	262	10.51	26.0	0.27	1.00	0.05	12.6	
Untreated Mill Room Sludge III (1 week composite)	370	10.55	11.1	0.22	0.14	0.06	0.3	0.04
Dry Overspray Solids (4 day composite)			68,200	11.4	159	0.2	1,685	

Table 27.39   Analysis of a metal preparation process waste stream

Component	Concentration (mg/l)
Biochemical Oxygen Demand, 5 day	16
Total Suspended Solids	388
Oil and Grease	6
Phosphorus	10
Sulfite	1
Sulfate	156
Chromium (Total)	0.02
Copper	0.07
Lead	0.13
Manganese	0.15
Nickel	0.86

into the bottom of the first extractor. The solvent flows out of the top of the first extractor and into the bottom of the second, and so on, until about 15,000 gallons of solvent have been pumped into 10 tons of pine chips. The spent chips are used for fuel, and the crude extract solution of rosin, solvent, and terpene oils, known as "drop solution," is pumped to tanks where water sprays wash out color bodies. A heavy, dark resin (nigre) separates from the wash water and is drained off. The remaining solution is evaporated to recover the solvent, and a series of successive equilibrium distillations remove the terpene oils, leaving a marketable residue known as wood rosin. The terpene oils are further fractionated by batch distillation into

turpentine, pine oil, dipentene, and other terpene derivatives. Black [1] presents a simplified flow diagram of these processes (Fig. 27.24).

For each ton of pine stumps, 100 to 600 gallons of extraction solvent and 2000 to 5000 gallons of water are used. Actually the solvent is recovered and reused (see Fig. 27.24) so that only about one gallon of solvent is wasted per ton of chips. Secondary refining processes are carried out by some naval-stores plants, in which the rosins and oils obtained from primary processing are utilized as raw materials. Black [1] presents the characteristics of primary-processed wastes of wood naval stores (Table 27.40) and also gives representative values of character and volume of secondary-processed wastes (Table 27.41).

A few remedial measures to handle naval-stores wastes have been proposed, although little, if any, waste treatment is currently practiced. The suggested methods include elimination (process change or stopping of waste discharge), by-product recovery (such as terpene hydrate), equalization of flow, recirculation and reuse, and waste treatment where necessary; pilot-plant results have shown that trickling filtration produces an 83 per cent BOD reduction.

References: Naval-Stores Wastes

1. Black, H. H., and V. A. Minch, "Industrial waste guide, wood naval stores," *Sewage Ind. Wastes* **25**, 462 (1953).
2. "1951 Industrial wastes forum," *Sewage Ind. Wastes* **24**, 869 (1952).

3. Palmer, R. C., "Producing naval stores from waste pine wood," *Chem. Met. Eng.* **37**, 140, 289, and 422 (1930).

4. Shantz, J. L., and T. Marvin, "Waste utilization," *Ind. Eng. Chem.* **31**, 585 (1939).

SPECIAL-MATERIALS INDUSTRIES

The major special-materials industries are glue manufacturing, wood preserving, candle manufacturing, and plywood making.

27.17 Animal-Glue Manufacturing Wastes

Sources of wastes. The manufacturing process may be

divided into three areas; the mill house, where stock is made amenable to glue extraction by washing and milling; the cook house, where the stock is cooked and the light liquor is extracted; and the drying rooms, where the light liquor is condensed and dried to the final product (Fig. 27.25).

The bulk of water-borne wastes come from the mill house and include lime from soaking, acids, wash water, dirt, hair, and miscellaneous solids brought in with the stock. Three types of stock (raw material) are used—green fleshings, hide stock, and chrome-split stock—and processing is somewhat different for each. Fleshings, the substance which joins animal skins to the carcass, is the most perishable raw mate-

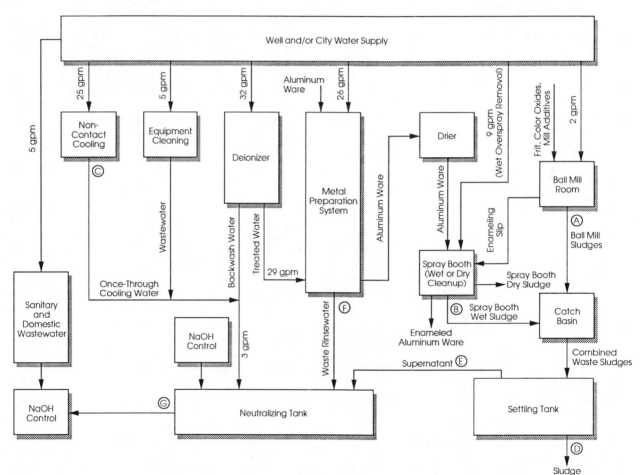

Figure 27.23 Estimate of water usage in porcelain enameling on aluminum flow diagram.

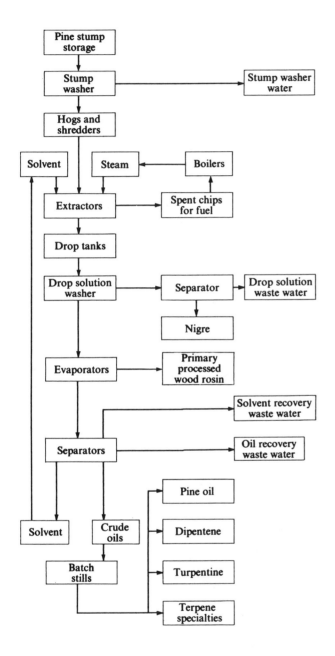

Fig. 27.24 Flow diagram for wood naval-stores plant, using naphtha as the solvent for the primary extraction process. (After Black [1].)

rial. They are washed for approximately 12 hours, acidulated, then soaked, before being brought to the cook house. Hides are first desalted and cleaned by milling for 6 hours, then limed and soaked for 80 to 90 days in large vats. After this soaking period they are washed free of lime, acidulated, and sent to the cook house. Chrome splits are first chopped up mechanically and limed for approximately 5 hours. The lime is washed out, sulfuric acid is added, and the washing begins. After the acid has been washed out, the stock soaks in dilute acid for approximately 8 hours before magnesite $[Mg(CO_3)_2]$ is added. After a final hour of soaking the stock is sent to the cook house.

Cook-house operations consist of loading the prepared stock into large kettles and cooking it with steam at temperatures ranging from 160° to 200°F. After about 6 hours the light liquor is drained off. The process is repeated three more times, each one yielding a lower grade of glue. The material left after the final cook cycle is called tankage. This is further treated to yield products such as grease, hair, and fertilizer.

The wastes generated have been shown to be amenable to biological treatment; methods such as foam separation, flotation, and sedimentation seem valid primary processes. Table 27.42 gives the unit waste loads recently found by the author in an animal-glue manufacturing plant.

This reference has been published since 1971

1. LaGrega, M. D., R. W. Klippel, and N. L. Nemerow, "An industrial waste case history—the animal glue industry," *Presented at 5th Mid-Atlantic Ind. Waste Conf., Drexel University* (1971).

Table 27.42 Summary of unit waste loads of an animal-glue manufacturing plant.*

Stock	Waste, gal	BOD, lb	Suspended solids. lb	COD, lb
Fleshings	50.5	2.5	4.25	4.8
Hide	54.7	0.58	1.92	1.42
Chrome stock	51	0.28	0.40	0.65

*All quantities are those produced per pound of glue.

Table 27.40 Representative values of character and volume of primary process naval-stores manufacturing wastes. (After Black [1].)

Primary process waste source	Discharge, gal*		Oxygen consumed (dichromate), ppm		Total solids, ppm		pH range		Phenols, ppm		BOD, 5-day, 20°C				Suspended solids				Population equivalent‡	
											ppm		lb*		ppm		lb*			
	N†	B†	N	B	N	B	N	B	N	B	N	B	N	B	N	B	N	B	N	B
Stump washer	350–3500		5700–16,400		11,000–20,000		4.5–6.7		0		50–1500		0.05–9.63		1220–16,900		2.34–105		0.3–58	
Drop solution wash	180		5200		1900		4.5–6.1				3130		4.34		150		0.22		29	
Solvent recovery	730	150	1100	2,800	285	150	6.1–7.2	3.3–3.5	0	0	550	2,100	3.5	2.5		16		0.63	20	16
Oil recovery	250	1170	4000	1,360	310	30	4.3–6.9	4.1–5.2	0	0	2170	60	5.7	0.58		65			34	4
Furfuraldehyde recovery	12			14,500		150		3.4–3.5				10,700		1.07		8				6

*Per ton of stumps.
†N, naphtha extraction; B, benzene extraction.
‡Per ton of stumps, based on 0.167 lb of 5-day BOD per capita per day.

Table 27.41 Representative values of character and volume of secondary process naval-stores manufacturing wastes. (After Black [1].)

Secondary process waste source	Unit	Discharge, gal*	Oxygen consumed (dichromate), ppm	Total solids, ppm	pH range	Phenols, ppm	BOD		Suspended solids		Population equivalent*†
							ppm	lb*	ppm	lb*	
Pale rosin refining											
Solvent recovery	1 ton stumps	490			3.6 +	0	820	3.35	12	0.05	20
Alcohol recovery		485			5.8–6.2	0	385	1.55	24	0.10	9.30
Polymerization											
Solvent recovery	1000 lb rosin	2100	430	1,530	1.7–3.2		75	1.33	28	0.5	8
Acid recovery		750	30		2.5–3.5		17	0.11			0.6
Disproportionation											
Condenser water	1000 lb rosin	3900	590	150	5.9–6.5	{2.0}	260	8.45	41	1.33	50
Alkaline wash water		140	79,000	14,600	8.5–11.9		38,100	44.4	3,350	3.90	265
Hydrogenation Waste rosin oils	1000 lb rosin	1	960,000	410,000	12.2 +		382,000	3.2	26,500	0.22	19
Isomerization (wash water)	1000 gal pine oil	500	17,000	128,000	12.2 +	1970	14,800	49.3	75	0.25	295
Synthesization (condenser water)	1000 lb rosin	6000	830	415	9.5–10.5	0.2	270	13.5	130	6.5	80

*Per unit.
†Based on 0.167 lb 5-day BOD per capita per day.

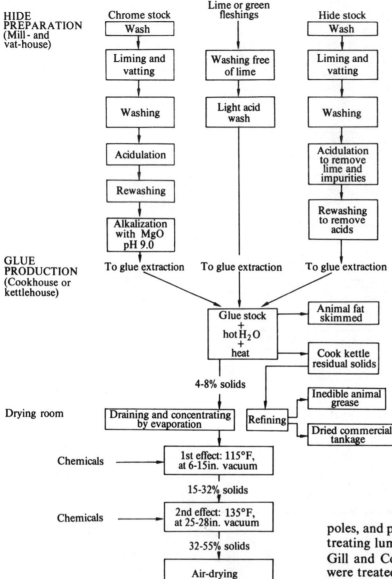

HIDE
PREPARATION
(Mill- and
vat-house)

GLUE
PRODUCTION
(Cookhouse or
kettlehouse)

Drying room

Chemicals

Chemicals

Fig. 27.25 Glue manufacturing process flow chart.

27.18 Wood-Preservation Wastes

There are about 400 wood-preserving plants in the United States. Almost all use pressure processes to preserve wood with either creosote or pentachlorophenol or both. Both preservatives are increasing in total use, the creosote chiefly to preserve crossties,

poles, and piling and pentachlorophenol primarily for treating lumber, poles, and cross-arms. According to Gill and Corey [2], 262,000,000 cubic feet of wood were treated with preservatives and fire retardants in 1965 in the United States.

Origin of wastes. Air-dried timber (or sometimes properly conditioned green wood) is first steamed at about 20 psi for up to 12 hours while the condensate is continually removed. The second step involves a relatively short period of evacuation before the pressure is raised with air to 30 to 90 psi. During this latter period the retort is filled with either creosote or pentachlorophenol oil. Air is vented while the preservative is added until all the air has been displaced by the chemical. A maximum pressure of about 200 psi is used for a period of 2 to 8 hours. As the pressure is then released, air trapped in the wood bubbles out, bringing any excess preservative with it. The preservative is usually too valuable to waste and is returned to the storage tank. A final vacuum is applied to remove any excess oil from the surface of the wood.

The condensate removed during the first steaming process appears to be the major source of waste, since it contains dissolved preservative left on the walls of the retort in previous cycles. An attempt is usually made to recover preservatives by pumping the condensate to sumps for differential gravity separation. Creosote, being denser than water, is drawn off the bottom of the sump and returned to storage tanks or directly to the retort. Further recovery is obtained by using additional settling tanks in series. Pentachlorophenol oil, being lighter than water, is removed from the top of the sump and returned to storage tanks.

The water and material extracted from the wood during steaming (the other major source of waste) contain a considerable portion of contaminating matter and are usually treated similarly to the steam condensates. After all the differential separation has taken place, the residual water-bearing wastes are discharged to the appropriate watercourse.

Characteristics of the waste. Middlebrooks [3] gives the following characteristics of the total waste:

Constituent	Raw waste concentration (mg/liter)
COD	11,500–19,600
BOD	2800–5000
Total solids	6340
Total volatile solids	5730
Phenols	17–85
pH	5.2
Suspended solids	1420
Volatile suspended solids	1100
COD from raw waste filtrate	7080
Nitrogen (total) as N	89
Nitrogen (NH_3) as N	32
Nitrogen (organic) as N	57
Phosphates (total)	< 5.0

Although Middlebrooks does not give unit volumes of waste, he notes a volume of about 3000 gallons per day. He recognizes that "although the waste water from the wood preservation process is highly contaminated, the volume is relatively small."

Treatment of wastes. Both Middlebrooks [3] and Gaudy *et al.* [1] found that this waste was amenable to biological treatment. The former also found that flocculation with lime achieved a significant reduction in COD and BOD. Chlorination is relatively ineffective in reducing the COD. However, chemical flocculation followed by chlorination reduced the COD of the waste by 85 per cent and reduced the phenol concentration to less than 0.5 mg/liter. Gaudy found considerable success with oxidation-pond treatment.

References: Wood-Preservation Wastes

1. Gaudy, A. G., Jr., R. Scudder, M. M. Neely, and J. J. Perot, "Studies on the treatment of wood preserving wastes," Presented at Symposium on Selected Papers in Water and Wastewater Technology, 55th Meeting of the American Institute of Chemical Engineers, Houston, Texas, Feb. 7–11, 1965.
2. Gill, T., and E. A. Corey, *Wood Preservation Statistics—1965*, Forest Service, U.S. Department of Agriculture, Washington, D.C. (1966).
3. Middlebrooks, E. J., "Wastes from the preservation of wood," *J. Sanit. Eng. Div. Am. Soc. Civil Engrs.* **SA1**, No. 5785, 41 (1968).

The following references have been published since 1968.

1. Gaudy, A. F., R. Scudder, M. M. Neeley, and J. J. Perot, "Studies on the treatment of wood preservation wastes, *Proc. 64th Mtg Amer. Wood Preservation Assoc.*, 196 (1968).
2. Middlebrooks, E. and E. Pearson, "Wastes from preservation of wood," *Proc. Ind. Waste Conf. 23rd*, Purdue University, 213 (1968).
3. Stump, V. L. and F. Agardy, "Investigation of a wood curing waste," *J. San. Eng. Div. ASCE* **95** (SA6), Proc. Paper 6953 (1969).
4. Vonfrank, and J. C. Eck, "Water pollution control in the wood preservation industry," *Proc. 65th Mtg Amer. Wood Preservation Assoc.*, 151 (1969).

27.19 Candle-Manufacturing Wastes

The candle industry has moved with the times.* From the old, white, hand-made product has evolved the modern, colored, mass-produced, long-lasting, and dripless candle. Quality candles always incorporate stearic acid, although the most important single ingredient in candle manufacturing is still paraffin wax. The two types of candle are named after these ingredients. The paraffin candle is the less expensive one; it melts easily and burns down rapidly. Because they also drip, paraffin candles are used for the multicolor drip effects popular today. The stearic-acid candle burns longer, stays cleaner, and does not drip, because of its 20 to 30 per cent stearic-acid content (the remainder being mainly paraffin wax).

Whatever their composition, candles are made in only two ways, either dipped or molded. In the dipping method, racks of suspended wicks are repeatedly dipped into a vat of molten wax. With each dip a new layer is built onto the candle until it is complete. In the molding method, molten wax is poured into molds around already-positioned wicks. The wax ingredients are carefully controlled to ensure that the candles shrink after cooling and do not stick to the molds. High-quality candles are finally dipped into a bath of Cenwax G (hydrogenated castor oil) to give them a hard shell. When the candle is lit, this shell collects the molten wax and prevents dripping.

---

*"Lighting the way," *Wallace and Tiernan Topics* **20**, 1 (1966).

It is reported* that little waste is evolved from this industry, because any spilled wax is simply recovered, melted, and reused. However, when a candle manufacturer makes his own stearic acid (in an "integrated" plant) there is a possibility of some waste being produced. Stearic acid is made from the high-pressure steam-splitting of tallow, with glycerine and oleic acid separated and sold as by-products. The only published treatment for wastes from an "integrated" candle-manufacturing plant is anaerobic digestion.

24.19 Plywood-Plant Glue Wastes

Plywood plants are divided into two parts: the *green end*, designed to store, debark, veneer, and prepare the wood laminates for sheet formation; and the *glue end*, where the laminates are glued, pressed, trimmed, sanded, and packaged for shipment. The complete plant processes are shown schematically in Fig. 27.26.

Major wastes from these plants originate from the glue kettles and the glue spreaders. Bodien [1] reports that a typical plant, producing 100 million square feet of plywood per year (one-half interior and one-half exterior grade), makes about 11 batches of interior glue and 9 batches of exterior glue a day (equivalent to 400,000 and 350,000 pounds, respectively, per month). Washing the glue-mixing equipment produces a small volume of highly concentrated waste. A typical plant

---

*"Lighting the way," *Wallace and Tiernan Topics* **20**, 1 (1966).

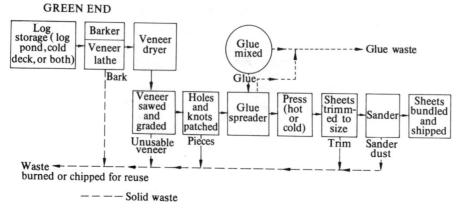

Fig. 27.26 Plywood plant flow diagram. (After Bodien [1968].)

Table 27.43 Typical plywood-plant waste characteristics. (After Bodien.†)

Plant no.	1966 production, ft² (⅜-inch basis)	Average discharge, gpm	pH	COD, mg/l	Total PO₄, mg/l	Total Kjeldahl nitrogen, mg/l	Phenol, mg/l	Suspended solids, ppm.	Total solids, ppm
						Waste-water analyses after settling			
1	100,000,000	18.2							
2	135,000,000	30.2	11.6	1814	15	110	1667	148	1627
3	100,000,000	21.6	9.4	1917	9	64	1790	356	1814
4	70,000,000	54.0	10.8	1621	12	3	222	330	1120

†Bodien, D. G., *Progress Report, Plywood Plant Glue Waste Disposal*, Report no. PR-2. United States Department of Interior, Technical Projects Branch, Northwest Region, Pacific Northwest Water Laboratory, Corvallis, Oregon, February 1968.

also has six periods of use of glue-spreaders per day; the spreaders are usually washed down once a shift when interior glue is used and at least once a day for exterior glue. The difference in frequency is due to the relatively short pot life of interior glues (6 to 8 hours).

Exterior glues contain a furfural extraction of corncobs and oat hulls, as well as wheat flour, phenolic formaldehyde resin, and caustic soda and ash. Interior glues contain dried blood, soya flour, lime, caustic soda, and sodium silicate. In addition, formaldehyde and pentachlorophenol may be added to produce a glue toxic to insect pests. Typical waste characteristics, derived from Bodien's study, are shown in Table 27.43.

Treatment. Most plywood plants practice settling, either in ponds or septic tanks. This treatment removes some of the glue solids and wood chips, but achieves little reduction in phenols, phosphates, or total Kjeldahl nitrogen. Some plants discharge their glue wastes to municipal treatment systems, where high pH, glue solids, and wood chips have been known to cause difficulty. Because of the high organic content of the relatively low-volume glue wastes, properly controlled incineration offers a potential solution to glue-waste disposal.

The following references have been published since 1968.

1. Bodien, D. G., "Progress report, plywood plant glue waste disposal," *U.S. Dept. of int., Technical Projects Branch, Pacific Northwest Water Lab., Corvallis*, Oregon, Report No. PR-2 (1968).

2. Boyoston, J. R., "Plywood and sawmill liquid waste disposal," *Forest Products J.* **21**, 58 (1971).

3. EPA, "Development document for effluent; Limitation guidelines and new source performance statistics for plywood and wood preservation," *EPA Rpt. #440/1–73/023* (1973).

4. Haskell, H. H., "Handling phenolic resin adhesive wash water in southern plywood plants," *Forest Prod. J.* **21**, 64 (1971).

27.21 Cement Industry

Origin. The following processes are required in the manufacture of cement:

1. *Quarrying raw materials.* Lime is primary ingredient, but materials such as sand, clay, shale, iron ore, and blast-furnace slag are also used in the manufacturing of cement.

2. *Raw grinding.* This process involves the crushing, pulverizing and mixing of raw materials so that they can be acted upon by the intense heat of the kilns.

3. *Initial blending.* During this process the raw material is blended to get a mixture of the right composition for a product of uniform quality.

4. *Fine grinding.* This process involves using either one of the two different methods:
   (a) *Wet process:* Raw materials which are very moist are ground with water and fed into the kiln in a slurry.

(b) *Dry process*: Materials are dried, ground, and fed to the kiln in a dry state.

5. *Kiln.* The finely ground mixtures are taken from storage tanks and fed into the kiln where, subjected to temperatures of nearly 3000°F, the materials reach a point of incipient fusion and form hard, marble-sized balls called clinker. Clinker coming from the kiln is rapidly cooled by air.

6. *Finish grinding.* Clinker, mixed with a small amount of gypsum to regulate the setting time, is ground into a fine powder. This powder is cooled (generally by a water jacket heat exchange) and transported within the plant by pneumatic pumps, the air to which is supplied by water-cooled compressors.

These pollutants arise from cement-plant use of water for the following:

1. *Cooling*—gives rise to thermal pollution.
2. *Raw material washing and beneficiation*—produces high pH and alkalinity, total dissolved and suspended solids.
3. *Process-slurry water*—a problem only in the event of spillage
4. *Dust control*—uses wet scrubbers to collect kiln dust from effluent gases. Also, sprays roads to prevent dust caused by truck traffic (contributes to the run-off problem).

5. *Dust leaching*—nine plants in the United States have a leaching system for recovery and reuse of collected kiln dust. In leaching, dry dust is mixed into a slurry, placed into a clarifier for settling, the underflow of which is returned to the kiln. The overflow containing high pH, alkalinity, suspended solids, dissolved solids, potassium, and sulfate is discharged. This constitutes the most severe water pollution problem in the industry.

6. *Dust disposal*—collected dust is mixed into a slurry and fed into a pond for solid settling. The settled solids are not recovered and the overflow (leachate) is discharged.

The manufacture of cement is a continuous process which is normally interrupted only to reline the kilns.

The largest volumes of water used in cement manufacturing are essentially nonpolluting. Process water is evaporated and most cooling water is not contaminated except for the increased temperature.

Characteristics. The EPA has divided the industry into three subcategories: nonleaching plants, leaching plants, and materials storage piles runoff. The water pollutants associated with the cement industry and their present loadings from nonleaching and leaching subcategories are as follows:

Pollutant	Units loading/product	Leaching plants		Nonleaching plants	
		Minimum	Maximum	Minimum	Maximum
pH	lb/ton	6.0	12.0	6.0	12.3
Total dissolved solids	lb/ton	0.11	26.11	0	15.74
Total suspended solids	lb/ton	0	8.99	0	14.67
Alkalinity	lb/ton	0	8.02	0	7.73
Potassium	lb/ton	0.36	22.58	0	2.42
Sulfate	lb/ton	1.23	31.35	0	3.24
Temperature rise	°F	0	19.8	0	30.6

Of the 154 nonleaching plants in the United States, 23 per cent are presently achieving essentially no discharge of pollutants (less than 0.01 lb./ton of product not including runoff). The remaining 77 per cent can achieve the same by 1977 with existing state-of-the-art technology.

Treatment. For plants in the leaching subcategory, the main control and treatment methods involve segregation of dust-contact streams and neutralization with stack gases followed by sedimentation with recycling and reuse of wastewater. The devices employed include: (a) cooling towers or ponds to reduce the temperature of cooling process waters; (b) settling ponds to reduce the concentration of suspended solids; (c) containment ponds to dispose of waste kiln dust; and (d) clarifiers to separate solids in dust-leaching operations.

Pollution of waterways caused by storage piles runoff and dust contamination can be reduced by locating storage piles where storm waters would be contained, paving areas used by vehicles; and frequently building ditches around the plant area draining to a holding sump.

The treatment technology currently practiced can adequately control pH, alkalinity, and suspended solids, but not the dissolved solids. Handling the dissolved solids will require the adaptation of additional treatment technology by the industry. The technology that appears most promising for concentration of leachate is electrodialysis which has been successfully applied in Japan for more than a decade to the concentration of sea water for the recovery of salt. If electrodialysis were used, the concentrated stream could be easily evaporated and the salts in the dilute stream would be low enough to be recycled to the leading system.

As stated earlier, very few operations in the manufacturing of cement add pollutants to the water used. For the most part, with the exception of leaching systems, pollution results from practices that allow materials to come in contact with the water. Pollutant levels can be greatly reduced or even eliminated by instituting "good-housekeeping" practices or by more extensive reuse and recycling of contaminated waters.

References

1. "Cement Manufacturing point source category," *U.S. Environ. Prot. Agency*, Pub. No. 440/1–74–005–a, January 1974.
2. Daugherty, K. E. "Review of cement industry pollution control," *Amer. Ceramic Soc. Bull.* **54**, 189 (1975).
3. The Portland Cement Association, "The making of Portland Cement," Pub. No. E108, Chicago, Ill.
4. The Portland Cement Association, "The drama of cement making," Pub. No. E103–15M–6–54 (1953).

27.22 Wood Furniture and Fixture Manufacturing

Origin. The production process is generally a dry process. However, wet spray booths and laundry facilities do produce liquid wastes. The EPA has concluded that this industry can achieve a level of waste control to eliminate discharge of process wastewater into navigable waters by July 1, 1977. This can be achieved using available technology.

Characteristics. The primary pollutants come from the finishing process. The use of bleaching, straining, sealing, and topcoating agents that are removed in the spray booths are the major pollutants.

Treatment. Acceptable treatment methods include evaporation ponds, spray irrigation, burning with boiler fuel, discharge to municipal treatment or hauling to a landfill.

References

1. U.S. Environmental Protection Agency, "Effluent limitations guidelines and performance standards for wood furniture and fixture manufacturing," *EPA* 440/1–74/033–a, November 1974.
2. U.S. Environmental Protection Agency, "Economic analysis of proposed effluent guidelines for wood furniture and fixture manufacturing," *EPA* 230/1–74–057, October 1974.

27.23 Asbestos Wastes

Origin. Asbestos is mined from the earth, crushed to 5/16 of an inch or less, sorted, and sent to the manufacturing plant. During this process the waste asbestos tailings are stockpiled and then disposed of. If preventive measures are not taken at this time, rainwater will flow through the tailings, thereby increasing the pH and total amount of suspended solids in the runoff.

Asbestos can be processed by two different methods. A dry process can be used in which all necessary cleaning is done with forced air; this process creates a dust problem, however, as the asbestos tailings are blown into the air and no liquid waste is developed. The other method is a wet process in which water is used for cleansing the crushed ore. The water is then filtered out and either recycled or treated and discharged.

Characteristics. The wastes from asbestos mining and manufacturing are generally high in BOD, COD, pH, and suspended solids. It has been found that once asbestos tailings enter the human body cancer may develop. Although it takes from 20 to 30 years for cancer to develop much research is presently being done to develop a higher quality of treatment for this industry.

Treatment. The treatment method presently used at most asbestos mines is to construct drainage ditches and dams to prevent rainwater from flowing through the waste asbestos tailings and to collect that which does and treat it with sulfuric acid to lower the pH before allowing it to reach the stream. At asbestos manufacturing plants the wastewater is treated with sulfuric acid and retained in settling/percolating ponds which are periodically dredged; then the sludge is dried and landfilled.

Since the discovery of the carcinogenic effects of asbestos tailings in potable water supplies, the EPA has been pushing for stricter laws to prevent this from affecting society. The best method developed for removal of the asbestos fibers appears to be the use of chemical coagulation with iron salts and polyelectrolytes followed by filtration. This method results in better than 99.8 per cent removal of fibers from water containing 12×10^6 fibers per liter.

References

1. Anonymous, "Crisis in Silver Bay," *Time* **102**, 96 October 22, 1973.
2. Anonymous, "Woodsreef Asbestos," *Mining Mag.* **128** (No. 6) (1973).
3. Harris, R. H., and E. M. Brecher "Is the water safe to drink?", *Consumer Reports* **39**, 438 (1974).
4. Lawrence, J., H. M. Tosine, T. W. S. Pang, and H. W. Z. Zimmerman, "Asbestos-like fibers in drinking waters could be a potential health hazard," *Water Res.* **9**, 397 (1974).
5. "Reports," *Sci.* **185**, No. 4154, 853.

27.24 Paint and Printing Ink Wastes

Paint manufacturing consists of the process of mixing pigments with a suitable vehicle and grinding them to a satisfactory fineness, color, and consistency. Oils, resins, solvents, plasticizers, pigments, extenders, and dyes are all used in paint manufacturing and all contribute to pollution problems.

Paint includes a variety of materials such as enamels, lacquers, varnishes, undercoats, surfacers, primers, sealers, fillers, stoppers, etc. Paints contain pigments and a second ingredient, a binder. Pigments have decorative and protective properties, while the binder (a resin or polymer) holds the pigment particles together and binds them to the surface.

Characteristics and treatment. *Vapor wastes* are generated mainly from reactors, mixing and grinding, and filling lines. The four most commonly used types of equipment for vapor control are exhaust systems, scrubbers, incinerators, and condensers. Other less frequently used methods of control are carbon absorption, water absorption, combination scrubbers and incinerators, and diluted extraction.

Liquid wastes are generated by reactor condenser by-products, caustic and other water-based washes, solvent-based washes, solvent-based rejected material, water-based rejected material, and washes containing heavy metals. Settling tanks, ponds, and basins are by far the most common type of liquid-waste-control equipment, but distillation and incineration are also widely used. Other types of equipment in use are: filtration, solvent-oil scrubbers, water scrubbers, septic tanks, neutralization and filtration, multipass sedimentation, three-compartment classification pits, tertiary waste treatment (for water-based waste, exact type of treatment not given) and coagulation. Oxidation, reduction, neutralization, and filtration are used on water-based wastes containing heavy metals.

Disposal of solid wastes from liquid waste clean-up is also a problem. Most companies depend on commercial truck haulers, landfills, or dumps. This type of disposal will not be allowed in the future, and disposal of this type of waste presents the most difficult problem facing the paint industry today.

The following references have been published since 1956.

1. Batelle Labs, "Fluidized bed incineration of selected carbonceous industrial wastes," *Water Poll. Contr. Res. Series*, U.S. EPA 12120 FYF (1972).
2. Briscoe, R. V., "Treatment of electrophoretic painting effluents," *Trans. Inst. Met. Finish* **50**, 199–206 (1972).
3. EPA, "Inorganic chemicals industry profile," *EPA* Prog. No. 12020 EJI (1971).
4. EPA, "Major inorganic products," *EPA* 440/1–73/007 (1973).
5. Fader, S. W., "Barging industrial liquid wastes to sea," *J. WPCF* **44**, 314 (1972).
6. Koerner, H., "Processing Latex wastewater," *Ger. Offen Pat.*, #2,021,826 (cl.C.02c) (1971; Appl.) (1970).
7. Lungren, H., "Air flotation purifies waste water from latex polymer manufacture," *Chem Eng. Symp. Prog. Series* **65**, 191–195 (1969) .
8. Scott, W. H., "System for the removal of latex from wastewater," *U.S. Pat.*, # 3,726,794 (Cl.210–104; Bold) (1973); Appl. 102,015 (1970).
9. Shreve, R. N., "Paint, varnish, lacquer and allied industries," *The Chemical Process Industries* (2nd ed.), New York: McGraw-Hill, 494–527 (1956).
10. Zimmerman, R. L., "Environmental improvement from a product standpoint," *J. Paint Tech.* **45**, 58–61 (1973).

Printing ink

The most active studies of printing inks are in the deinking field. Deinking involves the use of a special system for the removal of coated and printed material from paper to form a useable pulp. This pulp is then used in the various paper stock including tissue, fine paper, boxboard, foodboard, and newsprint. The deinking process involves the following stages: (1) pulping or defibering, (2) cleaning and screening, (3) washing out contaminants, (4) dewatering, and (5) bleaching. The washing stage yields the wastes we are interested in.

In the washing stage the stock is cleaned of all dispersed inks, clays, and chemicals. There are five methods used in the washing stage: (1) vacuum cylinder, (2) American disc filter, (3) Lancaster washer, (4) Sidehill washer and, (5) floatation. The washing in the first four methods allows stock flow to pass over an open surface that permits the removal of water and chemicals by drainage through the stock. The fifth method washes the stock by flotation of the inks on air bubbles; the ink goes to the surface of the stock and is mechanically removed.

Characteristics. The effluent from the washing stage contains lost fiber, clay, coating material, ink particles, and deinking chemicals (such as sodium hydroxide, sodium silicate, sodium carbonate, sodium hypochlorite). This waste is characterized by high BOD, high suspended solids which may not be present in pure ink manufacturing as contrasted to ink removal from paper, and high pH.

Treatment. The following general treatment methods are applied to deinking wastes:

1. *Plain sedimentation* This method uses settling tanks with provision for sludge handling and disposal. It also uses lagoons and settling basins.
2. *Upflow clarification* Wastewater is introduced at the bottom of the settling tank; advantage is thereby taken of filtration of the waste through the sludge blanket.
3. *Coagulation* Best results have been obtained by chemical coagulation with lime, sulfuric acid, alum, sodium aluminate plus alum, activated silica plus alum, activated silica plus lime, and magnesium salts.
4. *Biological aeration* Since the bulk of the BOD of deinking waste is in dissolved form, the biological aeration method was developed. In this process the overflow from primary sedimentation is aerated in the presence of biological floc, followed by secondary settling for removal and recirculation of the floc. Supplemental nitrogen is added for nutrient purposes.

Sludge from these processes is presently used as a low-grade fill material.

References: Printing ink wastes

1. Anonymous, "Wisconsin tissue effluent plant pioneers European process here," *Paper Trade J.* **158**, 32, (1974).
2. Bathe, O., "Effect of anti-pollution laws on the industrial paint market," *Polymers, Paint, and Color J.* **163**, 711 (1973).
3. Bowers, D., "Control of liquid wastes in printing ink manufacturing plants," *Am. Ink Maker* **51**, 16 (1973).

4. Broadbent, D., "Energy conservation and pollution control in the paint finishing industry," *Prod. Fin.*, 8 (1974).

5. Cysin, H., "Environmental problems of the chemical industry," *J. of Oil and Color Chemists Assn.* **56**, 515 (1973).

6. Gaugush, F., "Update of air and water regulations for environmental protection," *J. of Paint Tech.* **45**, 70 (1973).

7. Gove, G. and J. McKeown, "Wastes from waste paper utilization," *TAPPI*, Report No. 45, 57 (1972).

8. Keszthelyi, S., "Foams; fundamental characteristics, prevention, and destruction" *J. of Paint Tech.* **46**, 31 (1974).

9. Martens, D. W. and R. W. Gordon, "Toxicity and treatment of deinking wastes containing detergents," *Int. Pacific Salmon Fisheries Comm. Report*, No. 25 (1974).

10. Sugaya, T., "Treatment system handles flexo ink and starch waste," *Paperboard Packaging* **58**, 28 (1973).

11. Tackett, R., "State of the art of waste disposal in the coatings industry," *J. of Paint Tech.* **46**, 63 (1974).

12. VanSoest, R. and T. W. Wright, "Sludge handling in a deinking mill," *Internat. Symp. of Pulp and Paper Process Control*, 79, April 1973.

13. Williams, R., "Latex waste treatment," *J. of Paint Tech.* **46**, 46 (1974).

14. Zimmerman, R. L., "Environmental improvement from a product standpoint," *J. of Paint Tech.* **45**, 58 (1973).

CHAPTER 28

CHEMICAL INDUSTRIES

Once again we must draw a very thin, and often questionable, distinction between the chemical industries described in this chapter and the materials industries of the previous chapter. Undoubtedly, there are some industries which could be classified in either chapter. In general, however, the chemical industry encompasses smaller plants which produce basic chemicals and raw materials to be used by other manufacturers, whereas the materials industry consists of larger plants producing materials for direct public use.

Chemical wastes are produced by plants that manufacture acids, bases, detergents, cornstarch, powder and explosives, insecticides and fungicides, fertilizers, silicones, plastics, resins, synthetics, and other substances which are often used as raw materials for further manufacturing processes. Chemical processes vary greatly, according to the nature of the substance being manufactured; they include chemical reactions at high temperatures and pressures with or without catalyst and flux, separation of a liquid from a gas, solid, or other liquids, and so forth. Some of the manufacturing methods are sedimentation, flotation, evaporation and distillation, washing, filtering, electrolysis, burning, centrifugal separating, absorption, crystallization, and screening.

Chemical wastes include acids, bases, toxic materials, and matter high in BOD, color, and inflammability (phosphorus) and low in suspended solids. Often chemical wastes require neutralization, as in the manufacture of silicones, smokeless powder, TNT, insecticides, and herbicides which are acid in character. Many can be treated by some biological-oxidation method such as trickling filters, activated sludge, or lagooning. It has been found that if cornstarch waste is mixed with an equal volume of domestic sewage, it can be treated by either activated sludge or trickling filters.

Coagulation is necessary for some wastes, such as those containing phosphorus. Wastes with a complex molecular structure are often separated by some physical process. Detergents, which present the problem of foaming in aeration tanks, have been found amenable to biological degradation, although some of the so-called "hard" syndets are resistant to this. Lake studies have shown that the advent of detergents in domestic sewage has increased the residual soluble inorganic phosphorus by as much as 100 per cent. This increase was directly related to frequent nuisance blooms of blue-green algae.

28.1 Acid Wastes

Acid wastes may be discharged by any of the plants mentioned in the apparel, materials, and chemical industries. Since most states have laws relating to stream standards which require that the pH of a receiving stream be maintained between 6.0 and 9.0, acid wastes usually cannot be allowed to flow untreated into our watercourses. In this chapter we are especially concerned with acids arising from chemical plants producing primary raw materials such as dyes, explosives, pharmaceuticals, and silicone resins. The most important are the dilute wastes of hydrochloric, sulfuric, and sometimes nitric acid. Because of the varied uses of acids, the origin of any one of these acid wastes may bear little or no resemblance to the origin of another.

Regardless of the degree of acidity, the main method of treatment of acid wastes is neutralization (see Chapter 8). Gehm [1] describes a method of neutralizing acid wastes by means of an upflow limestone bed, which handled wastes with up to 10,000 ppm of mineral acidity in a bed capable of receiving 0.1 mgd of waste (Fig. 28.1). Shugart [2] describes a process utilizing lime for automatically neutralizing

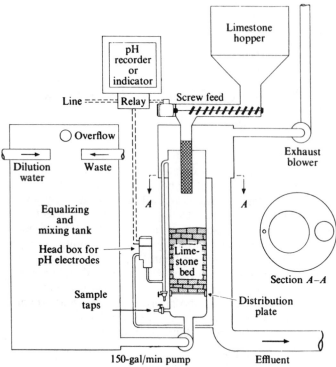

Fig. 28.1 Proposed design of an up-flow neutralizing bed, capable of handling 100,000 gal/day of nitro-cellulose waste containing from 10,000 to 15,000 ppm mineral acidity. (After Gehm [1]).

acid citrus wastes, employing Beckman pH electrodes for control (Fig. 28.2).

As mentioned in Chapter 27, oil refineries use sulfuric acid for desulfurization, improvement of color, refining of lubricating oils, as a catalyst in alkylation, and for other miscellaneous purposes. Two processes are commonly used by oil refineries for disposal of sulfuric-acid wastes: spray-burning and indirect combustion. Spray-burning involves spraying waste acid into a hot combustion chamber (1700 to 2000° F) with small amounts of excess air added to oxidize hydrocarbons. Sulfur is converted to SO_2 and hydrocarbons to CO_2 and H_2O, the hot gases are cooled and dried, and the SO_2 is absorbed to make new sulfuric acid. The principal reaction of the second method, indirect combustion, is the reduction of the sulfuric acid in the sludge by the hydrocarbons which are present. Granular by-product coke is recirculated through a mixer, acid sludge is added to this circulating stream, and heat is applied in the decomposing chamber. Flow diagrams of these processes are presented in Fig. 28.3.

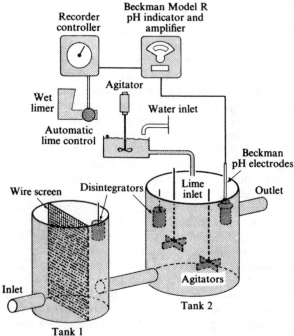

Fig. 28.2 Diagram showing equipment layout for treatment of orange and lemon wastes [3].

Dickerson and Brooks [4] describe a process using lime neutralization for a mixed nitric and sulfuric acid waste from a nitrocellulose manufacturing plant. Although the acids varied widely in volume, concentration, and ratio, neutralization was accomplished effectively in a multiple-unit reaction chamber provided with two-point pH-controlled addition of dolomitic lime slurry.

Tully [5] also describes an upflow limestone bed acid-neutralizing unit for treating a mixture of hydrochloric and sulfuric acids in varying concentrations—wastes from the manufacture of certain resins. The wastes were diluted until they reached a concentration of less than 1 per cent and then passed upward through a 3-foot expanded limestone bed at an average rate of 20 to 30 gallons per minute per square foot of bed area. The effluent pH averaged 4.6, and the 1958 operating costs were about

$0.49 per ton of 1 per cent acid neutralized. The installation is shown in Figs. 28.4 and 28.5.

References: Acid Wastes

1. Gehm, H. W., "Up-flow neutralization of acid wastes," *Chem. Met. Eng.* 51, 124 (1944).
2. Shugart, P.L. "Automatic pH control replaces manual operation for acid waste treatment," *Public Works Mag.* 85, 67, (1962).
3. "Shortcut to neutral water (ion exchange)" Editorial, *Chemical Week* 90, 69 (1962).
4. Dickerson, B. W., and R. H. Brooks, "Neutralization of acid wastes," *Ind. Eng. Chem.* 42, 599 (1950).
5. Tully, T.J. "Waste-acid neutralization" *Sewage Ind. Wastes* 30, 1385 (1958).
6. Temple, K.L. and A.H. Colmer, "the formation of acid lime drainage," Proc. 6th Industrial Waste Conf., Purdue Univ., Feb. 1951.

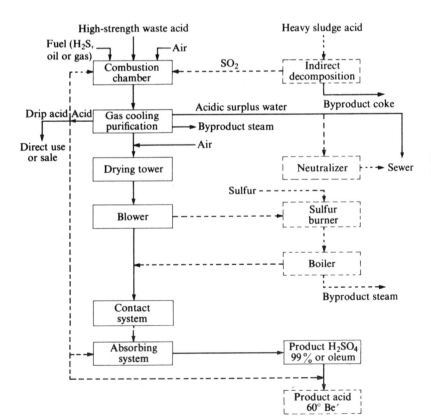

Fig. 28.3 Flow diagram showing (solid lines) basic elements in the spray-burning type of acid recovery plant, (dotted lines) supplementary features which may be used, depending on particular requirements, and the necessary features present in a combination of the spray-burning and indirect-combustion processes. (Courtesy T. R. Harris, Monsanto Chemical Co.)

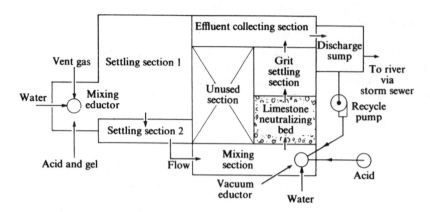

Fig. 28.4 Schematic plan and flow diagram of existing acid neutralization system [6].

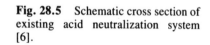

Fig. 28.5 Schematic cross section of existing acid neutralization system [6].

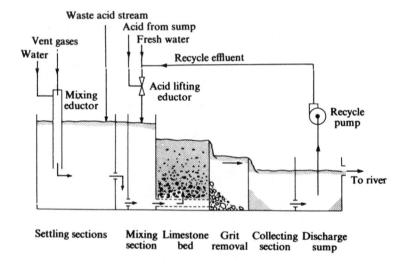

28.2 Cornstarch-Industry Wastes

Although this industry might be included under food processing in Chapter 26, it is described here because cornstarch products are so widely used in the chemical and materials field. This industry processes corn to produce starch, oil, and feed. A bushel of corn weighs on average about 56 pounds [19] and by the wet-milling process this yields about 32 pounds of pearl starch (used in the manufacture of textiles), 1.6 pounds of oil, and 13 to 14 pounds of feed. The composition of the corn kernel is approximately as follows: carbohydrates, 80 per cent; protein, 10 per cent; oil (fat), 4.5 per cent; fiber, 3.5 per cent; and minerals, 2 per cent.

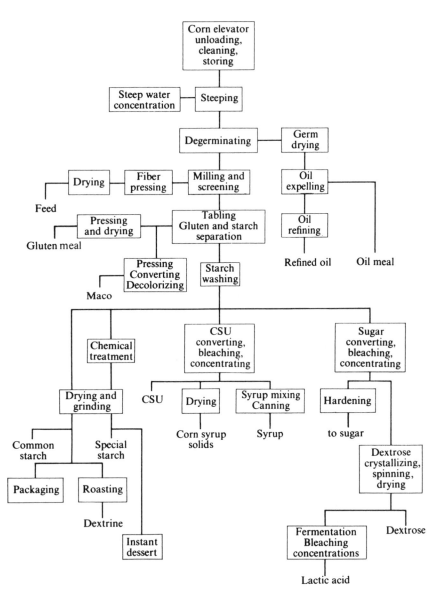

Fig. 28.6 Flowsheet of wet-milling process for corn. (After Van Patten and McIntosh [24].)

In the early 1930s industry began a waste-water reuse program which reduced plant losses to less than 0.5 per cent of the dry corn raw material. This has become known as the "bottled-up" system.

Clean corn is steeped in a dilute solution of sulfuric acid in order to loosen the hull, soften the gluten, and dissolve minerals or organic matter in the kernel. Next the corn is ground to free, but not crush, the germ; the ground corn is mixed with water and placed in settling tanks. When the germs float to the top, they are skimmed off, and oil is pressed or extracted from them. The kernel residues are ground finely, to separate the soluble starch and gluten from the fiber and hull, which are known as "grits and bran" and are used as feed additives. The starch is separated from the gluten by settling, centrifuging, and countercurrent washing. The gluten is added to feed, and the starch is filtered and washed on vacuum filters and dried. Then it is either marketed as starch, or modified starch, or hydrolyzed into corn syrup or corn sugar. Both the feed-producing and starch-manufacturing processes evolve a process-water waste containing about 3 per cent of the corn in soluble form. Van Patten and McIntosh [24] present a flow sheet for wet milling of corn (Fig. 28.6), and Hatfield [19] presents a flow sheet of both the wet-milling process and "bottling-up" reuse of process waters in cornstarch manufacture (Fig. 28.7).

Even before waste-treatment practices were utilized

in the industry, the "bottling-up" process was common, since it was introduced to abate stream pollution, and it can now be considered as an actual part of the cornstarch-plant process. The wastes from this industry consist of residues and leaks from the reuse processes. The bottling-up process has been described [19] as: (1) recirculation of process water, (2) evaporation of a portion of this recirculated water as steep water, and (3) addition of all dried organic residues to the gluten to make an improved cattle feed.

Major wastes from the cornstarch plants are: (1) volatile organics entrained in the evaporator condensate, (2) syrup from final wastes, and (3) wastes from bottling-up processes—arising primarily because of an imbalance between the amount of fresh

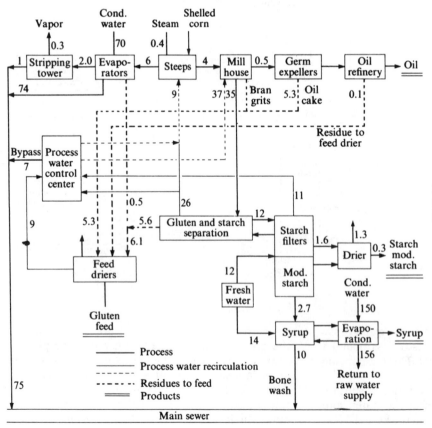

Fig. 28.7 Flow sheet showing cornstarch production. (After Hatfield [19].)

water added, the amount of recirculated water, and the amount of steep water taken to the evaporators. Hatfield [19] lists the population equivalents of these wastes (Table 28.1). For each bushel of corn processed, 40 gallons of water are used directly in the process and 100 to 200 gallons per bushel for other purposes, although much of this is reused [21].

Table 28.1 Population equivalents of cornstarch wastes [19].

Waste source	Population equivalent*
Steep-water evaporators	30,000
Light bone wash	8,000
Cleanup, etc.	12,000
Total	50,000

*Per 50,000-bushel grind.

Van Patten and McIntosh [24] describe waste reduction at the American Maize Products Company, which manufactures starches, sugars, syrups, and fermentation products from corn. In 1951, the company spent about $850,000 to replace equipment and install modifications in production procedures that would reduce its waste BOD load by 92 per cent. Since the residual cornstarch-plant wastes generally contain organic solubles from whole-kernel corn and corn syrup, they are particularly amenable to biological treatment, especially when mixed with domestic sewage. The wastes are normally quite hot, and all factors directly attributable to this characteristic must be considered; for example, oxygen solubility is lowered, but at the same time digestion is enhanced. Settling is hindered by the heat of the wastes, and sewer lines tend to clog as the starch wastes cool. Equalization, to effect both cooling and homogeneity of the wastes, should thus be practiced prior to biological treatment, either at the industrial-plant site or in the municipal treatment plant.

Hatfield *et al.* [9] used trickling filters with a 5:1 recirculation ratio to effect a 90 per cent BOD reduction, with an influent BOD concentration of 1400 ppm or less and a loading of about 4 pounds of BOD per

cubic yard per day. Control of pH and addition of nitrogen and phosphorus were necessary.

References: Cornstarch-Industry Wastes

1. Adinoff, J., "Disposal of organic chemical wastes to underground formations," *Ind. Wastes* **1**, 40 (1955).
2. Douglass, I. B., "By-products and waste in potato processing," in Proceedings of 15th Industrial Waste Conference, Purdue University, May 1960, p. 99.
3. Fergason, A. A., "Bacterial utilization of potato starch wastes," in Proceedings of 15th Industrial Waste Conference, Purdue University, May 1960, p. 258.
4. Greeley, S. A., and W. D. Hatfield, "The sewage disposal works of Decatur, Ill.," *Trans. Am. Soc. Civil Engrs.* **94**, 544 (1930).
5. Greenfield, R. E., *et al.*, "Cornstarch waste treatment with sewage, Decatur, Ill.," *Sewage Ind. Wastes* **19**, 951 (1947).
6. Greenfield, R. E., G. N. Cornell, and W. D. Hatfield, "Cornstarch wastes," in Proceedings of 3rd Industrial Waste Conference, Purdue University, May 1947, p. 360.
7. Greenfield, R. E., G. N. Cornell, and W. D. Hatfield, "Industrial wastes: cornstarch processes," *Ind. Eng. Chem.* **39**, 583 (1947).
8. Gurnham, C. F. (ed.), *Principles of Industrial Waste Treatment*, John Wiley & Sons, New York (1955), p. 375.
9. Hatfield, R., E. R. Strong, F. Heinsohn, H. Powell, and T. G. Stone, "Treatment of wastes from a corn industry by pilot-plant trickling filters," *Sewage Ind. Wastes* **28**, 1240 (1956).
10. Hatfield, W. D., "Operation of the pre-aeration plant at Decatur, Ill.," *Sewage Works J.* **3**, 621 (1931).
11. Hatfield, W. D., "Special test for corn starch wastes," *Sewage Ind. Wastes* **22**, 1381 (1951).
12. Hatfield, W. D., "Cornstarch processes," in *Industrial Wastes, Their Disposal and Treatment*, ed. W. Rudolfs, Reinhold Publishing Corp., New York (1953), p. 132.
13. Haupt, H., "By-product recovery from starch wastes," *Sewage Ind. Wastes* **8**, 350 (1936).
14. Hussman, W., "Starch waste treatment experiments," *Sewage Ind. Wastes* **6**, 342 (1934).

15. Mohlman, F. W., "Treatment of packing-house, tannery, and corn-products wastes," *Ind. Eng. Chem.* **18**, 1076 (1926).

16. Mohlman, F. W., "Utilization and disposal of industrial wastes," in Proceedings of 1st Industrial Waste Conference, Purdue University, November 1944, p. 43.

17. Mohlman, F. W., and A. J. Beck, "Disposal of industrial wastes," *Ind. Eng. Chem.* **21**, 205 (1929).

18. Pulfrey, A. L., R. W. Kerr, and H. R. Reintjes, "Wet milling of corn," *Ind. Eng. Chem.* **32**, 1483 (1940).

19. Rudolfs, W., *Industrial Waste Treatment*, Reinhold Publishing Corp., New York (1953), p. 132.

20. Sjostrom, O. A., "Treatment of waste water from a starch and glucose factory," *Ind. Eng. Chem.* **3**, 100 (1911).

21. Corn Industry Research Foundation, *The Story of Corn and Its Products*, New York (1952).

22. Van Patten, E. M., and G. H. McIntosh, "Waste savings at American Maize Products Company," in Proceedings of 6th Industrial Waste Conference, Purdue University, February 1951, p. 344.

23. Van Patten, E. M., and G. H. McIntosh, "Corn products manufacture, waste load reduction," *Sewage Ind. Wastes* **24**, 1443 (1952).

24. Van Patten, E. M., and McIntosh, G. H., "Liquid industrial wastes, corn products manufacture," *Ind. Eng. Chem.* **44**, 483 (1952).

25. Wagner, T. B., "Disposal of starch factory wastes," *Ind. Eng. Chem.* **3**, 99 (1911).

Suggested Additional Reading

The following references have been published since 1961.

1. Buzzell, J. C., A. L. Caron, S. J. Rychman, and O. J. Sproul, "Biological treatment of protein water from the manufacture of potato starch. I," *Water Sewage Works* **V-3**, 327 (1964).

2. Hemens, J., P. G. J. Meiring, and G. J. Stander, "Full-scale anaerobic digestion of effluents from the production of maize-starch," *Water Waste Treat. J.* **9**, 16 (1962).

3. Ling, J. T., "Pilot investigation of starch-gluten waste treatment," in Proceedings of 16th Industrial Waste Conference, Purdue University Engineering Extension Series, no. 109, May 1961, p. 217.

4. Vennes, J. W., and E. G. Olmstead, "Stabilization of potato wastes," *Official Bulletin* (Official Organ of the North Dakota Water and Sewage Works Conference) **29**, 10 (1961).

The following references have been published since 1968.

1. Benik, F. K., "Application of reverse osmosis to waste streams in corn processing industry," *Water Poll. Contr., Ont.* **107**, 24–26; 47–49 (1969).

2. Maric, T., "Wastewaters from meat processing and starch plants and their purification," *Technika* **24**, 1485–1491 (1969).

3. Schwartz, J. H., "Potato starch factory waste effluents recovery," *J. Sci. of Food and Agric.* **23**, 977 (1972).

4. Segfried, C. F., "Purification and starch industry wastewater," *Eng. Bull. Purdue Univ., Eng. Ext. Srs.* **132**, 1103–9 (1969).

5. Willenbrink, R. V., "Control and treatment of corn and soybean wastes," *Water and Sewage Works* **114**, Ref. No. R198-R202 (1968).

28.3 Phosphate-Industry Wastes

Two billion years ago, when the earth was first formed, the molten rocks cooled and solidified into rocks containing small amounts of apatite, a tricalcium fluorophosphate mineral. Exposed to the elements, these rocks slowly weathered, were washed into streams, and eventually dissolved in the ocean. Species of sea life withdrew these minute forms of phosphorus, now combined with calcium, limestone, quartz, sand, and so forth, to built their shells and bodies. These various forms of sea life eventually died, settled to the bottom of the ocean, and formed thick layers of deposits containing phosphorus. Such a deposit is now being commercially mined mainly in Florida, which 10 million years ago was at the bottom of the ocean [20]. About 70 per cent of the world's supply of phosphate rock comes from an area in central Florida approximately 50 miles in diameter, centered in the small city of Bartow.

The phosphate rock in this region is found in the form of small pebbles embedded in a matrix of phosphatic sands and clays. These beds, sedimentary in origin, are overlaid with nonphosphate sands and lime rock of more recent origin. Although mining is seldom carried out below 60 feet, the phosphate rock is found at varying depths down to several hundred feet. The overburden is first stripped off with large drag lines and the exposed matrix is excavated in strips; this excavated matrix is dropped into a previously prepared pit and mixed with water from hydraulic guns that wash the rock into a pump sump. From there, the mixture of phosphate matrix and water is carried in pressure pipelines to a washer plant and the larger phosphate-rock particles are removed by screens, shaker tables, and hydraulic sizing cones. All the particles retained on a 200-mesh screen are recovered, by hydraulic sizing in large clarifiers and by a flotation process in which the phosphate particles are selectively coated (with a material such as tall or rosin oil) after conditioning with NaOH for pH control.

The nonphosphatic sands are removed by another flotation process in which, after dewatering and treatment with sulfuric acid, the silica material is selectively coated with an amine and floated off, thus concentrating the phosphate content in the tailings sufficiently for commercial purposes. The finished product is dried in kilns and either sold directly as fertilizer or further processed to superphosphate or triple phosphate. It may also be burned in electric furnaces to produce elemental phosphorus or phosphoric acid.

The electric-furnace conversion process (Fig. 25.8) is described by Horton et al. [10]. Phosphate rock is blended with coke, the reducing agent, and silica, which acts as a flux, to form the furnace charge. In the furnace, the charge is smelted by electrical energy to liberate phosphorus as a gas, leaving a dense slag— a mixture of calcium silicate and ferrophosphorus— which is tapped periodically. The furnace gases (with temperatures of 500 to 800°F) consist of phosphorus,

carbon monoxide, and various other gaseous impurities in small amounts. The gases are cleaned, usually by electrostatic precipitators, and condensed by water to a heavy liquid; after which the water and heavy liquid phosphorus are separated by sedimentation.

Wastes from the phosphate industry arise from (1) mining the rock and (2) processing the rock to elemental phosphorus and other pure chemicals. In mining, the main wastes originate from the washer plant, where phosphate rock is separated from the water solution, and the flotators, where phosphate particles are separated from the impurities retained on the screens. In processing the phosphate, the major source of water-borne waste is the condenser water bleed-off from the reduction furnace (see Fig. 28.8).

The wastes from phosphate-rock mining are high in volume. Flotation plants [5] commonly use and discharge approximately 30,000 gpm of waste, containing fine clays and colloidal slimes, as well as some tall oil (a resinous by-product from the manufacture of chemical wood pulp) or rosin oil from the flotators. The condenser water bleed-off from phosphorus refining varies in volume from 10 to 100 gpm and has as its most important ingredient the elemental phosphorus in colloidal form, which may ignite if allowed to dry out in ditches. Another significant component of the waste is fluorine, which is also present in the furnace gases. Horton [10] gives the general characteristics of phosphorus wastes (Table 28.2).

To treat these wastes, land areas are provided for tailings, storage, and settling of slime. These methods,

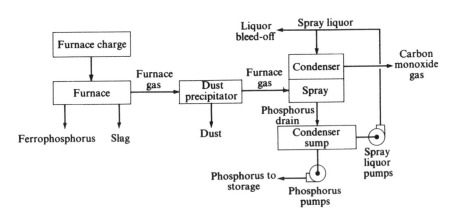

Fig. 28.8 Electric furnace for conversion of phosphate rock into phosphorus [10].

Table 28.2   Phosphorus wastes [10].

Characteristic	Amount or value
pH	1.5–2.0
Temperature	120–150°F
Elemental phosphorus	400–2500 ppm
Total suspended solids	1000–5000 ppm
Fluorine	500–2000 ppm
Silica	300–700 ppm
P_2O_5	600–900 ppm
Reducing substances as (I_2)	40–50 ppm
Ionic charge	Predominantly positive (+)

and the use of modern flotators, comprise the bulk of phosphate-rock waste-treatment measures. However, it is also common to use mechanical clarifiers for removing sand tailings and to store waste water in lagoons prior to reuse [36].

Among the treatment or disposal methods for phosphorus-refining plants that have been investigated or attempted are not only the above-mentioned lagooning, oxidizing, and settling (with or without prior chemical coagulation), but also filtering and centrifuging. Horton *et al.* [10] state that coagulation and settling appear to offer the best solution. They present their pilot-plant results in schematic form (Fig. 28.9) and state that concentrations of 40-fold (25 per cent solids) were obtained by a simple coagulation and sedimentation process.

References: Phosphate-Industry Wastes

1. Bixler, G. H., J. Work, and R. M. Lattig, "Elemental phosphorous," *Ind. Eng. Chem.* **48**, 2 (1956).
2. Bowles, O., "Florida phosphate holds attention of American Institute of Mining Engineers," *Rock Products* **53**, 164 (1950).
3. Breton, E., Jr., and N. H. Waggaman, "Calcium fluoride," U.S. Patent no. 2410043, 29 October 1946.

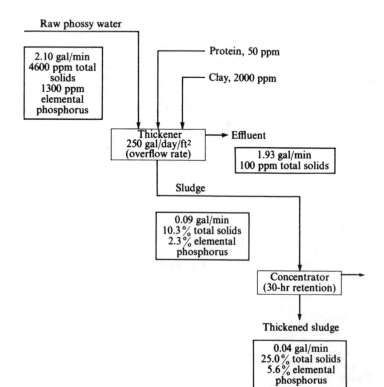

Fig. 28.9   Summary of pilot plant for the treatment of wastes from phosphorus refining. (After Horton *et al.* [10].)

4. Bridgers, G. L., J. W. Moore, and H. M. McLeod, Jr., "Phosphatic animal-feed supplement," *Ind. Eng. Chem.* **41**, 1391 (1949).

5. Fuller, R. B., "The position of the pebble phosphate industry in stream sanitation," in Proceedings of 1st Annual Public Health Engineering Conference, University of Florida, November 1948; Bulletin no. 26, April 1949.

6. Fuller, R. B., "The position of the pebble phosphate industry of Florida in stream sanitation as of November, 1949," in Proceedings of 5th Industrial Waste Conference, Purdue University, November 1949.

7. Fuller, R. B., "Phosphate industry position in Florida," *Sewage Ind. Wastes* **23**, 700 (1951).

8. Hall, J. P., and N. A. Hodges, "Recovery of phosphate fines," U.S. Patent no. 2113727, 12 April 1938.

9. Hignett, T. P., and M. R. Siegel, "Recovery of fluorine from stack gases," *Ind. Eng. Chem.* **41**, 2493 (1949).

10. Horton, J. P., J. D. Molley, and H. C. Bays, "Processing of phosphorus furnace wastes," *Sewage Ind. Wastes* **28**, 70 (1956).

11. Lea, W. L., and G. A. Rohlich, "Removal of phosphate from treated sewage," *Sewage Ind. Wastes* **26**, 261 (1954).

12. Lenhart, W. B., "Latest recovery methods highlight Florida phosphate plant," *Rock Products* **54**, 74 (1951).

13. Lenhart, W. B., "Developments in producing of phosphates," *Rock Products* **55**, 80 (1952).

14. McCarthy, J. A., and W. E. Cassidy, "Viability of bacteria in the presence of phosphates," *J. New England Water Works Assoc.* **57**, 287 (1943).

15. Maynard, P., "Lightweight aggregate from phosphate slimes," *Eng. Mining J.* **152**, 92 (1951).

16. Murdock, H. R., "Industrial waste," *Ind. Eng. Chem.* **43**, 89*A* (1951).

17. Murdock, H. R., "Industrial wastes," *Ind. Eng. Chem.* **44**, 115*A* (1952).

18. Orago, A., "Three new steps in treating Florida phosphate rock," *Eng. Mining J.* **151**, 78 (1950).

19. Owen, R., "Removal of phosphorus from sewage effluent with lime," *Sewage Ind. Wastes* **25**, 548 (1953).

20. "Phosphate, the servant of mankind," *Oil Power* **26**, 3 (1951).

21. State Water Pollution Control Board, "Phosphates," in *Water Quality Criteria*, Publication no. 3, Sacramento, California (1952), p. 322.

22. Rudolfs, W., "Phosphates in sewage and sludge treatment," *Sewage Ind. Wastes* **19**, 43 (1947).

23. Sawyer, C. N., "Some new aspects of phosphates in relation to lake fertilization," *Sewage Ind. Wastes* **24**, 768 (1952).

24. Seyfried, W. R., "Recovery of values from phosphate rock," U.S. Patent no. 2152364, 28 March 1939.

25. Specht, R. C., "Effect of waste disposal of the pebble phosphate rock industry in Florida on conditions of receiving streams," *Mining Eng.* **187**, 779 (1950).

26. Specht, R. C., "Phosphate waste studies," *Florida Univ. Eng. Exp. Sta. Bull.* **32**, February 1950.

27. Specht, R. C., "Disposal of wastes from the phosphate industry," *J. Water Pollution Control Federation* **32**, 969 (1960).

28. Specht, R. C., and W. E. Herron, Jr., "Lightweight aggregate from phosphate slimes," *Rock Products* **53**, 96 (1950).

29. Swainson, S. J., "Washing and concentrating Florida pebble phosphate," *Mining Met.* **25**, 454 (1944).

30. "Symposium on phosphates and phosphorus," *Ind. Eng. Chem.* **44**, 1519 (1952).

31. Thompson, D., "Ultrasonic coagulation of phosphate tailing," *Bull. V. Polytech. Inst. Eng. Exp. Sta. Ser.* **75**, 5 (1950).

32. "Twelfth Purdue Industrial Waste Conference," *Ind. Wastes* **3**, 48 (1957).

33. Waggaman, W. H., and R. E. Bell, "Factors affecting development, Western Phosphate," *Ind. Eng. Chem.* **42**, 269 (1950).

34. Waggaman, W. H., and E. R. Ruhlman, "Conservation problems of the phosphate industry," *Ind. Eng. Chem.* **48**, 360 (1956).

35. Wakefield, J. W., "Semi-tropical industrial waste problems," in Proceedings of 7th Industrial Waste Conference, Purdue University, May 1952, p. 503.

36. Wakefield, J. W., "As Florida grows, so does its industrial waste problem," *Wastes Eng.* **24**, 495 (1953).

37. Williams, D. E., F. L. MacLeod, E. Murrell, and H. Patrick, "Animal feeding test," *Ind. Eng. Chem.* **41**, 1396 (1948).

38. Wright, D. M., "Dewatering materials such as Florida phosphate rock," U.S. Patent no. 2158169, 16 May 1939.

Suggested Additional Reading

The following references have been published since 1962.

1. Chodak, M. E., "Chemical industry solids disposal problems," *J. Proc. Inst. Sewage Purif.*, Part 5 (1962), p. 431.

2. Jones, W. E., and R. L. Olmsted, "Waste disposal at a phosphoric acid and ammonium phosphate fertilizer plant," Purdue University Engineering Extension Series, Bulletin no. 112, 1962, p. 198.

3. Knopsack-Griesheim, A. G., "Treating waste waters obtained in the production of phosphorus," British Patent no. 888085, 24 January 1962.

4. Patton, V. D., "Phosphate mining and water resources," *Ind. Water Wastes* **8**, 24 (1963).

The following references have been published since 1970.

1. Anonymous, "Alum may combat algae," *Sci. News Letter*, 370 May 29, 1971.

2. Anonymous, "Enzyme inhibition used as assay for phosphates," *Env. Sci. and Tech.* **4**, 883 (1970).

3. Anonymous, "Fairfax goes for AWT," *Water and Wastes Eng.* **10**, 38–39 (1973).

4. Anonymous, "Instrumentation for environmental monitoring," University of California, Berkeley Lab., *Water* **2**, Rpt. LBL-1 (1973).

5. Anonymous, "North Carolina's dustless terminal," *World Parts* **35**, 12–13 (1973).

6. Anonymous, "Michigan removes phosphorus," *Env. Sci. and Tech.* **5**, 297 (1971).

7. Anonymous, "Modified activated sludge process takes out phosphates," *Env. Sci. and Tech.* **5**, 15 (1971).

8. Anonymous, "New acid route gets the test," *Chem. Wk.* **109**, 20 (1971).

9. Anonymous, "Phosphorus still the culprit," *Sci. News* **100**, 248 (1971).

10. Anonymous, "Phosphorus in fresh water and the marine environment," *Int'l Assoc. on Water Pollut. Research in Water Research* **7**, 1–2 (1973).

11. Anonymous, "Profitable phosphate removal claimed," *Detergents and Specialties* **7**, 48 (1970).

12. Anonymous, "Spectrum," *Env. Mag.* **13**, 27 (1971).

13. Anonymous, "Spectrum," *Env. Mag.* **13**, 28 (1971).

14. Bahls, L. L., "Diatom community response to primary wastewater effluent," *J. WPCF* **45**, 134–144 (1973).

15. Brodersen, K. and I. Larsen, "Production of elemental phosphorus from sludge produced by precipitation of municipal wastewater with aluminum salts," *Denmark Atomenergikommissionen Risoc*, Rpt Risol-M-1492 (1972).

16. Chung, S. K., "Wastewater treatment properties of Fe (II) and Fe (III) compounds," *Water and Sew. Works* **120**, 54–58 (1973).

17. Clesceri, M. I., "Phosphorus removal from wastewaters; a cost analysis," *Water and Sew. Works* **120**, 82–86; 88–91 (1973).

18. EPA, "Economic analysis of proposed effluent guidelines; The industrial phosphate industry," *EPA Office of Planning and Evaluation*, EPA-30/1-73-021 (1973).

19. Evers, R. H., "Advanced wastewater treatment techniques for removing nitrogen and phosphorus," *Water and Sew. Works*, Ref. No. R15-R16, R18-R19, R134, R136-R137 (1973).

20. Grigoropoulos, S. G., R. C. Vedden, and D. W. Max, "Fate of aluminum-precipitated phosphorus in activated sludge and anaerobic digestion," *J. WPCF* **43**, 2366 (1971).

21. Gunopolos, A. and F. I. Vilen, "Process evaluation—phosphorus removal," *J. WPCF* **43**, 1975 (1971).

22. Hetling, L. J. and I. G. Carcich, "Phosphorus in wastewater," *Water and Sew. Works* **120**, 59–62 (1973).

23. Hou, P. H., "Complementary/role of Fe (III), sulfate and calcium in precipitation of phosphate from solution," *Env. Letters* **5**, 115–136 (1973).

24. Jenkins, D., A. B. Menor, and J. F. Ferguson, "Recent studies of calcium phosphate precipitation in wastewaters," *Applications of New Concepts of Physical-Chemical Wastewater Treatment, Int'l Conf.; Papers: Eckenfelder, W. W. and L. K. Cecil*, Elmsford, New York: Pergamon Press 211–230 (1972).

25. Kappe, D. S. and S. E. Kappe, "Algal growth exciters," *Water and Sewage Works* **118**, 245–248 (1971).

26. LeClaire, B. P., "Selected references on phosphorus removal," *Canada Water Pollut. Contr. Directorate, Env. Protection Service Rpt. Series, Economic and Technical Review*, Rpt EPS 3-WP-73-2 (1973).

27. McAchran, G. E. and R. D. Hogue, "Phosphate removal from municipal sewage," *Water and Sew. Works* **118**, 36–39 (1971).

28. McCaull, J., "Building a shorter life." *Env. Mag.* **13**, 8 (1971).

29. Meyer, C. L., "De invloed van josfaat op de toxiciteit van koper voor een clg," *TNO Nieuws* **27**, 468–473 (1972).

30. Michel, F. M., "Nitrogen and phosphorus uptakes by 'chlorella pyrenoidosa' in sewage treatment processes," *Water and Sew. Works* **120**, 76–79 (1973).

31. Morgan, W. E., and E. G. Gruh, "An investigation of phosphorus removal mechanisms in activated sludge systems," *Gov't Rpts Announcements* **73**, 112 (1973).

32. Nelson, D. W., and J. V. Mannering, "Nitrogen and

phosphorus composition of surface runoff as affected by tillage method," *J. of Env. Quality* **2**, 292–295 (1972).

33. Ripley, P. G., and G. L. Lamb, "Filtration of effluent from a biological chemical system," *Water and Sewage Works* **120**, 66–69 (1973).

34. Van Vuuren, L. R. J., "Combined physical, chemical and biological treatment of effluents," *Inst. of Water Pollut. Contr., S. African Branch Bienniel Conf. in Water Pollut. Contr.* **72**, 227–230 (1973).

35. Weiss, C. M., "Algal response to detergent phosphate levels," *Water Pollut. Contr. Fed. J.* **45**, 480–489 (1973).

28.4 Soap- and Detergent-Industry Wastes

The soap and detergent industry produces relatively small volumes of liquid wastes directly, but causes great public concern when its products are discharged after use in homes and factories.

In soap manufacturing, the waste waters discharge into trap tanks on skimming basins, where floatable fatty acids are recovered. The recovered fatty acids may not only pay all costs of operation, but also amortize the treatment-plant investment. Gibbs [9] successfully treated soap-plant wastes by flotation with fine air bubbles for a retention period of 40 minutes. The floated sludge was skimmed into a receiving tank, from which it was periodically pumped back to the soap factory for reprocessing or recovery.

Detergents are a class of surface-active compounds used as cleansing agents. They can be grouped as anionic, cationic, and nonionic agents. The synthetic detergents are effective in wide pH ranges and do not form insoluble precipitates in hard water, as many soaps do. Use of synthetic organic detergents in place of, and in addition to, soap has increased at a rapid rate, the production in the United States growing from about 28 million pounds in 1941 to more than 2 billion pounds in 1955 [32]. These synthetic compounds are used not only in households, but also increasingly in the textile, cosmetic, pharmaceutical, metal, paint, leather, paper, and rubber industries, because of their properties of dispersing, wetting, and emulsifying [20].

Many difficulties are reportedly being caused by these residual, water-soluble detergents, such as: (1) interference with oxygen transfer in activated-sludge treatment and in receiving streams, (2) excessive frothing, (3) toxicity to fresh-water game fish, and (4) difficulty of removal at water-treatment plants. The main concern of the public appears to be the presence of these compounds in water supplies, which implies not only that we may be forced to drink a small amount of detergent in our water, but also that other pollution, of a more dangerous type, could conceivably find its way into the water supply. Bogan [2] found that, although all detergents (Table 28.3) decompose somewhat by biological attack, the rate of decomposition is related to chemical structure. Branching in the alkyl group of the alkyl-aryl detergent types causes a definite retarding of oxidation of the alkyl-benzene-sulfonate detergents. Susceptibility of a nonionic substance to biochemical oxidation, other things being equal, decreases as the size of the polyoxyethylene hydrophilic group increases.

Research is currently being carried out on the replacement of polyphosphates in detergents with organic acids to reduce the algae problem. Studies have shown that both alkyl-benzene-sulfonate and linear alkyl-sulfate can be removed by percolation through sand–silt-type soils.

References: Soap- and Detergent-Industry Wastes

1. Bell, C. E., "Assimilation of hydrocarbons by microorganisms," *Advan. Enzymol.* **10**, 443 (1950).

2. Bogan, R. H., "The biochemical oxidation of synthetic detergents," in Proceedings of 10th Industrial Waste Conference, Purdue University, 1955, p. 231.

3. Bogan, R. H., and C. N. Sawyer, "Biochemical degradation of synthetic detergents," *Sewage Ind. Wastes* **26**, 1069 (1954).

4. Borden, L., and P. C. F. Isaac, "Effects of synthetic detergents on the biological stabilization of sewage," *Surveyor* **115**, 915 (1956).

5. Committee Report, Association of Soap and Glycerine Producers, "Determinations of orthophosphate, hydrolyzable phosphate and total phosphate," *J. Am. Water Works Assoc.* **50**, 1563 (1958).

6. Degens, P. N., Jr., H. Van der Zee, and J. D. Kommer, "Effects of synthetic detergents on the settling of suspended solids," *Sewage Ind. Wastes* **26**, 1081 (1954).

7. Downing, A. L., and L. J. Scragg, "The effect of synthetic detergents on the rate of aeration in diffused-

Table 28.3   Biochemical oxidation of the principal detergent types. (After Bogan [2].)

Synthetic detergent, class and type	5-day 20°C BOD				BOD, Warburg method (acclimated seed)	
	Sewage seed		Acclimated seed			
	ppm × 10³	%*	ppm × 10³	%*	ppm × 10³	%*
Anionic						
Akyl benzene sulfonate						
n-dodecyl	237	10.0	1046	44.2	300	12.7
Keryl	0	0	535	21.6	240	10.1
Tetrapropene	0	0	81	3.4	40	1.7
Alkyl sulfate						
n-dodecyl	1307	57.2	1362	59.7	930	40.7
Dupanol C	1250	4.8	1330	57.4	920	45.2
Sulfonated ester						
Igepon AP-78	1315	59.8	1460	66.5	990	45.0
Sulfonated amide						
Igepon T-77	1443	52.0	1558	56.0	1560	56.0
Nonionic						
Alkyl phenoxy						
polyoxyethylene						
Neutronyx 600	0	0	116	5.4	254	11.8
Igepal CA 630	132	6.1	124	5.7	440	20.4
Rohm and Hass						
OPE-5	310	13.1	390	16.4	170	7.2
Polyethoxy amide						
Ethomid HT/15	996	47.5	880	42.0	412	19.5
Ethomid HT/60	29	1.5	310	15.9	16	8.2
Polyethoxy ester						
Ethofat C/15	1000	46.5	880	42.0	640	29.1
Ethofat C/60	220	11.8	240	12.8	160	8.5
Polyglycol ethers						
Pluronic F68	124	6.0	20	1.1	67	3.6

*Based on theoretical amount of oxygen required for complete conversion to CO_2 and H_2O.

air activated sludge plants," *Water Waste Treat. J.* **7**, 102 (1958).

8. Fairing, J. D., and F. R. Short, "Spectrophotometric determination of alkyl benzene sulfonate detergents in surface water and sewage," *Anal. Chem.* **28**, 1827 (1956).

9. Gibbs, F. S., "The removal of fatty acids and soaps from soap-manufacturing waste waters," in Proceedings of 5th Industrial Waste Conference, Purdue University, 1949, p. 400.

10. Gibbs, F. S., "Soap manufacturing wastes, removal of fatty acids and soaps," *Sewage Ind. Wastes* **23**, 700 (1951).

11. House, R., "Analytical development work for detergent ABS determination in waste waters," *Sewage Ind. Wastes* **29**, 1225 (1957).

12. Lammana, C., and M. F. Mallett, *Basic Bacteriology*, Williams & Wilkins Co., Baltimore (1953).

13. McGauhey, P. H., and S. A. Klein, "Removal of ABS by sewage treatment," *Sewage Ind. Wastes* **31**, 877 (1959).

14. McKinney, R. E., "Syndets and waste disposal," *Sewage Ind. Wastes* **29**, 654 (1957).

15. McKinney, R. E., and E. J. Donovan, "Bacterial degradation of ABS," *Sewage Ind. Wastes* **31**, 690 (1959).

16. McKinney, R. E., and J. M. Symons, "Bacterial degradation of ABS," *Sewage Ind. Wastes* **31**, 549 (1959).

17. Maloney, G. W., and W. D. Sheets, "Detergent builders and BOD," *Sewage Ind. Wastes* **29**, 263 (1957).

18. Manganelli, R., H. Heukelekian, and C. N. Henderson, "Persistence and effect of alkyl aryl sulfonate in sludge digestion," in Proceedings of 15th Industrial Waste Conference, Purdue University, May 1960, p. 199.

19. Munro, L. A., and M. Yatabe, "Frothing of synthetic sewages," *Sewage Ind. Wastes* **29**, 883 (1957).

20. Niven, W. W., *Fundamentals of Detergency*, Reinhold Publishing Corp., New York (1950).

21. Oldham, L. W., "Investigations into the effects of a non-ionic synthetic detergent on biological percolating filters," *J. Inst. Sewage Purif.*, Part 2 (1958), p. 136.

22. Porter, J. R., *Bacterial Chemistry and Physiology*, John Wiley & Sons, New York (1946).

23. Raybould, R. D., and L. H. Thompson, "Tide and santomesse in the sewage works," *Surveyor* **115**, 41 (1956).

24. Roberts, F. W., "The removal of anionic syndets by biological purification processes—observations at Luton and Letchworth," *Water Waste Treat. J.* **6**, 302 (1957).

25. Roberts, F. W., and G. R. Lawson, "Some determinations of the synthetic detergent content of sewage sludge," *Water Waste Treat. J.* **7**, 14 (1958).

26. Sawyer, C. N., "Effects of synthetic detergents on sewage treatment processes," *Sewage Ind. Wastes* **30**, 757 (1958).

27. Sawyer, C. N., R. H. Bogan, and J. R. Simpson, "Biochemical behavior of synthetic detergents," *Ind. Eng. Chem.* **48**, 236 (1956).

28. Sawyer, C. N., and D. W. Ryckman, "Anionic synthetic detergents and water supply problems," *J. Am. Water Works Assoc.* **49**, 480 (1957).

29. Sheets, W. D., and G. W. Maloney, "Synthetic detergents and the BOD test," *Sewage Ind. Wastes* **28**, 10 (1956).

30. Stanier, R. Y., "Problems of bacterial oxidative metabolism," *Bacterial. Rev.* **14**, 179 (1950).

31. Soap & Detergent Association, *Synthetic Detergents in Perspective*, compiled by the Association, 295 Madison Ave., New York, N.Y. (1962).

32. Truesdale, G. A., "Foaming of liquids containing synthetic detergents," *Water Waste Treat. J.* **7**, 108 (1958).

33. U.S. Tariff Commission Reports, "Synthetic organic chemicals, U.S. production and sales," 2nd Series, Government Printing Office, Washington, D.C.

34. Weaver, P. J., "Determination of trace amounts of alkyl benzene sulfonates in water," *Anal. Chem.* **28**, 1922 (1956).

Suggested Additional Reading

The following references have been published since 1962.

1. Abrams, I. M., and S. M. Lewon, "Removal of ABS from water by chloride cycle—anion exchange," *J. Am. Water Works Assoc.* **54**, 43 (1962).

2. Anderson, D. A., "Growth responses of certain bacteria to ABS and other surfactants," in Proceedings of 19th Industrial Waste Conference, Purdue University, 1964, p. 592.

3. "Biodegradability of detergents: a story about surfactants," *Chem. Eng. News* **41**, 102 (1963); *Textile Technicians' Dig.* **20**, 2272 (1963).

4. "Armour solves problems of waste water treatment," *Soap Chem. Specialties* **43**, 47 (1967).

5. Barth, E. F., and M. B. Ettinger, "Anionic detergents in waste water received by municipal treatment plants," *J. Water Pollution Control Federation* **39**, 815 (1967).

6. Basu, A. K., "Treatment of effluents from the manufacture of soap and hydrogenated vegetable oil," *J. Water Pollution Control Federation* **39**, 1653 (1967).

7. Bock, K. J., "Biological investigations of detergents," *Textil-Rundschau* **17**, 87 (1962); *Chem. Abstr.* **56**, 15303 (1962).

8. Buescher, C. A., and D. W. Rychman, "Reduction of foaming of ABS by ozonation," in Proceedings of 16th Industrial Waste Conference, Purdue University, May 1961, p. 251.

9. Burgess, S. G., and L. B. Wood, "Removal and disposal of synthetic detergents in sewage effluents," *J. Proc. Inst. Sewage Purif.* (1962), p. 158.

10. Chadwell, J. H., "Soap plant waste treatment," in Proceedings of 13th Ontario Industrial Waste Conference, June 1966, p. 231.

11. Eden, G. E., *et al.*, "The destruction of ABS in sewage treatment processes," *Water Sewage Works* **108**, 275 (1961).

12. Eldridge, E. F., "Irrigation as a source of water pollution," *J. Water Pollution Control Federation* **35**, 614 (1963).

13. Eisenhauer, H. R., "Chemical removal of ABS from wastewater effluents," *J. Water Pollution Control Federation* **37**, 1567 (1965).

14. Eldib, I. A., "Testing biodegradability of detergents," *Textile Technicians' Dig.* **20**, 4372 (1963).

15. El'kun, D. I., and D. B. Maeve, "Waste waters from production of synthetic detergents as a source for acetic acid and its homologs," *Inst. Lesokhim. Prom.* **14**, 85 (1961).

16. Evans, F. L., and D. W. Rychman, "Ozonated treatment of wastes containing ABS," in Proceedings of 18th Industrial Waste Conference, Purdue University, 1963, p. 141.

17. Feng, T. H., "Exploration of sludge adsorption of syndets," *Water Sewage Works* **109**, 183 (1962).

18. Gates, W. E., and J. A. Borchardt, "Nitrogen and phosphorus extraction from domestic waste water treatment plant effluent by controlled algal culture," *J. Water Pollution Control Federation* **36**, 443 (1964).

19. Hlavsa, E., "Treating waste water containing synthetic detergents," Czechoslovak Patent no. 105679, 1962.

20. "Ion exchange systems may remove detergents wastes," *Eng.* **32**, 369 (1961).

21. Jackson, D. F., "The effects of algae on water quality," in Proceedings of 1st Annual Water Quality Research Symposium, Albany, N.Y., New York State Health Department, *FGB*-20, 1964.

22. Isaac P. C. G., and S. H. Jenkins, "A laboratory investigation of the breakdown of some of the newer synthetic detergents in sewage treatment," *J. Proc. Inst. Sewage Purif.*, part 3, p. 314 (1960).

23. Jenkins, S. H., N. Harkness, A. Lennon, and K. James, "The biological oxidation of synthetic detergent in recirculating filters," *Water Res.* **1**, 31 (1967).

24. Johnson, W. K., and G. J. Schroepfer, "Nitrogen removal by nitrification and denitrification," *J. Water Pollution Control Federation* **36**, 1015 (1964).

25. Kaufmann, H. P., and F. Malz, "The adsorptive precipitation and the skimming of detergents from aqueous solutions," *Fette, Seifen, Anstrichmittel* **62**, 1024 (1960); *Chem. Abstr.* **55**, 9739 (1961).

26. Koefer, C. E., "The syndet problem after five years of progress," *Public Works Mag.* **95**, 82 (1964).

27. Klein, S. A., and P. H. McGauhey, "Detergent removal by surface stripping," *J. Water Pollution Control Federation* **35**, 100 (1963).

28. Klein, S. A., "Effect of ABS on digester performance," *Water Sewage Works* **109**, 373 (1962).

29. Kling, W., "Surface action agents in effluents," *J. Textile Inst. Proc. Abstr.* **54**, A140 (1963).

30. Knesta, V., "Elimination of synthetic detergents from industrial waste waiters," *Chem. Prumysl* **13**, 281 (1963).

31. "Knocking out ABS by biodegradation wastes," *Eng.* **34**, 41 (1963).

32. Kucharski, J., "Production of synthetic detergents and problems in industrial waste disposal," *Chemik* **15**, 94 (1962).

33. Kumke, G. W., and C. E. Renn, "LAS removal across an institutional trickling filter," *Am. Oil Chem. Soc. J.* **43**, 92 (1966).

34. Lashen, E. S., and K. A. Booman, "Biodegradability and treatability of alkylphenol ethoxylates," Proceedings 22nd Industrial Waste Conference, Purdue University, Eng. Ext. Series 129, p. 211, 1967.

35. Lockwood, J. C., "Detergent industry and clean waters," *Soap Chem. Specialties* **41**, 67 (1965).

36. Ludzack, F. J., and M. B. Ettinger, "Controlling operation to minimize activated sludge effluent nitrogen," *J. Water Pollution Control Federation* **34**, 920 (1962).

37. Manay, K. H., W. E. Gates, J. D. Eye, and P. K. Deb, "The adsorption kinetics of ABS on fly ash," in Proceedings of 19th Industrial Waste Conference, Purdue University, 1964, p. 146.

38. Mackenthum, K. M., "A review of algae lake weeds and nutrients," *J. Water Pollution Control Federation* **34**, 1077 (1962).

39. Mackenthum, K. M., "Public Health Service research on algae and aquatic nuisances," in Proceedings of 1st Annual Water Quality Research Symposium, Albany, N.Y., New York State Health Department, 20 February 1964.

40. Maloney, T. E., "Detergent phosphorus effect on algae," *J. Water Pollution Control Federation* **38**, 38 (1966).

41. McGauhey, P. H., and S. A. Klein, "Degradable pollutants—a study of new detergents," unpublished data.

42. McGauhey, P. H., and S. A. Klein, "Travel of synthetic detergents with percolating water," in Proceedings of 19th Industrial Waste Conference, Purdue University, 1964, p. 1.

43. Metzgen, A., "Washing and scanning agents in waste water," *Spinner Weber Textilveredl.* **78**, 541 (1960).
44. Nemerow, N. L., and M. C. Rand, "Algal nutrient removal from domestic waste waters," in Proceedings of 1st Annual Water Quality Research Symposium, Albany, N.Y., New York State Health Department, 20 February 1964.
45. O'Neill, R. D., "Exotic chemicals," in Proceedings of 1st Annual Water Quality Research Symposium, Albany, N.Y., New York State Health Department, 20 February 1964.
46. Perlman, J. L., "Detergent biodegradability," *Textile Technicians' Dig.* **20**, 5583 (1963).
47. "Phosphates may lose detergent markets," *Chem. Eng. News* **45**, 18 (1967).
48. Pitter, P., and J. Chudoba, "Removing soaps by coagulation," *Vodni Hospodarstvi* **12**, 164 (1962).
49. Renn, E. E., W. A. Kline, and G. Orgel, "Destruction of linear alkylate sulfonates in biological waste treatment by field test," *J. Water Pollution Control Federation* **36**, 864 (1964).
50. Reymonds, T. D., "Pollutional effects of agricultural insecticides and synthetic detergents," *Water Sewage Works* **109**, 352 (1962).
51. Samples, W. R., "Removal of ABS from waste-water effluent," *J. Water Pollution Control Federation* **34**, 1070 (1962).
52. Sawyer, C. N., "Causes, effects and control of aquatic growth," *J. Water Pollution Control Federation* **34**, 279 (1962).
53. Schoen, H. M., *et al.*, "Foam separation," *Ind. Water Wastes* **6**, 71 (1961).
54. Sengupta, A. K., and W. W. Pipes, "Foam fractionation–the effect of salts and low molecular weight organics on ABS removal," in Proceedings of 19th Industrial Waste Conference, Purdue University, 1964, p. 81.
55. Soap & Detergent Association Technical Advisory Council, *Synthetic Detergents in Perspective*, New York (1962).
56. Spade, J. F., "Treatment methods for laundry wastes," *Water Sewage Works* **109**, 110 (1962).
57. Sweeney, W. A., "Note on straight chain ABS removal by adsorption during activated sludge treatment."
58. Swisher, R. D., "Biodegradation of ABS in relation to chemical structure," *J. Water Pollution Control Federation* **35**, 877 (1963).
59. Swisher, R. D., "Chemical mechanism of straight chain ABS biodegradation," *Textile Technicians' Dig.* **20**, 4374 (1963).
60. Swisher, R. D., "LAS major development in detergents," *Chem. Eng. Progr.*, 41 (1964).
61. Swisher, R. D., "Transient intermediate in the biodegradation of straight chain ABS," *J. Water Pollution Control Federation* **35**, 1557 (1963).
62. Symons, J. M., and L. A. Del Valle-Rivera, "Metabolism of organic sulfonates by activated sludge," in Proceedings of 16th Industrial Waste Conference, Purdue University, May 1961.
63. "Symposium-field evaluation of LAS and ABS treatability," in Proceedings of 20th Industrial Waste Conference, Purdue University, 1965, p. 724.
64. Tatsumi, C., and M. Nakiagawa, "Production of fatty microorganisms. IV: Production of fat yeast from soap waste liquor," *Hakko Kyokaishi* **19**, 396 (1961).
65. Waymon, C. H., and J. B. Robertson, "Adsorption of ABS on soil minerals," in Proceedings of 18th Industrial Waste Conference, Purdue University, 1963, p. 253.
66. Waymon, C. H., J. B. Robertson, and C. W. Hall, "Biodegradation of surfactants under aerobic and anaerobic conditions," in Proceedings of 18th Industrial Waste Conference, Purdue University, May 1963, p. 578.

The following references have been published since 1970.

1. Anonymous, "Don't count NTA out as a detergent phosphate requirement," *Env. Sci. and Tech.* **5**, 747 (1971).
2. Anonymous, "Ecolo-G, Bohack Syndets seized by U.S. Marshals," *Deterg. and Spec.* **8**, 46 (1971).
3. Anonymous, "New syndet base," *Deterg. and Spec.* **8**, 44 (1971).
4. Anonymous, "Nonionics are set for cleanup," *Chem. Wk.* **109**, 35–39 (1971).
5. Anonymous, "Phosphates are back in favor," *Chem. Wk.*, **109**, 20 (1971).
6. Anonymous, "Phosphate, NTA substitute offered by GAF," *Deterg. and Spec.* **8**, 37 (1971).
7. Anonymous, "Purex boosts nonionics," *Chem. Wk.* **109**, 39 (1971).
8. Anonymous, "Return to soap is feasible," *Sci. News.* **101**, 11 (1972).
9. Anonymous, "Silicates cut alkalinity," *Chem. Wk.* **109**, 41 (1971).

10. Anonymous, "Soapers square off," *Chem. Wk.* **109**, 25 (1971).
11. Anonymous, "Spectrum," *Env. Mag.* **13**, 24 (1971).
12. Anonymous, "Spectrum," *Env. Mag.* **13**, 26 (1971).
13. Anonymous, "Syndet materials," *Deterg. and Spec.* **8**, 56 (1971).
14. Christie, A. E., "Trisodium nitrilotriacetate and algae," *Water and Sew. Works*, 58–59 February 1970.
15. Segalos, H. A., "The great phosphate hoax," *Deterg. and Spec.* **8**, 18 (1971).
16. Semling, H. V., Jr., "International water symposium," *Deterg. and Spec.* **7**, 52 (1970).
17. Walther, H. J., and D. Knaack, "Elimination of ABS and alkanesulfate in the oxidation ditch," *Wasser-Wirt-Wassertech* **20**, 306; *Chem. Abs.* **74**, 67456z (1971).

28.5 Explosives-Industry Wastes

The explosives industry is concerned with three major processes servicing the public and private interests of our society: (1) manufacture of TNT, (2) manufacture of smokeless powders, and (3) manufacture of small-arms ammunition. Although this industry is particularly active during wartime, the peacetime uses of guns, ammunition, and explosives are also significant, because of hunting for sport, mining operations, construction projects which require blasting, and the manufacture of such items as fireworks and cap pistols.

In the manufacture of TNT (trinitrotoluene), toluene is mixed with a solution of nitric and sulfuric acids, under proper temperature conditions, and nitrate groups are gradually added, one by one, until the product is primarily trinitrotoluene, which is then washed free of residual acid, crystallized, and purified with sodium sulfite. The impure beta and gamma trinitrotoluenes are washed out of the alpha trinitrotoluenes as soluble sulfonates, after which the purified product is finally remelted, flaked, and packaged.

In the manufacture of smokeless powder, purified cotton, known as cotton linters, is treated with a mixture of nitric and sulfuric acids, producing cellulose nitrate (guncotton). Various methods of purifying this product are used, including boiling, macerating, and washing; it is then mixed with ether–alcohol and a stabilizer to render it colloidal. The powder is granulated by pressing it through steel dies, the solvent is recovered, and the powder dried and blended for shipping.

In the manufacture of small-arms ammunition, the brass case is made first, then the projectile, after which the percussion cap is filled, and finally all the parts, including the smokeless powder, are assembled.

Table 28.4 Quantities of TNT wastes [12].

Characteristic	Waste per 100,000 lb of explosive produced (TNT and DNT)			
	Plant A	Plant B	Plant C	Average
Flow, million gallons	1.17	1.08		1.12
Free mineral acid as H_2SO_4, lb	2,070	1,210	3140	2140
Sulfates, lb	5,560	5,450	2840	4620
NH_3 nitrogen, lb	49.7		27.2	38.5
NO_2 nitrogen, lb	140	179	60	116
NO_3 nitrogen, lb	1,062		302	684
Oxygen consumed, lb	8,360	4,990	1055	4800
Total solids				
Volatile, lb	9,460	6,180	6440	7360
Ash, lb	12,240	10,220	4980	9150
Suspended solids				
Volatile, lb	200	118	170	163
Ash, lb	1,380	130	6	505

The major wastes originate from making the cartridge cases and projectile jackets; this process consists of extruding and annealing metals, pickling them in acids, washing them with detergents, and lubricating the dies. Some wastes also come from the lead shop, where lead ingots are drawn into wire and shaped into projectiles.

The characteristics of TNT wastes are that they are generally clear, highly colored, strongly acid, have a high percentage of volatile solids, a "chemical" odor and an acid taste, and are quite resistant to alteration after they reach the receiving waters. The acid wash waters from the washing after nitration and the sulfite-purification wash water (red water) are the two major wastes. The nitration wash water is acid and yellow, while the red water is alkaline and so intensely red that it appears almost black. Smith and Walker [12] presents typical results of two TNT-plant wastes (Tables 28.4 and 28.5).

The four principal smokeless-powder wastes are (1) acid residue remaining after nitrating the cotton and purifying the product, (2) guncotton lost in water in which cellulose nitrate is boiled, (3) ether-alcohol lost from the solvent-recovery system, and (4) aniline

Table 28.5   Average of analytical results of TNT wastes [12].*

Plant	pH	Color	Odor	Acidity		Oxygen consumed	SO$_4$	Nitrogen			Total solids		Suspended	
				Methyl red	Phenolphthalein			NH$_2$	NO$_2$	NO$_3$	Volatile	Ash	Volatile	Ash
A	2.4	7,100	70	291	485	795	672	5.3	15	107	1004	1273	22	144
B	2.7	6,300	16	134	178	551	604		20		686	1123	14	15
Average A and B	2.6	6,700	43	212	332	673	638	5.3	18	107	850	1198	18	80
C(no cooling water)	1.2	34,000	11	3230	3460	1057	2923	2.8	62	310	5490	4685	17	0

*All results are given in parts per million, except for the pH, color, and odor concentration.

Table 28.6   Average analytical results for wastes from three powder plants [12].*

Plant	pH	Color	Odor	Acidity		5-day BOD	Oxygen consumed	SO$_4$	Nitrogen		Total solids		Suspended solids		Soap hardness
				Methyl red	Phenolphthalein				NO$_2$	NO$_3$	Volatile	Ash	Volatile	Ash	
A	< 1.6			1860	1990	49.1	76.2	1280	2.7	530			29	24	
B	< 1.6	53	21	1540	1610	47.6	99.8	1100	1.9	470	687	354	54	136	241
C(combined flow)		52		2820	2970	62.9	91.4	1512	2.4	650	2645	229	29	13	368
C(pyro-cotton)	1.1	36	4	4250	4400	52.3	106.0	2265	2.6	970	3930	346	37	7	526
C(finish)	8.2	104	29			83.9	62.0	1	2.3	9	69	122	13	26	50

*All results are given in parts per million, except for the pH, color, and odor concentration.

from the manufacture of diphenylamine. One estimate [12] of the quantities of waste from manufacturing 100,000 pounds of powder gives 89,500 pounds of mixed acids, 2500 pounds of alcohol, and 125 pounds of cotton, all in a total of 8.3 million gallons of waste water. Smith and Walker [12] also present the average analytical results from three powder plants (Table 28.6) which produce a large volume of strongly acid liquid waste, containing high concentra-

tions of sulfates and nitrates.

The predominant characteristics of small-arms ammunition wastes are that they are turbid, greenish-gray in color, and may have an oily or soapy odor, contain copper and zinc from acid pickling baths, and grease from the cutting oils and soaps. The average waste characteristics for three typical plants [12] are given in Tables 28.7 and 28.8, which show that a flow of 36,000 gallons contains 100 pounds of

Table 28.7   Average analytical results, waste flows from three small-arms ammunition plants [12].*

Plant	Acidity (methyl red)	Alkalinity (methyl orange)	Grease	Copper	5-day BOD	Oxygen consumed	Total solids			Suspended solids			Sulfate
							Total	Volatile	Ash	Total	Volatile	Ash	
A		213	184	31	141	67	1577	445	1132	276	205	71	244
B	6		419	22	138	103	1092	505	587	408	372	36	397
C	60		502	86	295	163	1753	675	1078	408	374	34	647
Average			368	46	191	111	1474	542	932	364	317	47	309

*All results are given as parts per million.

Table 28.8   Waste products per 100,000 rounds of ammunition [12].*

Product and plant	Million gallon flow	Alkalinity	Mineral acid	SO₄	Grease	Copper	BOD	Oxygen consumed	Total solids		Suspended solids	
									Volatile	Ash	Volatile	Ash
0.50-caliber ammunition												
A	0.0623	131		265	284	29.7	159	64	31.4	504	252	116
B	0.0975	20		212	475	16.3	144	137	567	474	442	39.7
Average	0.0799	76		239	380	23.0	152	67	299	489	347	78
0.30-caliber ammunition												
A	0.0252	58		37	46	7.9	42	22	114	141	69	19
B	0.0242		9.40	73	40	5.4	17	14	58	120	34	3
Average	0.0242			55	43	6.7	30	18	86	131	52	11
Combined output												
A	0.0417	74		84	64	10.5	49	23	155	392	71	24.7
B	0.0408		3.0	104	144	7.5	47	40	174	198	129	12.2
C	0.0258		12.9	140	109	18.6	64	35	146	232	81	7.4
Average	0.0361			76	106	12.2	53	33	158	274	94	14.8

*All results are given in pounds.

Table 28.9 Average analytical results, treatment plants for ammunition wastes [12].*

Sampling	pH	Acidity		Alka-linity	Grease	Copper	Total solids			Suspended solids			5-day BOD	Odor	
		Methyl red	Phenol-phthalein				Total	Vola-tile	Ash	Total	Vola-tile	Ash		Concen-tration	Type
Raw waste	3.5	60	239		709	86	1763	684	1079	413	379	34	298	174	Oily
After grease removal	3.6	55	211		54	82								55	Oily
Influent (final settling)	6.9			333		64				867	358	509		48	Soapy
Final effluent	6.9			43	8	9	1510	302	1208	25	13	12	47	14	Soapy

*All results are given in parts per million, except for pH and odor.

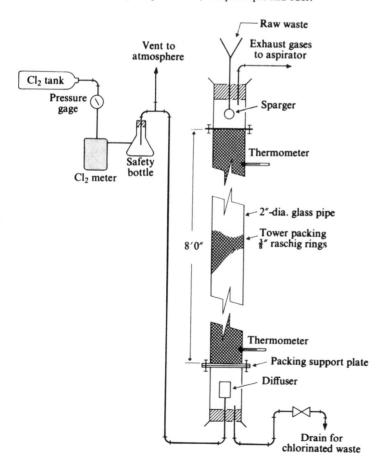

Fig. 28.10 Chlorination tower for treating TNT wastes. (After Edwards and Ingram [2].)

grease, 12 pounds of copper, 95 pounds of volatile suspended solids, and a 5-day BOD population equivalent of 300 people for each 100,000 rounds of mixed ammunition.

Smith and Walker [12] describe a plant treating ammunition wastes by means of grease flotation and chemical precipitation (Table 28.9), which removed 84 per cent of the BOD and 94 per cent of the suspended solids. The pH was also raised from 3.5 to 6.9 and the odor reduced about 90 per cent.

Dickerson [1] concluded that coagulation and sedimentation are not effective methods of treatment for powder-manufacturing wastes and that chlorination, though effective, is too costly. He did state, however, that these wastes are amenable to decomposition by both aeration and biological oxidation.

Ruchhoft et al. [7] concluded that black garden soil was best for filtering TNT wastes and that such soils can remove an amount of TNT up to about 0.1 per cent of their weight. Dosing and mixing activated carbon into the waste were more effective in removing TNT than filtration through carbon.

More recently, Edwards and Ingram [2] stated that the color of TNT wastes is very resistant to biological and chemical attack, since reducing substances such as SO_2, $FeSO_4$, $Na_2S_4O_7$, and hydrosulfites are not effective and adsorption, on materials such as clays and activated carbon, may remove much color, but is impractical because of cost. It is true that all the color can be removed by ion exchange, but short cycles and difficulty in regeneration make this method uneconomical. Oxidizing agents seem to be most promising, but only chlorine can remove color at a reasonable cost. For example, fresh TNT stellite waste can be chlorinated directly in a column (Fig. 28.10) and more than 90 per cent of the color can be removed if one uses about 9000 ppm of chlorine—a considerable dosage.

References: Explosives-Industry Wastes

1. Dickerson, B. W., "Treatment of powder plant wastes," in Proceedings of 6th Industrial Waste Conference, Purdue University, February 1951, p. 30.
2. Edwards, G., and W. T. Ingram, "The removal of color from TNT wastes," J. Sanit. Eng. Div. Am. Soc. Civil Engrs. 81, separate no. 645 (1955).
3. Edwards, G. P., and W. T. Ingram, "TNT wastes, description of treatment methods," Sewage Ind. Wastes 26, 1484 (1954).
4. Mohlman, F. W., "Industrial wastes in wartime," Sewage Works J. 15, 1164 (1943).
5. Morris, R. L., and J. D. Dougherty, "Infrared identification of degraded TNT wastes in streams and shallow wells," in Proceedings of 15th Industrial Waste Conference, Purdue University, May 1960, p. 281.
6. Ruchhoft, C. C., et al., "TNT wastes, color reactions and disposal procedure," Sewage Ind. Wastes 18, 339 (1946).
7. Ruchhoft, C. C., M. LeBosquet, Jr., and W. G. Meckler, "TNT wastes from shell-loading plants," Ind. Eng. Chem. 37, 937 (1945).
8. Ruchhoft, C. C., and W. G. Meckler, "Colored TNT derivative and alpha TNT in colored aqueous alpha-TNT solutions," Ind. Eng. Chem. (analytical edition) 17, 430 (1945).
9. Ruchhoft, C. C., and W. G. Meckler, "TNT waste treatment studies," Sewage Ind. Wastes 18, 779 (1946).
10. Ruchhoft, C. C., and F. J. Noms, "Estimation of ammonium picrate in wastes from bomb and shell loading plants," Ind. Eng. Chem. (analytical edition) 18, 480 (1946).
11. Schott, S., C. C. Ruchhoft, and S. Megregian, "TNT waste," Ind. Eng. Chem. 35, 1122 (1943).
12. Smith, R. S., and W. W. Walker, "Surveys of liquid wastes from munitions manufacturing," U.S. Public Health Service, Reprint no. 2508, Public Health Reports nos. 58, 194, 1365, and 1393.
13. Wilkinson, R., "Treatment and disposal of waste waters containing picric acid and dinitrophenol," Ind. Chem., Jan.–Feb. 1951.
14. Wilkinson, R. W., "Shell-filling plant waste treatment and disposal," Sewage Ind. Wastes 20, 590 (1948).

Suggested Additional Reading

The following references have been published since 1958.
1. Madera, Solin, Vucka, "The biochemical reduction of TNT—the course and intermediary substances of reduction of 2,4,6-TNT," Sb. Vysoke Skoly Chem. Technol. Praze, Oddil Fak. Technol. Paliv. Vody 3, 129 (1959).

2. Schuster, G., "Possible biochemical decomposition of hard-to-handle constituents of individual waste water," *Math.-Naturw. Reihe* **11**, 179 (1962).

3. Sercu, C., "Chemical plant waste incinerator," *Natl. Fire Protect. Quart.* **55**, 90 (1961).

4. Solin, V., and M. Kustka, "The treatment of waste waters containing TNT by sprinkling on ashes," *Sb. Vysoke Skoly Chem. Technol. Praze, Oddil Fak. Technol. Paliv. Vody* **2**, 247 (1958).

5. U.S. Department of Commerce, "Air and stream pollution control: preliminary survey of thermal methods for TNT red water disposal," Pamphlet no. 556, Washington, D.C., Office of Technical Service (1961), p. 810.

6. Chakraborty, A. K., "Wastes and effluents in the chemical industries," *Technologyl, Sindri* **3**, 72–74 (1966).

7. Lever, N. A., "Disposal of nitrogenous liquid effluent from Modderfontein Dynamite factory," *Proc. 21st Ind. Waste Conf., Purdue Univ. Eng. Exten. Series* **121**, 902–925 (1966).

8. Siele, V. I. and C. Ribando, "Lab study of an ext'n synthetic technology for the elimination of pollution from Mahon fog filter waters at TNT plants," *U.S. Clearinghouse Fed. Sci. Tech. Info.* No. 724114, 29 pp. (1971).

9. Freeman, D. J. "Continuous Fixation and Removal of Explosive Wastes from Pink Water using surfactant Technology." Proc. 40 Purdue Ind. Waste Conf., 1985, P. 659.

10. Kuo, C. J. and R. C. Ahlert, "Catalytic Oxidation of Munitious Wastewater". Proc. 38 Purdue Ind. Waste Conf., 1983, P. 377.

11. Mannebach, R. A. et al., "Munitious Manufacturing Wastewater Treatment: A Gross-Rosts Facility-Case History," Proc. 37 Purdue Ind. Waste Conf., 1982, P. 213.

12. Freeman, D. J. and Q.A. Colitti, "Removal of Explosives from Load-Assemble-Pack Wastewater (Pink Water) Using Surfactant Technology," Proc. 36. Purdue Ind. Waste Conf., 1981, P. 383.

13. Smith, L.L., "Evaluation of an Anaerobic Rotating Surface System for Treatment of a Munition Wastewater Containing Organic and Inorganic Nitrates," Proc. 37 Purdue Ind. Waste Conf., 1979, P. 628.

14. Semmens, M. J. et al. "Treatment of an RDX-TNT Waste from a Munitous Factory," Proc. 39 Purdue Ind. Waste Conf., 1984, P. 837.

15. Shelby, S. E. et al., "A Case Study for the Treatment of an Explosives Wastewater from an Army Ammunition Plant," Proc. 39 Purdue Ind. Waste Conf., 1984, P. 821.

28.6 Formaldehyde Wastes

This one chemical, formaldehyde, warrants individual consideration in this chapter for several reasons. First, it is used in numerous industries such as plastics, leather, and antibiotics. Second, by virtue of its special chemical nature, it has long been considered an effective antiseptic agent, and treatment of a biologically inhibiting agent presents difficult or, at least, unique problems. The toxicity problem is illustrated in the work of Gellman and Heukelekian [4], in which they proved that a formaldehyde concentration somewhere between 130 and 175 ppm was lethal to bacteria in sewage (Table 28.10) and that even in small concentrations a lag period occurs, owing to the action of sublethal doses of formaldehyde on the biological flora in sewage. Another valuable conclusion evolved from their research, which they summarized as follows: "By a process of adaptation and selection of bacteria, the oxidizable concentration of formal-

Table 28.10 Effects of formaldehyde concentrations in sewage on oxidation of formaldehyde. (After Gellman and Heukelekian [4].)

Days of oxidation	Formaldehyde concentration, ppm			
	45	90	130	175
1	0	0	0	0
2	15	0	0	0
3	21	72	43	0
$3\frac{1}{4}$	23	81	92	0
4	25	89	112	0
7	28	90	135	0

dehyde was increased from 135 ppm to 1750 ppm." This means that even a toxic chemical can serve as food for bacteria, *provided* that the bacterial species is carefully acclimated to the food.

Dickerson [2] reported at about the same time on a synthetic-resin plant producing a total BOD load of 2000 pounds per day, with individual BOD values ranging from 300 to 10,000 ppm, and formaldehyde concentrations up to 5000 ppm (90 per cent of the total BOD load) in a volume of about half a million gallons of water. He found that a high-rate trickling filter can be used for treatment of formaldehyde, organic oils, and organic acids and can provide satisfactory reduction, as shown in Table 28.11. He advised holding a uniform pH, but said that adjustment is not necessary as long as the pH stays somewhere between 4.5 and 8.5. His report also showed that toxicity is only relative and that long periods of pilot-plant operation are necessary to ascertain the correct degree of oxidation required [2].

In later studies, Dickerson *et al.* [3] report on a complete treatment plant for formaldehyde wastes involving aeration and activated sludge along with trickling filtration, for treating about 4700 pounds

Table 28.11 Trickling filtration of formaldehyde waste. (After Dickerson [2].)

Concentration of formaldehyde, ppm	Removal of formaldehyde, lb/yd³	Removal efficiency, %
110	1.12	23
184	1.25	16
266	1.75	15
300	3.10	23
360	3.45	28

of BOD, with a removal efficiency of approximately 90 per cent (see Fig. 28.11). Operating costs (1952–1954) were estimated at about 2 cents per pound of BOD treated.

References: Formaldehyde Wastes

1. Dickerson, B. W., "A high-rate trickling filter pilot plant for certain chemical wastes," *Sewage Works J.* **21**, 685 (1949).
2. Dickerson, B. W., "High-rate trickling filter operation on formaldehyde wastes," *Sewage Ind. Wastes* **22**, 536 (1950).
3. Dickerson, B. W., C. J. Campbell, and M. Stankard, "Further operating experiences in biological purification of formaldehyde wastes," in Proceedings of 9th Industrial Waste Conference, Purdue University, May 1954, p. 331.
4. Gellman, I., and H. Heukelekian, "Biological oxidation of formaldehyde," *Sewage Ind. Wastes* **22**, 1321 (1950).
5. Ragan, J. L., and R. H. Maurea, "Industrial waste disposal by solar evaporation," *Ind. Water Wastes* **8**, 37 (1963).
6. Sakharnov, A. V., *et al.*, "Purification of waste waters from formaldehyde and phenol in phenol-formaldehyde resin production," U.S.S.R. Patent no. 141814, Oct. 16, 1961; application December 1960.
7. "Treatment of trade water containing formaldehyde," *Effluent Water Treat. J.* **3**, 88 (1963).
8. Waldemeyer, T., "Treatment of formaldehyde wastes by activated sludge," *Surveyor* **111**, 445 (1952).

The following references have been published since 1966.

1. Aadegeest, M., "Purification of wastes from phenol-formaldehyde resin production," *Ger. Offen.* #2,054,753 (Cl. CO2c) (1971); *Appl.* (1969).
2. Neely, W. B., "The adaptation of *aerobacten aerogenes* to the stress of sub lethal doses of formaldehyde," *J. Gen Microbiol.*, **45**, 187–197 (1966).

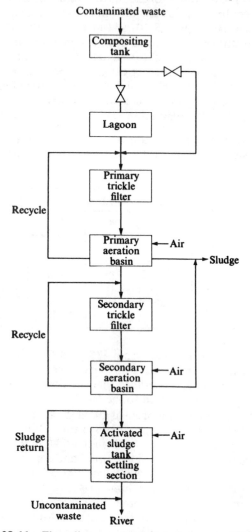

Contaminated waste

Compositing tank

Lagoon

Primary trickle filter

Primary aeration basin — Air

Recycle

Sludge

Secondary trickle filter

Recycle

Secondary aeration basin — Air

Sludge return

Activated sludge tank — Air

Settling section

Uncontaminated waste   River

Fig. 28.11 Flow diagram of complete treatment plant for formaldehyde wastes [3].

3. Rogouskaya, C., M. Lazarava, and L. Kostina, "The influence of increase temperatures on the biocoenosis of activated sludge and the intensity of decomposition of organic compounds," *Lab. Studies*, USSR.

28.7 Pesticide Wastes

The most troublesome waste from the production of chemicals used to make insecticides, herbicides, and pesticides is that arising from production of 2,4-dichlorophenoxyacetic acid (2,4-D). The chemical actually reaching the waste stream is dichlorophenol (DCP).

Mills [7] originally treated this waste by passing the waste water through a filter bed of activated carbon, at a cost of about $6 per pound of dichlorophenol removed, but the method was soon abandoned because of the cost. The alkaline chlorination process then came into favor, which gives 95 to 98 per cent destruction of dichlorophenol, although the effluent still contains about 25 ppm of dichlorophenol and the cost is somewhat more than $1 per pound of dichlorophenol destroyed. In his search for an economical method of removing the remaining DCP, Mills concluded that, although it is a phenolic compound, it cannot be biologically oxidized under the same conditions as phenol.

Another major insecticide, DDT, is manufactured by the reaction of monochlorobenzene and chloral alcohol in the presence of H_2SO_4 containing 20 per cent free SO_3. The crude DDT is purified either by washing with large volumes of water or by neutralization with Na_2CO_3 after draining off the spent acid. The wastes contain a great deal of acid; for example, the neutralization process results in 500 gallons of waste acid per ton of DDT manufactured. The acid waste contains 55 per cent H_2SO_4, 20 per cent ethyl hydrogen sulfate, and 20 per cent chlorobenzene sulfuric acid. In addition, this method produces about 800 gallons of wash water per ton of DDT, containing 2 to 6 per cent acid, and 90 gallons of wash water from centrifuges. Other wastes arise from the manufacture of chloral alcohol [4]. The only waste-treatment method used has been dumping the wastes into sufficient dilution water, such as the ocean. Treatment by municipalities in biological plants is possible, but is greatly hampered by the toxicity of most of the constituents of the wastes.

Increased interest is being shown in the effect of new organic insecticides on fish and wildlife. Rachel Carson's book aimed at the layman, *Silent Spring*, which first appeared in the *New Yorker* magazine in June 1962 and was published in book form by Houghton Mifflin, is an example of current public interest in this subject. Aldrin, toxaphene, rotenone, and dieldrin are a few of the more toxic sprays recently reported to be responsible for large-scale destruction of aquatic life in many streams and farm ponds [3, 5, 6, 10, 15]. These insecticides have been found to be lethal even in very small concentrations. For example, Burdick [1] found that 0.05 to 0.40 ppm of

Table 28.12 Sanitary characteristics of raw and treated parathion wastes. (After Stutz [10].)

Characteristic	Raw wastes*	Plant effluent*
COD	3,000	100
Total solids	27,000	18,000
Volatile solids	25%	0.1%
pH	2.0	7.0
Acidity	3,000	
Sodium	6,000	
Chlorides	7,000	
Phosphates	250	
Nitrogen	20	
Sulfates	3,000	
Parathion		< 0.1
p-Nitrophenol		< 1.0

*All results are given in mg/liter unless otherwise indicated.

rotenone was lethal to brown trout. The toxicity of these chemicals varies according to the type of fish and increases with an increase in the temperature of the water.

At the present time, little treatment of these wastes (other than those described by Mills) is being practiced. However, there is increasing concern over keeping these chemicals out of the watercourses.

Stutz [1966, 10] reports successful biological oxidation of parathion wastes after nine years of experimentation. He characterized these wastes as very strong, highly mineralized, and acid in nature. Liquid wastes are prechlorinated, lagooned, neutralized with limestone, adjusted for pH control with soda ash, treated by activated sludge for 5 to 9 days at a suspended-solids concentration of 18,000 ppm in mixed liquor, clarified, and discharged to the city sewer. Influent and effluent analyses are given in Table 28.12.

References: Pesticide Wastes

1. Burdick, G. E., H. J. Dean, and E. J. Harris, "Toxicity of emulsifiable rotenone to various species of fish," *New York Fish Game J.*, January 1955, p. 36.
2. Chanin, G., and R. P. Dempster, "A complex chemical waste," *Ind. Wastes* **3**, 155 (1958).
3. Doudoroff, P., M. Katz, and C. M. Tarzwell, "Toxicity of some organic insecticides to fish," *Sewage Ind. Wastes* **25**, 840 (1953).
4. Grindley, J., "Effluent disposal in DDT manufacture," *Ind. Chem.* **26**, November 1950.
5. Hoffman C. H., and A. T. Drooz, "Effects of a C-47 airplane application of DDT on fish food organisms in two Pennsylvania watersheds," *Am. Midland Naturalist* **50**, 175 (1953).
6. Ingram, W. M., and C. M. Tarzwell, "Selected bibliography of publications relating to undesirable effects upon aquatic life by algicides, insecticides, and weedicides," Publication no. 400, U.S. Public Health Service (1954).
7. Mills, R. E., "Development of design criteria for biological treatment of an industrial effluent containing 2,4-D waste water," in Proceedings of 14th Industrial Waste Conference, Purdue University, May 1959, p. 340.
8. "Report on New Jersey Sewage and Industrial Wastes Association meeting, industrial waste problems," *Ind. Wastes* **3**, 72 (1958).
9. Tarzwell, C. M., "Disposal of toxic wastes," *Ind. Wastes* **3**, 48 (1958).
10. Tarzwell, C. M., and C. Henderson, "Toxicity of dieldrin to fish," *Trans. Am. Fisheries Soc.* **86**, (1956).
11. Warrick, L. F., "Blitz on insects creates water problems," in Proceedings of 6th Industrial Waste Conference, Purdue University, February 1951, p. 455.
12. Warrick, L. F., "Fish kills by leaching of insecticides," *Sewage Ind. Wastes* **24**, 924 (1952).
13. Weiss, C. M., "Response of fish to sub-lethal exposures of organic phosphorus insecticides," *Sewage Ind. Wastes* **31**, 580 (1959).
14. Wilson, I. S., "The Monsanto plant for the treatment of chemical wastes," *J. Inst. Sewage Purif.*, Midland Branch, 18 March 1954.
15. Young, L. A., and H. P. Nicholson, "Stream pollution resulting from the use of organic insecticides," *Progressive Fish Culturist* **13**, 193 (1951).

Suggested Additional Reading

The following references have been published since 1961.

1. Buescher, C. A., and J. H. Dougherty, "Chemical oxidation of selected organic pesticides," *J. Water Pollution Control Federation* **36**, 1005 (1966).
2. Coley, G., and C. N. Stutz, "Treatment of parathion wastes and other organics," *J. Water Pollution Control Federation* **38**, 1345 (1966).
3. Lutin, P. A., and J. J. Cibulka, "Oxidation of selected carcinogenic compounds by activated sludge," Purdue University Engineering Extension Series, Bulletin no. 118, 1965, p. 131.
4. Morris, J. C., and W. J. Weber, Jr., "Adsorption of biochemically resistant material from solution," U.S. Public Health Service, Environmental Health Series Water Supply and Pollution Control, 999-WP-33, March 1966, p. 108.
5. Oshina, I. A., and N. K. Tyurina, "Clarification of 2,4,D production by adsorption," *Chem. Abstr.* **63**, 12860 (1965).
6. Pitter, P., and F. Tucek, "Influence of waste water from chlorophenol production on biological purification," *Chem. Abstr.* **65**, 8556 (1966).
7. Pitter, P., and J. Chudoba, "Purification of waste effluents from the industrial production of the fungicides Kaptan and Faltan," *Chem. Abstr.* **63**, 12864 (1965).
8. Randall, C. W., and R. A. Lauderdale, "Biodegradation of malathion," *J. Sanit. Eng. Div. Am. Soc. Civil Engrs.* **93**, 145 (1967).
9. Riklis, S. G., and A. R. Perkins, "The hydrolysis of chlorobenzenesulfonic acid, a by-product of DDT manufacture," *Khim. Prom.*, 1961, p. 461.
10. Stutz, C. N., "Treating parathion wastes," *Chem. Eng. Progr.* **62**, 82 (1966).
11. Teasley, J. I., "Identification of cholinesterase inhibiting compound from industrial effluent," *Environ. Sci. Technol.* **1**, 411 (1967).
12. Wilroy, R. D., "Industrial wastes from scouring rug wools and the removal of dieldrin," Purdue University Engineering Extension Series, Bulletin no. 115, 1963, p. 413.
13. Winar, R. M., "The disposal of waste water underground," *Ind. Water Eng.* **4**, 21 (1967).
14. Woodland, R. G., M. C. Mall and R. R. Russell, "Process for disposal of chlorinated organic residues," *J. Air Pollution Control Assoc.* **15**, 56 (1965); *Chem. Abstr.* **62**, 12890 (1965).

The following references have been published since 1968.

1. Anonymous, "Evaluation of accumulation, translocation, and degradation of pesticides at land wastewater disposal sites," *Comm. Bus. Daily* **27**, May 25 (1973).

2. Anonymous, "Treatment of effluent from manufacture of chlorinated pesticides with a synthetic, polymeric adsorbent, Amberlite XAD-4," *Env. Sci. and Tech.* **7**, 138–141 (1973).

3. Armstrong, N. E. and P. N. Storrs, "Biological effects of waste discharges on coastal receiving waters," *Citation No. 71-6TC0499* (1969).

4. Atkins, P. R., "The pesticide manufacturing industry— current waste treatment and disposal practices," *U.S. Nat'l Techn. Info. Service, Gov't Rpts. Announcements* **72**, 146 (1972).

5. Bulla, E. D. and E. Edgerley, "Photochemical degradation of refractory organic compounds," *J. WPCF*, **40**, 546 (1968).

6. Canter, L., D. Nance and D. Rowe, "Effects of pesticides on wastewater," *Water and Sew. Works* **116**, 230 (1969).

7. Chau, A. S., "Analysis of chlorinated hydrocarbon pesticides in waters and wastewaters, methods in use in water quality division laboratories," *Ottawa: Canada Dept. of the Environment, Inland Waters Branch, Water Quality Div.*, 61 (1972).

8. Eichelberger and J. J. Lichtenberg, "Carbon absorption for recovery of organic pesticides," *J. AWWA* **63**, 25 (1971).

9. Gabovich, R. D. and I. L. Kurinnoi, "Ozonization of water containing petroleum products, aromatic hydrocarbons and organic pesticides," *Gig. Naselennykli Mest.*, 31–5 (1967).

10. Haan, C. T., "Movement of pesticide by runoff and erosion," *ASAE Paper 70–706* (1970).

11. Hindin, E. and P. J. Bennett, "Organic compounds removed by reverse osmosis," *Water and Sew. Works.*, **116**, 466 (1969).

12. Huang, J. C. and C. S. Liao, "Adsorption of pesticides by clay minerals," *ASCE J. Sanitary Eng.*, SA5 (1970).

13. King, P., "Removal of trace organics from water by adsorption on coal," *OWRR Project A-015-VA(2)* (1970).

14. Lawless, E. W., "The pollution potential in pesticide manufacturing," *EPA Tech. Studies Rpt. TS007204* (1972).

15. Leigh, G. M., "Degradation of selected chlorinated hydrocarbon insecticides," *J. WPCF*, **41**, R450 (1969).

16. Lue-Hing, E. and S. D. Brady, "Biological treatment of organic phosphorus pesticide waste waters," *23rd Ind. Waste Conf., Purdue University* (1968).

17. Pheiffer, T. H., Donnelly, D. K. and D. A. Possehl, "Water quality conditions in the Chesapeake Bay system," *U.S. EPA, Region III, Annapolis Field Office*, Tech. Rpt. No. 55 (1972).

18. Pitter, P. and J. Chudoba, "Biological treatment of pesticide waste water containing colloidal sulfur," *Chem. Ind. (London)* **52**, 1846 (1968).

19. Reimann, K., "Behavior and toxicology of pesticides in wastes and sewage," *Muenchnes Beitr. Abwasser-, Fisch,- Flussbiol* **16**, 200 (1969).

20. Stockton, D. L., "Reduction of pollution at the source for organic chemicals and pharmaceutical plants by decentralized treatment," *Ind. Proc. Design for Pollut. Contr., Proc., Charleston, WV, AICHE Workshop*, **4**, 50–54 (1972).

21. Wershaw, R., P. Burcar, and M. Goldberg, "Interactions of pesticides with natural organic material," *Env. Sci. and Tech.* **3**, 271 (1969).

22. Whaley, H., G. Lee, R. Jeffrey, and E. Mitchell, "Thermal destruction of DDT in an oil carrier," *Canadian Dept. of Energy, Mines Resources*, Res. Rpt. 225 (1970).

28.8 Plastic and Resin Wastes

Plastics and resins are chainlike structures known chemically as polymers. All polymers are synthesized by one or more of the following processes: bulk, solution, emulsion, and suspension. A typical production reaction requires the addition of a free radical initiator and modifiers to the monomer, the building block of the polymer. This polymerization process creates relatively little water-borne waste, compared with other chemical manufacturing processes. In most cases, the preliminary step—the synthesis of the monomer—creates considerably more waste than the production of the polymer from the monomer.

The U.S. Department of the Interior Profile [1] separated these chemical industries into nine subdivisions, which represent over 85 per cent of all plastic production: (1) cellulosics, (2) vinyls, (3) styrenes, (4) polyolefins, (5) acrylics, (6) polyesters and alkyds, (7) urea and melamine resins, (8) phenolics, and (9) miscellaneous resins. This profile reports that the current total production of 14.25×10^9 pounds per year generates 113×10^6 pounds of water-borne waste, expressed as 5-day BOD. Since that is

Table 28.13   Production of plastics and resins.

Division	Production, 10^9 lb/year		Increase, %	Percentage of total production in 1967
	1962	1967		
Cellulosics	0.48	0.50	4	3.5
Vinyls	1.55	2.80	80	19.5
Styrenes	1.25	2.25	80	16.0
Polyolefins	2.00	4.35	82	30.5
Acrylics	0.18	0.29	61	2.0
Polyesters and alkyds	0.68	1.10	62	7.5
Urea and melamines	0.49	0.66	35	4.5
Phenolics	0.69	1.05	52	7.5
Miscellaneous resins	0.65	1.25	93	9.0
Total	7.97	14.25	78	100.0

roughly equivalent to the waste discharged by 1,800,000 persons, we cannot overlook this industry as a major contributor to the organic waste loads on our streams and lakes. It is further reported that of the BOD generated, 55 per cent is removed by treatment-plant systems, while the remaining 45 per cent (51×10^6 pounds per year) is discharged to watercourses. The assumption is that 1 pound (dry weight) of industrial waste generates 0.75 pounds of BOD. From Table 28.13 it can be seen that the polyolefins, vinyls, and styrenes make up about two-thirds of all production, while all divisions except the cellulosics have made substantial production gains during the 5-year period from 1962 to 1967.

Each subdivision of this industry manufactures its product in a different manner from any other. It would be very difficult for the author to present, or the reader absorb, all the details of each subdivision's processes and wastes. However, brief descriptions are given to help the reader in understanding the industry.

1. Cellulosics. Cellulosics, which are plastic materials produced from cellulose, range from regenerated cellulose, or cellophane, to the nitrocellulose in gun-cotton. Purified wood pulp is the main raw material. In the United States, purified cellulose (the major product) is made by the xanthate process; in Europe the cuprammonium process is widely used. In the xanthate process cellulose is treated in a solution of NaOH and CS_2. The resultant cellulose xanthate solution is coagulated, and cellulose is regenerated in the form of a continuous film by acidification (Fig. 28.12).

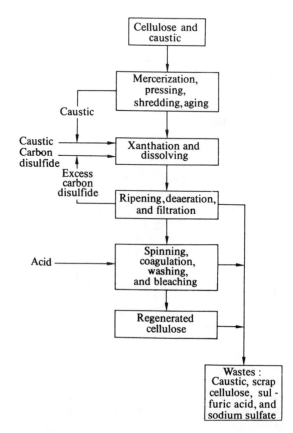

Fig. 28.12   Flow diagram of the production of regenerated cellulose.

The wastes contain much biodegradable cellulosic materials, sulfates, and heavy metals. The level of waste generated has been reported to range from 0.015 to 0.10 pounds per pound of product manufactured. The bulk of the remainder of the waste is reported to be H_2SO_4, $NaSO_4$, and heavy metals.

Neutralization of acids with caustic is a common practice. In some operations, such as biological treatment, CS_2 may be toxic. Scrap cellophane from production processes is also a waste that requires disposal; usually it is buried or incinerated.

2. Vinyl resins.

The mono- and copolymers of vinyl chloride are among the oldest and most versatile thermoplastic resins. Production is essentially in batch operations, and suspension polymerization is the most widely used process in terms of the variety and quantity of products; it accounts for 85 to 90 per cent of the total vinyl resins produced. In this method, the monomer is dispersed as small droplets in a stabilized suspending medium consisting of water and 0.01 to 0.50 per cent by weight of the suspending agents, such as polyvinyl alcohol, gelatin, and cellulose ethers. The suspension is then heated in a reactor in the presence of catalysts such as benzoyl, lauroyl, and tert.-butyl peroxides in order to initiate polymerization. When polymerization is complete, the polymer suspension is taken to a blowdown tank or stripper, where residual unreacted monomer is recovered. The stripped polymer is transferred to a blend tank and mixed with sufficient other batches to form a lot. Finally, the polymer slurry mix is pumped to centrifuges, where it is washed and dewatered, and dried in rotary driers. A simplified flow diagram of the suspension process for polyvinyl chloride is shown in Fig. 28.13.

The effluent from the centrifugation step contains most of the contaminants from these plants—suspending agents, surface-active agents, catalysts, small amounts of unreacted monomer, and significant amounts of very fine particles of the polymer product. It has been reported for an average plant, producing 100 million pounds of vinyl resins a year, that the waste contains 1 million pounds of BOD, 150,000 pounds of suspended solids and 100 to 200 $\times$ 10^6 gallons of waste.

Primary-sedimentation and activated-sludge treatments have been reported to produce less than 1 per cent reduction in BOD or COD and 98 per cent removal of suspended solids and 89 per cent removal of BOD and COD and 98 per cent removal of suspended solids, respectively.

3. Polystyrene resins and copolymers.

Polystyrene's combination of physical properties and ease of processing (by injection molding and extrusion) makes it a unique thermoplastic. The crystal-clear product has excellent thermal and dimensional stability, high flexural and tensile strength, and good electrical properties.

The fundamental manufacturing process for polystyrene resins and copolymers is a batch process that uses a combination of both bulk (mass polymerization) and suspension methods. The styrene monomer, or mixtures of monomers, is purified by distillation or caustic washing to remove inhibitors. The purified raw materials, together with an initiator,

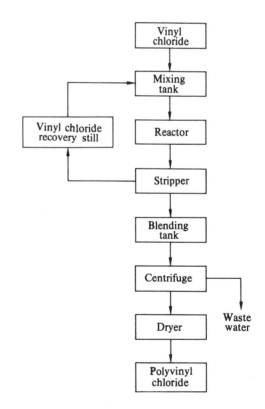

Fig. 28.13 Flow chart for polyvinyl chloride production.

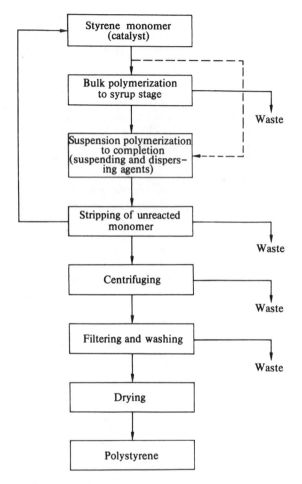

Fig. 28.14   Flow chart for polystyrene production.

are charged into stainless-steel or aluminum poly-merization vessels, which are jacketed for heating and cooling and contain agitators. Polymerization of the monomer is carried out at about 90°C to ap-proximately 30 per cent conversion, at which stage the reaction mass is syrupy. During this prepoly-merization step, water is used only as a heat-exchange medium. Since it does not come into contact with the product and is therefore not contaminated, it can be recirculated. The prepolymer, or partially polymeriz-ed mass, is then transferred to suspension-poly-merization reactors containing water and proprietary suspending and dispersing agents. The reactors are

usually jacketed, and the contents stirred in stainless-steel vessels. The syrupy mass is broken up into drop-lets by means of the stirrer and held in suspension in the aqueous phase. Temperature is a critical variable in the further polymerization of the product. After completion of polymerization, the polymer suspension is sent to a blowdown tank where any un-reacted monomer is stripped. The stripped batch is centrifuged, and the polymer product is filtered, washed, and dewatered. A flow chart for this process is shown in Fig. 28.14.

Reaction water (suspension medium) and wash water are the two significant sources of waste water in the production of polystyrene. Some cooling water is lost through evaporation; however, the amount lost is insignificant, compared with the primary sources of water waste. Approximately 1.5 gallons of water, other than cooling water, is used for each pound of polymer product. The pollutional character of the effluent is slight, because of the small quantities of additives (catalyst and suspending agents), used in suspension polymerization; and the low reaction-medium temperatures required (120 to 180°F). The catalysts are generally of the peroxide type; the suspending agents may be methyl or ethyl cellulose, polyacrylic acids, polyvinyl alcohol, and numerous other naturally occurring materials such as gelatin, starches, gums, casein, zein, and alginates. Inorganic materials such as calcium carbonate, calcium phos-phate, talc, clays, and silicates may also be present in effluent reaction water.

No plants employing typical technology have waste-treatment facilities. It is estimated that over 90 per cent of the waste is discharged to municipal sewers.

4. Polyolefins (polyethylenes). The polyethylenes pro-duced today run the gamut of molecular weight from waxes of a few thousand to polyethylenes of several million molecules. In addition to this range in molec-ular weights, an equally wide range in stiffness is available. In decreasing order of utilization, poly-ethylene is used for: film and sheet, injection mold-ing, blow-molded bottles, cable insulation, coatings, pipe, and all other uses. Since this represents about 30 per cent of the plastic and resin industry, its pro-duction is significant.

One fundamental process for manufacturing poly-ethylene is the high-pressure method. The final

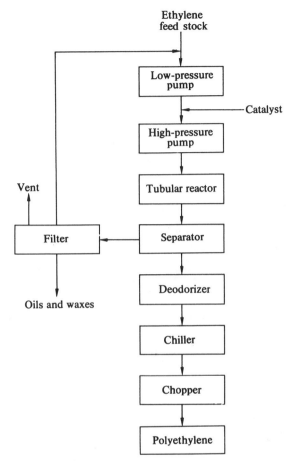

Fig. 28.15 Tubular-reactor process for low-density polyethylene production.

ethylene), of which the most common type is the Phillips process, which uses a supported catalyst of chrome and alumina. Figures 28.15 and 28.16 depict the high-pressure and low-pressure methods of producing polyethylene.

Low-density polyethylene processes create no significant water wastes. Water contacts the product only at the chiller-chopper step. Analysis of a typical highly recirculated chill water revealed a very low total organic carbon of 0.4 ppm. The high-density polyethylene processes also produce no significant water wastes. Typical process waste waters contain a BOD of less than 10 ppm. Potential hazards that might generate water-borne wastes are improper

product is often called low-density polyethylene. A high-purity ethylene stream is elevated to a suitable pressure and passed through a reactor (of a tubular or an autoclave design) in the presence of free radical initiators. The resulting polymer-monomer mixture is separated by pressure reduction into a monomer-rich and a polymer-rich stream. The monomer-rich stream can be cleaned and recycled as feed to the reactor or used in another process. The polymer-rich stream is usually further concentrated by a second separation step and then extruded into ribbons or strands for pelletizing. Another process is the low-pressure method (producing high-density poly-

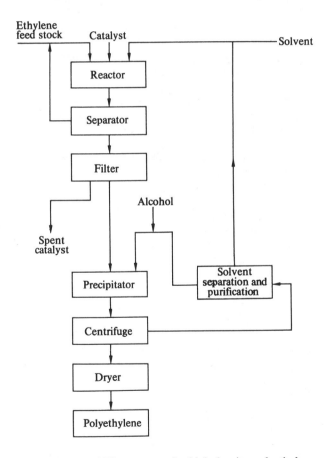

Fig. 28.16 Phillips process for high-density polyethylene production.

operation, spills, and washdown of equipment and facilities.

5. Acrylics. Acrylic resins are made by three processes: bulk, solution, and emulsion polymerization. Bulk polymerization is used for cast sheets and molding and extruding powders. Solution polymerization is used to produce coatings for industrial sales, including automobile paints and fabric coatings. Emulsion polymerization is used mainly to produce coatings for trade sales, such as home paints. About 40 per cent of acrylic resins are made by this last process, which is a batch operation. The monomer is combined with the catalyst, water, and surfactant in a large vat. Polymerization and emulsification are carried out simultaneously. Some of the water is removed and additional surfactant added. Lumps are removed by either filtration or centrifugation, and the emulsion is placed in storage. The final product contains about 50 per cent acrylic resin. Figure 28.17 presents a flow diagram of the emulsion process.

The bulk and solution polymerization methods create very little waste. Virtually all wastes from the manufacture of acrylic resins are from emulsion polymerization. This process creates a concentrated water waste from washing the vat between batches and from lumps that are filtered or centrifuged out of the emulsion mix. The waste contains acrylic monomer, acrylic polymer, emulsifying agents, and catalyst; it is white, highly turbid, and has a high suspended-solids content. It is reported that for every pound of product there are 0.125 gallons of water and 0.0015 pounds of BOD in the waste.

Acrylic plants generally feed their waste into municipal systems. This has created a number of problems from the start and a variety of solutions. These range from a self-contained, closed-end, waste-treatment plant, in which the water is continuously recycled, to a wash-water treatment plant that also treats the waste for a municipality in a joint operation. One particular plant, where some waste-removal-efficiency data were available, reports an 85 per cent BOD removal by the activated-sludge treatment.

6. Polyester and alkyd resins. Alkyds and polyesters are characterized by great variation in their formulations, not only according to the class of resin produced, such as oil-modified polyesters (alkyds), unsatured polyesters (laminates amd molding com-

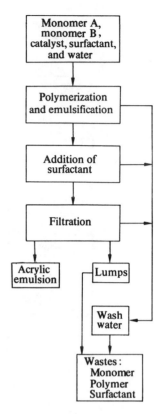

Fig. 28.17 Flow diagram for the production of acrylic emulsion.

pounds), and linear polyesters (films and fibers), but also within each group. The fundamental manufacturing process for alkyd and polyester resins is batch-type condensation polymerization of a dibasic acid and a polyfunctional alcohol. Polymerization in the presence of oils or fatty acids results in a complicated polymer known as an alkyd resin. The polymerization process is illustrated in Fig. 28.18.

The significant wastes associated with the production of alkyd and polyester resins are (1) unreacted volatile fractions of raw materials, which either appear in the withdrawn water of esterification and in the water used in scrubbers or are vented to the atmosphere, and (2) residue in kettles cleaned out with either caustic solutions or solvents.

Flotation and land disposal are the only two methods of waste treatment known to be utilized,

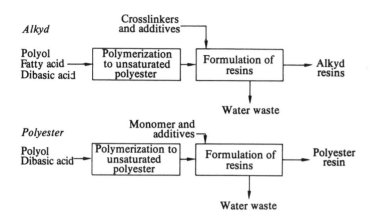

Fig. 28.18   Flow chart for polyester production.

other than discharge to a municipal sewer system. It is maintained that there is no adequate treatment process for waste water from polyester production.

7. Urea and melamine resins. Urea and melamine resins, which can be used interchangeably, compete with phenolics. Their superior tensile strength and modulus of rupture command a higher price than phenolics, and their electrical properties are outstanding. The fundamental manufacturing process for urea and melamine resins is batch-condensation polymerization of the urea or melamine with formalin (a 40 per cent solution of formaldehyde in water). The raw materials, urea or melamine and formalin, together with catalysts, miscellaneous additives, and modifiers of a proprietary nature, are charged into a jacketed reaction vessel and heated to initiate the reaction. Once initiated, the reaction is exothermic, and the heating steam is shut off. Cooling water is introduced into the jacket to control the reaction temperature. The mixture is refluxed until the proper degree of polymerization takes place. The resin is soluble in water, so no separation occurs as in the case of phenolic–formaldehyde condensation resins. The urea and melamine resins are vacuum-dehydrated until the solids content is 50 to 60 per cent and then are either sold as a solution or spray-dried and sold as a solid product. A flow diagram for the manufacturing processes is shown in Fig. 28.19.

The significant water wastes from the production of urea and melamine resins are: water introduced with

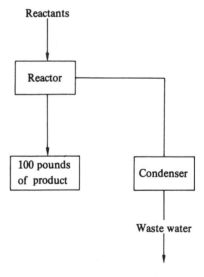

Fig. 28.19   Flow chart for the production of urea melamine, and phenolic resins. To produce 100 lb of urea resin, 75 lb of urea and 188 lb of formalin are put into the reactor; the waste water contains 133 lb H_2O from formalin, 43 lb H_2O from the reaction, 3.5 lb urea, and 3.5 lb formaldehyde. For 100 lb of melamine resin, the quantities are 52 lb melamine and 182 lb formalin; the waste water contains 109 lb H_2O from formalin, 21 lb H_2O from the reaction, 2 lb melamine, and 2 lb formaldehyde. For 100 lb of phenolic resin, the quantities are 92 lb of phenol, 73 lb of formalin, and 0.3 lb of catalyst; the waste water contains 44 lb H_2O from formalin, 17 lb H_2O from the reaction, 3 lb phenol, 1 lb formaldehyde, and 0.3 lb non-volatile matter, with a pH of 6.5.

the raw materials, water formed as a product of the condensation reaction, caustic solutions used for cleaning the reaction kettles, and blowdown from cooling towers. Quantities of waste can be computed from Fig. 28.19. The waste water discharged from resin plants has a temperature of about 85 to 90°F.

Treatment is carried out by lagooning and thermal incineration, as well as discharge to municipal plants. Lagoons are designed with no overflows, so that evaporation and seepage are the only means of volume reduction. Malodors have been observed with this method. Incineration can also lead to air pollution problems, depending upon the fuel used and the operating efficiency.

8. *Phenolic resins.* The phenol-derived resins are the oldest family of resins. A wide variety of products and uses make up the distribution pattern. The major uses are for inexpensive casting and plywood bonding. The fundamental manufacturing process for phenolic resins is batch-condensation polymerization of phenolics with formalin. The phenolics and formalin, together with catalysts and miscellaneous additives and modifiers, are charged into a jacketed reaction kettle and heated to initiate the reaction. Once initiated, the reduction is exothermic; heating is terminated, and cooling water is introduced into the jacket to control the reaction temperature. The mixture is refluxed until the contents separate into two, a heavy viscous resin layer and an aqueous one. At this point, a vacuum is applied and the temperature is raised to remove the water. The molten resin is drained into a pan where it solidifies on cooling. The manufacturing process is illustrated in Fig. 28.19.

The significant water wastes from the production of phenolic resins are: water introduced with the raw materials, water formed as a product of the condensation reaction, caustic solutions used for cleaning the reaction kettles, and blowdown from cooling towers. The process water resulting from resin production is about 7.3 gallons per 100 pounds of resin. About 600 gallons of cooling water are required to control the reaction: 42 gallons are discharged to the sewers as cooling-tower blowdown and 558 gallons are recirculated. The waste waters discharged from resin plants have a temperature of about 85 to 90°F.

Phenolic-resin wastes are treated by lagooning, phenol extraction, and thermal incineration, and also

are discharged to municipal sewage-treatment plants. Normally a single-stage phenol-extraction plant will remove about 96 per cent of the phenols and 100 per cent of the formaldehyde. One report of the character of waste from such plants is as follows:

Phenol	1,600 ppm
BOD	11,500 ppm
Chlorine demand	68 ppm
Total solids	500 ppm
Volatile solids	250 ppm
Total suspended solids	40 ppm
Volatile suspended solids	20 ppm
pH	6.4

Municipal sewage authorities indicate a high degree of accommodation for phenolic waste. Studies have shown that both phenolics and formaldehyde are biodegradable in conventional biological sewage-treatment processes, providing the concentration is maintained below toxic levels. The difficulties that do arise are due to fluctuation in pH. Therefore, combination of phenolics and municipal wastes is technically feasible, and pretreatment is generally not required.

9. *Miscellaneous resins.* The miscellaneous resins are chemically unrelated but have one common feature—low-volume production. They include polyurethanes, epoxy, acetal, polycarbonates, silicone, nylon 6, and coumarone-indene. Several of the resins generate no water wastes. Some specialty resins are manufactured by just a few companies. Accurate reports of treatment practices are not available.

References: Plastic and Resin Wastes

1. *The Cost of Clean Water*, Vol. III, Industrial Waste Profile no. 10: Plastic Materials and Resins, U.S. Department of the Interior, Washington, D.C. (1967).

2. Mayo Smith, W., *Manufacture of Plastics*, Vol. 1, Reinhold, New York (1964).

3. Raff, R. A., and J. B. Allison, *Polyethylene*, Interscience, New York (1956).

The following references were published since 1968.

1. Anonymous, "Environmental currents," *Env. Sci. and Tech.*, 272 (1970).

2. Anonymous, "Plastic wastes yield to pyrolysis," *Env. Sci. and Tech.*, 473 (1970).

3. Ardegust, M., "Purification of wastes from phenol-formaldehyde resin production," *Ger. Offen*, 2,054,753 (1971).

4. Bernadiner, M. N., A. P. Shurygin, and B. S. Esilevich, "Thermal decontamination of wastewaters from the product of nylon 66 fibers," *Khim. Volkna*, **4**, 67–70 (1969).

5. Chereshkevich, L. V. and D. D. Chegodaer, "Reuse of ftoroplast-4 wastes," *Plast. Massy*, USSR.

6. Dzurovcin, L. and J. Slioka, "Melting furnace for fusing wastes from the production of thermoplastic polymer fibers," *Czech* **132**, 140 (1969).

7. Forschung and Patsntverwertung, "Reusing granular or fibrous polyester wastes," *Fr.*, 1,566,868 (1969).

8. Huenecke, K., "Preventing phenol-containing wastewater in the production of phenol and cresol formaldehyde resins," *Ger. Abs.* 1,595,770 (1971).

9. Illing, G., "Reworking of man-made fibers, thermoplastic sheets, and expanded plastics to granulated injection molding and extrusion compounds," *Kunststoff technik* **8**, 196–8 (1969).

10. Izard, E. F. and W. C. Lindsey, "Recovery of diamines and tetracarboxylic acids from aromatic polyamides," *Fr. 1, #*509,269 (1968).

11. Komissarov, S. A., "Fibrous filler," *Prom. Obraztsy, Tovarnye Znaki* **46**, 77 (1969).

12. Komleva, T. B. and Z. I. Pokrovenko, "Organization of sanitary-hygenic control of the discharge and use of products made from polymeric materials," *Gig. Prime'n. Polm. Mater. Izdelii Nkh.*, No. 119–25 (1969).

13. Marinich, V. K. and O. V. Firosova, "Removal of a lubricant from wastewaters of kapron production," *Khim. Volkna*, **4**, 76–8 (1969).

14. Mathews, V., "Composition polymeryable to polyamides," *Brit.*, 1,201,652 (1970).

15. Petru, K., "Use of plastic production wastes," *Plast. Hnoty Kauc.* **6**, 165–170 (1969).

16. Pashkov, A. B. and N. M. Udovin, "Processing acid wastes from styrene copolymers chloromethylation," *Brit.*, 1,162,078 (1969).

17. Potts, J. F., "Continuous pyrolysis of plastic wastes," *Ind. Water Eng.* **7**, 32–35 (1970).

18. Schmidt, L. I., S. M. Gordon, and Nkitinai, "Purification of waste waters from the production of polystyrene-plastics," *Izobret, Prom. Obraztsy, Torarnye Anaki* **45**, 157 (1968).

19. Steinmetz, C. E. and W. J. Day, "Treatment of waste from polyester manufacturing operations," *Chem. Eng. Progr. Symp. Series* **65**, 188–190 (1969).

20. Zege, I. D. and A. U. Sakhernov, "Purification of industrial wastes in the production of butanolized phenol-formaldehyde resins," *Lakokrasoch. Meter. Ikh Primen.* 2, 70–72 (1971).

28.9 Fertilizer Industry Wastes

Origin. A study done by Wellman-Lord Inc. has divided the fertilizer industry into three main categories each with several subcategories, which are (1) fertilizer raw materials; (2) fertilizer intermediates; and (3) fertilizer products. Fertilizer raw materials will not be considered here as they are mining operations that obtain the elemental phosphorous, potash, and sulphur. (They are discussed in section 28.3). The other two categories are further divided as follows:

(A) Fertilizer intermediates:
   1. Sulphuric acid
   2. Phosphoric acid
   3. Nitric acid
(B) Fertilizer products:
   1. Solid fertilizers
     (a) N-fertilizers
       (1) Ammonium nitrate
       (2) Urea
       (3) Ammonium sulfate
     (b) P-fertilizers
       (1) Superphosphates
     (c) NP-fertilizers
       (1) Monoammonium phosphate
       (2) Diammonium phosphate
     (d) NPK-fertilizers
     (e) Blended fertilizers
   2. Liquid fertilizers
     a. Ammonia
     b. Liquid formations
     c. Slurry formations

Each of these products are produced in many ways and must be discussed separately.

Origin of Wastes

Sulphuric Acid Production. Sulphur dioxide is produced by the burning of elemental sulphur in a dry air steam at high temperatures. The hot SO_2 is then cooled and converted by catalytic action to SO_3. After passing the catalytic beds, the SO_3 gas is cooled and then forced countercurrent in a packed column reactor to 98–99 per cent H_2SO_4, where the SO_3 is readily hydrolized to H_2SO_4.

The reactions are essentially complete and consequently no acid is lost to an effluent. However, two problems do exist. In all of the reactions heat is released in considerable quantities and cooling facilities are required. With cooling facilities comes the inherent problems of wastewater disposal, which shall not be discussed here. The other problem to be considered with in plant is that of accidental spills. Technically this problem can be solved quite easily. A collection and holding system for the spills may be installed so that spilled material can be recycled back to the process. (Acid wastes are discussed in Section 28.1.)

Phosphoric Acid Production.   At the present time, most phosphoric acid is manufactured by the wet process. Generally, the method involves the use of a strong acid to dissolve the phosphate (P_2O_5) portion of the phosphate rock. The two most commonly used acids for this process are sulfuric and nitric, but hydrochloric acid has also been applied:

$$Ca_3(PO_4)_2 + HA + H_2O \rightarrow 3Ca(A) \cdot 2H_2O$$
$$+ 2H_3PO_4.$$

With each of the different acids distinct wastes result. Minor quantities of calcium, fluorine, iron, aluminum, silica, and uranium are found. When sulfuric acid is used, another major waste is the calcium sulfate (gypsum) produced in the acidulation of the rock. For every ton of P_2O_5 produced, five tons of gypsum must be removed.

In the nitric-acid process, soluble calcium nitrate is produced. This must be removed from the phosphoric-acid solution, and it can be done in two ways. Process *A* adds ammonia and carbon dioxide and converts $Ca(NO_3)_2$ to solid $CaCO_3$ and soluble ammonium nitrate. The entire mixture is then converted directly to solid fertilizer and consequently produces no effluent. Process *B* treats the acidulated mass thermally to remove some of the $Ca(NO_3)_2$ and converted solid $CaCO_3$, which is removed as waste. The remainder is solidified as a fertilizer product.

The hydrochloric-acid acidulation produces a soluble calcium chloride by-product that is separated from the phosphoric acid through liquid-liquid extraction. The solvent selectively separates the H_3PO_4 from the remaining mass and is then split from the H_3PO_4 by specific gravity difference leaving H_3PO_4, solvent, and impurities separate.

With the use of any of the acids, an appreciable quantity of fluorine is evolved as a gas. The release of the gas originates primarily from the attack tank with lesser quantities from the other equipment. The gas must be removed by water scrubbing and consequently the result is a wastewater to be dealt with. In a like manner, small quantities of NH_3 escape from ammoniator units and are scrubbed out with fluorine.

Concentrating and purification operations, following raw material production involve evaporation, solvent extraction, and sedimentation. Small quantities of acid and still more fluoride are wasted along with the water condensate and solid impurities.

Nitric Acid Production.   Nitric acid is produced by catalytic oxidation of ammonia with air. The process is carried out in the following three steps: (1) oxidation of ammonia to form nitric oxide: (2) oxidation of nitric oxide to form nitrogen peroxide; and (3) reaction of nitrogen peroxide with water to form nitric acid. The system is completely closed, recycling all the unreacted intermediates back into the process. The only wastes that are characteristic are lubrication oils for process machinery. Care must be taken to collect these before they enter a wastewater stream. Preventative measures should also be taken to prevent accidental spills.

Nitrogen Fertilizers.   Ammonia is the basis for all of these fertilizers and is by far the most significant, as far as its effects on wastewater characteristics. There are four methods of ammonia production; each has several small process differences. The most widely used method of ammonia production is steam reforming of natural gas.

In this process, natural gas and steam are passed through a suitable catalyst to form hydrogen, carbon monoxide, and carbon dioxide. Air is then added to supply the required nitrogen. Oxygen in the air, added to the partially reformed gas, burns to form water and to supply additional heat to complete the reforming reaction.

Natural gas usually contains small traces of sulfur

compounds that are harmful to the catalysts used in the ammonia production. These sulfur compounds are removed prior to reforming by activated carbon absorption or molecular sieving. The absorber or sieve is then reactivated by passing steam through the bed. The steam and contaminants are then vented to the atmosphere.

The reformed gas, containing CO, CO_2, and water vapor must be cleaned to remove these three gases since they are detrimental to the synthesis reaction. Carbon monoxide is reacted with steam over a suitable catalyst to produce additional hydrogen and CO_2. All constituents are gaseous and do not produce a fluid effluent. Carbon dioxide removal is accomplished in several ways. Scrubbing with various alkaline solutions such as mono-, di- or tri-ethynolamine, or other solutions can be done. Under high pressure the CO_2 reacts with these solutions at normal temperatures. By heating and reducing pressure, the scrubbing agent is recovered and the CO_2 is discharged as a gas. Process gas still contains small amounts of CO and CO_2, which are converted catalytically to methane for use in boilers.

Some producers remove carbon monoxide by scrubbing at high pressure with a copper-carbonate solution containing acetic or formic acid. The solution is regenerated by reducing the pressure and heating with steam. Ammonia water vapor and carbon monoxide are vented to the atmosphere. The solution is regenerated. The copper-liquor scrubbing is followed with either ammonia or sodium-hydroxide wash to remove the final traces of carbon dioxide. These final solutions cannot be regenerated and are constantly wasted.

Compression and synthesis of the ammonia produce no wastewater, since each step is a total recycle system.

The major problem with wastewater from this process is the boiler blowdown water. It is the largest contaminated water stream associated with the fertilizer producers.

Three other methods of ammonia production are presently in use; however, they amount to only about six per cent of the total United States production. In one method, compressed air and hydrogen are burned in a combustion furnace to produce the 3-to-1 hydrogen-nitrogen mixture. Heat is recovered. The CO_2 contained in the air is removed by sodium-hydroxide scrubbing. The sodium/carbonate solution is discharged as waste. Again cooling water and boiler water are a major problem.

The partial oxidation process is used to produce a very small portion of the ammonia. Liquid air is separated into liquid oxygen and liquid nitrogen. All but the liquid nitrogen are wasted as gas to the atmosphere. Natural gas and steam are burned and quenched to produce hydrogen, CO, CO_2, argon, and methane. The CO and steam are converted to CO_2 and hydrogen by passing them through an iron oxide catalyst. The CO_2 is then removed by the monoethynolamine method, producing no wastewater. Two liquid waste streams exist: (1) scrubbing atmospheric CO_2 with sodium hydroxide and (2) the excess quench water which contains dissolved CO, CO_2, methane, and argon.

Ammonia can also be produced with a coke oven gas. The gas is compressed; high molecular weight hydrocarbons are removed by condensation and are washed with light oils to remove benzine (as a by-product). The process gas is then scrubbed with CO_2 and H_2S. The ammonia solution is regenerated with steam and water washing and CO_2 and H_2S are vented to the atmosphere. A final wash with a sodium-hydroxide solution removes remaining CO_2 and this wash is wasted. Liquid nitrogen from an air-separation plant is used to supply nitrogen for the synthesis. The recovered methane is used in boiler heating and the ethylene as a by-product. The gas is then compressed and subjected to the conventional synthesis processes.

Ammonium Nitrate. Three processes are used to manufacture ammonium nitrate: (1) the Stengle process; (2) the Prill process; and (3) the crystalizer process.

The Stengle process. Ammonia vapor and 60 per cent nitric acid are reacted in a packed tower. The vaporized water and entrained nitrate are recovered by condensation. The condensed liquid is then recycled as a process feed. The hot melt is then dried with hot dry air to reduce the water content. The noncondensible gases, air and water vapor, are vented to the atmosphere. Water from the final condenser is wasted to the sewer. The dried ammonium nitrate is broken, granulated, screened, and bagged, with the fines recycled back to the melt.

The Prill process. Ammonia and nitric acid react in an agitated vessel and are concentrated, either by heat of reaction vaporization of water alone or with vacuum evaporation. The steam containing trace amounts of ammonium nitrate is vented to the atmosphere. Hot ammonium-nitrate melt is then prilled to produce the solid ammonium nitrate. The air is discharged to the atmosphere along with small amounts of ammonium nitrate dust. Any dust or particles caught are redissolved and then returned to the reactor.

The Crystallizer process. Crystallization production of ammonium nitrate is essentially the same as prilling except for the drying. Drying is done in a rotating drum with hot air.

Wastewater is generally not a problem here since all products can be collected and returned to the process.

Urea. Liquid ammonia and carbon dioxide are reacted at 160 to 200°C and 100–250 atmospheres to produce urea, ammonium carbamate and water. Heat is removed, and the carbamate is returned to the head of the process. The urea can either be left as a liquid or prilled to produce a solid.

No wastewater stream is inherent in this process since it is a complete recycle system. Spills should be controlled.

Ammonium Sulfate. Sulfuric acid and ammonia are reacted in a solution of mother liquid from the crystallizer centrifuge. The water evaporated by the heat of reaction is removed by vacuum ejectors and the crystals of ammonium sulfate are removed from the mother liquid by centrifugation. The crystals are dried and bagged, with the fines being returned to the process. Again, no wastewater stream is inherent in this process.

Normal Superphosphate. Sixty to seventy per cent H_2SO_4 and ground phosphate rock are rapidly mixed (in seconds). Because the reaction is highly exothermic, the mass solidifies rapidly, producing gases that are collected and water-scrubbed. After the mass has cooled for 1 or 2 hours, it is moved to a curing room for 2 to 3 weeks. After curing, it is sized, bagged, and shipped.

The gas is mainly flourine which, when scrubbed out, presents the wastewater treatment problem.

Triple Superphosphate. Two methods of triple superphosphate production are presently in use. The run-of-pile process involves mixing the raw materials of ground phosphate rock and 64–75 per cent H_3PO_4 acid. The reaction is similar to that in normal superphosphate, highly exothermic. Solidification in this process normally takes place on a slow-moving belt enroute to the curing area. After curing, the solids are sized and bagged.

The alternative process combines the ground phosphate rock and 56 per cent H_3PO_4 in an agitated vessel. The lower concentration of acid allows for better reaction completion and handling. The slurry is then dried and prepared for sale.

The raw-waste load is the same as that of normal superphosphate.

Monoammonium Phosphate (MAP). The most prominent process used in MAP production is the slurry process. Ammonia and H_3PO_4 in the 30 to 40 per cent P_2O_5 range are brought together in the reaction vessel with or without agitation. These materials react exothermically to produce a slurry which is distributed over recycled dried material. The undersized granules are separated and returned to the process.

Production of MAP by mixing the H_3PO_4 acid (concentration in 54–60 per cent P_4O_5 range) and NH_3 in a jet-reactor is also used. The exothermic reaction produces a MAP slurry and super-heated steam that are sprayed counter current to cool dry air. The solidified MAP particles are mechanically separated and removed to storage. Normally no wastewater effluent exists.

Diammonium Phosphate (DAP). Monoammonium phosphate is reacted further with ammonia to produce the diammonium phosphate granules which are sized and either returned to the process or bagged for shipment. Dust particles are scrubbed out and reused in the process. Any secondary scrubbing must be wasted and will present a problem.

NPK Fertilizers. Three methods are generally used in the production of NPK fertilizers. The first and simplest is that of blending dry fertilizers together to produce the desired composition. Since this is a dry process, no waste stream is inherent with proper inhouse practices.

A phosphoric acid based NPK fertilizer is produced

simply by coating potash particles with DAP. The wastes from this process are similar to those in the DAP manufacturing and present similar problems.

The third process is based on nitric acid. It is identical to the process described in phosphoric acid acidulation with the small modification of the addition of KCl to the later stages. The slurry is then dried and granulated presenting essentially the same waste problem as that of the phosphoric acid acidulation process.

Blended Fertilizers. The blending of fertilizers is done on a dry basis and, therefore, has no inherent waste stream. Care must be taken in maintaining plant conditions so that accidental spills may be handled dry and not washed away.

Liquid Fertilizers. Liquid fertilizers are simply raw products or combinations of raw products from basic manufacturing processes. Wastewaters characteristic to each of these were previously described.

Characteristics and Treatment

Disregarding treatment of boiler blowdown and cooling tower effluent, which is by far the largest wastewater problem in the industry, several treatment schemes have been used. Most approaches have been based on the origin of the product—that is, H_3PO_4 or NH_3.

The effluent streams can be characterized as either a phosphoric acid effluent or an ammonia effluent.

The phosphoric acid effluent is high in flouride concentration, low in pH, high in phosphate, and high in suspended solids. Standard practice has been to contain the water for reuse, allowing enough time for solids sedimentation. The problem is that fresh water addition may result in overflow of the retained water. The retention ponds should be carefully designed. If overflow is evident, it must be treated with lime and clarified.

A slightly more sophisticated method for discharging overflow water is with a two-stage liming process. The first lime treatment brings the pH up to 3 or 4 and reduces the fluoride concentration to 20–25 mg/l and the P concentration to 50–60 mg/l. The CaF_2 precipitate is settled out and the effluent is treated again with lime to raise the pH to 6 or 7. The F and P concentrations are reduced to about 10 mg/l. The water is clarified and released to a receiving stream.

Rabosky and Miller (1974) carried the precipitation process further. The two-stage fluoride removal process of lime precipitation and then alum-polyelectrolytes flocculation and sedimentation was found to be a workable method. Initial fluoride concentrations of 95 to 135 mg/l were successfully reduced to 1 to 2 mg/l using this approach.

Maximum precipitation of the fluoride with lime occurred at a pH greater than or equal to 12.0; however, optimum coagulation with the alum-polyelectrolyte occurred at a pH of about 6 or 7. Consequently, the pH had to be readjusted by the addition of concentrated H_2SO_4.

Recovery of the fluorine as fluosilicic acid (a very valuable product) was investigated by Malin (1972) in an actual fertilizer plant. The off gases from the phosphoric acid evaporators were condensed and concentrated and had a ready market in the aluminum and steel industries.

Bhattacharya, Grieves, and Romans (1972) precipitated equimolar concentrations of orthophosphate and fluorine with lanthinum (III) which could then be readily floated at acidic pH's with the addition of the strongly basic anionic surfactant, Sodium Lauryl-Sulfate. Better than 95 per cent floatation was achieved over pH ranges from 3.5 to 6.0 using 0.023 MOLE SLS/MOLE ORTHOPHOSPHATE + FLUORIDE. Maximum floatation of 98 per cent was obtained at a pH of 4.0 +, which produced a collapsed foam volume 10 per cent of the initial waste volume.

The other effluent type is characteristic of ammonia production and ammonia-containing products. Most of the contamination comes from ammonia production itself. It is characteristically high in ammonia from effluent gas-scrubbing and gas-cleaning operations and high in sodium hydroxide or carbonate from gas-cleaning processes.

Methods of treatment are now being investigated for specific wastes. Bingham (1972) experimented with the removal of NH_4NO_3 by ion exchange. Continuous flow removal of NH_4^+ with a strong acid cation resin was followed by a weak base resin in the hydroxide form to remove the anions. Both resins were regenerated and the backwashes were combined to form a recovered NH_4NO_3 that was recycled back into the plant processes. There was no effluent.

Siddiqi, Ratnaparkhi, and Agarwl (1971) used a submerged, fixed-film, biological reactor for nitrification of waste waters high in NH_4^+ and free of organics. They found almost complete bio-oxidation was achieved, which was four times more efficient than bio-filtration.

Air stripping of ammonia was investigated by Rao and Ramprasad (1971). They found that NH_4^+ removal was enhanced up to pH 11 and exponentially removed with temperature increase. The removal rate was also found to be independent of the initial concentration. Sodium hydroxide rather than lime was recommended to control pH because it was more economical and interferred less with aeration.

In all cases it is evident that future waste streams will change because of the increasingly-stiffer air-pollution regulations. This must be carefully considered in any waste parameter prediction.

References

1. Anonymous, *Development Document for Proposed Effluent Limitations Guidelines and New Source Performance Standards for the Basic Fertilizer Manufacturing Point Source Category*, November 1973.
2. Battille Memorial Inst., "Inorganic fertilizer and phosphate mining industries," *Water Pollut. and Contr.*, September 1971.
3. Bingham, E. C., "Water pollution problems at a nitrogen fertilizer plant," *Proc. 4th Annual Northeastern Regional Antipollution Conf.*, 104, July 1971.
4. Das, A. K. and J. A. Khan, "Nitrogen removal from fertilizer wastes," *Technology, Sindri* 2, 10–14 (1965).
5. Das, A. and J. A. Khan, "Nitrogen removal from fertilizer effluents II," *Technology, Sindri* 3, 41–45 (1966).
6. Desai, M. W. and M. S. Varde, "Effluents from fertilizer plant at Bombay," *Tech., Sindri* 3, 22–26 (1966).
7. Gartrell, F. E. and J. C. Barber, *Chem. Eng. Progr.* 62, 44–47 (1966).
8. James, G. R., "Stripping ammonium nitrate from vapors," *Chem. Eng. Progr.* 69, 79–82 (1973).
9. Koncz, E., "Nitrogenous fertilizers at Craiova," *Revue. Chim. Buc.* 17, 290–92 (1966).
10. Prasad, R. R. and B. K. Dutta, "Effluents at Sindri," *Tech., Sindri* 3 65–68 (1966).
11. Roy, A. K. and B. B. Roy, "Recovery of ammonium sulfate," *Tech., Sindri* 3, 77–80 (1966).
12. Sachett, W. J., "Collection and treatment systems," *U.S. Pat.* #3,499,731.
13. Saltnov, V. S., "Ammonium sulfate wastes," *Khim Prom.* 45, 235 (1969).
14. Sarbaev, A. N., "Urea wastes," *USSR*, Pat.#239863.
15. Sharma, R. D. and M. G. Garg, "Effluent disposal at Nangel," *Tech., Sindri* 3, 91–95 (1966).
16. Stocker, W. F. and W. M. Reiter, "Phosphoric acid manufacture," *Chem. Eng. Progr.* 70, 59 (1974).
17. Walters, C. T., "Waste disposal in fertilizer plants," *Anal. Instrum.* 7, 294–300 (1969).

Recent References

1. Anonymous "Treatment of effluents from African explosives and Chemical Industries Limited Modderfontein factory," *Water Pollut. Abs.* 45, 497 (1972).
2. Bhattacharya, D., R. B. Grieves, and J. D. Romans, "Precipate coflotation of orthophosphate and fluoride" *Proc. 27th Ind. Waste Conf., Purdue University*, 270 (1972).
3. Bingham, E. C., "Fertilizer maker stops nitrogen," *Water and Wastes Eng.* 9, F-5 (1972).
4. David, M. L., J. M. Malk, and C. C. Jones, "Economic impact of costs of proposed effluent limitation guidelines for the fertilizer industry," *USEPA*, 230/1-73-010, October 1973.
5. Landy, J. A., "Chromate removal at a Saudi Arabian fertilizer complex," *Water Pollut. Contr. Fed. J.* 43, 2242 (1971).
6. Malin, M. H. Jr., "Cleanup pays off for fertilizer plant," *Environ. Sci. and Tech.* 6, 400 (1972).
7. Rabosky, J. G. and J. P. Miller, Jr. "Fluoride removal by lime precipitation and alum and polyelectrolyte coagulation" *Proc. 29th Ind. Waste Conf., Purdue University*, 669 (1974).
8. Rao, A. V. S. P. and G. Ramprasad, "A study on air stripping of ammonia" *Water Pollut. Abs.* 44, 418 (1971).
9. Siddiqi, R. H., et al., "Nitrification in treatment of nitrogenous fertilizer industry wastewater," *Water Pollut. Abs.* 44, 418 (1971).
10. Wellman, Lord, Inc., "Study report, industrial waste studies program, Group 6: Fertilizers," *Water Quality Office EPA* Cont. No. 68-01-0029, July 1971.

28.10 Toxic Chemicals

The Federal Water Pollution Control Act Amendments of 1972 (see Chapter 2.3) in Section 307 (a) (1) directed the Environmental Protection Agency's

Administrator to publish (and from time to time revise thereafter) a list which includes any toxic pollutant or combination of such pollutants for which an effluent standard will be established. "The Administrator in publishing such list shall take into account the toxicity of the pollutant, its persistence, degradability, the usual or potential presence of the affected organisms in any waters, the importance of the affected organisms, and the nature and extent of the effect of the toxic pollutant on such organisms."

Mercury, a toxic chemical, is a naturally-occurring element on earth, although about 10,000 tons of it are extracted from cinnabar ore (HgS) each year in the United States. About one-half is released directly into the environment through industrial wastes. The accumulation of mercury in predatory fish, such as tuna and swordfish, constitutes the major danger to man. The principal industries discharging mercury are electric wire and equipment, paper processing, and caustic-soda manufacturing.

Many catastrophies, such as that in Minamata Bay, Japan, have resulted in death from eating fish contaminated with mercury. Industrial treatment of these wastes primarily involves recovery in-plant

and better housekeeping procedures.

USA Today depicts 7 billion pounds of toxics reaching our environment in 1987 (2, 1989). (See Fig. 28.20.) Most (3.9 billion) are discharged on or in the land while 2.6 billion spew into the air and the remaining 0.5 billion reach the waterways directly. All of these are considered hazardous and discussed by your authors in Chapter 32 extensively.

The National Wildlife Federation pinpointed 500 industrial plants as the nation's largest producers of toxic pollution (3, 1989). Twenty-four plants produced about 3.5 billion pounds of toxic wastes in 1987. A total of 118 plants reported releases of at least 10 million pounds and at least 2.6 million pounds were released by each of the 500 polluters. The 1987 toxic emissions (of the 327 hazardous chemicals) were submitted to the U.S. Government by 18,383 plants nationwide.

References

1. Libman, B. Y. and N. S. Fuks, "Purification of wastewaters from the production of toxic organic phosphorus compounds," *Zh. Vses. Khim. Obshchest* **12**, 651 (1967).

Fig. 28.20 (USA Today, 2, 1989)

2. *USA Today*, July 31, 1989, page 5B.
3. "Group names 500 top toxic polluters," *The Miami Herald*, August 11, 1989, page 7A.

28.11 Mortuary Science Wastes

Introduction. The practice of embalming in the United States was actually started during the Civil War, when Dr. Thomas Holmes was commissioned by the United States Government to devise a technique of "body preservation and disinfection" to be used on the Union soldiers killed during the war. Essentially, the Government was looking for a method of preservation and disinfection for those bodies which had to be transported great distances before reaching their final resting place. Obviously, there was at that time, great concern that the decomposition of these "unburied" bodies might lead to other health hazards.

From its inception, the major goal of the mortuary science industry has been to achieve 100 per cent disinfection and preservation of dead bodies. However, today we realize that such a goal is unrealistic. One-hundred per cent disinfection cannot be accomplished because the water used for making the embalming solution contains chemical and biological impurities and because it is virtually impossible to remove all of the body's fluids, even using the most sophisticated techniques available today. It has been estimated that at least 17 gallons of embalming solution would be required to displace most of the body's fluids.

The first embalming fluid, which was devised and used by Dr. Holmes, was bichlorate of mercury, an excellent disinfecting agent. Around 1900 a cooling process, which did not require the injection of fluids, came into use as the most effective method of disinfecting and preserving a body. The process consisted of simply storing the body on a block of ice. Today, the practice of disinfection and preservation is accomplished with organic reagents, the primary one being formaldehyde.

The term Formalin is actually the proper term for the embalming solution, as the formaldehyde is diluted with water to obtain the proper strength. Formalin, which is commercially prepared, contains materials such as (1) light metals, especially Al and Mg; (2) coagulents and anti-coagulants to obtain the desired chemical reaction in the body; (3) precipitants, which precipitate chemical elements (e.g., metals in both the body and the solution) so that no adverse reactions occur to inhibit the embalming process; and (4) sequestering agents, which are used in the separation of body tissues to develop a better "fix," or embalmed body. The exact chemical content of the formaldehyde solution is the trade secret of each company manufacturing it.

The practice of using heavy metals for disinfecting and perserving bodies, has been outlawed for legal and medical reasons. For example, if a casket is dug up or somehow disturbed (i.e., by earthquake or flood, etc.), the concentrated heavy metals would be exposed to the environment; or, if "foul play" was the suspected cause of death, these heavy metals would prohibit a chemical analysis of the remains. Thus, these heavy metals have been replaced with lighter metals such as Al and Mg.

Characteristics. There are essentially three possible sources of pollution to be dealt with in the mortuary science industry. The first involves crematories; incineration of the body might release toxic elements into the air. The chances of this occurring, however, appear remote, since the incineration process requires excess oxygen to ensure complete combustion. The ashes collected are usually placed in an urn and given back to the deceased's relatives.

The second and most obvious source of pollution from this industry involves the embalming process. The aspirated body fluids and excess embalming fluid are discharged into the sewer. These wastes have high BOD value, as well as bacteriological pollution potential. Note that when a dead body is transported from one state to another, stringent regulations must be met, especially if the deceased died of an infectious disease. However, the liquid wastes from these bodies are indiscriminantly discharged into sewers.

What happens if the cause of cancer in humans is ultimately traced to viral organisms? What are the possibilities that the viral organisms (e.g., in blood cancer or leukemia) discharged in these liquid wastes may survive and even multiply in sewerage? The transmission or pathenogenic organisms via natural waterways is well understood. Dr. Jonas Salk was a great proponent of the idea that cancer is caused by viral organisms which could be transmitted by our waterways.

The final source of pollution could come from the decomposition of bodies which have been buried for many years. Metal vaults are believed to remain intact for at least 40 years. Eventually, however, these vaults rust out and their contents are exposed to the soil environment. It is important to consider the diffusion rate of the elements contained in these vaults, the assimilative capacity of the soil, and most importantly, the effect of these elements have on the neighboring soil and aquatic environment. An underground stream passing directly under a cemetary could transmit of these elements to surface soils and waters.

Pine and other wooden vaults are known to consistently decompose and cave in after only 3 to 5 years. Many cemetaries, as well as some religious and state agencies, forbid the use of wooden vaults in particular areas because they decompose and cave in leaving cavities in the ground. This presents real pollution potential.

They have also recently discovered that concrete vaults are not as safe as was once thought. Concrete vaults that are reinforced with steel are generally as secure as metal vaults. Concrete vaults that are not reinforced with steel have been known to crack open (especially the concrete lid) and leak water (especially under adverse conditions such as abundance of melting snow and ice). Tree roots seeking nutrition are another cause of the cracking of concrete vaults. It is essential to remember that concrete is a porous material, and, although it is a better material than wood for vaults, it cannot compare to the quality of metal vaults.

To date, very little research has been done on the rate of decomposition of buried bodies, which is essentially dependent upon the amount of embalming received and the environmental factors in the vicinity of these vaults. It is well known that once a casket begins to leak water, due to its oxygen content, the body undergoes a saponification reaction called adiposcere and takes on a waxy appearance. The rate of decomposition thereafter depends upon many environmental factors.

Treatment. The only research that this industry appears to be engaged in at this time is in the formulation of new embalming solutions. Presently, other aldehydes are being experimented with to determine the degree of embalming acquired. Glutaraldehyde has been found to be an excellent substitute for formaldehyde, but is expensive. New amounts and types of light metals are also being introduced into these embalming solutions. With the refinement of the technique, these new compounds could also prove to be potential pollutants.

Remember that the mortuary science industry does not conduct its business on a day-to-day basis—the treatment of two bodies per week is considered above average. Thus it should be relatively easy to design a waste-treatment process for use in this industry; some sort of holding tank (anaerobic followed by aerobic) based on the average waste volume per week would suffice. This would give some degree of treatment to these wastes and would possibly destroy some of the microorganisms found in this waste.

A good example of an anti-corrosive material, which could be used in caskets and vaults to avoid future pollution from cemetaries, is aluminum. The use of biodegradable materials in embalming solutions is another good example of an improvement which could be made in this industry.

Suggested Readings in Mortuary Science Industry

History

Bowman, L. E., *The American Funeral Industry*, "A study in guilt, extravagance, and sublimity," introduced by Hary A. Overstreet, Washington: Public Affairs Press (1959).

Dincauze, D. F., "Crematory cemeteries in Eastern Massachusetts," *Peabody Museum*, Cambridge, Mass. (1968).

Iron, P. E., *The Funeral: Vestiqe or Value?* Nashville: Abingdon Press (1966).

Mitford, J., *The American Way Of Death*, New York: Simon & Schuster (1963).

"Undertakers and undertakings—Law & Legislation U.S., funeral homes: legal and business problems," Sam P. Douglass, Chairman, *N.Y. Practicing Law Institute* (1971).

Periodicals

The American Funeral Director, New York, New York.
Casket and Sunnyside, New York, New York.
Mortuary Management, Los Angeles, California.

Legislation

New York State Funeral Directors Manual, Albany, New York.

Abstract Literature

"Funeral Service Abstracts," published annually by the *National Assn. of Colleges of Mortuary Science,* Westport, Conn. 06880.

Pfohl, R. C. "Chemical Industries—The Embalming and Mortuary Science Industry," *A Report for Industrial Wastes Treatment Class—Syracuse University,* January 13, 1972.

28.12 Hospital and Laboratory Wastes

Literature describing hospital and laboratory wastes is not readily available. It is to be expected that the treatment problems posed by these wastes will be unique to the institution involved; therefore, a variety of treatment methods may be needed. Often, where chemicals are used in conjunction with biological contaminants, the chemicals alone may be sufficient to inactivate the biological component.

In one Hungarian hospital a system, where settling followed by a trickling filter is used for waste treatment, was evaluated (Csandy and Zsuzsanna, 1972) to determine whether peak discharges of antibiotics would inhibit the biological component of the filter. It was found that sewage dosed with up to 7 mg/l of streptomycin did not affect treatment, whereas 12 mg/l caused some inhibition of biological action. The effluent used in a South African hospital was found after routine disinfection to contain fewer pathogens than the municipal sewer into which it was being discharged (Grabow, 1972).

References

1. Csanady, M. and D. Zsuzsanna "Trickling filter experiment for purification of antibiotic-containing hospital sewage," *Water Res.* **6**, 1541 (1972).
2. Grabow, W. O. K. and E. M. Nupen, "Load of infectious microorganisms in the wastewater of two South African hospitals," *Water Res.* **6**, 1557 (1972).

28.13 Polychlorinated Biphenyls

Introduction.   Although it is only during the last few years that the public has been made aware of PCBs and their effects, they are the most widespread contaminant in the environment. There are traces of PCB everywhere on earth, but it is concentrated near industrialized areas because of its passage through sewage systems, fluvial transport, and leaching from landfill dumps. Manufactured in this country since 1929 by Monsanto (sole producer), PCBs have a wide range of industrial and domestic applications (see "Origins"). Over 400,000 tons were produced from 1948 to 1973. The first reported trace of PCBs in this country occurred in a study dealing with the sediments off the Santa Barbara coast in 1945.

The toxic effects of PCBs—ranging from death in the lower invertebrates to physiological disturbances in primates and humans, are known. In 1957 millions of chicks in the eastern and midwestern United States were killed by a minute impurity (chlorinated dibenzofuran) produced in the manufacture of PCBs. That same year, industrial workers in Germany who were in contact with PCBs developed inclusion cysts, comedons, pustules, and other chloracne cysts.

In 1967 over 1000 Japanese became ill when they ingested rice oil contaminated with PCBs. Symptoms included eye discharge, weakness, vomiting, intestinal disturbances, weight loss, and skin lesions. After five years the Japanese still showed symptoms of PCB poisoning. Yusho, a disease characterized by brown pigmentation of the skin, nails, and gums, was the most prevalent physiological disturbance in the Japanese newborn. Still births and decreased growth in Japanese boys was also reported.

Other reported physiological disturbance include the swelling of livers and jaundice in humans and primates, enzyme system disturbances, hyperplasia and dysplasia of gastric mucosa (eventual ulceration and hemorrhaging); growth inhibition in hampsters; decrease in immunosuppression in birds and mammals; inhibition of photosynthesis and growth of algae; decrease in reproduction of oysters and shrimp; killing of pelagic birds; premature births in sea-lions.

The long-term effects of PCBs are still unknown. However, PCBs in conjunction with other chemicals known to cause cancer, will have a synergistic effect and will abet a specific chemical in causing cancer at a much lower level than normal. Also, PCBs have the same effect with toxicity levels of other chemicals—that is, the level of the lethal dose is greatly reduced if PCBs are present.

PCBs are always found wherever pesticides are located because PCBs are formed as impurities during the manufacture of many chlororganic pesticides such as parathion, aldrin, dieldrin, eldrin, and DDT. It was recently discovered that DDE, a decomposition product of DDT, further breaks down into DDMU, which is commonly found as a vapor in our atmosphere. In the presence of ultraviolet light, DDMU decomposes into PCBs. Hence, our atmosphere is the global transporter of PCBs either in the vapor phase (air) or liquid phase (rain or snow). It is also the reason why every point on earth is contaminated with PCBs.

Although PCBs can be isolated via reverse osmosis, adsorption, or organic solvent extraction, the only known way to entirely destroy PCBs is through pyrolysis in a special industrial furnace at 1700–2400°F.

Origins. PCBs originate in transformers and power capacitors, hydraulic fluids, diffusion pump oil, heat transfer applications, plasticizers for many flexible products including printing plates, adhesives for manufacture of brake linings, clutch faces, and grinding wheels, manufacture of safety and acoustical glass, lamination of ceramics and metals, washable wall coverings and upholstering materials, adhesives for envelopes and tapes, coatings for ironing board covers, delustering agents for rayon, flame-proofer for synthetic yarns, waterproofing canvas, additives in primers, paints, and varnishes, film casting solutions, sealants for concrete and asphalt, manufacture of printing inks, paper manufacturing, lubricating and cutting oils, aluminum foil, tires, wire insulation, plastic coatings used for food containers, carbonless carbon paper, coatings for thermographic duplicating machines, Xerox toner, toilet soaps, barrier creams, degreasers, waxes for tool and die casting process, insulating tape, water treatment chemicals, coloring compounds, additives to zinc alloys, caulking compounds, manufacture of plastic bottles, by-product in manufacture of pesticides, disintegration or decomposition product of DDT, mothproofing compounds for clothing and carpeting, electric dishwasher detergent, household waxes and cleansing agents, and fluorescent light ballasts.

Note. Since 1971, the use of PCBs in the United States has been limited to the manufacture of transformers and high voltage capacitors. As of 1975, there isn't a substitute for their high dielectric properties, heat resistance (very little decomposition), and non-flammable characteristics. The manufacture, use, or import of PCBs is banned in Sweden (1970) and Japan (1972).

Estimates of Distributions in N. America

Landfill dumps = 300,000 tons
Sediments in rivers and Great Lakes = 20,000 tons
Sediments on Continental Shelf = 10,000 tons
Oceanic waters (N. Atlantic between 26°N and 63°N) = 20,000 metric tons upper 200 meters of water)

Characteristics. PCBs are chlorinated aromatic organic compounds, similar to DDT, which are very stable with no known BOD problem since they cannot be decomposed by bacterial, enzymic, or any other biological or ordinary chemical means. In the environment the half-life is not known because of the stability of the molecule; molecular weight varies according to the amount of chlorination. Solubility in water is very low with the solubility depending on the amount of chlorination—as the chlorination increases, the solubility decreases. PCBs are colorless until mixed with water, then a turbid solution (suspension) is formed. Compounds are very soluble in fats; hence, they are found in adipose tissue. PCBs are toxic (lethal) to lower forms of life and cause physiological problems in higher forms of life.

E.P.A. Standards. Maximum level acceptable in fish = 0.5 ppb, if used for eating purposes by humans.

Treatment
A. *Isolation Techniques*—PCBs can be extracted from water solutions using hexane; absorbed from solutions or vapors by activated charcoal or polymeric resins (Amberlite XAD-4 or XAD-7); or reverse osmosis can be used.
B. *Destruction of PCB molecule*—the only known method of destroying the PCB molecule is through use of pyrolysis in a special industrial furnace at a temperature of 2400°F. (Envirotech Corporation claims 99.9% destruction at a temperature of 1700°F, but EPA disputes the claim).

Common Names of PCBs. Phenoclor, Clophen A50, Therminol FR-L, Kaneclor-400, and Aroclor with following numbers (Aroclor 1016, 1221, 1232, 1242, 1248, 1254, 1260, 1262, 1268). With the Aroclor brand the last two numbers refer to the percentage of chlorination with the exception being #1016 which is 41% chlorinated.

References

1. Ahnoff, M. and B. Josefsson, "Clean up procedures for PCB analysis on river water extracts," *Bull. Environ. Contam. and Toxicology* **13**, 159 (1975).
2. Allen, J. R., "Biological effects of PCB's and PCT's on the subhuman primate," *Environ. Res.* **6**, 344 (1973).
3. Allen, J. R. and D. H. Norback, "PCB and PCT induced gastric mucosal hyperplasia in primates," *Sci.* **179**, 498 (1973).
4. Allen, J. R., "Tissue modification in monkeys as related to absorption, distribution, and excretion of PCBs," *Bull. Environ. Contam. and Toxicology* **2**, 86 March 1974.
5. Anonymous, "PCBs and the environment" *Nat. Tech. Info. Ser.* NTIS Report COM-72-10419 (1972).
6. Anonymous, "Sludge furnace eliminates PCB residuals," *Water and Sewage Works* **119**, 50 (1972).
7. Anonymous, "PCBs—Their use and control," *Org. for Econ. Cooperation and Development*, OECD Report 02046 (1973).
8. Anonymous, "Phenols in refinery wastewater can be oxidized with hydrogen peroxide," *Oil and Gas J.* **73** 84 (1975).
9. Bengtson, S. A. and A. Sodergen, "DDT and PCB residues in airborne fallout and animals in Iceland," *Ambio* **3**, 84 (1974).
10. Bache, C., *et al.*, "PCB residues, accumulation in Cayuga Lake Trout with age," *Sci.* **177**, 1191 (1972).
11. Bedford, J. W., "The use of polyurethane foam plugs for extraction of PCBs from natural waters," **12**, 622 (1974).
12. Berglund, F., "Levels of PCBs in foods," *Environ. Health Perspectives* No. 67 April 1972.
13. Biddleman, T. F. and C. E. Olney, "High volume collection of atmospheric PCBs," *Bull. of Environ. Contam. and Toxicology* **11**, 442 May 1974.
14. Bitman, J., *et al.*, "Biological effects of PCBs in rats and quail," *Environ. Health Perspectives*, No. 1, 145 April 1972.
15. Boyle, R. H., "Of PCB ppms from GE and a SNAFU from EPA and DEC," *Audubon* **77**, 127 (1975).
16. Boyle, R. H., "Those ubiquitous PCBs," *Audobon* **77**, 128 (1975).
17. Bradley, R. L. Jr., "PCBs in man's food—a review," *J. of Milk and Food Techn.* **36**, 155 (1973).
18. Burns, J. E., "Organochlorine pesticide and PCB in biopsied human adipose tissue—Texas 1968–72," *Pesticides Monitoring J.* **7**, 122 (1974).
19. Cecil, H. C., "PCB induced decrease in liver vitamin A in Japanese quail and rats," *Bull. of Environ. Contam. and Toxicology* **9**, 179 (1973).
20. Cherry, R. and L., "What's in the water we drink?," *New York Times Mag.* 38, December 8, 1974.
21. Choi, P. S. K., *et al.*, "Distribution of PCBs in an aerated biological oxidation wastewater treatment system," *Bull. of Environ. Contam. and Toxicology* **11**, 12 (1974).
22. Cole, D. and F. W. Plapp, "Inhibition of growth and photosynthesis in *Chlorella phrenoidosa* by a PCB and several insecticides," *Environ. Entomology* **3**, 217 (1974).
23. Crook, E. H., *et al.*, "Removal and recovery of phenols from industrial waste effluents with Amberlie XAD polymeric adsorbents," *Ind. Eng. Chem.* **14**, 113 (1975).
24. Crosby, D. G. and K. W. Moilanen, "Photodecomposition of chlorinated biphenyls," *Bull. of Environ. Contam. and Toxicology* **10**, 372 (1973).
25. Crossland, J. and K. P. Shea, "The hazard of impurities," *Environ.* **15**, 35 (1973).
26. Cutkomp, L. K., *et al.*, "The sensitivity of fish AT Pases to PCBs," *Environ. Health Perspectives*, No. 1, 165, April 1972.
27. Dahlgren, R. B., *et al.*, "PCBs: Their effects on penned pheasants," *Environ. Health Perspectives*, No. 1, 89, April 1972.
28. Delong, R. L. *et al.*, "Premature sea lion births and organochlorine residues," *Sci.* **181**, 1168 (1973).
29. Dogushi, M. and S. Fukano, "Residue levels of PCT, PCB, and DDT in human blood," *Bull. of Environ. Contam. and Toxicology* **13**, 57 (1975).
30. Dube, D. J., *et al.*, "PCBs in treatment plant effluent," *Water Poll. Contr. Fed.* **46**, 966 (1974).
31. Dunham, R., "Japan's dying—Are we next?," *Environ. Quality* **4**, 14 (1973).
32. Editorial, "Controversy continues over PCBs," *Chem. and Eng. News* **49**, 32 (1971).

33. Edwards, V. H. and P. F. Schubert, "Removal of 2,4D and other persistent organic molecules from water supplies by reverse osmosis," *Am. Water Works Assn.* **66**, 610 (1974).

34. Fine, S. D., "Final environmental impact statement rule-making on PCBs," *FDA Report*, December 18, 1972.

35. Fisher, N. and C. Wurster, "Individual and combined effects of temperature and PCBs on growth of three species of phytoplankton," *Environ. Pollu.* **5**, 205 (1973).

36. Fisher, N. S., "Chlorinated hydrocarbon pollutants and photosynthesis of marine phytoplankton: A reassessment," *Sci.* **189**, 463 (1975).

37. Fries, G. F., "PCB residues in milk of evironmentally and experimentally contaminated cows," *Environ. Health Perspectives* **1**, 55 (1972).

38. Fries, G. F., *et al.*, "Long-term studies of residue retention and excretion by cows fed a PCB (Aroclor 1254)," *J. Agricultural and Food Chemistry* **21**, 117 (1973).

39. Greichus, Y. A., *et al.*, "Insecticides, PCB, and mercury in wild cormorants, pelicans, their eggs, food, and environment," *Bull. of Environ. Contam. and Toxicology* **9**, 321 (1973).

40. Hammer, D. I., "PCB residues in the plasma and hair of refuse workers," *Environ. Health Perspectives*, No. 1, 83, April 1972.

41. Hammond, A. L., "Chemical pollution; PCBs," *Sci.* **175**, 155 (1972).

42. Harvey, G. R., *et al.*, "PCBs in North Atlantic ocean water," *Sci.* **180**, 644 (1973).

43. Harvey, G. R. and W. G. Steinhauer, "Atmospheric transport of PCBs to the North Atlantic," *Atmospheric Environ.* **8**, 777 (1974).

44. Harvey, G. R., *et al.*, "Decline of PCB concentrations in North Atlantic surface water," *Nature* **252**, 387 (1974).

45. Haque, R. and D. Schmedding, "A method of measuring the water solubility of hydrophobic chemicals: Solubility of five PCBs," *Bull. of Environ. Contam. and Toxicology* **14**, 13 (1975).

46. Hesselberg, R. J. and D. D. Scher, "PCB and p,p,DDE in blood of cachectic patients," *Bull. of Environ. Contam. and Toxicology* **11**, 202 (1974).

47. Hom, W. *et al.*, "Deposition of DDE and PCBs in dated sediments of the Santa Barbara Basin," *Sci.* **84**, 1197 (1974).

48. Hoopingarner, R., *et al.*, "PCB interaction with tissue culture cells," *Environ. Health Perspectives*, No. 1, 155, April 1972.

49. Huff, J. E. and J. S. Wasson, "Health hazards from chemical impurities: clorinated dibenzodiozins and chlorinated dibenzofurans," *International J. of Environmental Studies* **6**, 13 (1974).

50. Hutzinger, O., *et al.*, "Photochemical degradation of chlorobiphenyls (PCBs)," *Environ. Health Perspectives* **1**, 15 (1972).

51. Jan, J., *et al.*, "Excretion of some pure PCB isomers in milk of cows," *Bull. of Environ. Contam. and Toxicology* **13**, 313 (1975).

52. Jensen, S., "The PCB story," *Ambio* **1**, 123 (1972).

53. Jensen, S. and G. Sundstrom, "Structures and levels of most chlorobiphenyls in two technical PCB products and in human adipose tissue," *Ambio* **3**, 70 (1974).

54. Johnston, D. W., "PCBs in sea birds from Ascension Island, South Atlantic Ocean," *Bull. of Environ. Contam. and Toxicology* **10**, 368 (1973).

55. Keil, J. E., *et al.*, "PCBs (Acrclor 1242): effects of uptake on E. Coli growth," *Environ. Health Perspectives*, No. 1, 175, April 1972.

56. Kinter, W. B., *et al.*, "Studies on the mechanism of toxicity of DDT and PCBs-disruption of osmorefulation on marine fish," *Environ. Health Perspectives*, No. 1, 169, April 1972.

57. Kolbye, A. C., "Food exposures to PCBs," *Environ. Health Perspectives*, No. 1, 85, April 1972.

58. Kreitzer, J. F. and J. W. Spann, "Tests of pesticidal synergism with young pheasants and Japanese quail," *Bull. of Environ. Contam. and Toxicology* **9**, 250 (1973).

59. Kuratsune, M. and Y. Masuda, "PCBs in noncarbon copy paper," *Environ. Health Perspectives*, **1**, 61 (1972).

60. Lichtenstein, E. P., "PCBs and interactions with insecticides," *Environ. Health Perspectives*, **1**, 151 (1972).

61. Lied, A. J. and D. D. Bills, "Accumulation of dietary PCBs (Aroclor 1254) by rainbow trout," *J. Agr. and Food Chemistry*, **22**, 638 (1975).

62. Litterst, C. L. and E. J. VanLoon, "Teime course of introduction of microsomal enzymes following feeding with PCB," *Bull. of Environ. Contam. and Toxicology*, **11**, 206 (1974).

63. Martell, J. M., *et al.*, "PCBs in suburban watershed, Reston, Virginia," *Environ. Sci. and Techn.*, **9**, 872 (1975).

64. Marx, J. L., "Drinking water: another source of carcinogens?," *Sci.*, **186**, 809 (1974).

65. Maugh, T. H., II, "DDT: an unrecognized source of PCBs," *Sci.*, **180**, 578 (1973).

66. Maugh, T. H., II, "PCBs, still prevalent, but less of a problem," Sci., **178**, 388 (1972).

67. Masuda, Y., *et al.*, "Comparison of PCBs in Yusho patients and ordinary persons," *Bull. of Environ. Contam. and Toxicology*, **11**, 213 (1974).

68. Moore, S. A. Jr. and R. C. Harriss, "Effects of PCB on marine phytoplankton communities," *Nature* **240**, 356 (1972).

69. Mosser, J. E., "Interactions of PCBs, DDT, and DDE in a marine diatom," *Bull. of Environ. Contam. and Toxicology* **12**, 665 (1974).

70. Nisbet, I. C. T. and A. F. Serofim, "Rates and routes of transport of PCBs in the environment," *Environ. Health Perspectives* **1**, 21 (1972).

71. Norback, D. H. and J. R. Allen, "Chlorinated aromatic hydrocarbon induced modifications of the hepatic endoplasmic reticulum: concentrated membrane arrays," *Environ. Health Perspectives* **1**, 137 (1972).

72. Offner, H., *et al.*, "The effect of DDT, Lindane, and Aroclor 1254 on brain cell culture," *Environ. Physiology and Biochem.* **3**, 204 (1973).

73. Peakall, D. B., "PCB occurrence and biological effects," *Residue Rev.* **44**, 1 (1972).

74. Peakall, D. B., *et al.*, "Embryonic mortality and chromosomal alterations caused by Aroclor 1254 in ring doves," *Environ. Health Perspectives* **1**, 103 (1972).

75. Price, A. and R. L. Welch, "Occurrence of PCBs in humans," *Environ. Health Perspectives* **1**, 73 (1972).

76. Risebrough, R. W. and B. deLappe, "Accumulation of PCBs in ecosystems," *Environ. Health Perspectives* **1**, April 1972.

77. Sanders, H. O. and J. H. Chandler, "Biological magnification of a PCB (Arochlor 1254) from water by aquatic invertebrates," *Bull. of Environ. Contam. and Toxicology* **7**, 257 (1972).

78. Savage, E. P., *et al.*, "A search for PCBs in human milk," *Bull. of Environ. Contam. and Toxicology* **9**, 222 (1973).

79. Schoor, W. P., "Problems associated with low solubility compounds in aquatic toxicology tests: Theoretical model and solubility characteristics of Aroclor 1254 in water," *Water Res.* **9**, 937 (1975).

80. Staiff, D. C., *et al.*, "PCB emission from fluorescent lamp ballasts," *Bull. of Environ. Contam. and Toxicology* **12**, 455 (1974).

81. Stalling, D. L. and F. L. Mayer, Jr., "Toxicities of PCBs to fish and environmental residues," *Environ. Health Perspectives* **1**, 159 (1972).

82. Tinker, J., "PCBs at Maendy: Scare or disaster," *New Scientist* **58**, 760 (1973).

83. Trout, P. E., "PCB and the paper industry," *Environ. Health Perspectives* **1**, 63 (1972).

84. Uchiyama, M., "Co-carcinogenic effect of DDT and PCB feedings on methlcholanthrene-induced chemical carcinogenesis," *Bull. of Environ. Contam. and Toxicology* **12**, 687 (1974).

85. Umeda, G., "PCB poisoning in Japan," *Ambio* **1**, 132 (1972).

86. Uthe, J. F., *et al.*, "Field studies on the use of coated porous polyurethane plugs as indwelling monitors of organo-chlorine pesticides and PCB contents of streams," *Environ. Letters* **6**, 103 (1974).

87. Villeneuve, D. C., *et al.*, "Residues of PCBs and PCTs in Canadian and imported cheeses—Canada, 1972," *Pesticides Monitoring J.* **7**, 95 (1973).

88. Sonksen, M. K. and S. P. Busch, "Destruction of PCB contaminated fuel oil in an Aluminum Melting Furnace," Proc. 39 Purdue Ind. Waste Conf., 1984, p. 353.

89. Sonksen, M. K. and J. A. Lease, "Evaluation of Cement Dust Stabilization of PCB Contaminated Sludges," Proc. 37 Purdue Ind. Waste Conf., 1982, p. 405.

90. Fox, L.L. and N.J. Merrick, "Controlling Residual PCB in Wastewater Treatment through Conventional Means," Proc. 37 Purdue Ind. Waste Conf., 1982, p. 413.

28.14 Chloralkali Wastes

Process: Chlorine and caustic soda are produced in an electrolytic cell with a mercury cathode. In the mercury cell, continuously fed brine is partly decomposed in one compartment between a graphite anode and the moving mercury cathode, forming chlorine gas at the anode and sodium amalga at the cathode. The overall reaction is shown as:

$$2NaCl_{(aq.)} + Hg \rightarrow Cl_2 \uparrow + Na(Hg)$$ (See Figure 28.21.)

The sodium amalgam shown in the above reaction is usually decomposed with water to yield hydrogen gas and very pure caustic soda. A typical flow sheet of the entire process is shown in the following Figure 28.20 after MacMullin (2, 1950).

Characteristics: Wastes usually originate from the purification of the caustic soda and in lessening the clogging of the mercury cell diaphragm with a resultant voltage increase. The purification using soda ash and caustic soda removes calcium, iron, and magnesium in the form of sludge which is generally wasted. The other major effluent is the electrolysis plant effluent. The latter is high in total solids (mostly dissolved and inorganic and contains copper, chromium, iron, lead, zinc, manganese, nickel, tin, and relatively high amounts of mercury, about half of which is soluble.)

Modern chloralkali plants are based upon membrane technology, which saves about 25 percent of energy as compared to mercury-based cells (3, 1989). In addition, membrane cells produce caustic soda at 35 percent concentration (compared to 10 percent from diaphragm cells). But, most important from our environmental viewpoint, the mercury contamination (about 10 ppm) of the caustic soda is avoided. Energy conservation and environmental impact have dictated the use of membrane cells. Even the asbestos-containing diaphragms cause problems when disposing of the contaminated separators. For a clearer understanding of the three types of chloralkali production cells in use today, the reader is directed to Figure 28-22.

Treatment: In-plant control of mercury waste as well as reuse of effluents is recommended. Aeration and dechlorination may achieve 50% reduction in mercury as well. Adsorption of mercury on clays or activated carbon has been shown to reduce mercury to about one part per million (Hamza et al. 1, 1983)

References

1. Hamza, A., O.Elsefai, and H.Saleh, "Mercury Removal from a Chloralkali Plant in Egypt," Proc. of the 38th Purdue Univ. Ind. Waste Conf., p. 339, 1983.
2. MacMullin, R. B., "By-Products of Amalgam-Type Chlorine Cell" Chem. Eng. Progress, p. 440, (1950).
3. Wett, Ted, "The Edge in Electrolyzers," *Chem. Business* July/August '89, p. 41.

28.15 Organic Chemicals

Although organic chemicals are discussed in this Chapter in a number of sections, they represent a large enough division to be considered separately in this Section. The organic chemicals, plastics, and synthetic fibers (OCPSF) industry contains many large, diverse and highly complex plants. Most industrial facilities in this class produce these chemicals as their primary

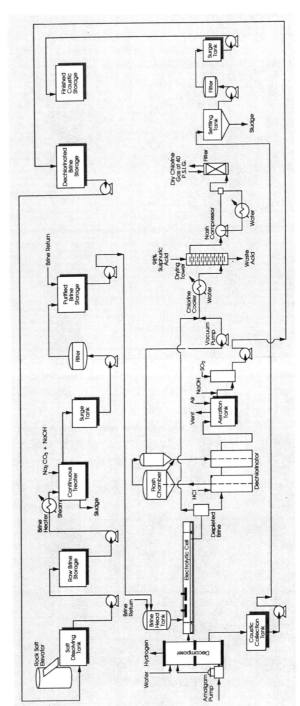

Fig. 28.21 Typical Flow Sheet Amalgam Cell Process for Chlorine. By-product shown is caustic. (After MacMullin, 1950)

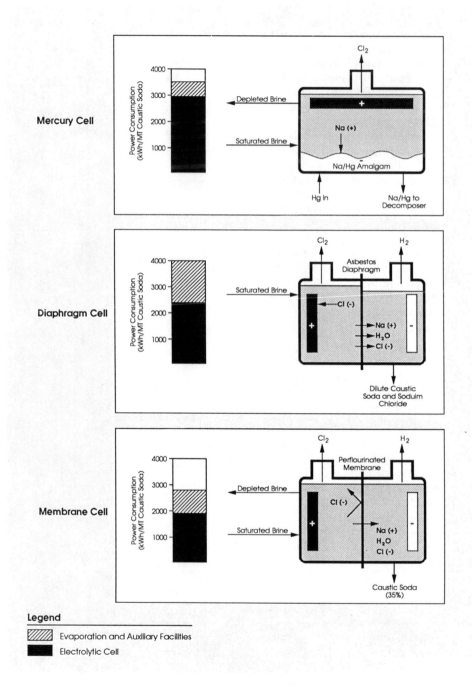

Fig. 28.22   (3, 1989)

function. The remainder are considered secondary plants and include chemical-producing industries such as pharmaceuticals and others found in this and other Chapters. Although over 25,000 different organic chemicals, plastics, and synthetic fibers are manufactured, less than half of these products are produced in excess of 1,000 pounds per year. (E.P.A. 1, 1987).

Different products are made by varying the raw materials, the chemical reaction conditions, and the chemical engineering unit processes. The products being manufactured at a single large chemical plant can vary on a weekly or even daily basis. Thus, a single plant may simultaneously produce many different products using a variety of continuous and batch operations, and the product mix may change on a weekly or daily basis.

Origin

The organic chemicals industry began with the isolation and commercial production of aromatic hydrocarbons (e.g., benzene and toluene and phenolics from coal tar). As more organic compounds possessing valuable properties were identified, commercial production methods for these compounds became desirable. The early products of the chemical industry were dyes, explosives, and pharmaceuticals. The economic incentive to recover and use by-products was a driving force behind the growing synthetic chemicals industry.

The plastics and synthetic fibers industry began later as an outgrowth of the organic chemicals industry. The first commercial polymers, rayon and bakelite, were produced in the early 1900's from feedstocks manufactured by the organic chemicals industry.

Chemicals derived from coal were the principal feedstocks of the early industry, although ethanol, derived from fermentation, was the source of some alipatic compounds. Changing the source of industry feedstocks to less expensive petroleum derivatives lowered prices and opened new markets for organic chemicals, plastics, and synthetic fibers during the 1920's and 1930's. By World War II, the modern organic chemicals and plastics and synthetic fiber industries based on petro-chemicals were firmly established in the United States.

In 1989, the organic chemicals, plastics and synthetic fibers (OCPSF) industry includes production facilities of two distinct types: those whose primary function is chemical synthesis, and those that recover organic chemicals as by-products from unrelated manufacturing operations such as coke plants (steel production) and pulp mills (paper production) described in Chapter 27. The majority of the plants in this industry are plants that process chemical precursors (raw materials) into a wide variety of products for virtually every industrial and consumer market.

Approximately 90 percent (by weight) of the precursors, the primary feedstocks for all of the industry's thousands of products, are derived from petroleum and natural gas. The remaining 10 percent is supplied by plants that recover organic chemicals from coal tar condensates generated by coke production.

E.P.A. relates all the industries listed in the Standard Industrial Classification (SIC) Manual as being matched to production of organic chemicals, plastics, or synthetic fibers and is shown in Figure 28.23.

Wastewater Characteristics

As a result of the wide variety and complexity of raw materials and processes used and of products manufactured in the OCPSF industry, an exceptionally wide variety of pollutants are found in the wastewaters of this industry. This includes conventional pollutants (pH, BOD_5, TSS, and oil and grease); an unusually wide variety of toxic priority pollutants (both metals and organic compounds); and a large number of nonconventional pollutants. Many of the toxic and nonconventional pollutants are organic compounds produced by the industry for sale. Others are created by the industry as by-products of their production operations.

EPA promulgated new source performance standards (NSPS) on the basis of the best available demonstrated technology. NSPS are established for conventional pollutants (BOD_5, TSS, and pH) on the basis of BPT model treatment technology. Priority pollutant limits are based on BAT model treatment technology. These are shown in Table 28.14.

Wastewater Treatment

To control the wide variety of pollutants discharged by the OCPSF industry, OCPSF plants use a broad range of in-plant controls, process modifications, and end-of-pipe treatment techniques. Most plants have implemented programs that combine elements of both in-plant control and end-of-pipe wastewater treatment. The configuration of controls and technologies differs from plant to plant, corresponding to the differing

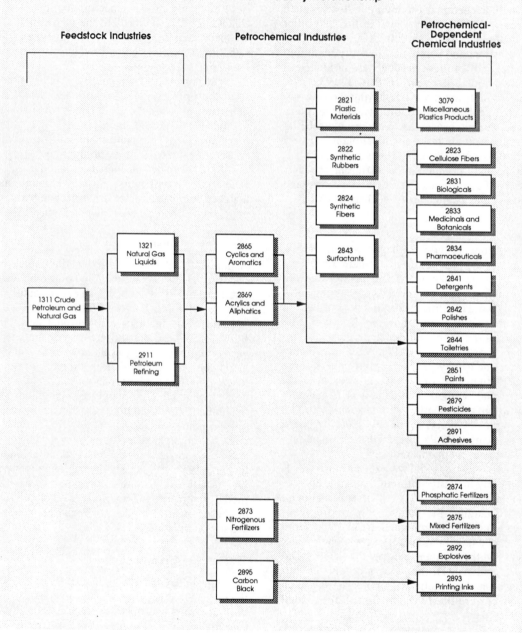

Fig. 28.23   (After EPA, 1, 1987) Relationships Among the SIC Codes Related to the Production of Organic Chemicals, Plastics, and Synthetic Fibers

Table 28-14 (After EPA,1,1987).   Bat Effluent Limitations and NSPS for the End-of-Pipe
Biological Treatment Subcategory

Pollutant Number	Pollutant Name	SAT Effluent Limitations and NSPS[1]	
		Maximum for Any One Day	Maximum for Monthly Average
1	Acenaphthene	59	22
3	Acrylonitrile	242	96
4	Benzene	136	37
6	Carbon Tetrachloride	38	18
7	Chlorobenzene	28	15
8	1,2,4-Trichlorobenzene	140	68
9	Hexachlorobenzene	28	15
10	1,2-Dichloroethane	211	68
11	1,1,1-Trichloroethane	54	21
12	Hexachloroethane	54	21
13	1-1-Dichloroethane	59	22
14	1,1,2-Trichloroethane	54	21
16	Chloroethane	268	104
23	Chloroform	46	21
24	2-Chlorophenol	98	31
25	1,2-Dichlorobenzene	163	77
26	1,3-Dichlorobenzene	44	31
27	1,4-Dichlorobenzene	28	15
29	1,2-Dichloroethylene	25	16
30	1,2-Trans-dichloroethylene	54	21
31	2,4-Dichlorophenol	112	39
32	1,2-Dichloropropane	230	153
33	1,3-Dichloropropene	44	29
34	2,4-Dimethylphenol	36	18
35	2,4-Dinitrotoluene	285	113
36	2,6-Dinitrotoluene	641	255
38	Ethylbenzene	108	32
39	Fluoranthene	68	25
42	Bis(2-Chloroisopropyl)ether	757	301
44	Methylene Chloride	89	40
45	Methyl Chloride	190	86
52	Hexachlorobutadiene	49	20
55	Naphthalene	59	22
56	Nitrobenzene	68	27
57	2-Nitrophenol	69	41
58	4-Nitrophenol	124	72
59	2,4-Dinitrophenol	123	71
60	4,6-Dinitro-o-cresol	277	78
65	Phenol	26	15
66	Bis(2-ethylhexyl)phthalate	279	103
68	Di-n-butyl phthalate	57	27
70	Diethyl phthalate	203	81
71	Dimethyl phthalate	47	19
72	Benzo(a)anthracene	59	22
73	Benzo(a)pyrene	61	23

Continued on next page.

Continued from previous page.

74	3,4-Benzofluoranthene	61	23
75	Benzo(k)fluoranthene	59	22
76	Chrysene	59	22
77	Acenaphthylene	59	22
78	Anthracene	59	22
80	Fluorene	59	22
81	Phenanthrene	59	22
84	Pyrene	67	25
85	Tetrachloroethylene	56	22
86	Toluene	80	26
87	Trichloroethylene	54	21
88	Vinyl Chloride	268	104
119	Total Chromium[2]	2,770	1,110
119	Total Copper[2]	3,380	1,450
121	Total Cyanide[3]	1,200	420
122	Total Land[2]	690	320
124	Total Nickel[2]	3,980	1,690
128	Total Zinc[2,4]	2,610	1,050

[1]All units are micrograms per liter.
[2]Metals limitations apply only to noncomplexed metal-bearing waste streams, including those listed in Table X-4. Discharges of chromium, copper, lead, nickel, and zinc from "complexed metal-bearing process wastewater," listed in Table X-5, are not subject to these limitations.
[3]Cyanide limitations apply only to cyanide-bearing waste streams, including those listed in Table X-3.
[4]Total zinc limitations and standards for rayon fiber manufacture by the viscose process and acrylic fiber manufacture by the zinc chloride/solvent process are 6,796 ug/1 and 3,325 ug/1 for Maximum for Any One Day and Maximum for Montly Average, respectively.

mixes of products manufactured by different facilities. In general, direct discharges treat their wastes more extensively than indirect dischargers.

The predominant end-of-pipe control technology for direct dischargers in the OCPSF industry is biological treatment. The chief forms of biological treatment are activated sludge and aerated lagoons. Other systems, such as extended aeration and trickling filters, are also used, but less extensively. All of these systems reduce biochemical oxygen demand (BOD_5) and total suspended solids (TSS) loadings, and in many instances, incidentally remove toxic and nonconventional pollutants. Biological systems biodegrade some of the organic pollutants, remove bio-refractory organics and metals by sorption into the sludge, and strip some volatile organic compounds (VOCs) into the air. Well-designed biological treatment systems generally incorporate secondary clarification unit operations to ensure adequate control of solids.

Other end-of-pipe treatment technologies used in the OCPSF industry include neutralization, equalization, polishing ponds, filtration, and carbon adsorption. While most direct dischargers use these physical/chemical technologies in conjunction with end-of-pipe biological treatment, many direct dischargers use only physical/chemical treatment.

In-plant control measures employed at OCPSF plants include water reduction and reuse techniques, chemical substitution, and process changes. Techniques to reduce water use include the elimination of water use where practicable, and the reuse and recycling of certain streams, such as reactor and floor washwater, surface runoff, scrubber effluent, and vacuum seal discharges. Chemical substitution is utilized to replace process chemicals possessing highly toxic or refractory properties with others that are less toxic or more amenable to treatment. Process changes include various measures that reduce water use, waste discharges, and/or waste loadings while improving process efficiency. Replacement of barometric condensers with surface condensers, replacement of steam jet ejectors with vacuum pumps, recovery of product or by-product by steam stripping, distillation, solvent extraction or recycle, oil-water separation, and carbon adsorption, and the addition of spill control systems are examples of process changes that have been successfully employed in the OCPSF industry to reduce pollutant loadings while improving process efficiencies.

Another type of control widely used in the OCPSF industry is physical/chemical in-plant control. This treatment technology is generally used selectively on certain process wastewaters to recover products or process solvents, to reduce loadings that may impair the operation of the biological system, or to remove certain pollutants that are not treated sufficiently by the biological system. In-plant technologies widely used in the OCPSF industry include sedimentation/clarification, coagulation, flocculation, equalization, neutralization, oil-water separation, steam stripping, distillation, and dissolved air flotation.

Some OCPSF plants also use physical/chemical treatment after biological treatment. Such treatment is usually intended to reduce solids loadings that are discharged from biological treatment systems. The most common post-biological treatment unit operations are polishing ponds and multimedia filtration. These unit operations are sometimes used in lieu of secondary clarification or to improve upon substandard biological treatment systems. A few plants also use activated carbon after biological treatment as a final "polishing" step.

References

1. Environmental Protection Agency, "Development Document for Effluent Limitations Guidelines and Standards for the Organic Chemicals, Plastics and Synthetic Fibers," PB 88-171335 EPA 440/1-87/009, October 1987.

2. Kremer, J. G. et al., "Regulation of Toxic Organics in Ind. Sewer Discharges at the Sanitation Districts of Los Angeles County," Proc. 42 Purdue Ind. Waste Conf., 1987, P. 347.

3. Kincannon, D. F. et al., "Volatilization of Organics in Activated Sludge Reactors," Proc. 41 Purdue Ind. Waste Conf., 1986, P. 132.

4. Arendt, E. J. et al., "Case History: Rubber Tubing Products Plant Wastewater and RCRA Compliance," Proc. 41 Purdue Ind. Waste Conf., 1986, P. 352.

5. Soderberg, R. W. and R. E. Bockrath, "Treatability of diverse waste streams in the PACT activated carbon biological process," Proc. 39 Purdue Ind. Waste Conf., 1984, P. 121.

6. Powell, R. W. et al., "Wastewater Treatment Facilities for Exxon's New Corporate Research Facilities," Proc. 39 Purdue Ind. Waste Conf., 1984, P. 139.

7. Miner, R. P., "Stauffers' Utilization of Waste Products as Fuel," Proc. 39 Purdue Ind. Waste Conf., 1984, P. 147

8. Marston, K. R. and F. E. Woodard, "Treatment of High Strength Wastewater Containing Organic Solvents," Proc. 39 Purdue Ind. Waste Conf., 1984, P. 735.

9. Anderson, G. A. et al., "Biological Treatability Study of Wastewater from a Nylon Fibers and Plastics Facility," Proc. 38 Purdue Ind. Waste Conf., 1983, P. 201.

10. Kincannon, D. F. et al., "Predicting Treatability of Multiple Organic Priority Pollutant Wastewaters from Single-Pollutant Treatability Studies," Proc. 37 Purdue Ind. Waste Conf., 1982, P. 641.

11. Shaul, G. M. et al., "Treatment of Dye and Pigment Processing Wastewater by the Activated Sludge Process," Proc. 37 Purdue Ind. Waste Conf., 1982, P. 677.

12. Stover, E. L., "Biological Treatability of Specific Organic Compounds Found in Chemical Industry Wastewaters," Proc. 36 Purdue Ind. Waste Conf., 1981, P. 1.

13. Wykpisz, A. C., "Ultrafiltration System for latex Paint Wastewater Treatment," Proc. 35 Purdue Ind. Waste Conf., 1980, P. 416.

14. Van den Berg, L. and C. P. Lentz, "Performance and Stability of the Anaerobic Contact Process as Affected by Waste Composition, Inoculation, and SRT," Proc. 35 Purdue Ind. Waste Conf., 1980, P. 496.

CHAPTER 29

ENERGY INDUSTRIES

Many industrial activities that are seldom considered as contributing to stream pollution may nonetheless, alone or in combination with others, create pollution problems of considerable magnitude. Among these is the generation of electrical power by steam, which is carried out by central utility plants located throughout the country. We shall describe in this chapter the wastes that arise in the operations of industries involved, either directly or indirectly, in the production of power for public and/or private users. All these industries have one trait in common—the use of large volumes of water, mainly for cooling purposes.

Another difficult waste problem arises in the mining of coal, a material frequently used for energy production. Although the problems of acid mine drainage and coal-preparation wastes (the main sources of pollution in mining operations) could have been considered in Chapter 27 on "Materials Industries," they will be treated here, because coal is used almost exclusively for the production of power, whereas oil, as shown previously, has a multitude of other uses.

Steam is also produced by the heat generated by nuclear fission, so that nuclear wastes rightfully belong in this chapter. However, since the operation of nuclear power plants gives rise to a unique and extremely significant disposal problem, the discussion of radioactive power-plant wastes will be postponed to Chapter 30, which presents a unified treatment of the most important disposal problems arising in connection with radioactive wastes regardless of their origin.

29.1 Steam Power Plants

The operation of steam power plants involves the generation of heat from coal, oil, or other fuel to produce steam from exceptionally pure water. The steam is used to drive turbines, which in turn are coupled to generators. After driving the turbine, the steam is condensed and reused as boiler-water feed. Some steam is lost, and hence make-up water is required to balance the water cycle. Fresh water is generally used to cool the steam, but this water is heated by its contact with the steam and is therefore discharged to the receiving steam at an elevated temperature. The nature and extent of the problems connected with discharge of cooling waters vary, depending on the location, the availability, and the type of waterway into which the liquids may be discharged. When the location of generating stations permits disposal of wastes into salt water, the problems are minimized, but not completely eliminated, since the discharge of certain types of contaminated waste waters into coastal waters is prohibited by federal or state statutes.

Thermal pollution. Davidson and Bradshaw [1967,1] report that in coal-fired steam plants 6000 Btu's of heat must be dissipated by means of cooling water in heat-exchangers for every kilowatt-hour of electricity generated. The cooling-water discharges are often from 11° to 17°F higher than the temperature of the water in the stream. Cooling towers are now being required to prevent thermal pollution.

Cadwallader [1965, 2] reports that a single-pass condenser will limit the temperature use to 11°F while a two-pass condenser will give a temperature use of 16°F. Condenser design is dependent on the quantity of cooling water available and the permissible temperature of water returning to the watercourse. He predicts the cooling-water requirements

Table 29.1 Total water circulated through condensers.* (After Cadwallader [1965, 2].)

Year	Rivers	Lakes	Brackish and sea water	Cooling towers	Artificial reservoirs or ponds	Total
1959	12,428	3,254	7,820	2,755	556	26,813
1970	23,842	8,168	16,455	6,635	2185	57,285
1980	42,754	15,260	28,015	15,487	4893	106,409

*Data presented as billions of gallons circulated annually.

in the United States for the period 1959 to 1980 (Table 29.1). The combined increase in demand for river water, lake water, brackish, and sea water for cooling purposes will be approximately 300 per cent, whereas, for the same period, it is predicted that water circulated through cooling towers will increase almost 500 per cent and the use of artificial cooling reservoirs and ponds will increase about 800 per cent.

Skanks [1966, 3] reports that, for the steel industry, mechanical-draft cooling towers provide more efficient and less expensive cooling water than spray ponds and atmospheric towers. Berg [1963, 4] makes a cost evaluation of cooling towers and recommends that, if the costs of once-through and conservation methods differ by only about 30 per cent, a thorough study of cooling-water needs is justified.

Wastes other than cooling waters. Powell [5] presents ten classes of waste materials other than heated cooling water that must be disposed of by power stations:

1) hot, concentrated water salines from boiler and evaporator blowdown;
2) acid and alkaline chemical solutions used in cleaning power-plant equipment;
3) water from blowdown of cooling towers, containing minerals in high concentrations;
4) waste water from washing stack gases;
5) acid and caustic solutions resulting from the regeneration of ion-exchange softeners and demineralization water-treatment plants which are used for conditioning feed water for boilers and evaporators;
6) hot alkaline water from blowing down chemical softening plants;
7) acid water drainage from coal storage;
8) drainage from cinder and ash dumps;
9) sanitary sewage;
10) oil, greases, and miscellaneous solid and liquid wastes.

All boilers must be cleaned both before and during operation. Acid and alkaline solutions as well as special detergents are used as cleaning agents, and are periodically discharged as waste. Powell [5] gives the typical dosage of chemicals used for boiling out a new unit with a capacity of 15,000 gallons of water at the operating level (Table 29.2). Since many boilers are cleaned two or more times each year to remove scale from tubes and boiler-drum surfaces, central stations with many banks of boilers may have a frequently recurring acid-waste problem. Neutralization procedures, such as described in Chapter 25 for acid-waste treatment, are sometimes required for these steam-plant wastes. In normal operation, many boilers are blown down daily, or at least weekly, to eliminate precipitated sludge from the mud drum. This process constitutes a major share of the wastes from power plants.

Table 29.2 Boiler compounds for cleaning new boilers. (After Powell [5].)

Chemical	Dosage
Trisodium phosphate (Na_3PO_4)	600 lb
Sodium carbonate (Na_2CO_3)	250 lb
Sodium hydroxide (NaOH)	250 lb
Sodium sulfite (Na_2SO_3)	38 lb
Sodium nitrate ($NaNO_3$)	160 lb
Detergent	10 gal

Forbes [1967,6] describes pollutants from cooling tower blowdowns and makes suggestions for the control of microbiological as well as chemical agents that contribute to pollution.

Cadwallader [1965, 2] points out that steam-electric generating technology has rapidly advanced since World War II, steam pressures have jumped from 1200 to 3500 psi, and boiler-water quality demands have changed from tolerating contamination in parts per million to parts per billion. Demineralizing ion-exchange resins require acid and alkali regeneration every few days. He suggests treating by neutralization and then controlled release at low rates. Very high-pressure boilers must be cleaned on the average of once a year. The chemical cleaning solutions most commonly used are hydrochloric acid, acetic acid, potassium bromate, ammonia, corrosion inhibitors, detergents, and phosphates. He suggests discharge to an equalization and neutralization basin and slow, controlled-rate discharge into the condenser cooling water as it leaves the station.

Eckenfelder [1966, 7] describes industrial powerhouse water as having the following concentrations (in milligrams per liter): COD, 66; BOD, 10; suspended solids, 50; and condenser water as: COD, 59; BOD, 21; suspended solids, 24.

Rice [1966, 8] proposes using boilers and cooling towers as waste evaporators. He shows that it would cost $121.26 per million lb ($1.01/1000 gallons) to treat boiler-feed water for complete reuse in low-pressure boilers, whereas complete treatment of discharge water would cost $207.33 per million lb ($1.73/1000 gallons). Other cost comparisons are given by Rice.

Hoppe [1964] [9] describes the chemical treatment of cooling-tower water, listing the chemicals used and the waste problem they create, if any. Controlled bleed-off of cooling-tower water will help the waste problem. He suggests regulating bleed-off with a conductivity meter, with the number of cycles of cooling water varying from six to two depending on the hardness of the make-up water.

Some cooling processes require that the water serving as cooling agent be cooled itself prior to use. This precooling is not usually a prerequisite for the condensation of steam from boilers, but cooling waters for air conditioning or other uses within a plant may have to be precooled. Eventually the cooling water will be discharged or it will be reused to a point where it must be treated before further use. According to Riley [10], cooling mechanisms are usually made of wood and designed to produce intimate contact between a down-flowing stream of water and an up-flowing stream of air. The wooden structures contain fans, which move a current of air up through the cooling tower, and pumps, which lift the water to the top of the tower to supply the down-flowing stream of water. (The circulation of these waters through an industrial plant poses two major problems: the loss of water and the excessive quantities of chemicals required for treating these waters.) Cooling towers are primarily used in the absence of an adequate water supply which would make it possible to accomplish the cooling in one single pass (a procedure used by most power plants). The quality of this water must be controlled, for protection against slime and algae growths, corrosion, infiltration of foreign contaminants, and wood deterioration.

Fly ash. Another water-pollution problem is presented by fly ash, a solid waste which results from the use of pulverized coal. Fly ash weighs about 30 lb/ft^3 and constitutes about 10 per cent of the coal burned. These wastes are sometimes collected and deposited in, or on the banks of, races or receiving streams, to be carried away at high water. Jacobs [11] estimates that in 1948 public utilities in the United States discharged a total of about 3,000,000 tons of fly ash. For each carload of coal used per day, the storage area for fly ash amounts to about 2.1 acre·ft per year.

A common method of fly-ash removal from steam power-plant flue gases is by use of a wet scrubber, an electrostatic precipitator, or cyclone separators. In all cases, the fly ash may be transported by pumping in a water slurry. The common method of separation is settling in a decanting basin. Eckenfelder [1966, 7] describes an industrial power plant with waste water containing 6750 mg/liter of fly-ash and suggests lagooning as the method of treatment. An August 1969 article in *Power* reported that a dyke of a fly-ash settling pond of the 700 megawatt Appalachian Power Company station at Carbo, Virginia, broke, releasing a large quantity of alkaline liquor to the Clinch River, with the result that all aquatic life was killed for 125 miles downstream.

A new method of SO_2 removal from coal-fired steam power stations utilizes powdered limestone or dolomite that is injected into the combustion chamber. It reacts with the SO_2 in the hot gas stream to form calcium sulfate ($CaSO_4$) or magnesium sulfate ($MgSO_4$). For a coal with 3 per cent sulfur, the ratio of coal to limestone would be 10:1. A 500-megawatt plant burning 5000 tons of coal per day would produce 465 tons of fly ash per day. Removal of the $CaSO_4$ or $MgSO_4$ from the solution is difficult and researchers suggest reuse of the supernatant liquor from the decanting basin in the wet scrubber. A wet scrubber using a sodium-carbonate solution removes 98 per cent of the particulate matter and 91 per cent of the SO_2. The cost of a 25-megawatt pilot plant is $10 per kilowatt and the operating cost is $1.17 per ton of coal. The main problem with the system is disposal of large quantities of sludge.

Waste treatment. Bennett [11] describes the procedure of disposing of fly ash as follows. Cyclone separators separate the fly ash from gases, water is added to the collected fly ash, and the slurry is pumped to settling or decanting basins. The ash settles, and the water is decanted to the river. Powell [5] disposed of blowdown wastes by constructing a 14-mile sewer line from the plant to the county sewage system. Since the waste was highly acid and alkaline, caused corrosion of equipment, and interfered with the efficiency of biological-treatment units, all waste waters were discharged first into a holding tank and then released to the sewer at a uniform and low rate of flow. Little evidence is currently available of the treatment of other power-plant wastes.

Bender [1967, 12] reports that 20 million tons of fly ash were produced in the United States in 1966, of which 1.25 million tons (about 6 per cent) were put to a useful application. The 45 million tons of fly ash available in 1980 will contain 5 million tons of pure aluminum metal, 8 million tons of iron pellets with 65 per cent iron, 21 million tons of silica and sand, and lesser quantities of oxides of titanium, potassium, sodium, and phosphorus, also uranium and germanium.

Fly-ash may be used as a pozzolana ingredient in Portland cement, for making lightweight aggregate, as a mineral filler in asphalt pavement, as a cement replacement in concrete, as a pozzolana in soil stabilization, as a fill for land development, as a grout in oil wells, as a soil conditioner in agriculture, for the making of bricks, as a filter aid in the vacuum filtration of industrial sludges, as a "choking" material for large, heavy slag roadway base courses, in asphalt roofing and siding materials and as a coagulant in sewage treatment.

Waste characteristics. Gibsen [10] defines "boiler blowdown" as the water containing impurities in considerable concentration which is removed from the boiler circulatory system in order to maintain concentration control. In Gibsen's study, the water removed had a pH of around 11 and total-solids content of about 6000 ppm; however, these characteristics vary with the type of boiler, the operating pressure, and the type of boiler feed used. The blowdown water usually contains some other agents, such as antifoam materials or some type of high-phosphate organic agent. Powell [24] presents a formula and diagrams to show that the concentrations reached may be sufficiently high to constitute a disposal problem of considerable magnitude (Figs. 29.1 and 29.2). He also presents a detailed analysis of raw and concentrated tower-blowdown water (Table 29.3). Fly

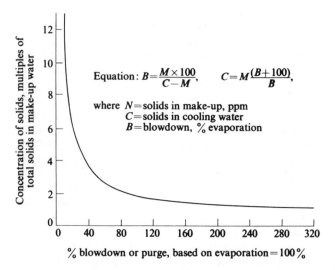

Equation: $B=\dfrac{M\times100}{C-M}$, $C=M\dfrac{(B+100)}{B}$,

where N = solids in make-up, ppm
C = solids in cooling water
B = blowdown, % evaporation

Fig. 29.1 Concentration of solids in water circulated over cooling tower as function of total losses due to windage, leakage, and blowdown. (After Powell [5].)

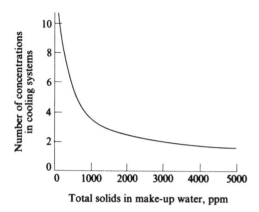

Fig. 29.2 Estimated allowable concentration of make-up water in cooling system for various amounts of total solids to make-up water. Chart based on use of acid when necessary and maintaining a Langelier index of 0.6 ± 0.1. (After Powell [5].)

Table 29.3 Analyses of raw make-up water and concentrated circulating water over cooling towers after corrective treatment with sulfuric acid.

Constituents	As	Make-up water*	Circu-lating water*
Calcium	Ca	84.0	345.0
Magnesium	Mg	30.0	123.0
Sodium	Na	107.0	438.0
Chloride	Cl	85.0	348.0
Sulfate	SO_4	417.5	1710.0
Nitrate	NO_3	0.2	0.8
Bicarbonate	HCO_3	16.0	65.0
Silica	SiO_2	8.3	35.0
Fluoride		0.5	2.0
Boron		0.1	0.4
Total		748.80	3067.2
Hardness, as calcium carbonate			
Calcium		210.0	863.0
Magnesium		124.0	505.0
Total hardness		334.0	1368.0
Methyl orange alkalinity		13.0	53.0
No. concentrations			4.1

*All results are given as parts per million.

ash contains carbon (1 to 60 per cent by weight), iron, aluminum, calcium, magnesium, silica, sulfur, titanium, and phosphorus. The particle-size distribution depends on the manner in which the fuel is burned. Normally, with ordinary stokers, 20 to 40 per cent of the ash has a diameter of less than 10 microns, and 80 to 90 per cent of the ash has a diameter of less than 200 microns.

Effects of power-plant wastes on receiving waters. The present author [1968, 13] has summarized some of the major effects on receiving streams of wastes from steam-generating plants, powered by either nuclear or fossil fuel. These effects are as follows:

1. Increase in temperature, which
   a) lowers the amount of oxygen which can be dissolved in water;
   b) increases bacterial as well as aquatic vertebrate activity, which rapidly diminishes already lowered oxygen resources;
   c) increases the growth rate of microscopic plant life and fish;
   d) may cause fish to hatch earlier in spring, ahead of the availability of organisms on which they feed;
   e) increases sensitivity of aquatic life to toxic elements;
   f) decreases value of water for drinking-water use;
   g) may kill small aquatic crustaceans with the sudden temperature rise as they pass through condensers.

2. Addition of salts to water from boiler compounds (mainly phosphates, carbonates, sulfates, and certain organic compounds), which can
   a) stimulate algae growth;
   b) decrease evaporation rates slightly;
   c) increase hardness of water;
   d) make water more corrosive to boats and home water equipment.

3. Addition of disinfectants, such as chlorine and copper sulfate, to decrease slime formation in cooling waters, which
   a) adds color and taste to receiving water;
   b) reduces bacteria population level.

4. Possible addition of radioactive matter to receiving water, which

  a) may concentrate and thus increase in fish and human food;

  b) may introduce trace amounts in water supply which may be harmful in the long term to people drinking water.

5. Reduction of the deeper, cold-water (hypolimnion) layer of water in a lake which serves as a cooling-water supply, which may

  a) cause mixing of upper layer (epilimnion) and lower layer (hypolimnion), with an overall decrease or increase in oxygen depending on whether the initial concentration of oxygen in the deep hypolimnion layer is withdrawn, heated, and subjected to some aeration before being spread out on the surface;

  b) withdraw stored-up algal nutrients (salts such as phosphates and nitrates) and put them into the epilimnion, where they become readily available for algae and other plant growth;

  c) withdraw accumulated dead, organic matter from the bottom and remove a certain amount of it by entrainment on the pump suction inlet screens (some passing the screens may be oxidized by the chlorine compounds added by plant to control slime growth).

In general, there are five methods of solving thermal pollution problems posed by power plants:

1) management of the resource, by dilution, dispersion, increasing turbulence to increase aeration, and cooling, as by using cooling-water storage ponds;

2) improving efficiency of thermal electric plants, as by the use of closed-circuit (evaporative) cooling;

3) utilization of waste heat, as in process heating, desalting water, heating building, etc.

4) disposal of waste heat to atmosphere, in spray cooling towers or in diversion channels;

5) using new methods of electric power generation, such as non-steam-driven turbines or air-cooled condensers.

References: Steam Power Plants

1. Davidson, B. and R. W. Bradshaw, "Thermal pollution of water systems," *Environ. Sci. Technol.* 1, 619 (1967).

2. Cadwallader, W.W., "Industrial wastewater control," Power, ed., C. F. Gurnham Academic Press, New York, 1965.

3. Shanks, R. I., "Water conservation in heavy industry," *Steel Times* 193, 682 (1966); *Eng. Index*, June 1967, p. 247.

4. Berg, B.R., W. Lane, and I. E. Larson, "Wateruse and related costs with cooling towers," Illinois State Water Survey, Circular No. 86, 1963 Eng. Index, 1965, page 2619.

5. Powell, S. T., "Power generating stations can develop stream pollution problems," *Ind. Eng. Chem.* 46, 112A (1954).

6. Forbes, M. C., "Cooling towers not unmixed blessing in pollution control," *Oil, Gas J.* 65, 88 (1967); *Eng. Index*, December 1967, p. 292.

7. Eckenfelder, W. W., Jr., *Industrial Water Pollution Control*, McGraw-Hill Book Co., New York (1966) p. 25.

8. Rice, J. K. "Water management reduces waste and recovers water in plant effluents," *Chem. Eng.*, 26 September 1966, p. 125.

9. Hoppe, T. C., "Industrial cooling water treatment for minimum pollution from blowdown," *Proc. Am. Power Conf.* 28, 719 (1966); *Chem. Abstr.* 67, 25252 (1967).

10. Hedgepeth, L. L., "1957 industrial wastes forum—solving the cooling tower blowdown pollution problem," *Sewage Ind. Wastes* 30, 539 (1958).

11. Jacobs, H. L., "Fly ash disposal," *Sewage Ind. Wastes* 22, 1207 (1950).

12. Bender, R.J. "Fly ash utilization makes slow progress," Power III 116. (1967).

13. Nemerow, N. L., "Some major effects of nuclear powered steam generating plant wastes on receiving lakes and streams," Paper presented at New York State Society of Professional Engineers's Meeting, Ithaca, N.Y., 15 November 1968.

29.2 Scrubber Wastes

Origin. Wet scrubber control and removal of particulate and gaseous discharges from stacks convert an initial air pollution problem into a water pollution problem. Liquid wastes are created as scrubbing liquid passes through a gaseous flow, entrapping contaminants by interception, impingement, diffusion, or agglomeration. A wet scrubber's effluent water is either recirculated after appropriate treatment; discharged into a municipal sewer system, provided the municipal plant has adequate capacity to facilitate additional wastewater flows generated by a wet scrubber; or treated and discharged into a receiving stream.

Character. The stack-air discharges from fossil-fueled combustion plants that are to be controlled include particulate matter and sulfur oxides. Impingement of water droplets upon particulates entraps and contains them for ultimate disposal. Sulfur oxides are removed from the flue gas by an absorbent such as an alkaline lime substance. Recent experimentation has tried to separate the lime slurry action from the scrubber, thereby eliminating deposition of solids in the scrubber, scaling, or plugging of the mistor. This operation circulates a "clear solution" through the scrubber, with subsequent passage through a lime bath. Treatment of the reactant depends upon the absorbent, such as ammonia, sodium or organic nitrogen compounds employed in the clear solution. Research continues to establish disposal or chemical recovery techniques for the reactants, as well as to determine the efficiency of the split operation (otherwise known as "double alkali process").

Treatment. Scrubber water is subject to solids removal and pH adjustment before recirculation through the system. Solids removal may be achieved by use of a centrifuge, a continuous gravity settler, or merely a settling pond. Selection of the system to be used depends upon the quantity, concentration and character of solids. In addition, the system efficiencies are weighed against capital and operating costs and income from recovered chemicals.

In addition to solids removal and pH adjustment, chemical treatment may be necessary if the scrubber water is discharged into a municipal sewage system or receiving stream. The type and degree of treatment would depend upon (1) the type and capacity of the municipal treatment operation or (2) the nature and best usage of the stream.

As with most treatment operations, ultimate solids disposal must be accounted for. As stated previously, chemical recovery techniques have been considered. Presently, the great quantities of solid to be disposed of limit the feasibility of landfilling. However, the trend is towards using spent limestone quarries or coal mines as disposal sites.

References

1. Accortt, J., Plymley, A. L., and J. R. Martin, "Fine particulate removal and SO_2 absorption with a two stage wet scrubber," *J. Air Poll. Contr. Assn.*, No. 10 (1974).
2. Battelle Memorial Institute, "Applicability of organic liquids to the development of new processes for removing SO_2 from flue gases," March 1969.
3. Corey, R., *Principles and Practices of Incineration*, New York: Wiley (1969).
4. Cross, F. and R. W. Ross, "Effluent water from incinerator flue-gas scrubbers," *Proc. 1968 Nat, Ubcuberatir Conf.*, *ASME*, New York, May 5, 1968.
5. Drake, E. R., G. Hooker, and S. C. Stowe, "Sulphur dioxide recovery methods," *Dow Chemical*, Michigan, January 30, 1945.
6. Eckert, J. S. and R. Stringle, "Performance of wet scrubbers on liquid and solid particulate matter," *J. Air Poll. Contr. Assn.* **24**, No. 10 (1974).
7. Hollinden, G. A. and N. Kaplan, "Status of application of lime-limestone wet scrubbing processes to power plants," Recent Advances in Air Poll. Contr., *A.I.Ch.E.* **70**, (1974).
8. Mannen, L. W., R. E. West, and F. Kreith, "Removal of SO_2 from low sulphur coal combustion gases by limestone scrubbing," *J. Air Poll. Contr. Assn.* **24**, No. 1 (1974).
9. Noyes Data Corp., *Air and gas cleanup equipment*, Park Ridge, New Jersey (1972).
10. Slack, A. V., "Control of SO_2 emissions," *Power Plants and Clean Air—the State of the Air Power Eng. Soc.*, New York, Proc. Winter Meeting, January 1973.
11. Sproull, W. T., *Air pollution and its control*, New York: Exposition Press (1970).

29.3 The Coal Industry

There are two major wastes associated with the production of coal: coal-preparation wastes (coal washeries) and coal-mine drainage wastes (acid mine drainage). Since these wastes and their subsequent effects and treatment are so different, they are usually considered as separate categories; however, in the present context, they will be considered as two different wastes associated with one industry.

Origin of Wastes

Coal-preparation wastes. After coal is mined and brought to the surface, it is processed by "breakers," or preparation plants, where the impurities are removed. Coal solids from a diameter of $3\frac{1}{4}$ inches down to $\frac{3}{64}$ inch are now used by the public; they are described, in order of decreasing size, as egg, stove, chestnut, pea, buckwheat, rice, barley, and No. 4 and No. 5 buckwheat. Both Rickert [1] and Parton [2] describe in considerable detail the series of crushing, screening, classifying, and washing processes involved in the cleaning of coal and the separating of the various sizes. During the cleaning operation, the fine coal becomes suspended in the large quantity of water required for the cleaning. One plant [1] discharges 9000 gpm, approximately 4 per cent of its volume being solids.

Acid mine-drainage wastes. Acid mine-drainage wastes result from the passage of water through mines where iron disulfides, usually pyrites, are exposed to the oxidizing action of air, water, and bacteria. Coal and adjacent rock strata buried in the earth contain sulfur in the form of various compounds. In the process of mining, the sulfuritic materials are uncovered and exposed to air and moisture, with the result that the sulfide oxidizes to ferrous sulfate (copperas) and sulfuric acid, according to the equation

$$2FeS_2 + 7O_2 + 2H_2O \rightarrow 2FeSO_4 + 2H_2SO_4. \quad (1)$$

As the mine is drained, the oxides form additional sulfuric acid in the water. Water finds its way into a strip mine primarily through surface runoff during periods of rainfall. (Ash [3] found that up to 97 per cent of the rainfall on an extensively mined area was pumped out as mine drainage.)

Since the ferrous sulfate in the presence of sulfuric acid is quite resistant to oxidation, the leaching action of the mine water carries off most of the iron and sulfur in this form. After a time, however, there is further oxidation by atmospheric oxygen and from the products of the reaction in Eq. (1), we have

$$2FeSO_4 + O + 2H_2SO_4 \rightarrow Fe_2(SO_4)_3 + H_2O + H_2SO_4. \quad (2)$$

When reactions (1) and (2) are considered to be occurring simultaneously, we have

$$4FeS_2 + 15O_2 + 2H_2O \rightarrow 2Fe_2(SO_4)_3 + 2H_2SO_4. \quad (3)$$

When the sulfuric acid concentration is diluted by receiving streams, the ferric sulfate hydrolyzes as follows:

$$Fe_2(SO_4)_3 + 6H_2O \rightarrow 2Fe(OH)_3 + 3H_2SO_4. \quad (4)$$

It is apparent that all the sulfur which, at the start, was present in the ground as an insoluble sulfate is now dissolved in the receiving stream as sulfuric acid.

There is also much evidence that bacterial activity plays an important role in acid formation [4, 5] 6.]. The sulfur-oxidizing bacterium, *thiobacillus thiooxidans*, an autotrophic bacterium, uses inorganic sulfur, thiosulfate, or tetrathionate for food and manufactures sulfate, which is eventually converted to sulfuric acid. It thus appears that the formation of acid mine water is both a chemical and a biological reaction and may occur in almost any mine where oxidizable sulfur comes into contact with air and water.

Character of Wastes

Coal-preparation wastes. Richert [1] presents typical size and ash analyses of the solids which are discharged from a coal-cleaning plant (Table 29.4). These data indicate that 7.4 per cent of the solids are retained on a 28-mesh sieve (openings of 0.023 inch). The refuse content increases progressively with the increase in the size of the particles and the proportion of clay present. The solids consist of coal, shale, clay, sandstone, bone, and bony coal. In addition to clay, relatively small amounts of other minerals are also present, the more common ones being calcite, gypsum, kaolin, and pyrite. The clays and shales break down into fine particles when subjected to the action of water or when immersed in it, and thus are responsible for the large quantities of semicolloidal particles present in suspension in the washing water. Hall [1965, 7] reports that coal-washing water has as its principal pollutant suspended solids and may contain calcium and magnesium sulfates and iron. It is sometimes acid but is usually kept alkaline to mini-

Table 29.4 Hydroclassifier operating data. (After Richert [1].)

Discharge from cleaning plant	Feed	Underflow	Overflow
Waste water, gpm	9000	560	8400
Solids, %	4.1	35.0	1.0
Solids, tons/hr	91.0	60.0	31.0
Recovery, % waste water	100.0	67.0	33.0
Ash, %	32.3	25.4	42.9

Mesh	Screen analysis, washout		
+10	0.2	0.2	0
+14	0.2	0.3	0
+20	2.4	2.0	0
+28	4.6	5.0	0
+35	7.6	10.0	0.2
+48	9.4	14.5	0.2
+65	9.6	15.0	0.2
+100	10.8	10.5	0.2
+150	8.6	14.5	0.4
+200	7.0	9.0	1.8
+270	2.6	4.0	2.0
−207	37.2	6.0	95.0

Table 29.5 Analysis of two different mine drainage waste waters [8].

Mine drainage analyses	Mine A	Mine B
Drainage flow, gpm	2500	900
pH	3.7	6.2
Free acid, ppm	124	4
Total acid, ppm	466	13
SiO_2, ppm	14	9.6
Al, ppm	17	1.9
Fe, ppm	22*	0.2
Mn, ppm	10	4.6
Ca, ppm	95	34
Mg, ppm	55	12
SO_4, ppm	746	172
Cl, ppm	9	2.4

*Apparently, a considerable amount of iron originally in solution has been precipitated out in suspended form.

mize corrosion of processing equipment. He describes the black solid discharge from coal-preparation plants as finely divided clay, black shale, and other minerals.

Acid mine-drainage wastes. The concentration of acid in mine waters varies widely, from less than 100 ppm to nearly 50,000 ppm, with typical values of about 100 to 6000 ppm of H_2SO_4, 10 to 1500 ppm of $FeSO_4$, 0 to 350 ppm of $Al_2(SO_4)_3$, and 0 to 250 ppm of $MnSO_4$. Mine drainage presents numerous anomalies, and the acidity of the drainage cannot be predicted with certainty. For example, a study of two mines located within a few miles of each other showed that the drainage from the mine whose coal contained 3 per cent of sulfur was 200,000 gal/day of water, with an alkalinity of 170 ppm, whereas the other mine, whose coal had a sulfur content of 2.6 per cent, discharged 130,000 gal/day of water with an acidity of 30,000 ppm [9].

The Bureau of Mines [4] also provides data illustrating the variability of mine-drainage waters (Table 29.5). Mine A has a highly acid drainage,

containing relatively large quantities of iron, other metals, and sulfates, while the drainage of mine B exhibits primarily the characteristics of typical ground water. Hinkle [10, 11] provides representative analyses of mine waters from a West Virginia area that is badly polluted with acid from the soft-coal country (Table 29.6).

Acidic stream water to be used for industrial or domestic purposes will require treatment that will vary depending on the eventual use. Acidity can be overcome by the addition of alkaline materials, but the resultant water is usually "hard" or contains an excess of alkaline materials. Although this water may be softened or treated for the excess of alkaline compounds, the problem is not always eliminated, because an amount of "foaming" and priming occurs as a result of the high chemical content [12]. Table 29.7 shows the characteristics of a Pennsylvania surface-water stream that receives acid mine drainage. From this table the reader may note the following: low pH; high specific conductance; high concentration of iron, aluminum, manganese, and magnesium; high sulfate concentration; high dissolved solids; high degree of hardness, most of which is of noncarbonate origin, which makes its removal by chemicals more expensive; and high acidity. All these observations serve to emphasize the great polluting potential of acid mine waters and the subsequent difficulty for any users of stream water receiving such waste waters.

Table 29.6 Analysis of some West Virginia mine water. (Hinkle [10])

Analysis*	Source of water			
	Pittsburgh seam	Sewickley seam	Upper Freeport	Roof drips, Pittsburgh
pH	2.2	7.7	4.3	8.3
Acidity				
Phenolphthalein	3880	12	40	8
Methyl orange	1420	−440 (alk)	10	−800 (alk)
Methyl red	2420	−408 (alk)	30	−
Total solids	9362	2466	733	3105
Organic and volatile solids	4616	270	69	149
Fixed mineral matter	5746	2196	664	2956
Insoluble matter	−	25	47	14
SiO_2	78	19	42	7
Fe_2O_3	1520	3	8	35
Al_2O_3	231	7	12	4
Mn_2O_3	+	5	0	147
CaO	589	121	156	199
MgO	268	53	60	24
Na_2O	754	829	24	1130
Sulfates as SO_3	4760	987	369	1315
Chloride	28	24	7	6

*All results are given as ppm, except for the pH.

Treatment of Coal-industry Wastes

Coal-preparation wastes. Parton [2] describes three methods of treatment or developments that have helped to reduce the amount of solids reaching the streams:

1) installation of settling and impounding facilities for collecting the fine-size anthracite (silt) in the waste water discharged from the wet washing of coal;
2) increased demand for fine coal, formerly discarded as refuse, which reduces the quantity of solid reaching the waste;
3) adoption of more efficient methods of cleaning the ultrafine sizes, e.g. froth flotation was developed as a method of solving this problem.

Richert and Bishop [1] show a sketch of the treatment plant at the Tamaqua Colliery, which has made use of all three developments (Fig. 29.3). The froth-flotation process is used to clean the coal sludge; i.e., the mineral is floated to the surface by an artificially induced froth. Since it is important that the bubbles be strong enough to support the weight of the floated coal, pine oil—an organic polar compound capable of forming a stable froth on the surface—is used as the frother. Operating results for the primary and secondary (scavenger) flotation units are given in Tables 29.8 and 29.9 respectively.

Sometimes rivers must be dredged to reclaim the "culm," or deposited river coal. Van Ness [13] describes an operation for recovering riverborne anthracite from the Susquehanna River. When the reclaimed coal is cleaned, it becomes steam-size coal, containing 16 per cent ash and 18 per cent moisture and supplying 10,000 Btu's of heat per pound.

Coal-mining wastes. Since 1962, twelve methods of coal-mining wastes treatment have received most consideration.

Table 29.7 Chemical analyses* of a Pennsylvania stream† containing acid mine drainage

Mean discharge, ft³/sec	Temperature, °F	Color	pH	Specific conductance at 25°C, μmhos	Silica	Aluminum	Iron	Manganese	Calcium	Magnesium	Combined sodium and potassium	Bicarbonate	Sulfate	Chloride	Fluoride	Nitrate	Dissolved solids	Hardness as $CaCO_3$ Total	Hardness as $CaCO_3$ Noncarbonate	Total acidity as H_2SO_4
40	48	2	3.70	1840	23	77	0.83	15	139	94	16	0	1170	12	0.3	1.0	1750	1200	1200	580
73.2	58	1	3.60	1430	17	37	0.57	7.2	117	74	13	0	798	7.0	0.1	0.2	1230	660	660	322
59.3	52	2	3.30	1430								0	829		0.1			844	844	308
54.2	66	1	3.70	1690								0	980		0.4			900	900	384
25.1	67	1	3.30	1670	13							0	996		0.2			716	716	536
29.2	67	1	3.25	1340	15	37	1.2	7.2	1000	67	3.0	0	744	7.0	0.1	0.4	1120	779	779	312
41.7	62	4	3.20	1870	21							0	1140	8.0	0.0	0.5		1010	1010	
32.4	62	2	2.60	1930								0	1300	12	0.0	0.1		587	587	690
43.6	60	1	3.50	1470	14	34	0.85	5.7	144	86	6.6	0	899	6.0	0.0	3.5	1430	931	931	248
30.4	68	7	3.30	1590	14							0	1090	4.0				650	650	496
22.3		0	3.15	1940	25	64	1.6	8.2	167	107	1.2	0	1190	21	2.0	1.1	1690	1270	1270	436

*All results are given in parts per million unless otherwise specified.
†Panther Creek at Tamaqua, Pa.

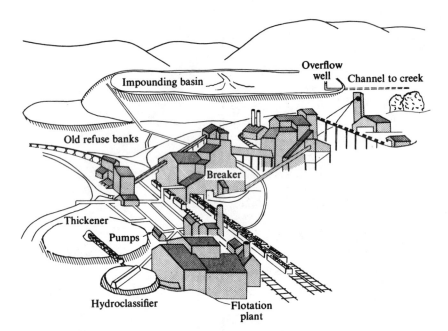

Fig. 29.3 Waste-water treatment at Tamaqua Colliery [1].

1. Thickener. The first step in clarification of coal-washery waste waters is to concentrate the material with hydraulic cyclones or thickeners. The clarified water is returned to the coal-washing circuit, while the high-solids underflow from the thickener may be passed through a drum-type filter or pumped to settling ponds. Clarified water from the pond may be returned to the coal-preparation circuit or discharged to a stream. Many clarification systems are overloaded and the immediate and economic solution is to minimize the amount of water necessary for proper coal preparation. In some cases the fine coals recovered from washery water have economic value but normally such water is disposed of in a permanent impoundment or a specially constructed pit.

2. Settling ponds. Deane [1966] states that waste water from a strip-mine coal-cleaning plant should be cycled through two settling ponds in series and the effluent from the second pond should be reused. Woodley [1967] advocates using large slurry ponds for wash water with reuse. Abandoned strip-mining excavations are excellent slurry ponds. Hummer [1965] describes minus 58-mesh refuse disposal in storage ponds and dam design. Cooley [1944] reports that Virginia limits discharge from coal-preparation plants to 2 mg/liter of settleable solids and

no discoloration. Several hours of sedimentation in a lagoon usually effects removal.

3. Coagulation. Hrebacka [1965] reports that the new mechanical mining techniques produce a great amount of fines (small unusable coal dust) and an increased amount of sludge. He discusses sedimentation with polyelectrolytes and sludge disposal. Fomenko *et al.* [1965] show that clarifier efficiency may be markedly increased by determining the optimum water-slurry system. Rozgaj [1965] found that organic and inorganic coagulants were unsatisfactory when used alone, but combinations of the two gave good settling in 30 minutes. The inorganic coagulants were H_2SO_4, $Ca(OH)_2$, $Al_2(SO_4)_3$, $CaCl_2$, and NaOH. The organic coagulants (polyacrylamides) were Separan 2610, Separan NP1020, and Aerofloc 550. Bochkaren and Baryshnikov [1963] and Olfert [1965] also found polyacrylamide to be an effective coagulant. Green [1966] used a mixture of clay and polyelectrolyte blended in ratios from 10:1 to 2:1 respectively in concentrations of 0.25 to 15 mg/liter to achieve settling of finely divided solids. Johnson [1964] found that free silica and hydrated lime (1 to 5 lb/ton of suspended solids) achieved good coagulation. Kumanomedo and Ogeshima [1966] used alum and either cationic, anionic, or nonionic polymers to

Table 29.8 Operating results, primary flotation circuit [1].

	Feed	Coal	Refuse
Gal/min	1000	280	720
Solids, %	10.0	40.6	9.6
Solids, tons/hr	52.0	33.5	18.5
Recovery, %	100.0	64.5	35.5
Ash, %	30.0	13.0	60.8

Screen and ash analyses

Mesh size	Wash-out	Ash, %	Wash-out	Ash, %	Wash-out	Ash, %
+28	3.7	11.3	5.7	8.8	1.5	14.2
+48	20.7	18.2	21.4	9.2	10.9	32.9
+100	34.0	27.9	34.0	12.8	26.5	60.5
+200	23.2	33.4	24.7	13.9	24.8	65.8
−200	18.4	46.4	14.2	19.5	36.3	70.0
Total	100.0		100.0		100.0	

Table 29.9 Operating results, scavenger flotation circuit [1].

	Feed	Coal	Refuse
Gal/min	360.0	25.0	335.0
Solids, %	15.4	41.5	11.7
Solids, tons/hr	15.2	3.1	12.1
Recovery, %	29.2	6.0	23.2
Ash, %	61.4	24.1	70.7

Screen and ash analyses

Mesh size	Wash-out	Ash, %	Wash-out	Ash, %	Wash-out	Ash, %
+28	3.5	16.9	2.5	14.9	3.5	16.4
+48	14.7	46.4	17.8	15.6	14.2	48.4
+100	31.2	66.1	31.9	24.6	32.2	73.7
+200	28.1	66.6	29.9	26.3	26.7	78.1
−200	22.5	72.1	17.9	30.6	23.4	79.4
Total	100.0		100.0		100.0	

coagulate coal-washery wastes. Day [1965] found that water-soluble polymeric chemicals aid solids settling of coal tailings and clay and effectively produce a reusable water.

4. Froth flotation. Hamilton et al. [1967] summarize improved treatment of coal fines by froth flotation and discuss flotation reagents, bubble contact, cell design, and kinetics studies. Adamson [1965] describes fines treatment by flocculation or froth flotation, depending on the character of the coal and catering to the fraction under 0.5 mm in a 400 to 600 ton/hr coal-cleaning plant. Mistrek [1966] reports using a product of olefin hydroformylation in amounts of 1 to 2.5 kg/ton as a flotation agent. Negulescu et al. [1964] report that a flotation agent rendered coal-washing water too high in oily matter and phenols for discharge and instead it was effectively reused. Roe [1964] advocates recycling flotation agents to decrease dosages required and recommends pressure rather than vacuum filtration of the froth-flotation tailings. Bucklen and Smith [1964] attain 97 per cent coal recovery in an air-lift froth-flotation cyclone. Fernèy [1964] reports on reclamation of fine coal from settling ponds using cyclone froth flotation or centrifuging. Florin [1964] describes drying and sintering

of froth-flotation tailings in a shaft furnace. Sorokin and Tsiperovich [1964] found that methylcyclohexane was an outstanding reagent in coal flotation. Flotation-concentrated coal fines are in one case sent to a dryer. The dryer exhaust goes to a cyclone unit and an impingement wet scrubber to remove ultra-fines.

5. Thickeners and hydrocyclones. Smidt [1967] reports on clarifying and thickening of coal slurry in a new Dorr-type thickener. Visman [1967] describes a multicyclone which produces semi-solid lumps from coal slurries, which may be further compacted into pellets with vibratory screens.

6. Centrifuges. Smidt [1965] reports on the use of a shell, solid-bowl centrifuge, and flocculation agents to dewater flotation refuse mechanically and so to reduce water consumption by 70 per cent. Iwaski et al. [1967] describe a new vacuum and stream centrifuge used to dewater fine suspensions. Llewellyn [1964] found that a continuous solid-bowl centrifuge is practical and economical in dewatering froth-flotation concentrates. Bruk [1965] defined working and construction parameters for coal-washing centrifuges. Eveson and Pickin [1966] use polyelectrolyte coagulant aids for settling bentonite and shale and centrifuged the sludge to produce a cake with 36 per cent moisture.

7. Filtration. Berger [1965] studied continuous rotary vacuum filtration combined with froth flotation for fines separation. Heertjes [1964] studied parameters affecting filtration including weight of particles, their concentration, resistance to filtration, and the use of filter aids. Washburn [1963] describes a closed circuit plant which handles 5000 to 7000 tons per shift. Flotation and flocculants are not used. Underflow from the thickener at 50 per cent solids is dewatered on two 1000-ft² dish filters. Hill *et al.* [1962] discuss use of the filter press to dewater froth-flotation tailings. They suggest using a maximum chamber size of 60 in.² and a ram-type pump for filling and bringing the press up to pressure.

8. Burial. May and Berg [1966] report that segregation and burial of bone and rider coal wastes from strip mining, which contain very acid materials, limit bad effects.

9. "Gob" piles. Davison and Jefferies report that the stability of pit waste heaps is influenced by vegetative cover which depends on nutrition factors. The recovery of quality coal from old anthracite tailings banks eliminates regional eyesores, acid drainage, fills abandoned strip mines, and provides jobs for coal workers [Coal Age **70**, 1965]. To recover coal from spoil banks, they must be 19 per cent coal by weight [*Tech. Digests* **9**, 1967].

10. Atomized water. Brady *et al.* [1967] experimented with removing fines of minus 60-mesh anthracite from a slurry by atomizing water and removing mist in an air current. This process recovers 67 to 79 per cent.

11. Dry cleaning. The use of an electropneumatic concentrating separator to clean all coal particles greater than $\frac{5}{8}$ inch without water would eliminate stream pollution and cleaning of sludge ponds [*Coal Age* **71**, 1967]. Thoma and Geppert [1967] describe a dry-cleaning process for use at the start of the coal-preparation process.

12. Trough separation. Tanaka *et al.* [1966] found that a trough separation unit was effective in separating shale and sandstone in coal refuse.

In a discussion of the problems associated with the treatment of acid mine drainage, Hert [14] recommended the following five control measures:

1) drainage control and diversion of water, to prevent water from entering the mining area, and rapid removal of any water present;
2) proper disposal of sulfur-bearing materials, to ensure that none of the "gob" (sulfuric refuse) comes in contact with water;
3) elimination of slug effects of pumping, i.e., to equalize loading to treatment plants by distributed pumping;
4) sealing terminal activities, actually a process of sealing up abandoned mines to prevent water from entering the sulfur-bearing soil;
5) treatment of mine drainage and in certain circumstances chemical treatment of controlled quantities of drainage from workings, to protect water quality.

Hert also points out that experience has indicated how mine waste-water drainage can be prevented to a great degree by proper advance planning; for example, the control of surface drainage, by protective measures such as barrier dams and so forth, should be worked out prior to opening a mine. After closing down operations, proper measures should be taken to prevent drainage, such as covering stretches of gob roads with packed clay material.

Neutralization is always a possibility, but has thus far been found too costly and presents many difficulties from an operating standpoint. The problem of collecting water for treatment is especially difficult in strip mines, where volumes and flow channels are constantly changing. Theoretically, one ton of either soda ash or limestone, or $\frac{3}{4}$ ton of hydrated lime, is required to neutralize one ton of H_2SO_4 in acid mine water. Braley [15] presents some results of a pilot-plant neutralization treatment. He concluded that, although neutralization is theoretically and physically possible, it is impractical.

After reviewing the research and plant studies, the author concludes that the formation of acid mine water can be prevented by observing the following general rules: keep water out, keep drainage moving, segregate sulfuritic materials, and neutralize acid pools. These four objectives can be realized by the following means. The major method of keeping surface water out is the construction of drainage ditches along the bank above the high wall of the stream. Underground water can be collected in a sump and pumped outside the stripping area. Proper grading keeps water moving so that it will not have a long contact time with the sulfuritic material. The segregation of sulfuritic material, in order to prevent it from contributing to acid mine drainage, can be achieved by keeping floors clean, burying acid-forming refuse, and not leaving coal exposed. A proper job of backfilling or grading the land so that it will shed water will permanently stop or reduce both the formation of acid and the total amount of drainage discharged.

Since 1962, fifteen major methods of preventing and treating acid mine drainage have been proposed.

1. Reducing water inflow. Hall [1965, 7] feels that the most effective method is to minimize contact (both in time and quantity) between water and acid-producing materials, by control of water flow. Flowing streams and surface run-off may be diverted around and away from mining operations by means of high-wall diversion ditches, drains, and conduits to carry flowing waters through or around the mining operation, and by rechanneling or diverting streams away from the mine. Water that does enter should be removed from the mine as quickly as possible. In underground mining, contact between flowing water and acid-producing materials can be minimized by sealing off the surface of the earth above the mine to close cracks, fissures, sinkholes, and other openings when they can be detected, by picking up water as close as possible to the point of entry in the mine, and by conducting it through and out of the mine either in closed conduits or in ditches or sewers that prevent further contact of the water with acid-producing materials.

Conrad [1966] stresses improved mining practices and mine layout to prevent and reduce water flow in mines. Deane [1966] says that water encountered in strip-mine operations should be intercepted by a high wall and drained to diversion ditches. Dye may be used to locate underground seepage [*Coal Age* **69**, 1965].

2. Flooding abandoned mines. Hall [1965, 7] states that flooding abandoned mines is effective in excluding oxygen when the mine is below drainage level and the water does not jeopardize active mines.

Deane [1966] suggests that the final strip-mine cut be covered with water. Water eliminates air and prevents oxidation and acid formation. Several flooded pits in southern Indiana went from pH 2.8 to 5.8 and these waters were used for municipal supplies. Woodley [1967] also proposed flooding Indiana strip mines.

3. Lagooning of acid mine water. Hall [1965, 7] suggests that surface flows from strip mines in areas where annual rainfall and evaporation are equal may be impounded with no discharge to streams. Prolonged lagooning will cause a reduction in the suspended ferric hydroxide and the ferrous sulfate will be oxidized and precipitate out, These lagoons eventually fill with deposits and must be cleaned out or a new lagoon constructed.

Campbell *et al.* [1964] report variable recovery from acid mine drainage in shallow strip-mine lakes. Some in Missouri took 45 years. Alkaline farm run-off and location of "gob" piles were important.

4. Proportioning of acid drainage to streams. Hall [1965, 7] points out that equalization basins and the proportioning of discharge with stream flow helps to minimize downstream effects.

5. Acid mine water used for coal washing. Dillon [1967] describes a coal-cleaning plant where acid mine water is used to wash coal. The raw coal has 30 per cent $CaCO_3$ and $MgCO_3$. The plant treats 600 tons/hr of raw coal using an average of 225 gpm of mine water (about 23 gals/ton). Mine water at pH 3 with 4340 ppm acidity as $CaCO_3$ and 551 ppm Fe (approximately 50 per cent ferrous) is discharged as effluent from the process with a pH of 6.7 to 7.1 and 0 to 1 ppm Fe. Lovell and Reese [1966] studied the use of acid mine water for coal washing and found it amenable in the froth-flotation, flocculation, and agglomeration processes of coal preparation.

6. Neutralization. Deul and Mihok [1967] experimented with neutralization using limestone and lime. They found that waters with low to moderate Fe could be treated by mixing with limestone, which produced a pH of 7 to 8 and < 7 ppm Fe. Waters high in ferrous Fe required a reaction time of less than 15 minutes and second-stage treatment with lime was necessary for short reaction times. They found that the sludges compacted well.

Gerard [1966] in a pilot-plant study shows that lime-neutralization, aeration, sedimentation, and dewatering produces an effluent containing less than 5 ppm Fe and with a pH of 7 to 7.5 but the process is expensive and has high operating costs. Matasov [1967] found that aeration helped the lime-neutralization of H_2SO_4 by promoting the formation of heavier particles of Fe $(OH)_3$ instead of $Fe^-(OH)_2$. Conrad [1967] proposes using an automatic lime feeder with continuous mixing controlled by a pH analyzer. Lunney [1964] stresses strict pH control in lime-neutralization to comply with effluent standards and to minimize the neutralizing agents, undesirable sludges, and plant facilities required. Denby [1965] describes lime-neutralization and sludge removal. Girard and Kaplan [1967] report the costs of lime-neutralization as ranging from $0.07 to $1.09 per 1000 gal or 5.2 cents to $3.25 per ton of coal produced. Hall [1965, 178] reports that neutralization with lime or related alkalis is expensive and waters containing iron must be intimately mixed to prevent deactivation of the alkali particles with an iron-hydroxide precipitate covering.

The Pennsylvania State Health Department has awarded a 55 million dollar contract to neutralize acid mine water from abandoned coal mines near Wilkes Barre [*Power* 112, 1968].

There is available an automatic limer, giving flows from $\frac{1}{2}$ to 100 gpm, which requires no attention other than periodic filling of the lime hopper. Water enters the unit through a 4-inch pipe and is discharged into an overshot water wheel that drives a stirrer in a lime-feed tank and operates a vibrating device to prevent blockage of the lime-feed chamber.

7. Sealing of abandoned mines. Hall [1965] feels that sealing a mine to prevent the entry of air is difficult or impossible because of the permeability of the overlying strata caused by cracking during settlement. Porges *et al.* [1966] also discuss mine sealing. Moebs [1966] reports that the U.S. Bureau of Mines is sealing an abandoned mine with a highly acid drainage, on a trial basis.

8. Covering strip mines and "gob" piles with earth. Woodley [1967] suggests that coal-haulage roads made of acid-producing materials should be covered. He notes that pollution from "gob" piles is extensive.

In one 18-acre site with 30,000 gpd seepage, the following ranges in water quality were measured over a two-year period: pH, 2.8 to 3.6; methyl orange acidity, 284 to 7200 mg/liter; phenolphthalein acidity, 2450 to 12,600 mg/liter; sulfates, 7900 to 30,000 mg/liter; Fe, 760 to 4240 mg/liter; and Mg, 40 to 50 mg/liter. Covering the site with two feet of soil has not decreased either the acidity or the seepage. Hall [1965] states that covering acid-forming materials with earth in strip mines is effective if the materials and earth are compacted to a sufficient density. Deane [1966] suggests that "gob" piles should be covered with soil and planted or, better yet, the spoil should be returned to the active or a nearby inactive mine. He suggests controlling silting by leaving the strip-mine area ungraded to get higher percolation rates and better vegetative growth.

9. Deep-well disposal. Linden and Stefanko [1966] describe subsurface disposal and point out that this method requires a knowledge of the geology of the area and of the chemical and physical properties of the water. The problem of chemical precipitation in the disposal formation must be considered. Dutcher [1966] reports on deep-well disposal in sandstone.

10. Aeration. Rhodes [1967] has patented a process of extended aeration with metallic Fe, which neutralizes the H_2SO_4 and forms insoluble Fe sulfates which are removed by sedimentation. He suggests using metallic Fe from old automobile bodies.

11. Ion exchange. Pollio and Kunin [1967] found that an anion-exchange resin, Amberlite, IRA-68, followed by aeration and clarification was efficient in treating 3000 ppm acidity.

12. Activated sludge. Klappach [1965] found that activated sludge provides adequate treatment.

13. Radiation. Steinberg *et al.* [1968] propose removal of Fe (II) by neutralization with limestone followed by gamma-radiation from Co^{60}. At pH 5.7, with aeration, Fe (II) went from 409 to less than 1 ppm in a short time, since gamma-radiation acts as a catalyst and crystalline precipitate is formed. The process also removes organic pollutants since gamma-radiation oxidizes the material to CO_2 and H_2O. The cost range is $0.05 to $0.25 per 1000 gal using Co^{60} or Cs^{137} as the radiation source.

14. Flash distillation. Westinghouse has a 5-mgd demineralization plant using flash distillation to make "ultra-pure" water from acid mine water [*Environ. Sci. Technol.* **1,** 600, 1967].

15. Modern coal mining. Hall [1965, 7] points out that acid mine drainage from modern coal mining is decreasing, because strip mines are becoming depleted and underground mines are in deeper seams where less water is encountered. The present underground full-seam mining techniques are less likely to cause acid drainage. The mine roof is deliberately collapsed after the coal is removed and subsidence occurs quickly with less void space, causing less breakage of acid-forming materials and generally compacting the earth above the mine. Today most acid-forming materials are brought to the surface with the coal, are separated in the coal-cleaning plant, and properly disposed of in coal-refuse piles.

1. Rickert, E. E., and W. T. Bishop, "Wash water treatment and fine coal recovery," *Ind. Eng. Chem.* 42, 626 (1950).

2. Jones, D. C., "Acid mine water, its control reduces stream pollution," *Mechanization* Part I, 15, 10 (Oct.

3. Ash, S. H. "Acid mine drainage problems" U.S. Bureau of Mines No. 508

4. Colmer, A. R., and M. E. Hinkle, "The role of microorganisms in acid mine drainage: a preliminary report," *Science* 106, 253 (1947).

5. Temple, K. L, and A. R. Colmer, "The autotrophic oxidation of iron by a new bacterium: Thiobacillus ferrooxidans," *J. Bacteriol.* 62, 605 (1951).

6. Temple, K. L., and A. R. Colmer, "The formation of acid mine drainage," *Mining Eng.* 3, 1090 (1951).

7. Hall, E. P., "Coal mining," in *Industrial Waste Water Control*, Academic Press, New York (1965), p. 169.

8. U.S. Bureau of Mines, "Acid mine water in the anthracite region of Pennyslvania," Technical Paper no. 710.

9. Hoffert, J. R., "Acid mine drainage," *Ind. Eng. Chem.* 39, 642 (1947).

10. Hinkle, M. E., and W. A. Koehler, "Investigations of coal mine drainage," West Virginia University, Engineering Experiment Station, Bureau of Coal

11. Hinkle, M. E., and W. A. Koehler, "The action of certain microorganisms in acid mine water," American Institute of Mining and Metallurgical Engineers, Technical Publication no. 2381, 1948.

12. Leitch, R. D., "Acid mine drainage in Western Pennsylvania," U.S. Bureau of Mines, Reports of Investigations, no. 2889, September 1928.

13. Van Ness, B., Jr. "Recovery of river coal," *Ind. Wastes* 1, 232 (1956).

14. Hert, O. H., "Practical control measures to reduce acid mine drainage," in Proceedings of 13th Industrial Waste Conference, Purdue University, May 1958, p. 18

15. Braley, S.A. "A pilot plant Study of the neutralization of acid drainage from bituminous coal mines" Penn. State Health Dept. 1951.

CHAPTER 30

RADIOACTIVE WASTES

Radioactive waste materials are the unwanted by-products formed during nuclear fission, which is induced by bombarding the nuclei of heavy elements (those with a mass number greater than 230) with neutrons. The resulting fission products, isotopes of approximately 30 elements, have mass numbers in the range of 72 to 162, are for the most part solids, and emit beta particles, together with electromagnetic radiation (gamma rays). The emission in turn produces further changes in the identity of the isotopes. This so-called "decay" proceeds according to the laws of monomolecular theory and is measured in terms of half-lives (the time required for half of the radioactive atoms to disentegrate), which vary from seconds to milleniums.

Unlike the alpha and beta emissions, the gamma rays ionize only a little and hence are exceedingly penetrating. Since these radiations are capable of altering the atoms of matter through which they pass and have a cumulative effect in nature, they may cause irreparable damage to living tissue.

Fission is the basic process used in the operation of nuclear reactors (piles), in the production of plutonium for military purposes, or in the conversion of nuclear energy into electricity. Radioactive waste matters are produced regardless of whether fission is controlled, as in the reactor of a power station, or occurs explosively, as in the atom bomb. The chemical separation of fissioned products and their conversion to nuclear fuel elements are the most important sources of radioactive wastes, in terms of level of activity and frequency of occurrence. To eliminate any radiation hazard to man, the disposal of these wastes must follow one of these two general principles: (1) concentrate and contain (applicable to highly active wastes) or (2) dilute and disperse (suitable for large volumes of wastes exhibiting a relatively low level of radioactivity).

To help the reader comprehend the unique problems that have to be solved by industries handling radioactive materials, we shall digress momentarily to present some basic quantitative information on radiation.

The unit frequently encountered in studies of radioactivity is the curie,* which is defined as the quantity of radioactive matter giving 3.70×10^{10} disintegrations per second. For mixed radioisotopes, the maximum permissible concentration (MPC) of radioactive matter in water is set by the Atomic Energy Commission at 10^{-7} microcurie/milliliter, or one tenbillionth of a curie per liter. Since it is expected that by 1980 the activity of fission products produced each year will amount to approximately 100 billion curies, the reader will readily comprehend the magnitude of the problem posed by the treatment of radioactive wastes.

Industries using radioactive matter must accept the responsibility for protecting our biological environment. Waste, whether solid, liquid, or gaseous, which may eventually come in contact with human beings must be rendered harmless to man's biological system. Unfortunately, this is not easily accomplished. Whereas industrial wastes lend themselves to stabilization by chemical, physical, or biological processes, radioactive wastes do not respond sufficiently to any of these methods of treatment. Only time can render radioactive wastes inactive, and in

---

*Until 1948, the curie was officially defined as the quantity of radium emanation (radon) in equilibrium with one gram of radium.

some cases hundreds of years must pass before the wastes are safe for discharge to the environment. Thus, at present, storage appears to be the only way of successfully solving the disposal problem. But if storage, particularly in liquid form, should continue to be the only means of effective disposal, the accumulated solutions of wastes could amount to 200 million gallons by 1980 and more than 2000 million gallons by the year 2000. Although many improved storage systems have been developed through research, only a few systems have proved both economical and practical.

As mentioned before, radioactive wastes may be liquid, gaseous, or solid; furthermore, they may exhibit high or low levels of activity. In the following sections, we shall discuss the origin and disposal of the three types of waste, as well as the different methods of treatment that may be indicated, depending on the amounts of radioactive materials present.

30.1 Origin of Wastes

The most frequent sources of radioactive wastes are the following.

Processing of uranium ores (in mining plants in the southwestern United States) produces considerable volumes of alpha emitters, mainly radium. Storage of these wastes will usually remove suspended radioactive materials.

Laundering of contaminated clothes usually produces large volumes of water containing decreasing concentrations of radioactive materials as the laundry progresses through a series of cycles. The initial wash water generally requires treatment.

Research-laboratory wastes contain all the materials prevalent in chemical, metallurgical, and biological research operations. Some of the wastes are in the form of stable isotopes of the radioactive chemicals. An acceptable method of disposal mixes the particular radioisotope with a quantity of stable isotopes of the same element in a certain ratio. This reduces the danger of contamination, since any living organism will take up the various isotopes of an element in the proportion in which they are present.

Hospitals contribute radioactive wastes from diagnostic and therapeutic uses. Iodine-131 and phosphorus-32 are the radioisotopes which predominate in hospital wastes. Fortunately these possess short half-lives, and simple detention tanks can render them inactive.

Fuel-element processing produces high-level wastes; however, the waste products of secondary operations may exhibit low levels of radioactivity and it is usually preferable to separate the two types. Fuel reprocessing, which also produces high-level wastes, is normally done only at AEC installations. Industry, therefore, will not be involved in this waste-disposal problem. However, the future policy of the government or the AEC may change, and private industry may be allowed to reprocess its own fuel.

Power-plant cooling waters may acquire some radioactivity. The contamination may be due to leaks in the piping caused by corrosion or to neutron bombardment of salts present in the cooling water. The water is stored for some time to permit decay and is then monitored before discharge to a river at a controlled rate.

30.2 Nuclear Power-Plant Wastes

We shall now describe in some detail the wastes encountered and the treatment needed in nuclear power plants and in nuclear fuel processing.

Falk [1] describes a boiling-water reactor which will produce steam at both 1000 psig and 500 psig (Figs. 30.1 and 30.2). The radioactive wastes encountered in reactor operations are to a large extent due to leakage, blowdown, maintenance, refueling, and other mechanisms. Falk states that corrosion products formed in the system (circulating reactor water is used as a source of heat) are the primary source of radioactive isotopes in the reactor water. Those formed outside the reactor vessel, particularly those in the circulating and feed water systems, may be carried into the vessel and become radioactive, along with corrosion products formed within the vessel. Hence it is mandatory that the water used for cooling purposes, as well as that which is used as a source of steam, be ultra-pure. Any salts or other impurities in the water may capture neutrons and become radioactive. Another potential source of radioisotopes in the reactor water is the fission products formed within the fuel elements. It may be possible, but fortunately

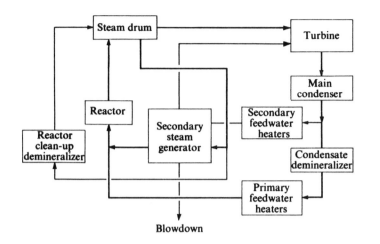

Fig. 30.1 Schematic flow diagram of power plant cycle. (After Falk [1].)

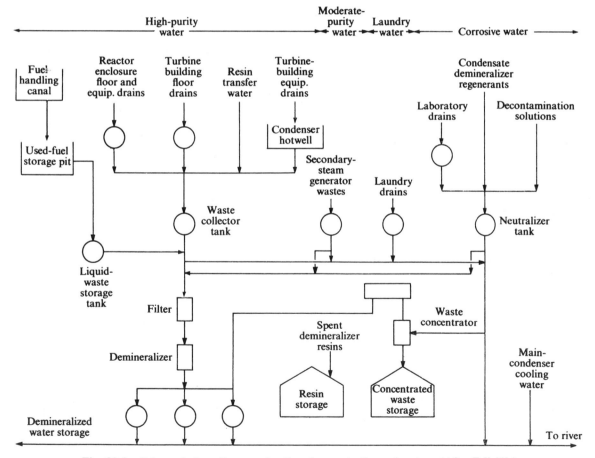

Fig. 30.2 Schematic flow diagram of radioactive waste disposal system. (After Falk [1].)

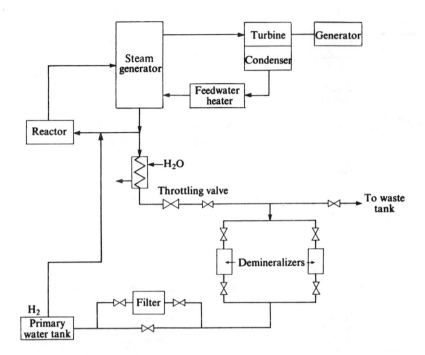

Fig. 30.3 Schematic diagram showing by-pass purification using mixed-bed demineralizers required to maintain high-purity water in a primary system. (After Medin [2].)

not probable, that holes in the cladding of some of the fuel elements could occur, with subsequent release of some fission products to the reactor water. The quantity of radioactive isotopes in the reactor water thus depends on corrosion rates, frequency of failure of fuel-element cladding, and rate of removal by condensate and reactor clean-up demineralizers.

The possible presence of radioactive isotopes in the water necessitates waste-treatment precautions. Medin [2] describes a pressurized-water package power reactor that the army has recently installed at Fort Belvoir, Virginia. In the primary system, great care is taken to maintain the water at a high level of purity to minimize build-up of excessive radioactivity caused by either impurities or corrosion products. No primary water is wasted, but a portion is removed, purified, and recirculated. The purification is accomplished with demineralizers (mixed ion exchangers), a micrometallic filter, and a hold-up tank (Fig. 30.3). Because of the danger of stress corrosion, it is essential that the boiler water contain neither oxygen nor chlorides in any perceptible concentration. Raw water is deaerated and evaporated to reduce the content of chloride and oxygen, which should not exceed 0.3 and 0.03 ppm, respectively.

Table 30.1   Characteristics of solvent extraction processes. (After Blomeke [3].)

Process	Application	Solvent	Salting agent	Approximate volume of untreated high-activity waste*
Purex	Pu, natural	TBP in hydrocarbon	HNO_3	990 gal/metric ton U
Redox	Enriched uranium	Hexone	$Al(NO_3)_3$	1000 gal/metric ton U
Hexone-25	U-Al alloys	TBP in hydrocarbon	$Al(NO_3)_3$	700 liters/kg enriched U
TBP-25	Th-U^{233}	TBP in hydrocarbon	$HNO_3 + Al(NO_3)_3$	670 liters/kg enriched U
Thorex			$Al(NO_3)_3$	1360 gal/metric ton Th

*Volumes refer to the wastes as they come from the extraction column.

30.3 Fuel-processing Wastes

Eight tons of liquid waste result per year from the typical average-size, nuclear reactor. Irradiated reactor fuels are chemically processed to reclaim the unburned nuclear fuel and recover the transmutation products, such as Pu^{239} or U^{233}, from mixtures of the fission products and inert components of the fuel [3]. Several processes in current use are described by Blomeke et al. [3] and are listed in Table 30.1. All involve solvent extraction of the type illustrated in Fig. 30.4.

Generally speaking, all fuel-processing methods are based on similar principles. Solid fuels, which have been stored for about three to four months, are dissolved in HNO_3 and fed to the extraction column (Fig. 30.4), where uranium and plutonium are extracted with the particular solvent chosen. To remove traces of fission products, the extracted uranium and plutonium are scrubbed by an aqueous salt solution added at the top of the extraction column. The waste solutions, which contain more than 99.9 per cent of

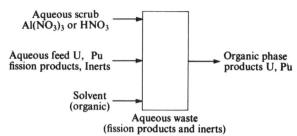

Fig. 30.4 Solvent extraction separation column. (After Blomeke et al. [3].)

the total fission products and inerts in the feed and scrub, leave at the bottom of the column. The actual concentration of both inert and radioactive elements found in fuel-processing wastes depends on the operating characteristics of the particular chemical plant and the particular extraction process used. Blomeke et al. [3] give the characteristics of the high-activity wastes produced by the five common extraction processes (Table 30.2). In addition to the concentration of inert chemicals, all the wastes con-

Table 30.2   Characteristics of current high-activity wastes. (After Blomeke [3].)

Characteristic*	Purex	Redox	Hexone 25	TBP- 25	Thorex
H, M	0.93	− 0.3	− 0.2	1.33	− 0.05
Al, M		1.08	1.6	1.63	0.62
Na, M		0.23			
NH$_4$, M			1.4		
Hg, M			0.01	0.01	0.01
NO$_3$, M	0.93	3.05	6.0	6.2	1.8
F, M					0.039
Cr$_2$O$_7$, M		0.06			
NH$_2$SO$_3$, M				0.04	
Fe, Ni, Cr, gm/liter	< 1	< 1	< 1	< 1	< 1
SiO$_2$			< 1	< 1	
PO$_4$, SO$_4$					< 1
Volume, untreated	990 gal/ton U	1000 gal/ton U	700 liters/kg U	670 liters/kg U	1360 gal/ton Th
Specific gravity	1.03	1.16	1.25	1.25	1.10
Boiling point, °C	101	108	105	105	101
Freezing point, °C	− 3	− 18	− 24	− 24	− 15
Specific heat	0.97	0.78	0.7	0.7	0.85
Volume after evaporation	60 gal/ton U	490 gal/ton U	510 liters/kg U	500 liters/kg U	380 gal/ton Th
Volume after neutralization	80 gal/ton U	830 gal/ton U	860 liters/kg U	840 liters/kg U	640 gal/ton Th

*Chemical composition is exclusive of fission products and heavy elements.

tain fission products and lesser amounts of uranium, plutonium, and other heavy elements. The presence of these heavy elements is attributable mainly to process losses, which usually approximate 0.1 per cent. The wastes obtained by the Purex process [4] are mainly solutions of fission products in nitric acid and their physical properties are essentially those of 1 M HNO$_3$. Trace constituents of iron, nickel, and chromium are present, from corrosion of stainless-steel equipment. These wastes lend themselves quite readily to concentration by evaporation. The fission product ruthenium begins to oxidize as the waste is concentrated. All other extraction-process wastes contain various amounts of aluminum. The Redox waste [5] contains some sodium and some dichromate. Hexone-25 and TBP-25 wastes [6] contain

mercury, and the Thorex waste [7] has both mercury and fluoride. The trace constituents in these wastes include those from products of corrosion, as in the Purex wastes, and also small amounts of SiO$_2$ and other impurities originally present in the fuel.

Blomeke [3] has also tabulated the characteristics of three high-activity power-reactor fuel-processing wastes (Tables 30.3 and 30.4). These wastes, in addition to being radioactive, are corrosive and possess peculiar chemical properties. None of them can be neutralized without precipitation of dissolved salts. For every megawatt·day of irradiation, about 1.1 gm of fission products (total stable and radioactive) is formed. Consequently, in most of the untreated wastes, the total content of fission products, consisting of isotopes of some 37 elements having atomic

Table 30.3 Characteristics of high-activity power reactor fuel-processing wastes. (After Blomeke [3].)

Characteristic*	Zirconium, TBP-25	Stainless steel, TBP-25	Darex, TBP-SS-28
H, M	1.0	3.4	2.0
Al, M	0.75		
Zr, M	0.55		
Fe, M		0.07	0.18
Cr, M		0.02	0.05
Ni, M		0.007	0.018
NO$_3$, M	2.3	2.7	2.3
F, M	3.2		
SO$_4$, M		0.5	
Cr$_2$O$_7$, M	0.01		
Fe, Ni, Cr, gm/liter	< 1		
Cl, gm/liter			< 1
Mn, P, Si, gm/liter		< 1	< 1
Volume, untreated	330 liters/kg U	330 liters/kg U	1760 gal/ton U
Specific gravity	1.2	1.1	1.2
Boiling point, °C	101	106	107
Freezing point, °C	Metastable < 25°C	− 6	− 22
Specific heat		0.84	0.75
Viscosity, cp	2	1.2	1.3
Volume after evaporation			300 gal/ton U (80 gm stainless steel/liter)

*Chemical composition is exclusive of fission products and heavy elements.

Table 30.4   Power reactor fuel-recovery processes. (After Blomeke [3].)

Process	Fuel	Process description
Aqueous HF – TBP-25	High Zr, enriched U	HF dissolution; Zr and F complexing with Al; oxidation of Pu with Cr_2O_7; extraction with TBP
Sulfuric acid – TBP-25	Stainless steel, enriched U	H_2SO_4 dissolution of stainless steel; HNO_3 dissolution of U; extraction with TBP
Darex – Purex	Stainless steel, natural U	Dilute aqua regia dissolution; distillation of HCl; extraction with TBP

weights near that of zinc, amounts to 1 gm/liter. Mixtures of fission products are characterized by the great amount of heat generated.

Culler [8] also lists the physical, chemical, and radiochemical natures of high-level fuel-processing wastes (Table 30.5), and Silverman [9] tabulates some values of inert and radioactive aerosols arising from ore processing (Table 30.6).

Table 30.5   Physical, chemical, and radiochemical characteristics of high-level fuel-processing wastes. (After Culler [8].)

Characteristic	Natural U-Pu		Enriched U-235, diluent-salted
	Salted	HNO_3-salted	
Physical			
Boiling point, °C	120	112	103
Specific gravity	1.18	1.24	1.23
Concentration possible	3	3	2
U^{235} burned (unneutralized), gal/gm	2.5	0.5	5.0
U^{235} burned (neutralized with NaOH), gal/gm	10.0	1.5–2.0	20.0
Chemical			
Acidity, N	− 0.3 (basic)	8.0	− 0.2
Total salts, M			
Unneutralized	1.2	0.2	2.0
Neutralized	6.0	8.2	Acid storage
Solids stability	Unstable basic	Stable	Unstable basic
Radiochemical			
Curies/gal of acid	80	400	2000
Curies/gal of base	20	200	Acid storage
Lead shielding for 1 cm³ acid (in.)*	3.5	4	4
Lead shielding for 500 gal/acid (in.)*	11.5	12	12
Watts/gal of acid	0.3	1.2	5.4

*Inches of lead shielding required for protection.

Table 30.6   Particle size and distribution of some inert and radioactive aerosols. (After Blatz [9].)

Source[*]	Aerosol	Geometric[*] mean diameter, Mg	Mass median[*] diameter, $M'g$	Standard geometric diameter, σ_g
a	Atmospheric dust from 14 U.S. cities, average	0.54 μ	0.97 μ	1.56
b	Atmospheric dust as measured by electron microscope (Knolls Atomic Power Laboratory)	0.028	1.12	3.05
c	Beryllium fluoride fume, BeF_2, from furnace-pouring operation, 10 ft from furnace	0.36	2.3	2.2
d	Iron oxide fume from open-hearth furnace			
	Before waste-heat boiler	0.047	0.65	2.55
	After waste-heat boiler	0.057	0.82	2.60
e	Fission-product source pilot plant			
	Radioactive material measured by light microscope	0.47	2.25	2.17
	Radioactive material measured by electron microscope	0.014	0.14	2.39
	Nonradioactive material measured by light microscope	0.42	11.30	2.86
f	Uranium oxide fume produced by burning scrap	0.12	8.11	3.29
g	Separations process stack effluent during dissolving‡	0.2	3.1	2.6
h	Sodium oxide from burning Na metal§	0.04	0.17	2.0
i	Air-borne dust from incinerator ash disposal	0.43	101	3.86
j	Fume from ferrosilicon electric furnace during tapping	0.43	2.77	2.2

†Conversion to mass or geometric size based on mathematical conversion $\log Mg = \log M'g - 6.91 \log 2\sigma_g$.

*The following references were used:

a. J. E. Ives, *et. al.*, *Atmospheric Pollution of American Cities for the Years 1931 to 1933*, Public Health Bulletin no. 224, U.S. Government Printing Office, Washington, D.C. (1936).

b. J. J. Fitzgerald and C. G. Detwiler, "Collection Efficiency of Air Cleaning and Air Sampling Media," *Am. Ind. Hyg. Assoc. Quart.* **16**, 123 (1955).

c. A. J. Vorwald (ed.), *Pneumoconiosis*, p. 378, Hoeber, New York (1950).

d. C. E. Billings, W. D. Small, and L. Silverman, "Pilot-plant Studies of Continuous Slag-wool Filter for Open-hearth Fume," *J. Air Pollution Control Assoc.* **5**, 159 (1955).

e. J. J. Fitzgerald and C. G. Detwiler, "Size Distribution of Particles Produced by Fission Product Source Pilot Plant," *KAPL* 1232, General Electric Co., Schenectady, N.Y., 1954.

f. E. W. Conners, Jr., and D. P. O'Neil, "Efficiency Studies of a High Efficiency, High Temperature Filter against Freshly Generated Uranium Oxide Fume," *ANL* 5453, Argonne National Laboratory, Lemont, Ill., June 1954.

g. J. J. Fitzgerald, "Evaluation of *KAPL* Separations Process Stack Effluent," *KAPL* 1015, General Electric Co., Schenectady, N.Y., 1952.

h. R. C. Lumatainen and W. J. Mechan, "Removal of Halogens, Carbon Dioxide and Aerosols from Air in a Spray Tower," *ANL* 5429, Argonne National Laboratory, Lemont, Ill., February 1955.

i. W. H. Megonnell, J. H. Ludwig, and L. Silverman, "Dust Exposures during Ash Removal from Incinerators," *Arch. Ind. Health*, **15**, 215 (1957).

j. L. Silverman and R. A. Davidson, "Electric Furnace Ferro-silicon Fume Collection," *J. Metals* (*Trans. AIME*) **203**, 1327 (1955).

‡Obtained by light microscopy; electron microscopy showed a mean size of 0.05.

§Light-field microscopy gave a mean size of 0.2μ (magnification $\times$ 900).

30.4 Treatment of Radioactive Wastes

Any discussion of the treatment of radioactive wastes must be preceded by some information about the extent to which the waste must be treated. The waste engineer is, at this point, well acquainted with the tolerance limits of streams and municipal treatment plants with respect to "usual" types of industrial waste. Because radiation is a unique phenomenon with which the engineer is not too familiar, our attention will be directed to tolerance values pertaining to radioactive wastes. The main threat to human beings arises by contamination (in excess of the critical tolerance value) of the water used for human consumption or recreation, the fish harvested as food, or the edible plants which, irrigated with the contaminated water, have absorbed some of the radioactive substances [10]. The International Commission on Radiation Protection [11] has established limits of maximum permissible concentration (MPC) for the United Nations, and the National Committee for Radiation Protection and Measurement has established similar values for the United States. The latter group has set the MPC for the general population at five roentgens over a 30-year period. According to Eliassen [10], one roentgen (r) is the quantity of radiation which causes one gram of tissue to absorb 97 ergs of energy.

Because of their sensitivity to radiation, the reproductive organs are most seriously affected by exposure to radioactivity. Beadle [12] estimates that the dosage mentioned above will produce 33 mutations (births of abnormal children) per 10,000 births. The MPC's in air and in domestic water supply for most radioelements are listed in *Handbook 52*, from the National Bureau of Standards [13]. *Handbook 59* [14], a supplement to *Handbook 52*, contains data concerning the allowable internal exposure for radiation workers and advises a safety factor of ten for the general population outside the controlled area. The AEC provides the following regulations [14] concerning the disposal of radioactive wastes by discharge into sanitary sewage systems and burial in soil.

20.303 *Disposal by release into sanitary sewerage systems.* No licensee shall discharge licensed material into a sanitary sewerage system unless:
(a) It is readily soluble or dispersible in water; and
(b) The quantity of any licensed or other radioactive material released into the system by the licensee in any one day does not exceed the larger of subparagraphs (1) or (2) of this paragraph:
   (1) The quantity which, if diluted by the average daily quantity of sewage released into the sewer by the licensee, will result in an average concentration equal to the limits specified in Appendix B [Table 30.8] Table I, Column 2 of this part; or
   (2) Ten times the quantity of such material specified in Appendix C [Table 30.9] of this part; and
(c) The quantity of any licensed or other radioactive material released in any one month, if diluted by the average monthly quantity of water released by the licensee, will not result in an average concentration exceeding the limits specified in Appendix B [Table 30.8] Table I, Column 2 of this part; and
(d) The gross quantity of licensed and other radioactive material released into the sewerage system by the licensee does not exceed one curie per year. Excreta from individuals undergoing medical diagnosis or therapy with radioactive material shall be exempt from any limitations contained in this section.

20.304 *Disposal by burial in soil.* No licensee shall dispose of licensed material by burial in soil unless:
(a) The total quantity of licensed and other radioactive materials buried at any one location and time does not exceed, at the time of burial, 1000 times the amount specified in Appendix C [Table 30.9] of this part; and
(b) Burial is at a minimum depth of four feet; and
(c) Successive burials are separated by distances of at least six feet and not more than 12 burials are made in any year.

The *Federal Register* specifies items of records, exceptions, and enforcement, and reprints the Appendixes A, B, and C, containing maximum permissible limits referred to in AEC regulations 20.303 and 20.304 (Tables 30.7 through 30.9). Falk [1] also presents a table of MPC values in water for several typical radioisotopes and adds for comparison the equivalent concentrations expressed in more conventional units (Table 30.10). Lundgren [15] concludes that the doses of radiation allowed in sewers by the AEC will have relatively little effect on bacteria and other microorganisms, or on the functioning of their enzyme systems in waste-disposal plants. Considering the structural organization of disposal plants, the amount of radiation present, and the dosage of radiation that will affect microorganisms, we see that no immediate problem will arise due to hazards of radiation.

Table 30.7   Permissible weekly dose of radiation. (From *Federal Register*, Jan. 29, 1957).

Conditions of exposure		Dose in critical organs (mrem)			
Parts of body	Radiation	Skin, at basal layer of epidermis	Blood-forming organs	Gonads	Lens of eye
Whole body	Any radiation with half-value layer greater than 1 mm of soft tissue	600*	300*	300*	300*
Whole body	Any radiation with half-value layer less than 1 mm of soft tissue	1500	300	300	300
Hands and forearms, or feet and ankles, or head and neck	Any radiation	1500†			

*For exposures of the whole body to X- or gamma rays up to 3 mev, this condition may be assumed to be met if the "air dose" does not exceed 300 mr, provided the dose to the gonads does not exceed 300 mrem. "Air dose" means that the dose is measured by an appropriate instrument in air in the region of highest dosage rate to be occupied by an individual, without the presence of the human body or other absorbing and scattering material.

†Exposure of these limited portions of the body under these conditions does not alter the total weekly dose of 300 mrem permitted to the blood-forming organs in the main portion of the body, to the gonads, or to the lens of the eye.

Treatment of liquid radioactive wastes. Two main procedures have been used for the disposal of liquid wastes: concentration and storage, with subsequent burial, or dilution and dispersal, with subsequent discharge to sewers or streams. A typical treatment procedure is illustrated in Fig. 30.5, which shows in schematic fashion the handling of all liquid wastes. It is interesting to note that the ultimate disposition of all liquid wastes is accomplished by either burying in the earth or dumping in the ocean. Although the ocean was once considered adequate to handle all the radioactive materials we could manufacture, it has recently been proved woefully inadequate as a safe receptacle for the radioactivity we will be discharging by the year 2000.

For example, the AEC, despite recent findings that ocean disposal areas for radioactive wastes have so far been safe, warns that extreme care must be taken in the future, in particular with respect to possible damage to the containers. For that reason, the AEC refused at one time to grant that portion of a license to the Industrial Waste Corporation of Houston, Texas, which would have permitted the organization to use sea disposal for radioactive wastes from hospitals, research facilities, and industry.

A combination of dilution and dispersal is the usual method of disposing of liquid wastes of low activity.

Dilution may have to be preceded by temporary storage to allow sufficient time for the reduction of radioactivity by natural decay. Temporary storage is usually restricted to short-lived isotopes (Na^{22}, P^{32}, I^{131}). Liquid wastes of very low activity are held for some time to permit decay and are then discharged directly to the watercourse. Generally, wastes having activity concentrations greater than 20 $\mu c/cm^3$ are considered too radioactive for temporary storage. There are practical limits to the amount of radioactive waste that can be treated by dilution and discharged. Lieberman [16] recommends the following limits of radioactivity discharged per million gallons of sewage effluent: (1) 1 millicurie of Sr^{90} or Pu^{210}, (2) 100 millicuries of any radioactive material having a half-life of less than 30 days, (3) 10 millicuries of any other radioactive material.

Low-level wastes are considered to be almost exclusively by-products. They are activation and fission products that are usually trapped on ion-exchange resins or filters and are solidified by encasement in concrete or asphalt for burial. Also included as low-level wastes are contaminated paper, clothing, and tools.

Low-level wastes are buried at six federal or state-owned sites in the states of Kentucky, Nevada, South Carolina, Illinois, New York, and Washington. These

Table 30.8 Permissible concentrations in air and water above natural background. (From Code of Federal Regulations Title 10. Revised Jan. 1, 1971.)

Element (atomic number)	Isotope[1]		Table I		Table II	
			Column 1	Column 2	Column 1	Column 2
			Air (μc/ml)	Water (μc/ml)	Air (μc/ml)	Water (μc/ml)
Actinium (89)	Ac 227	S	2×10^{-12}	6×10^{-5}	8×10^{-14}	2×10^{-4}
		I	3×10^{-11}	9×10^{-3}	9×10^{-13}	3×10^{-1}
	Ac 228	S	8×10^{-8}	3×10^{-3}	3×10^{-9}	9×10^{-4}
		I	2×10^{-8}	3×10^{-3}	6×10^{-10}	9×10^{-4}
Americium (95)	Am 241	S	6×10^{-12}	1×10^{-4}	2×10^{-13}	4×10^{-6}
		I	1×10^{-10}	8×10^{-4}	4×10^{-12}	2×10^{-5}
	Am 242m	S	6×10^{-12}	1×10^{-4}	2×10^{-13}	4×10^{-6}
		I	3×10^{-10}	3×10^{-3}	9×10^{-12}	9×10^{-5}
	Am 242	S	4×10^{-8}	4×10^{-3}	1×10^{-9}	1×10^{-4}
		I	5×10^{-8}	4×10^{-3}	2×10^{-9}	1×10^{-4}
	Am 243	S	6×10^{-12}	1×10^{-4}	2×10^{-13}	4×10^{-6}
		I	1×10^{-10}	8×10^{-4}	4×10^{-12}	3×10^{-5}
	Am 244	S	4×10^{-6}	1×10^{-1}	1×10^{-7}	5×10^{-3}
		I	2×10^{-5}	1×10^{-1}	8×10^{-7}	5×10^{-3}
Antimony	Sb 122	S	2×10^{-7}	8×10^{-4}	6×10^{-9}	3×10^{-5}
		I	1×10^{-7}	8×10^{-4}	5×10^{-9}	3×10^{-3}
	Sb 124	S	2×10^{-7}	7×10^{-7}	5×10^{-9}	2×10^{-5}
		I	2×10^{-8}	7×10^{-4}	7×10^{-10}	2×10^{-4}
	Sb 125	S	5×10^{-7}	3×10^{-3}	2×10^{-8}	1×10^{-5}
		I	3×10^{-8}	3×10^{-3}	9×10^{-10}	1×10^{-4}
Argon (18)	A 37	Sub[2]	6×10^{-3}	—	1×10^{-4}	—
	A 41	Sub	2×10^{-6}	—	4×10^{-8}	—
Arsenic (33)	As 73	S	2×10^{-6}	1×10^{-2}	7×10^{-8}	5×10^{-4}
		I	4×10^{-7}	1×10^{-2}	1×10^{-8}	5×10^{-4}
	As 74	S	3×10^{-7}	2×10^{-3}	1×10^{-8}	5×10^{-5}
		I	1×10^{-7}	2×10^{-3}	4×10^{-9}	5×10^{-5}
	As 76	S	1×10^{-7}	6×10^{-4}	4×10^{-9}	2×10^{-5}
		I	1×10^{-7}	6×10^{-4}	3×10^{-9}	2×10^{-5}
	As 77	S	5×10^{-7}	2×10^{-3}	2×10^{-8}	8×10^{-8}
		I	4×10^{-7}	2×10^{-3}	1×10^{-8}	8×10^{-8}
Astatine (85)	At 211	S	7×10^{-9}	5×10^{-3}	2×10^{-10}	2×10^{-6}
		I	3×10^{-8}	2×10^{-3}	1×10^{-9}	7×10^{-5}
Barium (56)	Ba 131	S	1×10^{-6}	5×10^{-3}	4×10^{-8}	2×10^{-4}
		I	4×10^{-7}	5×10^{-3}	1×10^{-8}	2×10^{-4}
	Ba 140	S	1×10^{-7}	8×10^{-4}	4×10^{-9}	3×10^{-3}
		I	4×10^{-8}	7×10^{-4}	1×10^{-9}	2×10^{-5}
Berkelium (97)	Bk 249	S	9×10^{-10}	2×10^{-2}	3×10^{-11}	6×10^{-4}
		I	1×10^{-7}	2×10^{-2}	4×10^{-9}	6×10^{-4}
	Bk 250	S	1×10^{-7}	6×10^{-3}	5×10^{-9}	2×10^{-4}
		I	1×10^{-6}	6×10^{-3}	4×10^{-8}	2×10^{-4}
Beryllium (4)	Be 7	S	6×10^{-6}	5×10^{-2}	2×10^{-7}	2×10^{-3}
		I	1×10^{-6}	5×10^{-2}	4×10^{-8}	2×10^{-3}
Bismuth (83)	Bi 206	S	2×10^{-7}	1×10^{-3}	6×10^{-9}	4×10^{-5}
		I	1×10^{-7}	1×10^{-3}	5×10^{-9}	4×10^{-5}

Table 30.8   (*continued*)

Element (atomic number)	Isotope[1]		Table I		Table II	
			Column 1	Column 2	Column 1	Column 2
			Air (μc/ml)	Water (μc/ml)	Air (μc/ml)	Water (μc/ml)
	Bi 207	S	2×10^{-7}	2×10^{-3}	6×10^{-9}	6×10^{-5}
		I	1×10^{-8}	2×10^{-3}	5×10^{-10}	6×10^{-5}
	Bi 210	S	6×10^{-9}	1×10^{-3}	2×10^{-10}	4×10^{-5}
		I	6×10^{-9}	1×10^{-3}	2×10^{-10}	4×10^{-5}
	Bi 212	S	1×10^{-7}	1×10^{-2}	3×10^{-9}	4×10^{-4}
		I	2×10^{-7}	1×10^{-2}	7×10^{-9}	4×10^{-4}
Bromine (35)	Br 82	S	1×10^{-6}	8×10^{-3}	4×10^{-8}	3×10^{-4}
		I	2×10^{-7}	1×10^{-3}	6×10^{-9}	4×10^{-5}
Cadmium (48)	Cd 109	S	5×10^{-8}	5×10^{-3}	2×10^{-9}	2×10^{-4}
		I	7×10^{-8}	5×10^{-3}	3×10^{-9}	2×10^{-5}
	Cd 115m	S	4×10^{-8}	7×10^{-4}	1×10^{-9}	3×10^{-5}
		I	4×10^{-8}	7×10^{-4}	1×10^{-9}	3×10^{-5}
	Cd 115	S	2×10^{-7}	1×10^{-3}	8×10^{-9}	3×10^{-5}
		I	2×10^{-7}	1×10^{-3}	6×10^{-9}	4×10^{-5}
Calcium (20)	Ca 45	S	3×10^{-8}	3×10^{-4}	1×10^{-9}	9×10^{-8}
		I	1×10^{-7}	5×10^{-3}	4×10^{-9}	2×10^{-4}
	Ca 47	S	2×10^{-7}	1×10^{-3}	6×10^{-9}	5×10^{-5}
		I	2×10^{-7}	1×10^{-3}	6×10^{-9}	3×10^{-5}
Californium (98)	Cf 249	S	2×10^{-12}	1×10^{-4}	5×10^{-14}	4×10^{-6}
		I	1×10^{-10}	7×10^{-4}	3×10^{-12}	2×10^{-8}
	Cf 250	S	5×10^{-12}	4×10^{-4}	2×10^{-13}	1×10^{-5}
		I	1×10^{-10}	7×10^{-4}	3×10^{-12}	3×10^{-5}
	Cf 251	S	2×10^{-12}	1×10^{-4}	6×10^{-14}	4×10^{-6}
		I	1×10^{-10}	8×10^{-4}	3×10^{-12}	3×10^{-5}
	Cf 252	S	2×10^{-11}	7×10^{-4}	7×10^{-13}	2×10^{-5}
		I	1×10^{-10}	7×10^{-4}	4×10^{-12}	2×10^{-5}
	Cf 253	S	8×10^{-10}	4×10^{-3}	3×10^{-11}	1×10^{-4}
		I	8×10^{-10}	4×10^{-3}	3×10^{-11}	1×10^{-4}
	Cf 254	S	5×10^{-12}	4×10^{-6}	2×10^{-12}	1×10^{-7}
		I	5×10^{-12}	4×10^{-6}	2×10^{-12}	1×10^{-7}
Carbon (6)	C 14	S	4×10^{-6}	2×10^{-2}	1×10^{-7}	8×10^{-4}
	(CO_2)	Sub	5×10^{-5}	—	1×10^{-6}	—
Cerium (58)	Ce 141	S	4×10^{-7}	3×10^{-3}	2×10^{-8}	9×10^{-5}
		I	2×10^{-7}	3×10^{-3}	5×10^{-9}	9×10^{-6}
	Ce 143	S	3×10^{-7}	1×10^{-3}	9×10^{-9}	4×10^{-5}
		I	2×10^{-7}	1×10^{-3}	7×10^{-9}	4×10^{-5}
	Ce 144	S	1×10^{-8}	3×10^{-4}	3×10^{-10}	1×10^{-5}
		I	6×10^{-9}	3×10^{-4}	2×10^{-10}	1×10^{-5}
Cesium (55)	Cs 131	S	1×10^{-5}	7×10^{-2}	4×10^{-7}	2×10^{-3}
		I	3×10^{-6}	3×10^{-2}	1×10^{-7}	9×10^{-4}
	Cs 134m	S	4×10^{-5}	2×10^{-1}	1×10^{-6}	6×10^{-3}
		I	6×10^{-6}	3×10^{-2}	2×10^{-7}	1×10^{-3}
	Cs 134	S	4×10^{-8}	3×10^{-4}	1×10^{-9}	9×10^{-6}
		I	1×10^{-8}	1×10^{-3}	4×10^{-10}	4×10^{-5}

Table 30.8 (*continued*)

Element (atomic number)	Isotope[1]		Table I		Table II	
			Column 1	Column 2	Column 1	Column 2
			Air ($\mu c/ml$)	Water ($\mu c/ml$)	Air ($\mu c/ml$)	Water ($\mu c/ml$)
	Cs 135	S	5×10^{-7}	3×10^{-3}	2×10^{-8}	1×10^{-4}
		I	9×10^{-8}	7×10^{-3}	3×10^{-9}	2×10^{-4}
	Cs 136	S	4×10^{-7}	2×10^{-3}	1×10^{-8}	9×10^{-5}
		I	2×10^{-7}	2×10^{-3}	6×10^{-9}	6×10^{-5}
	Cs 137	S	6×10^{-8}	4×10^{-4}	2×10^{-9}	2×10^{-5}
		I	1×10^{-8}	1×10^{-3}	5×10^{-10}	4×10^{-5}
Chlorine (17)	Cl 36	S	4×10^{-7}	2×10^{-3}	1×10^{-8}	8×10^{-5}
		I	2×10^{-8}	2×10^{-3}	8×10^{-10}	6×10^{-5}
	Cl 38	S	3×10^{-6}	1×10^{-2}	9×10^{-8}	4×10^{-4}
		I	2×10^{-6}	1×10^{-2}	7×10^{-8}	4×10^{-4}
Chromium (24)	Cr 51	S	1×10^{-5}	5×10^{-2}	4×10^{-7}	2×10^{-3}
		I	2×10^{-6}	5×10^{-2}	8×10^{-8}	2×10^{-5}
Cobalt (27)	Co 57	S	3×10^{-6}	2×10^{-2}	1×10^{-7}	5×10^{-4}
		I	2×10^{-7}	1×10^{-2}	6×10^{-9}	4×10^{-4}
	Co 58m	S	2×10^{-5}	8×10^{-2}	6×10^{-7}	3×10^{-3}
		I	9×10^{-6}	6×10^{-2}	3×10^{-7}	2×10^{-3}
	Co 58	S	8×10^{-7}	4×10^{-3}	3×10^{-8}	1×10^{-4}
		I	5×10^{-8}	3×10^{-3}	2×10^{-9}	9×10^{-5}
	Co 60	S	3×10^{-7}	1×10^{-3}	1×10^{-8}	5×10^{-5}
		I	9×10^{-9}	1×10^{-3}	3×10^{-10}	3×10^{-3}
Copper (29)	Cu 64	S	2×10^{-6}	1×10^{-2}	7×10^{-8}	3×10^{-4}
		I	1×10^{-6}	6×10^{-3}	4×10^{-8}	2×10^{-4}
Curium (96)	Cm 242	S	1×10^{-10}	7×10^{-4}	4×10^{-12}	2×10^{-5}
		I	2×10^{-10}	7×10^{-4}	6×10^{-12}	3×10^{-5}
	Cm 243	S	6×10^{-12}	1×10^{-4}	2×10^{-13}	5×10^{-6}
		I	1×10^{-10}	7×10^{-4}	3×10^{-12}	2×10^{-5}
	Cm 244a	S	9×10^{-12}	2×10^{-4}	3×10^{-13}	7×10^{-6}
		I	1×10^{-10}	8×10^{-4}	3×10^{-12}	3×10^{-5}
	Cm 245	S	5×10^{-12}	1×10^{-4}	2×10^{-13}	4×10^{-6}
		I	1×10^{-10}	8×10^{-4}	4×10^{-12}	3×10^{-5}
	Cm 246	S	5×10^{-12}	1×10^{-4}	2×10^{-13}	4×10^{-6}
		I	1×10^{-10}	8×10^{-4}	4×10^{-12}	3×10^{-3}
	Cm 247	S	5×10^{-12}	1×10^{-4}	2×10^{-13}	4×10^{-9}
		I	1×10^{-10}	6×10^{-4}	4×10^{-12}	2×10^{-5}
	Cm 248	S	6×10^{-13}	1×10^{-5}	2×10^{-14}	4×10^{-7}
		I	1×10^{-11}	4×10^{-5}	4×10^{-13}	1×10^{-6}
	Cm 249	S	1×10^{-5}	6×10^{-2}	4×10^{-7}	2×10^{-3}
		I	1×10^{-5}	6×10^{-2}	4×10^{-7}	2×10^{-3}
Dysprosium (66)	Dy 165	S	3×10^{-6}	1×10^{-2}	9×10^{-8}	4×10^{-4}
		I	2×10^{-6}	1×10^{-2}	7×10^{-8}	4×10^{-4}
	Dy 166	S	2×10^{-7}	1×10^{-3}	8×10^{-9}	4×10^{-3}
		I	2×10^{-7}	1×10^{-3}	7×10^{-9}	4×10^{-5}
Einsteinium (99)	Es 253	S	8×10^{-10}	7×10^{-4}	3×10^{-11}	2×10^{-5}
		I	6×10^{-10}	7×10^{-4}	2×10^{-11}	2×10^{-5}

Table 30.8 (*continued*)

Element (atomic number)	Isotope[1]		Table I		Table II	
			Column 1	Column 2	Column 1	Column 2
			Air (μc/ml)	Water (μc/ml)	Air (μc/ml)	Water (μc/ml)
	Es 254m	S	5×10^{-9}	5×10^{-4}	2×10^{-10}	2×10^{-5}
		I	6×10^{-9}	5×10^{-4}	2×10^{-10}	2×10^{-5}
	Es 254	S	2×10^{-11}	4×10^{-4}	6×10^{-13}	1×10^{-5}
		I	1×10^{-10}	4×10^{-4}	4×10^{-12}	1×10^{-5}
	Es 255	S	5×10^{-10}	8×10^{-4}	2×10^{-11}	3×10^{-5}
		I	4×10^{-10}	8×10^{-4}	1×10^{-11}	3×10^{-5}
Erbium (68)	Er 169	S	6×10^{-7}	3×10^{-3}	2×10^{-8}	9×10^{-5}
		I	4×10^{-7}	3×10^{-3}	1×10^{-8}	9×10^{-5}
	Er 171	S	7×10^{-7}	3×10^{-3}	2×10^{-8}	1×10^{-5}
		I	6×10^{-7}	3×10^{-8}	2×10^{-8}	1×10^{-4}
Europium (63)	Eu 152 (T/2 = 9.2 hrs)	S	4×10^{-7}	2×10^{-3}	1×10^{-8}	6×10^{-5}
		I	3×10^{-7}	2×10^{-3}	1×10^{-8}	6×10^{-5}
	Eu 152 (T/2 = 13 yrs)	S	1×10^{-8}	2×10^{-8}	4×10^{-10}	8×10^{-5}
		I	2×10^{-8}	2×10^{-3}	6×10^{-10}	8×10^{-5}
	Eu 154	S	4×10^{-9}	6×10^{-4}	1×10^{-10}	2×10^{-5}
		I	7×10^{-9}	6×10^{-4}	2×10^{-10}	2×10^{-5}
	Eu 155	S	9×10^{-8}	6×10^{-3}	3×10^{-9}	2×10^{-4}
		I	7×10^{-8}	6×10^{-3}	3×10^{-9}	2×10^{-4}
Fermium (100)	Fm 254	S	6×10^{-8}	4×10^{-3}	2×10^{-9}	1×10^{-4}
		I	7×10^{-8}	4×10^{-3}	2×10^{-9}	1×10^{-4}
	Fm 255	S	2×10^{-8}	1×10^{-3}	6×10^{-10}	3×10^{-5}
		I	1×10^{-8}	1×10^{-8}	4×10^{-10}	3×10^{-5}
	Fm 256	S	3×10^{-9}	3×10^{-5}	1×10^{-10}	9×10^{-7}
		I	2×10^{-9}	3×10^{-5}	6×10^{-11}	9×10^{-7}
Fluorine (9)	F 18	S	5×10^{-6}	2×10^{-2}	2×10^{-7}	8×10^{-4}
		I	3×10^{-6}	1×10^{-2}	9×10^{-8}	5×10^{-4}
Gadolinium (64)	Gd 153	S	2×10^{-7}	6×10^{-3}	8×10^{-9}	2×10^{-4}
		I	9×10^{-8}	6×10^{-3}	3×10^{-9}	2×10^{-4}
	Gd 159	S	5×10^{-7}	2×10^{-3}	2×10^{-8}	8×10^{-5}
		I	4×10^{-7}	2×10^{-3}	1×10^{-8}	8×10^{-5}
Gallium (31)	Ga 72	S	2×10^{-7}	1×10^{-3}	8×10^{-9}	4×10^{-5}
		I	2×10^{-7}	1×10^{-3}	6×10^{-9}	4×10^{-5}
Germanium (32)	Ge 71	S	1×10^{-5}	5×10^{-2}	4×10^{-7}	2×10^{-3}
		I	6×10^{-6}	5×10^{-2}	2×10^{-7}	2×10^{-3}
Gold (79)	Au 196	S	1×10^{-6}	5×10^{-3}	4×10^{-8}	2×10^{-4}
		I	6×10^{-7}	4×10^{-3}	2×10^{-8}	1×10^{-4}
	Au 198	S	3×10^{-7}	2×10^{-3}	1×10^{-8}	5×10^{-5}
		I	2×10^{-7}	1×10^{-3}	8×10^{-9}	5×10^{-5}
	Au 199	S	1×10^{-6}	5×10^{-3}	4×10^{-8}	2×10^{-4}
		I	8×10^{-7}	4×10^{-3}	3×10^{-8}	2×10^{-4}
Hafnium (72)	Hf 181	S	4×10^{-8}	2×10^{-3}	1×10^{-9}	7×10^{-5}
		I	7×10^{-8}	2×10^{-3}	3×10^{-9}	7×10^{-5}
Holmium (67)	Ho 166	S	2×10^{-7}	9×10^{-4}	7×10^{-9}	3×10^{-5}
		I	2×10^{-7}	9×10^{-4}	6×10^{-9}	3×10^{-5}

Table 30.8 (*continued*)

Element (atomic number)	Isotope[1]		Table I		Table II	
			Column 1	Column 2	Column 1	Column 2
			Air (μc/ml)	Water (μc/ml)	Air (μc/ml)	Water (μc/ml)
Hydrogen (1)	H3	S	5×10^{-6}	1×10^{-1}	2×10^{-7}	3×10^{-3}
		I	5×10^{-6}	1×10^{-1}	2×10^{-7}	3×10^{-3}
		Sub	2×10^{-3}	—	4×10^{-5}	—
Indium (49)	In 113m	S	8×10^{-6}	4×10^{-2}	3×10^{-7}	1×10^{-3}
		I	7×10^{-6}	4×10^{-2}	2×10^{-7}	1×10^{-3}
	In 114m	S	1×10^{-7}	5×10^{-4}	4×10^{-9}	2×10^{-5}
		I	2×10^{-8}	5×10^{-4}	7×10^{-10}	2×10^{-5}
	In 115m	S	2×10^{-6}	1×10^{-2}	8×10^{-8}	4×10^{-4}
		I	2×10^{-6}	1×10^{-2}	6×10^{-8}	4×10^{-4}
	In 115	S	2×10^{-7}	3×10^{-3}	9×10^{-9}	9×10^{-5}
		I	3×10^{-8}	3×10^{-3}	1×10^{-9}	9×10^{-5}
Iodine (53)	I 125	S	5×10^{-9}	4×10^{-5}	8×10^{-11}	2×10^{-7}
		I	2×10^{-7}	6×10^{-3}	6×10^{-9}	2×10^{-4}
	I 126	S	8×10^{-9}	5×10^{-5}	9×10^{-11}	3×10^{-7}
		I	3×10^{-7}	3×10^{-3}	1×10^{-8}	9×10^{-5}
	I 129	S	2×10^{-9}	1×10^{-5}	2×10^{-11}	6×10^{-8}
		I	7×10^{-8}	6×10^{-3}	2×10^{-9}	2×10^{-4}
	I 131	S	9×10^{-9}	6×10^{-5}	1×10^{-10}	3×10^{-7}
		I	3×10^{-7}	2×10^{-3}	1×10^{-8}	6×10^{-5}
	I 132	S	2×10^{-7}	2×10^{-3}	3×10^{-9}	8×10^{-6}
		I	9×10^{-7}	5×10^{-3}	3×10^{-8}	2×10^{-4}
	I 133	S	3×10^{-8}	2×10^{-4}	4×10^{-10}	1×10^{-5}
		I	2×10^{-7}	1×10^{-3}	7×10^{-9}	4×10^{-5}
	I 134	S	5×10^{-7}	4×10^{-3}	6×10^{-9}	2×10^{-5}
		I	3×10^{-6}	2×10^{-2}	1×10^{-7}	6×10^{-4}
	I 135	S	1×10^{-7}	7×10^{-4}	1×10^{-9}	4×10^{-6}
		I	4×10^{-7}	2×10^{-3}	1×10^{-8}	7×10^{-5}
Iridium (77)	Ir 190	S	1×10^{-6}	6×10^{-3}	4×10^{-8}	2×10^{-4}
		I	4×10^{-7}	5×10^{-3}	1×10^{-8}	2×10^{-4}
	Ir 192	S	1×10^{-7}	1×10^{-3}	4×10^{-9}	4×10^{-5}
		I	3×10^{-8}	1×10^{-3}	9×10^{-10}	4×10^{-5}
	Ir 194	S	2×10^{-7}	1×10^{-3}	8×10^{-9}	3×10^{-5}
		I	2×10^{-7}	9×10^{-4}	5×10^{-9}	3×10^{-5}
Ieon (26)	Fe 55	S	9×10^{-7}	2×10^{-2}	3×10^{-8}	8×10^{-4}
		I	1×10^{-6}	7×10^{-2}	3×10^{-8}	2×10^{-3}
	Fe 59	S	1×10^{-7}	2×10^{-3}	5×10^{-9}	6×10^{-5}
		I	5×10^{-8}	2×10^{-3}	2×10^{-9}	5×10^{-5}
Krypton[2] (36)	Kr 85m	Sub	6×10^{-6}	—	1×10^{-7}	—
	Kr 85	Sub	1×10^{-5}	—	3×10^{-7}	—
	Kr 87	Sub	1×10^{-6}	—	2×10^{-8}	—
	Kr 88	Sub	1×10^{-6}	—	2×10^{-8}	—
Lanthanum (57)	La 140	S	2×10^{-7}	7×10^{-4}	5×10^{-9}	2×10^{-5}
		I	1×10^{-7}	7×10^{-4}	4×10^{-9}	2×10^{-5}

Table 30.8   (*continued*)

Element (atomic number)	Isotope[1]		Table I		Table II	
			Column 1	Column 2	Column 1	Column 2
			Air (μc/ml)	Water (μc/ml)	Air (μc/ml)	Water (μc/ml)
Lead (82)	Pb 203	S	3×10^{-6}	1×10^{-2}	9×10^{-8}	4×10^{-4}
		I	2×10^{-6}	1×10^{-2}	6×10^{-8}	4×10^{-4}
	Pb 210	S	1×10^{-10}	4×10^{-6}	4×10^{-12}	1×10^{-7}
		I	2×10^{-10}	5×10^{-3}	8×10^{-12}	2×10^{-4}
	Pb 212	S	2×10^{-8}	6×10^{-4}	6×10^{-10}	2×10^{-5}
		I	2×10^{-8}	5×10^{-4}	7×10^{-10}	2×10^{-5}
Lutetium (71)	Lu 177	S	6×10^{-7}	3×10^{-3}	2×10^{-8}	1×10^{-4}
		I	5×10^{-7}	3×10^{-3}	2×10^{-8}	1×10^{-4}
Manganese (25)	Mn 52	S	2×10^{-7}	1×10^{-3}	7×10^{-9}	3×10^{-5}
		I	1×10^{-7}	9×10^{-4}	5×10^{-9}	3×10^{-5}
	Mn 54	S	4×10^{-7}	4×10^{-3}	1×10^{-9}	1×10^{-4}
		I	4×10^{-8}	3×10^{-3}	1×10^{-9}	1×10^{-4}
	Mn 56	S	8×10^{-7}	4×10^{-3}	3×10^{-8}	1×10^{-4}
		I	5×10^{-7}	3×10^{-3}	2×10^{-8}	1×10^{-4}
Mercury (80)	Hg 197m	S	7×10^{-7}	6×10^{-3}	3×10^{-8}	2×10^{-4}
		I	8×10^{-7}	5×10^{-3}	3×10^{-8}	2×10^{-4}
	Hg 197	S	1×10^{-6}	9×10^{-3}	4×10^{-8}	3×10^{-4}
		I	3×10^{-6}	1×10^{-2}	9×10^{-8}	5×10^{-4}
	Hg 203	S	7×10^{-8}	5×10^{-4}	2×10^{-9}	2×10^{-5}
		I	1×10^{-7}	3×10^{-3}	4×10^{-9}	1×10^{-4}
Molybdenum (42)	Mo 99	S	7×10^{-7}	5×10^{-3}	3×10^{-8}	2×10^{-4}
		I	2×10^{-7}	1×10^{-3}	7×10^{-9}	4×10^{-5}
Neodymium (60)	Nd 144	S	8×10^{-11}	2×10^{-3}	3×10^{-12}	7×10^{-5}
		I	3×10^{-10}	2×10^{-3}	1×10^{-11}	8×10^{-5}
	Nd 147	S	4×10^{-7}	2×10^{-3}	1×10^{-8}	6×10^{-5}
		I	2×10^{-7}	2×10^{-3}	8×10^{-9}	6×10^{-5}
	Nd 149	S	2×10^{-6}	8×10^{-3}	6×10^{-8}	3×10^{-4}
		I	1×10^{-6}	8×10^{-3}	5×10^{-8}	3×10^{-4}
Neptunium (93)	Np 237	S	4×10^{-12}	9×10^{-5}	1×10^{-13}	3×10^{-6}
		I	1×10^{-10}	9×10^{-4}	4×10^{-12}	3×10^{-5}
	Np 239	S	8×10^{-7}	4×10^{-3}	3×10^{-8}	1×10^{-4}
		I	7×10^{-7}	4×10^{-3}	2×10^{-8}	1×10^{-4}
Nickel (28)	Ni 59	S	5×10^{-7}	6×10^{-3}	2×10^{-8}	2×10^{-4}
		I	8×10^{-7}	6×10^{-2}	3×10^{-8}	2×10^{-3}
	Ni 63	S	6×10^{-8}	8×10^{-4}	2×10^{-9}	3×10^{-5}
		I	3×10^{-7}	2×10^{-2}	1×10^{-8}	7×10^{-4}
	Ni 65	S	9×10^{-7}	4×10^{-3}	3×10^{-8}	1×10^{-4}
		I	5×10^{-7}	3×10^{-3}	2×10^{-8}	1×10^{-4}
Niobium (Columbium) (41)	Nb 93m	S	1×10^{-7}	1×10^{-2}	4×10^{-9}	4×10^{-4}
		I	2×10^{-7}	1×10^{-2}	5×10^{-9}	4×10^{-4}
	Nb 95	S	5×10^{-7}	3×10^{-3}	2×10^{-8}	1×10^{-4}
		I	1×10^{-7}	3×10^{-3}	3×10^{-9}	1×10^{-4}
	Nb 97	S	6×10^{-6}	3×10^{-2}	2×10^{-7}	9×10^{-4}
		I	5×10^{-6}	3×10^{-2}	2×10^{-7}	9×10^{-4}

Table 30.8 (*continued*)

Element (atomic number)	Isotope[1]		Table I Column 1 Air ($\mu c/ml$)	Table I Column 2 Water ($\mu c/ml$)	Table II Column 1 Air ($\mu c/ml$)	Table II Column 2 Water ($\mu c/ml$)
Osmium (76)	Os 185	S	5×10^{-7}	2×10^{-3}	2×10^{-8}	7×10^{-5}
		I	5×10^{-8}	2×10^{-3}	2×10^{-9}	7×10^{-5}
	Os 191m	S	2×10^{-5}	7×10^{-2}	6×10^{-7}	3×10^{-3}
		I	9×10^{-6}	7×10^{-2}	3×10^{-7}	2×10^{-3}
	Os 191	S	1×10^{-6}	5×10^{-3}	4×10^{-8}	2×10^{-4}
		I	4×10^{-7}	5×10^{-3}	1×10^{-8}	2×10^{-4}
	Os 193	S	4×10^{-7}	2×10^{-3}	1×10^{-8}	6×10^{-5}
		I	3×10^{-7}	2×10^{-3}	9×10^{-9}	6×10^{-5}
Palladium (46)	Pd 103	S	1×10^{-6}	1×10^{-2}	5×10^{-8}	3×10^{-4}
		I	7×10^{-7}	8×10^{-3}	3×10^{-8}	3×10^{-4}
	Pd 109	S	6×10^{-7}	3×10^{-3}	2×10^{-8}	9×10^{-5}
		I	4×10^{-7}	2×10^{-3}	1×10^{-8}	7×10^{-5}
Phosphorus (15)	P 32	S	7×10^{-8}	5×10^{-4}	2×10^{-9}	2×10^{-5}
		I	8×10^{-8}	7×10^{-4}	3×10^{-9}	2×10^{-5}
Platinum (78)	Pt 191	S	8×10^{-7}	4×10^{-3}	3×10^{-8}	1×10^{-4}
		I	6×10^{-7}	3×10^{-3}	2×10^{-8}	1×10^{-4}
	Pt 193m	S	7×10^{-6}	3×10^{-2}	2×10^{-7}	1×10^{-3}
		I	5×10^{-6}	3×10^{-2}	2×10^{-7}	1×10^{-3}
	Pt 197m	S	6×10^{-6}	3×10^{-2}	2×10^{-7}	1×10^{-3}
		I	5×10^{-6}	3×10^{-2}	2×10^{-7}	9×10^{-4}
	Pt 197	S	8×10^{-7}	4×10^{-3}	3×10^{-8}	1×10^{-4}
		I	6×10^{-7}	3×10^{-3}	2×10^{-8}	1×10^{-4}
Plutonium (94)	Pu 238	S	2×10^{-12}	1×10^{-4}	7×10^{-14}	5×10^{-6}
		I	3×10^{-11}	8×10^{-4}	1×10^{-12}	3×10^{-5}
	Pu 239	S	2×10^{-12}	1×10^{-4}	6×10^{-14}	5×10^{-6}
		I	4×10^{-11}	8×10^{-4}	1×10^{-12}	3×10^{-5}
	Pu 240	S	2×10^{-12}	1×10^{-4}	6×10^{-14}	5×10^{-6}
		I	4×10^{-11}	8×10^{-4}	1×10^{-12}	3×10^{-6}
	Pu 241	S	9×10^{-11}	7×10^{-3}	3×10^{-12}	2×10^{-4}
		I	4×10^{-8}	4×10^{-2}	1×10^{-9}	1×10^{-3}
	Pu 242	S	2×10^{-12}	1×10^{-4}	6×10^{-14}	5×10^{-6}
		I	4×10^{-11}	9×10^{-4}	1×10^{-12}	3×10^{-5}
	Pu 243	S	2×10^{-6}	1×10^{-2}	6×10^{-8}	3×10^{-4}
		I	2×10^{-6}	1×10^{-2}	8×10^{-8}	3×10^{-4}
	Pu 244	S	2×10^{-12}	1×10^{-4}	6×10^{-14}	4×10^{-6}
		I	3×10^{-11}	3×10^{-4}	1×10^{-12}	1×10^{-5}
Polonium (84)	Po 210	S	5×10^{-19}	2×10^{-5}	2×10^{-11}	7×10^{-7}
		I	2×10^{-19}	8×10^{-4}	7×10^{-12}	3×10^{-5}
Potassium (19)	K 42	S	2×10^{-6}	9×10^{-3}	7×10^{-8}	3×10^{-4}
		I	1×10^{-7}	6×10^{-4}	4×10^{-9}	2×10^{-5}
Praseodymium (59)	Pr 142	S	2×10^{-7}	9×10^{-4}	7×10^{-9}	3×10^{-5}
		I	2×10^{-7}	9×10^{-4}	5×10^{-9}	3×10^{-5}
	Pr 143	S	3×10^{-7}	1×10^{-3}	1×10^{-8}	5×10^{-5}
		I	2×10^{-7}	1×10^{-3}	6×10^{-9}	5×10^{-5}

Table 30.8   (*continued*)

Element (atomic number)	Isotope[1]		Table I		Table II	
			Column 1	Column 2	Column 1	Column 2
			Air (μc/ml)	Water (μc/ml)	Air (μc/ml)	Water (μc/ml)
Promethium (61)	Pm 147	S	6×10^{-8}	6×10^{-3}	2×10^{-9}	2×10^{-4}
		I	1×10^{-7}	6×10^{-3}	3×10^{-9}	2×10^{-4}
	Pm 149	S	3×10^{-7}	1×10^{-3}	1×10^{-8}	4×10^{-5}
		I	2×10^{-7}	1×10^{-3}	8×10^{-9}	4×10^{-5}
Protoactinium (91)	Pa 230	S	2×10^{-9}	7×10^{-3}	6×10^{-11}	2×10^{-4}
		I	8×10^{-19}	7×10^{-3}	3×10^{-11}	2×10^{-4}
	Pa 231	S	1×10^{-12}	3×10^{-5}	4×10^{-14}	9×10^{-7}
		I	1×10^{-10}	8×10^{-4}	4×10^{-12}	2×10^{-5}
	Pa 233	S	6×10^{-7}	4×10^{-3}	2×10^{-8}	1×10^{-4}
		I	2×10^{-7}	3×10^{-3}	6×10^{-9}	1×10^{-4}
Radium (88)	Ra 223	S	2×10^{-9}	2×10^{-5}	6×10^{-11}	7×10^{-7}
		I	2×10^{-10}	1×10^{-4}	8×10^{-12}	4×10^{-6}
	Ra 224	S	5×10^{-9}	7×10^{-5}	2×10^{-10}	2×10^{-6}
		I	7×10^{-10}	2×10^{-4}	2×10^{-11}	5×10^{-6}
	Ra 226	S	3×10^{-11}	4×10^{-7}	3×10^{-12}	3×10^{-8}
		I	5×10^{-11}	9×10^{-4}	2×10^{-12}	3×10^{-5}
	Ra 228	S	7×10^{-11}	8×10^{-8}	2×10^{-12}	3×10^{-8}
		I	4×10^{-11}	7×10^{-4}	1×10^{-12}	3×10^{-5}
Randon (86)	Rn 220	S	3×10^{-7}	—	1×10^{-8}	—
		I	—	—	—	—
	Rn 222	S	1×10^{-7}	—	3×10^{-9}	—
Rhenium (75)	Re 183	S	3×10^{-6}	2×10^{-2}	9×10^{-8}	6×10^{-4}
		I	2×10^{-7}	8×10^{-3}	5×10^{-9}	3×10^{-4}
	Re 186	S	6×10^{-7}	3×10^{-3}	2×10^{-8}	9×10^{-5}
		I	2×10^{-7}	1×10^{-3}	8×10^{-9}	5×10^{-5}
	Re 187	S	9×10^{-6}	7×10^{-2}	3×10^{-7}	3×10^{-3}
		I	5×10^{-7}	4×10^{-2}	2×10^{-8}	2×10^{-3}
	Re 188	S	4×10^{-7}	2×10^{-3}	1×10^{-8}	6×10^{-5}
		I	2×10^{-7}	9×10^{-4}	6×10^{-9}	3×10^{-5}
Rhodium (45)	Rh 103m	S	8×10^{-5}	4×10^{-1}	3×10^{-6}	1×10^{-2}
		I	6×10^{-5}	3×10^{-1}	2×10^{-6}	1×10^{-2}
	Rh 105	S	8×10^{-7}	4×10^{-3}	3×10^{-8}	1×10^{-4}
		I	5×10^{-7}	3×10^{-3}	2×10^{-8}	1×10^{-4}
Rubidium (47)	Rb 86	S	3×10^{-7}	2×10^{-3}	1×10^{-8}	7×10^{-5}
		I	7×10^{-8}	7×10^{-4}	2×10^{-9}	2×10^{-5}
	Rb 87	S	5×10^{-7}	3×10^{-3}	2×10^{-8}	1×10^{-4}
		I	7×10^{-8}	5×10^{-3}	2×10^{-9}	2×10^{-4}
Ruthenium (44)	Ru 97	S	2×10^{-6}	1×10^{-2}	8×10^{-8}	4×10^{-4}
		I	2×10^{-6}	1×10^{-2}	6×10^{-8}	3×10^{-4}
	Ru 103	S	5×10^{-7}	2×10^{-3}	2×10^{-8}	8×10^{-5}
		I	8×10^{-8}	2×10^{-3}	3×10^{-9}	8×10^{-5}
	Ru 105	S	7×10^{-7}	3×10^{-3}	2×10^{-8}	1×10^{-4}
		I	5×10^{-7}	3×10^{-3}	2×10^{-8}	1×10^{-4}
	Ru 106	S	8×10^{-8}	4×10^{-4}	3×10^{-9}	1×10^{-5}
		I	6×10^{-9}	3×10^{-4}	2×10^{-10}	1×10^{-5}

Table 30.8   *(continued)*

Element (atomic number)	Isotope[1]		Table I		Table II	
			Column 1	Column 2	Column 1	Column 2
			Air (μc/ml)	Water (μc/ml)	Air (μc/ml)	Water (μc/ml)
Samarium (62)	Sm 147	S	7×10^{-11}	2×10^{-3}	2×10^{-12}	6×10^{-5}
		I	3×10^{-10}	2×10^{-3}	9×10^{-12}	7×10^{-5}
	Sm 151	S	6×10^{-8}	1×10^{-2}	2×10^{-9}	4×10^{-4}
		I	1×10^{-7}	1×10^{-2}	5×10^{-9}	4×10^{-4}
	Sm 153	S	5×10^{-7}	2×10^{-3}	2×10^{-8}	8×10^{-5}
		I	4×10^{-7}	2×10^{-3}	1×10^{-8}	8×10^{-5}
Scandium (21)	Sc 46	S	2×10^{-7}	1×10^{-3}	8×10^{-9}	4×10^{-5}
		I	2×10^{-8}	1×10^{-3}	8×10^{-10}	4×10^{-5}
	Sc 47	S	6×10^{-7}	3×10^{-3}	2×10^{-8}	9×10^{-5}
		I	5×10^{-7}	3×10^{-3}	2×10^{-8}	9×10^{-5}
	Sc 48	S	2×10^{-7}	8×10^{-4}	6×10^{-9}	3×10^{-5}
		I	1×10^{-7}	8×10^{-4}	5×10^{-9}	3×10^{-5}
Selenium (34)	Se 75	S	1×10^{-6}	9×10^{-3}	4×10^{-8}	3×10^{-4}
		I	1×10^{-7}	8×10^{-3}	4×10^{-9}	3×10^{-4}
Silicon (14)	Si 31	S	6×10^{-6}	3×10^{-2}	2×10^{-7}	9×10^{-4}
		I	1×10^{-6}	6×10^{-3}	3×10^{-8}	2×10^{-4}
Silver (37)	Ag 105	S	6×10^{-7}	3×10^{-3}	2×10^{-8}	1×10^{-4}
		I	8×10^{-8}	3×10^{-3}	3×10^{-9}	1×10^{-4}
	Ag 110m	S	2×10^{-7}	9×10^{-4}	7×10^{-9}	3×10^{-5}
		I	1×10^{-8}	9×10^{-4}	3×10^{-10}	3×10^{-5}
	Ag 111	S	3×10^{-7}	1×10^{-3}	1×10^{-8}	4×10^{-5}
		I	2×10^{-7}	1×10^{-3}	8×10^{-9}	4×10^{-5}
Sodium (11)	Na 22	S	2×10^{-7}	1×10^{-8}	6×10^{-9}	4×10^{-5}
		I	9×10^{-9}	9×10^{-4}	3×10^{-10}	3×10^{-5}
	Na 24	S	1×10^{-6}	6×10^{-3}	4×10^{-8}	2×10^{-4}
		I	1×10^{-7}	8×10^{-4}	5×10^{-9}	3×10^{-5}
Strontium (38)	Sr 85m	S	4×10^{-5}	2×10^{-1}	1×10^{-6}	7×10^{-3}
		I	3×10^{-5}	2×10^{-1}	1×10^{-6}	7×10^{-3}
	Sr 85	S	2×10^{-7}	3×10^{-3}	8×10^{-9}	1×10^{-4}
		I	1×10^{-7}	5×10^{-3}	4×10^{-9}	2×10^{-4}
	Sr 89	S	3×10^{-8}	3×10^{-4}	3×10^{-10}	3×10^{-6}
		I	4×10^{-8}	8×10^{-4}	1×10^{-9}	3×10^{-5}
	Sr 90	S	1×10^{-9}	1×10^{-5}	3×10^{-11}	3×10^{-7}
		I	5×10^{-9}	1×10^{-3}	2×10^{-10}	4×10^{-5}
	Sr 91	S	4×10^{-7}	2×10^{-3}	2×10^{-8}	7×10^{-5}
		I	3×10^{-7}	1×10^{-3}	9×10^{-9}	5×10^{-5}
	Sr 92	S	4×10^{-7}	2×10^{-3}	2×10^{-8}	7×10^{-5}
		I	3×10^{-7}	2×10^{-3}	1×10^{-8}	6×10^{-5}
Sulfur (16)	S 35	S	3×10^{-7}	2×10^{-3}	9×10^{-9}	6×10^{-5}
		I	3×10^{-7}	8×10^{-3}	9×10^{-9}	3×10^{-4}
Tantalum (73)	Ta 182	S	4×10^{-8}	1×10^{-3}	1×10^{-9}	4×10^{-5}
		I	2×10^{-8}	1×10^{-3}	7×10^{-10}	4×10^{-5}
Technetium (43)	Tc 96m	S	8×10^{-5}	4×10^{-1}	3×10^{-6}	1×10^{-2}
		I	3×10^{-5}	3×10^{-1}	1×10^{-6}	1×10^{-2}

Table 30.8   (*continued*)

Element (atomic number)	Isotope[1]		Table I		Table II	
			Column 1	Column 2	Column 1	Column 2
			Air (μc/ml)	Water (μc/ml)	Air (μc/ml)	Water (μc/ml)
	Tc 96	S	6×10^{-7}	3×10^{-3}	2×10^{-8}	1×10^{-4}
		I	2×10^{-7}	1×10^{-3}	8×10^{-9}	5×10^{-5}
	Tc 97m	S	2×10^{-6}	1×10^{-2}	8×10^{-8}	4×10^{-4}
		I	2×10^{-7}	5×10^{-3}	5×10^{-9}	2×10^{-4}
	Tc 97	S	1×10^{-5}	5×10^{-2}	4×10^{-7}	2×10^{-3}
		I	3×10^{-7}	2×10^{-2}	1×10^{-8}	8×10^{-4}
	Tc 99m	S	4×10^{-5}	2×10^{-1}	1×10^{-6}	6×10^{-8}
		I	1×10^{-5}	8×10^{-2}	5×10^{-7}	3×10^{-8}
	Tc 99	S	2×10^{-6}	1×10^{-2}	7×10^{-8}	3×10^{-4}
		I	6×10^{-8}	5×10^{-8}	2×10^{-9}	2×10^{-4}
Tellurium (52)	Te 125m	S	4×10^{-7}	5×10^{-3}	1×10^{-8}	2×10^{-4}
		I	1×10^{-7}	3×10^{-3}	4×10^{-9}	1×10^{-4}
	Te 127m	S	1×10^{-7}	2×10^{-3}	5×10^{-9}	6×10^{-5}
		I	4×10^{-8}	2×10^{-3}	1×10^{-9}	5×10^{-5}
	Te 127	S	2×10^{-6}	8×10^{-3}	6×10^{-8}	3×10^{-4}
		I	9×10^{-7}	5×10^{-3}	3×10^{-8}	2×10^{-4}
	Te 129m	S	8×10^{-8}	1×10^{-3}	3×10^{-9}	3×10^{-5}
		I	3×10^{-8}	6×10^{-4}	1×10^{-9}	2×10^{-5}
	Te 129	S	5×10^{-6}	2×10^{-2}	2×10^{-7}	8×10^{-4}
		I	4×10^{-6}	2×10^{-2}	1×10^{-7}	8×10^{-4}
	Te 131m	S	4×10^{-7}	2×10^{-3}	1×10^{-8}	6×10^{-5}
		I	2×10^{-7}	1×10^{-3}	6×10^{-9}	4×10^{-5}
	Te 132	S	2×10^{-7}	9×10^{-4}	7×10^{-9}	3×10^{-5}
		I	1×10^{-7}	6×10^{-4}	4×10^{-9}	2×10^{-5}
Terbium (65)	Tb 160	S	1×10^{-7}	1×10^{-3}	3×10^{-9}	4×10^{-5}
		I	3×10^{-8}	1×10^{-3}	1×10^{-9}	4×10^{-5}
Thallium (81)	Tl 200	S	3×10^{-6}	1×10^{-2}	9×10^{-8}	4×10^{-4}
		I	1×10^{-6}	7×10^{-3}	4×10^{-8}	2×10^{-4}
	Tl 201	S	2×10^{-6}	9×10^{-3}	7×10^{-8}	3×10^{-4}
		I	9×10^{-7}	5×10^{-3}	3×10^{-8}	2×10^{-4}
	Tl 202	S	8×10^{-7}	4×10^{-3}	3×10^{-8}	1×10^{-4}
		I	2×10^{-7}	2×10^{-3}	8×10^{-9}	7×10^{-5}
	Tl 204	S	6×10^{-7}	3×10^{-3}	2×10^{-8}	1×10^{-4}
		I	3×10^{-8}	2×10^{-3}	9×10^{-10}	6×10^{-5}
Thorium (90)	Th 228	S	9×10^{-12}	2×10^{-4}	3×10^{-13}	7×10^{-6}
		I	6×10^{-12}	4×10^{-4}	2×10^{-13}	10^{-5}
	Th 230	S	2×10^{-12}	5×10^{-5}	8×10^{-14}	2×10^{-6}
		I	10^{-11}	9×10^{-4}	3×10^{-13}	3×10^{-5}
	Th 232	S	3×10^{-11}	5×10^{-5}	10^{-12}	2×10^{-6}
		I	3×10^{-11}	10^{-3}	10^{-12}	4×10^{-5}
	Th natural	S	3×10^{-11}	3×10^{-5}	10^{-12}	10^{-6}
		I	3×10^{-11}	3×10^{-4}	10^{-12}	10^{-5}
	Th 234	S	6×10^{-8}	5×10^{-4}	2×10^{-9}	2×10^{-5}
		I	3×10^{-8}	5×10^{-4}	10^{-9}	2×10^{-5}

Table 30.8 (*continued*)

Element (atomic number)	Isotope[1]		Table I		Table II	
			Column 1	Column 2	Column 1	Column 2
			Air ($\mu c/ml$)	Water ($\mu c/ml$)	Air ($\mu c/ml$)	Water ($\mu c/ml$)
Thulium (69)	Tm 170	S	4×10^{-8}	1×10^{-3}	1×10^{-9}	5×10^{-5}
		I	3×10^{-8}	1×10^{-3}	1×10^{-9}	5×10^{-5}
	Tm 171	S	1×10^{-7}	1×10^{-2}	4×10^{-9}	5×10^{-4}
		I	2×10^{-7}	1×10^{-2}	8×10^{-9}	5×10^{-4}
Tin (50)	Sn 113	S	4×10^{-7}	2×10^{-3}	1×10^{-8}	9×10^{-5}
		I	5×10^{-8}	2×10^{-3}	2×10^{-9}	8×10^{-5}
	Sn 125	S	1×10^{-7}	5×10^{-4}	4×10^{-9}	2×10^{-5}
		I	8×10^{-8}	5×10^{-4}	3×10^{-9}	2×10^{-5}
Tungsten (Wolfram) (74)	W 181	S	2×10^{-6}	1×10^{-2}	8×10^{-8}	4×10^{-4}
		I	1×10^{-7}	1×10^{-2}	4×10^{-9}	3×10^{-4}
	W 185	S	8×10^{-7}	4×10^{-3}	3×10^{-8}	1×10^{-4}
		I	1×10^{-7}	3×10^{-3}	4×10^{-9}	1×10^{-4}
	W 187	S	4×10^{-7}	2×10^{-3}	2×10^{-8}	7×10^{-5}
		I	3×10^{-7}	2×10^{-3}	1×10^{-8}	6×10^{-5}
Uranium (92)	U 230	S	3×10^{-10}	1×10^{-4}	1×10^{-11}	5×10^{-6}
		I	1×10^{-10}	1×10^{-4}	4×10^{-12}	5×10^{-6}
	U 232	S	1×10^{-10}	8×10^{-4}	3×10^{-12}	3×10^{-5}
		I	3×10^{-11}	8×10^{-4}	9×10^{-13}	3×10^{-5}
U 233	U 233	S	5×10^{-10}	9×10^{-4}	2×10^{-11}	3×10^{-5}
		I	1×10^{-10}	9×10^{-4}	4×10^{-12}	3×10^{-5}
	U 234	S	6×10^{-10}	9×10^{-4}	2×10^{-11}	3×10^{-5}
	U 235	I	1×10^{-19}	9×10^{-4}	4×10^{-12}	3×10^{-5}
	U 235	S	5×10^{-10}	8×10^{-4}	2×10^{-11}	3×10^{-5}
		I	1×10^{-10}	8×10^{-4}	4×10^{-12}	3×10^{-5}
	U 236	S	6×10^{-10}	1×10^{-3}	2×10^{-11}	3×10^{-5}
		I	1×10^{-10}	1×10^{-3}	4×10^{-12}	3×10^{-5}
	U 238	S	7×10^{-11}	1×10^{-3}	3×10^{-12}	4×10^{-5}
		I	1×10^{-10}	1×10^{-3}	5×10^{-12}	4×10^{-5}
	U 240	S	2×10^{-7}	1×10^{-3}	8×10^{-9}	3×10^{-5}
		I	2×10^{-7}	1×10^{-3}	6×10^{-9}	3×10^{-5}
	U-natural	S	7×10^{-11}	5×10^{-4}	3×10^{-12}	2×10^{-5}
		I	6×10^{-11}	5×10^{-4}	2×10^{-12}	2×10^{-5}
Vanadium (23)	V 48	S	2×10^{-7}	9×10^{-4}	6×10^{-9}	3×10^{-5}
		I	6×10^{-8}	8×10^{-4}	2×10^{-9}	3×10^{-5}
Xenon (54)	Xe 131m	Sub	2×10^{-5}	—	4×10^{-7}	—
	Xe 133	Sub	1×10^{-5}	—	3×10^{-7}	—
	Xe 133m	Sub	1×10^{-5}	—	3×10^{-7}	—
	Xe 135	Sub	4×10^{-6}	—	1×10^{-7}	—
Ytterbium (70)	Yb 175	S	7×10^{-7}	3×10^{-3}	2×10^{-8}	1×10^{-4}
		I	6×10^{-7}	3×10^{-3}	2×10^{-8}	1×10^{-4}
Yttrium (39)	Y 90	S	1×10^{-7}	6×10^{-4}	4×10^{-9}	2×10^{-5}
		I	1×10^{-7}	6×10^{-4}	3×10^{-9}	2×10^{-5}
	Y 91m	S	2×10^{-5}	1×10^{-1}	8×10^{-7}	3×10^{-3}
		I	2×10^{-5}	1×10^{-1}	6×10^{-7}	3×10^{-3}

Table 30.8   (*continued*)

Element (atomic number)	Isotope[1]		Table I		Table II	
			Column 1	Column 2	Column 1	Column 2
			Air (μc/ml)	Water (μc/ml)	Air (μc/ml)	Water (μc/ml)
	Y 91	S	4×10^{-8}	8×10^{-4}	1×10^{-9}	3×10^{-5}
		I	3×10^{-8}	8×10^{-4}	1×10^{-9}	3×10^{-5}
	Y 92	S	4×10^{-7}	2×10^{-3}	1×10^{-8}	6×10^{-5}
		I	3×10^{-7}	2×10^{-3}	1×10^{-8}	6×10^{-5}
	Y 93	S	2×10^{-7}	8×10^{-4}	6×10^{-9}	3×10^{-5}
		I	1×10^{-7}	8×10^{-4}	5×10^{-9}	3×10^{-5}
Zinc (30)	Zn 65	S	1×10^{-7}	3×10^{-3}	4×10^{-9}	1×10^{-4}
		I	6×10^{-8}	5×10^{-3}	2×10^{-9}	2×10^{-4}
	Zn 69m	S	4×10^{-7}	2×10^{-3}	1×10^{-8}	7×10^{-5}
		I	3×10^{-7}	2×10^{-3}	1×10^{-8}	6×10^{-5}
	Zn 69	S	7×10^{-6}	5×10^{-2}	2×10^{-7}	2×10^{-3}
		I	9×10^{-6}	5×10^{-2}	3×10^{-7}	2×10^{-3}
Zirconium (40)	Zr 93	S	1×10^{-7}	2×10^{-2}	4×10^{-9}	8×10^{-4}
		I	3×10^{-7}	2×10^{-2}	1×10^{-8}	8×10^{-4}
	Zr 95	S	1×10^{-7}	2×10^{-3}	4×10^{-9}	6×10^{-5}
		I	3×10^{-8}	2×10^{-3}	1×10^{-9}	6×10^{-5}
	Zr 97	S	1×10^{-7}	5×10^{-4}	4×10^{-9}	2×10^{-5}
		I	9×10^{-8}	5×10^{-4}	3×10^{-9}	2×10^{-5}
Any single radionuclide not listed above with decay mode other than alpha emission or spontaneous fission and with radioactive half-life less than 2 hours.		Sub	1×10^{-6}	—	3×10^{-8}	—
Any single radionuclide not listed above with decay mode other than alpha emission or spontaneous fission and with radioactive half-life greater than 2 hours.	—	—	3×10^{-9}	9×10^{-5}	1×10^{-10}	3×10^{-6}
Any single radionuclide not listed above, which decays by alpha emission or spontaneous fission.	—	—	6×10^{-13}	4×10^{-7}	2×10^{-14}	3×10^{-8}

[1] Soluble (S); Insoluble (I).

[2] "Sub" means that values given are for submersion in a semispherical infinite cloud of airborne material.

Note: In any case where there is a mixture in air or water of more than one radionuclide, the limiting values for purposes of this Appendix should be determined as follows:

1. If the identity and concentration of each radionuclide in the mixture are known, the limiting values should be derived as follows: Determine, for each radionuclide in the mixture, the ratio between the quantity present in the mixture and the limit otherwise established in Appendix B for the specific radionuclide when not in a mixture. The sum of such ratios for all the radionuclides in the mixture may not exceed "1" (i.e., "unity").

(Footnotes for Table 30.8 continued)

Example: If radionuclides A, B, and C are present in concentrations C_A, C_B, and C_C, and if the applicable MPC's, are MPC_A, and MPC_B, and MPC_C respectively, then the concentrations shall be limited so that the following relationship exists:

$$\frac{C_A}{MPC_A} + \frac{C_B}{MPC_B} + \frac{C_C}{MPC_C} \leq 1$$

2. If either the identity or the concentration of any radionuclide in the mixture is not known, the limiting values for purposes of Appendix B shall be:

a. For purposes of Table I, Col. 1–6 $\times$ 10^{-13}
b. For purposes of Table I, Col. 2–4 $\times$ 10^{-7}
c. For purposes of Table II, Col. 1–2 $\times$ 10^{-14}
d. For purposes of Table II, Col. 2–3 $\times$ 10^{-8}

3. If any of the conditions specified below are met, the corresponding values specified below may be used in lieu of those specified in paragraph 2 above.

a. If the identity of each radionuclide in the mixture is known but the concentration of one or more of the radionuclides in the mixture is not known the concentration limit for the mixture is the limit specified in Appendix "B" for the radionuclide in the mixture having the lowest concentration limit; or

b. If the identity of each radionuclide in the mixture is not known, but it is known that certain radionuclides specified in Appendix "B" are not present in the mixture, the concentration limit for the mixture is the lowest concentration limit specified in Appendix "B" for any radionuclide which is not known to be absent from the mixture.

Table 30.9 Permissible quantities of radioactive material. (From Code of Federal Regulations Title 10 revised as of Jan. 1, 1971.)

Material	Micro-curies	Material	Micro-curies	Material	Micro-curies
Americium-241	.01	Hafnium-181	10	Platinum-197	100
Antimony-122	100	Holmium-166	100	Plutonium-239	.01
Antimony-124	10	Hydrogen-3	1000	Polonium-210	0.1
Antimony-125	10	Indium-113m	100	Potassium-42	10
Arsenic-73	100	Indium-114m	10	Praseodymium-142	100
Arsenic-74	10	Indium-115m	100	Praseodymium-143	100
Arsenic-76	10	Indium-115	10	Promethium-147	10
Arsenic-77	100	Iodine-125	1	Promethium-149	10
Barium-131	10	Iodine-126	1	Radium-226	.01
Barium-140	10	Iodine-129	0.1	Rhenium-186	100
Bismuth-210	1	Iodine-131	1	Rhenium-188	100
Bromine-82	10	Iodine-132	10	Rhodium-103m	100
Cadmium-109	10	Iodine-133	1	Rhodium-105	100
Cadmium-115m	10	Iodine-134	10	Rubidium-86	10
Cadmium-115	100	Iodine-135	10	Rubidium-87	10
Calcium-45	10	Iridium-192	10	Ruthenium-97	100
Calcium-47	10	Iridium-194	100	Ruthenium-103	10
Carbon-14	100	Iron-55	100	Ruthenium-105	10
Cerium-141	100	Iron-59	10	Ruthenium-106	1
Cerium-143	100	Krypton-85	100	Ramarium-151	10
Cerium-144	1	Krypton-87	10	Ramarium-153	100
Cesium-131	1000	Lanthanum-140	10	Ramadium-46	10
Cesium-134m	100	Lutetium-177	100	Ramadium-47	100
Cesium-134	1	Manganese-52	10	Scandium-48	10
Cesium-135	10	Manganese-54	10	Selenium-75	10
Cesium-136	10	Manganese-56	10	Silicon-31	100
Cesium-137	10	Mercury-197m	100	Silver-105	10
Chlorine-36	10	Mercury-197	100	Silver-110m	1
Chlorine-38	10	Mercury-203	10	Silver-111	100
Chromium-51	1000	Molybdenum-99	100	Sodium-24	10
Cobalt-58m	10	Neodymium-147	100	Strontium-85	10
Cobalt-58	10	Neodymium-149	100	Strontium-89	1
Cobalt-60	1	Nickel-59	100	Strontium-90	0.1
Copper-64	100	Nickel-63	10	Strontium-91	10
Dysprosium-165	10	Nickel-65	100	Strontium-92	10
Dysprosium-166	100	Niobium-93m	10	Sulphur-35	100
Erbium-169	100	Niobium-95	10	Tantalum-182	10
Erbium-171	100	Niobium-97	10	Technetium-96	10
Europium-152 9.2 h	100	Osmium-185	10	Technetium-97m	100
Europium-152 13 yr	1	Osmium-191m	100	Technetium-97	100
Europium-154	1	Osmium-191	100	Technetium-99m	100
Europium-155	10	Osmium-193	100	Technetium-99	10
Fluorine-18	1000	Palladium-103	100	Tellurium-125m	10
Gadolinium-153	10	Palladium-109	100	Tellurium-127m	10
Gadolinium-159	100	Phosphorus-32	10	Tellurium-127	100
Gallium-72	10	Platinum-191	100	Tellurium-129m	10
Germanium-71	100	Platinum-193m	100	Tellurium-129	100
Gold-198	100	Platinum-193	100	Tellurium-131m	10
Gold-199	100	Platinum-197m	100	Tellurium-132	10

Table 30.9   (*continued*)

Material	Micro-curies	Material	Micro-curies	Material	Micro-curies
Terbium-160	10	Uranium-233	.01	Zirconium-93	10
Thallium-200	100	Uranium-234—Uranium-235	.01	Zirconium-95	10
Thallium-201	100	Vandium-48	10	Zirconium-97	10
Thallum-202	100	Xenon-131m	1000	Any alpha emitting	
Thallium-204	10	Xenon-133	100	radionuclide not listed	
Thorium (natural)	50	Xenon-135	100	above or mixtures of	
Thulium-170	10	Ytterbium-175	100	alpha emitters of	
Thulium-171	10	Yttrium-90	10	unknown composition	.01
Tin-113	10	Yttrium-91	10	Any radionuclide other than	
Tin-125	10	Yttrium-92	100	alpha emitting	
Tungsten-181	10	Yttrium-93	100	radionuclides, not listed	
Tungsten-185	10	Zinc-65	10	above or mixtures of beta	
Tungsten-187	100	Zinc-69m	100	emitters of unknown	
Uranium (natural)	50	Zinc-69	1000	composition	.1

Table 30.10   (After Falk [1].)

Radioisotope	Half-life	Maximum permissible concentration in H_2O, $\mu c/cm^3$	Equivalent concentration, gm/cm^3
Mn^{56}	2.6 hr	3×10^{-3}	1.4×10^{-16}
Cu^{64}	12.8 hr	5×10^{-3}	1.3×10^{-15}
Fe^{59}	45 days	10^{-4}	2.0×10^{-14}
Zr^{95}	65 days	6×10^{-4}	2.8×10^{-14}
Cs^{137}	33 yr	2×10^{-3}	2.6×10^{-11}
Sr^{90}	19.9 yr	8×10^{-7}	4.0×10^{-15}
Unknown mixture		10^{-7}	10^{-15} to 10^{-16}

sites are estimated to have the capacity to handle low-level wastes until approximately 1995. The sites are managed by private companies and regulated by the states. Recent federal announcements indicate more stringent qualifications for low-level storage of the transuranic wastes. It appears that shipments to the disposal sites of low-level wastes will be limited to 10 nanocuries (10^{-9} curie) per gram of transuranic elements, a level not easily ascertained because of the difficulty of obtaining representative samples.

High-level wastes are generated during fuel reprocessing. They contain separated fission products and small amounts of transuranic elements. Normally, the two most significant fission products are Sr^{90} (half life, 28 years) and Cs^{137} (half life, 30 years). Recently, attention has also been focused on I^{129} (half life, 1.6 years), which is present in very small quantities.

It is common practice to treat solid wastes, as well as high-level radioactive liquid wastes, by concentration and burial. A small volume of waste is less difficult to monitor, package, transport, and dispose of than a larger volume with the same activity. However, the cost of concentration must not exceed that saved by the concentration process. From Fig. 30.5, the three most widely used methods of concentration are coprecipitation, evaporation, and ion exchange.

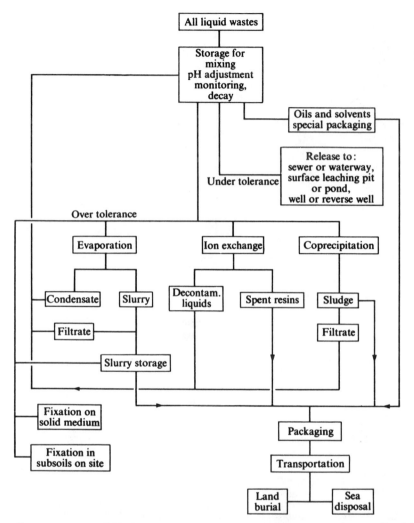

Fig. 30.5 Stages in processing liquid radioactive waste.

Coprecipitation. This is the precipitation of an otherwise soluble (because of its low concentration) material along with an insoluble precipitate. The insoluble precipitate acts as a "scavenger" or "carrier" to co-precipitate radioactive ions from solution. Usually the pH is adjusted to an alkaline condition prior to precipitation to form metal hydroxide flocs, which aid in the scavenging process. After the supernatant settles, the sludge is removed, packaged for shipment, and disposed of by burial. Further concentration, however, may be effected by drying the sludge prior to burial. Coprecipitation is known to yield decontamination factors of 200 to 1000, which is less than

that realized by evaporation; however, the process is much less expensive.

Evaporation. This is at present the most widely used concentration process for water solutions containing a mixture of wastes of low-level activity. The wastes are boiled, the resultant vapor is condensed, and the condensate is subsequently released to the sewer, provided there is no significant radioactivity. The high-activity concentrate is usually a sludge that solidifies upon cooling and is then transferred to polyethylene-lined 55-gallon drums for ultimate burial. Foaming and priming with resultant carry-over occur and present operational difficulties. Although evaporation is expensive, it usually is very effective and provides decontamination factors of 100,000 to 1,000,000.

Ion exchange. In this process the radioactive ions in the waste are exchanged for stable ions of the exchanger. Cation exchange is the principal ion-exchange process used to concentrate radioactive waste. Both synthetic resins and Montmorillonite clay serve as exchangers. This method is customarily reserved for small volumes of solutions containing low concentrations of solids and exhibiting low levels of radioactivity. With these wastes, the exchangers may be used for long periods before regeneration becomes necessary.

Eliassen [1964, 17] reports that high-level aluminum-bearing fission-product wastes can be incorporated into ceramic glazes (glasses) by the addition of silica and fluxes to calcined wastes and firing the mixture at 1450°C. A suggested flowsheet is presented in Fig. 30.6.

Treatment of solid radioactive wastes. Solid wastes may be disposed of by burial, by incineration if combustible, or by remelting if metallic. Typical stages in processing solid wastes are presented in Fig. 30.7.

Burial. The AEC operates several burial areas in this country. In selecting these sites, the AEC has been guided by one imperative consideration, namely, to reduce environmental hazards below maximum tolerance levels. Constant surveillance should be possible, and the land should be forbidden to the public for generations to come. Burial grounds, as established by the AEC, should: (1) be not less than 10 acres in area; (2) be accessible; (3) have greater than 15, and preferably greater than 20, feet of unconsolidated sedimentary overburden on the bedrock; (4) the overburden should be sufficiently coherent that the vertical, or nearly vertical, walls of a burial trench will stand up for short periods of time; (5) areas should not be located directly upstream from a ground-water course, from existing or potential plant sites, or nearby populated areas.

Radioactive wastes are usually buried in narrow, deep trenches excavated by means of a back hoe and backfilled with several feet of earth cover compacted by bulldozers. Burial in containers is, as one would expect, expensive, and one must also consider the cost

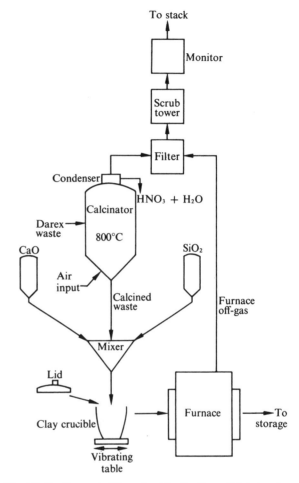

Fig. 30.6 Suggested flow sheet for fixation of high-level radioactive waste in glass. (After Eliassen [17].)

of excavating the trenches. At present, a major portion of the disposal problem is connected with the ultimate disposal of low-volume, high-level liquid wastes resulting from the chemical processing of spent reactor-fuel elements. Since the effective life of the fission-product mixtures making up these wastes is on the order of several hundred years, and since the future portends great volumes of wastes, the present practice of storing the wastes in tanks cannot be considered a permanent solution to the problem. Rich [168] predicts that for at least 100 years mined-out salt mines will provide enough storage space for anticipated waste production. The states of New

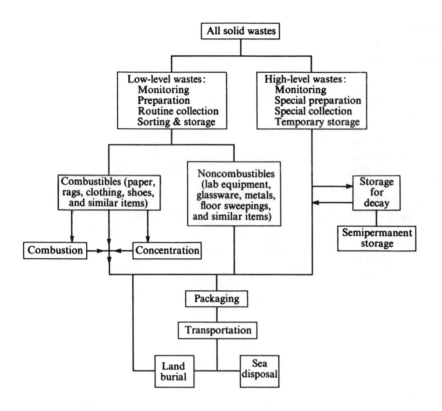

Fig. 30.7 Stages in processing solid radioactive waste.

York, Michigan, Ohio, and Kansas are ideally suited to this type of disposal. Rich and Rodger [18] conclude that high-level wastes can be concentrated and converted into solids at a cost of no more than 10 cents per gallon. Containers can be fabricated (probably) out of $\frac{1}{8}$-inch aluminum bronze at a cost of $50 per cubic foot of capacity. The cost of storage in mined-out salt deposits is estimated at $1 per square foot of floor space. Rich concludes that if one could succeed in developing an economical method of concentrating and fixing the radioactive waste in a nonleachable solid, a system of isolation would thus be available for the ultimate, safe, and economical disposal of high-level wastes.

According to recent reports, storage of radioactive wastes in glass appears to be an effective, safe, and inexpensive method of atomic-waste disposal. The British Atomic Energy Authority at Harwell has developed a method by which radioactive waste materials may be converted into insoluble glasslike particles. A mixture of silica and borax is added to a nitric-acid

solution of the concentrated liquid wastes. The solution is then treated until the liquid evaporates, and the remainder becomes red hot, sinters, and melts. On cooling, the meltage solidifies to a glass of predetermined composition, containing from 20 to 30 per cent of waste oxides.

The solidified wastes would be vitrified in silicate glass, which has excellent nonleaching qualities. A 1000 megawatt nuclear plant will burn about 1 ton of U^{235} annually. The solidified high-level wastes from this plant, after the reprocessing of the spent fuel, would be about 100 cubic feet annually, including the inert material for a 2:1 dilution.

Incineration. The burning of combustible wastes has not been adopted on any large scale, but many experiments have been carried out. The by-products of active-waste incineration (stack gases and ashes) are also radioactive. The problem of radioactive gases will probably become more acute as atomic-energy installations increase. This was brought home dramatically when an overheated nuclear reactor at Windscale (Cumberland, England) spewed radioactive iodine[131] over an area of 200 square miles on October 10, 1957. As a consequence, all milk produced in the area was declared unsafe, barred from shipment, and dumped.

Filters, electrostatic precipitators, scrubbers, settling chambers, inertial collectors, or wet collectors can be used in the stacks to reduce or eliminate air contamination. Persons handling radioactive ashes must be careful to avoid inhalation or direct exposure to concentrated radioactive material. Volume reduction by incineration may exceed a ratio of 50:1, but the problem of contamination control, as well as high initial and operating costs, discourage the use of this method of solid-waste treatment.

Remelting. Contaminated iron and steel remelted in an induction furnace produces a slag which carries off a major portion of uranium originating from the processing of natural uranium ores. The content of radioactive uranium in the melted steel is reduced by a factor of 40 to 1.

Treatment of gaseous radioactive wastes. Radioactive atmospheric contaminants can be divided into two main categories, gases and particulate matter. These wastes result mainly from mining and ore processing, and also from incineration of contaminated combustibles. The principal method of controlling such radioactive materials at source has been the application of local exhaust ventilation. The principles of local exhaust-hood design apply equally well to the handling of radioactive contaminants. Features of ventilation for radioactive laboratories (Silverman [9]) which may differ from those governing ventilation in general are:

1. Sufficient air supply must be provided to equal that exhausted by the hoods.
2. No recirculation of air shall be permitted in ventilating radioactive laboratories.
3. It is essential to maintain the laboratories in which active materials are handled at a slight negative pressure with respect to adjacent non-active chemical or physical laboratories, corridors, or offices.
4. Supply air should always be taken from the outside or from uncontaminated areas.
5. Adequate by-passes, crossovers, and auxiliary equipment should be installed to provide ex-

haust-air flow in the event of shutdown or power failures.
6. For highly contaminated working areas, a parallel system for stand-by or emergency operation is essential in case of power or exhauster failure.
7. Converging velocities from exhaust ducts should range from 2000 ft/min for gases to 5000 ft/min for particulates of heavy metals.
8. Air- and gas-cleaning equipment should meet design decontamination factors and be readily accessible or removable without exposing maintenance personnel to undue radiation danger. The equipment should also be provided with efficient monitoring devices, as well as continuous resistance- or pressure-loss indicators or recorders.

From an economic standpoint, the disposal of radioactive gases should, wherever permissible, utilize the atmospheric dispersion and dilution obtaining for tall stacks. Other procedures which are sometimes used are: (1) containment by ventilating the gases to a detention chamber, where they are stored for an adequate period to permit decay before release to the atmosphere; (2) scrubbing and adsorption at normal or extremely subnormal temperatures; (3) adsorption on special materials; (4) reaction with solid materials; and (5) combustion and special reactions.

Gases created by incineration or combustion of solid wastes are not a significant problem since these processes do not create large amounts of radioactive gases. Where gases have been involved, and dilution by the atmosphere is not adequate, ethanolamine scrubbing towers for C^{14} or S^{35} have been used. For the removal of particulates created by nuclear-energy operations, inertial collectors, high-efficiency filters, or electrostatic precipitators are used.

30.5 Cost of Radioactive-Waste Treatment

Eliassen *et al.* [9] present a breakdown of costs for radioactive waste treatment in various locations:

Large-diameter cyclone	$0.10–0.25 per cfm
Inertial scrubbers (power-driven)	$0.15–0.25 per cfm
High-efficiency cellulose asbestos filters	$0.04–0.06 per cfm
Single-stage electrostatic precipitator	$0.50–2.00 per cfm

Cost for off-site disposal of solid wastes from Rocky Flats, Colo, facility, April 1954–Sept. 1955	$2.95 per hundred-weight $1.02 per ft³ $0.08 per ton-mile* (750 mi)
Estimated cost of burial of solid radioactive waste, National Reactor Testing Station, 1955	$9.11 per cubic yard
Estimated cost (capital) for burial of concentrated liquid wastes in trenches	$0.37–2.00 per gallon
Total evaporation costs (Argonne National Laboratory)	$0.12 per gallon
Total evaporation costs (Knolls Atomic Power Laboratory)	$0.138 per gallon
Total annual operating cost of Los Alamos waste-treatment plant	$5.86 per 1000 gallons
Initial capital cost of Los Alamos waste-treatment plant, plus laboratory facilities	$350,000

References: Radioactive Wastes

1. Falk, C. F., "Radioactive liquid waste disposal from the Dresden nuclear power station," Nuclear Engineering and Science Conference, Reprint no. 102, Session XVI, March 1958.
2. Medin, A. L., "Army package power reactor water treatment and waste disposal," *Ind. Eng. Chem.* 50, 989 (1958).
3. Blomeke, J. O., E. D. Arnold, and A. K. Gresky, "Characteristics of reactor fuel process wastes," Nuclear Engineering and Science Conferences, Preprint no. 44, Session XVI, March 1958.
4. Irish, E. R., and W. H. Reas, "The Purex process—a solvent extraction reprocessing method for irradiated uranium," Symposium on the Reprocessing of Irradiated Fuel, Brussels, U.S. Atomic Energy Commission, *TID*-7534, May 1957, p. 83.
5. Lawroski, S., and M. Levenson, "Redox process—a solvent extraction reprocessing method for irradiated uranium," U.S. Atomic Energy Commission, *TID*-7534, May 1957, p. 45.
6. Stevenson, C. E., "Solvent extraction processes for enriched uranium," U.S. Atomic Energy Commission, *TID*-7534, May 1957, p. 152.
7. Bruce, F. R., "The Thorex process" U.S. Atomic Energy Comm., TID-7534, May 1957, page 180.
8. Culler, F. L., "Notes on fission product wastes from proposed power reactors," U.S. Atomic Energy Commission, Report no. *CF* 55-4-25, Oak Ridge National Laboratory, 1955.
9. Blatz, H. (ed.), *Radiation Hygiene Handbook*, McGraw-Hill Book Co., New York (1959).
10. Eliassen, R., "Atomic wastes disposal: an international problem," Paper presented at Conference on Biological Waste Treatment, Manhattan College, April 1960.
11. Federal Register of Atomic Energy Commission, *Standards for Protection against Radiation*, Part 20, Title 10, Code Federal Regulations, February 1957.
12. Beadle, G. W., "Ionizing radiation and the citizen," *Sci. Am.* 201, 219 (1959).
13. "Maximum permissible amounts of radioisotopes in the human body, and maximum permissible concentrations in air and water," in *United States National Bureau of Standards Handbook*, no. 52, U.S. Government Printing Office, Washington, D.C. (1953).
14. "Maximum permissible radiation exposures to man," Insert to accompany *United States National Bureau of Standards Handbook*, no. 59, U.S. Government Printing Office, Washington, D.C. (1957).
15. Lundgren, D., "Effect of radiation on cells and their bacterial enzymes," Paper presented at Syracuse University Civil Engineering Seminar, 1960.
16. Lieberman, J. A., "Engineering aspects of the disposal of radioactive wastes from peacetime power reactors," *Ind. Wastes* 1, 278 (1956).
17. Eliassen, R., "Disposal of high level radioactive wastes," *J. Water Pollution Control Federation* 36, 201 (1964).
18. Rich, C. G., "The disposal of high-level radioactive wastes in salt formations," in Proceedings of 13th Industrial Waste Conference, Purdue University, May 1958, p. 581.

CHAPTER 31

NON-POINT SOURCE POLLUTION

31.1 Introduction

A non-point source of pollution can be defined as any source of water pollution not coming directly or continually from a pipe. Generally characterized as area wide sources with multiple, diffuse discharges, these sources are often tied to rainfall runoff and therefore take on the same variant nature. Due to this lack of a distinct or continuous effluent which can be routinely sampled or collected for characterization, treatment, or monitoring, non-point sources lack suitability for regulation by the effluent limitations approach. Hence regulations must center on sources control, which requires the establishment of a cause and effect relationship between an activity and the detrimental consequences of the ensuing pollutant discharge. For many cases, this is much easier said than done.

It is convenient to establish two broad categories (that is, rural and urban) for non-point discharges; subdivide these categories further when dealing with a specific problem.

Potential sources of pollution from the rural environment include wildlife wastes, leached plant residues, fertilizers, herbicides and pesticides, nutrient and organic matter eroded with soil, inorganic sediments from the weathering of rocks and soil, acid mine drainage, farm animal wastes, and excess drainage from rural septic systems. Since surface water runoff is the major transporting mechanism, topography and vegetation exert a major influence on both the quality and quantity of this runoff.

31.2 Land Non-point Wastes

Subcategories of rural lands with the potential of contributing to the problem of non-point pollution include forest and range land, fields used for crop production, animal lands (pastures and barnyards), and land used for mining purposes.

Subcategories of urban contaminants may include street litter, ice-control chemicals, rubber, decaying vegetation, domestic pet wastes, fallout of combustion products from automobile and industrial stack emissions, lawn chemicals, pesticides and herbicides, heavy metals, asbestos fibers from auto brake linings, oil, grease, gasoline, paint, dust, dirt, and any other substance (natural, household, commercial, or industrial product) that finds its way into a gutter or storm sewer. The major portion of these contaminants consists of dirt and dust particles, the greatest amount of which is found within six inches of a curb. Pollutant accumulation per unit length of gutter is a function of the time between street cleanings. Therefore, the major factor controlling the quality of urban discharge is runoff. Due to the impervious nature of the greater part of urban lands which increases the volume of runoff for a given rainfall event, urban lands have a higher yield of pollutants than rural areas.

The following summarizes major non-point sources, their pollutional characteristics, and the methods used to control such discharges.

Precipitation contains dirt, dust, combustion products (sulfur and nitrogen oxides), and anything else that is emitted into the atmosphere. Rainfall contaminants are a function of local land-use and air-circulation patterns. Prevention of emissions is the only feasible control.

Contributions from forest and range lands include sediments and wildlife and vegetation decay products. When undisturbed, these lands serve as an excellent indicator of natural runoff quality. However, any disturbance of the land cover, such as logging, causes increased runoff volumes and larger pollutant discharges. Sediments are the major pollutant arising from logging activities. Other possible contaminants include pesticides and herbicides, fertilizers, and any other chemical applied in connection with silvicultural activities.

In general, minimizing disturbances to the land surface will minimize erosion and pollutant production. Erosion-control practices which will reduce sediment production during logging activities include proper design and maintenance of logging roads and skid trail, rapid reforestation of cut-over areas, avoidance of clear cutting, and use of alternative logging procedures. Maintenance of the proper land area to animals ratio on range lands will prevent overgrazing and allow natural regeneration of grasses. Reduce chemical losses by applying only needed amounts at the proper time.

Agricultural croplands discharge sediments, excess fertilizers, nutrients and organics from applied manures, pesticides and herbicides, and leached plant residues. Since surface-water runoff is the carrying medium for these, traditional soil conservation practices that control runoff, such as contour farming, strip cropping, terracing, and crop rotation are the most effective means of controlling pollutant discharges. Again, proper timing and correct amounts are important in reducing chemical losses. Manures should be incorporated immediately into soil to prevent runoff losses. In addition, it should not be spread on frozen ground when runoff potential is high. Plowing plant residues under will prevent any leaching losses.

Accumulation of animal wastes in barnyards and pastures can result in the discharge of organic matter and its decay products, as well as any inorganic salts present from feed supplements, via contact with surface runoff. Animals with access to streams passing through pastures may discharge wastes directly to the streams while watering.

Mining activities result in sediment and acid drainage discharges from spoils and overburden piles at abandoned mines. Control of surface drainage, which prevents the formation of acids and erosion, is the easiest way to control the problem on abandoned mining sites. In new mines, modern mining and reclaimation techniques should be instituted to prevent the problem from arising. Collection and treatment of the mines drainage may become necessary in some cases; this may be accomplished by limestone neutralization, ion exchange, electrodialysis, or ion oxidation.

Though dirt and dust are urban lands' major contaminants, this runoff may include anything. Mechanical street sweeping reduces unwanted discharge. Public education concerning the hazards of unwanted discharge to gutters and storm sewers will help to lesson its impact on urban drainage.

References

1. Aleti, A., S. Y. Chiu, and A. D. McElroy, "Methods for identifying and evaluating the nature and extent of non-point sources of pollutants from agriculture," *Proc. 1974 Cornell Agr. Waste Man. Conf.*, Cornell University, Ithica, New York, 10 (1974).
2. Biggar, J. W. and Corey, R. B. "Agricultural drainage and eutrophication," *Eutrophication: Causes, Consequences, Correctives*, Nat. Acad. Sc. Wash, D.C., 404–445 (1969).
3. Field, R. and E. J. Struzeski, "Management and control of combined sewer overflows," *J. Water Poll. Control Fed.* **44** 1393 (1972).
4. Grurek, W. J. and J. G. Broyan, "A natural nonpoint phosphate input to small streams," *Proc. 1974 Cornell Agr. Wastes Man. Conf.*, Cornell Univ. Ithica, New York, 39 (1974).
5. Klausner, S. D., *et al.*, "Surface runoff of soluble nitrogen and phosphorus under two systems of soil management," *J. Environ. Qual.* **3**, 42 (1974).
6. Kunishi, H. M., *et al.*, "Phosphate movement from an agricultural watershed during two rainfall periods," *J. Agr. and Food Chem.* **20**, 900 (1972).
7. Lin, Shundar, "Nonpoint rural sources of water pollution," *Circular 111 Illinois State Water Survey*, Urbana, Illinois (1972).
8. Murphy, T. A., "Nonpoint source control to meet

water quality goals," *Proc. 1974 Cornell Agr. Wastes Man. Conf. 2*, Cornell University, Ithica, New York, (1974).

9. Office of Air and Water Programs, "Identifications and control of pollution from salt water intrusion," EPA-430/9-73-013, *U.S. Govt. Printing Office*, Wash. D.C. (1973).

10. Office of Air and Water Programs, "Ground water pollution from subsurface excavation," *U.S. Govt. Printing Office*, Wash. D.C., EPA 430/9-93-012 (1973).

11. Office of Air and Water Programs, "Processes, procedures, and methods to control pollution resulting from silvicultural activities," *U.S. Govt. Printing Office*, EPA 430/9-73-010, Washington, D.C. (1973).

12. Office of Air and Water Programs, "Processes, procedures, and methods to control pollution from mining activities," EPA 430/9-73-011, *U.S. Govt. Printing Office*, Washington, D.C. (1973).

13. Office of Air and Water Programs, "Processes, procedures and methods to control pollution resulting from all construction activity," EPA 430/9-73-007, *U.S. Govt. Printing Office*, Washington, D.C. (1973).

14. Romkens, M. J. M. and D. W. Nelson, "Phosporus relationships in runoff from fertilized soils," *J. Environ. Quality* 3, 10 (1974).

15. Schuman, G. E. and R. E. Burwell, "Precipitation Nitrogen contributions relative to surface runoff discharges," *J. Environ. Qual.* 3, 366 (1974).

16. Taylor, A. W., and W. M. Edwards, and E. C. Simpson, "Nutrients in streams draining woodland and farmland near Coshocton, Ohio," *Water Resources Res.* 7, 81 (1971).

17. Timmons, D. R. and R. F. Holt, "Leaching of crop residues as a source of nutrients in surface runoff waters," *Water Resources Res.* 6, 1367 (1970).

18. USEPA, "Selected urban storm water abstracts," EPA, R2-72-127 (1972).

19. USEPA, "Combined sewer overflow seminar papers," EPA, 670/2-73-077 (1973).

20. Weidner, R. B., *et al.*, "Rural runoff as a factor in stream pollution," *J. Water Poll. Contr. Fed.* 41, 377 (1969).

31.3 Urban Non-point Street Wastes

Origin and characteristics. The urban runoff generated during a period of rain or appreciable snow melt can create three types of discharges: (1) combined sewer overflows, (2) storm drainage in separate systems, and (3) overflows from infiltrated sanitary sewers. Storms and snow melts have a pronounced affect on overflow quality. Contents of concern in urban runoff and sewer overflow are solids, infectious bacteria, dissolved organics and inorganics, nutrients, heavy metals, and pesticides. Organic content can average one-half the strength of raw sewage, but is in much greater quantity. Separate system analyses have shown solids concentration to equal that of untreated sewage, five-day BOD's to equal secondary effluent, and bacterial counts two to four times lower than untreated sewage, but two to four times greater than the limits for water contact recreation.

Highways and streets are major sources of contaminant material in urban runoff. These contaminant materials have been shown to be ice-control chemicals, material washed from vehicles, and other inorganic mineral-like materials. A great portion of the pollutional potential existing with these materials seems to be associated with the fine-solids fraction. Based on a 1974 estimate annual usage is 9 to 10 million tons of sodium chloride, 0.3 million tons of calcium chloride, and 11 million tons of abrasives.

Catch basins have been shown to provide little or no effect in removing the important fraction of fines in the contaminant material, but improved street cleaning practices can reduce pollution load potential.

Treatment. Where to attack the problems (at the source, along the way, at the end, or a combination of these) has been considered in recent studies. At the source contaminants may be limited by more efficient street cleaning practices, control of pesticides, and minimization of erosion. Substitute forms of deicing such as in slab melting systems, mobile thermal melters, substitute compounds for NaCl, hydrophobic substances in pavements, and pavements that release energy have been studied but do not seem feasible at this time due to their costs as compared to sodium chloride as a deicing agent. Along the way in-line storage or backwater impoundments have been considered to have the following advantages: (1) an opportunity for equalization is created; (2) they are simple to design and operate; (3) they are responsive to storm changes; (4) they are unaffected by quality and flow changes, and (5) they can be operated in combination with a dry-weather treatment plant. The following are disadvantages of impoundments: (1) they are large; (2) they are costly; (3) they depend on treatment for dewatering and solids disposal. Other forms of treatment include solids removal by sedimentation, dissolved air flotation, screening, and filtration. These treatment processes, however, are insensitive to flow variation and are not effectively utilized during storm periods.

References

1. Anonymous, "Chemical spreader automation," *Better Roads* (1973).
2. Anonymous, "Major considerations in use of deicing chemicals," *Better Roads* (1974).
3. Broecker, Schwartz, Sloan, and Ancona, "Road salt as an urban tracer," *Proc. Street Salting-Urban Water Quality Workshop*, SUNY College of Forestry, Syracuse, New York, July 1971.
4. Bubeck, Diment, Deck, Baldwin, and Liton, "Runoff of deicing salt; effect on Irondequoit Bay, Rochester, New York," *Proc. Street Salting-Urban Water Quality Workshop*, SUNY College of Forestry, Syracuse, New York, July 1971.
5. Clary, A. C., "Deicing: A public works director's dilemma," *Proc. Street Salting-Urban Water Quality Workshop*, SUNY College of Forestry, Syracuse, New York, July 1971.
6. "Erosion-control value of fiberglass," *Rural and Urban Roads*, February 1974.
7. Freestone, F., "Runoff of oils from rural roads treated to suppress dust," *Environ. Prot. Tech. Ser.*, EPA-R2-72-054, October 1972.
8. Hanes, R., L. Zelazny, and R. Blaser, "Effects of deicing salts on water quality and biota," *Highway Research Board*, Report No. 91 (1970).
9. Hawkins, R., "Street salting and water quality in Meadowbrook New York," *Proc. Street Salting-Urban Water Quality Workshop*, SUNY College of Forestry, Syracuse, New York, July 1971.
10. Hutchinson, F. E., "The effect of highway salt on water quality in selected Maine rivers," *Proc. Street Salting-urban Water Quality Workshop*, SUNY College of Forestry, Syracuse, New York, July 1971.
11. Judd, J. H., "Effect on urban salt runoff on lake stratification," *Ibid*, No. 10.
12. Kunkle, S. H., "Effects of road salt on a Vermont stream," *Ibid*, No. 10.
13. Lincoln, H., "Sensible salting success," *Better Roads*, March 1974.
14. Patrick, R., "Effects of channelization on the aquatic life of streams," *Highway Research Board*, Special Report No. 138.
15. Sartor, J. and G. Boyd, "Water quality improvement through control of road surface runoff," *Water Poll. Contr. in Rural Areas*, Chapter 5, Sec. 21, Jewell and Swan (1974).
16. Sharp, R. W., "Road salt as a polluting element," *Ibid*, No. 10.
17. Struzeski, E. J., "Environmental impact of highway deicing," *Ibid*, No. 10.
18. Wood, F. O., "Salt—the universal deicing agent," *Ibid*, No. 10.

31.4   Leachate from Landfill Sites

Origin.   During recent years leachate production has steadily gained recognition as a potential water contaminant. Its formation is a result of infiltration and percolation of rainfall, groundwater, runoff, or flood water into and through an existing or abandoned solid waste landfill site. Upon saturation of the site, the volume of water migrating from the site balances that which infiltrates. It is difficult to predict leachate characteristics due to the variable nature of the constituents of landfill. However, trends are quite obvious upon examination of landfills that are exclusively designated for municipal solid wastes or solid waste ash residue.

Character.   Municipal solid waste typically consists of paper and fibrous material (64 per cent), food wastes (12 per cent), metallics (8 per cent), glass and ceramics (6 per cent), and moisture (20 per cent). Although the moisture content initiates decomposition of readily available organic waste, the high percentage of fibrous products is responsible for the occurrence of moisture-absorption of by-product water generated by decomposition. However, if a landfill becomes saturated by infiltrating water, further decomposition will occur. The rapid initial decomposition rate of refuse will consume free oxygen and further decomposition occurs under anaerobic conditions.

The decomposition of the organic fraction will yield soluble organic substances measurable as BOD, COD, volatile dissolved solids, ammonia, and organic nitrogen. The decomposition process, particularly anaerobic, is typified by the production of volatile acids due to acid fermentation and the subsequent reduction of the acids to methane and carbon dioxide during methane fermentation. Upon conversion of the volatile acids pH neutrality is approached.

In addition, chemical analyses of refuse-derived leachate indicates the presence of heavy metal cations (e.g. calcium and iron) and alkaline metals (e.g. sodium or potassium). Their notable presence results from the increased solubility of metallic compounds in an acidic environment.

Anions present include bicarbonates, sulfates, and chlorides. Also, the combination of appropriate soluble ions and carbon dioxide contribute to hardness measured as calcium carbonate. Staining of water or ground surfaces is indicative of leaching soluble iron.

In contrast to refuse leachate, ash residue leachate is characterized by an alkaline pH. Hence, primarily salts of the alkali metals (i.e. sodium) are leached. The anions, such as nitrates, chloride and sulfates, leach rapidly and completely. Heavy metals such as iron, calcium and chromium are present in insoluble forms.

Another difference in character is the slight BOD and COD associated with ash as opposed to the substantial BOD of refuse-derived leachate with its noted organic content. Suspended and dissolved solids also appear in ash and refuse leachates.

Concerning both leachates, the release of contaminants is directly proportional to the amount of leachate produced and thus the amount of moisture added. Furthermore, leachate strength is a function of its stability and the rate at which it occurs.

Treatment. There are two means of minimizing the effect of leachate upon groundwater: (1) control of source and receptor of infiltrating fluids, and (2) treatment of leachate regardless of fluid source and receptor. The latter method would in most cases be more costly in terms of capital expenditure and operation. Further, the cost range would vary, depending on the climatic and geological conditions affecting a landfill, the constituents of the fill, and the type of treatment selected. Leachate treatment

is generally recommended if initial leachate control methods are ineffective or cannot be practiced due to particular site conditions (e.g., geologic or proximity to flood areas) or if, due to lack of understanding of the need for leachate control or financial limitations during initial operation of landfill, leachate control is not practiced.

The control of leachate formation by minimizing water infiltration can be achieved by intercepting groundwater flow in drainage ditches, as well as by compacting the contents of the fill and providing cover material and grading. With the movement of fluids through soil, based upon soil porosity, size and shape of pores, and pressure gradient, use of a cover soil with a minimal permeability coefficient would limit infiltration. Recently, impermeable synthetic materials have been used to isolate the landfill contents from exterior water movement. Accordingly, proper soil or material selection (such as colloidal clay rather than silty sand) limits infiltration of ground-water flow and percolation of moisture within the refuse through the soil.

In addition, it is important to note the natural attenuating capability of soil, particularly as a filter and/or adsorbent of suspended solids in the leachate. Construction of a trench several feet away from and surrounding a fill serves as a natural treatment by filtration, dilution, and dispersion of contaminated water the percolates through the fill. The accumulated water may be treated as described in the following paragraphs.

The treatment of leachate from landfills is greatly dependent upon the nature of its constituents; namely, municipal waste which is subject to organic decomposition and ash residue which is essentially devoid of organics. The treatments discussed below would apply, in particular, to municipal waste, although application would be appropriate when a fill is used concurrently for ash and municipal wastes or if the organic content of the ash is high due to incomplete burnout of the material.

The type of treatments which have been evaluated and, in some cases, implemented as a viable treatment are:

1. Recirculation of leachate through landfill site
   (a) by injection
   (b) by spray irrigation

2. Biological
  (a) Aerobic
  (b) Anaerobic
  (c) With or without tertiary treatment
3. Physical–chemical

The concept of leachate recirculation, simulating accelerated landfill age, and thus leachate stabilization, had been gaining popularity. The recirculation of leachate through a landfill enhances the biological activity within a fill, thus increasing the rate of biological stabilization of organic material in the wastebed. Consequently, leachate neutrality is achieved in considerably less time under conditions of recirculation than without.

An element necessary for biological stabilization is the availability of nutrients such as nitrogen, phosphorous. The seeding of fresh leachate with recirculated leachate provides a continuous source of nutrients for ultimate decomposition of organic matter. Installation of gas vents will disperse into the atmosphere methane and carbon dioxide which may accumulate in landfill pockets or fissures.

Although treatment by recirculation is effective in minimizing leachate effect upon groundwaters (or streams), the mechanism of such a treatment must be further developed.

Biological treatment, either aerobic or anaerobic, and tertiary treatment would be an appropriate means of refuse leachate control. A typical scheme could be an aerated lagoon with subsequent filtration and activated carbon adsorption.

Successful use of physical-chemical treatment has been mixed. Lab tests indicate a fair removal of contaminants, while a full-scale treatment plant reports that the treated effluent is suitable for potable use. It is apparent from the data that the strength of the influent was not very potent and could easily be purified by coagulation, sedimentation, pH neutralization, and chlorination. There is only one such plant operating in the United States; obviously, implementation of additional plants and analyses of their effluent will provide further insight into the treatment's efficiency and monetary cost for such efficiency.

Ash leachate is essentially devoid of BOD, COD, and heavy metals. As previously stated, alkali salts are detectable, and represent a source of water contamination. Generally, the maximum salt content has not been shown to be greater than an average of one-third of that found in ocean water. Thus the effect of the flow of leachate should be based upon the existing character and best usage of the receiving stream or groundwater flow.

In addition, the leachate has a desirable (slightly alkaline) pH for disposal into groundwater, streams, or municipal treatment plants. Treatment that may be necessary could be solids filtration, which could be achieved by forming a soil filtration zone between the landfill and receptor.

References

1. Apgar, M. A. and D. Langmuir, "Ground water pollution potential of a landfill above the water table," *Ground Water* 9, No. 76 (1971).
2. Boyle, W. C. and R. K. Ham, "Biological treatability of landfill leachate," *J. Water Poll. Contr. Fed.* 5, 46 (1974).
3. California State Water Pollution Control Board, "Report on the investigation of leaching of ash dumps," Sacramento, California, *State Water Quality Contr. Board*, Pub. No. 2 (1952).
4. California State Water Pollution Control Board, "In-situ investigation of movements of gases produced from decomposing refuse," *State Water Quality Contr. Board*, Sacramento, California, Pub. No. 31 (1965).
5. Chamberlain, G. M., "Changes ahead for sanitary landfill designers, operators," *Am. City*, August 1975.
6. Crusberg, T. C., "Chemical and biological characteristics of leachate," *Proc. 1973 Second Ann. Conf., Worcester Polytechnic Inst.*, January 9, 1973.
7. Dowell: Division of Dow Chemical Company, "Dowell Soil Sealant Systems," Midland, Michigan.
8. Garland, G. and D. C. Mosher, "Leachate effects of improper land disposal," Office of Solid Waste Management for USEPA, *Waste Age*, March 1975.
9. Hassen, A. A., "Effects of sanitary landfill on quality of ground water—general background and current study," *1st Ann. Solid Waste Symp.*, Pasadena, California, May 5, 1971.
10. Hill. A. D., "Pave old gravel pit—Town gets sanitary landfill," *Asphalt Inst., Roads and Streets*, August 1973.
11. Hughes, G. M., R. A. Landon, and R. N. Farrolden, "Hydrology of solid waste disposal sites in northeastern Illinois," *Ill. State Geol. Survey* (1971).

12. Lenard, J. F. and M. Dilaj, "Stop leachate problems," *Water and Wastes Eng.*, (1975).

13. Martin Associates, "Abandoned limestone quarry recreated into showplace landfill operation," King of Prussia, Pennsylvania (1972).

14. Metro Water Treatment Corporation, "Treatment of leachate from sanitary landfills by independent physical-chemical methods," *Metro Systems Div.*, Lansdale, Pennsylvania (1975).

15. Pohland, F. G. and P. Maye, III, "Landfill stabilization with leachate recycle," Georgia, EPA, *3rd Annual Environ. Eng. & Science Conf.*, Louisville, Kenya, March 5–6 1973.

16. Pollard, W. S., "Proposed design of treatment system for Cortland County landfill leachate," unpublished report, *Cornell University*, August 14, 1972.

17. Qasim, S. R. and J. C. Burchinal, "Leaching from simulated landfills," *J. Water Poll. Contr. Fed.*, March 1970.

18. Rovers, F. A., G. J. Farguhar, and J. P. Nunan, "Landfill contaminant flux-surface and subsurface behavior," *21st Ontario Ind. Waste Conf.*, June 23–26 1974.

19. Salvato, J., W. G. Wilkie, and B. Mead, "Sanitary landfill-leaching prevention and control," *J. Water Poll Control Fed.* **43**, No. 10 (1971).

20. Vardy, P., "Design, environmental management, and economic considerations for sanitary landfills," *Waste Age*, January/February 1974.

21. Zanoni, A. E. and A. A. Funaroli, "Potential for ground water pollution from the land disposal of solid wastes," Marquette University, Milwaukee, Wisconsin, *CRC Critical Reviews in Environ. Contr.*, May 1973.

CHAPTER 32

HAZARDOUS WASTES

32.1 Definitions

Generators of waste materials can ascertain whether their waste is hazardous by either referring to a published list or by complying with the EPA definition of hazardous waste.

The United States Environmental Protection Agency defines the characteristics of a hazardous waste in terms of ignitability, corrosivity, reactivity, and Extractive Procedure toxicity. These are mentioned by Nemerow (Industrial Solid Wastes (1) 1983) in Table 10.1 and discussed briefly in this text in the following four sections of this chapter.

Many hazardous wastes contain toxic materials which can be classed in more than one category. In these cases the waste will be grouped in the category of major significance. The reader must understand, however, that certain constituents of a waste may also be important and from another category.

The legal United States definition of a hazardous waste is extremely complicated and even controversial. At the time of writing this book, an interested person is urged to refer to the Environmental Protection Agency Regulations for Identifying Hazardous Wastes—40–CFR261; 45–FR33119, May 19, 1980 and revised July 1, 1983 and ammended on ten different occasions from then until July 25, 1985.

The reader can ascertain the status of a certain waste by referring to the following four figures published by the Bureau of National Affairs, Inc. Washington, D.C. 20037. The waste must first receive a legal definition of a RCRA solid waste (See Fig. 32.1). It can then be classified as hazardous and handled as such by referring to Figure 32.2. Regulations covering hazardous wastes are shown in Figure 32.3 and for those subject to control under Subtitle C of the EPA Regulations in Figure 32.4.

32.2 Dangers

Once a decision has been made to manufacture a potentially toxic chemical or product, the door is open to catastrophic release of either to the environment. Careful control must be maintained by the industrial manufacturer and product users from the very beginning. The dangers of toxic wastes being released to the environment are real and originate from three sources: (1) production of the toxic chemical, (2) use of the toxic chemical in subsequent manufacturing a product, and (3) release of a toxic product (or other chemical) from the use or wastage of the product. In addition, there is always the potential hazard of escape of any of these toxic materials during transportation from any point to another location for use or disposal. In reality all of these dangerous omissions or releases of toxic material to the environment occur accidentally. No one in his right mind purposely intends to cause these catastrophies. But, accidents occur frequently enough to be considered a usual occurrence no matter how infrequently they happen. The only reasonable prevention—other than non-use of hazardous materials—is to design fail-safe devices to operate when the accidents occur.

32.3 Scope of Problems and Costs of Solutions

Pearce (2) (UNEP, 1983, pg. 57) reports that in the industrialized nations, at least, by far the largest proportion of hazardous waste generated, is recycled or reclaimed by industry. He gives as an example Great Britain, where about ten million tons per year are recovered by industry while only 3.4 million tons are rejected for disposal.

United States Environmental Protection Agency determined (1980) that costs of hazardous waste disposal

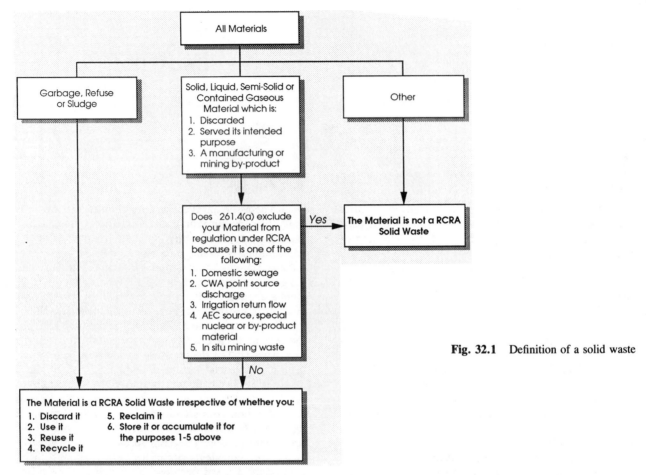

Fig. 32.1 Definition of a solid waste

were major factors in attaining final solutions to the problem. Although costs vary widely, E.P.A. gave ranges for various treatments as shown in Table 32.1.

Table 32.1 Cost of Hazardous Waste Disposal Practices (U.S.E.P.A., 1980)

Technology	$/Metric Ton
Land spreading	2-25
Surface impoundment	14-180
Chemical fixation	5-500
Secure chemical landfill	50-400
Incineration (land-based)	75-2,000

A more recent E.P.A. rough general cost comparison for hazardous waste treatment (Assessment of Incineration—1985 March) gives the following:

Type of Waste Management	Type or Form of Waste	Price 1981 $/metric Tonne
Landfill	Drummed	168-240
	Bulk	55-83
Deep Well Injection	Oily wastewater	16-40
	Toxic rinse water	132-164
	Cyanides/heavy metals, and highly toxic waste	66-791
Land Incineration	Liquids	53-237
	Solids and highly toxic liquids	395-791

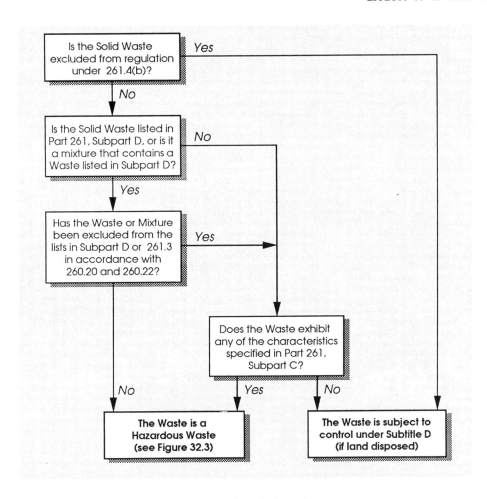

Fig. 32.2 Definition of a hazardous waste

32.4 Effects on Environment

Hazardous wastes injure the air, water, and land environments drastically. They often effect human beings directly and most certainly indirectly. The deterioration of the environment is extensive and often irreversible—or reversible at heavy cost of both money and manpower. They are caused by wastes which burn, corrode, react violently, are toxic, are acutely hazardous to health and welfare of society.

It is becoming increasingly evident that traces of hazardous chemicals, including suspected carcinogens, are getting into the food chain via surface water supplies. People are exposed through drinking water, air, and food such as fish originating in rivers and lakes.

The toxic pollutants find their way into surface water supplies from waste treatment plants, coal burning, refuse incineration, and atmospheric fallout. An example of this dangerous effect is that occurring in the U.S. Great Lakes basin (Great Lakes Study 3, 1985) in which 1,065 hazardous or potentially hazardous substances were identified in the lakes, with the highest concentration in the southern edges. Besides the recreational and commercial fishing which go on in these lakes, many large cities, such as Syracuse, New York, take in raw water from them (Lake Ontario) for drinking water purposes. Drinking waters must now be carefully examined and analyzed for potentially hazardous chemicals.

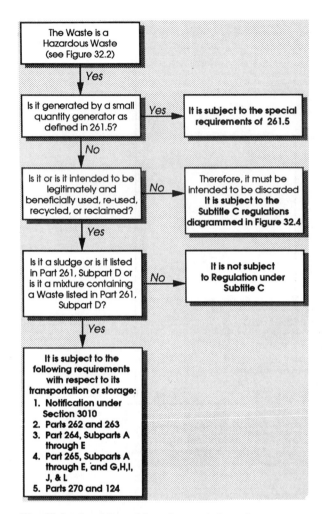

Fig. 32.3 Special provisions for certain hazardous waste [Amended by 48 FR 14153. April 1, 1983]

32.5 Philosophies of Solutions

Problems caused by hazardous wastes must be solved by (1) not producing them in the first place, (2) by collecting and reusing every bit of them in some other product not effecting the environment, or (3) by destroying them in such ways that there will either be nothing left or their residues will be compatible with and harmless to the environment.

Any of these solutions cost money to instigate and/or take ingenuity and research to implement. They cannot

usually be solved in ways heretofore practiced by municipalities and industries in rather conventional manners.

One of the major objectives of this book will be to explore some potentially feasible means of solving what is now a major problem of our society—that of disposing of our hazardous wastes.

Developed countries such as the United States, West Germany, France, etc. tend to locate chemical plants which utilize or produce hazardous chemicals in less developed countries in order to minimize production and distribution costs. A lesson rapidly being learned by developed countries is that these lower production costs cannot be gleaned at the expense of the local environment. Hazards in developing countries are just as dangerous and intolerable from these plants as they are in developed nations. It was originally thought possible by developed nations to export environmental damages—and also acceptable to lesser developed countries in order to gain the economic consequences. The Bhopal incident accentuates the fact that this kind of thinking on anyone's part no longer exists.

32.6 Varied Types of Industrial Hazardous Wastes

In 1974 the U.S. Environmental Protection Agency listed 15 major industries which discharge wastes containing hazardous substances. They indicated the specific hazardous materials in each waste as shown in Table 32.3A.

Sundaresan et al. (UNEP pg. 70, 1983) groups industries producing toxic and hazardous wastes into two categories:

(1) conventional solid wastes generated in large quantities and stored near the factory such as phosphatic fertilizers and thermal power plants

(2) other semi-solid and solid residues including liquid wastes resulting from the organic and inorganic chemical industries. These wastes are not voluminous but some are highly toxic and potentially hazardous.
Yakowitz (107, 1988) presents a list of 14 types of hazardous wastes (Table 32.2B). It appears to be the most comprehensive breakdown of specific, general characteristics.

The American Chemical Society is especially concerned with the various types of hazardous wastes. It published the following Table 32.3A of types of these wastes of concern to its members (Hazardous Waste Management (4) 1984).

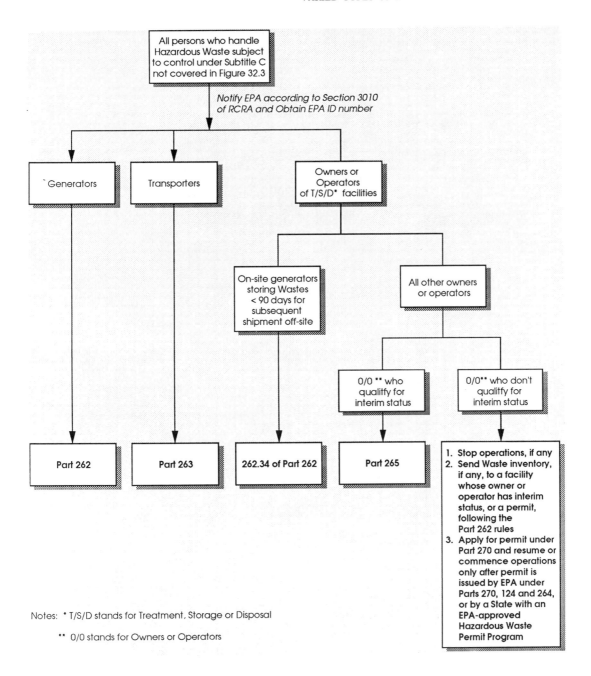

Fig. 32.4   Regulations for hazardous waste not covered in
diagram 3 [Amended by 48 FR 14153. April 1, 1983]

Table 32.2A   Hazardous Waste Industries

Industry	Hazardous Substances
1. Mining and Metallurgy	As,Dd,Cr,Cu,Cn,Pb,Hg,Se,Zn
2. Paint and Dye	Cd,Cr,Cu,Cn,Pb,Hg,organics,Se
3. Pesticide	As,Cl-hydrocarbons,Cn,Pb,Hg,organics,Zn
4. Electrical and Electronic	Cu, " " " " Se
5. Printing and Duplicating	As,Cr,Cu,Pb,organics,Se
6. Electroplating-Metal Finishing	Cd,Cr,Cn,Cu,Zn
7. Chemical Manufacturing	Cl-hydrocarbons,Cr,Cu,Pb,Hg,organics
8. Explosives	As,Cu,Pb,Hg
9. Rubber and Plastics m	Cl-hydrocarbons,Cn,Hg,organics,Zn
10. Battery	Cd,Pb,Ag,Zn
11. Pharmaceutical	As,Hg,organics
12. Textile	Cr,Cu,organics
13. Petroleum and Coal	As,Cl-hydrocarbons,Pb
14. Pulp and Paper	Hg,organics
15. Leather	Cr,organics

32.7 Ignitable Hazardous Wastes

The Environmental Protection Agency (5) (1980) classifies these ignitable types of hazardous wastes in four categories as follows:

1. A liquid that has a flash point less than 60°C. (140°F.). Exemption—aqueous solution with less than 24% alcohol.
2. A waste that is not a liquid but is capable under standard temperature and pressure of causing fire through friction, absorption of moisture or spontaneous chemical changes, and, when ignited, burns so vigorously and persistently that it creates a hazard.
3. An ignitable compressed gas
4. An oxidizer

From the above definitions, such ignitable hazardous wastes can originate from all three physical states: gases, liquids, and solids.

32.8 Waste Oils

Waste oils not only are objectionable aesthetically in any of our three environments (air, water, or land), but also are classified as toxic or hazardous because they are ignitable. Burning by itself is a hazard to our lives and property. Further, the products of combustion can also be toxic to our breathing. For all these reasons, oils must be treated and ultimately disposed of so as to prevent toxicity to any environment.

Automotive crankcase used waste oil constitutes a large proportion of waste oils. Collection, recovery, and repurification of waste oils represents the ideal solution to the disposal problems. However, economics must justify such methods. When oil prices rise, justification for recovery is enhanced. When damage costs to the environment are included in the economics—and added to the current price of new oil—justification for recovery is also improved. For evaluating recovery of waste oils, the physico-chemical analysis of average waste oils is necessary. A typical analysis is given in the following Table 32.3B. The main obstacle to recovery is that of collection. The use of strategically-located collection stations where individuals and garages can bring their oils will enhance recovery. Other deterrents to recovery include additives in the oils, environmental laws on reprocessing, and removal of tax incentive laws for reuse of waste oils.

Other possibilities for treatment include: (1) mixing waste oils with lighter new fuel oil for burning in power plants, and (2) incorporating with municipal solid wastes and tree and bush cuttings or beach littoral collections for composting. No actual data is currently available on either of these methods.

32.9 Spent Oil Emulsions

These wastes originate mainly from metal fabricating and power plants where mixtures of oils and waters are used to cool moving parts. Since these oils are contaminated with water and grit or cuttings, they can be centrifuged or skimmed for concentration prior to transportation away from the site of origin. Lubricating oils can also be heated, coagulant added, and centrifuged prior to reuse in the same plant.

32.10 Other Oily Wastes

Other waste oils originate from ship discharges—either accidental or planned. Oil water mixtures from bilges are usually discharged to sea when the bilge water level gets too high. Sometimes the bilge water is coagulated and filtered before discarding the water. In that case a residual sludge remains to be disposed of. In proper proportions and with good premixing and feeding devices, it may be burned in special furnaces, even on board ship. Usually it is held in a collection tank and pumped out at an onshore station. Such stations can handle larger quantities of oily sludges more effectively and economically. Furnaces or incinerators can be designed specifically to burn these wastes.

Table 32.2B   List of Hazardous Characteristics (After Yakowitz, 107, 1988)

Code Number	Characteristics
H1*	Explosive An explosive substance is a solid or liquid substance (or mixture of subtances) which is in itself capable by chemical reaction of producing gas at such a temperature and pressure and at such a speed as to cause damage to the surroundings.
H2*	Oxidizing Substances which, while in themselves are not necessarily combustible, may, generally by yielding oxygen, cause or contribute to the combustion of other materials. (Organic substances which contain the bivalent-0-0-structure are thermally unstable substances which may undergo exothermic self-accelerating decomposition.)
H3*	Inflammable The word "flammable" has the same meaning as "inflammable." Inflammable liquids are liquids, or mixtures of liquids, or liquids containing solids in solution or suspension (for example, paints, varnishes, lacquers, etc. but not including substances otherwise classified on account of their dangerous characteristics) which give off an inflammable vapour at temperatures of not more than 60.5°C, closed-cup test, or not more than 65.6°C, open-cup test. (Since the results of open-cup tests and of closed-cup tests are not strictly comparable and even individual results by the same test are often variable, regulations varying from the above figures to make allowances for such differences would be within the spirit of this definition.) Inflammable solids are solids, other than those classed as explosives, which under conditions encountered are readily combustible, or may cause or contribute to fire through friction.
H4**	Irritating Non-corrosive substances and preparations which, through immediate, prolonged or repeated contact with the skin or mucous membrane, can cause inflammation.
H5**	Harmful Substances and preparations which, if they are inhaled or ingested or if they penetrate the skin, may involve limited health risks.
H6*	Toxic Substances and preparations which, if they are inhaled or ingested or if they penetrate the skin, may involve serious, acute or chronic health risks and even death.

Table 32.2B   (*Continued*)

Code Number	Characteristics
H7**	Carcinogenic Substances and preparations which, if they are inhaled or ingested or if they penetrate the skin, may induce cancer in man or increase the incidence (a).
H8*	Corrosive Substances which, by chemical action, will cause severe damage when in contact with living tissue, or, in the case of leakage, will materially damage, or even destroy, other items or a means of transport; they may also cause other hazards.
H9*	Infectious Substances containing viable micro-organisms or their toxins which are known, or suspected, to cause disease in animals or humans.
H10*	Liberation of flammable gases in contact with water Substances which, by interaction with water, are liable to become spontaneously inflammable or to give off inflammable gases in dangerous quantities.
H11	Liberation of corrosive fumes in contact with air or water.
H12	Liberation of toxic gases in contact with air or water.
H13	Capable, by any means, after disposal, of yielding another material, e.g., leachate, which possesses any of the characteristics listed above.
H14	Ecotoxic Substances which, if released, present or may present immediate or delayed adverse impacts to the environment by means of bioaccumulation and/or toxic effects upon biotic systems.

*Definition taken from *Transport of Dangerous Goods*, Recommendations of the United Nations Committee of Experts on the Transport of Dangerous Goods, Third Revised Edition, United Nations, New York, 1985.
**Definition taken from Article 2 of the European Communities Council Directive of 18th September 1979 amending for the sixth time Directive 67/548/EEC on the approximation of the laws, regulations and administrative provisions relating to the classification, packaging and labelling of dangerous substances [Directive 79/831/EEC.]
(a) Guidance with regard to this characteristic may be obtained by consulting the lists of known and strongly suspected carcinogens published periodically by the International Agency for Research on Cancer.

Onshore oil sludge storage tanks can also be used to receive tank cleaning wastes. These wastes are usually more dilute, containing mixtures of condensed steam and oil.

Power stations both stationary and moving generally utilize centrifugation to purify their fuel oil before burning in order to remove the last traces of water. The latter contaminant may reduce the boiler efficiency. It is possible to use these or other standby centrifuges to recover oil from tank cleanings or even bilge waters. The recovered oil could be burned in the power plant's boiler or repurified, if necessary, and sold or reused for lubricating purposes.

32.11 Oily Shipboard Waste or Refinery Sludges

Waste oils can be mixed with soils and decomposed by naturally adapted soil bacteria. Dotson et al. (Yard spreading (6) 1974) report that bacteria of the genus Pseudomonas grew most rapidly in a soil-oil mixture. These bacteria continued to grow until this food source was exhausted. Despite this finding Dotson et al. blame lack of information on fullscale studies as the reason for not using this disposal method. A mixed population such as that of soils is more efficient than a single species or genus. Petroleum hydrocarbons vary in susceptibility to decomposition. High molecular weight, viscosity, and crystallinity are properties that inhibit biological oxidation and decomposition. Straight chain medium molecular weight hydrocarbons such as kerosene and light motor oils oxidize readily, while aromatic types are more resistant. Some evidence of oil decomposition in soil is provided by the oxidation of the asphalt coatings of underground pipes and the disappearances of oil leakages in lakes and shore areas attributed to microbial assimilation.

Table 32.3A RCRA-Regulated Hazardous Wastes

The following list, taken from the EPA survey, provides estimates of the percentage of generators producing specific types of RCRA-regulated hazardous wastes during 1981. The sum of the total exceeds 100% because a generator may have produced more than one type of waste, and a given waste stream may have been a mixture that consequently was reported under multiple characteristics.

Waste Group Generated	% of Establishments Generating
Spent solvents-halogenated and nonhalogenated	51.0%
Ignitable wastes*	43.4%
Corrosive wastes*	33.4%
Spilled, discarded, or off-specification commercial chemical products or manufacturing chemical intermediates	28.8%
EO toxic wastes*	27.8%
Electroplating and coating wastewater treatment sludges and cyanide-bearing solutions and sludges	16.4%
Statutory hazardous wastes (i.e., not listed or regulated as hazardous wastes by EPA or state)	12.2%
Listed industry wastes from specific sources	10.2%
Acutely hazardous wastes	10.2%
Reactive wastes*	7.1%

*These waste groups include wastes that, while not specifically listed in EPA's list of hazardous wastes, exhibit one of the hazardous characteristics, such as high flammability, high or low pH levels, toxicity, and volatility or violent reactive tendencies with other substances.

32.12 Corrosive Hazardous Wastes

EPA (5) (1980) considers corrosive any wastes falling into either of the following two classes:

1. An aqueous waste that has a pH less than or equal to 2, or greater than or equal to 12.5.

2. A liquid that corrodes steel at a rate greater than 6.35 (0.250 inches) per year at 55 °C.

These corrosive wastes could be either very acid or very alkaline and thereby hazardous to any environment into which they were emitted.

32.13 Acid Wastes

Acidic wastes may be discharged by any plant manu-facturing materials, apparel and chemicals. Most important and hazardous are those coming from chemical plants such as those producing dyes, explosives, pharmaceuticals and silicone resins. The major ones are dilute wastes from hydrochloric, sulfuric, and sometimes nitric acid. Regardless of the degree of acidity or the origin of acid wastes, the main method of treatment is neutralization.

32.14 Alkaline Wastes

Alkaline wastes are considered corrosive at pH values greater than 12.5 and hence classified as hazardous. As such they must be treated properly before release to the environment. Like acidic wastes they too must be neutralized as the authors describe in Chapter 8.

32.15 Reactive Industrial Wastes

EPA (5) (1980) rather broadly defines reactive wastes into eight distinct categories:

1. Normally unstable, readily undergoes violent change without detonating.
2. Reacts violently with water.
3. Forms potentially explosive mixtures with water.
4. When mixed with water, generates toxic gases, vapors, or fumes in dangerous quantities.

Table 32.3B Typical Waste Automotive Oil Composition

Variable	Value
Gravity, °API	24.6
Viscosity at 100°F	53.3 Centistokes
Viscosity at 210°C	9.18 "
Flash point	215°F (C.O.C. Flash)
Water (by distillation)	4.4% volume
BS&W	0.6 by volume
Sulfur	0.34 weight %
Ash, sulfated	1.81 " "
lead	1.11 " "
Calcium	0.17 " "
Zinc	0.08 " "
Phosphorous	0.09 " "
Barium	568 ppm
Iron	356 "
Vanadium	5 "

5. Cyanide or sulfide bearing waste when exposed to pH conditions between 2 and 12.5 can generate toxic gases, vapors, or fumes in dangerous quantities.
6. Capable of detonation or explosive decompostion or reaction at standard temperature or pressure.
8. Forbidden explosives, class A or class B explosives.

32.16 Wastes Containing Cyanides and Isocyanides

These wastes are generally recognized by the public as being very 'reactive' and thereby classified as hazardous. They are classified as generating toxic gases, vapors, or fumes in dangerous quantities. They are used and released from many of our chemical industries and metal plating plants. These compounds aid in solubility and plating of protective metals or machined parts varying from typewriters to silverware. They also can serve as a basic starting chemical for nitrogenous fertilizers. However, when they are allowed to reach the environment without treatment and in sufficient concentration, they are deadly.

Other chemicals containing cyanides are also used and occasionally released to the environment. Such a situation occurred accidentally at Bhopal, India, in December 1984 when a tank of methyl isocyanate leaked the gas and killed over 2,000 persons. The chemical,

MIC, has a formula of CH_3NCO or

$$HC - N = C$$

with H above and H, O below, is reported by Sax (Dangerous Properties—1979) to be highly dangerous when exposed to heat, flame or oxidizers. It is highly irritating to the skin, eyes, and mucous membrane and can cause pulmonary edema and can be absorbed via the skin.

This chemical is an important starting compound for the manufacture of nitrogenous fertilizers. Since it exists as a liquid and boils at the relatively low temperature of 39.1° celsius, it must be kept quiescent and cool until used or oxidized to degrade the cyanide.

32.17 Sulfide Residues

These residues or sludges may result from the natural or artificial precipitation of sulfur bearing wastes from a variety of chemical wastes such as sulfur dyes from textile plants or acid sludges from oil refinery processing.

32.18 Oil Refinery Sludges

Spent caustic wastes from catalytic polymerization and alkylate washes in oil refineries contain considerable amounts of sulfides. Acidification of these caustic wastes (with H_2SO_4) removes some of the objectionable compounds. The acid sludges may be used as a source of fuel or to produce by-products such as oils, tars, asphalts, resins, fatty acids and chemicals. Some refineries recover sulfuric acid from the acid sludges for their own use. Without reuse or recovery, these acid sludges still contain high amounts of precipitated sulfides which, in turn, could be released to the environment if the pH or other conditions caused the leaching of the sludge.

32.19 Trihalomethanes

Organic compounds reaching drinking water supplies mainly from waste cleaning solutions contain toxic chemicals such as trichloroethylene, tetrachloroethylene, methyl chloride, vinyl chloride, carbon tetrachloride. These chemicals are all relatively volatile— reverting from the liquid to the gaseous state readily— and can be removed from the water supply by either granular activated carbon adsorption or air stripping.

Halogenated methane or ethane compounds are highly reactive, unstable, and can release toxic gases. They are, therefore, considered hazardous. Such compounds can be released in their manufacture in the petrochemical industry or in the resulting solvent use in dry cleaning establishments.

Major methods of treatment include prevention of their release by collection tanks for accidental overflow spills and rejects. Most of these compounds are colorless liquids with ethereal odors, relatively low boiling points and high vapor pressures, and very irritating to the conjunctiva and hence dangerous to the health of any breathing human or animal.

32.20 E P Toxic Industrial Wastes

Toxicity to humans and animals as described in this chapter and in chapter 28.10 can affect different organs of the body. For example, carcinogenic toxicity causes cancer in living cells and tissues while genotoxicity causes genetic damage to these same cells and tissues. On the other hand, neurotoxicity causes toxic effects only on the nervous system, nephrotoxicity on the kidneys and hepatoxicity on the liver. It is not the object of this book to discuss details of toxicity, but only to recognize that there are various types and each may be caused by many different chemicals at different concentrations for different periods of exposure.

EPA (5) (1980) considers a solid waste toxic if, when extracted by the EP method (Nemerow, Industrial Solid Wastes, 1983, page 116) (1), the leachate contains ions of constituents equal to or greater than 100 times the Primary Drinking Water Standard. If the waste contains less than 0.5% filterable solids, then the filtrate is considered the extract.

32.21 Mercury–Containing Wastes

Caustic chlorine plants use mercury cells and waste mercury. Sundaresan et al. (7) (UNEP 73, 1983) report a plant producing 60 tons of alkali per day and discharges 1400 m³ of wastewater with a pH of about 11 and from 3 to 5 mg/l of mercury. Brine sludges are also reported to contain from 18 to 20 mg Hg/l. Leaching of mercury into water supplies with subsequant biomagnification causes serious diseases of the nervous system to humans eating fish from such waters.

Mercury, a toxic chemical, is also a naturally–occurring element on earth. However, about 10,000 tons of it are extracted from HgS, cinnebar ore, in the United States each year. Of this about half is released directly into the environments through the discharge of industrial wastes.

The accumulation of mercury in predatory fish, such as tuna and swordfish, constitutes a major danger to humans eating the fish. Many catastrophies, such as that in Minimata Bay, Japan have resulted in death from eating fish contaminated with mercury. Other industries besides the chlor–alkaline plants discharging mercury–laden wastes are electric wire and equipment and paper processing.

Treatment of mercurous wastes by industry involves primarily recovery within the plant and better housekeeping practises to eliminate wastes.

32.22 Metal Containing Sludge

Biocycle (Hazardous Waste Landfill (8) 1983) describes the satisfactory use of chrome–bearing sewage sludges as compost for stabilizing embankments of berms, drainage ditches, and roadsides.

Many varied types of industrial and municipal–originated sludges contain metals which constitute a hazard to the environment receiving them.

Municipalities contribute two main sludge-type wastes which contain metals: (1) wastewater (sewage) sludge, and (2) refuse solids or sludge. The concentrations of metals in both of these municipal sludges depends upon the types of industries within the collection system and the habits and practises of these plants and people in the district.

Domestic sewage sludge often contains various amounts of heavy metals. The latter originates in wastes discharged mainly from small electroplating shops and garages located within the municipal limits. The quantities are difficult to control even though many cities have enacted ordinances limiting allowable concentrations of metallic ions. Generally, no more than 1 ppm of copper, cyanide, or chromium and 2 to 5 ppm of zinc or nickel should be allowed in sewage plant influents. If the quantities exceed these values, concentrations in the sludges resulting from sedimentation or biological treatment contain excessive amounts of metals which hinder anaerobic digestion processes.

Chemical precipitation at elevated pH values aid in the removal of heavy metals by precipitation of the hydroxide or carbonate. Refer to chapter 27.6 for treatment details.

32.23 Other Inorganic Wastes

In general, the presence and removal of inorganic/dissolved minerals from wastewaters have been given relatively little attention by environmental engineers. But, with the passage of the TOSCA legislation of 1976 and RCRA legislation of 1976, these minerals which are deemed toxic and the type and concentration are given serious consideration for treatment and removal.

Chlorides, phosphates, nitrates, sulfates and certain metals are examples of the more common and significant inorganic dissolved solids. Among the methods employed mainly for removing inorganic matter from wastes are: (1) evaporation, (2) dialysis, (3) ion exchange, (4) algae, (5) reverse osmosis, (6) chemical precipitation, and (7) certain oxidation-reduction reactions. The authors describe these treatments in Chapter 12.

32.24 Pickling Liquors

Before applying the final finish to steel products, the manufacturer must remove dirt, grease, and especially iron oxide scale which accumulates on the metal before and during fabrication. Normally this is carried out by plunging the steel material into dilute sulfuric (about 20% by weight) acid. The process, known in the trade as "pickling," produces a hazardous waste called pickling liquor containing primarily ferrous sulfate.

32.25 Galvanizing and Metal Plating Wastes

The authors present a good description of the origin, characteristics, and treatment of these wastes in Chapter 27.5 and 27.6. After metals have been fabricated into the appropriate sizes and shapes to meet customer specifications, they are finished to final product requirements. Finishing is accomplished by stripping, removal of undesirable oxides, cleaning, and plating. In plating, the metal to be plated acts as the cathode while the plating chemical metal in solution acts as the anode. The overall liquid wastes are relatively small in volume but are extremely toxic as confirmed by the EP toxicity test. Most significant toxic metals are chromium, zinc, copper, nickel, and tin. Acids, cyanides, alkaline cleaners, grease and oil are also found in these wastes and must be considered when planning treatment of metals.

32.26 Graphic and Photographic Waste Liquors

Wastewater from large scale film developing and printing operation contains "spent" solutions of developer and fixer which carry thiosulfates and silver. The solutions are usually alkaline with various amounts and types of organic reducing agents along with the silver. The silver metal is toxic in certain concentrations and, therefore, qualifies this waste as a toxic one.

Treatment to eliminate the toxicity involves removal and recovery of valuable silver metal. The authors described in Chapter 27.3 three methods generally used for silver recovery: metallic replacement, electrolysis, and precipitation.

32.27 Salts

Salts are generally considered non-toxic primarily because in most water resources such as lakes, streams, and groundwater their concentration is still relatively low. However, salt concentrations of over 2,000 to 3,000 parts per million are generally toxic to plants. Salts gradually build up in receiving waters, especially those being reused for irrigating croplands where rainfall is low. Here salts may never flush out of the root zone of the crops. Where drainage from these croplands is hampered by underlying clay layers, salt pockets in the groundwater will build up.

Citing figures by the United Nations Food and Agricultural Organization (FAO), Earthscan reports that today "about 120 million hectares (300 million acres)—half of the world's irrigated land—suffer from reduction of crop yields due to salinization" (Salt of the Earth (9) 1984).

One way of controlling the toxicity of salts in groundwater caused by continual reuse of irrigation waters with inadequate land drainage is planting of alfalfa or in certain cases some grasses. Alfalfa is an extremely thirsty and deep-rooted perennial. It will soak up the high-salty water table and cause it to lower to below the normal crop root level. After a number of seasons or years, the original crops of barley or wheat can be replanted without the danger of salt toxicity. In the meantime, the alfalfa or grasses can be harvested for other uses.

When certain industries such as pickle processing and textile finishing discharge heavily salted wastes, they may increase the salt concentration in receiving waters to toxic levels. In such cases, these wastes should be segregated and concentrated further by solar evaporation or membrane filtration, and the salt then recovered and reused by industry.

Without improved salt removal from industrial wastewaters, our water resources will eventually become salt-toxic. When and if that occurs, civilization as we have known it will begin to deteriorate and dry up.

A possible solution is to develop more salt tolerant crops as foods. Hodges and O'Leary describe (Helping Crops (10) 1985) plants known as halophytes that grow like weeds along seacoasts and estuaries. One of them, salicornia, a plant that grows to a foot or two in height looks like a bunch of green pencils. The plant produces oil high in polyunsaturates at a greater yield than soy beans, a high protein food product, and a principal source of vegetable oil. They also found two varieties of another crop, atriplex, yield as much animal feed as alfalfa and can be harvested several times a year. They report much work progressing on developing hybrid strains of salt-tolerant wheat, rice, barley, and tomatoes.

32.28 Phosphatic Fertilizer Sludges and Ore Extraction Wastes

Wastes from the phosphate industry arise from (1) mining the rock and (2) processing the rock to elemental phosphorous and other pure chemicals. In mining the main wastes originate from the washer plant when the rock is separated from the water solution, and the flotators, where phosphate particles are separated from the impurities retained on the screens. In processing the phosphate, the major source of water-borne waste is the condenser water, bleed-off from the reduction furnace.

The wastes are high in volume and contain fine clays and colloidal slimes as well as some resinous oil from the flotators. Both mining and processing wastes contain small concentrations of radon emitted from uranium rock associated with the phosphate rock. Flotation wastes contain some resinous oils. Mining slimes also contain other heavy metals. Toxic constituents are classed as radioactivity, oils, and heavy metals. (Refer to Chapter 28.3.)

The major hazard of these sludges comes from leaching of the toxic contaminants into water supplies or exposing humans to radioactivity from reused gypsum as building material.

32.29 Ore Extraction Wastes

Copper, lead, and zinc can be extracted from their ores and concentrated to obtain pure metals. In copper processing about four tons of concentrated ore produces a ton of pure copper metal. Slag, which develops as a waste from the metal smelting, contains silicon, iron, and magnesium, as well as 6 to 8 percent zinc, 0.5 percent manganese, 0.4 percent copper and 1.7 percent sulfur. Reused spent acid from the electrolyte is occasionally wasted and contains small amounts of toxic metals considered hazardous to the environment. Refer to Chapter 27.5.

32.30 Power Plants

The operation of steam power plants generally involves the generation of heat from coal, oil, or uranium fuel to produce steam with exceptionally pure water. The steam is used to drive turbines which in turn are coupled to generators.

Toxicity of residuals arising from the production of electricity in the above manner originates in the following:

(1) hot, concentrated water salines from boiler and evaporator blowdown. These may include various metals and hexavalent chromium used as a corrosion inhibitor. These are hazardous as described under Chapter 32.1.
(2) acid and alkaline chemical solutions used in cleaning power plant equipment. These are hazardous as described under Chapter 3.
(3) acid water drainage from coal storage and ash ponds. These also are hazardous as described in Chapter 3.
(4) leakage or discharge of nuclear-fired power plant cooling waters. These are hazardous as described in Chapter 6.
(5) fuel (nuclear) reprocessing plant wastes. These are also hazardous as described in Chapters 32.3 and 32.4.

In Chapters 29 and 30, the authors characterize this industry and its processes in great detail.

32.31 Nitrogenous Fertilizers

In all these fertilizers, ammonia is the basis and by far the most significant compound in the wastewaters. In the fertilizer effluents from the production of ammonium phosphates, a considerable amount of fluorine gas is driven off or elemental and fluorosilicious acid fluoride is found in the condensates. This element and traces of arsenic are the main hazardous wastes. They are toxic as classified in Chapter 32.4. The major method of treatment of the toxic components has been storage of the wastewaters in lagoons from which they are equalized and proportioned to discharge over longer periods of time, coagulation of the fluoride with lime and/or alum–polyelectrolytes, or recovery of the fluorine directly as fluorosilicious acid.

32.32 Wood Preserving Wastes

Almost all of the 400 United States wood preserving plants use pressure processes to preserve wood (for long term exposure to soil, water, and air) with either creosote or pentachlorophenol or both. Air dried timber is first teamed at about 20 psi for up to 12 hours while the condensate is continuously removed. Following evacuation and raising the pressure with air to 30–90 psi, the retort is filled with the creosote or pentachlorophenol oil. When all the air has been expelled by the preservative, pressure is maintained at about 200 psi for 2–8 hours. Entrapped air and preservative bubble out after the pressure is released. The preservative escaping is recovered and returned to a storage tank for reuse. Excess oil is removed by a vacuum prior to steaming. This condensate is the source of the major portion of toxic waste. Refer to Chapter 27.22.

32.33 Acute Hazardous Industrial Wastes

This category of hazardous wastes includes industrial wastes which are a danger to human beings and animals in very small concentrations when released into the environment. They affect the health of all living beings in an insidious manner—usually invisible and undetectable without sophisticated sampling and measuring equipment. Such wastes include radioactive, biological, and asbestos–laden air and water.

32.34 Radioactive Wastes

Major origins of these wastes are from the use of nuclear energy by power plants and hospitals. Natural fallout from nuclear explosions is also an ever present possibility and has occurred in the past. In power plants radioactive wastes can be released in contaminated cooling waters, leaks and subsequent wash-downs. In addition, fuel elements need reprocessing to renew the uranium rods periodically. Acid reprocessing wastes containing high levels of radioactivity can be released either at the power plant site or at designated national reprocessing sites. For example, a tank filled with radioactive gas ruptured at an uranium reprocessing plant on January 4, 1986 in Gore, Oklahoma. The container was being heated at the time (Worker killed, dozens hurt—1986) when it burst, releasing about 14,000 pounds of slightly radioactive uranium hexafluoride gas which breaks down into toxic hydrogen fluoride and low level radioactive uranyl fluoride particles. One person died of acid burns and 14 were hospitalized for skin and respiratory system exposures.

American homes that may be contaminated with radon gas could be causing as many as 30,000 lung cancer deaths per year. The gas rises to the land surface from any source of uranium and can travel horizontally for miles before penetrating cracks in house foundations. If the air within the building is not exchanged continually with outside air, the radon concentration can build up to levels which are dangerous.

Since these wastes cause genetic changes, usually in the form of some type of cancer in humans and animals coming in contact with them, proper disposal is mandatory. The authors describe these wastes and their treatment in some detail in Chapter 30. Generally, they are either concentrated and contained or diluted and dispersed. In the United States the plans are for using the former method on all high level radioactive wastes and burying them in governmentally-selected and monitored, safe, deep underground sites. Low level radioactive wastes are retained until safe levels of radioactivity are reached or are diluted with other wastes and discharged to sewers or streams.

Radioactive wastes may be released to the air, land, or water environments from many varied sources. Some of these include hospitals, power plants, and various industrial laboratories. Its ultimate site for disposal determines—in many cases—the hazard to the environment.

One unusual source of radioactive contamination occurred in several loads of steel construction bars shipped to Arizona from a Mexican plant. The steel was contaminated with Cobalt 60, a man-made radioactive material produced in nuclear reactors. It is also used to treat cancer tumors and in gauging devices. In this case (Miami Herald—1984) the steel bars were already in place in several construction sites in Arizona when the radioactivity was discovered. Some bars contained up to 350 millerems per hour; the permissable dosage for U.S. workers in the nuclear industry is 5,000 millerems per year.

As another step in using underground burying of these wastes, the Department of Energy recently (Areas in 7 states (12) 1986) named 12 areas in seven states as prime candidates for the nation's second nuclear waste dump, tentatively scheduled for operation shortly after the year 2000. The sites proposed are given in the following Table 32.4.

Three sites in Nevada, Texas, and Washington states have already been chosen for the first nuclear waste depository scheduled for construction beginning in about 1996. Important criteria besides the political and social acceptability of these disposal sites are (1) sufficient bedrock formations, ability to withstand earthquakes,

Table 32.4   Seven Proposed Waste Storage Sites

1. Georgia: Lamar, Monroe and Upison counties, 214 square miles.
2. Maine: Hancock, Penobscot and Washington counties, 52 square miles; Androscoggin, Cumberland and Oxford counties, 385 square miles.
3. Minnesota: Marshall, Pennington, Polk and Red Lake counties, 300 square miles; Norman and Polk counties, 113 square miles; Benton, Mille Lacs, Morrison, and Sherburne counties 397 square miles.
4. New Hampshire: Cheshire, Hillsborough, Merrimack and Sullivan counties, 78 square miles.
5. North Carolina: Franklin, Johnson and Wake counties, 142 square miles; Buncombe, Haywood and Madison counties, 105 square miles.
6. Virginia: Bedford, 209 square miles; Halifax and Pittsylvania counties, 307 square miles.
7. Wisconsin: Langlade, Menominee, Marathon, Oconto, Portage, and Shawano counties, 1,094 square miles.

and the proper composition to prevent any contamination and subsequent escape of groundwater.

As of 1986, about 10,000 metric tons of spent fuel from 98 nuclear power generators are stored in water pools near the reactors.

The nuclear industry is still protected from some excess environmental liability by the Price-Anderson Act of 1957. This legislation was intended to spur investment in commercial nuclear power. Lately there has been public pressure not to renew this Act (Nuclear industry(13)1986). The main argument now against the provisions of the Act is that it also encourages recklessness and negligence which may lead to catastrophic disasters.

Under the provisions of the Nuclear Waste Policy Act of 1982, the Public Service Electric and Gas Company of New Jersey, along with other operators of nuclear plants, has signed fuel disposal contracts for its nuclear plants with the U.S. Department of Energy. These contracts require the federal government to ultimately take title and provide necessary services to transport, package and place the spent fuel in underground repositories (PSE&G Annual Report(14)1984). Utilities are required to pay a fee of one mill per kilowatthour of nuclear energy produced to fund the disposal program.

In the case of an "all-out" nuclear war or an acciden-tal drop of a nuclear bomb, the environment (at least) would become a gross hazard to all living beings. Chemist John Birks (paper presented at A. C. S. meeting, Miami Beach, May 1985) calculates that approximately 200 million tons of chemical-laden smoke would pour into the atmosphere after a wide scale nuclear exchange. This smoke, Birks says, would include carcinogenic asbestos fibres released from burning ceiling tiles and asphalt shingles, large amounts of poisonous CO's, NO's, and CH_4 from forest fires, and millions of tons of noxious chemicals from the combustion of rubber, plastics, petroleum products and industrial chemicals.

32.35   Biological Wastes

Wastes from decomposing organic foods or bodies of humans or animals represent a hazard to those people who come into direct contact with them. The dangers are largely biological—that is the diseases associated with decomposing organic matter. The infectious diseases can be carried by flies, mosquitos and rodents. Nemerow (Industrial Solid Wastes (1) 1984) lists many of these diseases by these carriers. Our main concern in this text is hazardous wastes of a biological nature contributed in hospital wastes. Also of recent concern during the 1980's are with substances that interfere with reproduction (genotoxins). Certain hazardous chemicals can cause low sperm counts, structural sperm abnormalities, testicular cancer, and birth defects (Toxics and Male Infertility (15) 1985). Some examples of these chemicals given include:

Dibromochloropropane(DBCP)—temporarily   sterilized all the men who handled it during manufacturing

Dioxin—caused severe reproductive deformities in mice and monkeys
    —caused birth defects in Vietnamese children exposed to Agent Orange
    —caused disease to American soldiers serving in Vietnam exposed to Agent Orange

2,4,5T—caused high rates of miscarriage in women near sprayed areas

DBCP, Kepone and Dioxin—caused sterility and other genotoxic effects in mice and rodents and probably humans

Ten antibiotics including penicillin and tetracycline—suppress sperm production temporarily

Tagamet(Stress-related drug)—reduced sperm counts by 43%

In addition, radioactivity and x-rays have long been known to cause both sterility and structural abnormalities in sperm.

32.36 Hospital Wastes

Although wastes from hospitals include domestic, contaminated and special wastes, we are concerned primarily with the pathological waste component. This biological waste is hazardous due to contamination with pathogens. Tubercular lungs are particularly hazardous due to the potential for airborne release of pathogens. The organism, mycobacterium tuberculosis, has been found especially prevalent and dangerous in these wastes as have the bacteria of the pasteurella, brucella and psittacosis groups as well as certain viruses. Surgical and autopsy wastes contain pathogen and present dangers if not handled properly. Certain of the other contaminated waste may also be classed as hazardous since they contain blood, pus, and sputum which can infect the air with many microbiological agents. Some special waste also contain radioactivity such as C^{14}, Cr^{51}, Au^{198}, I^{151}, Fe^{50}, P^{32}, and Na^{24}. These must be segregated and handled as suggested in Chapter 28.12. The pathological and contaminated wastes usually comprise less than 10 percent of the total hospital wastes.

Incineration is the safest method of disposal of these hazardous hospital wastes (Nemerow, Industrial Solid Wastes (1)1984). However, since this is costly unless done collectively and often includes dangers and malpractices, such systems must be designed and operated with care. Separation of these wastes from other hospital wastes, transportation, and storage must precede proper incineration.

One incidence of improper practise was recently reported and serves to illustrate some of the problems involved with disposing of hospital hazardous wastes (Incinerator belches foul smoke (16) 1985).

A malodorous pall hangs over an area of northwest Dade County, Florida. The source of the odor is the Metro Waste Services Inc. where leak-proof metal containers of infectious hospital waste are trucked in over a gravel path, hauled up a concrete ramp, and fed into a smoky incinerator that even its owner admits is a rust-eaten "piece of junk." The state and county officials have cited the operator claiming it was not burning the waste completely and that contaminated runoff was seeping into the ground. The two major concerns—even when the incinerator is repaired—are (1) the smoke itself, when it hangs close to the ground and penetrates the building and surrounding area is obnoxious and (2) the potential that live organisms could exist in both the smoke and in the drainage from the stored waste. This situation is typical of many incinerator operators who attempt to burn these wastes on a small scale.

Although high temperature incineration is conceded by most environmentalists to be the preferred method of hospital waste disposal, inefficient or incomplete combustion of halogenated plastics, bactericides, and hazardous pharmaceutic may yield toxic dioxin, HC1, and chlorine gas. Doyle et al. (The Smoldering question (17) 1985) claim that alkaline scrubbing systems can effectively neutralize and remove acid gas and most toxic air contaminants. Hospital wastes contain about 20–30 percent plastics (contrasted with about 5 percent for municipal wastes), and generally average 7500 to 10,000 BTU/pound of heating value (compared to 5,000 BTU/pound for municipal refuse). There are many hazardous materials in these wastes which must be considered when incinerating them. Infectious components (about 10 percent of the total) can usually be destroyed completely by proper incineration. However, burning does not destroy inorganic constituants like mercury, some cytotoxic agents used in chemotherapy and anti-neoplastic agents. Although dioxin and hydrochloric acid are the major toxic components following incineration, furans and PCB's are also present. Hospital incinerators are usually relatively inefficient in that they operate at lower temperatures and contain shorter stacks than the larger commercial or industrial ones.

In addition to hospitals, we now are aware of laboratories, doctor and dentist offices, and small medical treatment centers which contribute potentially infectuous wastes. These include needles, syringes, gauze, dressing gowns, I.V. tubings and used blood storage bags. As of 1988, 31 states now have laws requiring hospitals, at least, to disinfect (steam sterilize or incinerate) such waste before they bury it in landfills. However, both methods have limitations and hence may not be satisfactory. Further, tracing illegally dumped wastes of these types is very difficult. More illegally-dumped medical wastes occur today because of the banning of small and scattered amounts in conventional landfills, and burning is costly. However, some companies are currently working on small disinfection units for individual medical waste suppliers and also on incinerators capable of burning these wastes without creating air contamination from gases resulting from plastic burning.

32.37 Biological Wastes and Other Contaminents

There were 4,742 reports of salmonella infection in Illinois, Michigan, Indiana, Wisconsin and Iowa according to Ms. Weidel, legal counsel for the Illinois State Inspector General's office during April 1985. At least three deaths were also linked to the outbreak. Salmonella is an infection of the gastrointestinal tract that usually causes cramps, vomiting, diarrhea and fever. It causes death in one of every 1,000 cases, and the ones most at risk are the elderly and children. It is usually caused by contaminated food—in the above situation it was traced to contaminated milk from Hillfarm Dairy in suburban Melrose Park, Illinois. Normally, most dairies use pasteurization, a high–temperature heating process, to kill salmonella as well as other bacteria. Among the hypotheses put forward to explain this Illinois outbreak are (1) incomplete pasteurization; (2) contamination introduced after pasteurization; (3) heat-resistant salmonella bacterium; and (4) sabotage.

Viruses growing in polluted waters have been proven to be hazardous to human health even though not ingested directly in drinking water. For example, eating non–cooked or even poorly–cooked shellfish represents a clear risk to our health. DuPont, of the University of Texas Health Science Centers, says (Eating raw shellfish (18) 1986) that "reports appearing with alarming frequency in the U.S., Europe, and Australia" demonstrate a "clear risk" of gastrointestinal illness and hepatitus-A infection from eating raw or steamed clams and oysters contaminated with viruses or bacteria apparently originating in polluted water. A recent study reported in the New England Journal of Medicine documents 103 outbreaks in which 1,017 people became ill during 8 months in 1982 and concluded that the illness usually was associated with raw clams or oysters but that there was an "unexpected high attack rate" among people who ate steamed clams that had not been cooked sufficiently.

Nausea, vomiting, diarrhea or abdominal cramps were major ill effects. Closing polluted waters to the taking of shellfish represents the most effective preventative of this health hazard, but often closing occurs "after the fact" and too late to prevent all illnesses. Many shellfish lovers resist ample cooking required to kill viruses.

32.38 Asbestos–Laden Wastes

Asbestos dust can originate in municipal refuse from discarded auto brake linings and all kinds of old building demolition wastes. It can also come from a variety of common industries mostly associated with the building trades. Sax (Dangerous Properties (19) 1979) gives an excellent table of potential sources of origin of asbestos material. He also includes a table of non-hazardous asbestos wastes so that workers and users of commodities do not get the idea that all products containing asbestos are dangerous. Your authors have taken the liberty of rearranging this information into a table containing both hazardous and non-hazardous uses of asbestos.

The health hazards relating to asbestos involve contracting the disease of asbestosis by breathing in mass doses of asbestos dust over long periods of time. This can lead to the development of cancer of the lungs which is especially aggravated by smoking at the same time. Exposure to the blue (crocidolite) asbestos fibre can also lead to cancer of the chest or stomach walls known as "mesothelioma".

Treatment usually follows prevention techniques such as the use of inhalation masks, highly–efficient exhaust systems, and moisture laden air systems during working or disassemblying of lagging and insulation.

Treatment of the wastes primarily has been by sedimentation and/or filtration of the fibres and landfill disposal. Landfilling is recommended only during wet–type operations which eliminate dust. The authors report using chemical coagulation treatment in Chapter 27.23 with iron salts or polyelectrolytes followed by filtration to remove better than 99.8% of fibres from water containing 12×10^6 fibres per liter.

To avoid treatment problems, the EPA has proposed the elimination of the use of asbestos in making (1) cement pipe, (2) roofing felt, (3) flooring felt, and (4) floor tile. Its use in asbestos lining in brakes, however, so far has not been included in the elimination list (Ban

Table 32.5   Sources of Asbestos Wastes (adapted from Sax, 1979)

Hazardous sources of asbestos usage	Non hazardous sources of asbestos usage
1. Thermal lagging and delagging (special risk from 'blue' asbestos mainly from steel and power plants	1. Fillers which are combined with other ingredients, mainly in linoleum, floor tiles, rubber paints plastics, adhesives, roofing, and motor assembly.
2. Furnace insulation, mainly from heavy industry	2. Grinding in assembly of brake and clutch parts.
3. Heat and sound insulation, mainly from locomotive, railroad car, boiler, chemical plants, and gas works mfg.	3. Asbestos washers and gaskets
4. Asbestos-cement sheets and boards of insulation, mainly from building trade plants.	

on asbestos—1986). There is still strong criticism for such banning of asbestos by the U.S. industries and the Canadian government. They contend that the "lethal practices of the past have been rectified, and that current applications of asbestos do not pose unacceptable health risks." We must point out that about 85% of asbestos used today in the U.S. comes from Canada. The U.S. industries use this asbestos currently to make those products which bind the fibres tightly so that they cannot be dislodged easily to reach the environment. The arguments for and against their present usage is expected to continue for a long period.

32.39 Toxic Industrial Wastes

While hazardous materials are those that are included in all hazard classes, toxic materials are those that cause toxic effects. Therefore all toxic materials are hazardous, but not all hazardous materials are classed as toxic. The reader may observe that these definitions

are also verified in this book. While Chapter 32 deals with hazardous materials only, Chapter 28.10 describes toxic materials and their treatment. Special treatments of toxic materials may also be found in the following section.

Several of the newer chemical industrial plants manufacture organic chemicals for sale to producers of other products discharge some of these wastes to the environment. Many are extremely toxic to aquatic life, animal life, or humans in very small quantities. Most of these chemicals were either unheard of or seldom used ten to twenty years ago. Few of them were measurable in small concentrations with laboratory scientific equipment until the last ten years. Even now we are uncertain of the long-term potential toxic effects of many of these modern chemicals. In chapter 32.9 the limits of some of the more common of the recent organic chemicals are given. In this section we discuss the production effects, and treatment of such chemicals as are contained in organic laden wastes, as well as pesticides, pharmaceuticals, and laboratory chemicals.

32.40 Organic–Laden Wastes

Chemical plants containing organic chemicals such as PCB, phenol, formaldehyde, spent solvents, halogens, organic sulfurs and nitrogens, paints and varnish residues, organic acids, dyes and pigments, explosive and defoliant chemicals, are considered in this section. They are so diverse as to defy classification except that they are all organic and potentially toxic.

32.41 PCB–Laden Oil/PCB–Containing Wastes

Polychlorinated byphenols originate in transformers and power capacitors, hydraulic fluids, diffusion pump oil, heat transfer applications, plasticizers for many flexible products including printing plates and numerous other products. The major properties of heat resistance without decomposition and non-flammability are hard to duplicate in other substitute chemicals. Despite this, their use has been banned in the United States for all but transformers and high voltage capacitors.

Reclass 50, a mobile operation conducted by a two man crew can bring a high-voltage transformer from 60,000 to less than 50 ppm faster than any other system (Graham (20) 1985). In this process the unit is put through three soaking steps over a period of seven to 15 months, depending on the transformer size, PCB concentration, and operating temperature. Except for a short period, the transformer remains in service. First, the original askarel is drained and replaced with two flushings of TF-1, a proprietary material, which in turn is replaced with a final soak of silicone. The last step is filling with Carbide's silicone dielectric fluid, which will be used for the remainder of the transformer's service life.

P C B's are a group of oily, colorless, organic liquids in the same chemical family as the pesticide DDT. They have been linked to birth defects, reproductive failures, liver problems, skin lesions, cancers and tumors.

Before 1972 PCB's were used in all non-drying microscopic immersion oils since they possess great stability and optical properties still said to be unmatched today.

Cargille (PCB's in perspective, 1983), whose company makes PCB-containing chemicals, recommends to its customers that used and unused chemicals be sent back to Cargille for reuse. However, the new labelling regulation of EPA "forcing disposal" is hindering this solution.

Fox and Merrick (Toxic and Hazardous Wastes (21) 1983) meet PCB discharge standards of 1.0 ppb by the following three operations and then conventional wastewater treatment:

1—eliminating their discharge into the storm and sanitary sewer system by source control.

2—reroute PCB contaminated wastewater to the existing combined industrial-sanitary conventional treatment plant.

3—flushing contaminated systems and storing PCB contaminated oil and debris in tanks prior to disposal.

Treatment of PCB's has been primarily by banning their production or use. Where found they must be recovered and incinerated, pyrolyzed, or reused.

Sunohio developed a PCBX process which chemically destroys PCB's in transformer mineral oil. It is described in some detail in Figure 32.5.

Following the Fullers Earth Dual Filters shown in Figure 32.5, the treated fluid flows through a fourth stage filter, a vacuum degasser, and a final fifth stage submicron filter. As the treated fluid flows through the sequence of filters, all the spent reagent and reaction by-products are removed yielding a clean, reusable fluid with a PCB concentration of less than 2 ppm. The

reaction by-products, spent reagents, and exhausted filter mediums are collected in approved containers, solidified, and disposed of in an appropriate landfill.

According to Sunohio, because PCBX treatment reclassifies transformers and allows for continued use of the oil, capital expenditures are avoided. Faulted transformers can be rebuilt after PCBX treatment instead of disposing of them. PCB oil leaks and spills that often result in high cleanup costs, operation disruption, extreme adverse public disclaim, and potential EPA fines all can be avoided by proper PCB treatment.

Another method of detoxifying PCB wastes-especially those of sludges such as collect in lake or river bottoms—is the fluidized bed combustion process. American Toxic Waste Disposal, Inc. of Waukegan, Illinois uses this system for the PCB muds dredged from Lake Michigan (The Amicus Journal (22) 1985). It involves grinding the sediments into fine particles and forcing hot air (1200 F) through them on a perforated plate of bed until the PCB's vaporize. A cyclone separator and scrubbers then isolate the PCB gases which are converted to liquid form. Then the PCB's can either be incinerated or adsorbed in activated carbon filters for later disposal in a relatively smaller space than that required for the untreated waste muds. One study, however, found it required two-thirds of a barrel of fuel oil to burn each cubic yard of sediment. This represents a significant cost to the reclaimer/disposer.

32.42 Phenols and Formaldehydes

Phenol is an acutely toxic chemical containing the benzene ring (C_6H_5OH). In acute poisoning the main effect is on the central nervous system. Also, absorption from skin contact of phenol may be very rapid and death results from collapse within 30 minutes to several hours. In industry we encounter phenols mainly being used or produced in integrated steel mills, synthetic textile mills, and resin (plastic) manufacturing.

Phenol and formalin are used as starting chemicals for manufacturing the phenolic-derived resins. The phenolics and formalin, together with catalysts, are fed into an insulated and heated reaction tank. After reacting, draining following water evaporation, and cooling the resin is used for product manufacturing.

Cooling waters, condensates, blowdown wastes result in about 7 pounds of wastewater per 100 pounds of resin. These wastes usually contain high phenol con-

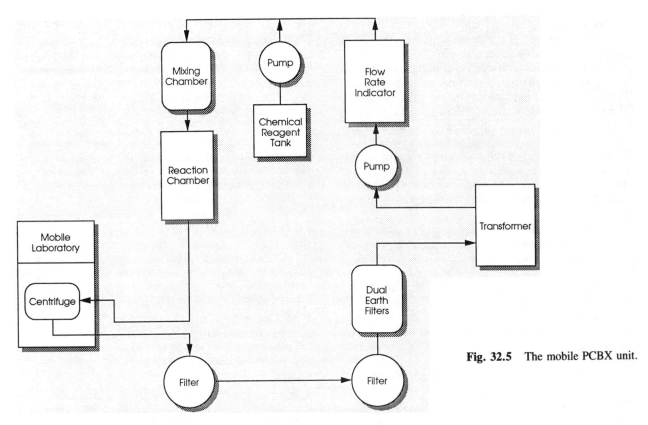

Fig. 32.5 The mobile PCBX unit.

tents (over 1000 ppm) and high BOD (over 10,000 ppm).

Treatment has been by phenol extraction, lagooning, thermal incineration, and sometimes by discharge (and dilution) into municipal sewage systems.

Phenolic wastes, although bactericidal, have been degraded biologically under the proper environmental conditions usually involving bacterial adaptation periods. For example, Zoltek (Bacteria Digesting Toxic Wastes (23) 1984) took a groundwater already contaminated with phenols and terpenes allegedly from a nearby plant which treats wool with coal tar creosote and contracted for pumping it to the sewage treatment plant. His laboratory results showed the positive action of municipal sewage in degrading these compounds during biological treatment.

32.43 Spent Solvents

Organic solvents are used in many chemical industries for assisting in production. Dry cleaning and paint in-

dustries are typical major users of solvents. Solvents used in paint, lacquer, and varnish manufacturing include ketones, aromatics, aliphatics, alcohols, glyco ethers, glycol ester ethers, glycols, glycol esters, terpines, etc. The solvents aid in keeping the pigments dissolved and are eventually evaporated either during the manufacturing process or during the product use. Waste solvents discharged following condensation are potentially toxic to receiving waters.

Solvents of the chloromethanes are found in the wide spread dry cleaning industry establishments as well as in dewaxing of oils, refrigerant preparation, and medicine formulations. Most are made by chloringating methane in the presence of oxygen and high velocities under a temperature of about 700° F with product recirculation for heating. The exact product obtained can be varied somewhat to satisfy market demands. Where these chemicals are used in sufficient quantities to warrant environmental concerns, they must be collected and treated (usually by an outside contractor) and not discharged untreated to the local air, ground, or water.

Treatment is usually by concentration by adsorption and/or incineration.

Dedert offers a "Supersorbon" process to recover solvents (Supersorbon–Process—1985). Hazardous solvents are recovered from the contaminated air in various chemical plants by drawing the air through activated carbon and regenerating the carbon by backflashing with steam and subsequent decanting off the condensed water as shown in the following schematic, Figure 32.6. Solvents recovered are toluene, naphtha, lactol spirits and similar blends of aromatic, aliphatic and parafinnic hydrocarbon solvents. Acetates, alcohols, ketones and chlorinated hydrocarbons can be recovered with added steps such as distillation and chemical drying. Claims are made of over 90 percent yield of solvent. Some operational data indicates that 2–4 pounds of steam, .07–.20 of electricity, 3–5 gallons of cooling water are required per pound of solvent.

In addition, from 1–2 pounds of activated carbon must be replaced for each ton of recovered solvent.

Another source of the solvent turpentine originates from the manufacture of naval stores. In this industry, shredded wood chips, usually pine, are loaded into a battery of extractors into which live steam and another solvent—usually naptha, permeate an added and maintained at about 80 psig (pounds per square inch gage). Most of the naptha is separated from the turpentine,

pine oil, and rosin in a concentrating evaporator. Some turpentine escapes condensation and some other is lost during purification of the pine oil and rosin residues.

Still another source of solvents such as xylene, benzene, toluene napthalene as well as acetylene, etheylene, propylene and butene is the petrochemical industry. These solvent chemicals are actually derived from coal, oil, or natural gas and are precursors of finished products such as rubber, plastics, and fibres.

By distilling, compressing, fractionating, and reforming natural gas, petroleum, and coal under a variety of specific operating conditions (such as time, temperature and pressure), these organic chemicals (solvents) are obtained. Inadvertently, some solvent is lost to the environment during the conversion or in the use of them to produce final useful products. Most of these solvents are known to be toxic—usually carcinogenic—in certain concentrations.

32.44 Organic Wastes Containing Halogens, Sulfur, and Nitrogen

A variety of chemical and manufacturing industries produce wastes containing organic halogens, sulfur, and nitrogen. Sometimes they are discharged with other contaminants and difficult to separate. For example, the explosive manufacturing plants which produce

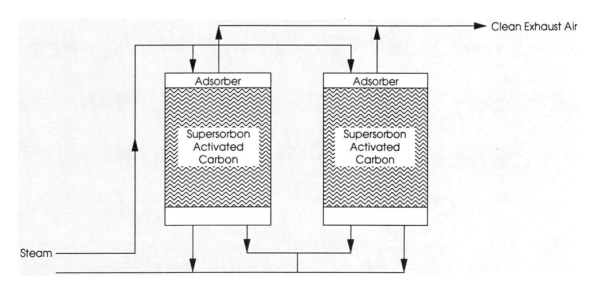

Fig. 32.6 Supersorbon Solvent Recovery

TNT wastes, also highly acid and colored, textile finishing plants which may use sulfur or azo dyes and produce wastes which are colored alkaline, and replete with suspended and colloidal solids. Organic halogens are also produced by chemical manufacturers by direct chlorination of lower aliphatic hydrocarbons. They also can be produced by disinfection with halogen, usually chlorine, or organic waste residues. For example, chlorobenzene, an important intermediate for sulfur colors, and a solvent used to make other products such as aniline, phenol, chloronitrobenzene, etc., is produced by passing dry chlorine through benzene in iron vessels using ferric chloride as a catalyst and keeping the reaction at 140°F. The same compound could be produced with less efficiency as a result of disinfection of a benzene–containing wastewater. In both cases, some chlorobenzene would be wasted into the environment and contribute to the toxicity of the medium

32.45 Special Treatments for Hazardous Wastes

In this section we report, describe, and evaluate the many types of treatment systems currently being used or suggested for counteracting the negative impact of hazardous wastes in the environment. Although we cover the conventional physical, chemical, and biological systems, we do not go into basics very deeply since Nemerow (Industrial Water Pollution 24, 1987) and others already provide this information. On the other hand, we cover more extensively newer, more sophisticated treatments which are directed specifically for handling hazardous wastes. For example, waste solidification, regional exchanges, plasmolysis, and recovery and reuse are explained with greater care because of their implications in hazardous wastes field. Examples are given of treatment of specific hazardous wastes by appropriate methods whenever possible.

32.46 General Treatment

Hazardous wastes are incompatable not only with the environments, but also with other more common wastes of industry. Because of the direct and insidious dangers of hazardous wastes, they must be considered separately as infringements upon our normal industrial practises and upon every citizen's right to live and use environmental amenities. Those industrial wastes which burn, corrode, over-react, or are toxic will be considered as hazardous and given special priority in treatment preference and type.

Pearce (25, page 57, UNEP, 1983) states that of the 3.4 million tons of hazardous wastes produced by industry in the United Kingdom just less than 80% are landfilled, while less than 12% are disposed of in the ocean; 3% incinerated, and the remaining 5–6% are treated in other ways.

United States EPA (26, 1980) determined that only 10% of the treatment methods used for hazardous wastes are "acceptable." These are detailed in the following Table 32.6.

Van Noordwyk et al. (27, 1980) classified hazardous wastes by quantity generated and disposal method used. Organic chemicals (SIC 2861, 2865, and 2869) were categorized in the following Table 32.7.

Nemerow reported (Industrial Wastes 28, 1984) that amounts of commercial hazardous wastes are collected and treated in the following ways, Table 32.8.

In a National Research Council staff digest (Reducing Hazardous Waste Generation 29, 1985), they find the RCRA law of 1976 defines as hazardous certain wastes because they may (a) cause or significantly contribute to an increase in mortality or an increase in serious irreversible, or incapacitating reversible, illness; or (b) pose a substantial present or potential hazard to human health or the environment when improperly treated, stored, transported, disposed of, or otherwise managed. They arrived at four general principles that should govern efforts to reduce the generation of hazardous waste.

1. No single approach to encouraging waste reduction will be most effective in all circumstances.

Table 32.6 Hazardous Waste Disposal in the United States (USEPA, 1980)

Disposal Method	Percent of total
Acceptable	
Controlled incineration	6
Secure landfills	2
Recovered	2
Unacceptable	
Unlined surface impoundments	48
Land disposal	30
Uncontrolled incineration	10
Other	2

Table 32.7

Method of Treatment	Quantities* (M's/year) Onsite	Offsite
Landfill	483,000	113,000
Incineration	2,250,000	51,000
Controlled	699,000	—
Uncontrolled	1,550,000	—
Deep Well	6,540,000	—
Biological treatment or lagoons	565,000	—
Recovery	267,000	
Total	10,100,000	164,000

*1977 data-wet basis

Table 32.8 Commercial Hazardous Waste Disposal Methods

Waste Management	Capacity Wet, 1000 Tons Year 1981	Volume Received Wet, 1000 Tons
Landfill	37,372	1,965
Land treatment and		
Solar evaporation	1,400	282
Chemical treatment	1,305	734
Deep well injection	1,095	475
Resource recovery	341	83
Incineration	102	80
Total handled by 9		
commercial waste		
treatment firms	41,615	3,610

2. Reductions in the generation of hazardous industrial wastes can be expected to occur through a series of loosely defined and overlapping phases. It is desirable to reduce the generation of hazardous wastes.

3. The costs of alternative methods of waste disposal should reflect the social costs of protecting public health and the environment.

4. Regulation will continue to play a crucial and central role in the overall waste management effort, but future waste reduction is more likely to be fostered by non-regulatory methods, such as information dissemination programs and economic incentives.

Currently, there are many new processes in various stages of development for treating and destroying all types of hazardous wastes. The processes are compiled from response to two national solicitations for new hazardous waste treatment ideas, from several literature reviews, and through contact with experts in the field (Assessment of Incineration 30, March; 1985). These process are summarized in Chart 3, Table 32-10.

Torricelli informs us that the Office of Technology Assessment cited 26 systems or products that could effectively and permanently destroy toxic sites (Lets Destroy Wastes, 31, 1985). Among these he chose to bring out four: chemical detoxification, biodegradation, encapsulation, and supercritical water oxidation.

32.47 Segregation

Segregation involves the removal of hazardous indus-

trial wastes from other process wastes for subsequent special handling. Segregated hazardous waste can either be treated by the industrial plant at its own facility or picked up by a service company for transporting to a central, specially designed treatment plant for hazardous waste.

Rollins Environmental Services located in Logan Township New Jersey is generally conceded to be the largest waste processing plant in New Jersey (Civil Engineering 32, 1979). Rollins requires information of what the waste is and for the company to segregate and label wastes for pickup at its plant site. If wastes are not segregated and instead mixed, serious problems can occur. For example, a mixture of chlorinated hydrocarbons and metal sludge waste will create the problem of corroding clay or plastic liners in landfills (due to the chlorinated hydrocarbons) or releasing metals to the atmosphere if incinerated (due to heavy metals in sludge).

32.48 Location of Facilities

The problem of gaining acceptance for a site for any sort of treatment, including storage, has always been a difficult one for conventional waste treatment. The situation becomes even more sensitive and one faces public objections even at the mere mention of "toxic" or "hazardous" waste. An ideal site where no people live in the area and yet is situated near enough to suppliers does not exist in reality. Any site will take a great deal of time, patience, and tact in gaining its acceptance for

Table 32.9   Chart B Emerging Alternative Technologies (From Assessment of Incineration—1985)

Technology and Description	Suitable Wastes	Capacity Per Unit of Time	Cost Per Unit Of Waste	Environmental Data	Commercial Status
WET AIR OXIDATION: —process of oxidizing organic compounds in water at temperatures of 350° to 650° F.	very dilute organic and inorganic aqueous waste except highly refractory organics	10 gallons per minute	5 to 10 cents a gallon for a 10 gallon system costs for larger system not available	limited data available EPA currently evaluating a unit on cyanide, sulfides and nonhalogenated waste	technology currently available widespread use expected in next 2 years on aqueous waste
MOLTEN GLASS INCINERATION: —High temperature in furnace (23000°F) destroys organic waste streams —combustion gasses pass through ceramic filters —glass slag encapsulates inorganic residues	any combustible waste degree of halogenation not a consideration scrubbers required for HCL	existing glass manufacturing process 100-21,000 lbs per hour	cost figures not available	data not available testing needed ceramic tiles and residue encapsulation in slag indicate minimal impact	not reported
SUPERCRITICAL WATER: —inorganics, insoluble in supercritical water, are removed from waste streams —organics rapidly oxidized	aqueous waste streams with high levels of inorganics and toxic organics treatment of highly halogenated material not yet demonstrated	currently treats 1000-2000 gallons per day	cost figures not available	bench-scale test on various wastes indicates DREs of 98.5%-99.8%	commercial scale unites will be available in 1-3 years

Table 32.9   (*Continued*)

Technology and Description	Suitable Wastes	Capacity Per Unit of Time	Cost Per Unit Of Waste	Environmental Data	Commercial Status
HIGH TEMPERATURE ELECTRIC REACTOR: —two companies: Thagard Reseach Huber Corporation	process initially designed for solid waste	75-125 lbs of solids per hour	cost information not available	DREs far in excess of the 99.99% RCRA requirement	mid-summer 1985
—vertical reactor heated by electrodes implanted in the walls to pyrolize organic wastes	also suitable for liquid refractory waste streams	no figures available for liquids, but it is assumed throughput would be less	Huber claims cost comparable to conventional incineration		
MOLTEN SALT REACTOR: —burning and scrubbing (900°C) oxidize carbons of organic matter to carbon dioxide and water —byproducts (phosphorous, arsenic, sulfer, halogens) retained in melt as inorganic salt	designed for solid and liquid waste especially applicable to highly toxic and halogenated combustible waste with low percentage of ash	pilot scale facility processes 80 to 200 lbs per hour	cost data not available	no emissions organic salts are the only biproduct DREs for organics, pesticides & chemical warfare agents 99.99% to 99.99999%	currently available for commercial use but none operating on that scale to date
PLASMA ARC: —uses the high temperature of plasma (50,000F) to destroy hazardous waste —all hardware in 45 foot mobile trailer	highly toxic liquid waste streams degree of halogenation not a consideration	600 lbs. per hour in commercially sized unit	cost data not available	current demonstration in New York state will provide info on DREs and emissions	commercial scale unit to be operating in 1 to 3 years

use as location for storing, treating, and/or disposal of hazardous waste.

32.49 Treatment Required-Risk Evaluation

The type and extent of hazardous waste treatment required for protection of the environment is related to the amount of risk one is prepared to accept. An acceptable risk depends not only on the cost of protection involved, but also upon who will be required to pay for the environmental protection. Such risk analysis becomes simpler if the cost is borne directly by the one who will benefit from the treatment protection. In matters pertaining to hazardous waste treatment, however, it is more likely that industry will be required to pay for protecting some environment and its users quite far removed from the industry itself. Economists refer to such situations as external diseconomies.

Even when governments are forced to grant funds for such expenditures to protect certain people specifically involved, they are somewhat reluctant to do so. This is because their protectorate is widely based and often does not respond with a single voice in a unanimous and powerful manner. In other words, benefactors of governmental actions and expenditures for hazardous waste are difficult to identify as well as to quantify the effects.

For years the most acceptable practise of risk analysis and evaluation has been the use of cost-benefit relationships. That is, there must be a dollar benefit for every dollar spent to protect the environment from the adverse effects of hazardous wastes. With these wastes costs are not only huge, but also difficult to predict reliably. Benefits are even more troublesome to quantify because they usually relate to human health, the quality of life, and even a value of human life itself. Costs and benefits often can be manipulated by the person or group presenting the case for or against remedial action depending upon personal goals.

One device reported by Douglas Ginsburg (Tangled Rules—1985), administrator for information and regulator affairs at the Office of Management and Budget, is to "try to guide and question and kibitz the process in order to get as much risk reduction as possible for the expenditure." This procedure is laudatory, but is still highly subjective and will vary from one person to another and one case to another. What is needed is a dependable, consistent objective method. Perhaps it might be better to state publicly that we will seek the maximum risk prevention that the public will buy. Or that any environmental risk is too great to accept where hazardous wastes and human beings are involved. We have almost done this where the operation of nuclear power plants are concerned.

However, as Shabecoff points out (Tangled rules 33, 1985), "there is no consensus on what society is willing to pay to avoid risk." For example, a life saved has been valued at various amounts ranging from $400,000 to $7.5 million. Much lower life values have been set where masses of lives are involved. For example, each life lost in the Bhopal accident has been valued at only $30,000 by a preliminary offer of settlement.

The Administrator of EPA (1985), Lee Thomas, is quoted as suggesting, "we are a long way from the point at which decisions can or should be made by mechanistic application of any acceptable risk or cost "criterion." However, your authors still believe that the public would benefit more from an established, open, scientific process for risk evaluation even if it was not perfect and always open to some debate. At least its parameters would be known and its values could be seen and evaluated by both violators and recipients of hazardous wastes effects.

Dr. Steven Kelman, of Harvard University, believes that "we should hold the government responsible not for a risk-free society, but for a higher standard of behavior than people hold themselves" (Tangled rules 33, 1985). This is probably because in an individual's case his chances of death or disease are less than those of governments who represent more people. Unfortunately, governments seldom make decisions and laws in a vacuum and are usually influenced largely by individuals and their lobbyists. Therefore, government's decisions may appear to be similar to those of certain individuals or groups rather than society as a whole. This fact is yet another reason and argument for objective risk-benefit procedures rather than subjective ones.

32.50 Land Treatment

The use of vacant land as receptacles for hazardous waste is recommended only when the land can be upgraded from its previous use and when no additional adverse environmental consequence results. Qualifying lands should be far away from underground water supplies and residential communities where odors, percolate or contact with hazardous gases, liquids, or drying solids might take place. Since these ideal conditions seldom are found, we suggest underlying the entire land area with an impermeable membrane to prevent leachate from entering the water supply. Leachate should be collected and treated especially for heavy metals and toxic organics such as those originating from pesticides.

Land reuse for parking facilities, one-storied warehouse or parks are examples of upgraded uses following proper disposal of hazardous wastes.

Whenever possible, hazardous wastes—usually after some form of pretreatment—are returned to lands which are compatable with the residues. For example, many lands are low-lying and unusable in its present condition. Such lands could be used as receptacles for such wastes. Sundaresan (7, UNEP 71, 1983) states that DDT wastes are neutralized with lime and the lime sludge dumped in low-lying areas of land.

The Japanese are mixing solid wastes, excavated earth, gravel, and sludge as well as ash from incineration plants to form a source of new land (Land-poor but trash rich 34, 1986). They haul the mixture to Tokyo Bay where it is placed and has already served as real estate for Haneda Airport and Tokyo Disneyland, an oil terminal port, a power plant, an apartment complex with 4,500 units, an industrial park and even a sewage treatment plant. Of course, the competition for space has caused obstacles with other users of the bay. Shellfish have been reported to have been killed by the change in environment caused by the newly created land. But the pressures upon their society for "living room" has dictated reuse of solid wastes and a change in the environment.

32.51 Thermal Treatment

Heating hazardous wastes to some elevated temperature can bring about changes in the physical, chemical, and biological nature of some wastes so as to render them innocuous. This treatment is not to be confused with complete burning in the presence of excess air as in incineration.

Perhaps one of the most promising and utilized thermal process is pyrolysis. Pyrolysis involves burning of residue at about 900°C. in the absence of air. The process has been used in the chemical industry and often called "destruction distillation." It has been used for treating municipal refuse with the production of several byproducts such as light oil, gas, ammonium sulfate, and tar as well as solid waste residue. The gas energy is usually more than ample to supply the heat for pyrolysis. Whether pyrolysis of a hazardous waste will be effective in destroying the toxicity is a question that has not been fully resolved.

Should an industrial hazardous waste be organic and easily decomposable at relatively low (less than 900°C.) temperatures and yielding non-toxic products, pyrolysis may be a reasonably effective treatment process.

One form of thermal treatment is that of plasma technology. It developed from knowledge obtained from our space flight research which required superelevated laboratory temperature to simulate spaceship nosecone reentry conditions. Electrical energy (high voltage) is converted with high (85–90%) efficiency to heat energy. One such process of plasma technology developed by SKF Steel Company (Herlitz, 1983) recovers iron, zinc, and lead from steelmill baghouse dust. Without previous sintering, Herlitz reports, the fine material is pneumatically injected with coal powder into the lower part of a coke-filled shaft furnace provided with plasma generators where the material is injected. The oxides are instantaneously reduced, and liquid and gaseous metals are formed. The gases are extracted in a normal condenser. One such plant to produce 97% liquid iron, zinc, and lead from 80,000 tons of this dust per year was commissioned in Sweden in mid 1984. The process appears feasible for other solid wastes containing otherwise hazardous waste metals.

Another form of thermal treatment described by Worthy (35, 1982) in pilot plant stage is the "molten salt process." He describes a reactor tank which contains a constantly moving bed of molten soldium carbonate. The movement is maintained by a constant supply of wastes and air being fed beneath the melt surface which is maintained at temperatures of between 750°C. and 1,000°C. Destruction efficiencies are reported from bench scale studies to be over 99.99% for several organic hazardous wastes including toxic chemicals used in warfare, pesticides, and PCB's. This process is being developed by Rockwell International.

Worthy (35, 1982) also reveals that the Thagard Research Corporation has reported developing a high-temperature fluid wall type reactor. It consists of a tubular core of porous refractory carbon which is externally heated by carbon electrodes to give off ample radiant energy.

This energy activates reactants (wastes) which descend through the tube. Radiant energy provides rapid and immediate heating of the contaminating reactants. Gas of an inert type (usually nitrogen) circulates around the porous core which reduces contact of the reactants with the wall and prolongs the life of the core. Temperatures above 2200°C. are reported in the reactor.

Energy, Incorporated (35, 1982) is developing a fluidized bed turbulence to promote combustion of hazardous organic contaminants. It is based upon the principle that in supercritical water (374°C. and 218 atmospheres) certain insoluble organic compounds become highly soluble and complex organics are converted to low-molecular weight compounds. When these resultant products are reacted with oxygen, hydrogen, and carbon, they are rapidly oxidized and halogen, phosporous, sulfur and metals form insoluble salts and settle out. A non-hazardous stream of supercritical water (500°C. and 252 atmospheres) is used directly for process heat or indirectly to drive turbines for generating power. In bench scale studies such toxic organics as chlorobenzenes, DDT and PCB, as well as chlorinated ethanes have been destroyed.

Enelco/Von Roll uses a process of thermal destruction for disposing of industrial wastes (Thermal Systems 36, 1984). It is essentially a modernization of older incineration systems under highly-controlled conditions. Control of the waste combustion process involves accounting for the waste heat content and oxidation rate of each waste and then scheduling the mixing and feed rate of the waste such that the combustion heat release is reasonably constant. Pumpable waste is injected by its system at a predetermined rate resulting in a constant release of heat. Batch fed waste, injected at predetermined regular intervals, releases additional heat and because of the nature of batch feed process, this heat release will not be constant. Of the 23 VonRoll industrial waste disposal plants in operation, four are regional hazardous waste disposal plants. One serves Nyborg, Denmark and has the capacity to handle more than 60,000 metric tons of solvent, chemical by products, resins, paint and varnish residues, waste oil, plastics, halogenated hydrocarbons, and inorganic sludges and waste waters totalling 1,200 metric tons per week. The waste can be fed directly in steel drums. Kiln discharges of molten slag and flue gases generate steam to provide the majority of the heating requirements of the town of Nyborg. Recoverable materials are recovered and sold.

Dupont Company has provided a hazardous waste service since 1975 by thermal decontamination. It uses a furnace temperature of 600–750°C. and is claimed by DuPont to be the largest furnace in the world. Three sources of residual waste from this process shown in the following Figure 32.7 include residual furnace noncombustibles (carted away to a smelter), baghouse solids to landfills, and stack gases discharged to upper air.

32.52 Incineration

Incineration is a high-temperature burning process whereby combustible wastes are reduced to inert residues (ashes). It has been concurred that incineration provides an economic, nuisance-free, clean method of ultimately disposing of municipal refuse. However, gases and residual ashes remain as potential sources of pollution. In addition, when hazardous chemicals or materials are also incinerated, their combustion products must be evaluated to be assured that the toxicity has been removed by the burning.

Incineration takes place in a furnace consisting of an enclosed, refractory-lined structure equipped with grates and supplied with excess amounts of air in which the burning takes place. See Chapter 14.8 for more details. Temperatures in the furnace approximate 1,750°F. The combustion chamber is an enclosed, refractory-lined structure, sometimes combined with the furnace, in which more complete burning of residual flyash material and gaseous materials occurs. Following this is a subsidence chamber, a large, insulated chamber allowing the combustion gases to expand and reduce their velocity during which particles can settle prior to gas emission to the chimney (stack).

Flyash should be prevented from reaching outside air environment by settling and/or filtration and sometimes scrubbers. This collected ash can be combined with the residual grate-ash for final landfill or reuse in making other materials—such as bricks or cement blocks. The amount of residual ash depends primarily upon the composition of the original hazardous waste, exit gas velocity, and fly ash removal systems. Some of the fly ash may contain unburned organic matter and must be tested further for residual toxicity before ultimate dis-

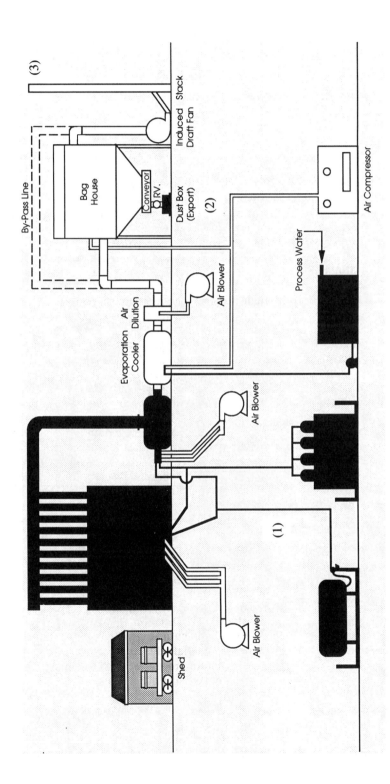

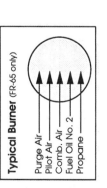

Fig. 32.7   Thermal decontamination process flow diagram.

charge. If residual toxicity remains, furnace temperatures may have to be increased, gas velocities reduced, or detention time increased in the combustion chamber, or all three used to reduce the discharge of toxic materials.

Final capital costs as well as operating costs may be relatively high as compared to other techniques. Therefore, design and operating conditions must be carefully controlled. Generally, we recommend both laboratory and pilot plant experimental burning to obtain proper design and operating procedures.

For coastal producers of toxic wastes land incinerators may be dangerous to inhabitants of those areas. In these cases incineration on ships on the ocean can serve to eliminate potential land air pollution.

Waste Management Company (Chicago, Illinois) has converted a cargo ship, the Vulcanus, to an ocean-going incinerator. The Company proposes to burn 3.6 million gallons of highly toxic PCB wastes on the high seas in the Gulf of Mexico.

Ship incineration has been purported to disperse acid wastes directly into the ocean, thereby avoiding costly scrubbing. Water, carbon dioxide, and vaporous hydrochloric acid falls out into the ocean within a short distance from the ship. Safe collection and transportation of these toxic wastes to the ship could be a major problem. Any accident involving the incinerator-ship during its travel to the burning site could result in major environmental damage.

Le Roy (25, 1983) reports incinerator capacity in France for burning 260,000 tonnes per year of hazardous waste including phenolated water, hydrocarbons, and chlorinated waste. He acknowledges relatively high costs for incineration of these wastes because of (1) technological complexity, (2) operating difficulties, and the (3) need to purify gas and smoke.

Freeman (Hazardous Wastes Incineration 37, 1981) limit hazardous waste incinerators to two types; liquid injection and rotary kiln systems. In the former, liquid wastes are fed along with support fuel and air for combustion into the incinerator which is maintained at 820°C. to 1600°C. Similar temperatures are also used in the latter rotary kilns which handle large volumes of both solid and liquid wastes. Selection of the proper type of incineration and for the proper reasons is imperative. Freeman presents tables of advantages and disadvantages of the two types of incinerators. Your authors summarize these in the following Table 32-11.

Noland (Incineration 38, 1984) demonstrated that a "transportable" incineration system could be disassembled, transported approximately 1000 miles, be reassembled, and fully operational within two weeks. He reports destruction of a "pink water" explosive plant waste by incineration of more than 99.99% (based upon primary kiln ash analyses), no detectable explosive wastes in the stack gas, and stack emissions were in compliance with all Federal and State regulations (including SO_2, HC1, NO_x, CO, and particulates).

The U.S.E.P.A. concluded in a recent study (Assessment of Incineration 30, 1985) that (1) incineration, whether at sea or on land, is a valuable and environmentally sound treatment option for destroying liquid hazardous wastes, particularly when compared to land disposal options now available and (2) there is no clear preference for ocean or land incineration in terms of risks to human health and the environment.

Incineration at sea represents a potential solution to counteract the objections of people because of environmental land pollution. On November 27, 1985, the EPA gave tentative approval to a hazardous waste burning trial aboard ship at sea (EPA clears test 39, 1985 and EPA to Permit waste burning 39, 1985). The test of burning about 700,000 gallons of PCB's and perhaps dioxin will take place in the spring of 1986 aboard a Chemical Waste Management ship in the Atlantic Ocean, 140 miles east of the mouth of Delaware Bay. The Company has provided $60 million in financial guarantees to cover a possible accident during the test burn. The entire test burn will cost about $1.8 million. The wastes will be transported by railroad from a landfill in Emille, Alabama and loaded aboard the ship in Philadelphia. Therefore the hazards of storage and transportation described in Section 10 of this chapter prevail also in this potential method of disposal.

Cheremisinoff (Thermal treatment, 102, 1988) recommends that "high temperature processes for specific applications should be particularly considered where available land is scarce, stringent requirements for land disposal exist, destruction of toxic materials is required, or the potential exists for energy recovery. He lists seven potential advantages and considerations of high temperature processes: (1) Maximum volume reduction which reduces ultimate disposal requirements. (2) Detoxification which destroys or reduces toxics. (3) Energy recovery from combustion of waste products. (4) Costs which are generally higher than for other disposal alternates. These are hardly advantages of high temperature treatment, but seem valid to your authors.

Table 32.10   Incineration

Rotary Kiln Liquid Injection	
Advantages	Disadvantages
Can be used for a wide variety of both liquids and solids (mixed or separate).	High capital costs for installation especially when using low feed rates.
No problem when melt occurs.	Careful operation in order not to damage refractory.
Drums or bulk containers with various feeds are used.	Airborne gases or particles may exit prior to complete oxidation.
Good mixing and air for solids.	Spherical or cylindrical solids may pass through faster and avoid complete burning.
Continuous ash removal does not interfere with burning.	
No moving parts in kiln.	Excess air required due to leaks which lowers fuel efficiency.
Wet scrubber can be added.	
Rotational speed of kiln can be varied to control residence time in burning.	If drying grates are used prior to kiln they may become plugged with heat.
No preheating, mixing needed.	High particulate loadings.
Can be operated at very high temps. of 1400°C to insure destruction of toxic chemicals.	Relatively low thermal efficiency.
	Liquid must be able to be atomized.
Able to handle a broad range of liquids.	Liquid must be heated sufficiently or supplemental fuel added.
No ash removal needed from incinerator bottom.	Must be capable of complete combustion without flames hitting refractory.
Can handle small amounts of liquids.	
Responds fast to waste temp. changes.	Liquid waste may clog burner nozzles.
Almost no moving parts.	Requires sophisticated instrumentation.
Low maintenance.	

Note: Pros and Cons of two main types of incinerators (adapted from Freeman (1981))

(5) Operating problems create high maintenance requirements—once again, not an advantage but an important consideration. (6) Personnel required are highly skilled and experienced. (7) Environmental impacts can results if air and solid effluents are not included in the treatment scheme.

The IIT Research Institute offers a new hazardous waste treatment process which essentially uses high-temperature as the treatment vehicle. It is accomplished by inserting tubular electrodes into organic-laden landfills. Radio-frequency energy charges the electrodes to heat the soil to 200°F.–1000°F. The vaporized hazardous pollutants are collected in the same tubular pipes and evacuated to a separate vapor treatment system.

32.53 Chemical, Physical and/or Biological Treatment

These treatments are designed purposely to render the industrial hazardous wastes free from toxic chemicals and materials.

Chemical treatment aims at removing smaller elements of toxic nature, generally dissolved or colloidal solids. In cases where removal of toxic dissolved organic solids are the objective, chemicals used are powerful oxidizing agents such as chlorine, potassium dichromate, or ozone. The oxidation proceeds rapidly, depending upon the concentration of organic matter and amount and contact time of oxidant used. End products are usually harmless gases such as carbon dioxide, nitrogen and water. This type of treatment is costly because of the continuous consumption of oxidizing chemicals and detention time involved. The advantage of using this process and especially chlorination is that any harmful bacteria and/or viruses present are also killed. Photolysis is another form of chemical oxidation which destroys organic matter to carbon dioxide and water. Certain heavy metal oxides, notably zinc oxide and titanium dioxide, are photocatalytic and in the presence of dissolved oxygen matter and beach sand when irradiated will decompose the organics by about 75% in 3 days.

Chemical treatment aimed at removing the larger colloidal solids is known as chemical coagulation. This is a process of destabilizing the colloids, aggregating them, and binding them together for ease of sedimentation. It involves the use of chemical flocs which absorb, entrap, or otherwise coalesce suspended matter that is too finely divided or small to be settled alone. The chemicals most widely used are alum ($Al_2(SO_4)_3$ and ferric sulfate ($Fe_2(SO_4)_3$. The reader is directed to Chapter 11 for theories of this treatment.

Physical treatment of hazardous materials in industrial wastes include (a) settling, (b) filtration, (c) adsorption or absorption, and (d) flotation. Please refer to Chapter 10 for theories of these treatments.

The removal of dissolved organic matter by *biological* means has long proven successful in the treatment of domestic sewage and certain industrial wastes. Adapted cultures of microorganisms are placed in contact with dissolved organic matter of these wastes under the proper environmental conditions of oxygen, temperature and pH. Organic matter is adsorbed and decomposed by the bacteria involved to yield carbon dioxide and water and some cell growth matter. The cell growth matter is later settled out and a portion returned to the process for biological enhancement, the remainder wasted and treated as sludge for ultimate disposal.

Some hazardous chemicals such as phenol from coking plants or synthetic textile mills can be degraded by certain adapted microorganisms to yield harmless and products. Efficiency of biological degradation depends upon adaptation of the microorganisms, contact time, oxygen available, and the proper environmental conditions. In actual practise we use two types of biological treatment processes; one circulates the microorganisms in a basin with oxygen and the wastes while the other passes the waste over a fixed surface containing the microorganisms and allowing permeation of air at the surfaces as well. There are many variations of each process in existence today. The reader can refer to Chapter 13 for these variations.

Zoltek (Bacteria Digesting Toxic Wastes 23, 1984) successfully used biological treatment at an existing sewage treatment plant to digest groundwater contaminated with phenolic compounds from an old pine tar plant.

Flathman and Githens (In-Situ Biological Treatment—1985) cleaned up a soil and groundwater spill of isopropanol, acetone, and tetrahydrofuran by a combination of biological and physical techniques. These resulted in 90% removal of IPA and THF within three weeks. Acetone, an intermediate oxidation product of IPA metabolism, was removed by the end of the sixth week. The spill originated from several buried tanks which leaked contents into a 12 foot basin of sand and pea gravel.

32.54 Neutralization

Because excessive acid (low pH) or alkaline (high pH) industrial wastes are considered toxic neutralization before discharge into the environment is a must. In ad-

dition, neutralization is usually considered desirable prior to most forms of biological treatment should the wastes contain large quantities of biodegradable organic matter as well. The authors, in Chapter 8, list and describe in some detail eight major methods of neutralization of either acid or alkaline wastes. Your authors suggest defining the titration curve of the potentially toxic industrial waste. This will allow the polluter to determine the exact quantity of neutralizing agent required to obtain any amount of neutralization obtained. Then the polluter is in a position to decide whether what amount of neutralization is cost effective. In order to do this he must also determine his damage costs involved in non-neutralization of the wastes.

32.55 Underground Injection

In locations where the subterranean environment is suitable, it is possible to inject potentially hazardous wastes under the ground. Deep well injection has been used to dispose of organic solutions from refinery petrochemical, chemical, paper and pharmaceutical plants. Ultimate fate of the hazardous waste should be ascertained prior to selecting this method of disposal. The acceptable waste should be free from suspended or clogging solids and material which will react adversely with the substrata into which it is being injected. The suitable underground environment should be permeable to a point of final residence and isolated there from any potential water resource. Chapter 13.12 describes these details.

In order to determine the suitability of the underground, a well is drilled and various depth core samples are analyzed for specific characteristics relating to its permeability and reactivity with the waste. Tests will also confirm the injection pressure required at various waste flow rates. Underground soils may be treated to improve permeability to reduce the injection pressure required.

The discharger must be aware of the dangers of (1) contaminating potable water supplies by lateral migration of the hazardous waste or even by vertical migration through open subsurfaces or mechanically-failed soil structures and of (2) causing movement of geological faults leading to potential earthquakes in the area. In this case the waste acts as a lubricant between two adjacent underground slabs of faulted rock causing them to move apart more easily.

32.56 Landfarming

The filling in of farming land with stabilized hazardous waste sludge is sometimes used as a disposal/utilization technique. Only in cases where the harvestable cover crop is not harmed by the sludge or is not used as a consumptive food material is this method acceptable. These constraints limit the use of this method for most hazardous wastes. However, in situations where flammable or ignitable wastes are involved, this method may be feasible. In this case the oils and other associated organic matter may be decomposed by soil bacteria using the matter for food and enhancing the surface cover crop growth. Bacterial/viral-contaminated waste sludge may also be landfarmed in certain instances where the cover crop will not be harvested in the near future. This may give ample time for the bacteria/virus to become innactivated. Stabilized and/or solidified hazardous waste may also be landfarmed in certain instances where the waste serves primarily as a soil stabilizer and a preventer of leaching into groundwater.

One type of industrial waste being landfilled is that from the construction industry. It is common practise to bury construction demolition and other debris such as damaged tree limbs, damaged plywood, wood forms, old concrete blocks and bricks, etc. These materials may slowly be broken down under the ground without interfering with the use of the surface crops growing.

A land cultivation report (U.S.E.P.A. Land Cultivation—40, 1978) states that 3 percent of all industrial waste can be disposed of by land cultivation. They found that the amount of industrial solid waste disposed of by this process was limited by soil texture, drainage, permeability, and the waste pH, bulk density, soluble salt and metal concentration, and nitrogen and phosphorous contents. Costs for this process vary widely from two to eighteen dollars per cubic meter of industrial waste. Transportation costs must be added to this amount.

Bahorsky (Land Application Treatment 41, 1983) concludes that land application of textile wastes can be less expensive in capital extended aeration systems, less expensive to operate, requires less training, produce no sludge problems, be significantly less energy intensive, and give consistently high-quality effluents. However, he makes no recommendation that the land be reused subsequently for farming. It does seem possible to farm on this land after a certain period of waste treatment.

32.57 Disposal in Natural Storage Areas

A commonly recommended solution to the disposal of hazardous wastes is to return them to their natural origin or to areas of the land compatable with them. For example, coal overburden as well as flyash and unburned carbon could be returned to coal mines before renovating the natural land.

Sundaresan et al. (7, UNEP 72, 1983) recommend returning neutralized, precipitate, and dewatered lime sludge from titanium dioxide wastes to abandoned ilemenite mines.

32.58 Permanent Landfills

In some cases hazardous wastes are disposed of either with or without some form of pretreatment into the land for permanent storage. Naturally many precautions must be taken in using this technique. Most important of these include the security of the waste in the landfill—it must not be reached from the surface of the land nor must it escape into surface or underground water supplies.

One such use is described by Sundaresan et al. (7, UNEP 72, 1983) for disposing of carbamate pesticide wastes. After neutralization, as a pretreatment, the wastewater is solar-evaporated in polyethylene lined ponds. The resulting lime sludge from neutralization is disposed of as landfill. The authors state that since the sludge may contain carbamates "adequate safeguards have to be taken." Presumably this means sealing the sludge in a permanent landfill.

LeRoy (25, 1983) acknowledges that since all hazardous wastes cannot be incinerated or detoxicated, some with low toxic content can be disposed of in landfill provided certain conditions are observed. Wastes are permitted in landfills based upon their quantity, solubility, toxicity, and concentration. In particular, sites for industrial waste where disposal is allowed in France must be impermeable, so that the waste and substances leached from it are adequately confined. He reports 14 such landfills in France at special disposal charges ranging from 80 to 500 francs per tonne.

In the United Kingdom Peter Pearce reports (7, UNEP, 57, 1983) that of the 3.4 tons of hazardous wastes generated per year and not recovered about 80% is landfilled. The remainder is either disposed of in oceans or by incineration or other treatment. About the same percentage generated is landfilled in the United

States as of 1987 (42, Pizzuto, 1981). Pearce points out some advantages for using permanent landfills for ultimate disposal of hazardous wastes. Adjacent land areas sometimes can be used by an industry to avoid costly transportation problems. Further, incineration, the major alternative treatment of hazardous wastes, costs about 2 to 3 times that of landfilling. Incineration has the disadvantage of not being suitable for many types of wastes and of generating toxic gaseous products and unburned particles. Efficient removal of heavy metals from the exhaust and damage to the incinerator structure also pose problems. Pearce (UNEP, 7, 58, 1983) lists the following six constraints in selecting a permanent landfill site for the ultimate resting place for hazardous wastes.

(1) Institutional constraints—usually legal and ownership of land.
(2) Geographical considerations—existing land and water use, topography, climate, hydrology, location of waste sources, population distribution, and transportation routes.
(3) Geological considerations—proximity to known fault zones or sink holes, potential for ground shifting causing landslides, etc.
(4) Waste characteristics—property and volume of wastes determine the design and storage capacity of a landfill area.
(5) Management priorities—such as financing during the beginning phases, operating and maintaining the landfill after closing the fill.
(6) Environmental and social considerations—ecology, water resources, archaeological resources, socio-political and socio-economic factors.

Hazardous wastes can usually be controlled landfilled unless they are known to be radioactive, highly flammable, explosive, excessively toxic, odorous or corrosive (DOE, 43, 1978). Non-compatible or highly leachable wastes must be given special consideration in designing safe fills. Sax (44, 1979) warns that we should watch out for the compatibility of each hazardous waste with others which may cause spontaneous reaction in the fill especially with water and acids. In such cases, pretreatment or segregation of different wastes may be required.

The question of how permanent are landfills is of importance. For example, liners preventing leaching of potentially toxic chemicals from the fill. It is too early to predict the longevity of various liner materials for landfills. With that in mind, it is preferable and safer to design these facilities for some leachate to protect water resources.

In New Jersey storage tanks containing volatile liquids have been painted white and covered with floating covers. Lower temperatures thus created yield lower evaporation rates and ensures no loss of toxic organics to the atmosphere.

Piasecki (45, 1983) pictorially depicts what can go wrong with permanently storing hazardous wastes in a landfill. They include (1) burrowing gophers attacking the landfill cover, (2) freezing temperatures shrinking and tearing the landfill liner, (3) mineral acid and solvents mixing and triggering a chemical fire, and (4) chemicals corroding waste collection pipes or weight of waste may crush these same pipes, (5) debris may clog perforated collection outlet pipes, (6) the landfill's protective cover can be breached by erosion, by new construction on the site, or by settling after bulk solids compact and barrels disintegrate further down in the fill. If the cover splits at the surface, rain water can reach the waste and overload the leachate collection system or cause the entire landfill to overflow.

Sachdev et al. (46, Use of Fly Ash—1983) investigated the use of fly ash with or without additives as a liner material for landfill disposal sites. They found the following parameters significant for use of fly ash as a liner material. They aimed for a hydraulic conductivity of 10^{-5} cm/sec. or less and a leachate with a minimal impact on groundwater.

Boiler Type—The type of boiler and boiler temperature affect the physical and chemical characteristics of the fly ash. Fly ash from a pulverized boiler is generally finer than from a cyclone or stoker fired boiler. The finer fly ash has more surface area and is therefore more reactive.

Coal Source—The coal source may affect fly ash characteristics. There are differences between western coal fly ashes and eastern coal fly ashes. Western coal fly ashes have higher lime content and higher pH than fly ashes from eastern coal. The higher lime content (greater than 10 percent) fly ashes usually exhibit self-hardening properties.

Particulate Control System—Electrostatic precipitators and bag houses generally collect a finer fly ash than mechanical collectors such as cyclone units.

The finer fly ashes have more surface area and are generally more reactive.

Fly Ash Handling Mode—Wet and dry fly ash handling have different effects on fly ash reactivity [3]. Wet handling reduces the reactivity of fly ash.

Loss-on-Ignition (LOI)—LOI indicates the amount of unburnt carbon in the fly ash. High LOI (and therefore high carbon content) in fly ash inhibits pozzolanic activity. The concrete industry specifies a maximum LOI value of 6 percent in fly ash for use in concrete manufacturing [4].

Particle Size—The particle size of fly ash is important. Finer fly ash offers more surface area for reaction and generally provides lower permeability than coarser material. A well-graded fly ash is preferred over a uniform particle size fly ash since it would generally exhibit smaller void spaces and therefore lower permeability. Fly ash utilized as a mineral admixture in portland cement concrete must have a minimum of 66 percent by weight passing through a No. 325 sieve [4].

pH—A higher pH (pH of 10 or more) is required to promote the precipitation of calcium silicate which is the source of bonding in the pozzolanic process.

Lime Content—Lime is the source of calcium which reacts in the pozzolanic reaction. It is the "free" lime in the fly ash which is available for reaction. Lime can be added if fly ash has low lime content. A fly ash with lime content greater than 10 percent normally exhibits self-hardening characteristics.

Amorphous Silica Content—The amorphous silica is the non-crystalline portion which reacts with calcium in the pozzolanic reaction. A fly ash with low amorphous silica would therefore have low reactivity. The amorphous silica content would generally increase with increasing boiler temperature.

However, the United States government officials report that one-third of the toxic waste landfills might be forced to close because of results obtained by underground monitoring and by new insurance requirements (U.S. Waste Deadline 51, 1985). Industry appears to be moving away from dumping toxic material into landfills and toward the incineration and treatment of hazardous wastes. In fact, 1,100 toxic landfills—almost two-thirds of the nation's total—have in fact closed because they cannot comply with federal environmental rules (Toxic sites close 52, 1985).

32.59 Ocean Dumping

The dumping of hazardous industrial wastes into ocean areas may be practised sometimes safely because of the following attributes of oceans: (1) they tend to dissolve and disperse the wastes three dimensionally—as contrasted to the surface use only of land resources; (2) oceans are mostly out-of-contact with man except for shipping and transportation-two rather "lowly" uses of water resources; (3) some pollutants may settle thousands of feet to ocean bottoms where presumably they will cause man no harm; (4) the ocean is already considered a mixture of a multitude of chemical elements and compounds; and (5) the ocean food chains are much longer than those we experience with on-land food chains and hence man is placed farther away from hazardous chemicals.

Oceans should be used only for wastes which when diluted fully will offer no threat to marine life—especially that kind which serves as food for man. The oceans should provide enough time before reuse by man for degradation of the hazardous industrial waste. Nemerow gives (Chapter 12, 11, Stream, Lake, Estuary, and Ocean Pollution 1985) a detailed description of the effects of some common ocean pollutants on the viability of the ocean resources. It is important to keep in mind that the aquatic ecosystem of oceans has a finite assimilative capacity for a particular contaminant without significant deleterious effects. The assimilative capacity of any particular ocean area is determined by physical processes such as mixing currents, geomorphology, types of sediments, types of water chemistry and biology.

The National Research Council (Ocean Disposal Systems 53, 1984) recommended that although oceans are suitable for discharge of industrial and municipal wastes, "certain xenobiotics (man-mad organics) are so persistent that they cannot be removed in treatment plants or diffused to safe limits in the ocean."

32.60 Secure Burial, Reuse, Chemical Fixation

Prouty et al. (Sludge Amended (54) 1983) used a containment method of incorporating toxic sludges into bricks during their manufacture. They claim savings in energy, since bricks are lighter, more porous, and enhanced insulation quality. Bricks are produced by mixing finely ground clay with water, forming it into the desired shape, drying, and finally burning it. They tested metal laden sludges and found a portion of the metals volatized during brick burning, but the remaining metals were bound in the brick and will not leach from it.

Wright and Caretsky (In Situ Treatment (49) 1981) give seven techniques for chemical fixation of toxic wastes. *Cement-based fixation* is a process which mixes toxic wastes in a slurry of water and cement. When the concrete hardens it can be used for various building purposes. Sometimes in surface coatings of asphalt, vinyl, or emulsion mixtures to increase the strength and decrease the permeabilities of the concrete. Radioactive wastes and heavy metal sludges have been "fixed" in this manner. *Lime-based fixation* involves mixing the waste with lime and fine-grained silicious waste material such as fly ash, blast furnace slag, or cement kiln dust. *Encapsulation fixation* process encloses previously bonded waste in a covering of non-reactive, inert material. Usually the toxic waste is mixed with a thermosetting plastic, placed in a mold and heated to fuse into a hard block. Many hazardous industrial wastes originated from the extraction, use and alteration of toxic elements found in native underground sites in a form similar to that in which they were found. The concept appears reasonable. However, several obstacles must be overcome. First, the volume, weight, or mass of hazardous materials must be reduced to a point where it would be economically feasible to return them underground. Second, the exact location to where they are returned must be suitable. That is to say, it must be available, owned by the disposer, and not likely to cause environmental problems to potential underground resources such as water, air or land. Thirdly, methods must be found to solidify and contain the hazardous wastes so that they will not migrate away from the burial site once they are placed underground. These are not insurmountable obstacles; however, they must be faced and settled before solidification becomes a viable solution for the safe disposal of hazardous wastes. *Self cementing fixation* is used with waste sludge containing large amounts of $CaSO_4$ or $CaSO_3$. Small portions are dewatered and calcined under controlled conditions and remixed with the entire waste sludge and additives as necessary. The product is a hard, plaster-like, relatively impermeable mass. This process is especially useful for sludges from power plant stack gas scrubbers. *Classification process* mixed

the toxic waste with sand before fusing into glass. This is used mainly for radioactive wastes. *Thermoplastic fixation* dries, heats, and mixes the toxic waste with a heated plastic material such as paraffin, bitumen, and polyethylene, cooled and resalting solid matter stored indefinitely. Although this process is mainly used for radioactive wastes, it may be used for certain organic solvents, oxidizing salts such as nitrates, chlorates or perchlorates, and dehydratable salts.

Organic polymer fixation produces a spongy, rather than a solid mass, by blending the waste with a prepolymer and catalyst. This is done at room temperature and usually in batch basis rather than a continuous process. It is also used for industrial radioactive wastes. Francis (Landfilled wastes—1984) found it much more economical and practical to treat a toxic metal, landfilled sludge in place than to remove it for treatment and subsequent alternative disposal. Essentially he found, that after detailed analysis of the landfilled sludge, nickel as the hydroxide was the major toxic contaminant. His plan was based upon the fact that nickel hydroxide solubility in water can be effectively controlled by keeping the pH at 10.2 its point of minimum solubility. The landfill sludge site was covered with a layer of finely ground calcium carbonate to neutralize acid rain before it reached the waste. Then the site was covered with mounded, compacted, and graded earth to divert rainwater and runoff from the waste. Finally a 12 inch layer of gravel followed by a 6 inch top soil layer was added on the earthen cover. The top soil is seeded to provide vegetation necessary to prevent erosion.

His system of containment although costly was estimated at about 10% of that necessary to remove sludge and to treat it other ways.

ENRECO (Project Reports (55) 1985) has successfully completed the solidification phase of over 50 sites during 1984. The company makes a careful analysis of solidification agents and then uses a unique injector to add and mix the solidification agents with the waste. Some agents which have been used include fly ash for oil field drilling fluids, kiln dust for oil field skim pits, and carbon for PCB impounded waste sludge. In the last case the solidified sludge was removed and transported to an acceptable landfill. In other cases the solidified mass was buried on site. Photographs of successful operations of 1—mixing, 2—solidification, 3—removing and 4—burying with land renewal.

Jones and associates point out that many hazardous wastes, especially metal plating wastes, may contain constituents which could interfere with the binding process of solidification with Type I Portland Cement and fly ash. Some include oil and grease, light weight oil, phenol, sulfates, strong base, pesticides, degreaser, lead, copper, and zinc. (Factors affecting Stabilization/Solidification—1985.)

32.61 Exportation

Since the beginning of time, man has developed the attitude that if he can give his problem to someone else, he is relieved of any further responsibility. Or "out of sight, out of mind." Therefore, it comes as no surprise to observe industry expounding the philosophy of contracting for services to carry away hazardous wastes to a site distant from the plant where it originated. The assumption is made in these cases that it is far less expensive to pay some agent to assume the responsibility for providing the service than to assume the liability and costs yourself. It is also presumed that the disposal of the hazardous waste is safe and effective at the distant site.

Industry should shoulder the responsibility (before exporting hazardous waste) of determining not only whether this method is more economical, but also whether it is effective and safe ultimately.

In June of 1983 a novel plan to export toxic PCB wastes from Florida to Honduras was uncovered (Export plan 56, 1983). The plan was foiled and no criminal charges were filed. But EPA and state officials agreed that this export plan underlined growing fears that strict regulations on toxic waste disposal in the United States could lead some American firms to try to dump their toxic materials abroad.

In March of 1989 a treaty between more than 100 countries was unanimously adopted restricting shipment of hazardous waste across borders (Conference Backs Curbs on Export—106, 1989). The pact requires waste exporters to notify and receive consent from receiving countries before shipping the waste. In addition, the treaty requires countries exporting waste and those receiving it to insure that the waste is ultimately discarded in an environmentally sound manner. The treaty prohibits shipments of hazardous waste to nations that have banned waste imports. It also requires all cross-border waste shipments be packaged and la-

beled properly and that all companies transporting or disposing of hazardous wastes be authorized to handle wastes. Among the wastes considered hazardous are clinical wastes from hospitals, wastes from pharmaceutical factories, PCB's, compounds containing mercury or lead, and wastes from the production or use of dyes, paints, and wood-preserving chemicals. The Treaty is a beginning and its nature will require additions, clarification, refining, and strengthening as time goes on.

32.62 Recovery and Reuse

Recovery and reuse of hazardous waste represents the ultimate in environmental and economical efficiency. If found feasible, such a solution prevents environmental damages and provides some monetary return to the generator of hazardous waste. Not to be overlooked are the indirect benefits of recovery and reuse to society. Most of these benefits are associated with a slower decrease in the world's natural resources—a so-called slowing down of entropy buildup. In addition, consumer prices for products made with recovered materials should be lower in the long run.

Therefore the goal of all industry and its environmental engineers should be that of recovering and reusing all wastes including especially hazardous ones. In chapter 15 we present a novel system for doing this with all industrial production. In chapters 6 and 7, some general principles of reuse and described. Here, we present only some examples of effective recovery and reuse of certain hazardous industrial wastes.

An important consideration in this chapter should be the question of whether the business of recovery and reuse of many chemical elements and compounds considered toxic to the environment represents a beneficial service to society independent from its profitability. If so, one could expect that it would be reasonable for public subsidization to some extent of such ventures. If not, then the business of recovery and reuse must stand solely on its own economic viability similar to that of other profit-making industries.

Prior to recovery of hazardous wastes one should attempt to reduce the amount of these wastes to a minimum. In that connection the Jacobs Engineering Group proposes four major incentives for what they refer to as "Waste Minimization." (The EPA Manual for Waste Minimization—103, 1988).

(1) Economics
   A-Landfill disposal cost increases
   B-Costly alternative treatment technologies
   C-Savings in raw material and manufacturing costs
(2) Regulations
   A-Certification of a waste management program on the hazardous waste manifest
   B-Biennial waste management program reporting
   C-Land disposal restrictions and bans
   D-Increasing permitting requirements for waste handling and treatment
(3) Liability
   A-Potential reduction in generator liability for environmental problems at both onsite and off-site treatment, storage, and disposal facilities
   B-Potential reduction in liability for worker safety
(4) Public Image and Environmental Concern
   A-Improved image in the community and from employees
   B-Concern for improving the environment.

32.63 Copper Recovery from Plating and/or Engraving Baths

When a hazardous industrial waste is composed almost wholly of copper as a contaminant, it usually can be recovered in quite pure form as the hydroxide. Such may be the case in electronic printing board wastes. Even when the wastewater is contaminated slightly with other metals such as iron or zinc, copper may be precipitated and then purified, if necessary, before recovery and reuse.

Copper hydroxide is quite insoluble at an optimum pH of between 7 and 9. For recovery and reuse NaOH is preferable to Ca(OH)$_2$ for raising the ph which contaminates the recovered sludge. The pH should be kept below 9 and preferably above 8 in order to prevent resolublization of the copper hydroxide precipitate. Other contaminating metals especially zinc and chromium, will be incompletely precipitated at this pH.

Another method of recovering copper from mixed metal wastes is that of depositing copper on brass chips in the presence of chloide. This process of copper recovery is used primarily for treating brass mill wastes where chromium, copper, and zinc are common impurities found in the effluent.

Cation exchangers may also be used on copper wastes such as those from the electronic industry in making printed circuits. Copper is present in these wastes as the sole contaminant. However, only analysis of the hazardous wastewater will reveal whether either of these latter two recovery methods is preferable to copper precipitation as hydroxide.

Cartwright describes a complete metal plating reclamation and waste treatment system designed, installed, and operated in Bloomington, Minnesota (Innovative Technology 57, 1984). The latest treatment technologies are detailed including reverse osmosis, two bed deionization, mixed bed deionization, activated carbon adsorption, ultrafiltration, ozonation, slant tube clarification and sludge dewatering. Over 90% of the rinse water from the electroplating line is purified back to 18 meg ohm quality and recycled; ultra pure makeup water is supplied from city sources and toxic wastes are precipitated as an insoluble sludge. Nickel-iron, nickel, copper, and gold metals are involved in preparing 10 cm diameter wafers of these metals on a special ceramic substrate. Cartwright describes the concept of the reclamation and minimal discharge as the "wave of the future."

32.64 Spent Caustic Soda Regeneration

Certain industrial wastes such as textile kiering and pulp mill cooking liquors contain spent sodium hydroxide in relatively high concentration and in relatively uncontaminated form. In these cases as well as in other special situations, caustic soda may be recovered and reused in the same industry or sold to other plants for other uses.

Dialysis or other variations of membrane separation is a major method used in recovering pure caustic soda. The caustic permeates the membrane and dissolves in water usually much easier and faster than any other contaminants contained in the waste. The authors give theories and rate reactions for typical dialysis systems in Chapter 12.2. The quantity of sodium hydroxide diffusing through the membrane depends upon the time, the area of the dialyzing surface, the mean concentration difference, and the temperature. Naturally, the membrane pore size and degree of clogging will control the flux rate (permeability) of the caustic soda.

Evaporation is also used to concentrate caustic soda for recovery and reuse especially when heat energy is relatively inexpensive and when impure caustic can be reused satisfactorily.

32.65 Mercury Recovery

Toxic mercury compounds occur in several industrial wastes, mainly in the alkali-chlorine plants using mercury cells. Small amounts of mercury escape these plants, end up in watercourses, and are picked up and biomagnified in fish from where they reenter the human food chain again. Because of the extreme toxicity of mercury to the human nervous system, it is very vital for industry to make certain none is released. All forms of mercury apparently revert to the specially toxic and dangerous methyl mercury form. Since mercury is found in very small amounts in the wastes from several plants using mercury cells or thermometers, it is seldom practical to design and install large, complicated treatment systems. Rather, it is more realistic to separate out by gravitational means (mercury is almost 14 times as dense as water) small quantities of waste mercury at the origin of its escape from the process or use and return them to reuse in the plant.

32.66 Cadmium Recovery

Cadmium accumulation in humans has been linked with hypertension, emphysema and bronchitis. Much cadmium wastes have been attributed to automobile battery manufacturers. The major treatment techniques for removing trace amounts of cadmium include chemical precipitation, cementation, reverse osmosis, ion exchange, chelating resins, and foam separation.

The nickel-cadmium cell is an excellent storage life battery. It is a secondary cell type; that is, it depends upon chemical reactions that are reversible by electric energy and therefore do not need chemical replacements. The cell uses nickel hydroxide as the positive electrode and dadmium-cadmium hydroxide as the negative electrode as the electrochemically active materials. Potassium hydroxide is the electrolyte in an aqueous solution. When these used batteries are delivered to chemical recovery plants, they are stored, washed, and valuable nickel and cadmium recovered.

Drainings, washings, and unusable metal parts of the batteries must be kept from polluting the soil and associated groundwater during the recovery process.

32.67 Silver Recovery from Plating and Photographic Wastes

Silver is both too toxic and too valuable to allow any to be discharged to waste. Two major industries use and waste sufficient silver to make it imperative for them to recover it: silver plating and photographic developing.

Silver can be precipitated as the chloride, sulfide, hydroxide, or insoluble silver. These are effected by the addition of sodium dulfide, sodium hydroxide and sodium borohydride ($NaBH_4$), a powerful reducing agent. In some instances ferric chloride is added to enhance silver solids separation. Any other heavy metal contaminants will also be precipitated and must be removed from the solids before recovering the silver.

With a lithographic film waste, Cook et al. (58, Case Histories, 1979) reports the borohydride system gave the best overall results based on residual metal levels, cost, and suitability of the precipitated silver for further recovery. His chemical cost was $0.50 per troy ounce of silver removed. Nemerow reports (Theories and Practices of Industrial 59, 1963) that recovering silver from a silverware plating plant waste by precipitating it as the chloride yielded silver concentrations in the effluent of less than 3.5 ppm from an original value of up to 250 ppm.

In addition to precipitation, your authors report in Chapter 27 that silver can also be recovered by metallic replacement and electrolysis. In the case of photographic wastes containing silver, spent hypo solutions are brought into contact with a metal surface such as steel stampings, zinc, or copper. After more than 2/3 of the silver is recovered, the metal is removed from the tanks, dried and sold to the refiner.

In the electrolysis method, both an anode and cathode are placed in the waste silver and an electric current passed through. Silver will plate out on the cathode. 32 square feet of cathodic area are required for treating 50 gallons of waste containing 19 ounces of silver. After 24 hours of air agitation, 98 percent of the silver is removed. About 600 troy ounces of silver can be recovered before the cathode has to be desilvered or replaced.

32.68 Waste Oil Recovery

Waste oils are found in industrial discharges from many varied types of plants and originate from uses such as lubricating, fuel, hydraulic circuits, and refining. As might be expected from such widespread uses, the qualities of the oily residues are vastly different.

Residual oils from transformers, turbines, and some hydraulic systems are sometimes referred to as "bright oils." These oils can be recovered and directly reused for purposes not demanding the highest quality (purity) oils. An example of such secondary reuse is in stripping moulds in foundries.

Waste oils from motor cooling, mill hardening and rolling and drawing of steel, and other metal lubricants and cooling are called "black oils." These oils can be recovered and regenerated for reuse by re-refining by specialized companies. It may be dangerous practice to incinerate these black oils rather than regenerate them because of the impurities such as lead given off during burning. In addition, the wasteful, uneconomical burning is eliminated by regeneration and reuse.

Fitzpatrick (Machine Shop Scrap, 60, 1985) discusses companies which turn, bore, or mill metal. A product of these operations is always some form of oil-contaminated scrap metal. The scrap has little or no value and, in addition, it is a nuisance in its present contaminated form. Fitzpatrick claims that a properly designed and sized system will reclaim valuable oil, upgrade the value of metal scrap and reduce handling and hauling costs. He also claims that these systems return investment capital at a rate comparable to other capital equipment. Scrap chips must first be collected by some units placed beneath the various machines, raised to crushers to reduce the scrap size, discharging the scrap into containers for storage for resale and the reclaimed oil to cleaning by settling and filtration for reuse as friction reducing and cooling agents.

Another location in the steel mill where oil may be recovered and reused is the quench water. As steel is quenched, the metal oxide and the oil combine to form a grimy scale. The scale accumulates into an oily, dirty sludge. One company reports using a Barrett Sludge Extractor (Sludge extractor 61, 1985) to recover most of the quench oil from the sludge and save about $30,000 per year in oil purchases. The Extractor operates like a centrifuge to spin the wet sludge at high speed and compact the sludge while driving the oil towards the center of the unit. From there the oil is collected, filtered and recovered for reuse while the dry sludge remaining is disposed of as required by local regulations.

Your authors report in Chapter 27.11 two major types of fuel oil wastes: diesel train refueling oil and auto and truck waste crank-case oils. When recovering these oils for reuse, they are generally heated, coagulant is added, and the resulting oily slurry centrifuged to remove the impurities. The centrifugate is then re-sold to oil refiners and/or distributors for reuse.

32.69 Solvents Recovery

Solvents are universally used by many industries both in the gaseous and liquid phase. Examples of such industries are dry cleaning, printing, paints and varnish, rubber, and metal degreasing. Besides objectionable odors, some solvents are photochemically reactive. Fluorocarbon solvents effect the upper layer (ozone) of the atmosphere. The recovery and reuse of waste solvents can be compared in philosophical and economical terms to that of waste oils. Solvents must be regenerated to separate them from their impurities. Regeneration is carried out by either steam stripping or rectification. Since regeneration equipment is expensive mainly from a capital investment standpoint, regional or collective facilities for several industries are recommended.

The Miami Herald (page 18, Oct. 27, 1983) reports of the Gold Coast Oil Company which distilled and recycled chemical wastes, subsequently closed due to pressure from Dade County, Florida, but responsible legally for any residual toxic wastes. The company left behind about 2,500 fifty-five gallon drums of unidentified waste, unrecyclable residues and solvents. It was gooey, sticky, and discolored. The drums leaked and the material tested for "alarmingly high concentrations" of heavy metals. The lesson here is that even recycling plants may run into environmental problems resulting from additional residual wastes.

Vara International of Vero Beach, Florida is one of a number of companies which specialize in solvent recovery. It has used a special pelletized activated carbon to adsorb solvents for over 40 years. The choice of carbon is very important; special concern should be uniform particle size, low ash content, high retentivity and strength.

32.70 Water Recovery and Reuse from Industrial Laundries

Industrial laundry wastewater, although not generally considered hazardous, may contain priority pollutants such as heavy metals. At the least it is a complex and variable mixture of high concentrations of organic materials and suspended solids. Van Gils et al. give a recent consolidation of analyses of typical industrial laundry wastes (Future of Water Reuse 62, 1984) in Table 32.11.

They reported that lime coagulation, settling, high-rate ultrafiltration, and fixed-bed carbon adsorption

Table 32.11 (After Van Gils et al. 1984) Typical Industrial Laundry Wastewater Constituent Concentration

BOD	1,300 mg/l
COD	5,000 "
Susp. Solids	1,000 "
Oil and Grease	1,100 "
Lead	4.5 "
Zinc	3.0 "
Copper	1.7 "
Chromium	0.88 "
Nickel	0.29 "
Chloroform	3.3 "
Benzene	2.5 "
Perchlorethylene	9.1 "
Toluene	5.2 "

proved effective in providing a reusable water at an operating cost of $3.80 per 1000 gallons of wastewater processed. No mention is made of the disposition of the contaminants removed. Presumably, they would be landfilled or burned during regeneration of the carbon.

Space is often a limiting factor in smaller laundries for normal physical-chemical treatment required. However, with increasing size of industrial laundries and more automated adsorption-filtration systems now encountered, these treatments and the reuse value of the water are more feasible.

High gradient magnetic separators consisting of a filter bed packed with a fibrous ferromagnetic material such as stainless steel or wool screens and magnetized by a strong external magnetic field which surrounds the bed have been described and recommended as useful for treating laundry wastewater (Oberteuffer 63, 1973) and (Delatour 64, 1973). This treatment is said to be extremely fast, with filtration rates as much as two magnitudes higher than conventional sedimentation-filtration processes. Once again contaminants removed must be disposed of ultimately by burning or landfilling. Other treatments for laundry wastes can be found in Chapter 25.7.

32.71 Dye Recovery and Reuse from Textile Wastes

Textile dye wastes were reported by your author as far back as 1952 (Textile Dye Wastes 65, 1952) as being pollutional in nature and varying in treatability depending upon their chemical nature. In 1957 Nemerow de-

scribed and classed commercial dyes according to their structure (Color in Industrial Wastes 66, 1957). It was at this time that it was proposed to alter the color of a major class of textile dyes, the azo group, in order to remove the color contaminant characteristic. The azo chemical structure was altered by reduction with stannous chloride and salting with NaCl.

From that time to the present, Nemerow has been advocating recovering and reusing textile dyes. Support for this thesis was based upon the difficulty and high cost of color removal from these wastes as well as the economic value of the recoverable dyes.

In 1982 (Industrial Water Pollution Control in Hong Kong 67, 1982) Nemerow suggested to the textile industries the removal of dyes on fine filter material such as clay or fine weave fabric. The dyes can be concentrated up to 80% solids before evaporating with waste heat the remaining 10-15% water. In considering the economics of dye recovery by this or any other method, one must consider the avoidance of damage costs or other destructive treatment costs as negative costs of dye recovery.

Recently, Bergenthal et al. (Textile Dyebath-68, 1984) demonstrated the technical feasibility of batch dyebath reconstitution and reuse at a carpet mill. They overcam several technical problems which are common to a wide variety of textile mills such as:

1. selecting product styles and shades that include dyebath reuse
2. reforming dye recipes to use a single dye group to dye many shades
3. demonstrating the feasibility of dyebath reuse on a portion of the product while the remainder continued with normal production
4. producing high-quality product with recycled dyebaths
5. adapting dyebath reuse to conform to the mill's standard dyeing procedures.

They observed 25–50% reduction in both pollutant and water use. Economic benefits were termed attractive and especially when dyebath dilution with steam condensates and overflow cooling waters was eliminated. This latter finding lends even more credence to our recommendation that dry dyes should be recovered, stored, and reused by adjustments with fresh dyes at appropriate times in the production schedule.

Nemerow finds recently (104, 1988 U.S. Pakistan Joint Research) that textile dye wastes can be treated economically and effectively by ultrafiltration. Membrane filtration studies are in progress.

32.72 Detoxification

Hazardous wastes can be detoxified in-place or prior to disposal in another site by a variety of processes depending upon the type of waste and special situation. Any of these processes portends a residual waste for ultimate disposal. However, the residual waste would no longer possess toxic properties. Detoxification at the site of hazardous waste origin is highly recommended by the authors. Not only is the toxic waste generator better able to detoxify his own waste as it originates, but also he is protecting the final disposer from accidental spills or malfunctioning systems. Also, the generator is legally and morally responsible for making his own waste safe for further transportation and/or treatment and disposal.

Sunohio developed and is promoting a detoxification process (69, 1981) consisting of a mobile chemical plant on a tractor-trailer truck. It has been used to detoxify polycholorinated biphenyls (PCB's). A chemical agent strips chlorine atoms from insulating liquids and thereby removes its toxicity. The promoters claim that the detoxified insulating fluids can them be reused for the same purpose without decreasing its effectiveness.

Piasecki (45, 1983) reports of recent EPA studies of soaking toxic solvents and acids in hot baths of molten salt that will destroy the toxic chemicals. The wastes are injected into a pool containing sodium carbonate heated to around 1,650° Fahrenheit. The hazardous hydrocarbons are burned off and converted to CO_2 when oxygen is supplied or available while sulfur and chlorine react with the salt and end up in the ash residue of only less than one percent of the original volume. One advantage of this process is that it allegedly takes place at temperatures lower than those of conventional incinerators. This minimizes the production and evolution of nitrous oxides, precursors of acid rain. Rockwell International offers a unit which can detoxify one ton of waste per hour.

Piasecki describes the process as pumping the contaminated oil (PCB) into a chemical soup that contains metallic sodium. The sodium strips the chlorine from the PCB's and combines with it, creating a form of sodium chloride. The biphenyls are removed as a harmless

sticky residue, and the oil can be returned to the transformer, used as fuel oil, or dumped without special handling. The process can reduce PCB contamination from a level of 10,000 parts per million to less than 2 parts per million. Acurex Corporation is developing a truck-treatment process to repurify contaminated soils of this PCB and return the soil to its original site.

LeRoy (25, 1983) described a collective detoxification establishment to consist mainly of:

1. a workshop for removing cyanide by oxidation
2. a workshop for removing chromate
3. an acid-base neutralizing workshop

He adds that machining fluids are also processed by physico-chemical methods such as ultrafiltration and evaporation.

Hanson (70, 1984) describes a new air stripper which removes over 99% of total chloroform extractables from a well water supply. The air stripping system is costing the town of Hartland, Wisconsin about 4 cents per thousand gallons of water treated and strips around 500 parts per billion of TCE at 1,000 gallons per minute. Once again, however, such a treatment transfers the toxic material from the water to the air environment.

Wright and Caretsky (In Situ Treatment 49, 1981) include soil flushing, chemical detoxification, and microbial inoculations in treatment of waste dumps to detoxify them. Soil is flushed by applying water at the surface and collecting leachate with the use of shallow well points. Analysis of the leachates will indicate whether they must be treated further or may be discharged safely to watercourses. This is especially useful when soils and wastes are acidic and likely to leach out heavy metals. Chemical detoxification is done in the same way except some chemical agent is added to degrade or alter the toxic material so as to render it harmless. Seeding the soil with a microbial growth which can degrade the contaminant(s) is also used when a suitable amount of time is available and the toxic material can be biodegraded.

The technique of "plasma arc" detoxification provides the most complete destruction with the highest energy level possible short of atomic reactors. It is claimed to have been invented by a Canadian engineer, Tom Barton. He transmits a high energy between two electrodes which heats a short chamber of air to temperatures in the range of 45,000 degrees Fahrenheit, transforming any impurity in the chamber into a plasma. A plasma can be defined in this case as "ionized gas composed of electrons and positive ions in such relative numbers that the gaseous medium is essentially electrically neutral." Sometimes plasma is referred to as a "fourth state of matter."

Familiar forms of plasma-like matter are lightening bolts, fluorescent lights, and welding arcs. Barton feeds a toxic waste into the middle of the chamber where it is attacked by the plasma disintegrating the molecular bonds which are broken, and only carbon, hydrogen, oxygen, and chlorine atoms remain. He pumps the gases to a scrubber when the potential for formation and release of residual toxic gases exist. Barton's development of the plasma arc represents an improvement over the NASA implementation of the 60's in space reentry of vehicles. His electrodes are said to withstand the heat and use for longer periods than the older NASA ones. Several companies in addition to the U.S. Army are researching the plasma arc method for destruction of not only toxic wastes generated directly by industry, but also for renovation and purification of contaminated soils.

Two more conventional methods of detoxifying waters containing small quantities of volatile organic chemicals such as trichloroethylenes from dry cleaning wastes are (1) air stripping and (2) adsorption.

(1) Air Stripping

This is accomplished by pumping water down through a packed column through which air is being blown up and discharged above the column to the atmosphere. The air strips the volatile organics from the water which is collected beneath the column. Air to water ratios as well as their temperatures are important factors in determining removal efficiencies.

An illustration of air-stripping was reported by Wenck and Josephson (Hazardous Wastes Sprayed Away 71, 1984). They drew polluted leachate from under a toxic landfill to a point where it could be pumped 1400 feet, then sprayed in an adjustable arc at a rate of about 200 gallons per minute. The jet mixed air with water at a ratio of 8,000 to 1. The treatment process not only eliminated toxic organics from a downstream well water, but also volatilized the same during spraying in the air.

(2) Granular Activated Carbon Adsorption

In this treatment contaminated water is fed through enclosed tanks containing a packed bed of specific size and density of activated coal material. The latter holds the volatile organic contaminants while the purified water is usually discharged at the tank bottom devoid of contaminants. The bed area and contact time are two of the more important parameters effecting the degree and rate of removal of these volatile organics.

Nyer reports that Rockaway Borough used granular activated carbon to remove trichloroethane from its underground drinking water at an average increase in the monthly bill for each household of $3.00. The system handles 1½ MGD at an influent TCE concentration of 335 ppb. (Nyer 72, Groundwater Treatment Technology, 1985).

Nyer also reports in his new book (72, 1985) that the city of Acton, Massachusetts air-strips the volatile organics from its groundwater with a removal of 96–99% at a total cost of $0.053 per 1000 gallons. All organics were removed to less than 1 ppb. This naturally eliminates the possibility of toxic trihalomethanes from the drinking water following disinfection.

Oil Recovery Systems, Inc. uses an air stripping system to remove volatile organics from water (Air Stripping 73, 1985). They report excellent results in removing chlorinated solvents such as trichloroethane, trichloroethelene, and methylene chloride and petroleum hydrocarbons such as gasoline and fuel oils. However, it is significant to note that the system of air stripping is often used in conjunction with activated carbon adsorption.

Weston engineers (Clean-Up 74, 1984) claim to have eradicated the first Superfund site cleanup by a combination of containment and treatment. They installed a three-foot wide soil-bentonite slurry wall around a 20 acre dump site, thereby significantly curtailing exfiltrated water from the site. Then they constructed a full-scale 300 ppm treatment plant (based upon successful pilot plant results) scheduled for completion in April 1985. The treatment consisted of chemical precipitation of the inorganics followed by pressure filtration to remove any remaining metallic floc. The waste stream was then preheated to about 90°C. and passed through a stripping column. Air was passed countercurrently through the column and the off-gases burned in a vapor incinerator. The column effluent was divided, the larger portion discharged within the slurry wall while the smaller portion subjected to biological treatment by extended aeration and then groundwater recharge outside the wall. NH_4Cl and H_3PO_4 were added to enhance biological treatment. Weston estimates that it will take about one and three-quarters years at the 300 ppm rate to provide the two full flushes of the contaminated dump site necessary to remove 90% of the contaminants held in the soil water. Capital costs approximated $5 million with annual operating costs of about $1.4 million.

GDS, Inc. has developed a patented system for soil and groundwater detoxification which they claim is cost and performance effective (Bio-Reclamation 75, 1983). The system involves pumping contaminated groundwater into activating tanks where the microorganisms are enriched with compounds of phosphorous and ammonia, and sometimes iron, magnesium, and manganese salts. The treated groundwater is then settled, pumped in trenches for recirculation by reinjection into the original underground site. They claim decontamination of both soil and groundwater by this process. They admit that complete site cleanup may take years, but overall results in a savings of both time and money. A schematic view of the system is shown in the following Figure 32.8.

The U.S.E.P.A. reported (Fungus eats—77, 1985) the fungus Phanerochaete chrysosporium was used to degrade dioxins, DDT, benzopyrene, and two kinds of polychlorinated biphenyls, or PCB's. The EPA predicts that contaminated soil could be innovulated or mixed with the fungus grown on sawdust or wood chips. They admit, however, that it may take a lengthy period, perhaps years, to degrade large amounts of the chemicals, but claim it's a better alternative.

ATW/CALWELD has introduced a "Detoxifier" system that incorporates a mobile, in situ treatment system with advanced chemical and biochemical technologies (Detoxifier 76, 1985). The system is capable of chemically detoxifying or biochemically degrading aqueous toxic substances or contaminated soils, in place, to assure a much more economical and safe clean-up solution than traditional remedial procedures. Mounted on a heavy-duty crane (See Figure 32.9), the systems include an intrusion device which is powered down into the impounded waste or soil. Computer control and feedback determines the rate of descent and speed of the device which mixes, turbulizes, pulverizes, and ho-

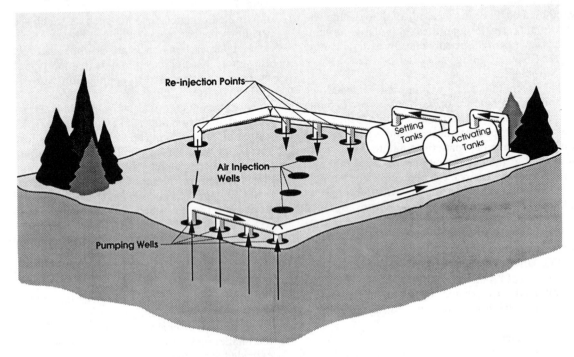

Fig. 32.8   Bioremediation system.

mogenizes the impounded contents while simultaneously selecting, feeding and integrating chemical reagents or biocatalysts and their nutrients with the

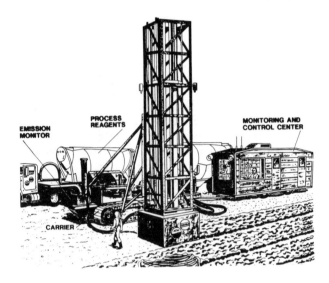

Fig. 32.9

conditioned contents. Oils and greases are biodegraded or transformed into non-soluble metallic soaps. The solidified mass will not reslurry according to the company. The company also states that any gases or vapors liberated during the process are collected and scrubbed.

NASA has shown that certain plants such as spider plants are effective in eliminating toxic gas concentrations of NO_2 and CO. Although this is perceived more as a treatment for indoor air, it could also be used under proper conditions for outdoor environments (NASA roots out—78, 1985).

Itvine and Busch reported in 1979 (Sequencing batch reactors 79,1979) the use of Sequencing Batch Reactors (SBR) as an excellent alternative to conventional activated sludge biological treatment for wastewater. More recently, Irvine (Enhanced Biological Treatment 80,1984) showed that the SBR can provide substantial saving in energy and costs by removing organic compounds found in hazardous wastes biologically, rather than with activated carbon. The SBR is a periodically operated, fill-and-draw reactor with five discrete pe-

riods in each cycle: fill, react, settle, draw and idle. Herzbrun now finds (Biological treatment of hazardous wastes 81, 1985) that total organic carbon (TOC) degradation averaged 76% and phenol degradation averaged 99% during the first month of bench scale operation. This type of treatment is intended to replace activated carbon adsorption treatment of leachates containing hazardous chemical compounds. Instead, activated carbon would be used only as a "polishing" device following SBR treatment.

Conversion of a hazardous-classified waste to a delisted one can sometimes diminish the risk to the generator of environmental suits. This may amount to considerable expenditures of both time and money in order to demonstrate to E.P.A. that treatment residuals do not have the characteristics for which they were originally listed as hazardous. In addition, building and paying for the amortization and operation expenses of a treatment system can be costly. Also, sometime in the future, the generator's treated and delisted waste might—once again—be listed as hazardous by EPA. In the long run, it may pay the industrial generator of a toxic waste to contract with an off-site disposal company for handling the waste. At least the industry will have a partner to share the liability of its environmental effect. Perhaps the off-site disposer can obtain delisting after treatment easier than the original industrial generator.

32.73 Regional Exchanges of Hazardous Wastes

The concept of regional exchanges or markets for transfering hazardous materials from one supplier to another user is a rather new and intriguing potential solution to the dilemma facing society. A regional center enhances the exchange rather than allowing the alternative disposal into the environment to occur. This process requires a great deal of communication, confidence, and consideration of all parties concerned. Advertisement of products materials, and chemicals available for exchange must be extensive but, at the same time, discrete. Sometimes buyers and sellers must remain anonymous. Management of the exchange must be competent and trustworthy, somewhat like a stock market exchange, but without the excessive product exposure to the general public. Certainly the objective of reuse without discharge is an admirable one. Only the procedure needs refinement, practice, and much experience. There is a similarity in the objectives of these exchanges with those of zero pollution attainment described in Chapter 15. The latter goal, however, is attained by all manufacturing of related products taking place in the same industrial complex.

LeRoy (25, 1983) reports that these exchanges have been set up in France by local initiative. Their main purpose is to create and develop new outlets for using waste and saving raw materials.

The National Conference on Waste Exchange (82, 1983) defines a waste exchange "as an operation that engages in transfer of either information concerning waste materials or the waste materials themselves." Waste exchanges were first organized in Europe where depletion of readily available natural resources and limited land disposal areas forced manufacturers to find alternative sources of raw materials. All of the foreign exchanges are information clearing houses and most of them are operated by trade organizations, primarily in the chemical industry. Material exchanges are the major innovation in the waste exchange concept developed in the United States from the mid 1970's.

Surplus materials, equipment, scrap metals, and discontinued products, as well as traditional wastes, may enter the waste exchange cycle. Industries are using the term, "investment recovery" to describe the entire process of recycling and reuse within some plants. Some firms have already investigated or instituted process modifications designed to allow for or enhance the reuse potential of by-products generated during normal manufacturing processes. Some modifications which have proven successful include: (1) substitution of reclaimed acid for typical new electro-plating acids, (2) source separation and segregation of various materials, (3) concentration and volume reduction, (4) altering raw materials specifications to allow substitution of lower quality inputs, (5) using intermediate reactions designed to modify waste stream components, (6) tightening process control to take advantage of by product (not waste) streams, and (7) educating plant management and workers on the benefits of resource reuse.

One deterrent to successful waste exchange programs is that of exposing private information to persons outside a particular industrial plant. The program must maintain an agreement of "confidentiality" to protect the proprietary interests of generators and to limit the direct identification of specific firms which are generating a particular material. Industry identifies "potential liability" as the primary reason for non-participation in waste transfer agreements.

Four requirements have been identified (National Conference 82, 1983) for industrial resource reuse and waste transfer:

(1) Participation in a waste exchange must be uncomplicated and cost-effective. The exchange itself must be reputable and reliable.
(2) Alternatives to conventional treatment and disposal methods which are presented by a waste exchange must be ethical and cost-effective.
(3) The exchange should have as wide an audience of potential users as possible and should have extensive contacts in the waste management field in order to be aware of all waste management options.
(4) The generator must know where and in what form his waste is being reused or disposed.

Many areas of cooperation between exchanges are possible including the following: (a) common data base shared on regional or national level, (b) trading of listings between exchanges for catalogue distribution, (c) network of regional contacts for information referral, and (d) licensing of exchanges to ensure maintenance of quality of service. However, industrial managers hesitate to participate in cooperative exchanges because of the potential liability for mismanagement of waste. Unfavorable economics, especially when affected by high transportation costs for hazardous wastes, may be another deterrent to cooperative exchanges.

It was reported by an Argonne National Laboratory (National Conference 82, 1983) Report that the amount of energy saved by one waste exchange over two and one-half years was 10×10^9 BTU. They calculated that a savings of 10^{12} BTU per year (the equivalent of 100,000 barrels of oil at 10^7 BTUs per barrel) would result if 50 exchanges as large or as effective as the one studied existed.

As with other liquid, gaseous, and solid wastes, there are generally no tax incentives, credits, or advantages for industries which recycle hazardous wastes. On the other hand, there are taxes levied for producing them. Exceptions are the states of New Jersey and California where tax advantages are given (or are being considered) for recycling waste.

One example of regional exchange of easte information is (SWIX 83, 1982) shown in Figures 32.10 and 32.11, which serves the Southern region of the U.S.A. It publishes and distributes to interested firms listings of materials available, materials wanted, and services available and wanted. Examples of listing forms are included in Figures 32.10 and 32.11. Materials include acid and alkalis, inorganic chemicals, metals, organic chemicals, solvents, oils and paints, paper, wood, plastic, rubber, glass, and miscellaneous. Services include recycling, equipment and supplies, consulting engineering, consulting legal and health-related, tank cleaning and lining, collection and transportation, storage, treatment and disposal, and miscellaneous.

32.74 Solar Evaporation

Some hazardous wastes, if sufficiently concentrated and of low volume, can be evaporated by natural sunlight to a very small volume for ultimate disposal or for permanent fixture at the same site. This method of treatment precludes ample sunlight, minimum rainfall and humidity, and adequate land for exposure to the sun's rays.

Sundaresan et al. (7, UNEP, 72, 1983) report solar evaporation used to dispose of quinophos, a highly toxic pesticide. The semi-solid sludge and the residue after solar evaporation are usually incinerated.

They also use solar evaporation ponds for concentrating urea plant spillages. The sludge after evaporation containing about 12% arsenic is stored in drums, sealed, and kept within the factor premises for final disposal (at unknown destination).

32.75 Landfilling with Leachate Treatment

The collection and subsequent treatment of leachates from hazardous waste landfills has become an acceptable alternative for disposing of these wastes. Collection systems must be designed to recover all the leachate and treatment systems must be designed and operated to remove all the hazardous constituents from the leachates. Leachate collection systems require the use of subterranean, impermeable barriers and drainage channels to central sump or reservoir sites. Treatments are concentrated mainly on removal of dissolved metals, minerals, and organic matter as described in earlier sections of this chapter.

Goldstein (84, See 5-3B) found metal plating leachates from landfills are sometimes acceptable to be pumped to publically owned sewage treatment plants (POTW).

Dupont reports the use of powdered activated carbon for the treatment of leachates from toxic waste dumps to remove these contaminants (DuPont 85, 1984).

32.76 Disposal into Publically Owned Wastewater Treatment Plants

Hazardous wastes may be discharged to municipal sewer systems if and when they are properly pretreated to remove and/or detoxify the hazardous components. Because of the cost savings of sending large volumes of wastewater to sewage systems, this method of treatment is especially interesting to industries. However, the hazardous component must first be removed or sufficiently diluted to be acceptable to the receiving municipality.

In keeping with the above, Coughlin et al. (7, Toxic and Hazardous Wastes, 1983) utilized a treatment technology for metallic wastes consisting of physicochemical processes that effectively change the hazardous soluble elements into recoverable non-soluble solids. The solids are further concentrated by a low pressure belt filter press that recovers 95% of the solids. The remaining solids are concentrated in an electroflotation system that results in a final treated effluent of 20 ppm of total suspended solids. The solid cake-like material is landfilled with no apparent hazard because of its relatively high pH. The water phase is then discharged to a publicly-owned treatment plant. This system represents a typical one for pretreatment of hazardous wastes prior to admission to POTP's.

32.77 Membrane Technology

The use of semi-permeable membranes as barriers for the removal of hazardous compounds or elements from reaching the external environment is gaining in popularity and acceptance.

Membrane openings or pores are so minute that considerable pressure must be applied to the waste to drive the fluid through them. In normal filtration, microfiltration, the pore sizes are relatively large and the waste solution to be filtered approaches the polymeric membrane perpendicularly as shown in Figure 32.12.

In another more common technique, the pore sizes are many times smaller and known as reverse osmosis or sometimes hyperfiltration and require more liquid

Fig. 32.10

Waste Management Services
Listing Form

**A separate form is required for each service listed.
Limit of two listings.**

1. Company Name: _____

2. Mailing Address: _____

   _____

   _____

3. Company Contact:_____   4. Title:_____

5. Signature: _____   6. Date:_____

7. Phone Number: (____)_____   8. SIC Code:_____

9. Fax Number: (_____)_____

LISTING INFORMATION:

(1) CLASSIFICATION:
   (Review all first, then select the **one** that best describes your firm's service)

   ❑   Recycling   ❑   Storage, Treatment, Disposal
   ❑   Equipment and Supplies   ❑   Emergcy Response/Clean-Up
   ❑   Environmental Consulting   ❑   Laboratory Analysis
   ❑   Legal and Health - Related Services ❑   Waste Minimization
   ❑   Tank Management   ❑   Miscellaneous
   ❑   Collection and Transportation

(2) DESCRIPTION OF SERVICE:
   (In 25 words or less, please describe your firm's service or product, keeping in mind
   what the reader of your listing may want to know. You may use the space provided
   below or type the listing on a separate piece of paper and attach it to this form.)

(3) LOCATION SERVED:
   (Give general area where service is available, e.g., Central Georgia, Southeastern U.S., etc.)

Send completed form to: **SWIX Clearinghouse**
                                 Post Office Box 960
                                 Tallahassee, FL   32302

Fig. 32.11

Material Available/Wanted
Listing Form

A separate form is required for each item listed.
Limit of two listings.

1. Company Name:_____ SIC Code #:_____

2. Mailing Address:_____

_____

3. Company Contact:_____ 4. Title:_____

5. Signature:_____ 6. Date:_____ 7. Phone: (___)_____

8. Fax Number: (___)_____

9. Check One Only: ❏ **MATERIAL AVAILABLE** ❏ **MATERIAL WANTED**

Classifications (review all first; then select **one** that best describes your material):

❏ ACIDS   ❏ OTHER ORGANIC CHEMICALS   ❏ WOOD AND PAPER
❏ ALKALIS   ❏ OILS AND WAXES   ❏ METALS AND
❏ OTHER INORGANIC CHEMICALS   ❏ PLASTICS AND RUBBER   METAL SLUDGES
❏ SOLVENTS   ❏ TEXTILES AND LEATHER   ❏ MISCELLANEOUS

10. Material to be listed (Main usable constituent, generic name):_____

11. The industrial process that generates this waste:_____

12. Main constituent and percentage:_____

13. Other constituents (including contaminants):_____

14. Percent by (check one): ❏ Volume   ❏ Wet Weight   ❏ Dry Weight

15. Physical State: ❏ Solution   ❏ Slurry   ❏ Sludge   ❏ Cake
     ❏ Aggregate   ❏ Solid   ❏ Dust   ❏ Gas

16. Miscellaneous information (e.g. pH, toxicity, reactivity, color, particle size, flash point, total solids, purchase date, manufacturer):

_____

17. Potential or intended use:_____

18. Packaging: ❏ Bulk   ❏ Drums   ❏ Pallets   ❏ Bales   ❏ Other:_____

19. Present Amount:_____ 20. Frequency: ❏ Continuous   ❏ Variable ❏ One Time

21. Quantity thereafter:_____ ❏ Pounds   ❏ Tons   ❏ Cubic Yards ❏ Gallons
     ❏ Kilograms   ❏ Cubic Meters   ❏ Liters   ❏ Other_____

per: ❏ Day   ❏ Week   ❏ Month   ❏ Quarter ❏ Year

22. Restrictions on amounts: ❏ None   ❏ Minimum   ❏ Maximum_____

23. Available to Interested Parties: ❏ Sample   ❏ Lab Analysis   ❏ Independent Analysis

24. For material wanted, acceptable geographic area (i.e. States, regions, countries):

_____

25. If necessary to speed communications, please check if your company's name, address and telephone number may be released.   ❏ YES   ❏ NO

Send completed form to: **SWIX Clearinghouse**
Post Office Box 960, Tallahassee, FL 32302

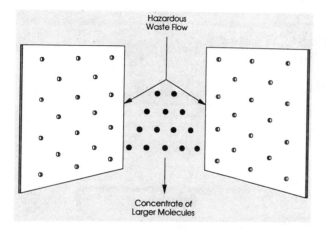

Hazardous
Waste Flow

Concentrate of
Larger Molecules

Fig. 32.12   Separation of hazardous wastes using membrane filtration.

pressure to force the smaller molecules through the pores of the membrane. The solution is passed over the surface of a specific semi-permeable membrane at a pressure in excess of the effective osmotic pressure of the feed solution.

In selecting the proper membrane to separate out hazardous contaminants from wastewater, several environmental parameters must be considered:

(1) pressure at which the membrane will be subjected. These vary from as low as 3 to 50 psig with microfiltration to as high as 200 to 1000 psig with reverse osmosis. At the higher pressures certain of the membranes may deteriorate or deform due to the mechanical strain put on them.

(2) temperature at which the polymeric membrane will be subjected. Most membranes will operate with varying efficiencies and lengths of life at temperatures between 0°C. and 85°C.

(3) pH range of operation of the membrane will effect its life. The value will be influenced by the hazardous waste solution pH.

(4) chemical compatibility of each membrane to the specific chemical makeup of the waste solution is important not only in determining the membrane choice, but also its durability under continued operation.

Membrane separation has proven useful for treating and recovering metal salts from plating bath wastes, re-

cycling lubricating oils and rinse waters from can-forming and other metal-working applications, concentrating wastes and conserving process waters in certain chemical processing plants, and in the future recovering pollutants from the large water users such as pulp and paper and textile industries. These applications will be enhanced by scarcity and increased prices of fresh water.

32.78 Electrochemical Treatment

One of the objections of chemical coagulation as a hazardous waste treatment device has been the difficulty in adjusting the exact dosages of coagulant required and pH needed to remove the metals. This results in higher operational costs for both chemicals and labor.

Electrochemical treatment eliminates or tends to overcome the objections of chemical coagulation. A direct current is conducted through a cell containing carbon steel electrodes. The current creates ferrous ions in the waste solution. The ferrous ion acts as a reducing agent, and in the case of the heavy metal chromium, results in its precipitation as the hydroxide. Other heavy metals such as As Cu, Ni, Zn, Pb and Sn are coprecipitated with the ferric hydroxide. It is claimed that no sulfates are formed as in normal chemical coagulation and that the effluent is acceptable for reuse in plating operations (Duffy 86, 1983). When using this bipolar cell (carbon-steel), one side is the anode and the other is the cathode. The ferrous ion produced at the anode and the hydroxide formed at the cathode react in the wastewater to form insoluble $(Fe(OH)_3$. Although no chemicals are added, the carbon steel electrodes are consumed, usually one to two pounds of iron per pound of heavy metal in the wastewater which is precipitated. Although the manufacturers admit that slightly more sludge is formed with the electrochemical treatment on a theoretical basis than with conventional chemical coagulation, in actual practice somewhat less sludge does result. Presumably this is because the exact stoichiometric amount of iron required actually goes into solution in electrochemical treatment. The exact dosage in coagulation, on the other hand, is difficult to control.

A typical cell operating at about 25 amps of dc current requires about five kilowatts of electricity per pound of heavy metal removed. Duffy reports the operational costs (which presumably includes both iron metal and electricity) would be about $1.00 per pound of heavy metal removed. He also claims that chelating compounds found in many plating wastewaters does not interfere with the treatment but is broken up by the iron. Although the effluent from electrochemical treatment may be non-hazardous and can be discharged safely to a receiving stream or sewage system, the sludge still remains to be disposed of (usually in landfills) and contains the toxic metals removed as hydroxides. A reduction of the pH of this sludge into the acid range will tend to liberate the metals into solution once again.

New Treatments

Since this field is so new and dynamic, many new treatment processes, in addition to those we have described, are being tested or researched. Susan Rasidel (105, '89) summarizes many of these in Table 32.12A.

32.79 Legal Problems of Industrial Hazardous Wastes

The authors have presented a tabular history of United States laws regulating toxic substances use and control of waste. It is interesting to note that these laws began as long ago as 1938 and continued to as recent as 1980. Of major concern to us in this book are the last four pieces of legislation summarized here for reference.

Date	Law	Regulating
1974, 1977	Safe Drinking Water Act	Drinking water Contaminants
1976	Resource Conservation and Recovery Act	Hazardous wastes
1976	Toxic Substances Control Act	Existing chemical hazards and new chemicals
1980	Superfund Act	Cleanup of spills of hazardous substance

The latest drinking water standards were presented by Nemerow(1974) in more specific terms. The recommended limits of any special elements and compounds are given in this reference.

The Resource Conservation and Recovery Act

(RCRA) of 1976 is presented and discussed by Nemerow(1, 1983 in Chapter 11, pages 123-132).

The Toxic Substances Control Act (TOSCA) of 1976, also discussed by Nemerow (1, 1983, page 102), severely restricts the handling of waste materials classified as toxic. Sixty five (65) elements, compounds, and substances are listed as toxic in Table 10.3 of that reference. In that table the author gives toxic limits for each of these chemicals for freshwater and saltwater aquatic life as well as for human life.

32.80 Superfund

In 1980 the Congress of the United States enacted legislation creating a "Superfund" to provide for cleanup of abandoned dumpsites and dangerous spills of toxic materials and to facilitate compensation of victims. The law imposes a tax on chemical manufacturing firms, the revenues from which are placed in a fund ($1.6 billion) to be utilized solely for these cleanups. Congress intended to force producers and disposers of these hazardous wastes to accept responsibility for their own wastes and for the fund to pay for any residual costs involved.

According to Grossman (7, page 138, 1983) the Superfund legislation required EPA to develop by June 1981 a National Contingency Plan to guide the search for and cleanup of dangerous sites and to prepare to respond to emergencies such as spills and explosions. A plan was finally proposed in March 1982. "The proposal is so vague as to provide no guarantee that Superfund resources and authority will be used to clean up any site. The plan implies that EPA cares more about saving money than cleaning up sites to protect human health."

Grossman reports that in its first use of Superfund authority, after a toxic dump site in Santa Fe Springs, California caught fire in July 1981, "top EPA officials quickly negotiated a private settlement with one of the responsible parties."

"The settlement limited the company's cleanup responsibility instead of requiring the cleanup to continue until the hazard was removed. It also committed EPA to testify on behalf of the company in any subsequent lawsuit against it arising from the dump and the fire."

It is alleged by EPA officials that the federal government will pay for 90% of the cleanup costs, while the individual states will pick up the remaining 10%. How-

Table 32.12A   Some New Treatments for Hazardous Waste, *Many Still on the Drawing Boards* (after Randle, 105, 89)

Company	On the market Use/Method	Costs
BioTrol Inc., Chaska MN A. Dale Pflug (612) 448 2515	Organics Bacteria attack contaminant molecules introduced in water, working on soil clean up.	Processing and capital: groundwater (100 ppm contamination), .5 to 2 cents/gal; soil, $50-$75 to $100-$125/ton.
Groundwater Technology Inc., Norwood, MA John Higley (800) 635 0053	Organics Small diameter Filter Scavenger, part of Filter Scavenger series, removes floating hydrocarbons.	Processing: electricity. Equipment: $10,000-$12,000/unit.
Micro-Bac International Inc., Austin, TX	Organics, assimilates heavy metals Bacteria metabolize toxic	Cost NA
Bob Billingsley (512) 837 1145 Ogden Environmental Services, San Diego	Organics Mobile circulating bed combustion incinerator.	Processing and service by Ogden: $100-$200/ton of soil.
Harold Diot (800) 876 4336 Permutit Co., Paramus, NJ Luis Rodguez (201) 967 6000	Heavy metals Sulfide precipitation process removes heavy metals from waste stream.	Processing: for existing system removing metals as hydroxides, at 100 gpm, $10-$30/day. Equipment: small batch unit, $50,000; large, over $50 million.
Resource Conservation Co., Bellevue, WA Douglas Austin (206) 828 2400	Organics, heavy metals Solvent extraction of contaminants from soil and sediment.	Processing and capital: $50-$150/cubic yard.
Ultrox International, Santa Ana, CA Jerome Berich (714) 545 5557	Organics Waste or groundwater exposed to ultraviolet light plus ozone or hydrogen peroxide.	Processing groundwater, low 15-20 cents/1000 gal, high, 5 cents/gal. Equipment: $60,000-$200,000 for 10 ppm TCE at 210 gpm.
Westinghouse Environmental Services, Madison, PA C. Keith Paulson (412) 722 5447	Organics, heavy metals Plasma, fired cupola remelts and recovers metals. Organics Pyroplasma destroys heat toxics in liquids. Organics Infrared conveyor belt drives heat from soil.	Service: $200-$300/ton. Service: less than $1/lb. Service: $200-$300/ton.
Zimpro/Passavant Rothschild, WI Bob Nicholson (715) 359 7211	Organics, some metals absorbed by carbon Biophysical system has powdered activated carbonic assist bacteria. Wet air oxidation burns high strength wastes.	Processing: $1/1000 gal. Equipment: smallest unit. $85,000; largest, $1 million. Processing: 6 cents/gal. Equipment: 10 gpm skid mounted unit, $2 million

Table 32.12A   (continued)

Company	Not yet *on the market* Used/Method	Costs
CF Systems Corp., Waltham, MA Tom Cody (617) 890 1200	Organics Liquified gas extraction uses gases as solvents to recover organic from refinery wastes including water, clay and oil.	Processing, operating and equipment: sludge, $75/cubic yard; water at 20 gpm, 20 cents/gal.
Freeze Technologies, Raleigh, NC Ken Hunt (919) 850 0600	Liquid-liquid separation process based on freezing.	Processing: NA. Equipment: $1 to $2 million.
EPA, Raritan Depot Edison, NJ Francine Everson (201) 548 8554	Organics Mobile carbon regenerator for water treatment systems. Organics, heavy metals Mobile soils washer separates fine soils and contaminants from sands and gravels. Organics, heavy metals Mobile in situ contaminant/treatment unit injects grout to contain spills; contaminants treated in place or removed for other treatment via flushing.	Costs NA Costs NA Costs NA
EPA, Cincinnati Alfred Kornel (513) 569 7421	Organics Potassium-Alkaline polyethylene glycolate reagent breaks halogen bond to chemically react with pollutants.	Processing: pilot unit (1.75 tons/batch), $400-$800/ton. Equipment: NA
IIT Research Institute, Chicago Gug Stresty (312) 567 4232	Organics Radiofrequency heating, in situ, brings contaminants and water vapor to surface, where collected for treatment.	Processing: $30-$60/ton. Equipment: 50-80 ton/day unit, $500,000.
Solar Energy Research Institute, Golden, CO John Thornton (303) 231 1269	Organics Contaminated water pumped from ground; solar energy system catalyst form free radicals to attack chemical bonds of pollutants.	Costs NA
Vertech Treatment Systems, Denver Nathan Chesley (303) 452 8800	Organics Aqueous phase oxidation of contaminated solution.	Processing: 10 gpm at 50,000 mg/liter chemical oxgen demand, 30 cents/gal; 50 gpm at same COD, 13 cents/gal. Equipment: 10 gpm unit, $3-$5.5 million.
Westinghouse Environmental Services, Madison, PA C. Keith Paulson (412) 722 5447	Organics, heavy metals verified Electric pyrolizer burns solids and sludges at extremely high temperatures.	Service: $200-$300/ton.

ever, when the property is government-owned, the federal share will be limited to 50%.

According to Skinner (7, UNEP page 69, 1983) EPA published the list of National Priorities Sites identifying 418 priority sites on December 20, 1982. Provisions were made to clean up these sites by any of the following options:

1. removing drums from the site
2. installation of a clay "cap" over the site
3. construction of ditches and dikes to control surface water
4. construct drains, liners, and grout curtains to control groundwater
5. provide an alternate water supply
6. relocate residents on a temporary or permanent basis

An important part of the Superfund program (according to Skinner) is to encourage voluntary cleanup by private industries and individuals when they are responsible for their releases. He reported (1983) that more than $121 million have been received from industry for clean-up. EPA also encourages state government involvement in the Superfund program. States and EPA may enter into a Cooperative Agreement in which a state takes a leading role in a remedial action. Federal money is transferred to the state and the state develops a workplan, schedule, and budget for the clean-up action. The state then contracts for any services it needs. EPA is responsible for monitoring the state's progress throughout the project.

In addition to the 10% state portion of the cleanup, the state must agree to maintain the site after remedial measures have been taken. Since CERCLA (Comprehensive Environmental Response, Compensation, and Liability Act) was passed in 1980 and up to September 1982, Superfund had agreed to allocations of $221 million (according to Skinner (7, 1983).

Krog, however, pleads for a revision of the Superfund Law as being lacking in both the original law and in Superfund II being considered in September of 1985 (Nations's toxic waste 87, 1985). She maintains that "toxic waste dumping doesn't generate money for anybody except attorneys." She points out that when an industry goes to court over a toxic waste dump site, the EPA must do likewise, which takes public funds out of the Superfund budget and transfers them to bank accounts for attorneys. And, according to her, this is not what Superfund was intended, but rather to clean up toxic dump sites.

Congressman Robert Torricelli, a member of the House Science and Technology Committee, points out the current problems and fallacies of the Superfund program (Let's Destroy Toxic Waste—31, 1985). He laments that our "20,000 toxic waste sites are not being cleaned up." "Hazardous dumps," he continues, "have merely been moved from one location to another, often creating new toxic sites in the process." He maintains that "the Superfund program promotes the storage of contaminated materials, not their permanent destruction." Although Torricelli does not say it, presumably this occurs because storage is still easier, less expensive and complicated than destruction of hazardous wastes. He contends that "despite the great intentions and accomplishments of the program, the result has been a lethal shell game that threatens to indefinitely prolong America's toxic legacy."

Some states have not been pleased by the progress made under Superfund. In New Jersey its Department of Environmental Protection attempts to block the sale of portions of property containing hazardous wastes until an entire site is cleaned up.

Governor Bob Graham of Florida signed into state law, July 1, 1983 (The Miami Herald 17A, July 2, 1983) a comprehensive environmental bill. The bill included a "provision to impose a 2 cents per barrel tax on pollutants once a state hazardous waste cleanup fund drops to $3 million." Also included in the bill was "a provision to allow homeowners and businesses with small amounts of hazardous wastes to bring it to a state collection site for free disposal." The main part of the new law included a one-time transfer of $11 million plus interest from an existing trust fund to help clean up hazardous waste sites in Florida (estimated to number about 200).

As an example of governments' intention to use new laws to prevent disasters from hazardous wastes, the Premier of the Province of Ontario, Canada (Statement by—89, July 2, 1985) said that "this government will ensure that environmental hazards do not eat away at the legacy we wish to leave our children. No longer will there be any question of who is responsible for preventing spills or cleaning them up. No longer will innocent victims be left without a route to compensation."

The U.S. Congress enacted the Resource Conservation Recovery Act (RCRA) in 1976 which charged the

E.P.A. with implementation. Among its statutes was that of operating a hazardous waste treatment, storage, or disposal facility. An operator must complete Part A and Part B of this permit. The former is merely an application giving a bare minimum of vital information. Part B permit involves an extensive study, survey, completion of forms, meetings and approvals before a 10 year permit can be issued.

Thorsen and Petura (90, Plan for your RCRA Part B Permit, 1983) give a schematic presentation of the steps needed to obtain Part B permit (Figure 32.13). The entire review of the Part B permit application is expected to take up to 6 months and when approved, contains very specific terms and conditions which the operator must follow.

In general, two types of applicant operator—each requiring different approach and consideration for permit—are involved: (1) industrial and (2) regional. The former is an individual plant primarily in business to make a profit and which produces a hazardous waste in the process. The latter is a collector, storer, and/or treator of hazardous wastes from a number of individual commercial plants. It, too, expects to make a profit, but as a direct result of the safe handling of the toxic waste. Thorsen and Petura visualize the decision of either type of applicant as a commitment of substantial time and money. They specify a Master Plan(not necessarily part of the application) which should describe:

1. types and approximate quantities of wastes to be handled at the site
2. activities and functions of the facility
3. the business related aspects (for internal use)
4. site development sequence and
5. strategy and resources required for implementation, including schedule, budget, and permit application sequence required to secure all necessary regulatory approvals and permits.

Because of the collective extensive hazardous waste management regulations, Thorsen and Petura recommend that "someone associated with the facility, with a substantial degree of regulatory familiarity, be actively involved with the team throughout the development of the permit application." They also point out that "the concerns and regulations of both federal and state agencies must be addressed."

In 1984, there were ten regional offices of the EPA to handle these Permit applications. They are shown geographically below for easy reference of the reader (Fig. 32.14).

Whitescaver and Clarke (7, Toxic and Hazardous Wastes, 1983) present toxic concentration limits for eight metals and eight toxic organics and illustrate the wide range of concentrations for the same toxic pollutant. This also shows that effluent guidelines were used seldomly but that 14 other techniques were employed, including negotiation and "best practical judgement" (BPJ).

Ms. Fisk, the Canadian Environmental Minister, announced on June 17, 1985 the new Regulation 309 under the Environmental Protection Act. This new regulation is aimed at controlling the process of handling industrial wastes from the generator to the receiver. It will clearly establish the responsibilities of the industries which create wastes, the haulers who carry them and the site operators who treat and dispose them. Recent toxic limits for air and water are given in Tables 32.12B and 32.13. The reader can use these values to become familiar with the hazardous chemicals and to develop an idea of the concentrations likely to result in serious illnesses and death.

Even OSHA (Occupational Safety and Health Administration) has become concerned and has pressured state labor departments to enact and enforce laws protecting plant workers from exposures to hazardous chemicals. In Florida, for example, on October 1, 1985 employers who handle any one of 1,400 chemicals listed by the state must:

1. Post notices informing employees of their rights to know about hazardous chemicals (and give a toll free telephone number to call for information).
2. Keep for 30 years material safety sheets for each toxic chemical on the list—giving their health hazards and pertinent facts about them.
3. Provide these information sheets to employees within 5 days of a request.
4. Train employees in the handling, hazards and emergency treatment for listed chemicals.
5. Notify the local fire department of chemicals used and stored.

A Water Pollution Control Federation workshop concerned with biomonitoring of toxic chemicals was summarized by Cairns (Biomonitoring for toxics—1985). He states that multi-species (rather than single

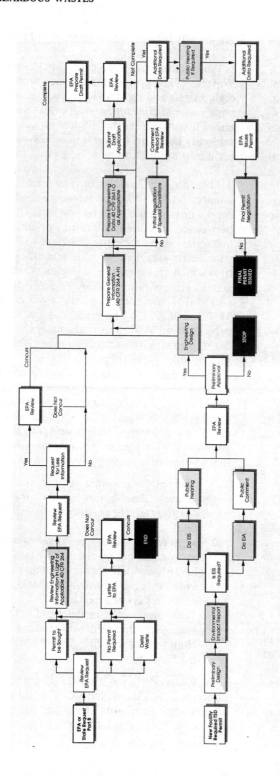

Fig. 32.13 RCRA TSD Part B permit application.

EPA Regions

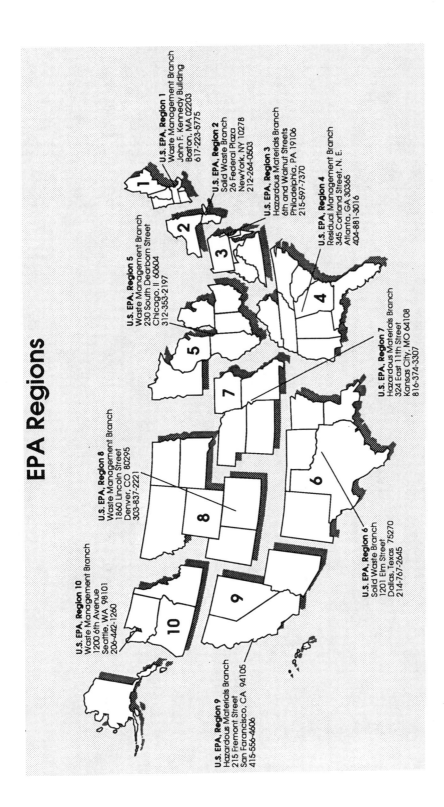

U.S. EPA, Region 1
Waste Management Branch
John F. Kennedy Building
Boston, MA 02203
617-223-5775

U.S. EPA, Region 2
Solid Waste Branch
26 Federal Plaza
New York, NY 10278
212-264-0503

U.S. EPA, Region 3
Hazardous Materials Branch
6th and Walnut Streets
Philadelphia, PA 19106
215-597-7370

U.S. EPA, Region 4
Residual Management Branch
345 Cortland Street, N. E.
Atlanta, GA 30365
404-881-3016

U.S. EPA, Region 5
Waste Management Branch
230 South Dearborn Street
Chicago, Il 60604
312-353-2197

U.S. EPA, Region 6
Solid Waste Branch
1201 Elm Street
Dallas, Texas 75270
214-767-2645

U.S. EPA, Region 7
Hazardous Materials Branch
324 East 11th Street
Kansas City, MO 64108
816-374-3307

U.S. EPA, Region 8
Waste Management Branch
1860 Lincoln Street
Denver, CO 80295
303-837-2221

U.S. EPA, Region 9
Hazardous Materials Branch
215 Fremont Street
San Francisco, CA 94105
415-556-4606

U.S. EPA, Region 10
Waste Management Branch
1200 6th Avenue
Seattle, WA 98101
206-442-1260

Fig. 32.14 U.S. EPA office locations.

Table 32.12B Water Quality Limits for Toxic Pollutants for Three Uses

Concentration of Toxic Material Considered Limit (ug/l)

Toxic Chemical	Freshwater Aquatic Life		Saltwater Aquatic Life		Human Health
	Acute	Chronic	Acute	Chronic	
1. Acenaphthene	1,700	—	970	710	20 (est.)
2. Acrolein	68	21	55	—	320
3. Acrylonitrile	7,550	—	not available	—	0.58–.006 lifetime
4. Aldrin-Dieldrin	.0019 (24 hr avg.) 2.5 maximum conc.	—	0.0019 (24 hr avg.) 0.71 (max. conc.)	—	.0071 ng/l – .71 ng/l
4A. Aldrin	3.0	—	1.3	—	.0074 ng/l – .74 ng/l
5. Antimony	9,000	1,600	not available	—	146
6. Arsenic	440	40	508	—	.22 ng/l – 22 ng/l
7. Asbestos	not available	—	not available	—	3,000–300,000 fibers/l
8. Benzene	5,300	—	5,100	700	.066–6.6
9. Benzidine	2,500	—	not available	—	.01 ng/l – 1.2 ng/l
10. Beryllium	130	5.3	not available	—	.37 ng/l – 37 ng/l
11. Cadmium	3.0 (100 ppm hardness) max. .025 (100 ppm hardness) avg	—	59 (maximum) 4.5 (avg.)	—	10
12. Carbon tetrachloride	35,200	—	500,000	—	.04–4.0
13. Chlorodane	2.4 max 0.0043 (24 hr. avg.)	—	.09 max .0040 (24 hr. avg.)	—	.046–4.6
14. Chlorinated benzenes	250	50 (fish 7.5 days)	160	129	hexachlorobenzene .072 ng/l – 7.2 ng/l tetrachlorobenzene 38 ug/l – 48 pentachlorobenzene 74–85 monochlorobenzene 488

15. Chlorinated Ethanes

	Freshwater Aquatic Life		Saltwater Aquatic Life		Human Health
	Acute	Chronic	Acute	Chronic	
1,2 dichloroethane	118,000	20,000	113,000	—	.094–9.4 (1,2 dichloroethane)
two trichloroethane	18,000	9,400	31,200	—	18.4 mg/l – 1.03 g/l
two tetrachloroethanes	9,320	2,400	9,020	—	.017–1.7 ug/l
pentachloroethane	7,240	1,100	390	261	—
hexachloroethane	980	540	940	—	.19–19 ug/l

Same as above

Table 32.12B *(Continued)*

| Toxic Chemical | Concentration of Toxic Material Considered Limit (ug/l) | | | | |
| | Freshwater Aquatic Life | | Saltwater Aquatic Life | | |
	Acute	Chronic	Acute	Chronic	Human Health
16. Chlorinated Napthalenes	1,600	—	7.5	—	not available
17. Chlorinated Phenols	30 to 500,000	970	440 to 29,000	—	0.1 (3 monochlorophenol) 0.1 (4 monochlorophenol) 0.04 (2,3 dichlorophenol) 0.5 (2,5 dichlorophenol) 0.2 (2,6 dichlorophenol) 0.3 (3,4 dichlorophenol) 1.0 (2,3,4,6 tetrachlorophenol) 2.6 mg/l (2,4,5 trichlorophenol) .12–12 ug/l (2,4,6 trichlorophenol) 1,800 ug/l (2 methyl 4 chlorophenol) 3,000 (3 methyl, 4 chlorophenol) 20 (3 methyl, 6 chlorophenol)
18. Chloroalkyl Ethers	238,000	—	not available	—	.00038 ng/l to .038 ng/l (for bischloromethyl ether) .003 ug/l – .3 ug/l (for bis 3 chloroethyl ether) 34.7 ug/l (for bis-2-chloroisopropyl ether)
19. Chloroform	28,900	1,240	not available	—	.019–1.9 ug/l
20. 2 Chlorophenol	4,380	2,000 (one fish species)	not available	—	.01 mg/l
21. Chromium	21 (max) .29 (avg. 24 hr.) 4700 ug/l (100 ppm trivalent hardness) 44 ug/l chronic toxicity	hexavalent	1260 (max) Cr^{vi} 18 (24 hr. avg.) 10,300 (Cr^{+3}) chronic toxicity	—	170 mg/l Cr^{iii} 50 ug/l (Cr^{vi})
22. Copper	5.6 (24 hr. avg.)	—	4.0 (24 hr. avg.)	—	1 mg/l (for taste and odor) none other available
23. Cyanide	22 ug/l (100 ppm hardness) max 32.5 (24 hr. avg.) 52 (max.)	—	23 (max) 2–30 mg/l	—	200 ug/l
24. DDT and Metabolites	1.1 max .001 (24 hr. avg. DDT) 0.6 acute toxicity TDE 1050 acute toxicity DDE	—	0.13 (max DDT) .001 (24 hr. avg.) 3.6 acute toxicity TDE 14 acute toxicity DDE	—	.0024 mg/l to .24 ng/l for DDT
25. Dichlorobenzenes	1,120	763	1970	—	400
26. Dichlorobenzidines	not available	—	not available	—	.00103–.103
27. Dichloroethylenes	11,600	—	224,000	—	.0033–.33
28. 2,4 Dichlorophenol	2,020	365	not available	—	.3 ng/l (for taste and odor) 3.09 mg/l (toxicity)

Table 32.12B *(Continued)*

Concentration of Toxic Material Considered Limit (ug/l)

Toxic Chemical	Freshwater Aquatic Life		Saltwater Aquatic Life		Human Health
	Acute	Chronic	Acute	Chronic	
29. Dichloropropanes Dichloropropenes	23,000	5,700	10,300	3,040	87 ug/l
30. 2,4 Dimethylphenol	2,120	—	not available	—	400 ug/l (taste and odor)
31. 2,4 Dinitrotoluene	330	220	590	370	.001–1.1 ug/l
32. 1,2-Diphenylhydrazine	270	—	not available	—	4–422 ng/l
33. Endosulfan	.056 (24 hr. avg.) .22 (maximum)	—	.0087 (24 hr. avg.) .034 (maximum)	—	74 ug/l
34. Endrin	.0023 (24 hr. avg.) .18 (maximum)	—	.0023 (24 hr. avg.) .037 (maximum)	—	1 ug/l
35. Ethylbenzene	32,000	—	430	—	1.4 mg/l
36. Fluoranthene	3,980	—	40	18	42 ug/l
37. Haloethers	360	122	not available	—	not available
38. Halomethanes	11,000	—	12,000	6,400	.019–1.9 ug/l
39. Haptachlor	.0038 (24 hr. avg.) .52 (maximum)	—	.0036 (24 hr. avg.) .053 (maximum)	—	.028–2.78 ng/l
40. Hexachlorobutadiene	90	9.3	32	—	.045–4.47 ug/l
41. Hexachlorocyclo-hexane (Lindane)	.080 (24 hr. avg.) 2.0 (maximum)	—	0.16	—	.92–9.2 ng/l
BHC	100	—	0.34	—	1.63–163 ng/l
42. Hexachlorocyclo-pentadiene	7.0	5.2	7.0	—	206 ug/l
43. Isophorone	117,000	—	12,900	—	1.0 ug/l (taste and odor)
44. Lead	3.8 ug/l (100 ppm hardness and 24 hr. avg.)	—	—	—	5.2 mg/l
45. Mercury	.00057 ug/l (24 hr. avg.) .0017 ug/l (maximum)	—	.025 (24 hr. avg.) 3.7 (maximum)	—	144 ng/l
46. Naphthalene	2,300	—	2,350	—	not available
47. Nickel	96 (100 ppm hardness and 24 hr. avg.) 1,800 (maximum)	620	7.1 (24 hr. avg.) 140 (maximum)	—	13.4
48. Nitrobenzene	27,000	—	6,680	—	19.8 mg/l
49. Nitrophenols	230	150	4,850	—	30 mg/l (taste and odor) 13.4 (for 2,4 dinitrocresol) 70 (for dinitrophenol)
50. Nitrosamines	5,850	—	3,300,000	—	.14–14 ng/l (for n-nitrosodimethylamine) .08–8.0 ng/l (for n-nitrosodiethylamine) .064–64 ng/l (for n-nitrosodi-n butylamine)

Table 32.12B (*Continued*)

Concentration of Toxic Material Considered Limit (ug/l)

Toxic Chemical	Freshwater Aquatic Life Acute	Freshwater Aquatic Life Chronic	Saltwater Aquatic Life Acute	Saltwater Aquatic Life Chronic	Human Health
51. Pentachlorophenol	55	3.2	53	34	490–49,000 ng/l (for n-nitrosodiphenylamine) 1.60–160 ng/l (for n-nitrosopyrolidine) 1.01 ng/l 30 ug/l (for taste and odor)
52. Phenol	10,200	2,560	5,800	—	3.5 mg/l 0.3 mg/l (for taste and odor)
53. Phthalate Esters	940	3	2,944	3.4	313 mg/l (dimethylphthalate) 350 mg/l (diethylphthalate) 34 mg/l (dibutylphthalate) 15 mg/l (di-2-ethyl-hexylphthalate)
54. Polychlorinated Biphenyls	0.014 (24 hr. avg.) 2.0 (maximum)	—	0.30 (24 hr. avg.) 10 (maximum)	—	.0079–.79 ng/l
55. Polynuclear Aromatic Hydrocarbons (PAH's)	not available	—	300	—	.28–28 ng/l
56. Selenium	35 (24 hr. avg.) 260 (maximum) 760 (inorganic scienate)	—	54 (24 hr. avg.) 410 (maximum)	—	10 ug/l
57. Silver	4.1 (maximum 100 ppm hardness)	0.12 (average chronic)	2.3	—	50 ug/l
58. Tetrachloroethylene	5,280	840	10,200	450	.08–8 ug/l
59. Thallium	1,400	40	2,130	—	13 ug/l
60. Toluene	17,500	—	6,300	5,000	14.3 mg/l
61. Toxaphene	.013 (24 hr. avg.) 1.6 (maximum)	—	.070 (maximum)	—	.07–7.1 ng/l
62. Trichloroethylene	45,000	21,900	2,000	—	.27–27 ug/l
63. Vinyl chloride	not available	—	not available	—	.20–20 ug/l
64. Zinc	47 ug/l (24 hr. avg.)	—	58 (24 hr. avg.)	—	5 mg/l

Table 32.13 Allowable Concentrations for Air Contaminants Resulting from Hazardous Waste Treatment, Storage and Disposal Emissions (E.P.A.)

Substance	ppm[a]	mg/M[3b]
Acetaldehyde	20	40
Acetic Acid	1	3
Acetic Anhydride	0.5	2
Acetone	100	200
Acetonitrile	4	7
Acetylene Dichloride (see 1,2-Dichloroethylene)		
Acetylene Tetrabromide	0.1	1
Acrolein	0.001	0.003
Acrylamide—Skin		0.03
Acrylonitrile—Skin	2	5
Aldrin—Skin		0.03
Allyl Alcohol—Skin	0.2	0.5
Allyl Chloride	0.1	0.3
C Allylglycidyl Ether (AGE)	1	5
Allyl Propyl Disulfide	0.2	1
2-Aminoethanol (see Ethanolarine)		
2-Aminopyridine	0.005	0.2
Ammonia	5	4
Ammonium Sulfamate (ammate)		2
n-Amyl Acetate	10	55
seo-Amyl Acetate	13	65
Aniline—Skin	0.5	2
Anisidine (o. p-isomers)-Skin		0.005
Antimony and compounds (as SB)		0.005
ANTU (alpha naphthyl thiourea)		0.03
Arsenic and compounds (as AS)		0.05
Arsine	0.005	0.02
Azinphos-Methyl—Skin		0.02
Barium (Soluble compounds)		0.05
Benzene	1	
p-Benzoquinone (see quinone)		
Benzoyl Peroxide		0.5
Benzyl Chloride	0.1	0.5
Biphenyl (see diphenyl)		
Bisphenol A (see diglycidyl ether)		
Boron Oxide		2
C Boron Trifluoride	0.1	0.3
Bromine	0.01	0.07
Bromoform—Skin	0.05	0.5
Butadiene (1,3-butadiene)	100	200
Butanethiol, see Butyl mercapran		
2-Butanone	20	60
2-Butoxy Ethanol (butyl cellosolve)—Skin	5	20
Butyl Acetate (n-butyl acetate)	15	70
seo-Butyl Acetate	20	95
tert-Butyl Acetate	20	95
Butyl Alcohol	10	30
sec-Butyl Alcohol	15	45
tert-Butyl Alcohol	10	30
C Butylamine—Skin	0.5	2
C tert-Butyl Chromate (as CrO_2)—Skin		0.01
n-Butyl Glycidyl Ether (BGE)	5	25
Butyl Mercaptan	1	4
p-tert-Butyltoluene	1	6
Cadmium Fume		0.01
Cadmium Dust		0.02
Calcium Arsenate		0.1
Calcium Oxide		0.5
Campher	0.2	
Carbaryl (sevin®)		0.5
Carbon Black		0.35
Carbon Disulfide	2	
Carbon Tetrachloride	1	
Chlordane—Skin		0.05
Chlorinated Camphene—Skin		0.05
Chlorinated Diphenyl Oxide		0.05
Chlorine	0.1	0.3
Chlorine Dioxide	0.01	0.02
C Chloride Trifluoride	0.01	0.04
C Chloroacetaldehyde	0.1	0.3
α-Chloroacetophenone (phenacylchloride)	0.005	0.03
Chlorobenzene (monochlorobenzene)	10	35
o-Chlorobenzyliden Malononitrile (OCBM)	20	100
2-Chloro1,3-Butadiene, see Chloroprene		
Chlorodiphenyl (42% chlorine)—Skin		0.1
Chlorodiphenyl (54% chlorine)—Skin		0.05
1-Chloro, 2,3-Epoxypropane (see epichlorhydrin)		
2-Chloroethanol (see ethylene chlorohydrin)		
Chloroethylene (see vinyl chloride)		
C Chloroform (trichloromethane)	5	25
1-Chloro-1-nitropropane	2	10
Chloropicrin	0.01	0.07
Chloroprene (2-chloro-1,3-butadiene)—Skin	3	9
Chromic Acid and Chromates		
Chromium, So. Chromic, Chromous Salts as Cr		0.05
Metal and Insol. Salts		0.1

Coal Tar Pitch Volatiles (benzene soluble fraction) Anthracene, BaP, Phenanthrene, Acridine, Chrysene, Pyrene		0.01
Copper Fume		0.01
Dusts and Mists		0.1
Cotton Dust (raw)		0.1
Crag® Herbicide		2
Cresol (all isomers)—Skin	0.5	2
Crotonaldehyde	0.2	0.6
Cumene—Skin	5	25
Cyanide (as CN)—Skin		0.5
Cyclohexane	30	100
Cyclohexanol	5	20
Cyclohexanone	5	20
Cyclohexene	30	100
Cyclopentadiene	8	20
2,4-D		1
DDT—Skin		0.1
DDVP (see Dichlorvos)		
Decaberane—Skin	0.005	0.03
Demetron®—Skin		0.01
Diacetone Alcohol (4-hydroxy-4-mathyl-2-pentanone)	5	25
1,2-Diaminoethane (see ethylenediamine)		
Diazomethane	0.02	0.04
Diborane	0.01	0.01
Dibutylphthalate		0.5
C o-Dichlorobenzene	5	30
p-Dichlorobenzene	8	45
Dichlorodifluoromethane	100	500
1,3-Dichloro-5,5-dimethyl Hydantoin		0.02
1,1-Dichloroethane	20	40
1,2-Dichloroethylene	20	80
C Dichloroethyl Ether—Skin	2	9
Dichloromethane (see methylenechloride)		
Dichloromonofluoromethane	100	400
C 1,1-Dichloro-1-Nitroethane	1	6
1,2-Dichloropropane, see propylenedichloride		
Dichlorotetrafluoroethane	100	700
Dichlorvos (DDVP)—Skin		0.1
Dieldrin—Skin		0.02
Diethylamine	3	8
Diethylamino (see ethyl ether)		
Difluorodibromomethane	10	90
C Diglycidyl Ether (DGE)	0.05	0.3
Dihydroxybenzene, see Hydroquinone		
Diisobutyl Ketone	5	30

Diisopropylamine—Skin	0.5	2
Dimethyl Acetamide—Skin	1	4
Dimethylamine	1	2
Dimethylaminobenzene (see xylidene)		
Dimethylaniline (N-dimethyl-aniline)—Skin	0.5	3.0
Dimethylbenzene (see xylene)		
Dimethyl 1,2-Dibromo-2,2-Dichloroethyl phosphate (dibrom)		0.3
Dimethylformamide—Skin	1	3
2,6-Dimethylheptanone (see diisobutyl ketone)		
1,1-Dimethylhydrazine—Skin	0.05	0.1
Dimethylphthalate		0.5
Dimethylsulfate—Skin	0.1	0.5
Dinitrobenzene (all isomers)—Skin		0.1
Dinitro-o-cresol—Skin		0.02
Dinitrotoluene—Skin	1	0.2
Dioxane (diethylene dioxide)—Skin	10	40
Diphenyl	0.02	0.1
Diphenylmethane Diixocyanate (see methylene biaphenyl isocyanate (MDI))		
Diporopylene Glycol Methyl Ether—Skin	10	60
Di-sec, Octyl Phthalate (Di-2-Ethylhexylphthlate)		0.5
Endrin—Skin		0.01
Epichlorohydrin—Skin	0.5	2
EPN—Skin		0.05
1,2-Epoxypropane (see propyleneoxide)		
1,3-Epoxy-1-Propanol (see glycidol)		
Ethanethiol (see ethylmercaptan)		
Ethanolamine	0.3	0.6
2-Ethoxyethanol—Skin	20	75
2-Ethoxyethylacetate (cellosolve acetate)—Skin	10	53
Ethyl Acetate	40	150
Ethyl Acrylate—Skin	3	10
Ethyl Alcohol (ethanol)	100	200
Ethylamine	1	2
Ethyl seo-amyl Ketone (5-heptanone)	5	25
Ethyl Benzene	10	45
Ethyl Bromide	20	90
Ethyl Butyl Ketone (3-heptanone)	5	25
Ethyl Chloride	100	300

Table 32.13   (*Continued*)

Substance	ppm[a]	mg/M[3b]
Ethyl Ether	40	120
Ethyl Formate	10	30
C Ethyl Mercaptan	1	3
Ethyl Silicate	10	85
Ethylene Chlorohydrin—Skin	0.5	2
Ethylenediamine	1	3
Ethylene Dibromide	2	
Ethylene Dichloride	5	
C Ethylene Glycol Dinitrate and/ or Nitroglycerin—Skin	0.02	0.1
Ethylene Glycol Monomethyl Ether Acetate (see methyl cellosolve acetate)		
Ethylene Imine—Skin	0.05	0.1
Ethylene Oxide	5	9
Ethylidine Chloride (see 1,1-dichloroethane)		
N-Ethylmorpholine—Skin	2	10
Ferbam		2
Ferrovanadium Dust		0.1
Fluoride as Dust		0.3
Fluoride (as F)		0.3
Fluorine	0.01	0.02
Fluorotrichloromethane	100	500
Formaldehyde	0.3	
Formic Acid	0.5	0.9
Furfural—Skin	0.5	2
Furfuryl Alcohol	5	20
Glycidol (2,3-epoxy-1-propanol)	5	15
Glycol Monoethyl Ether (see 2-ethoxyethanol)		
Halnium		0.05
Heptachlor—Skin		0.05
Heptane (*n*-heptane)	50	200
Hexachloroethane—Skin	0.1	1
Hexachlorocaphthalene—Skin		
Hexane (*n*-hexane)	50	180
2-Hexanone	10	40
Hexone (methyl isobutyl ketone)	10	40
see-Heryl Acetate	5	30
Hydrazine—Skin	0.1	0.1
Hydrogen Bromide	0.3	1
C Hydrogen Chloride	0.5	0.7
Hydrogen Cyanide—Skin	1	1
Hydropen Fluoride	0.3	
Hydrogen Peroxide (90%)	0.1	0.1
Hydrogen Selenide	0.005	0.02
Hydrogen Sulfide		
Hydroquinone		0.2
C Iodine	0.01	0.1

Iron Oxide Fume		1
Isoamyl Acetate	10	55
Isoamyl Alcohol	10	35
Isobutyl Acetate	15	70
Isobutyl Alcohol	10	30
Isophorone	3	14
Isopropyl Acetate	25	95
Isopropyl Alcohol	40	100
Isopropylamine	0.5	1
Isopropylether	50	210
Isopropyl Glycidyl Ether (IGE)	5	25
Ketene	0.05	0.09
Lead and its Inorganic Compounds		0.02
Lead Arsenate		0.015
Lindane—Skin		0.05
Lithium Hydride		0.002
LPG (liquefied petroleum gas)	100	180
Magnesium Oxide Fume		2
Malathion—Skin		2
Maleic Anhydride	0.03	0.1
C Manganese		0.5
Mercury		
Mesityl Oxide	3	10
Methanethiol (see methyl mercaptan)		
Methoxychlor		2
2-Methoxyethanol (see methyl cellosolve)		
Methyl Acetate	20	60
Methyl Acetylene (propyne)	100	165
Methyl Acetylene-Propadiene Mixture (MAPP)	100	180
Methyl Acrylate—Skin	1	4
Methyl-a (dimethoxymethane)	100	300
Methyl Alcohol (methanol)	20	25
Methylamine	1	1
Methyl Amyl Alcohol (see methyl isobutyl carbinol)		
Methyl (*n*-amyl) Ketone (see 2-heptanone)	10	50
C Methyl Bromide—Skin	2	8
Methyl Butyl Ketone (see 2-heptanone)		
Methyl ellosolve—Skin	3	8
Methyl Cellosolve Acetate—Skin	3	12
Methyl Chloride	10	
Methyl Chloroform	35	190
Methylcyclohexane	50	200
Methylcyclohexanol	10	50
o-methylcyclohexanone—Skin	10	45
Methyl Ethyl Ketone (MEK) (see 2-butanone)		
Methyl Formate	1-	25

Methyl Iodide—Skin	0.5	3	Petchloryl Fluoride	0.3	1
Methyl Isobutyl Ketone (see hexone)			Petroleum Distillates (naphtha)	30	200
			Phenol—Skin	0.5	2
Methyl Isocyanate—Skin	0.002	0.005	p-Phenylene Diamine—Skin		0.01
C Methyl Mercaptan	1	2	Phenyl Ether (vapor)	0.1	0.7
Methyl Methacrylate	10	40	Phenyl Ether-Biphenyl Mixture		
Methyl Propyl Ketone (see 2-pentznone)			(vapor)	0.1	0.7
			Phenylethylene (see styrene)		
C Methyl Styrene	10	50	Phenylglycidyl Ether (PGE)	1	6
C Methylene Bisphenyl Isocyanate			Phenylhydrazine—Skin	0.5	2
(MDI)	0.002	0.02	Phosdrin (mevinphos®)—Skin		0.01
Methylene Chloride	50		Phosgene (carbonyl chloride)	0.01	0.04
Molybdenum:			Phosphine	0.03	0.04
Soluble Compounds		0.5	Phosphoric Acid		0.1
Insoluble Compounds		2	Phosphorus (yellow)		0.01
Monomethyl Aniline—Skin	0.2	0.9	Phosphorus pentachloride		0.1
C Monomethyl Hydrazine—Skin	0.02	0.04	Phosphorus Pentasulfide		0.1
Morpholine—Skin	2	7	Phosphorus Trichloride	0.05	0.3
Naphtha (coaltar)	10	40	Phthalic Anhydride	0.2	1
Naphthalene	1	5	Picric Acid—Skin		0.01
Nickel Carbonyl	0.0001	0.0007	Pival® (2-pivalyl-1,3-indandione)		0.01
Nickel, Metal and Soluble Cmpds,			Platinum (soluble salts) as Pt		0.000?
as Ni		0.1	Propargyl Alcohol—Skin	0.1	
Nicotine—Skin		0.05	Propane	100	180
Nitric Acid	0.2	0.5	n-Propyl Acetate	20	85
Nitric Oxide	3	3	Propyl Alcohol	20	50
p-Nitroaniline—Skin	0.1	0.6	n-Propyl Nitrate	3	11
Nitrobenzene—Skin	0.1	0.5	Propylene Dichloride	8	35
p-Nitrochlorobenzene—Skin		0.1	Propylene Imine—Skin	0.2	0.5
Nitroethane	10	30	Propylene Oxide	10	25
Nitrogen Dioxide	0.5	0.9	Propyne (see methylacetylene)		
Nitrogen Trifluoride	1	3	Pyrethrum		0.5
Nitroglycerin—Skin	0.02	0.2	Pyridine	0.5	2
Nitromethane	10	25	Quinone	0.01	0.04
1-Nitropropane	3	9	RDX—Skin		0.2
2-Nitropropane	3	9	Rhodium, Metal Fume and Dusts,		
Nitrotoluene—Skin	0.5	3	as Rh Soluble Salts		0.0001
Nitrotrichloromethane (see			Ronnel		1
chloropicrin)			Rocenone (commercial)		0.5
Octachloronaphthalene—Skin		0.01	Selenium Compounds (as Se)		0.02
Octane	50	200	Selenium Hexafluoride	0.005	0.05
Oil Mist, Mineral		0.5	Silver, Metal and Soluble		
Organo (alkyl) Mercury		0.001	Compounds		0.001
Osmium Tetroxide		0.000	Sodium Fluoroacetate (1080)—		
Oxalic Acid		0.1	Skin		0.005
Oxygen Difluoride	0.005	0.01	Sodium Hydroxide		0.2
Paraquat—Skin		0.05	Stibine	0.01	0.05
Parathion—Skin		0.011	Scoddard Solvent	50	300
Pentaborane	0.0005	0.001	Strychnine		0.015
Pentachloronaphthalene—Skin		0.05	Styrene	10	
Pentane	100	300	Sulfur Hexafluoride	100	600
2-Pentanone	20	70	Sulfuric Acid		0.1
Perchloromethyl Mercaptan	0.01	0.08	Sulfur Monochloride	0.1	0.6
Perchloryl Fluoride	0.3	1	Sulfur Pentafluoride	0.003	0.03

Table 32.13 (*Continued*)

Substance	ppm[a]	mg/M[3b]
Sulfuryl Fluoride	0.5	2
Systox, see Demeton®		
2,4,5T		1
Tantalum		0.5
TEDP—Skin		0.02
C Terphenyls	0.1	0.9
1.1.1-Tetrachloro-2,2- difluoroethane	50	400
1.1.2,2-Tetrachloroethane—Skin	0.5	4
Tetrachloroethylene	10	
Tetrachloromethane (see carbon tetrachloride)		
Tetrachloronaphthalene—Skin		0.2
Tetraethyl Lead (as Pb)—Skin		0.008
Tetrahydrofuran	20	60
Tetramethyl Lead (as Pb)—Skin		0.007
Tetramethyl Succinonitrile—Skin	0.05	0.3
Tetranitromethane	0.1	0.8
Tetryl (2,4,6-trinitrophenyl- methylinitramine)—Skin		0.15
Thallium (soluble compounds)— Skin as Tl		0.01
Thiram		0.5
Tin (inorganic compounds, except oxides)		0.2
Tin (organic compounds)		0.01
Toluene	20	
C Toluene-2,4-diisocyanate	0.002	0.001
o-Toluidine—Skin	0.5	2
Toxaphene, see Chlorinated Camphene		
Tributyl Phosphate		0.5
1,1,1-Trichloroethane (see methylchloroform)		
1.1,2-Trichloroethane—Skin	1	5
Titaniumdioxide		2
Trichloroethylene	10	
Trichloromethane (see chloroform)		
Trichloronaphthalene—Skin		0.5
1,2,3-Trichloropropane	5	30
1,1,2-Trichloro 1,2,2- trifluoroethane	100	800
Triethylamine	3	10
Trifluoromonobromomethane	100	600
2,4,6-Trinitrophenol (see picric acid)		
2,4,6-Trinitrophenyl- methylnitramine (see tetryl)		
Trinitrotoluene—Skin		0.2

	ppm[a]	mg/M[3b]
Triorthocresyl Phosphate		0.01
Triphenyl Phosphate		0.3
Turpentine	10	60
Uranium (soluble compounds)		0.005
Uranium (insoluble compounds)		0.03
C Vanadium:		
V_2O_{25} Dust		0.08
V_2O_5 Fume		0.01
Vinyl Benzene (see styrene)		
Vinylcyanide (see acrylonitrile)		
Vinyl Toluene	10	50
Warfarin		0.01
Xylene (xylol)	10	45
Xylidine—Skin	0.5	3
Yttrium		0.1
Zinc Chloride Fume		0.1
Zinc Oxide Fume		0.5
Zirconium Compounds (as Zr)		0.5

[a]Parts of vapor or gas per million parts of contaminated air by volume at 25°C and 760 mm Hg pressure.
[b]Approximate milligrams of particulate per cubic meter of air.

test organism) are attractive in biomonitoring for toxics for the following reasons: they can give simultaneous data on toxicity and chemical fate; they are likely to use indigenous organisms; they can be cost effective; and they can be more representative of an ecosystem than single-species tests. Only organisms can integrate all the factors involved in the toxicity of an effluent; and as Cairns reminds us, "one of the primary reasons for desiring clean water is to protect the aquatic ecosystems." Often we are more concerned with the quality of the water for external or "out of stream" uses.

On November 8, 1984, the Solid and Hazardous Waste Amendments were added to RCRA as a reauthorization act. (The Weston Way—1985). This reauthorization mandates the EPA to selectively and aggressively prohibit the land disposal of specific hazardous wastes. It requires the EPA to require double liners and leachate collection systems for all landfill surface impoundments. In order to receive a RCRA permit, an owner of a landfill or impoundment will now be required to clean up or correct leaks of hazardous waste from the facility regardless of when the waste was managed. In 1986 the small generator exclusion will be lowered from one ton per month to 220 pounds per month. More generators, then, will be required to use RCRA disposal at costs of 15 to 30 times those of typical municipal landfill. When blending wastes with fuel oil and burning the mixture, the reauthorization requires EPA to tighten up its rules and dates for complying to abate abuses of the practice. Because of the need to use relatively large industrial boilers for blending and burning, some industries may decide to transport their wastes to central facilities with larger boilers. Governing and controlling by EPA of underground storage tanks will now include feedstock products and fuels. EPA is now required to develop new standards for both existing and new underground storage tanks in 1987. The amendments state that one of its intentions is to minimize the generation of hazardous wastes and the land disposal of hazardous waste by encouraging process substitution, materials recovery, and properly conducted surveys.

Bankruptcy apparently is no longer an escape route for an industry to avoid cleanup costs of hazardous waste dumps. The U.S. Supreme Court recently (A Clean Victory-92,1986) decided that bankrupt companies "retain an obligation to safeguard the public's safety after they go broke." In a 5 to 4 decision written by Justice Lewis F. Powell, Jr., "The Federal bankruptcy code is not meant to allow property to be abandoned in violation of state health laws," Justice Powell wrote. Although many more cases are expected to reach the Court, the public's interest appears to be a major concern in any decision involving hazardous wastes whether old and abandoned or current.

Our federal government is not exempt from problems of disposing of hazardous wastes legally. Federal agencies were found by a new congressional study (3 U.S. Agencies—93,1986) to have deposited tons of toxic wastes into a leaking California dump during 1985. Evidently the EPA had banned the use of the facility for "Superfund" waste. The bulk of the 8,300 tons of toxic wastes came from the Department of Defense. The General Accounting Office maintained, however, "that although EPA can mandate how it handles toxic wastes under its own jurisdiction, it lacks authority to force others to follow its policies relating to licensed dumps that are having environmental problems." The only way EPA can prevent federal agencies from using a particular disposal facility is to close it."

32.81 Industrial Insurance

Despite adequate precautions of waste planning, treatment, recovery and reuse industrial plants should protect themselves against eventual failure. Damages caused by accidental spills or unavoidable errors of discharge to the environment can be covered by adequate insurance.

In an editorial in the Wall Street Journal (1981), any company that stores, treats, or disposes of solvents, acids, alkalis or other hazardous materials is advised to have some insurance. The RCRA Act of 1976 holds industry responsible for a "cradle to the grave" accounting of hazardous materials produced by them. EPA is required by this Act by January of 1981 to ensure industry's coverage of $3 million for each case of long-term damage up to a maximum of $6 million a year, excluding legal defense costs. For sudden spills the required coverage is $1 million for each case and up to $2 million a year. Many insurance brokers are of the opinion that EPAs proposal limits are not sufficient to cover a major accident. In 1981 only three U.S. insurance companies were offering coverage for these damages. Therefore, many U.S. industries provide their own insurance for accidental spills.

Insurance problems have also carried over to the trucking industry, which is largely responsible for hauling away hazardous wastes. According to Giltenan (Pollution exclusion hit—94,1985) "the reluctance of the insurance industry to cover pollution liability is causing near panic among handlers of hazardous materials, and truckers are being especially hard hit." As a result of insurance costs and reluctance to insure, many truckers—especially marginal ones—are getting out of the hazardous materials business. Under Section 30 of the Motor Carriers Act of 1981, truckers are required to carry a minimum of $5 million insurance coverage for bodily, property, and environmental-restoration (pollution) liability. The spill area of liability has also become too costly for insurance carriers to continue underwriting this coverage. Some large truckers are paying the higher premiums and also carrying policies in excess of $5 million. Most of the higher insurance costs are blamed upon the increased tendency of courts to rule in favor of environmental pollution victims. Some of the higher costs are being passed on to chemical shippers making the costs of hazardous waste disposal even higher.

The Chemical Process Industry (CPI) notes the change in the insurance industry towards industrial environmental coverage (CPI Scrambles—95,1985). It feels that "inexpensive (pollution)insurance may never again be available, certainly not in the short term." Without adequate pollution, insurance the CPI faces troublesome times. How can it protect itself against liabilities associated with sudden and accidental mishaps, such as gas leaks, and non-sudden and gradual instance, like seepage from waste holding sites? Large companies are considering self-insurance to cover the dangers. Smaller companies are looking to trade association and broker assisted programs offering primary coverage for the major firm needs (worker's compensation, product liability, comprehensive general liability, and automobile liability/physical damage insurance). Insurance premiums, if obtainable, will be much more expensive, especially if pollution liability is clearly included. As a last resort, the federal government may enact legislation to assist the chemical industry. The exact nature of the assistance is not yet known, but it's suspected that it will be similar to that offered the nuclear industry and for those needing flood insurance.

Federation (WPCF) Committee Chairman Dan Hinricks recently revealed (Hazardous Waste Conference—96,1985) that because of the complex liability issues involved in hazardous waste clean-up, the insurance market has almost totally dried up. He said, "the need is there but if companies can't get insurance, they certainly aren't going to get into the business." He calls the insurance situation, "the single, most important concern in the hazardous waste industry."

32.82 Illegal Dumping

Jaffe (97,1983) asserts that "despite tougher controls on the disposal of toxic wastes, illegal dumping has not been curbed. Instead, the law enforcement officials say, waste haulers have found new ways to skirt the laws and new places to unload their poisonous cargo." States rely on a so-called "manifest system" which it forms for detailing the route and final location of all toxic wastes. In the manifest the kinds and quantity of waste chemicals are recorded. The hauler must include where the delivery is to be made and the disposer must tell the manner of final treatment. However, the illegal dumper's answer to the manifest system has been "creative accounting and forgery."

Many suits of illegal dumping have been filed and some decisions have already been rendered by the courts. It is not the purpose here to describe or elucidate the circumstances of each.

32.83 Storage of Hazardous Industrial Wastes

Hazardous wastes are generated by industry in small quantities generally. Whether they are exported for treatment and disposal or handled for treatment on the premises, they usually are stored temporarily. Storage is necessary in these cases in order to obtain sufficiently large quantities for efficient continual treatment. If these toxic wastes are collected and transported to some distant site for treatment, they often are also stored for some period of time at the distant site.

Once we accept storage as an integral step in the ultimate treatment of hazardous wastes, it becomes necessary to provide proper storage treatment. The following criteria should be integrated into the design and operation of hazardous waste storage.

1. limited access—entrance should be available only to previously certified and qualified plant personnel
2. proper ground location—storage should be located at a proper distance from living and working people. Consideration for wind direction and velocity
3. All storage ground area must be impermeable to spilled or leaked wastes. When spills or leaks occur they cannot be allowed to permeate into the ground but must be drained off the impermeable surface to a collection sump from which they are pumped to treatment or transfer.
4. Containers or tanks must be corrosion-proof and tight fitting.
5. Ample storage area-provide several times the anticipated maximum storage area in order account for breakdown in collection and treatment systems
6. Monitoring devices-leaks and fires into the surrounding air and on the ground out of the storage area should be monitored by continuously operated sensing equipment
7. Storage area leak plan-an action plan for catastrophe-type leak occurrence should be available, publicized, and readily available if such a hazard takes place.

The Chemical Conservation Corporation (Chem. Con.) has recently made available (New Transfer Waste Facility 98,1985) a new hazardous waste transfer (storage) facility to generators in central Florida. The property is 1.2 acres, with undeveloped area adjacent for expansion. It is enclosed by a 6 foot chain link fence, with a single security gate at the entrance. The truck area is graded concrete which will contain spills and minimize chemical absorption. With a capacity of 5,500 square feet for hazardous waste storage, the floor is heavily sealed concrete, divided into flammable and corrosive drum storage areas by a spill containment curb. Each area slopes to a center drain leading into one of two emerging 120 gallon waste containers located outside the building. Fire walls separate storage from personnel areas, and the storage electrical system is explosion proof. A hazardous waste sprinkler system protects the entire building with fire detection devices as well. These detection devices automatically signal the Orange County Fire Department and a portable telephone system insures constant communication. Detection devices on all entries to the storage building are linked to the Orange County Sheriff's Department.

Hazardous wastes stored here are transferred to ultimate disposal facilities for incineration, reclamation, neutralization, chemical fixation, or sanitary landfill. Currently, this storage (transfer) facility is handling 1.5 million gallons of hazardous waste per year for 350 industrial customers in 1500 drum quantities.

Safety Storage (Chemical Storage Containers 99, 1985) has designed and constructed containers to comply with both NFPA standards and OSHA regulations. Many local hazardous material storage ordinances require hazardous chemicals to be stored in secondary containment structures to prevent spills or leaks from contaminating groundwater. Their containers are made of 10-14 gauge ASTM-A569 steel, possess secondary spill capacity of 500 gallons, hold 30-55 gallon drums. The interiors are protected from chemical attack by a resistant epoxy coating. Outside demensions are 8'x8'x22'. Each unit is static-grounded to prevent ignition of flammable wastes by electrical discharge. Fire protection is provided with a water line and three fire sprinkler heads. Roofs are specially designed to handle explosions and floors are thick plywood underlaid with subfloors of 12 gauge epoxy-lined steel. Many other optional features for hazardous waste storage safety are available on an option basis. These containers may be leased to assist industry in conserving capital.

Stone advocates the use of selected, existing mined space as technically and economically feasible for permanent storage of untreatable wastes or the toxic end products of hazardous waste treatment (Update Storage of Hazardous Waste 100, 1985). Six advantages of this type of storage are given as:

1. In deep mined space the waste would be below drinking water aquifers
2. Isolation from the public and the surface ecology
3. If required, waste can be isolated from hydrological environment by encapsulation or containerization
4. Security can be readily maintained
5. In a sealed mine, no continuing maintenance will be required
6. If retrievability is desired, the mine could be used as a long-term underground warehouse.

Such treatment must protect all parts of the environment during construction, while adding waste, and for an indefinite period into the future. Storage underground utilizes the minimum of existing land surface. One possibility in mines is that of abandoned salt mines which are relatively free from people and industry. Although this practice has been used for storage of certain hazardous wastes for a number of years, one must be certain that the chemicals neither react with themselves nor the salt in the mine.

Storage of hazardous chemicals prior to and during use is a vital phase of material production. The reader can attest to this by reviewing some of the examples of accidents which have occurred recently from too little attention to this subject.

32.84 Transportation and Spill Prevention of Industrial Hazardous Wastes

Transporting hazardous wastes to an external site for storage, treatment, or disposal must be considered an integral part of its ultimate solution. It is usually done by truck or rail. Either mode of transportation involves hazards of accidental spills or wrecks which may release the wastes to an unsuspecting and unprepared environment. Trucks exhibit more flexibility in delivery as well as pickup. In the last decade it was estimated also that rail accidents were two or three times more prevalent than truck.

Although rail accidents are down, their severity is up (Chemicals on Rails 108, 1989). It is estimated that 4 billion tons of hazardous materials are shipped by rail annually, involving 250,000 shipments daily. The problem of moving hazardous chemicals by rail is not only that they may be explosive, poisonous, or corrosive, but also that trains move them through industrial areas and residential areas of cities. States and localities, along with the Federal Government, regulate chemical shipments by road and rail, but some confusion and complex compliance has resulted in burdens for carriers. Congress is examining possibilities of legislation allowing larger cities to approve of transportation routes and the Department of Transportation to select the safest routes.

Several states have incurred 10 or more railroad accidents involving toxic wastes [108]. In both modes of transportation, the waste generator must accept the responsibility of complete and correct disclosure of the chemical contents of the hazardous wastes. The more information about the composition and characteristics of the wastes to be transported, the safer will be the trip (in the case of an accident) to the disposal site. The transporter assumes the responsibility for safe delivery of the waste to the disposer. When the transporter accepts the waste, he should sign a verification statement with the generator that he is transporting a given volume of specific hazardous wastes. The same statement should be signed and given to the disposer once the wastes have arrived and been unloaded at the disposal site. Packaging of hazardous wastes should be done in conformance with the Department of Transportation regulations. Usually, liquid wastes are packaged in 55 gallon drums, sealed, and surrounded by vermiculite for insulation against the shock of collisions or pumped into tank cars supplied for the purpose by truckers or railroad haulers.

The truck fleet used by Chem Con (New Transfer Waste Facility 98,1985) travels over one million miles per year and has been accident-free for its three years of operation.

Nemerow discusses (Industrial Water Pollution 24,1987) the scavenger system that hauls, treats, reclaims, and disposes of a variety of industrial wastes. These wastes are usually small in volume and difficult and/or hazardous to dispose of by the producer in normal ways. The expenses of transportation may be considerable and determine in the long run whether exportation/treatment is a better alternative to on-site treatment.

32.85 Spills

As an example of protection against spills, the Proform Company offers a system for eliminating spills in areas where rail cars are loaded, unloaded, washed or fueled. They provide a structural fiberglass reinforced plastic track collector pan system to collect and isolate spills for treatment and disposal. The system components include collector pans that attach to rails and cross drains that connect with pre-installed underground pipes leading to a sump for treatment (Proform—101, 1985).

Jim Bradly, the Minister of the Environment for Ontario, Canada, announced on July 3, 1985 that "our initial priorities include regulation of the transportation of dangerous goods. . . ."

Bradley is new in this high position and recognized how vital it is to transport hazardous materials safely.

References

1. "Industrial Solid Wastes," A textbook by Nelson Leonard Nemerow, Chapter 10, Ballinger Publishing Company Cambridge, Mass., 1983.
2. "Industrial Hazardous Waste Management," Industrial and Environment Special Issue No.4,1983, UNEP, Paris, France.
3. "Great Lakes Study: Chemicals imperil food for millions," *The Miami Herald*, December 12, 1985, page 3C.
4. "Hazardous Waste Management," Dept. of Public Affairs, American Chemical Society, 1155 Sixteenth Street, N.W., Washington, D.C. 20036
5. Federal Register 1980, Washington, D.C., U.S.A. 261.21–261.24 EPA HW #D001–D003 and DO 17.
6. Dotson, G.K., Dean, R.B., Cooke, W.B. and Kennar, B.A., "Yard Spreading—A Conserving and Non-Polluting Method of Disposing of Oily Wastes," Waste Oil Report to Congress, Environmental Protection Agency, April 1974, page 151.
7. United Nations Environmental Protection, "Industrial Hazardous Waste Management," Industry and Environment-Special Issue No.4.,1983, Paris,France.
8. "Hazardous Waste Landfill Benefits from Compost," *Biocycle*, page 28, Sept.-Oct. 1983.
9. "Salt of the Earth," by Janet Raloff, *Science News* vol.126, 19, November 10,1984, page 298, Washington, D.C.
10. "Helping Crops Stand up to Salt," by Paul Raeburn, Industrial Technology, page 88, May 1985.
11. *Stream, Lake, Estuary, and Ocean Pollution*, A text by Nemerow, N.L., Van Nostrand Reinhold Publishing Company, New York City, 1985.
12. "Areas in 7 states eyed for nuclear dump site," *The Miami Herald*, January 17, 1986.
13. "Nuclear industry shouldn't escape liability fallout," An Editorial by Kathleen Welch (U.S. Public Interest Re-

search Group), *The Miami Herald*, January 9, 1986, page 31A.

14. Public Service Electric and Gas Co. 1984 Annual Report, page 14, Newark, New Jersey 07101.

15. "Toxics and Male Infertility," by M.Castleman, *Sierra Magazine* March/April 1985, page 49.

16. "Incinerator belches foul smoke," *The Miami Herald*, Section B, page 1, March 25, 1985.

17. Doyle, B.W., D.A. Drum, and J.D. Lauber, "The Smoldering question of hospital wastes," *Pollution Engineering* July 1985, page 85.

18. "Eating raw shellfish health risk, MD's say," *The Miami Herald*, March 13,1986, page 12A.

19. Sax, Irving, "Dangerous Properties of Industrial Materials," A text, Fifth Edition, Van Nostrand Reingold Pub.Co., New York City, 1979, page 824.

20. Graham, F.J., "PCB's Make a Slow Exit," *Chemical Business*, September 1985, page 72.

21. *Toxic and Hazardous Waste*, Proc.15th Mid-Atlantic Industrial Waste Conference, Edited by Lagrega, M.D. and L.K. Hendrian, June 26, 1983, Butterworth Publishers, Boston, Massachusetts.

22. *The Amicus Journal*, Spring 1985, page 25, Vol.6, No.4, MRDC, publication, New York City.

23. Zoltek, John, "Bacteria Digesting Toxic Wastes," The Overflow-Florida Water Pollution Control Operators Association, Nov-Dec. 1984, page 13.

24. Nemerow, N.L., "Industrial Water Pollution," Robert Krieger Publishing Company, Malabar, Florida, 1987.

25. LeRoy, E., "Processing hazardous waste in France," Industry and Environment-Industrial Hazardous Waste Management (UNEP Special Issue No.4., page 46,1983).

26. U.S.E.P.A., "Everybody's Problem, Hazardous Waste," Office of Water and Waste Management, Washington, D.C., 1980.

27. Van Noordwyck, H.J. et al., "Quantification of Municipal Disposal Methods for Industrially Generated Hazardous Wastes," Proc.6th Annual Research Symposium on Treatment of Hazardous Waste, USEPA-MERL, Cincinnati, Ohio, EPA/9-80-011.

28. "Industrial Wastes," Nemerow, Nelson, *Kirk-Othmer Encyclopedia of Chemical Technology, Vol.24, Third Edition*, John Wiley and Sons, page 228, 1984.

29. "Reducing Hazardous Waste Generation," A Digest of the Report by National Academy Press Washington, D.C., 1985.

30. "Assessment of Incineration as a Treatment Method for Liquid organic Hazardous Wastes," Summary and Conclusions, U.S.E.P.A., Office of Policy, Planning and Evaluation Washington, D.C. 20460, March 1985.

31. Torricelli, Robert G., "Let's Destroy Toxic Waste, Not Just Move It Around," An Editorial, *Wall Street Journal*, 9-26-'85, page 30.

32. Civil Engineering, "Waste Processing firms to play critical role in Hazardous Waste Management," Sept. 1979, (ASCE) page 85.

33. "Tangled rules on Chemical Hazards hamper U.S. Efforts to protect public," by Philip Shabecoff, *New York Times*, November 27, 1985, Page 13.

34. "Land-poor but trash-rich, the Japanese take over Bay," *The Miami Herald*, February 4, 1986, page 24.

35. Worthy, W., "Hazardous Waste:Treatment Technology Grows," Chemical Engineering News, March 8, 1982.

36. Thermal Systems for Waste and Refuse, ENELCO-Von Roll Environmental Elements Corp., Baltimore, Md. 21203 P.O. Box 1318.

37. H.Freeman, "Hazardous Wastes Incineration," Chapter 5, *Hazardous Wastes Management*, Edited by Peirce and Vesiland, page 59, Ann Arbor Science Pub.Inc., 1981.

38. "Incineration of Explosives Contaminated Soils," J.W. Noland, The Weston Way, Volume 10, Number 1, Winter 1984, page 3–6.

39. "EPA Clears Test on Incinerating Toxic Waste at Sea," by Robert Taylor, *Wall Street Journal* 11–27–'85, page 20, and "EPA to Permit Waste Burning off Jersey Coast," by Philip Shabecoff, *New York Times* 11–27–'85, page 16.

40. U.S.E.P.A., "Land Cultivation of Industrial and Municipal Solid Wastes," A State of the Art Study, EPA 600/2–78–1402 Washington, D.C.

41. "Land Application Treatment of Textile Knitting, Dyeing, and Finishing Wastewater," by Michael S. Bahorsky, Toxic and Hazardous Wastes Proc.15th Mid-Atlantic Industrial Waste Conf., page 8, 1983, Butterworth Publishers Woburn, Maine 01801.

42. Pizzuto, J.S., "Super-Fund-the U.S. Response," Symposium on Waste Disposal-the Challenge Madrid, Spain, International Association of Environmental Coordinators, Brussels, 1981.

43. Department of the Environment (DOE), "Cooperative Program on Research on the Behaviour of Hazardous Wastes in Landfill Sites," HMSO, London, England, 1978.

44. Sax, N.I., "Dangerous Properties of Industrial Materials," Van Nostrand Reinhold Co., New York City, 1979.

45. Piasecki, B., "Unfouling the Nest," *Science* Sept. 1983, Page 78.

46. Sachdev, D.R. et al., "Use of Fly Ash as a Liner Material for Utility Solid Waste Disposal Sites," 2nd Conf. on Municipal, Hazardous and Coal Waste Management, Pergammon Press, Coral Gable, Fla., Dec. 5–7,1983.

47. Boycem A.W., "Reconditioning of Dehydrated Sludge with Quicklime," Toxic and Hazardous Wastes, page 477, June 26, 1983.

48. Metry, A.A., M.F. Coia, M.H. Corbin, and A.L. Lenthe, "In Situ Closure of Sludge Lagoons," *Toxic and Hazardous Wastes*, page 538, 1983.

49. Wright, A.P. and S.D. Caretsky, "In Situ Treatment/ containment and Chemical Fixation," Chapter 6, *Hazardous Wastes Management*, Ann Arbour Science Pub. Inc., 1981.

50. Cheremisinoff, N.P., P.N. Cheremisinoff et al., *Industrial and Hazardous Wastes Impoundment*, Ann Arbor Science Pub.Co., Chapter 7, Siting Considerations; Chapter 9, Impoundment Facilities and Liners.

51. "U.S. Waste Deadline likely to shut down many dumps," *New York Times*, November 8, 1985, page 11.

52. "Toxic sites close rather than comply," *The Miami Herald*, December 7, 1985.

53. "Ocean Disposal Systems for Sewage Sludge and Effluent," National Academy Press, Washington, D.C., 1984.

54. Prouty, M.F., J. Alleman, and N. Berman, "Sludge Amended Brick Manufacture," *Toxic and Hazardous Wastes* page 492, June 26, 1983.

55. Project Reports—ENRECO,Inc., P.O. Box 9617, Amarillo, Texas 79105-9617.

56. "Export plan for wastes stirs fears," *Miami Herald* July 1, 1983, page 1G.

57. Cartwright, P.S., "Innovative Technology to treat toxic wastes from a thin film head manufacturing facility—A case history," Future of Water Reuse Proc.Vol.2, Water Reuse Symposium III, Aug.26–31, 1984, San Diego,Cal., page 778.

58. "Case Histories; Reviewing the use of sodium borohydride for control of heavy metal discharge in industrial wastewaters," M.M. Cook, J.A. Lander,Sr., and D.S. Littlehale, Proc.34th Purdue Univ.Industrial Waste Conf., May 8,1979, page 514.

59. *Theories and Practises of Industrial Wastes Treatment* by Nelson L. Nemerow, Addison Wesley Co., 1963, page 243.

60. Fitzpatrick, D.T., "Machine shop scrap processing and recycling," *Toxic and Hazardous Wastes* page 569, 1983.

61. "Sludge Extractor Reduces Quench Oil Use, Waste Load," *Pollution Equipment News* October 1985, page 93.

62. Future of Water Reuse-Water Reuse Symposium, G.J. Van Gils, M. Pirbazari, S.H.Kim, and J. Shorr, Volume 2, page 911, San Diego, Cal., August 26–31, 1984.

63. Oberteuffer, J.A., "High gradient magnetic separation," I.E.E.E. Trans. on Magnetics *9*,3, page 303, Sept.1973.

64. DeLatour, C., "Magnetic Separation in Water Pollution Control," E.E.E. Trans. On Magnetics *9*,3,page 314, Sept 1973.

65. "Textile dye wastes," by N.L. Nemerow, *Chemical Age* June 14,1952.

66. "Color in Industrial Wastes," by N.L. Nemerow, Journal of San.Eng.Div. of Amer. Soc. Civ. Engrs., Paper No.1180 *83*, SA1, Feb.1957.

67. "Industrial Water Pollution Control in Hong Kong," by N.L. Nemerow, UNIDO Report Vienna, Austria, July 5, 1982, DP/HOK/80/11–51/32.1J.

68. Bergenthal, J., "Textile Dyebath Reconstitution and Reuse," Future of Water Reuse Proc. Vol.2, Water Reuse Sumposium, Aug.24–26,1984, San Diego, Cal., page 840.

69. Sunohio, "New Process to Detoxify PCB's Approved by U.S.," *Miami Herald Newspaper* May 29, 1981.

70. Hanson, N., "Air Stripping is Effective in Removing TCE from Groundwater," Water/Engineering and Management, August 1984, page 18.

71. "Hazardous Wastes Sprayed Away," by N.C. Wench and P.D. Josephson, *Pollution Engineering 16*,9,Sept. 1984, page 33.

72. Nyer, E.K., *Groundwater Treatment Technology*, Van Nostrand Reinhold Publ. Co., New York City, 1985, page 149.

73. "Air Stripping—A Technical Data Bulleting," Oil Recovery Systems, Inc., Norwood, Maine 02062.

74. "Clean-Up of the Gilson Road Hazardous Waste Disposal Site, Nashua, New Hampshire," *The Weston Way* Volume 10, No. 3., Summer-Fall 1984, pages 3–7.

75. "Bio-Reclamation of Ground and Groundwater—Case History," Vidyut Jhaven and Alfred J. Mazzaca, Presented at 4th National Conf.On Management of Uncontrolled Hazardous Waste Sites, Oct 31–Nov.2. 1983.

76. "Detoxifier-Hazardous Waste Clean-up and Compliance," ATW/Calweld Inc. Santa Fe Springs,California 90670.

77. "Fungus eats away at toxins," *The Miami Herald* Page 1, June 15, 1985.

78. "NASA roots out earthy cure for polluted air in space stations," Paul Duke, *Wall Street Journal* August 28, 1985.

79. Irvine, R.L. and Busch, A.W., "Sequencing batch reactors-an overview," *Journ.Water Pollution Control Federation 51*, 235, (1979).

80. Irvine, R.L., et al., "Enhanced Biological Treatment of Leachates from Industrial Landfills," *Hazardous wastes* 1,123,(1984).

81. Herzbrun, P.A, R.L. Irvine, and K.C. Malinowski, "Biological treatment of hazardous waste in sequencing batch receptors," *Journ.Water Poll. Control Fed. 57*, 1163,Dec.1985.

82. *Proceedings of the National Conference on Waste Exchange*, Florida State University, Tallahassee, Florida, March 8–9,1983.

83. *The Southern Waste Information Exchange Catalog* (SWIX) Vol.11, No.1, Oct. 1982, P.O. Box 6487, Tallahassee, Fla. 32313.

84. "Landfilled Wastes treated in place," G.Z. Francis, *Pollution Engineering 16*,9, Sept.1984, page 37.

85. "Dupont to enter waste treatment," *The Miami Herald* June 5, 1984.

86. Duffy, J.G., "Electrochemical Removal of Heavy Metals from Wastewater," *Products Finishing*, August 1983, page 72.

87. "Nation's toxic waste is a killer topic—literally," An editorial by Kathleen Krog, *Miami Herald* 9–33–'85, page 22A.

89. Statement by the Honorable David R. Peterson, The First Session of the Thirty-Third Parliament of the Province of Ontario, July 2, 1985, page 29.

90. Thorsen, J.W. and J.C. Petura, "Plan for your RCRA Part B Permit," Clearwaters, Winter 1983, pages 12–15.

91. "Organic contamination:whistling past the graveyard," by Walter Weber,Jr., *Journ.Water Poll. Control Fed. 58* 1, 16, January 1986.

92. "A Clean Victory," an editorial, *The Miami Herald* February 3, 1986; "Ban on asbestos is unnecessary, industry, Canadians advise EPA," *The Miami Herald* January 24,1986, page 6A.

93. "3 U.S. agencies used banned toxic waste dump," *The Miami Herald*, January 1986.

94. Giltenan, E.F., "Pollution Exclusions Hit Truckers Hard," *Chemical Business* October 1985, page 38.

95. "CPI Scrambles for Pollution Insurance," *Chemical Business* November 1985, page 41.

96. "Hazardous waste conferences planned," Highlights (WPCF) 22, 12, 5, December 1985.

97. Jaffe, Mark, "Authorities struggle to curb illegal toxic waste dumping," *The Miami Herald*, Dec.1, 1983, page 13C.

98. "New Transfer Waste Facility Built in Central Florida," *The Florida Specifier;* page 20, April 1985.

99. *Chemical Storage Containers Safety*, P.O. Box 59037 San Jose, California 95159.

100. "Update-Storage of Hazardous Waste in Mined Space," R.B. Stone, Land Disposal of Hazardous Waste Proc.11th Annual Research Symposium EPA/600/9–85/013, April 1985, page 339.

101. "Proform-SFRP Track Collector Pan System," Proform Inc., 700 Terrace Lane, Paducah, KY. 42001.

102. "Thermal Treatment technologies for hazardous wastes," Cheremisinoff, P.N., *Pollution Engineering 20*,8, 50, Aug. 1988.

103. "The EPA Manual for Waste Minimization Opportunity Assessments," Jacobs Engineering Group, Pasadena, California, EPA/600/2–88–025, April 1988.

104. *U.S. Pakistan Joint Research on Advanced Renovation of Small Industry Wastes*, National Science Foundation Washington, D.C., Project INT–85–20198 (1988).

105. Randle, Susan, "Some New Treatments for Hazardous Waste," *Chemical Business*, February 1989, page 36.

106. "Conference Backs Curbs on Export of Toxic Waste," Steven Greenhouse, *New York Times*, March 23, 1989, page 1.

107. "Identifying,classifying and describing hazardous wastes," H.Yakowitz, UNEP Ind. & Env. Jan/Feb/Mar/1988, pg.6.

108. "Chemicals on Rails: A Growing Peril," John Cushman, Jr., *New York Times*, page 8, Aug. 2, 1989.

SECTION 306—NATIONAL STANDARDS OF PERFORMANCE

"SEC. 306. (a) For purposes of this section:

"(1) The term 'standard of performance' means a standard for the control of the discharge of pollutants which reflects the greatest degree of effluent reduction which the Administrator determines to be achievable through application of the best available demonstrated control technology, processes, operating methods, or other alternatives, including, where practicable, a standard permitting no discharge of pollutants.

"(2) The term 'new source' means any source, the construction of which is commenced after the publication of proposed regulations prescribing a standard of performance under this section which will be applicable to such source, if such standard is thereafter promulgated in accordance with this section.

"(3) The term 'source' means any building, structure, facility, or installation from which there is or may be the discharge of pollutants.

"(4) The term 'owner or operator' means any person who owns, leases, operates, controls, or supervises a source.

"(5) The term 'construction' means any placement, assembly, or installation of facilities or equipment (including contractual obligations to purchase such facilities or equipment) at the premises where such equipment will be used, including preparation work at such premises.

"(b) (1) (A) The administrator shall, within ninety days after the date of enactment of this title publish (and from time to time thereafter shall revise) a list of categories of sources, which shall, at the minimum, include:

"pulp and paper mills;
"paperboard, builders paper and board mills;
"meat product and rendering processing;
"dairy product processing;
"grain mills;
"canned and preserved fruits and vegetables processing;
"canned and preserved seafood processing;
"sugar processing;
"textile mills;
"cement manufacturing;
"feedlots;
"electroplating;
"organic chemicals manufacturing;
"inorganic chemicals manufacturing;
"plastic and synthetic materials manufacturing;
"soap and detergent manufacturing;
"fertilizer manufacturing;
"petroleum refining;
"iron and steel manufacturing;
"nonferrous metals manufacturing;
"phosphate manufacturing;
"steam electric powerplants;
"feroalloy manufacturing;
"leather tanning and finishing;
"glass and asbestos manufacturing;
"rubber processing; and
"timber products processing.

"(B) As soon as practicable, but in no case more than one year, after a category of sources is included in a list under subparagraph (A) of this paragraph, the Administrator shall propose and publish regulations establishing Federal standards of performance for new sources within such category. The Administrator shall afford interested persons an opportunity for written comment on such proposed regulations. After considering such comments, he shall promulgate, within one hundred and twenty days after publication of such proposed regulations, such standards with such adjustments as he deems appropriate. The Administrator shall, from time to time, as technology and alternatives change, revise such standards following the procedure required by this sub-section for promulgation of such standards. Standards of performance, or revisions thereof, shall become effective upon promulgation. In establishing or revising Federal standards of performance for new sources under this section, the Administrator shall take into consideration the cost of achieving such effluent reduction, and any non-water quality environmental impact and energy requirements.

"(2) The Administrator may distinguish among classes, types, and sizes within categories of new sources for the purpose of establishing such standards and shall consider the type of process employed (including whether batch or continuous).

"(3) The provisions of this section shall apply to any new source owned or operated by the United States.

"(c) Each State may develop and submit to the Administrator a procedure under State law for applying and enforcing standards of performance for new sources located in such State. If the Administrator finds that the procedure and the law of any State require the application and enforcement of standards of performance to at least the same extent as required by this section, such State is authorized to apply and enforce such standards of performance (except with respect to new sources owned or operated by the United States).

"(d) Notwithstanding any other provision of this Act, any point source the construction of which is commenced after the date of enactment of the Federal Water Pollution Control Act Amendments of 1972 and which is so constructed as to meet all applicable standards of performance shall not be subject to any more stringent standard of performance during a ten-year period beginning on the date of completion of such construction or during the period of depreciation or amortization of such facility for the purposes of section 167 or 169 (or both) of the Internal Revenue Code of 1954, whichever period ends first.

"(e) After the effective date of standards of performance promulgated under this section, it shall be unlawful for any owner or operator of any new source to operate such source in violation of any standard of performance applicable to such source.

SECTION 402—NATIONAL POLLUTANT DISCHARGE ELIMINATION SYSTEM

"SEC. 402. (a) (1) Except as provided in sections 318 and 404 of this Act, the Administrator may, after opportunity for public hearing, issue a permit for the discharge of any pollutant, or combination of pollutants, notwithstanding section 301 (a), upon condition that such discharge will meet either all applicable requirements under sections 301, 302, 306, 307, 308, and 403 of this Act, or prior to the taking of necessary implementing actions relating to all such requirements, such conditions as the Administrator determines are necessary to carry out the provisions of this Act.

"(2) The Administrator shall prescribe conditions for such permits to assure compliance with the requirements of paragraph (1) of this subsection, including conditions on data and information collection, reporting, and such other requirements as he deems appropriate.

"(3) The permit program of the Administrator under paragraph (1) of this subsection, and permits issued thereunder, shall be subject to the same terms, conditions and requirements as apply to a State permit program and permits issued thereunder under subsection (b) of this section.

"(4) All permits for discharges into the navigable waters issued pursuant to section 13 of the Act of March 3, 1899, shall be deemed to be permits issued under this title, and permits issued under this title shall be deemed to be permits issued under section 13 of the Act of March 3, 1899, and shall continue in in force and effect for their term unless revoked, modified, or suspended in accordance with the provisions of this Act.

"(5) No permit for a discharge into the navigable waters shall be issued under section 13 of the Act of March 3, 1899, after the date of enactment of this title. Each application for a permit under section 13 of the Act of March 3, 1899, pending on the date of enactment of this Act shall be deemed to be an application for a permit under this section. The Administrator shall authorize a State, which he determines has the capability of administering a permit program which will carry out the objective of this Act, to issue permits for discharges into the navigable waters within the jurisdiction of such State. The Administrator may exercise the authority granted him by the preceding sentence only during the period which begins on the date of enactment of this Act and ends either on the ninetieth day after the date of the first promulgation of guidelines required by section 304 (h) (2) of this Act, or the date of approval by the Administrator of a permit program for such State under subsection (b) of this section, whichever date first occurs, and no such authorization to a State shall extend beyond the last day of such period. Each such permit shall be subject to such conditions as the Administrator determines are necessary to carry out the provisions of this Act. No such permit shall issue if the Administrator objects to such issuance.

"(b) At any time after the promulgation of the guidelines required by subsection (h) (2) of section 304 of this Act, the Governor of each State desiring to administer its own permit program for discharges into navigable waters within its jurisdiction may submit to the Administrator a full and complete description of the program it proposes to establish and administer under State law or under an interstate compact. In addition, such State shall submit a statement from the attorney general (or the attorney for those State water pollution control agencies which have independent legal counsel), or from the chief legal officer in the case of an interstate agency, that the laws of such State, or the interstate compact, as the case may be, provide adequate authority to carry out the described program. The Administrator shall approve each such submitted program unless he determines that adequate authority does not exist:

"(1) To issue permits which—

"(A) apply, and insure compliance with, any applicable requirements of sections 301, 302, 306, 307, and 403;

"(B) are for fixed terms not exceeding five years; and

"(C) can be terminated or modified for cause including, but not limited to, the following:

"(i) violation of any condition of the permit;

"(ii) obtaining a permit by misrepresentation, or failure to disclose fully all relevant facts;

"(iii) change in any condition that requires either a temporary or permanent reduction or elimination of the permitted discharge;

"(D) control the disposal of pollutants into wells;

"(2) (A) To issue permits which apply, and insure compliance with, all applicable requirements of section 308 of this Act, or

"(B) To inspect, monitor, enter, and require reports to at least the same extent as required in section 308 of this Act;

"(3) To insure that the public, and any other State the waters of which may be affected, receive notice of each application for a permit and to provide an opportunity for public hearing before a ruling on each such application;

"(4) To insure that the Administrator receives notice of each application (including a copy thereof) for a permit;

"(5) To insure that any State (other than the permitting State) whose waters may be affected by the issuance of a permit may submit written recommendations to the permitting State (and the Administrator) with respect to any permit application and, if any part of such written recommendations are not accepted by the permitting State, that the permitting State will notify such affected State (and the Administrator) in writing of its failure to so accept such recommendations together with its reasons for so doing;

"(6) To insure that no permit will be issued if, in the judgment of the Secretary of the Army acting through the Chief of Engineers, after consultation with the Secretary of the department in which the Coast Guard is operating, anchorage and navigation of any of the navigable waters would be substantially impaired thereby; and

"(7) To abate violations of the permit or the permit program, including civil and criminal penalties and other ways and means of enforcement.

"(8) To insure that any permit for a discharge from a publicly owned treatment works includes conditions to require adequate notice to the permitting agency of (A) new introductions into such works of pollutants from any source which would be a new source as defined in section 306 if such source were discharging pollutants, (B) new introductions of pollutants into such works from a source which would be subject to section 301 if it were discharging such pollutants, or (C) a substantial change in volume or character of pollutants being introduced into such works by a source introducing pollutants into such works at the time of issuance of the permit. Such notice shall include information on the quality and quantity of effluent to be introduced into such treatment works and any anticipated impact of such change in the quantity or quality of effluent to be discharged from such publicly owned treatment works.

"(9) To insure that any individual user of any publicly owned treatment works will comply with sections 204(b), 307, and 308.

"(c) (1) Not later than ninety days after the date on which a State has submitted a program (or revision thereof) pursuant to subsection (b) of this section, the Administrator shall suspend the issuance of permits under subsection (a) of this section as to those navigable waters subject to such program unless he determines that the State permit program does not meet the requirements of subsection (b) of this section or does not conform to the guidelines issued under section 304(h) (2) of this Act. If the Administrator so determines, he shall notify the State of any revisions or modifications necessary to conform to such requirements or guidelines.

"(2) Any State permit program under this section shall at all times be in accordance with this section and guidelines promulgated pursuant to section 304 (h) (2) of this Act.

"(3) Whenever the Administrator determines after public hearing that a State is not administering a program approved under this section in accordance with requirements of this section, he shall so notify the State and, if appropriate corrective action is not taken within a reasonable time, not to exceed ninety days, the Administrator shall withdraw approval of such program. The Administrator shall not withdraw approval of any such program unless he shall first have notified the State, and made public, in writing, the reasons for such withdrawal.

"(d) (1) Each State shall transmit to the Administrator a copy of each permit application received by such State and provide notice to the Administrator of every action related to the consideration of such permit application, including each permit proposed to be issued by such State.

"(2) No permit shall issue (A) if the Administrator within ninety days of the date of his notification under subsection (b) (5) of this section objects in writing to the issuance of such permit, or (B) if the Administrator within ninety days of the date of transmittal of the proposed permit by the State objects in writing to the issuance of such permit as being outside the guidelines and requirements of this Act.

"(3) The Administrator may, as to any permit application, waive paragraph (2) of this subsection.

"(e) In accordance with guidelines promulgated pursuant to subsection (h) (2) of section 304 of this Act, the Administrator is authorized to waive the requirements of subsection (d) of this section at the time he approves a program pursuant to subsection (b) of this section for any category (including any class, type, or size within such category) of point sources within the State submitting such program.

"(f) The Administrator shall promulgate regulations establishing categories of point sources which he determines shall not be subject to the requirements of subsection (d) of this section in any State with a program approved pursuant to subsection (b) of this section. The Administrator may distinguish among classes, types, and sizes within any category of point sources.

"(g) Any permit issued under this section for the discharge of pollutants into the navigable waters from a vessel or other floating craft shall be subject to any applicable regulations promulgated by the Secretary of the department in which the Coast Guard is operating, establishing specifications for safe transportation, handling, carriage, storage, and stowage of pollutants.

"(h) In the event any condition of a permit for discharges from a treatment works (as defined in section 212 of this Act) which is publicly owned is violated, a State with a program approved under subsection (b) of this section or the Administrator, where no State program is approved, may proceed in a court of competent jurisdiction to restrict or prohibit the introduction of any pollutant into such treatment works by a source not utilizing such treatment works prior to the finding that such condition was violated.

"(i) Nothing in this section shall be construed to limit the authority of the Administrator to take action pursuant to section 309 of this Act.

"(j) A copy of each permit application and each permit issued under this section shall be available to the public. Such permit application or permit, or portion thereof, shall further be available on request for the purpose of reproduction.

"(k) Compliance with a permit issued pursuant to this section shall be deemed compliance, for purposes of sections 309 and 505, with sections 301, 302, 306, 307, and 403, except any standard imposed under section 307 for a toxic pollutant injurious to human health. Until December 31, 1974, in any case where a permit for discharge has been applied for pursuant to this section, but final administrative disposition of such application has not been made, such discharge shall not be a violation of (1) section 301, 306, or 402 of this Act, or (2) section 13 of the Act of March 3, 1899, unless the Administrator or other plaintiff proves that final administrative disposition of such application has not been made because of the failure of the applicant to furnish information reasonably required or requested in order to process the application. For the 180-day period beginning on the date of enactment of the Federal Water Pollution Control Act Amendments of 1972, in the case of any point source discharging any pollutant or combination of pollutants immediately prior to such date of enactment which source is not subject to section 13 of the Act of March 3, 1899, the discharge by such source shall not be a violation of this Act if such a source applies for a permit for discharge pursuant to this section within such 180-day period.

REFERENCE LIST OF USEPA EFFLUENT GUIDELINES DOCUMENTS

Category of EGD Industrial Studies	Subcategory	Document Number
Sugar Processing	a) Raw Cane Processing	EPA 440/1–75/044
	b) Beet	EPA 440/1–74/002b
Textile Mills	Textile Mills	EPA 440/1–82/022
Inorganic Chemicals Manufacturing	Inorganic Chemicals	EPA 440/1–82/007
Plastics & Synthetic Materials	Synthetic Polymers	EPA 440/1–74/036
Iron & Steel	Iron & Steel Volumes I thru VI	EPA 440/1–82/024
Steam Electric Powerplants	Steam Electric	EPA 440/1–82/029
Glass Manufacturing	a) Pressed and Blown Glass	EPA 440/1–75/034a
	b) Insulation Fiberglass	EPA 440/1–74/001b
	c) Flat Glass	EPA 440/1074/001c
Timber Products	A) Plywood & Wood	EPA 440/1–74/023a
	B) Timber Products	EPA 440/1–81/023
Pulp, Paper & Paperboard	Pulp, Paper & Paperboard and Builders' Paper & Board Mills	EPA 440/1–82/025

Category of EGD Industrial Studies	Subcategory	Document Number
Builders Paper & Board Mills	Builders Paper & Roofing	EPA 440/1–74/026
Metal Finishing	Metal Finishing	EPA 440/1–83/091
Coal Mining	Coal Mining	EPA 440/1–82/057
Pharmaceutical	Pharmaceutical	EPA 440/1–83/084
Ore Mining & Dressing	Ore Mining & Dressing	EPA 440/1–82/061
Paint Formulating	Paint Formulating	EPA 440/1–79/049
Concrete Products	Concrete Products	EPA 440/1-78/090
Gum and Wood	Gum and Wood	EPA 440/1–79/078
Carbon Black	Carbon Black	EPA 440/1–76/060
Battery Manufacturing	Battery Manufacturing	EPA 440/1–82/067
Foundries	Metal Molding	EPA 440/1–82/070
Coil Coating	Coil Coating	EPA 440/1–84/071
Porcelain	Porcelain	EPA 440/1–82/072
Aluminum Forming	Aluminum	EPA 440/1–82/073
Copper Forming	Copper Forming	EPA 440/1–84/074
Electronics	Electrical & Electronic Components	EPA 440/1–83/075
Nonferrous Metals	Nonferrous Metals	EPA 440/1–84/019

Category of EGD Industrial Studies	Subcategory	Document Number
Forming Dairy Products Processing	Forming Dairy Products Processing	EPA 440/1–74/021
Grain Mills	a) Grain Processing	EPA 440/1–74/028
	b) Animal Feed, Breakfast Cereal & Wheat	EPA 440/1–74/039
Canned & Preserved Fruits & Vegetables Processing	Citrus, Apple & Potatoes	EPA 440/1–74/027
Canned & Preserved Seafood Processing	Catfish, Crab, Shrimp	EPA 440/1–74/020
Cement Manufacturing	Cement Manufacturing	EPA 440/1–74/005
Feedlots	Feedlots	EPA 440/1–74/001
Electroplating	a) Copper, Nickel, Chrome & Zinc	EPA 440/1–74/003
	b) Electroplating Pretreatment	EPA 440/1–79/003
Organic Chemicals	Organic Chemicals & Plastics & Synthetic Fibers	EPA 440/1–83/009
Soaps & Detergents Manufacturing	Soaps & Detergents	EPA 440/1–74/018
Fertilizer Manufacturing	a) Basic fertilizer Chemicals	EPA 440/1–74/011

Category of EGD Industrial Studies	Subcategory	Document Number
	b) Formulated Fertilizer	EPA 440/1–75/042
Pertroleum Refining	Petroleum Refining	EPA 440/1–82/014
Nonferrous Metals	a) Bauxite Refining	EPA 440/1–74/091
	b) Primary Aluminum Smelting	EPA 440/1–74/019
	c) Secondary Aluminum Smelting	EPA 440/1–74/019
Phosphate Manufacturing	Phosphorus Derived Chemicals	EPA 440/1–74/006
Ferroalloy	Smelting & Slag Processing	EPA 440/1–74/008
Leather Tanning	Leather Tanning	EPA 440/1–82/016
Asbestos Manufacturing	Building, Construction & Paper	EPA 440/1–74/017
Rubber Processing	a) Tire & Synthetic	EPA 440/1–74/013
	b) Fabricated & Reclaimed Rubber	EPA 440/1–74/030
Timber Products Processing	A) Plywood & Wood	EPA 440/1–74/023
	B) Timber Products	EPA 440/1–81/023
Meat Products and Rendering	a) Red Meat Processing	EPA 440/1–74/012
	b) Renderer	EPA 440/1–74/031
Mineral Mining & Processing	Report to Congress —The Effects of	EPA 440/1–82/059

Category of EGD Industrial Studies	Subcategory	Document Number
	Discharges from Limestone Quarries on Water Quality	

Category of EGD Industrial Studies	Subcategory	Document Number
	and Aquatic Biota	
Pesticides	Pesticides	EPA 440/1–82/079
Plastic Processing	Plastic Molding & Forming	EPA 440/1–84/069